uvial System

Sediment Transfer through the Fluvial System

Edited by

VALENTIN GOLOSOV, VLADIMIR BELYAEV
Faculty of Geography, Moscow State University, Moscow, Russia

DES E. WALLING
Department of Geography, University of Exeter, Exeter, UK

Proceedings of the International Symposium held at Moscow, Russia from 2 to 6 August 2004. The symposium was jointly organized by the Faculty of Geography, Lomonosov Moscow State University and the International Comission on Continental Erosion (ICCE) of the International Association of Hydrological Sciences (IAHS); and co-sponsored by UNESCO and Moscow State University

IAHS Publication 288
in the IAHS Series of Proceedings and Reports

Published by the International Association of Hydrological Sciences 2004
IAHS Press, Centre for Ecology and Hydrology, Wallingford, Oxfordshire OX10 8BB, UK
IAHS Publication 288
ISBN 1-901502-67-8

British Library Cataloguing-in-Publication Data.
A catalogue record for this book is available from the British Library.

The papers included in this volume have been peer reviewed and some were extensively revised before publication.

IAHS is indebted to the employers of the Editors for the invaluable support and services provided that enabled them to carry out their task effectively and efficiently.

The papers were checked, formatted and assembled by Penny Farnell at IAHS Press, Wallingford, UK.

Publications in the series of Proceedings and Reports are available *only* from:
IAHS Press, Centre for Ecology and Hydrology, Wallingford, Oxfordshire OX10 8BB, UK
tel.: +44 1491 692442; fax: +44 1491 692448; e-mail: jilly@iahs.demon.co.uk

Printed in The Netherlands by Krips BV, Meppel.

Preface

Over the past 18 years, the International Commission on Continental Erosion (ICCE) of the International Association of Hydrological Sciences (IAHS) has organized a number of highly successful symposia dealing with various aspects of erosion and sedimentation. These have included:

The Symposium on Drainage Basin Sediment Delivery, Albuquerque, New Mexico, USA, 1986;
The Workshop on Erosion, Transport and Deposition Processes, Jerusalem, Israel, 1987;
The Symposium on Sediment Budgets, Porto Alegre, Brazil, 1988;
The Symposium on Sediment and the Environment, Baltimore, USA, 1989;
The Symposium on Erosion, Debris Flows and Environment in Mountain Regions, Chengdu, China, 1992;
The Symposium on Erosion and Sediment Transport Monitoring Programmes in River Basins, Olso, Norway, 1992;
The Symposium on Sediment Problems: Strategies for Monitoring Prediction and Control, Yokohama, Japan, 1993;
The Symposium on Variability in Stream Erosion and Sediment Transport, Canberra, Australia, 1994;
The Symposium on the Effects of Scale on the Interpretation and Management of Sediment and Water Quality, Boulder, USA, 1995;
The Symposium on Erosion and Sediment Yield: Global and Regional Perspectives, Exeter, UK, 1996;
The Symposium on Human Impact on Erosion and Sedimentation, Rabat, Morocco, 1997;
The Symposium on Modelling Soil Erosion, Sediment Transport and Closely Related Hydrological Processes, Vienna, Austria, 1998;
The Symposium on the Role of Erosion and Sediment Transfer in Nutrient and Contaminant Transfer, Waterloo, Canada, 2000;
The Workshop on Erosion and Sediment Transport Measurement in Rivers: Technological and Methodological Advances, Oslo, Norway, 2002;
The Symposium on the Structure, Function and Management Implications of Fluvial Sedimentary Systems, Alice Springs, Australia, 2002;
The Symposium on Erosion Prediction in Ungauged Basins: Integrating Methods and Techniques, Sapporo, Japan, 2003 .

The Moscow symposium on "Sediment Transfer through the Fluvial System" builds on, and extends, this coverage of key aspects of erosion and sedimentation, by focusing on the redistribution and storage of sediment within different parts of the fluvial system, as well as the fluvial system as a whole. Substantial progress has been achieved in this area in recent years, because the application of a range of different methods and techniques, many of them new, has provided detailed quantitative information on sediment behaviour and sediment transfer in small first order catchments, as well as in large river basins and their individual components (river channels, flood plains, etc.). An improved understanding of sediment transfer, storage and redistribution has many important practical applications, including soil conservation, catchment management, control of diffuse-source pollution, predicting and managing reservoir sedimentation, and the maintenance of irrigation systems and navigation channels, which link closely to the sustainable management of land, water and other natural resources. Co-sponsored by

UNESCO, the symposium represents a contribution to the International Sedimentation Initiative (ISI) of the UNESCO International Hydrological Programme (IHP VI).

The choice of Russia and Moscow as the venue for the symposium must be seen as particularly appropriate in view of the important contribution of Russian scientists to work in this field and the strong reputation of Moscow State University, which has undertaken much innovative work in this area. The Faculty of Geography of Moscow State University, which is the host of the Symposium, is the leading scientific centre in Russia and the former USSR for the study of erosion and sedimentation processes in different components of the fluvial system. The Laboratory for Soil Erosion and Fluvial Pro cesses (LSEFP), within the Faculty of Geography, was founded in 1969 by Professor N. I. Makkaveev. In his well-known monograph *River Channels and Erosion in River Basins*, published in 1955, Professor Makkaveev sought to emphasize that all fluvial processes are interdependent and that it is essential to take account of sediment transfer and redistribution within the wider river basin, when studying sediment transfer within the river channel, and *vice versa*. This theme will be key to many of the topics addressed by the symposium. Furthermore, Moscow State University will celebrate the 250th Anniversary of its foundation in early 2005 and it is appropriate and timely that participants from more than 22 countries will gather there to exchange findings and ideas.

The 61 papers published in this proceedings volume address the problems of understanding sediment redistribution and storage in different components of the fluvial system within many different regions of the world, including North and South America, Australia, Europe and Asia. They have been grouped into five themes, which cover the various pathways and linkages within the fluvial system, extending from small hollows on the slopes of first order basins to the outlets of large river systems. The relationships between erosion, sediment redistribution and storage and sediment transfer in headwater areas, are addressed by a series of papers included within the section dealing with *Sediment redistribution within small catchments in different environments*. The papers in the following section, which focuses on the *Interaction of rivers and river catchments in undisturbed and intensively cultivated basins*, explore the influence of natural and accelerated erosion on the slopes of a catchment on sediment transport within the river channel at different scales, ranging from small and medium-sized basins, through large basins and whole countries, to entire continents. The papers in the third section which deal with *Large river system functioning*, address the problems of interaction between channel morphology and sediment transport, for large rivers draining different environments and under different degrees of anthropogenic influence. The development of physically-based models provides a key input to understanding sediment transfer in fluvial systems and the papers in the fourth section, dealing with *Modelling of erosion and deposition processes*, document a wide range of models, ranging from models of soil detachment to models of meander development. Finally, the problems of understanding the transfer of sediment-associated pollutants through the different pathways and linkages of the fluvial system provide the focus for the final group of papers dealing with *Implications for nutrient and contaminant transfer*.

The editors would like to express their gratitude to UNESCO, IAHS and the Faculty of Geography, Moscow State University for their technical and financial support for the symposium. Thanks are also extended to everyone who helped with the preparation of this volume and the organization of the symposium more generally. In particular we thank Penny Farnell and Cate Gardner of IAHS Press for coordinating the publication of this volume.

Valentin Golosov, Vladimir Belyaev
Faculty of Geography, Moscow State University, Moscow 119992, Russia
Des E. Walling
Department of Geography, University of Exeter, Amory Building, Rennes Drive, Exeter EX4 4RJ, UK

Contents

2 Interaction of Rivers and River Catchments in Undisturbed and Intensively Cultivated Basins

3 Large River System Functioning

1 Sediment Redistribution Within Small Catchments in Different Environments

Climate-induced and local-scale erosion and sedimentation features in small catchments: Holocene history of two small valleys in Central Russia

YURI R. BELYAEV[1], ANDREY V. PANIN[1] & VLADIMIR R. BELYAEV[2]

1 *Department of Geomorphology and Palaeogeography, Moscow State University, GSP-2, Vorob'evy Gory, Moscow 119992, Russia*
panin@morpho.geogr.msu.su

2 *The Laboratory of Soil Erosion and Fluvial Processes, Moscow State University, GSP-2, Vorob'evy Gory, Moscow 119992, Russia*

Abstract Valleys draining small catchments in Central Russia demonstrate changing incision or aggradation tendencies in the second half of the Holocene. Such changes may reflect regional-scale climate dynamics. Alternatively, some may be only of local importance, resulting from catchment-scale events. A way to distinguish between these two causes is to compare the behaviour of several fluvial landforms within a relatively uniform territory. Two small valleys (catchment area 6.5 and 8.7 km^2) located close to each other were studied in detail in the field. Periods of high fluvial activity in the Middle Holocene (4.5–6 ka BP) and the last millennium, as well as a phase of stabilization during the Sub-Boreal and through to the middle of the Sub-Atlantic, are characteristic for both valleys and thus may be attributed to climate changes. Superimposed on these climate-induced tendencies are signals from local events such as forest fires and lateral shifts of recipient streams. The background for millennium-scale incision–infill cycles is the main tendency of incision over the entire Holocene controlled by a convex shape of long profile in both valleys and relatively erosion-resistant underlying material (the Mid-Quaternary glacial boulder clays).

Key words erosion; Holocene; Russian Plain; sedimentation; small catchments

INTRODUCTION

Investigations of the history of fluvial landform development are very important for better understanding the mechanisms and processes of their evolution. It gives us opportunity to verify our concepts of past environmental conditions, obtained by other palaeogeographical methods. Hence, it also moves us toward more precise long-range forecasts of future changes in drainage networks. Some of the events in a valley's history are characteristic for a certain catchment only, but others are of a regional scale. In order to distinguish between such events it is necessary to compare reconstructions for several valleys of the same region. Most such reconstructions for the Russian Plain have concentrated on small valleys of steppe and forest-steppe zones (Butakov, 1986; Sycheva, 1997; Panin *et al.*, 1998; Sycheva *et al.*, 1998; Hrutskiy *et al.*, 1998), whereas the forest zone has received much less attention. In view of this, for the study presented here we chose two small valleys situated in the south of the forest zone near the Moscow University research station Satino, where detailed investigations of geological and geomorphological structure have been conducted during the last 30 years (Komplexnyi analiz..., 1992; Rychagov, 1996; Panin *et al.*, 1999; Panin & Karevskaya, 2000; Belyaev *et al.*, 2003).

Fig. 1 Location of the case study valleys.

CASE STUDY SITES AND METHODS

The basins of the Cholokhovskaya stream and Yazvitsy stream are situated in the centre of the Russian Plain, about 110 km to the southwest of Moscow (Fig. 1). It is a marginal area of the Middle Pleistocene (Moscow age) glaciation. Modern landscape conditions and relief morphology both make this territory typical of the humid temperate centre of the Russian Plain. The two basins are similar in area (8.7 km^2: Cholokhovskaya, and 6.5 km^2: Yazvitsy) and in position in the drainage network. Both valleys have permanent streams at least in the lower parts.

The morphology of the valleys was studied using 1:5000–1:25 000 topographic maps and during field geomorphological mapping of key areas. Detailed investigations of the geological structure of the valleys were conducted along a number of cross-section profiles, at locations chosen to characterize morphologically different parts of valleys. Fossil organic matter provided a series of radiocarbon dates for alluvial and slope deposits (Table 1). Samples were analysed at the radiocarbon laboratories of the State Centre of Environmental Radiogeochemistry, National Academy of Science of Ukraine (Kiev), Moscow State University (MSU) and Institute of Geography, Russian Academy of Science (IGRAS). All dates cited here are in non-calibrated radiocarbon timescale.

Table 1 Radiocarbon dates of sediments from Cholokhovskaya and Yazvitsy valleys.

Sample	Profile	Pit	Depth (m)	Location	Material	Date
Dates from the Cholokhovskaya valley						
Ki-8411	b	1	0.45	Colluvial sheet base	Charcoal	1290±120
Ki-8412	b	2	0.57–0.58	Flood plain sediment	Charcoal	1940±120
Ki-8413	b	2	1.45	Flood plain sediment	Charcoal	3520±110
Ki-8414	b	4	1.05	Top of basal layer	Charcoal	5170±120
Ki-8415	d	12	0.40	Top of basal layer	Fossil wood	2470±120
Ki-8416	d	13	0.67	Basal layer	Fossil wood	3160±120
Ki-8417	d	16	0.44–0.45	Flood plain sediment	Charcoal	3860±120
Ki-8418	d	17	1.18	Top of basal layer	Charcoal	5125±110
Ki-8419	d	19a	0.51–0.54	Flood plain sediment above buried soil	Charcoal	4670±120
Ki-8420	d	19a	1.74–1.75	Top of basal layer	Charcoal	5830±120
Dates from the Yazvitsy valley						
Ki-5243	a	13	0.95	Colluvial sheet base	Charcoal	635±35
Ki-6441	e	31	1.05	Basal layer	Charcoal	4260±75
Ki-6442	e	31	1.05	Basal layer	Charcoal	4590±80
Ki-6443	e	28a	0.85	Colluvial sheet base	Charcoal	870±60
IGRAS-1643	i	14	0.6	Colluvial sheet base	Charcoal	740±180
MSU-1457	j	51	1.2	Colluvial sheet base	Charcoal	820±60
MSU-1474	e	31	1.05	Basal layer	Charcoal	4735±200
MSU-1475	e	28a	0.85	Colluvial sheet base	Charcoal	1100±60

MORPHOLOGICAL AND GEOLOGICAL STRUCTURE OF VALLEYS

The case study valleys appear to be very similar in both morphology and geological structure. They have convex longitudinal profiles typical for small valleys of this territory and strikingly similar features in their geological cross-sections (Figs 2 and 3).

Upper parts of valleys are incised in Late Pleistocene loessial loams which cover significant portions of interfluve areas in the region (Antonov & Rychagov, 2002). Further downstream, valleys begin to cut in to Middle Pleistocene glacial sediments (Dnieper and Moscow glaciation age boulder clays). The Cholokhovskaya stream valley is also cut into lenses of the Late Pleistocene alluvial-lacustrine sediment (Mikulino interglacial-age clays) in its lower reaches (Rychagov, 1996).

Cross-section profiles of the studied valleys are characterized by the presence of a complex of fluvial terraces, which are easily traced along both valleys (Figs 2 and 3), although their morphology and height above the thalweg vary noticeably. The terraces are composed of relatively thin (no more then 2.0–2.5 m thick) two-layered alluvial sediments. The thin lower layer (20–40 cm thick) is composed of variably sized coarse debris (up to boulders) within a clayey sand or sandy loam matrix, and is referred to as the basal layer. The upper layer is up to 1.0–1.5 m thick and is largely associated with a flood-plain-like sedimentation regime. It consists of finer material—sandy loams, silts or loamy sands—sometimes with prominent lamination. The composition of the alluvium (grain size, and the ratio between the thickness of the basal layer and the flood-plain alluvium) varies between different terraces within each cross-section, as well as within each terrace level along the valley. These changes provide important information about the dynamics of the effective channel-forming discharges and flood activity in the past.

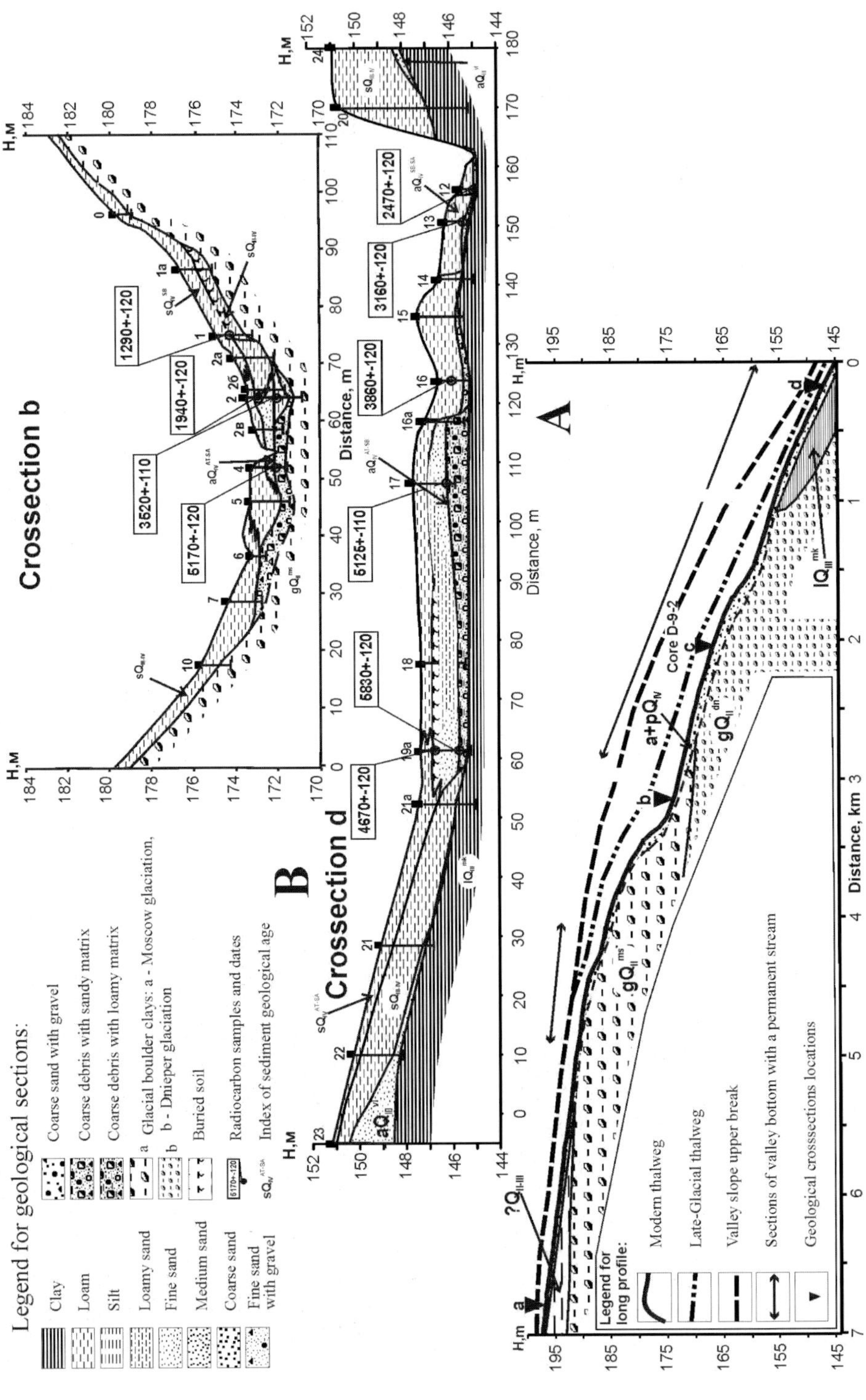

Fig. 2 Long profile (A) and geological cross-sections (B) of the Cholokhovskaya valley.

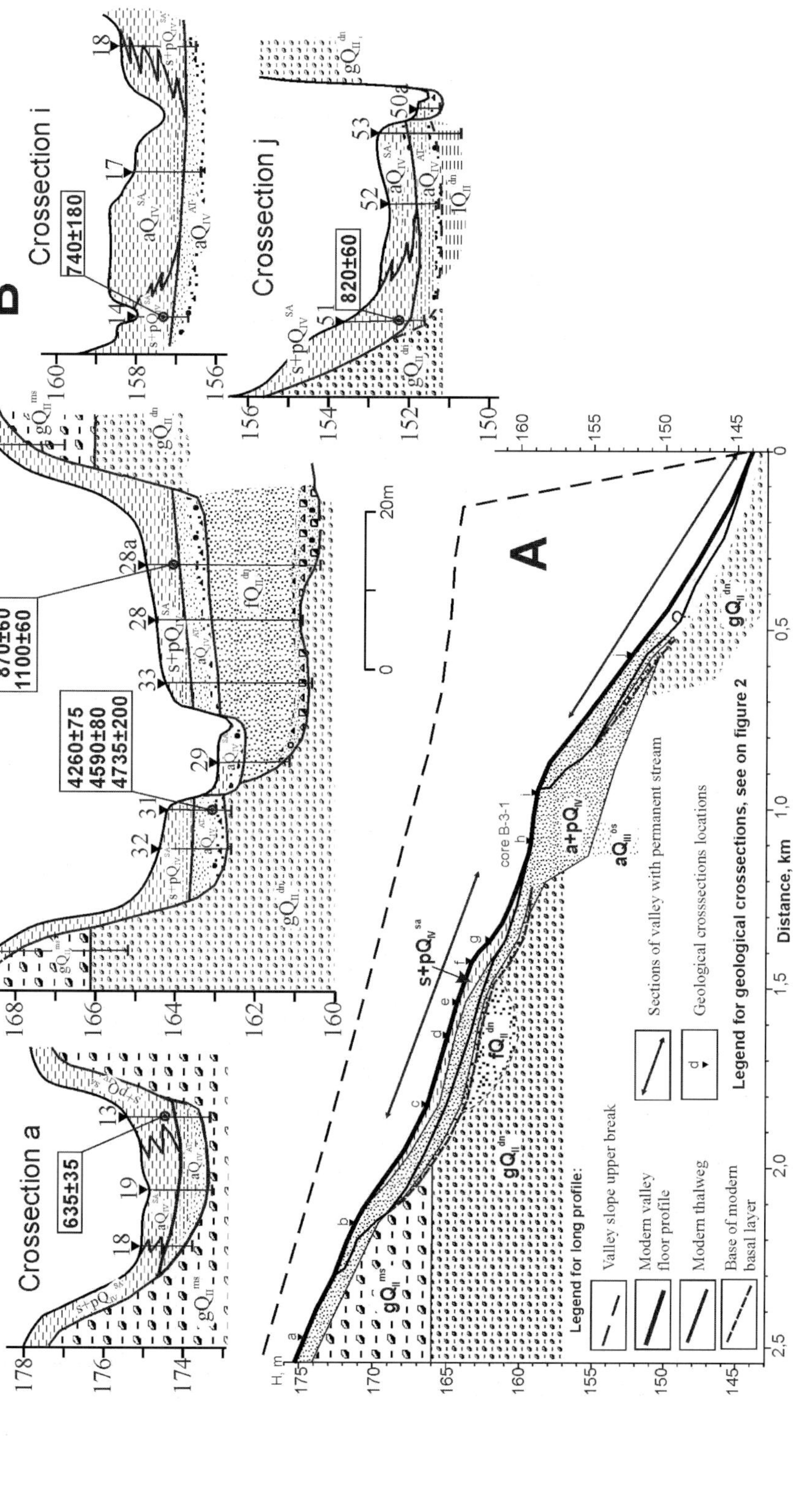

Fig. 3 Long profile (A) and geological cross-sections (B) of the Yazvitsy valley.

Detailed investigation of the geological structure of both valleys showed that there are no Late Quaternary (Valday glaciation age) or Early Holocene (Pre-Boreal and Boreal) sediments in their bottoms. The oldest Holocene alluvial material was found at the base of bottom infill sections in middle parts of both valleys, and in the highest terrace section in the lower part of the Cholokhovskaya stream (Figs 2 and 3). Dates from the top of the basal layer are 5830 ± 120 (Ki-8420), 5125 ± 110 (Ki-8418) and 5170 ± 120 (Ki-8414) from the Cholokhovskaya valley and 4260 ± 75 (Ki-6441), 4590 ± 80 (Ki-6442) and 4735 ± 200 (MSU-1474) from the Yazvitsy valley (Table 1). These dates allow us to associate these oldest alluvial sediments with the second half of the Atlantic period.

In the lower part of the Cholokhovskaya valley the alluvial unit is characterized by relatively large sized debris in the basal layer (up to in excess of 30 cm) and the increased thickness of this layer itself (up to 1 m) (Fig. 2(B), cross-section d). The upper layer of this unit consists of laminated sand, although in the Yazvitsy valley and in the middle part of the Cholokhovskaya valley it has a thinner, loamy composition. Alluvial units of younger age have the same two-layered structure, but they are generally thinner and have a normal ratio between basal and flood plain sediment thickness (i.e. the basal layer is substantially thinner than the overlying flood plain sediment layer). Importantly, the base of every younger alluvial unit lies deeper than that of the older alluvial unit in most of the studied cross-sections of the both valleys, except for upper parts and local zones of recent deposition. This suggests that incision is the main tendency in the history of this valley.

HOLOCENE HISTORY

An absence of Late Quaternary or Early Holocene alluvial sediments in the valley bottoms together with the radiocarbon dates for the oldest alluvial unit (Table 1) make it clear that, at least in the first half of the Holocene, incision was the main tendency in the development of the studied valleys. There may have been some minor aggradation episodes, but no evidence of that has been found, as lateral erosion was also active, widening valley floors and destroying all traces of any older sediment. Such an assumption is verified by other research concerning the intensification of erosion during the transition between glacial and interglacial periods (Sycheva, 1997). Intensive incision, starting in the Late Glacial, most likely occurred almost simultaneously along all the valleys; however, there was regressive erosion too.

For the Cholokhovskaya valley, the depth of the Early Holocene incision in the upper part was about 1.0–1.5 m, in the middle part 3.0–4.0 m and in the lower part, up to 2.0 m (Fig. 2(A)). Such a situation is typical when incision is caused by an increase of water runoff (Makkaveev, 1956). The size of the terrace that was formed and the sharpness of its upper boundaries prove that the valley floor at that period of time was much wider than today and that lateral erosion was very active. The relatively large size of debris in the basal layer and the width of the terrace complex lead us to a conclusion of highly effective channel-forming discharges and flood activity in the Cholokhovskaya and Yazvitsy valleys in the first half of the Holocene. Results of pollen analysis of the Atlantic age alluvium from Yazvitsy valley also confirm our assumption about higher precipitation levels during that period of time (Panin *et al.*, 1999).

In the lower part of the Cholokhovskaya valley intensive incision temporarily stopped about 5800 years ago (Table 1, Ki-8420). Dates and the geological cross-section (Fig. 2(B), cross-section d, Table 1, Ki-8420, Ki-8418) show that the period between 5800 and 5100

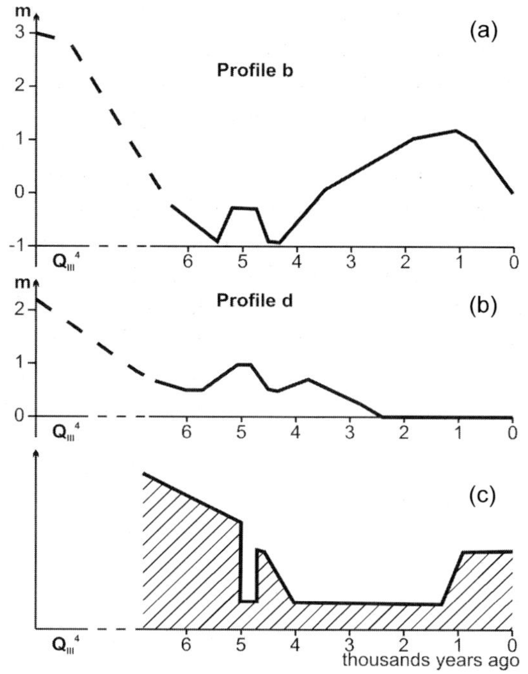

Fig. 4 Generalized tendencies of development (elevation of mean bed level) for (a) the middle reach and (b) the lower reach, and (c) qualitative changes in flood activity of the Cholokhovskaya stream.

years ago was marked by a slight decrease in flood activity (Fig. 4). Flood plain deposits became more clayey (Fig. 2(B), cross-section d, pits 17 and 18 in comparison to pit 19a) and horizontal channel deformation took place simultaneously with a slow and limited aggradation (Fig. 2(B), cross-section d).

Between 5100 and 4700 years ago (Table 1, Ki-8418, Ki-8419) flood activity and channel-forming discharges decreased dramatically (Fig. 4). Most of the floodplain in the lower part of the Cholokhovskaya stream became an unflooded terrace. On the surface of this terrace a fully stratified zonal soil profile was formed (Fig. 2(B), cross-section d).

About 4700 years ago flood activity increased again (Table 1, Ki-8419). The zonal soil was buried (Fig. 2(B), cross-section d) and incision in the lower part of the valley started again. At the beginning of this stage of incision the channel was braided, but most of its branches now exist only as oxbow depressions on the terrace surface. In the middle part of the Cholokhovskaya stream the Early Subboreal incision formed a narrow bottom cut, which is now buried by subsequent deposition (Fig. 2(B), cross-section b). However, the period of active erosion was very short. Dating of the alluvial infill of the bottom cut shows that it happened not later than 3900 years ago in the lower course and 3500 years ago in the middle course (Table 1, Ki-8413, Ki-8417). The same situation occurred in the Yazvitsy valley, where the period of intensive erosion is dated between 4200-4800 years ago (Fig. 3(B), Table 1, Ki-6441, Ki-6442, MSU-1474).

In the lower reach of the Cholokhovskaya stream a modern meandering belt has been formed since the Middle Sub-Boreal (about 3200 years ago). Simultaneously with meander shifts, continuous but slow incision has taken place. It was probably brought about by lateral shift or incision of the recipient stream, resulting in increased channel gradients in the Cholokhovskaya valley lower part. The basal layer here is very thin because receipt of coarse material from upstream was limited. The reason for that is the limited extent of this stage of incision which affected only a short part of the valley close to its mouth. At the same time, in the middle part fine sediment deposition took place, infilling a previously formed bottom cut (Fig. 2(B), cross-section b). The most appropriate explanation of this phenomenon is a decrease in flood activity.

Traces of a significant increase in flood activity—numerous dry channels and the main bottom cut, which still continues to incise or grow regressively in its upper part—are correlated with the Sub-Atlantic period. In the middle part of the Cholokhovskaya valley, the stream incised through a lens of sandy alluvium, the upper part of which is dated as 1940±120 years ago (Fig. 2(B), cross-section b, Table 1, Ki-8412). At the same time, the fact that dry channels situated on the surface of the currently unflooded terrace cut through young slope deposit sheets (Fig. 2(B), cross-section b, pit 6) is evidence that accumulation episodes and this most recent incision may be inter-related. Most probably, intensive incision and deepening of the valley started only after the last episode of slope-derived sediment accumulation. During that episode, thick sheets of slope-derived sediment partly covered valley floors. Abundant charcoal pieces were observed inside these sediment bodies. This suggests that forest fires were the most probable cause of this slope destabilization. Dates from different colluvial sheets are relatively close and vary between 1200 and 600 years ago (Table 1, Ki-5243, Ki-8411, Ki-6443, MSU-1475, MSU-1457, IGRAS-1643). Therefore, the last stage of incision in these two small valleys probably occurred during the last millennium and was caused by an increase of flood activity.

In recent centuries, human activity began to be a serious environmental factor within the study area. Human artefacts prove that agricultural activity already took place here in the 16th and 17th centuries. In the 18th century, the drainage basins of the Cholokhovskaya and Yazvitsy streams were partly cultivated. Later, cultivation ceased, arable fields were abandoned and secondary forests developed. All the artefacts found in fluvial sediments lay near the surface. Hence it is believed that, unlike the forest-steppe and steppe zone valleys (Golosov, 1998), human-induced accumulation in these two valleys was not very intensive.

DISCUSSION

Two main groups of factors are responsible for the trends revealed in the small valleys' development. The first is regional-scale climatic fluctuations, which have common reflections in both of the valleys. The second is a group of local catchment-scale events such as forest fires, human activities, processes of self-regulation in valleys and the influence of recipient streams. Comparison of the histories of the different valleys demonstrates that the most large-scale (in amplitude and duration) erosion or accumulation trends are similar for both valleys and occurred under a similar timeframe. This is verified by a striking similarity in morphology and geology in the two valleys investigated and especially in radiocarbon dates. The main climate-induced stages of the history of the small valleys are identified clearly (Fig. 4). The first is a stage of incision and high flood activity in the first half of the Holocene. The second is a stage of decreased flood activity, valley floor stabilization and limited aggradation in the Sub-Boreal and beginning of the Sub-Atlantic period, interrupted by a short increase of erosion 4700–3900 years ago. The last stage is of a recent increase of flood activity and erosion in the Late Sub-Atlantic. As additional support for such a conclusion, a similar Holocene history has been described for the Protva River, which is the main river of the territory (Panin & Karevskaya, 2000).

The most important of the local events that influenced the history of the two study valleys are forest fires. They caused intensive erosion in basins and accumulation of colluvial material on slopes and the valley bottoms. The influence of the recipient stream caused incision in the lower reach of the Cholokhovskaya stream in the Late Sub-Boreal. Internal fluvial system self-regulation processes (long profile adjustment via passage of regressive and transgressive waves of erosion and deposition) resulted in differences in the dates of interruption of the Early Holocene incision between middle and lower parts of the Cholokhovskaya valley. These waves of erosion and deposition can be clearly identified in the modern pattern of the valley floor morphology.

CONCLUSIONS

The main tendency in the Holocene history of the two studied small valleys is progressive incision. It is controlled by the convex shape of the long profiles and relatively erosion-resistant underlying material (the Mid-Quaternary glacial boulder clays). Three main long-term stages of the Holocene history of both valleys, induced by climatic fluctuations, are similar and can be summarized as follows:

- The Early Holocene – end of the Atlantic (about 5100 years ago). High flood activity. Intensive incision and lateral erosion. Wide valley floors were formed.
- Beginning of the Sub-Boreal – second half of the Sub-Atlantic. Decrease of runoff and channel-forming discharges. Stabilization of valley floor with limited aggradation in middle and upper parts. Between 4700 and 3900 years ago, short period of flood activity increase and incision.
- The Late Sub-Atlantic period (after 1300 years ago). Increase of flood activity and progressive incision.
- The main catchment-scale events influencing development of the study valleys during the Holocene are forest fires.

Acknowledgement This study has been supported by the Russian Foundation for Basic Research (project 03-05-64021). The authors thank Professor G. I. Rychagov (Department of Geography, MSU) for valuable comments and ideas.

REFERENCES.

Antonov S. I. & Rychagov, G. I. (2002) Mantle and slope formations in the middle Protva River basin (Pokrovno-sklonovye obrazovaniya basseina sredney Protvy) (in Russian with English summary). *Vestnik MGU, Series 5.* **4**, 39–44.

Butakov, G. P. (1986) *Pleistocenovyi periglyacial na vostoke Russkoy ravniny* (Pleistocene Periglacial in the east of the Russian Plain). Kazan University Publ., Kazan, Russia (in Russian).

Belyaev, Y. R., Panin, A. V. & Belyaev, V. R. (2003) Istoriya razvitiya balok centra Russkoy ravniny (na primere Cholokhovskoy balki, Satinskiy polygon MGU) (Evolution of balka's of the central Russian plain (the example of the Cholokhovskaya balka within the Satino polygon of MSU) (in Russian with English summary). *Vestnik MGU Series 5.* **5**, 55–63.

Hrutskiy, S. V., Semenov, O. P. & Kostsova, E. V. (1998) Processy reliefoobrazovaniya v periglyacialah pleistocena I sovremennye formy erosionnogo reliefa (Earth surface processes in Pleistocene periglacials and recent erosional forms) (in Russian with English summary). *Geomorphologia* **3**, 85–91.

Golosov, V. N. (1998) Redistribution of sediments within small river catchments in the agricultural zone of Russia. *Geomorphologie: relief, processes, environment.* **1**, 53–64.

Komplexnyi analiz chetvertichnyh otlozheniy Satinskogo uchebnogo poligona (Complex analysis of Quaternary sediments of the Satino key site) (1992). Moscow University Publ., Moscow, Russia (in Russian).

Makkaveev, N. I. (1956) Vliyanie stoka na prodolnyi profil reki (Influence of water runoff on longitudinal profile of the river). In: *Questions of Geography*, 199–205. Moscow-Leningrad, Russia (in Russian).

Panin, A. V. & Karevskaya, I. A. (2000) Istoriya formirovaniya poimy reki Protvy v predelah Satinskogo poligona MGU (History of the Protva river floodplain near the Satino key site) (in Russian with English summary). *Vestnik MGU. Series 5*. **4**, 52–62.

Panin, A. V., Karevskaya, I. A. & Markelov, M. V.(1999) Evoluciya doliny ruchiya Yazvitsy (bassein sredney Protvy) vo vtoroy polovine Golocena (Evolution of Yazvitsy stream valley (middle Protva basin) in the second half of the Holocene) (in Russian with English summary). *Vestnik MGU. Series 5*. **2**, 63–72.

Panin, A. V., Malayeva, E. M., Golosov, V. N., Ivanova, N. N. & Markelov, M. V. (1998) Geologo-geomorphologicheskoe stroenie i Holocenovaya istoriya razvitiya Berestovoy balki (Rostovskaya oblast). (Geologic-geomorphologic structure and development of berestovaya ravine (Rostov region) during the Holocene) (in Russian with English summary). *Geomorphologia* **4**, 70–85.

Rychagov, G. Y. (ed.) (1996) *Stroenie i istoriya razvitiya doliny reki Protvy* (Structure and history of Protva river valley) Moscow University Publ., Moscow, Russia (in Russian).

Sycheva, S. A. (1997) Evoluciya balochnoy sistemy v klimaticheskom ritme "oledenenie-mezhlednikovie-oledenenie" (Evolution of balka system during climatic rhythm "glaciation-interglaciation-glaciation") (in Russian with English summary). *Geomorphologia* **2**, 100–111.

Sycheva, S. A., Chichagova, O. A., Dayneko, E. K., Sulerzhitskiy, L. D. & Uzyanov, A. A. (1998) Etapy razvitiya erosii na Srednerusskoy vozvyshennosty v Golocene (Erosional stages on the Middle-Russian upland in Holocene) (in Russian with English summary). *Geomorphologia* **4**, 12–21.

Episodic sediment pulses generated by forested flood plain stripping: Bruces Creek, Nadgee State Forest, southeastern Australia

ASHLEY A. WEBB[1] & D. DRAGOVICH[2]

1 *State Forests of New South Wales, Research & Development Division, PO Box J19, Coffs Harbour Jetty, NSW 2450, Australia*
ashleyw@sf.nsw.gov.au

2 *Division of Geography F09, University of Sydney, NSW 2006, Australia*

Abstract Sedimentological, stratigraphic and radiocarbon analyses of flood plain sediment indicate that flood plain pockets of Bruces Creek, a sand-bed forest stream, were completely removed by a catastrophic flood around 1000 years BP. The present flood plain redeveloped by lateral migration prior to 575 years BP and was colonised by an emergent *Eucalyptus* forest and *Tristaniopsis laurina* (Water Gum) trees. Since 300 years BP, channel stability has been controlled by riparian vegetation and large woody debris (LWD), as indicated by obliquely accreted fine-grained sediment deposits on the channel banks. This work indicates that sediment transfer through small sand-bed forest systems is minimal during extended periods of vegetation-controlled channel and flood plain stability. However, episodic high-energy catastrophic floods can exceed thresholds of channel stability, totally destroy alluvial landforms and the flood plain forest, and result in the delivery of massive pulses of sediment to the channel.

Key words catastrophic erosion; catastrophic floods; flood plain stripping; riparian vegetation; sediment transfer; southeastern Australia

INTRODUCTION

Sediment transfer through small, forested drainage basins in southeastern Australia depends largely upon the relative influences of erosion by rainfall and flood events, and stabilization by vegetation. Recent Australian research has investigated the relative roles of high magnitude floods and natural riparian vegetation in controlling sediment stability at the channel and flood plain scale. On Wheeny Creek, a high energy bedrock-confined stream, Webb *et al.* (2002) found that vegetation and large woody debris (LWD) played a minor short-term (10^0 years) role in small-scale island and in-channel bench formation. However, in the longer term (10^1–10^3 years) it was concluded that the formation and destruction of fluvial landforms, and therefore the transfer and deposition of sediment, was flood-driven and flood-dominated (Webb *et al.*, 2002). In contrast, research on less confined and lower energy forest streams in southeastern Australia, such as Tonghi Creek (Webb, 2002; Webb & Erskine, 2003) and Thurra River (Brooks *et al.*, 2003), has demonstrated that channel and flood plain sediment storage can be controlled in the longer term by riparian vegetation. Webb & Erskine (2003), however, acknowledge that extreme floods have been responsible for the creation of alluvial cut-offs on Tonghi Creek around 400 years BP and for forested flood plain stripping around 1000 years BP.

The aim of this paper is to investigate the relative roles of riparian vegetation (and LWD) and extreme floods in controlling the late Holocene channel and flood plain stability of Bruces Creek, a partially bedrock- and terrace-confined forest stream, and to determine the consequences for sediment dynamics in such environments.

Study site

Bruces Creek is a 79 km^2 left bank tributary of the Wallagaraugh River, located in the southeast of New South Wales, Australia (Fig. 1) where mean annual rainfall is 975 mm. Drainage basin geology is dominated by Devonian granitoids of the Bega Batholith, Ordovician Adaminaby Group metasediments and Tertiary fluvial sediments. The study reach is 310 m long and is located in a riparian reserve in Nadgee State Forest where the drainage basin area is 68 km^2. In the study reach, Bruces Creek is a sand-bed stream that is continuously flanked by large but discontinuous pockets of flood plain (Webb, 2002). It is a medium energy stream with a mean bankfull specific stream power of 20.7 ± 1.9 Wm^{-2} (mean ± 2 standard errors). Regional flood frequency analyses indicate that bankfull discharge (22.7 ± 4.7 m^3 s^{-1}) has a return period of between 1.5 and 1.8 years on the annual maximum flood series and is approximately equal to the mean annual flood (23.6 m^3 s^{-1}) (Webb, 2002).

Riparian vegetation at Bruces Creek exhibits a distinct vertical zonation of species from the streambed to the flood plain. Shallow sections of the bed (riffles, runs and backwaters) are dominated by the macrophyte *Triglochin procerum* with occasional *Tristaniopsis laurina* (Water Gum), whereas pools are essentially free of vegetation, except for *T. laurina* attached

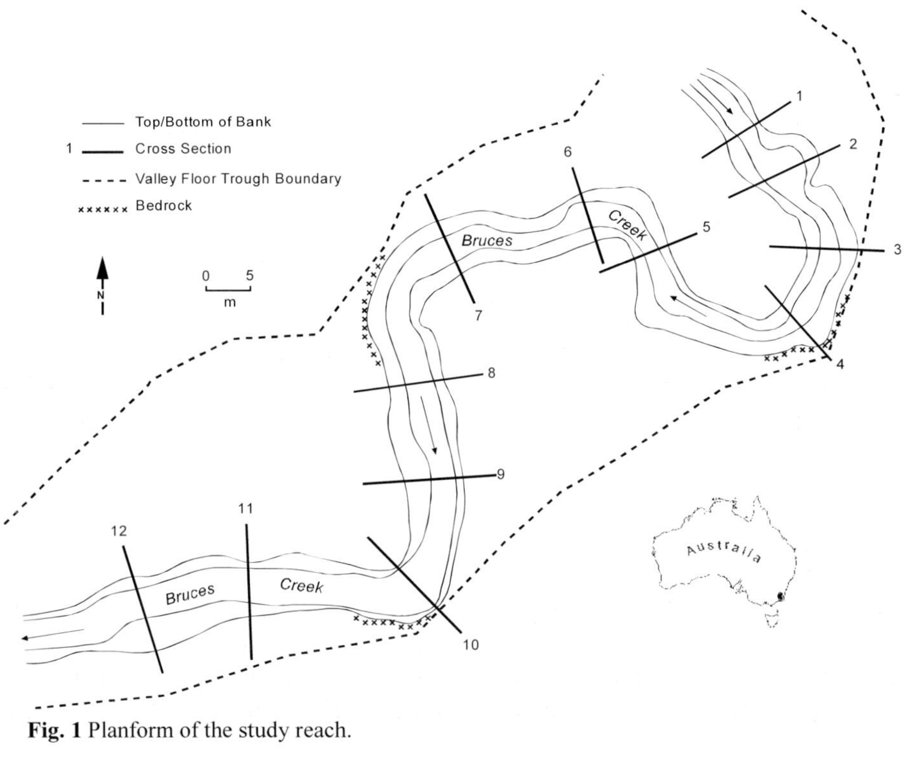

Fig. 1 Planform of the study reach.

to the lower stream bank. The bank vegetation is dominated by large *T. laurina* trees (up to 1.05 m diameter at breast height over bark—DBHOB) but also includes *Acacia mearnsii* and *Pittosporum undulatum* trees. Flood plain vegetation varies laterally away from the stream. *Tristaniopsis laurina* trees dominate next to the channel in association with species such as *A. mearnsii*, *P. undulatum* and *Babingtonia pluriflora*. Emergent trees occur on higher parts of the flood plain away from the immediate stream bank and include *Eucalyptus cypellocarpa* (maximum DBHOB 1.45 m), *E. elata*, *E. muelleriana* and *Angophora floribunda*. Further away from the channel banks on the slightly elevated remnant river terraces and footslopes, the density of riparian vegetation species declines while more sclerophyllous vegetation species become more dominant. The overstorey is dominated by eucalypts such as *E. muelleriana* and *E. cypellocarpa* along with hillslope species such as *Allocasuarina littoralis*.

Webb (2002) conducted a complete census of large woody debris (LWD) in the bed of the study reach. A total of 437 pieces of LWD was recorded at a mean density of 141 pieces per 100 m. The total loading of LWD was 751 $m^3\ ha^{-1}$. Both the loading and density are very high and plot close to the envelope curve for world rivers of similar catchment area (Gippel, 1995). The majority of LWD pieces are small (diameters 0.1–0.3 m). Despite being less abundant, medium size pieces (diameters 0.3–0.7 m) contribute the majority of the total LWD volume in Bruces Creek (Webb, 2002). *Tristaniopsis laurina* trees that grow within the channel banks contribute 16.9% of the total LWD volume in the study reach. Webb (2002) also found a high density of log-steps and debris dams in the study reach. Free-fall over log-steps accounted for 56% of the total head loss, and as a result of high bed roughness up to 3300 m^3 of sand are stored in the bed of the study reach (Webb, 2002).

METHODS

To establish the late Holocene chronology of erosional and depositional events at Bruces Creek, a sub-reach of the study site was selected for detailed investigation. This sub-reach comprised three pockets of flood plain that occur in the most upstream 130 m of the study reach at cross-sections 1–5 (Fig. 2). A combination of pits, trenches and auger holes was excavated. The sediments were described in the field (after Northcote, 1979) and representative samples subjected to particle size analyses in the laboratory to determine the flood plain stratigraphy and sedimentology. A total of 15 charcoal samples was collected for radiocarbon dating using the AMS method in order to establish ages for different landforms and erosional and depositional events. These data were combined with the biogeomorphic evidence presented above (after Webb, 2002) to reconstruct a chronology of late Holocene channel and flood plain evolution in the study reach.

Flood plain and bench stratigraphy

Pockets of flood plain have developed between the bedrock valley sides and river terrace remnants. The Pleistocene river terraces are characterized by a well-developed duplex soil comprising a loamy sand to loam fine sandy A_1 horizon with a bleached A_2 horizon overlying a mottled sandy medium clay to heavy clay B horizon (Fig. 3). The flood plain at

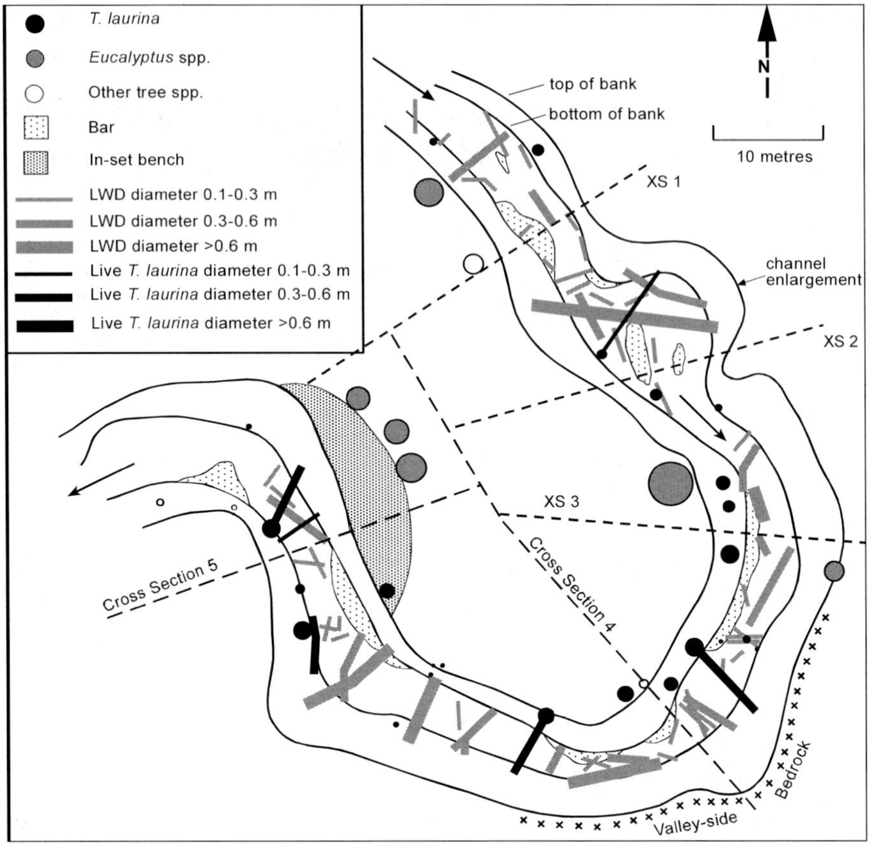

Fig. 2 Location of trees and large woody debris in a sub-reach of Bruces Creek. Circles represent tree diameters at breast height exaggerated by a factor of two.

cross-section 4 (Fig. 3(a)) comprises a series of coalescing point bar deposits that fine upwards from coarse sand to sandy loam. These sediments dip towards the channel and indicate that the dominant process of flood plain formation has been by lateral point bar accretion. They overlie a much coarser basal unit of slightly gravely coarse sand, and are capped by overbank deposits that fine laterally away from the channel (Fig. 3(a)). A unit of fine sandy loam has been plastered on the channel banks by the process of oblique accretion (Pickup & Warner, 1976) (Fig. 3(a)). These deposits have been induced by high roughness created by bank vegetation and LWD in the channel. Such deposition was observed in the field following a bankfull flood in April 2001. An inset bench has developed on the right bank of cross-section 5 and comprises a fining upwards sequence from coarse sand to fine sandy loam, overlying the basal unit of slightly gravely coarse sand (Fig. 3(b)).

Late Holocene chronology and formation of Bruces Creek

The five dates obtained from the stripped basal facies at cross-section 4 (Fig. 3(a)) do not become older with depth or with increasing distance from the present channel. The irregular

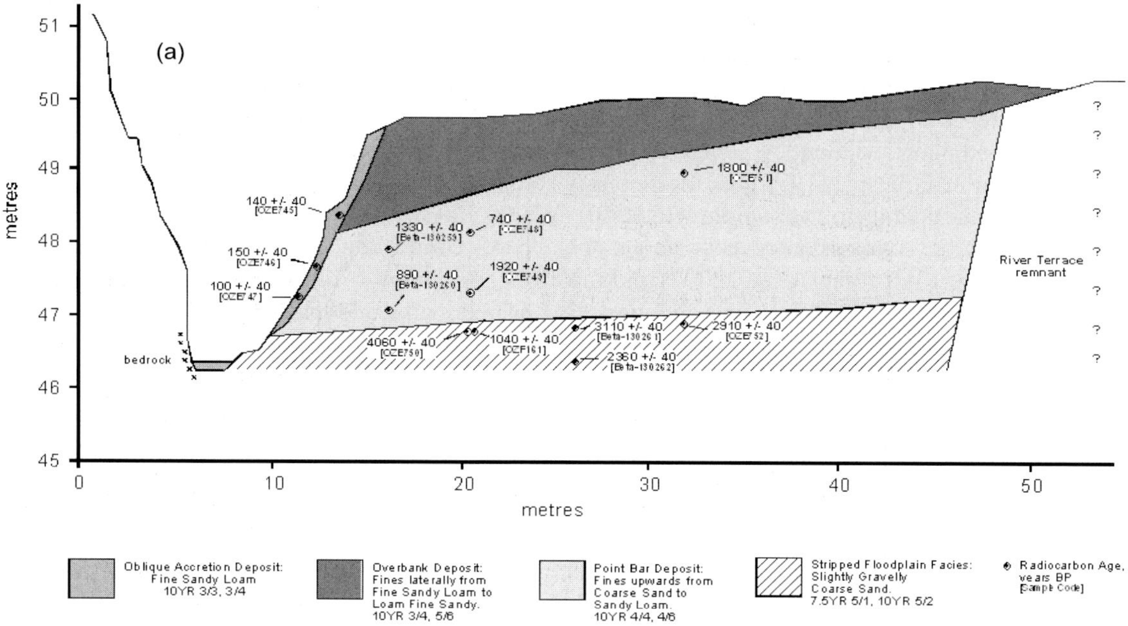

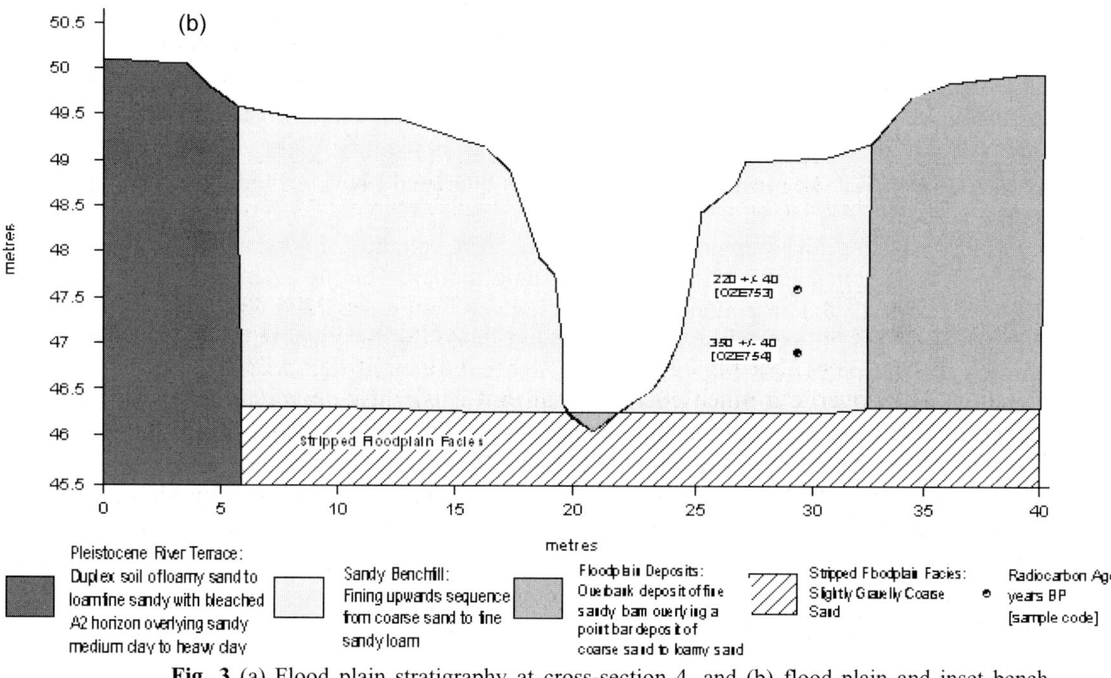

Fig. 3 (a) Flood plain stratigraphy at cross-section 4, and (b) flood plain and inset bench stratigraphy at cross-section 5.

age distribution of the dates, along with the coarse textured sediments, indicate reworking of the flood plain sediments and incorporation of older charcoal during substantial channel widening and flood plain stripping. It is plausible that a single catastrophic flood or a series of such events in rapid succession could effect large-scale flood plain erosion. The two adjacent charcoal samples from the stripped basal facies at cross-section 4 with an age difference of approximately 3000 radiocarbon years strongly supports substantial reworking of flood plain sediments. The age difference cannot simply be explained by sample size/mass (Blong & Gillespie, 1978), as the younger sample (OZF161) was in fact larger in diameter and mass than the other sample (OZE750) (Webb, 2002).

During catastrophic events, large pockets of flood plain can be completely removed (Nanson, 1986; Webb *et al.*, 2002) and the eroded sediment is often deposited in the channel and reworked downstream as a sediment slug or bed load wave (Erskine, 1994). Therefore, it is expected that a number of the charcoal samples dated from the coarse stripped flood plain facies would have been reworked from flood plain pockets located upstream of the study site. In this case, the youngest date is important for interpreting the timing of flood plain stripping. Given that the youngest sample has a radiocarbon age of 1040 ± 40 years BP, it is likely that the catastrophic event, or series of events, occurred no earlier than this time. The date, when calibrated to calendar years using the curve of Stuiver *et al.* (1998) yields a 2-sigma (95% probability) range of between 805 and 1050 years BP.

Five samples were dated from the laterally accreted coalescing point bar deposits at cross-section 4 (Fig. 3(a)) and although the AMS dates generally become younger closer to the channel, at least one date (Beta-130259) is older than the date obtained for charcoal located beneath it in the profile. As much of the material eroded during the catastrophic event(s) would have been subsequently reworked downstream, the arrangement of dates in the point bar deposits is expected. The dates suggest that following the catastrophic erosion of flood plain pockets around 1000 years BP, a new flood plain developed by lateral migration in the remaining void between the river terrace remnant and the bedrock valley side. Given the 2-sigma calibration (Stuiver *et al.*, 1998) of the youngest dates in the lateral accretion deposit (Fig. 3(a)), it is suggested that the flood plain formed approximately 1000 to 575 years BP.

Following the development of the present flood plain, *T. laurina* trees and associated riparian vegetation species colonized the channel banks. The presence of oblique accretion deposits suggests that the contemporary channel banks are relatively stable (Pickup & Warner, 1976). Calibration of the three dates obtained on charcoal excavated from the oblique accretion deposits (Fig. 3(a)) to calendar years (Stuiver *et al.*, 1998) suggests that the earliest time the banks became stable due to the influence of vegetation was 290 years BP. Thereafter, recruitment of LWD to the channel began after the trees reached maturity. The oldest radiocarbon date obtained from contemporary in-channel LWD was 160 ± 50 years BP (Webb, 2002). This date corresponds closely with the dates obtained for the oblique accretion deposits, supporting the hypothesis that channel and flood plain stability have only recently (10^2 years) been effected by riparian vegetation and associated LWD. Further evidence of recent vegetation colonization and LWD recruitment is provided by the fact that no LWD was found buried in the flood plain or bench sediments excavated at any of the cross-sections. Given that regional evidence indicates there have been no major changes in vegetation communities during the late Holocene (Kershaw *et al.*, 1991; Dodson, 1994), it is plausible that prior to catastrophic stripping, the former flood plain was forested. In this case, the lack of LWD buried in the flood plain supports the hypothesis that around 1000 years BP, the alluvial trough, including the flood plain, the flood plain forest and associated LWD, was cleaned out during a catastrophic flood or series of such events.

Since the channel was stabilized minor lateral channel adjustments have occurred. For example, channel widening has occurred at cross-section 2 (Fig. 2) where a very large debris dam has been outflanked and the channel locally enlarged by 100%. It is assumed that such localized channel expansions remain until the debris dam decays or is completely removed by a large flood. Inset benches then develop in the expansion. The inset bench at cross-section 5 (Fig. 3(b)) is likely to have been formed by this process. The radiocarbon dates obtained for charcoal incorporated in the sandy benchfill, when calibrated to calendar years (Stuiver *et al.*, 1998), suggest that the nucleus phase of bench formation began no earlier than 505 years BP. Given that Erskine & Livingstone (1999) demonstrated that in-channel benches are periodically reworked by floods, and that the two calibrated dates obtained in the bench are overlapping, it is likely that the bench formed around 300 years BP. Thus, channel adjustment at cross-section 5 occurred around the same time or just before vegetation-induced oblique accretion deposits were first deposited at cross-section 4.

CONCLUSIONS

In recent times (10^2 years) riparian vegetation and large woody debris have exerted a significant influence on channel and flood plain stability at Bruces Creek. Riparian trees, predominantly *T. laurina*, have been effective in binding the highly mobile sandy boundary sediments. The trees, along with a dense cover of shrubs and ground storey plants, have created high bank roughness that has induced the deposition of fine sandy sediments by oblique accretion. Recruitment of LWD to the channel has resulted in the formation of a large number of debris dams and log-steps, which dissipate considerable stream energy and therefore contribute to channel stability by reducing the competence of floods to erode the channel boundary. The high roughness in the channel created by LWD, furthermore, has induced the deposition and storage of large volumes of sand within the streambed. Apart from minor channel adjustments on a scale of hundreds of years (10^2 years), the majority of biogeomorphic evidence suggests that in partially confined, medium energy streams, such as Bruces Creek, short term channel and flood plain stability is controlled by riparian vegetation and associated LWD (Webb, 2002). However, the stratigraphic and radiocarbon evidence demonstrates that this is not the case in the long term. Catastrophic floods, like the one that occurred on Bruces Creek around 1000 years BP, are capable of exceeding thresholds of channel stability, totally destroying all alluvial landforms and completely removing the flood plain forest (Webb *et al.*, 2002; Erskine & Peacock, 2002). The results of this work indicate that sediment transfer through small sand-bed forest systems is minimal during the extended periods of channel and flood plain stability effected by riparian vegetation and LWD. However, episodic high-energy catastrophic floods can exceed thresholds of channel stability and result in the delivery of massive pulses of sediment to the channel.

Acknowledgements This research was funded by an Australian Postgraduate Award research scholarship, State Forests of NSW and two Australian Institute of Nuclear Science and Engineering grants. The authors are grateful for help provided by Wayne Erskine, Norm Webb, Natalie Marshall, Graham Lloyd and Chris Myers, with various aspects of this work.

REFERENCES

Brooks A. P., Brierley, G. J. & Millar, R. G. (2003) The long-term control of vegetation and woody debris on channel and floodplain evolution: insights from a paired catchment study in southeastern Australia. *Geomorphology* **51**, 7–29.

Blong, R. J. & Gillespie, R. (1978) Fluvially transported charcoal gives erroneous ^{14}C ages for recent deposits. *Nature* **271**, 739–741.

Dodson, J. R. (1994) Quaternary vegetation history. In: *Australian Vegetation* (ed. by R. H. Groves), 37–53. Cambridge University Press, London, UK.

Erskine, W. D. (1994) Sand slugs generated by catastrophic floods on the Goulburn River, New South Wales. In: *Variability in Stream Erosion and Sediment Transport* (ed. by L. J. Olive, R. J. Loughran & J. A. Kesby) (Proc. Symp., Canberra, December 1994), 143–151. IAHS Publ. 224. IAHS Press, Wallingford, UK.

Erskine, W. D. & Livingstone, E. A. (1999) In-channel benches: the role of floods in their formation and destruction on bedrock-confined rivers. In: *Varieties of Fluvial Forms* (ed. by A. Miller & A. Gupta), 445–475. John Wiley & Sons, Chichester, UK.

Erskine, W. D. & Peacock, C. T. (2002) Late Holocene floodplain development following a cataclysmic flood. In: *The Structure, Function and Management Implications of Fluvial Sedimentary Systems* (ed. by F. J. Dyer, M. C. Thoms & J. M. Olley) (Proc. Alice Springs Symp., September 2002), 177–184. IAHS Publ. 276. IAHS Press, Wallingford, UK.

Gippel, C. J. (1995) Environmental hydraulics of large woody debris in streams and rivers. *J. Environ. Engng* **121**, 388–395.

Kershaw, A. P., D'Costa, D. M., McEwen Mason, J. R. C. & Wagstaff, B. E. (1991) Palynological evidence for quaternary vegetation and environments of mainland southeastern Australia. *Quatern. Sci. Rev.* **10**, 391–404.

Nanson, G. C. (1986) Episodes of vertical accretion and catastrophic stripping: a model of disequilibrium flood-plain development. *Geol. Soc. Am. Bull.* **97**, 1467–1475.

Northcote, K. H. (1979) *A Factual Key for the Recognition of Australian Soils*. Rellim Technical Publications, Coffs Harbour, Australia.

Pickup, G. & Warner, R. F. (1976) Effects of hydrologic regime on magnitude and frequency of dominant discharge. *J. Hydrol.* **29**, 51–75.

Stuiver, M., Reimer, P. J., Bard, E., Beck, J. W., Burr, G. S., Hughen, K. A., Kromer, B., McCormac, J., Van der Plicht, J. & Spurk, M. (1998) Intcal 98 Radiocarbon age calibration, 24,000–0 cal BP. *Radiocarbon* **40**(3), 1041–1083.

Webb, A. A. (2002) Episodic erosion, riparian vegetation colonisation and the late Holocene stability of sand-bed, forest streams in southeastern Australia. PhD Thesis, University of Sydney, Australia.

Webb, A. A., Erskine, W. D. & Dragovich, D. (2002) Flood-driven formation and destruction of a forested floodplain and in-channel benches on a bedrock-confined stream: Wheeny Creek, southeast Australia. In: *The Structure, Function and Management Implications of Fluvial Sedimentary Systems* (ed. by F. J. Dyer, M. C. Thoms & J. M. Olley) (Proc. Alice Springs Symp., September 2002) 203–210. IAHS Publ. 276. IAHS Press, Wallingford, UK.

Webb, A. A. & Erskine, W. D. (2003) Distribution, recruitment, and geomorphic significance of large woody debris in an alluvial forest stream: Tonghi Creek, southeastern Australia. *Geomorphology* **51**, 109-126.

Temporal and spatial interactions of slope and catchment processes in the Central Spanish Pyrenees

JOSÉ M. GARCÍA-RUIZ, NOEMÍ LANA-RENAULT, SANTIAGO BEGUERÍA, BLAS VALERO-GARCÉS, TEODORO LASANTA, JOSÉ ARNÁEZ, JUAN I. LÓPEZ-MORENO, DAVID REGÜÉS & CARLOS MARTÍ-BONO

Instituto Pirenaico de Ecología, CSIC, Campus de Aula Dei, Apartado 202, ES-50080 Zaragoza, Spain
humberto@ipe.csic.es

Abstract Historical and present day changes in land use and plant cover explain the complex interactions assessed in the Central Spanish Pyrenees between geomorphic processes in hillslopes and channels. More intense erosion periods caused an enlargement of sediment source areas and an increase of sediment supply toward the rivers, characterized by braided channels. A sharp difference existed between the human-induced eroded areas and the rest of the territory, which shows moderate erosion rates. Farmland abandonment in the last few decades has reduced the sediment sources, thus decreasing both sediment delivery and streamflow, as a consequence of plant colonization of the old fields. Fluvial channels also show a decreasing torrential activity over the 20th century.

Key words Central Spanish Pyrenees; experimental basins and plots; geomorphic scales; hillslope–channel interactions; temporal scales

INTRODUCTION

Mountain areas, especially those located in Mediterranean environments, are characterized by a high interannual variability of rainfall and an extreme spatial heterogeneity of relief, plant cover and soil properties. There are several groups of factors to explain such a variability and diversity: (a) Mediterranean mountains are subject to the fluctuating activity of both the Azores Anticyclone and the Polar Front, causing severe contrasts in precipitation at different temporal scales (i.e. between dry and wet season or between dry and wet periods); (b) the relief is a result of the Alpine tectonics, resulting in a relatively young topography intensively dissected by the fluvial network. Deep valleys, steep gradients and a contrasting lithology explain the diversity of geomorphic landscapes, ultimately affected by moderate glacial activity at over 2000 m a.m.s.l.; (c) human disturbances have contributed to both soil erosion in most cases and soil conservation in others, and they have destroyed the original vegetation in many places and have created a mosaic landscape that, in addition, changes with time.

The Department of Soil Erosion and Land Use Changes (http://www.ipe.csic.es/erosion_hidrologia/home.htm) from the Pyrenean Institute of Ecology (Zaragoza, Spain) has studied the geomorphic evolution of the Central Spanish Pyrenees since 1987. Using combined field, experimental and laboratory approaches, it has accumulated a wealth of information on sediment yield and transport, temporal variability of sediment sources, historical evidence of soil erosion and extreme geomorphic events, and the relationships between land use/land cover changes and soil erosion. The purpose of this paper is to show the interactions between

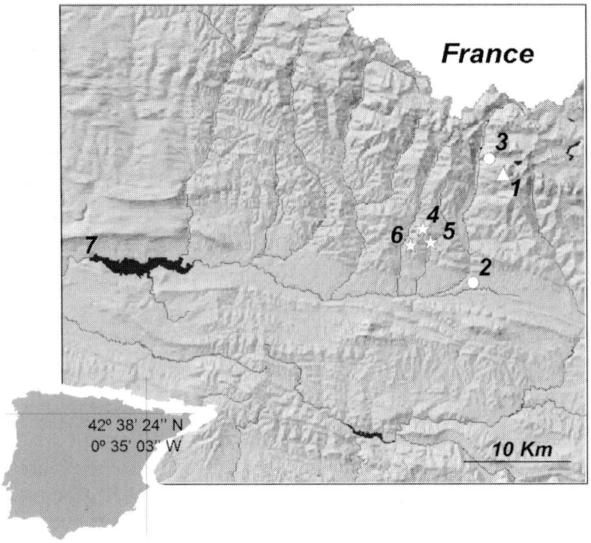

Fig. 1 The study area. (1) Collarada peak (2886 m); (2) Jaca; (3) Canfranc; (4) Aisa Valley Experimental Station; (5) Arnás catchment; (6) San Salvador catchment; (7) Yesa reservoir.

different temporal and spatial scales, in order to underline the processes that prevail at any scale considered.

THE STUDY AREA

The highest altitudes are located in the north of the study area (around 3000 m). Rivers run north–south across the Paleozoic area (mostly limestone and shale), the Inner Sierras (limestone and sandstone), the flysch sector and, finally, they cross the Inner Depression (marl, sandstone, claystone) and the Outer Sierras (limestone and sandstone) (Fig. 1).

Precipitation increases toward the north along the altitudinal gradient, and toward the west because of the Atlantic influence. The average annual precipitation in the northernmost sector of the Central Spanish Pyrenees reaches values above 1500 mm, and around 800 mm in the Inner Depression. The mean annual temperature decreases from north to south (8° in Canfranc and 11° in Jaca). The whole area is occasionally subject to very intense rainstorms, which can cause serious damage by flash floods.

Plant cover has been strongly impacted by human activities. As a result of the changes in land use during the 20th century, most cultivated fields have been abandoned, except in the Inner Depression and the valley bottoms (fluvial terraces), where cutting meadows and cereal crops prevail. Natural forests occur at any altitude in shady aspects and above 1400 m a.m.s.l in the sunny ones.

DATA AND METHODS

Studies on hillslope erosion have been undertaken at the Aisa Valley Experimental Station, where eight erosion plots under different traditional and modern land uses assess the effects of agricultural changes on runoff generation and soil erosion at a local scale (García-Ruiz *et al.*, 1995). The research includes the following land uses: cereal (barley), fallow land (alternating every 2 years with cereal), shifting agriculture (barley), abandoned cereal (now covered by dense herbaceous communities after 5 years of abandonment), abandoned shifting agriculture (with herbs now covering 60% of the soil after 4 years of abandonment), burnt plot (previously dense shrub cover), already re-established 2 years after a 1991 fire, meadow and dense shrub cover (representing the evolution of most of the hillslopes 30 years after farmland abandonment). The plots are 30 m^2 in size, located in a field with a 25% slope.

Basin-scale information on discharge and sediment transport has been collected from two different environments: The Arnás catchment (285 ha) was previously cultivated with cereals and abandoned 40 years ago. Now it is covered by shrubs with a variety of densities. The San Salvador catchment (136 ha) corresponds to a densely forest-covered, non-disturbed area. Both catchments are located at similar altitude, rock substratum and topography, and precipitation features are the same at any temporal scale.

In addition, remote sensing and Geographical Information Systems have been used to assess sediment sources at a larger scale (basins of more than 1000 km^2). Information is also available on reservoir siltation rates and sedimentological evolution of some of the Pyrenean reservoirs. The effects of changes in land cover and climate on water resources throughout the Central Spanish Pyrenees have also been analysed (Beguería *et al.*, 2003). Additional information on erosion during the last few millennia is available from proxy records as lacustrine sediment cores and analyses of slope screes and mass movements.

SOIL EROSION AND STREAMFLOW: FROM PLOTS TO LARGE BASINS

Runoff coefficients and soil erosion rates from plots are clearly higher in the cultivated plots, especially in the plot with shifting agriculture, followed by the fallow plot and the cereal plot. The lowest values are recorded from the plot with dense shrub cover, the meadow plot and the burnt (already re-colonized) plot (Fig. 2). In general, the erosion rates can be considered as moderate, ranging from 0.11 to 1.37 t ha^{-1} $year^{-1}$ in extreme cases. These results confirm the importance of old cultivated slopes as sediment sources in mountain areas, and the progressive decrease in soil erosion after farmland abandonment, especially when plant re-colonization reaches the stage of a mature shrub cover. The experimental data also demonstrate the importance of a dense shrub cover (probably similar to that of a forest) for soil conservation and the moderate response of overland flow during rainfall events. The actual landscape reflects very well the effects of shifting agriculture in steep slopes, frequent human-induced fires and grazing: areas with open scrubs on the sunny, convex slopes, affected by sheet wash erosion, soil stoniness, and relatively frequent occurrence of shallow landslides and debris flows, whereas concave areas are very stable.

At a small catchment scale, the hydrological and geomorphological response also depends on plant cover and land uses. Figure 3 shows, as an example, the hydrographs of the same rainfall and flood event for Arnás and San Salvador catchments. Arnás reacts almost immediately to every rainstorm event. This fast response is likely to be due to the presence of eroded areas close to the main channel, which are inherited from past farming activities and grazing. Nevertheless, the hydrological response tends to be, at an annual scale,

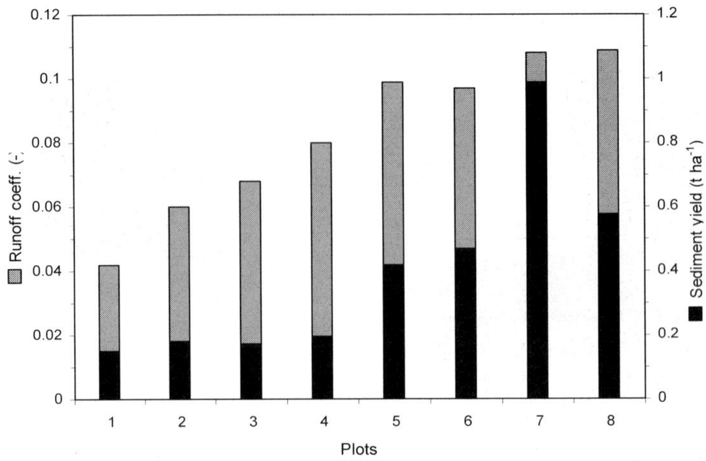

Fig. 2 Runoff and sediment yield from different land uses in the Aísa Valley Experimental Station. (1) shrub; (2) meadow; (3) burnt (old); (4) abandoned; (5) shifting agriculture; (6) cereal; (7) fallow; (8) burnt (recent).

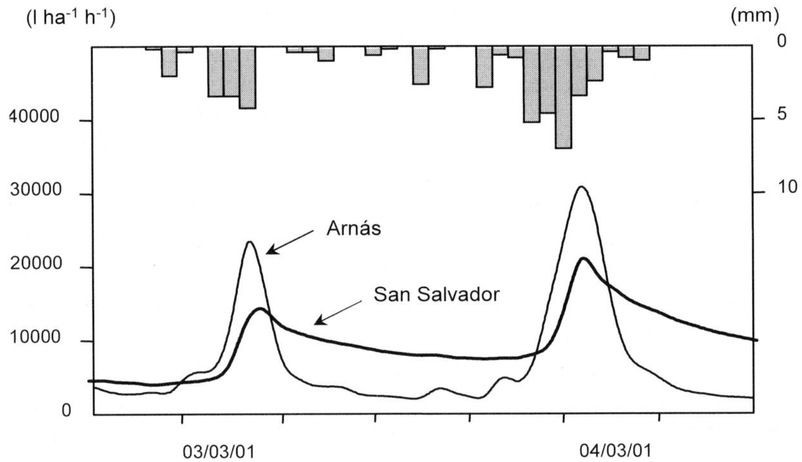

Fig. 3 Hydrographs from the same rainstorm event in the Arnás and San Salvador experimental catchments.

relatively complex: although the shape of the hydrograph suggests a prevailing Hortonian overland flow for the whole catchment, the correlation between rainfall and discharge is quite low (r^2 = 0.44), thus suggesting the importance of other factors, above all the antecedent soil humidity. Seeger *et al.* (2004) illustrate the occurrence of at least two situations during flood generation: (a) in the wet season the whole catchment is able to yield runoff, the hydrological response is fast and with a high peak flow; and (b) in the dry season, runoff is only supplied by the areas close to the channel, especially those affected by sheet wash erosion with open scrubs, and the peak flow is relatively low, even with intense and voluminous rainstorms. In San Salvador the hydrograph shows a very moderate response to rainfall events and a long recession curve after each event. In this case the role of

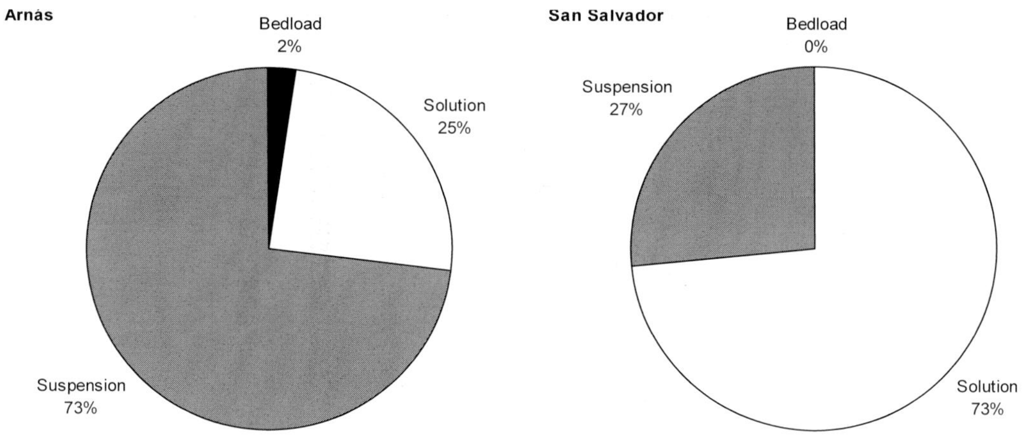

Fig. 4 Sediment balance in the Arnás and San Salvador experimental catchments.

interception and infiltration explains the moderate hydrological behaviour. Sediment transport also shows great differences between both catchments (Fig. 4). In Arnás 75% of the sediment outputs corresponds to suspended sediment, whereas 20% to solutes and 5% to bed load. In San Salvador 60% is carried as solutes and 40% as suspended sediment. Erosion rates are 2.04 and 1.87 t ha^{-1} $year^{-1}$, respectively, for the years 1999 and 2000.

At a large basin scale (regional scale: >1000 km^2) the use of remote sensing and models help to identify sediment sources. Beguería (2003) concludes that in the upper Aragón River basin, sediment sources are mainly related to: (a) the occurrence of marly outcrops, affected by gullying (extensive badlands) and located very close to the main channel, behaving as contributing areas during any rainstorm event; and (b) the past human activities, that caused large areas with sheet wash erosion on steep gradients close to the villages. These areas are not always well connected to the fluvial network, and they are contributing areas only during relatively infrequent rainstorm events.

At a large basin scale, sedimentation in the Yesa reservoir provides an erosion rate of 4.1 t ha^{-1} $year^{-1}$. Figure 5 shows how sediment delivery clearly increases as the area considered enlarges, due to the improvement between the interactions between slopes and channels in larger basins, and to the existence of very active sediment sources (bare marl outcrops) in the Aragón basin.

THE TEMPORAL APPROACH: EFFECTS OF LAND USE CHANGE

Investigation of soil erosion and streamflow through time reveals key periods of intense erosion mainly related to the removal of forest cover and farming of steep slopes.

The first evidence of human-induced plant cover disturbances is dated at about 4000 ^{14}C year BP. It is represented by the presence of charcoal and a small decrease in arboreal pollen percentages in some lake and slope records (Montserrat, 1992). The second dramatic erosion period occurred around the 12th century and represents a sudden and very significant lowering of the upper forest belt, in order to enlarge the summer pasture area. At present, the upper forest limit in many Pyrenean valleys reaches only 1600–1700 m, whereas the natural

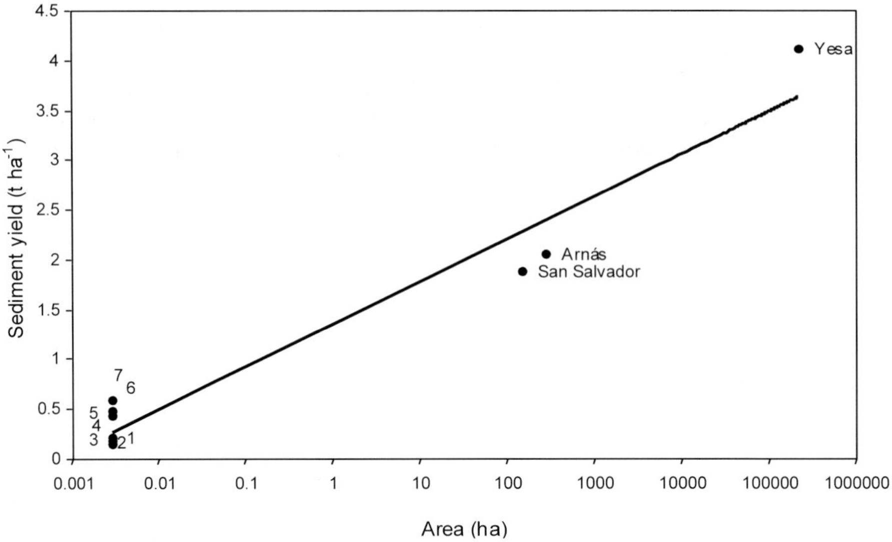

Fig. 5 Relations between area and sediment delivery in the Central Spanish Pyrenees. Plots data. (1) control area; (2) meadow; (3) burnt (old); (4) abandoned; (5) shifting agriculture; (6) cereal; (7) burnt (recent).

forest limit is at about 2200 m. The negative consequences of such a deforestation process were shallow translational landslides and dense rill networks, leaving large areas of the subalpine hillslopes with the bare substratum at the surface. Evidence of this erosion period has also been recorded in high mountain lake sediments (Montserrat, 1992), where increased charcoal percentages are associated with an intense period of soil erosion and a change in plant cover (substitution of trees by grasses).

Between the beginning of the 18th century and the first decades of the 20th century the population increased and all the farming possibilities in the Pyrenees were exploited. Thus, a new soil erosion period occurred, coinciding with the stage of maximum aggradation of the Ebro River delta in the Mediterranean. At the same time, alluvial plains were characterized by bare river bars and very unstable braided channels, suggesting a supply of large volumes of coarse sediment from the hillsides (García-Ruiz & Valero, 1998).

Since the middle of the 20th century farmland abandonment is the most important feature of Mediterranean mountains due to depopulation. This process has been very fast in such a manner that by 1970 all the fields in the hillslopes were already abandoned. As a consequence, the old fields and many overgrazed areas have been re-colonized with scrubs. In addition, government-induced re-afforestation affected many old cultivated areas. These changes in plant cover and land uses caused significant changes in runoff generation and soil erosion. Beguería *et al.* (2003) demonstrated decreasing streamflow in the central Spanish Pyrenees, regardless of climatic oscillations, between 1945 and 1995. The difference between the expected trends in discharge (based on climate) and the actual discharge implies that a non-climatic, time dependent factor has determined the decrease in discharge. Similarly, a decrease in the sedimentation rate has also been observed in large reservoirs in the last few decades (López-Moreno *et al.*, 2003). In relation to that, a clear reduction in sediment sources has been detected, as well as a partial stabilization of fluvial channels and alluvial fans (Gómez-Villar, 1996), now affected by entrenchment and plant colonization.

The study of the effects of different intensity rainfall events demonstrates that, in general, processes linked to rainfall events with return periods of less than 25 years can carry a lot of heterogeneous sediments, but the effects on the landscape are relatively minor. However, the less frequent events (>100-year return period) can introduce very important modification of the landscape, but since they affect a very localized area, they have limited regional importance (García-Ruiz *et al.*, 2002).

CONCLUSIONS

Temporal and spatial interactions in sediment sources, erosion and transport are very complex in the central Spanish Pyrenees, as in other Mediterranean mountains, due to the role of human activities in the triggering of erosion processes and in the evolution of the streamflow. Historical evidence demonstrates the existence of periods with higher sediment yield and transport, resulting in soil erosion, shallow landslides and, as a consequence, braided rivers. The Bronze and Iron Ages, the Lower Middle Age and the 17th–19th centuries seem to have been the most important erosion periods since the mid-Holocene. During those periods sharp differences existed between the human-induced eroded areas and the rest of the territory: the former, occupying around a half of the study area, were directly connected to the fluvial channels through natural and artificial drainage systems; the latter behave in a very moderate way, as the San Salvador catchment does at present, supplying clean waters, even during relatively infrequent rainstorm events.

Farmland abandonment during the second half of the 20th century has introduced very important hydrological and geomorphological changes. The bare, easily erodible marly and clayey outcrops are now the main sediment sources, whereas the old cultivated areas are colonized by scrubs and forests, and remain partly disconnected from the fluvial network.

Acknowledgements Funding for this research was provided by the following projects: HIDROESCALA (REN2000-1789-c04-01/GLO), PROHISEM (REN2001-2268-C02-01/HID) and PIRIHEROS (REN2003-08678/HID), all funded by CICYT, Spanish Ministry of Science and Technology.

REFERENCES

Beguería, S. (2003) Identificación y características de las fuentes de sedimento en áreas de montaña: erosión y transferencia de sedimento en la cuenca alta del río Aragón (Identification and characterization of sediment sources in mountain areas: erosion and sediment transfer in the Aragón river basin). PhD Thesis, University of Zaragoza, Zaragoza, Spain (in Spanish).

Beguería, S., López-Moreno, J. I., Lorente, A., Seeger, M. & García-Ruiz, J. M. (2003) Assessing the effect of climate oscillations and land use changes on streamflow in the Central Spanish Pyrenees. *Ambio* **32**, 283–286.

García-Ruiz, J. M. & Valero-Garcés, B. (1998) Historical geomorphic processes and human activities in the Central Spanish Pyrenees. *Mountain Res. & Develop.* **18**, 309–320.

García-Ruiz, J. M., Martí-Bono, C., Lorente, A. & Beguería, S. (2002) Geomorphological consequences of frequent and infrequent rainfall and hydrological events in Pyrenees mountains, Spain. *Mitigation and Adaptation Strategies for Global Change* **7**, 303–320.

García-Ruiz, J. M., Lasanta, T., Ortigosa, L., Ruiz-Flaño, P., Martí, C. & González, C. (1995) Sediment yield under different land uses in the Spanish Pyrenees. *Mountain Res. & Develop.* **15**, 229–240.

Gómez-Villar, A. (1996) *Conos aluviales en pequeñas cuencas torrenciales de montaña* (Alluvial fans in small mountain catchments). Geoforma Ediciones, Logroño, Spain (in Spanish).

López-Moreno, J. I., Beguería, S., Valero, B &, García-Ruiz, J. M. (2003) Intensidad de avenidas y aterramiento de embalses en el Pirineo Central español (Floods and reservoirs siltation intensity in the Central Spanish Pyrenees). *Ería*, **61**, 159–167 (in Spanish).

Montserrat, J. (1992) *Evolución glaciar y postglaciar del clima y la vegetación en la vertiente sur del Pirineo: estudio palinológioco* (Climate and vegetation glacial and postglacial evolution in the Pyrenean sunny aspects: Palinologic study). Instituto Pirenaico de Ecología, Zaragoza, Spain (in Spanish).

Seeger, M., Errea, M. P., Beguería, S., Arnáez, J., Martí, C. & García-Ruiz, J. M. (2004) Catchment soil moisture and rainfall characteristics as determinant factors for discharge/suspended sediment hysteretic loops in a small headwater catchment in the Spanish Pyrenees. *J. Hydrol.* (in press).

The role of pipe erosion and slopewash in sediment redistribution in small rainforest catchments, Sabah, Malaysia

A. M. SAYER[1], R. P. D. WALSH[1], M. A. CLARKE[2] & K. BIDIN[3]

1 *Department of Geography, University of Wales Swansea, Swansea SA2 8PP, UK*
ggsayer@swansea.ac.uk

2 *National Soil Resources Institute, Cranfield University, Silsoe, Bedfordshire MK45 4DT, UK*

3 *School of Science & Technology, Universiti Malaysia Sabah, Box no. 2073, 88999 Kota Kinabalu, Sabah, Malaysia*

Abstract The role of pipe erosion and slopewash in the redistribution of sediment in small rainforest catchments was investigated at sites in the Danum Valley Conservation Area in Sabah, Malaysian Borneo. Data loggers coupled to turbidity and flow depth sensors were installed in pipeflow and streamflow sites and the erosion bridge technique and overland flow traps were used to examine slopewash. The discharge and sediment responses from pipeflow and streamflow to nine storm events are presented. A single monitored pipe was found to contribute between 8 and 33% of stream stormflow and 3 to 61% of the stream sediment load in individual storm events. Overland flow, though comprising only a small proportion of rainfall, was found to be widespread and frequent, which may help to explain the comparatively high slopewash rates indicated by the erosion bridge results.

Key words data loggers; erosion bridge; pipeflow; rainforest; Sabah; sediment load; slopewash

INTRODUCTION

Quantitative piping research has concentrated on temperate areas (e.g. Gilman & Newson, 1980; Jones, 1981; Uchida *et al.*, 1999; Holden & Burt, 2002). In contrast, until recently piping in the humid tropics had received little attention, with the exception of a few qualitative (e.g. Baillie, 1975) or semi-quantitative (Walsh & Howells, 1988) studies. Most early studies suggested overland flow to be of little importance in tropical rainforest environments (Walsh, 1980). Recent work in Amazonia (e.g. Elsenbeer & Lack, 1996), southeast Asia (Douglas *et al.*, 1999; Sinun *et al.*, 1992) and northeast Australia (Elsenbeer *et al.*, 1994) has suggested the importance of both soil pipe systems and overland flow in forested humid tropical catchments and highlighted a need for systematic assessments of their roles in stream sediment and water budgets. This paper addresses this need by presenting results of an investigation into pipe erosion, slopewash and stream sediment transport in small rainforest catchments in the Danum Valley Conservation Area (DVCA) in Sabah, Malaysia.

STUDY AREA

The Danum Valley Conservation Area (DVCA) covers approximately 438 km^2 of primary lowland dipterocarp and montane rainforest in eastern Sabah (Fig. 1). Mean annual rainfall (1985–2003) at the Danum Valley Field Centre (<1 km from the study sites) was 2783 mm. Rainfall falls throughout the year, but with two wetter periods from October to January and

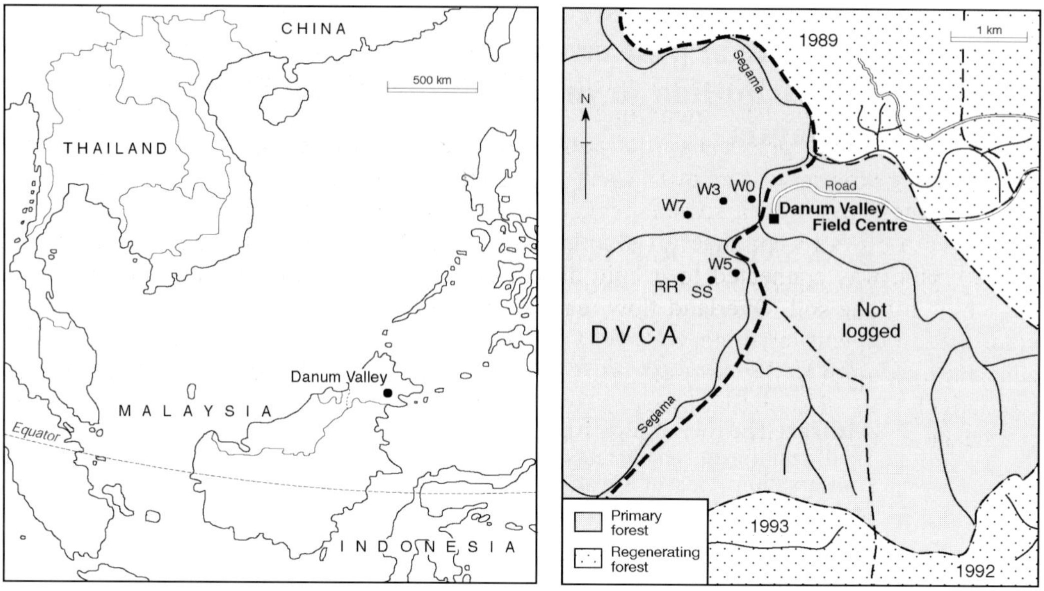

Fig. 1 Location map of study area and study sites.

May to June. The area is also susceptible to occasional ENSO-related droughts (Walsh & Newbery, 1999). The geology, which comprises Miocene mudstones, slumped breccias and tuffs, promotes local heterogeneity of soils, but Ultisols (equivalent to Alisols in the FAO-UNESCO system) are the most common soil group. These relatively deep soils (at least 1.5 m and often much deeper) are typically silt-loams, with a coarser topsoil (30–48% sand; 48–62% silt; 3–11% clay) and finer textured subsoil (10–19% sand; 66–73% silt; 15–22% clay) at depths of approximately 1 m.

METHODS

Pipeflow, slopewash and overland flow studies focussed at sites located in the extreme east of the DVCA close to the Danum Valley Field Centre (Fig. 1). Two small catchments known as W3 and W7 (13 203 and 5267 m^2 in size, respectively) were instrumented in November 2002 with weirs and CR10X Campbell Scientific data loggers programmed to continuously record the 15-min average output from 195-Analite turbidity probes and PDCR1830 Campbell Scientific pressure transducers. W3 has ephemeral surface streams in its upper part but is extensively piped in the lower part of the catchment whereas W7 is an entirely piped catchment. In W3 (the main focus of this paper) a streamflow site and a pipe (approx. 20 cm in diameter) that feeds into the stream were monitored.

Slopewash and erosional activity along slope profiles, around channel heads, and around collapsed pipe features were measured using the erosion bridge (or microprofiler) technique which involves the repeat measurement of the ground surface profile relative to a bar (bridge) which is mounted upon stakes at either end of a transect providing a stable datum. Two versions of the bridge were used: a 1.1 m wide bridge (Shakesby, 1993) at the pre-1997 sites and a 3 m wide bridge (Clarke *et al.*, 2002) at the sites established since 1997. In 1990, 29 erosion bridge transects were established between W3 and W0 on slopes around first-order channel heads; in 1997–1998 a further 40 slopewash sites were established down slope profiles at sites SS and RR; finally in 2001–2002 another 88 sites were installed at W0, W3, W5 and W7 around pipe collapse features and slope sites (see Fig. 1 for site locations). Erosion bridge transects were re-measured generally at 6-monthly or yearly intervals.

Overland flow was assessed using networks of simple overland flow traps. These comprised a 50 × 15 cm sheet of aluminium bent at the mid-point to form a v-shape. An outflow pipe was connected to a hole drilled at the bend and the "v" was inserted 3 cm vertically into the soil. Overland flow reaching the device was collected by a 500 ml vessel connected to the outflow pipe. Networks of traps were installed in 1999 along slope profiles adjacent to erosion bridge transects at RR, SS and between W3 and W0 (6, 4 and 4 traps, respectively). In 2002, networks were installed at W3 and W7 (6 and 5 traps, respectively). In each case overland flow was assessed for short monitoring periods.

RESULTS

Pipeflow hydrology, sediment dynamics and erosion

Tables 1 and 2 give summary details of pipeflow and streamflow discharge and sediment responses in W3 to nine rainfall events. Visual examination of the form of typical streamflow hydrographs (Fig. 2) defined the stormflow period to extend from the point when discharge first started to rise until the data-point three hours after peak discharge.

Overall 17.3% of streamflow during the nine events can be attributed to the single monitored pipe. The percentage of streamflow delivered by the monitored pipe tends to decline at higher rainfall intensities, but there are anomalies. Thus the smallest contribution of 7.7% arose when both maximum 15-minute and mean rainfall intensities were high (52 and 36.8 mm h^{-1}, respectively) during a small (18.4 mm) rainstorm, whereas the largest

Table 1 Discharge responses made by streamflow and pipeflow to eight rainfall events.

Date	Storm rainfall (mm)	Max 15-min rainfall intensity (mm h^{-1})	Mean rainfall intensity (mm h^{-1})	Antecedent rainfall (3 days)	Total stream Q (mm)	Total stream Q_s (mm)	Total pipe Q (mm)	Peak specific stream Q (mm h^{-1})	Peak specific pipe Q (mm h^{-1})	% stream Q supplied by monitored pipe
22 Dec. 2002	16.6	12.0	5.5	4.8	2.1	1.6	0.7	1.5	0.7	32.8
27 Dec. 2002	18.4	52.0	36.8	5.8	3.7	3.4	0.3	1.9	0.4	7.7
10 Jan. 2003	23.8	32.0	13.6	3.5	5.2	3.9	1.2	2.6	1.1	22.3
11 Jan. 2003	29.5	34.0	7.9	30.2	18.9	12.8	4.9	7.1	1.3	25.8
7 Feb. 2003	9.8	3.0	4.4	43.8	1.8	1.6	0.4	0.5	0.1	21.0
2 Mar. 2003	35.0	82.0	28.0	0.0	7.4	5.8	1.1	4.4	1.1	14.5
3 Mar. 2003	42.0	52.0	21.0	42.5	19.3	16.1	3.1	14.7	1.2	15.9
8 Jul. 2003	24.5	60.0	16.3	14.9	17.6	13.0	3.0	12.3	1.2	16.8
11 Jul. 2003	96.8	96.0	55.3	26.5	51.2	44.4	7.5	39.0	2.6	14.6
Total	296.4	†	†	†	127.2	102.6	21.9	†	†	†
Mean	†	47.0	21.0	†	†	†	†	9.3	1.1	17.3

Q, discharge; Q_s, stormflow discharge; †, not appropriate.

Table 2 Sediment responses made by streamflow and pipeflow to eight rainfall events.

Date	Total stream sediment load (kg)	Total pipe sedimen t load (kg)	% of stream sediment supplied by monitored pipe	Weighted mean streamflow SSC (mg l^{-1})	Weighted mean pipeflow SSC (mg l^{-1})	Peak streamfl ow SSC (mg l^{-1})	Peak pipeflow SSC (mg l^{-1})
22 Dec. 2002	1.3	0.8	61.0	103.9	193.6	284.0	495.4
27 Dec. 2002	2.1	0.4	16.5	93.7	200.0	287.4	387.8
10 Jan. 2003	5.1	1.5	28.9	159.2	207.6	599.2	443.2
11 Jan. 2003	12.5	4.0	31.8	108.8	134.2	245.8	477.9
7 Feb. 2003	0.5	0.1	16.6	48.5	38.4	65.8	55.4
2 Mar. 2003	13.7	1.7	12.4	305.8	261.7	2839.6	2655.6
3 Mar. 2003	27.5	3.6	13.0	234.3	192.3	609.3	1017.7
8 Jul. 2003	18.3	1.8	10.0	170.5	101.4	1226.0	557.4
11 Jul. 2003	440.8	11.5	2.6	1417.1	253.7	3320.1	858.0
Total	521.8	25.4	†	†	†	†	†
Mean	†	†	4.9	674.9	189.7	1053.0	772.0

SSC, suspended sediment concentration; †, not appropriate.

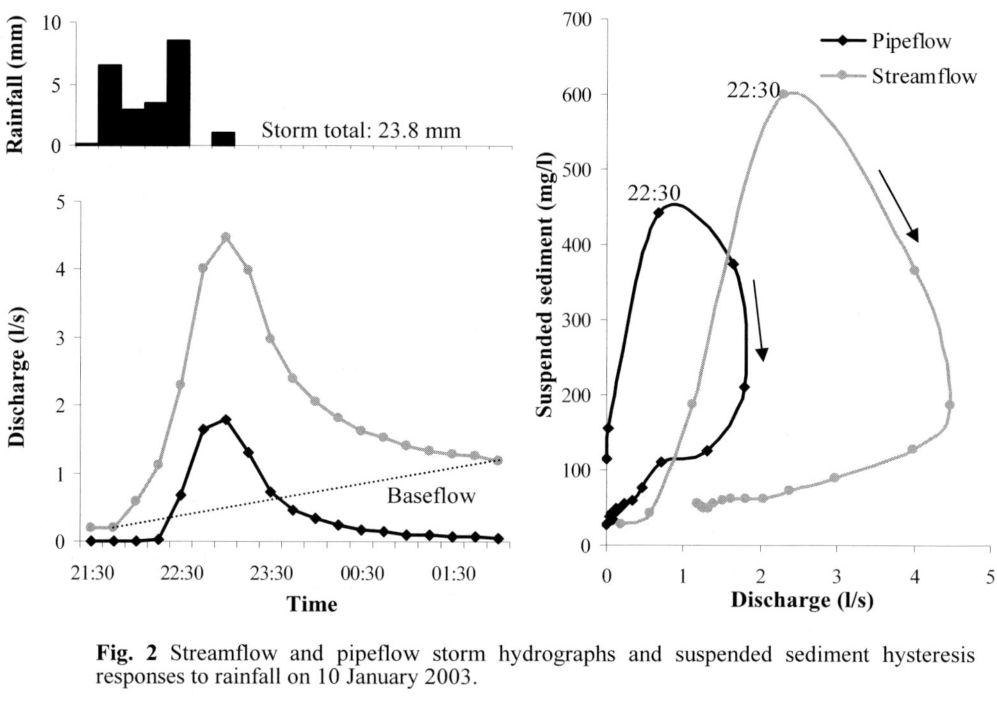

Fig. 2 Streamflow and pipeflow storm hydrographs and suspended sediment hysteresis responses to rainfall on 10 January 2003.

contribution of 32.8% arose from a similarly small storm (16.6 mm) but where maximum 15-min and mean rainfall intensities were much lower (12.0 and 5.5 mm h^{-1}, respectively). Total streamflow and pipeflow both increase with storm rainfall, but again anomalies indicate the influence of other variables. Figure 3 illustrates that responses are much lower after drier antecedent weather.

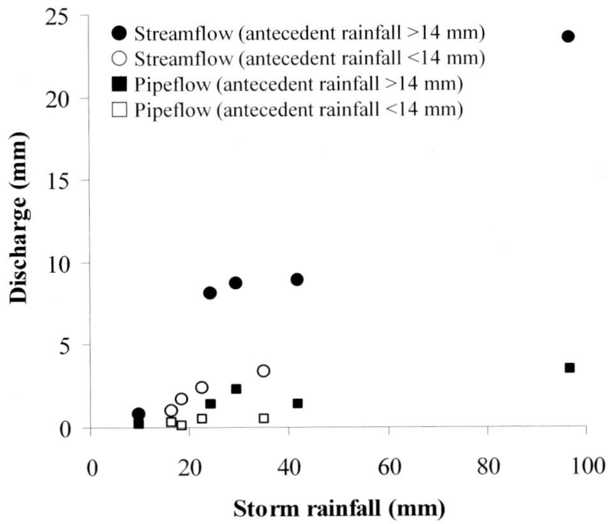

Fig. 3 Scatter plot showing the effect of antecedent rainfall (in the 3 days prior to the storm) on the relationship between storm rainfall and streamflow and pipeflow discharge.

Sediment transported by the monitored pipe accounted for 2.6–61% of the stream sediment budget in individual storm events (Table 2). Both streamflow and pipeflow sediment loads tend to increase with storm rainfall. The largest pipeflow sediment load of 11.5 kg (45% of the total pipe sediment discharge in the nine events) occurred in the 96 mm storm of 11 July 2003. Despite this response being three times larger than the next largest load, it accounted for only 2.6% of the stream sediment load of 440.8 kg, which in contrast, was 16 times greater than that of the next largest event.

The largest rainfall events do not necessarily result in the highest peak suspended sediment concentration (SSC). Although the 11 July 2003 storm resulted in the highest stream SSC (3320 mg l^{-1}), the peak pipeflow SSC response (858 mg l^{-1}) from this storm was exceeded in two smaller storms (35 and 42 mm; peak SSC 2656 and 1018 mg l^{-1}) when maximum 15-min rainfall intensity was exceptionally high (82 mm h^{-1} in the 35 mm storm) or there were very wet antecedent conditions.

Figure 2 illustrates the streamflow and pipeflow hydrographs and suspended sediment hysteresis responses in relation to the 23.8 mm storm of 10 January 2003. Response times of both the stream and pipe to rainfall are short (in this instance 90 min from the start of rainfall to peak discharge). Both the stream and the pipe exhibit clockwise hysteresis with SSC peaking before discharge. Discharge and sediment responses of the entirely piped W7 catchment, are very similar to those in W3.

Measurements of cross-sectional change between July 2002 and July 2003 at erosion bridge sites established in collapsed pipe sections and gullies downstream of pipe outlets at sites W0, W3, W5 and W7 all point to considerable erosional and depositional activity (Table 3). Many cross-sections show a combination of widening (erosion) and basal accretion (deposition) of a few centimetres over the period, but a few sites showed more dramatic change. Thus the headward erosion of one pipe window involved enlargement of a shallow surface depression into an incised 1.1 m deep and 30 cm wide gully over the monitoring period.

Overland flow and slopewash

Overland flow, although accounting for only a small percentage of rainfall, was widespread and frequent at all locations. During a monitoring period from late May to early July 1999, when 413 mm rain fell in 25 events, overland flow was most prolific at the high slope (mean 29°) RR site, where all six traps were found overflowing on three measurement occasions. At the intermediate slope (mean 23°) SS site, the results were more variable at the slope-top where one trap overflowed on two occasions and was 25–50% full on the other three but one trap did not record overland flow at all; the two mid-slope traps overflowed on every or all but one occasion. At the low slope (7°) W0 sites, traps on the mid-slope overflowed on each measurement occasion and traps at the slope-top sites varied from 25 to 100% full. During the monitoring period of 21–22 July 2002 when daily rainfalls were 19.8 and 17.5 mm, respectively, the relatively low-angle (5–18°) sites at W3 and W7 recorded overland flow in 10 of the 11 traps with catches of between 40 and 320 ml. Suspended sediment concentrations of overland flow recorded at W3 and W7 in July 2002 were very high (1953–8827 mg l^{-1}). Rates of erosion by slopewash at erosion bridge transects (Table 3) tend to increase sharply with slope angle. Thus mean surface lowering rates are much higher at the RR site (3.40 mm $year^{-1}$) than on the lower angle slopes at the two groups of W sites (0.36 and 0.55 mm $year^{-1}$). Much of the erosion at the long-term sites monitored since 1990 was recorded in the year (1995–1996) containing the largest rainstorm (Table 4).

Table 3 Slopewash erosion rates measured with the erosion bridge. For location of sites see Fig. 1.

Site	Period of measurement	n	Slope angle °		Erosion rate (mm $year^{-1}$)	
			Mean	Range	Mean	Range
W0–W3	June 1990–December 2002	12	18	6–31	–0.36	+0.6 to –1.9
RR	May 1999–January 2002	13	29	18–37	–3.40	+0.1 to –6.4
SS	May 1999–January 2002	11	23	17–28	–1.40	+1.3 to –5.5
W0/3/5/7	October 2002–June 2003	23	7	5–18	–0.55	+7.6 to –3.4

Table 4 Rainfall during the monitoring period.

Year	Annual rainfall (mm)	>50 mm falls	> 100 mm falls	Highest daily fall (mm)
1990	2729	10	1	135
1991	2609	10	1	114
1992	2366	8	0	77
1993	2501	8	0	92
1994	2978	7	3	123
1995	3294	11	0	99
1996	2989	6	1	163
1997	1918	3	0	93
1998	2139	6	2	124
1999	3382	15	0	90
2000	3501	13	3	140*
2001	3075	11	1	141
2002	2728	12	0	95

* 183 mm over a 20-h period straddling two rainfall days.

DISCUSSION AND CONCLUSIONS

Pipeflow is clearly an important contributor to the stormflow component of streamflow in the study area. The fact that the contribution of the monitored pipe to streamflow decreases at higher rainfall intensities may have two possible explanations. First, it may be that overland flow, which the traps showed to be frequent but volumetrically of limited importance in small–moderate events, becomes relatively more important at higher rainfall intensities. The exact mechanism (saturation or infiltration excess) by which overland flow is generated in the study area is unclear from previous studies (e.g. Sinun *et al.*, 1992; Bidin *et al.*, 1993). Alternatively, other pipes may be activated at higher intensities (which have been observed to occur) and the monitored pipe may have reached its maximum output. A large pipe outlet (approx. 50 cm) was witnessed to flow with hydrostatic pressure and with high turbidity during the storm of 8 July 2003 when 15-min max rainfall intensity was 60 mm h^{-1}.

Pipes clearly play a major role in supplying sediment to the stream, with the monitored pipe generating 10–61% of the stream's load in eight of the nine events. The tendency for the proportion of streamflow provided by the pipe to decline in more intense and larger rainstorms, with only 2.6% supplied in the largest event of 11 July 2003, is linked (as with storm runoff generation) to enhanced contributions from other pipes and perhaps also overland flow. The relative importance of the two remains, however, unclear as SSC of collected overland flow are of a similar order to peak pipe and stream SSC. The erosion bridge evidence both at slopewash and collapsed pipe sites records evidence of deposition as well as erosion and so not all the sediment mobilized by pipes and slopewash in a storm event reaches the gauging stations. It is therefore clear that pipes in the study area play a significant role in sediment redistribution and sediment load generation. Pipeflow and streamflow discharge and sediment responses to rainfall cannot, however, be explained solely by one variable. Total rainfall, rainfall intensity, antecedent weather conditions and associated changes in sources from pipes and overland flow have been highlighted here as important factors; future work will seek to elucidate the perhaps changing roles of these variables using a longer data set.

Acknowledgements The research was funded by a NERC studentship (NER/S/A/2001/06422), EU grant ERBIC18CT960102, a Dudley Stamp Memorial Fund grant from the Royal Society and by grants from the Royal Society's Southeast Asia Rainforest Research Programme. Grateful acknowledgement is also given to the BGRG for financially supporting conference attendance to present this paper. The authors would like to thank the Economic Planning Unit of the Prime Minister's Department of Malaysia and the Danum Valley Management Committee for research permission. Thanks also for the field assistance from Muhamad Jamil Hanapi and Johnny Larenus at the Danum Valley Field Centre and Dr Will Blake, University of Plymouth. This paper is number A/401 of the Royal Society's SE Asia Rain Forest Research Programme.

REFERENCES

Baillie, I. C. (1975) Piping as an erosion process in the uplands of Sarawak. *J. Tropical Geogr.* **41**, 9–15.

Bidin, K., Douglas, I. & Greer, T. (1993) Dynamic response of subsurface water levels in a zero-order tropical rainforest basin,

Sabah, Malaysia. In: *Hydrology of Warm Humid Regions*, (Proc. Yokohama Symp., July 1993), 491–496. IAHS Publ. 216. IAHS Press, Wallingford, UK.

Clarke, M. A., Walsh, R. P. D., Sinun, W., Larenus, J. & Hanapi, J. (2002) Soil erosion and surface roughness dynamics in relation to slope angle in primary and logged tropical rainforest, eastern Sabah. In: *Man and Soil in the Third Millennium* vol. II. (ed by J. L. Rubio, R. P. C. Morgan, S. Asins & V. Andreu), 1471–1484. Geoforma Ediciones, Logroño, Spain.

Douglas, I., Bidin, K., Balamurungan, G., Chappell, N. A., Walsh, R. P. D., Greer, T. & Sinun, W. (1999) The role of extreme events in the impacts of selective tropical forestry on erosion during harvesting and recovery phases at Danum Valley, Sabah. *Phil. Trans. Roy. Soc. Lond.* **B354**, 1749–1761.

Elsenbeer, H. & Lack, A. (1996) Hydrometric and hydrochemical evidence for fast flowpaths at La Cuenca, Western Amazonia. *J. Hydrol.* **180**, 237–250.

Elsenbeer, H., West, A. & Bonell M. (1994) Hydrologic pathways and stormflow hydrochemistry at South Creek, northeast Queensland. *J. Hydrol.* **162**, 1–21.

Gilman, K. & Newson, M. D. (1980) *Soil Pipes and Pipeflow: A Hydrological Study in Upland Wales.* BGRG Research Monograph 1, Norwich, UK.

Holden, J. & Burt, T. P. (2002) Piping and pipeflow in a deep peat catchment. *Catena* **48**, 163–199.

Jones, J. A. A. (1981) *The Nature of Soil Piping: A Review of Research.* BGRG Research Monograph 3, Norwich, UK.

Sinun, W., Wong, W. M., Douglas, I. & Spencer, T (1992) Throughfall, stemflow, overland flow and throughflow in the Ulu Segama rain forest, Sabah, Malaysia. *Phil. Trans. Roy. Soc. Lond.* **B335**, 389–395.

Shakesby, R. A. (1993) The soil erosion bridge: a device for micro-profiling soil surfaces. *Earth Surf. Processes Landf.* **18**, 823–827.

Uchida, T., Kosugi, K. & Mizuyama, T. (1999) Runoff characteristics of pipeflow and effects of pipeflow on rainfall-runoff phenomena in a mountainous watershed. *J. Hydrol.* **222**, 18–36.

Walsh, R. P. D. (1980) Runoff processes and models in the humid tropics. *Z. Geomorphol. Suppl. Bd* **36**, 176–202.

Walsh, R. P. D. & Howells, K. A. (1988) Soil pipes and their role in runoff generation and chemical denudation in a humid tropical catchment in Dominica. *Earth Surf. Processes Landf.* **13**, 9–17.

Walsh, R. P. D. & Newbery, D. M. (1999) The ecoclimatology of Danum, Sabah, in the context of the world's rainforest regions, with particular reference to dry periods and their impact. *Phil. Trans. Roy. Soc. London* **B354**, 1869–1883.

Change of fluvial sediment transport rates after a high magnitude debris flow event in a drainage basin in the Northern Limestone Alps, Germany

FLORIAN HAAS, TOBIAS HECKMANN, VOLKER WICHMANN & MICHAEL BECHT

Department of Physical Geography, Kath. University of Eichstätt-Ingolstadt, Osten 18, D-85072 Goettingen, Germany

florian.haas@geographie.uni-muenchen.de

Abstract This paper deals with the influence of debris flows on fluvial sediment transport in a drainage basin in the Northern Limestone Alps, Germany. A high magnitude rainstorm event in the year 2002 triggered several debris flows on slopes and in channels. Weekly quantification of the sediment transport in small channels in the study area Lahnenwiesgraben since the year 2000, show a drastic increase in sediment yield after the debris flow event due to extremely high sediment mobilization in the channel beds. The investigations show that erosion and deposition by debris flows also affect the long-term fluvial sediment budget. Debris flows change the state of the fluvial geomorphic system and induce prolonged high fluvial sediment transport. The composition of the transported sediment (organic content, granulometry) may also be modified. The interaction of geomorphic processes is an important issue in modelling sediment budgets and landform development.

Key words debris flow; fluvial erosion; Germany; Northern Limestone Alps; sediment budget; sediment cascades

INTRODUCTION

Debris flows are a common geomorphic process in mountainous geosystems. In spite of long recurrence intervals (low frequency–high magnitude events), they greatly contribute to the development of landforms (Zimmermann *et al.*, 1997) due to their high sediment yield.

A great deal of research has been undertaken to understand and model debris flows as a geomorphic process (e.g. Becht, 1995; Zimmermann *et al.*, 1997; Rieger, 1999; Hagg & Becht, 2000; Wichmann & Becht, 2003, 2004a,b), while their influence on fluvial processes, and thus the sediment budget of alpine drainage basins, has not yet been quantified. The significance of debris flows has been pointed out by various studies (e.g. Dietrich & Dunne, 1978; Caine & Swanson, 1989; Lehre 1981; Becht 1995), but a quantification of the changes to the sediment budget beyond short-term assessment has not been undertaken.

Within the scope of our research on fluvial and slope aquatic erosion processes in alpine geosystems, the amount of change caused to the fluvial system by debris flow events in the year 2002 has been investigated (mapping, quantification, modelling).

The investigation of the fluvial erosion processes forms part of the project *Sediment Cascades in Alpine Geosystems* (SEDAG), which aims at quantifying and modelling the sediment budget of mountain drainage basins. For this purpose, detailed studies of several geomorphic and hydrologic processes including soil creep, rockfall, landslides, debris flows, avalanches, aquatic and fluvial processes and the corresponding sediment storages are in progress (Heckmann *et al.*, 2002; Keller & Moser, 2002; Unbenannt, 2002; Schrott *et al.*, 2003; Wichmann & Becht, 2003).

STUDY AREA

The Lahnenwiesgraben drainage basin (16.7 km^2) is located in the Northern Limestone Alps ("Ammergebirge"; Bavaria) northwest of Garmisch-Partenkirchen (Fig. 1). It is drained by the Lahnenwiesgraben creek, which is tributary to the River Loisach. The "Hoher Kramer" in the southeast of the drainage basin is the highest peak with an elevation of 1985 m a.s.l. The maximum vertical distance in the drainage basin is 1279 m. Its geology consists of Triassic ("Hauptdolomit", "Plattenkalk" limestone, "Kössener" marls) and Jurassic series ("Aptychen" limestone and marls). The presence of glaciers during the last glaciations is indicated by several moraine deposits.

Most of the area is covered by vegetation consisting of forest (coniferous, mixed and *Pinus mug*; 75% of the drainage basin) and alpine meadows (18%).

METHOD

Fluvial and erosion processes on the slopes have been measured since the summer of 2000 using denudation gauges and sediment traps. Precipitation has been measured by two writing

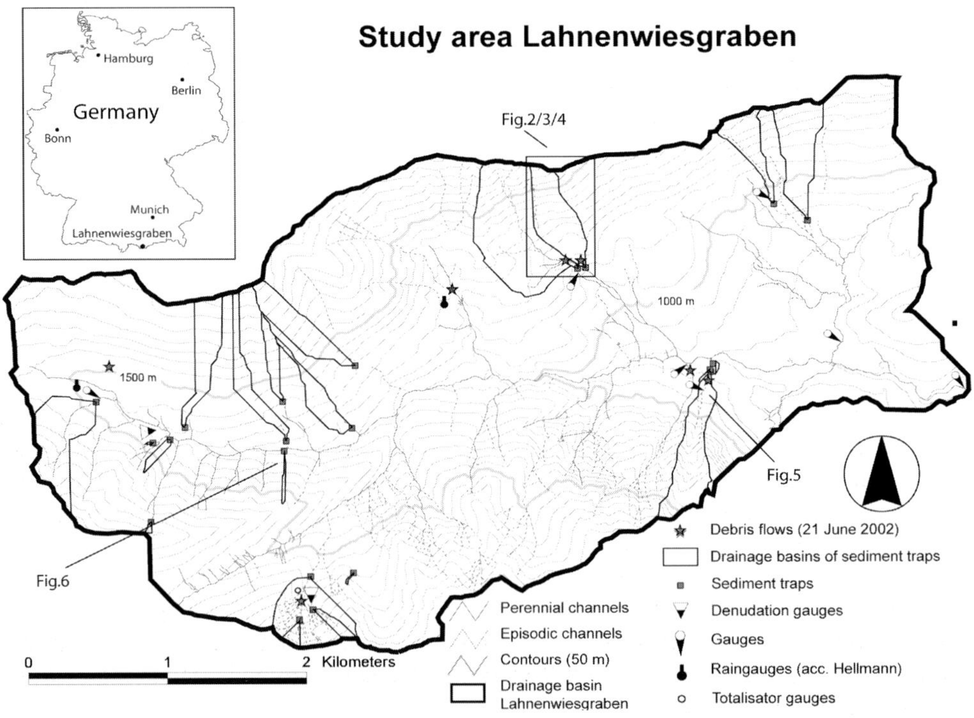

Fig. 1 Study area Lahnenwiesgraben.

raingauges (acc. Hellmann) and a totalisator gauge. In addition, runoff has been monitored by gauges in six creeks within the study area (Fig. 1).

Plastic troughs with a volume of 65 l installed in small channels are being used as sediment traps (Becht, 1995). Twenty-one sub-basins in the study area have been equipped with these sediment traps that are emptied weekly (Fig. 1). The sediment samples taken from the traps are weighed and analysed in the laboratory (granulometry and organic content).

The sub-basins have areas of between 1100 and 443 000 m^2 (determined by means of GIS analysis and mapping/terrestrial surveying). They have been chosen in order to represent a sufficient number of geofactor combinations (vegetation, geology, slope etc.) of the study area. On the basis of measured sediment transport rates and the geofactor combinations of the sub-basins, a regionalization of slope erosion for the whole study area is performed in order to establish a spatially distributed model.

The debris flow deposits of the year 2002 (see next section) that affected some sub-basins have been mapped and surveyed in order to determine process pathways and sediment yield.

HIGH MAGNITUDE RAINSTORM EVENT AND CONSEQUENCES

On 21 June 2002, two intense rainstorm events occurred in the Lahnenwiesgraben study area. The first event at 08:00 h with 46 mm/30 min precipitation triggered only slope type debris flows. The second event at 14:00 h with about 70 mm h^{-1} precipitation triggered several debris flows in channels all over the drainage basin (see Fig. 1).

The forest road and numerous installations (sediment traps, gauges) were destroyed by these channel-type debris flows. In one sub-basin ("Herrentischgraben"), a debris flow ran over a sediment trap and deposited a large quantity of sediment on the forest road below (Fig. 2). This event caused the mobilization and deposition of 300 t of sediment in the sub-basin, which corresponds to 1500 times the mass of the sediment transported by fluvial processes within the period July 2000–June 2002.

The sediment trap destroyed by the debris flow was reinstalled later. As fluvial transport was measured before the event, and measurement has been continued since then, the change of fluvial sediment transport rates caused by the debris flow can be quantified.

RESULTS

Figure 3 shows that fluvial sediment yield within the sub-basin "Herrentischgraben" has greatly increased after the debris flow, as soil and vegetation cover (shrubs and scrub) have been destroyed by the event. The debris flow incised the channel, leading to lateral erosion and local bank failure. This led to the mobilization of a great amount of sediment (Pleistocene deposits and weathered bedrock). In addition, the debris flow created deposits (debris flow cones, levées) now delivering material to fluvial erosion processes. These observations are corroborated by analyses of organic content and granulometry of the transported material. Figure 4 shows a remarkable coarsening after 21 June, which can be explained by the fact that the debris flow has exposed deeper and coarser layers of hillslope sediments to fluvial erosion.

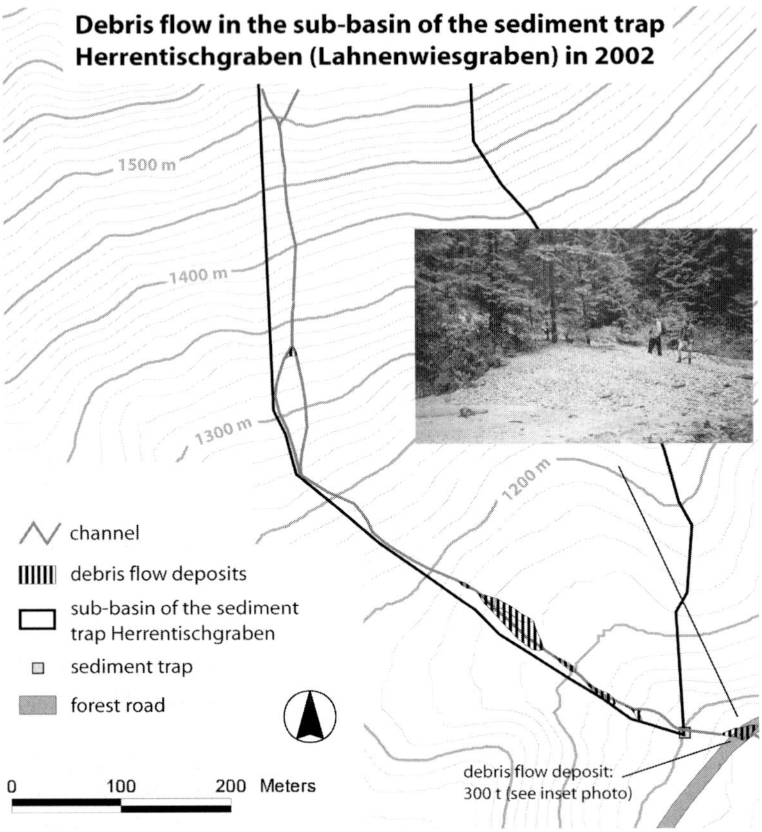

Fig. 2 Debris flow in the sub-basin of the sediment trap "Herrentischgraben" (Photo: V. Wichmann).

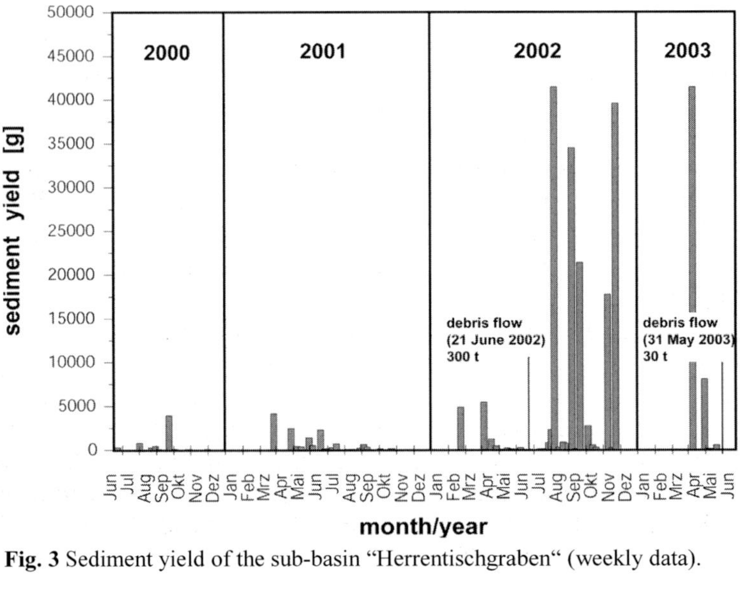

Fig. 3 Sediment yield of the sub-basin "Herrentischgraben" (weekly data).

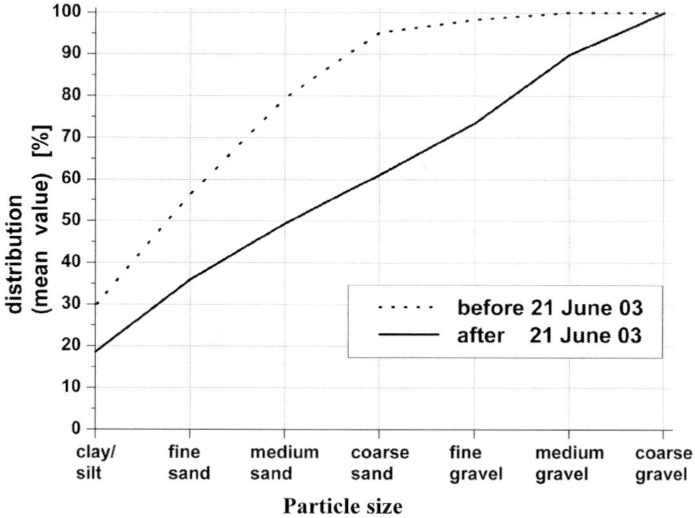

Fig. 4 Particle size distribution (mean value) of the transported material before and after the debris flow (Atterberg scale DIN4022; Leser, 1977).

The organic content of the sediment samples decreased from 15% before the event to about 8% after the event. This is due to the destruction and removal of vegetation cover and organic soil horizons from the surface.

The fact that important properties of the system have been changed by the debris flow not only led to increased fluvial sediment yield, but also influenced further debris flow activity. An event on 31 May 2003 in the same sub-basin was triggered by a precipitation event (45 mm/90 min) that would not have initiated (and did not initiate) a debris flow before the 2002 event. The sediment yield of the 2003 event (30 t), however, was an order of magnitude below the 2002 event.

Similar observations could be made in the "Königsstand" sub-basin, where an increase in fluvial sediment transport after a debris flow in 2002 could be measured too (Fig. 5). Contrary to the "Herrentischgraben" sub-basin, transport rates had started to decrease again soon (at the end of the year 2002), until another debris flow occurred on 31 May 2003.

The main difference between the two sub-basins is the channel size. The channel in the "Königsstand" sub-basin has a comparatively large cross section, while the cross section of the channel in the "Herrentischgraben" sub-basin is smaller and thus it is reacting more profoundly to debris flows (lateral erosion, bank failure, destruction of vegetation and/or soil cover). Increased sediment yield in the "Königsstand" sub-basin is mainly derived from sediments supplied to the channel by bank failure and levees. These sediment sources obviously are depleted after comparatively short time intervals. In the "Herrentischgraben" sub-basin, the profound change in the system has led to longer-reaching changes.

Sub-basins not affected by debris flows have been observed to react differently. For example, the sub-basin "Sulzgraben 2" showed increased fluvial transport (bed load) during the two rainstorms in 2002 and 2003, but no evidence of continued higher transport rates was found afterwards. This may corroborate our observation that long-term increased sediment yield in the sub-basins "Herrentischgraben" and "Königsstand" has indeed been caused by the debris flow events of 21 June 2002.

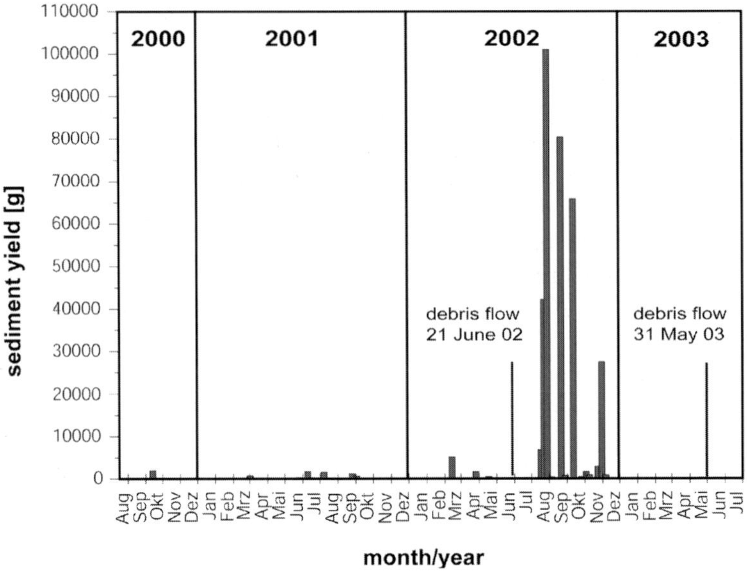

Fig. 5 Sediment yield of the sub-basin "Königsstand" (weekly data).

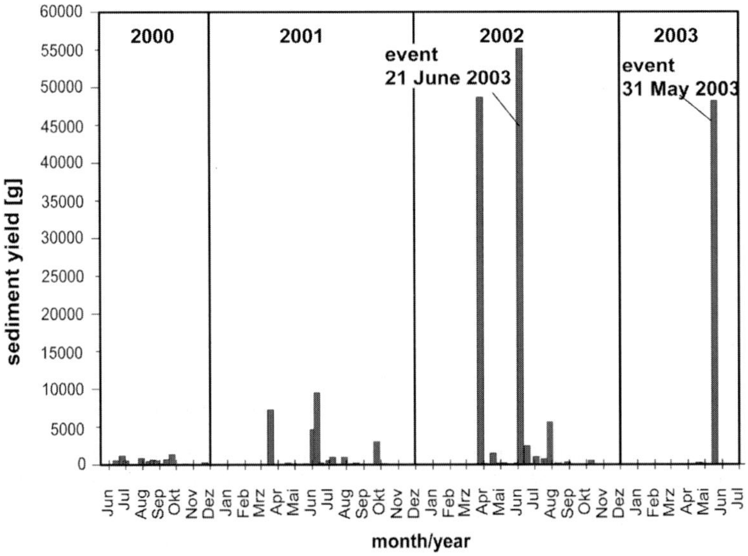

Fig. 6 Sediment yield of the sub-basin "Sulzgraben 2" (weekly data).

CONCLUSION

The results clearly indicate that debris flows are a very important geomorphic process contributing greatly to the sediment budget (Dietrich & Dunne, 1978; Lehre, 1981; Zimmermann *et al.*, 1997), especially to the fluvial sediment budget. According to our measurements, they can have a significant impact on drainage basins beyond the time of their occurrence, causing a change in the system that may lead to increased fluvial erosion and transport for years. Increased sediment availability also results in lower precipitation thresholds for the initiation of new debris flows.

At this stage, a period of increased geomorphic activity due to debris flows is observed in some sub-basins of the study area, the duration of which can not be readily assessed. In order to model sediment yield of fluvial processes and landform development, the influence of such low frequency–high magnitude events has to be taken into account. If debris flows are not taken into account, the long-term contribution of fluvial processes to the sediment budget may be greatly underestimated. Moreover, deriving transport rates from short-term observations is prone to errors and long-term observations are necessary for a better understanding of sediment budgets in alpine drainage basins.

The investigations described above are being continued within the SEDAG project in order to allow conclusions on future behaviour and the relaxation time of the system after debris flow impacts. The considerations concerning the interaction of geomorphic processes are incorporated in our approach to modelling landform development in the Lahnenwiesgraben study area.

Acknowledgement This study is funded by the German Science Foundation (DFG, Bonn), which is gratefully acknowledged by the authors.

REFERENCES

Becht, M. (1995) Untersuchungen zur aktuellen Reliefentwicklung in alpinen Einzugsgebieten (Investigations of the current landform development in alpine drainage basins). Münchener Geographische Abhandlungen, A47. Munich, Germany (in German).

Caine, N. & Swanson, F. J. (1989) Geomorphic coupling of hillslope and channel systems in two small mountain basins. *Z. Geomorph. N. F.* **33**(2), 189–203.

Dietrich, W. E. & Dunne, T. (1978) Sediment budget for a small catchment in mountainous terrain. *Z. Geomorph. N. F.* Suppl. **29**, 191–206.

Heckmann, T., Wichmann, V. & Becht, M. (2002) Quantifying sediment transport by avalanches in the Bavarian Alps—first results. *Z. Geomorph. N. F.* Suppl. **127**, 137–152.

Keller, D. & Moser, M. (2002) Assessments of field methods for rock fall and soil slip modelling. *Z. Geomorph. N. F.* Suppl. **127**, 127–135.

Lehre, A. K. (1981) Sediment budget of a small California Coast Range drainage basin near San Francisco. In: *Erosion and Sediment Transport in Pacific Rim Steeplands* (ed. by T. R. H. Davies & A. J. Pearce) (Proc. Christchurch Symp., January 1981), 123–139. IAHS Publ. 132. IAHS Press, Wallingford, UK.

Leser, H. (1977) *Feld- und Labormethoden in der Geomorphologie* (Field and laboratory methods in Geomorphology). Berlin, New York, USA (in German).

Rieger, D. (1999) Bewertung der naturräumlichen Rahmenbedingungen für die Entstehung von Hangmuren—Möglichkeiten zur Modellierung des Murpotentials. (Assessment of the natural conditions leading to slope type debris flows—possibilities for modelling the debris flow potential). Münchener Geographische Abhandlungen, A51. Munich, Germany (in German).

Schrott, L., Hufschmidt, G., Hankammer, M., Hoffmann, T. & Dikau, R. (2003) Spatial distribution of sediment storage types and quantification of valley fill deposits in an Alpine basin, Reintal, Bavarian Alps, Germany. *Geomorphol.* **55**, 45–63.

Unbenannt, M. (2002) Fluvial sediment transport dynamics in small alpine rivers—first results from two Upper Bavarian catchments. *Z. Geomorph. N. F.* Suppl. **127**, 197–212.

Wichmann, V. & Becht, M. (2003) Modelling of geomorphic processes in an alpine catchment. In: *GeoDynamics* (ed. by P. Atkinson, G. Foody, S. Darby & F. Wu) (Conf. Proc. GeoComputation 2003, Southampton).

Wichmann, V. & Becht, M. (2004a) Modellierung geomorphologischer Prozesse zur Abschätzung von Gefahrenpotentialen (Modelling of geomorphic processes for natural hazard zonation). *Zeitschrift für Geomorphologie N. F.* Suppl.-Bd (in German) (in press).

Wichmann, V. & Becht, M. (2004b) Spatial modelling of debris flows in an alpine drainage basin. In: *Sediment Transfer through the Fluvial System* (Proc. Moscow Symp., August 2004), 370–376. IAHS Publ. 288. IAHS Press, Wallingford, UK (*this issue*).

Zimmermann, M., Mani, P., Gamma, P., Gsteiger, P., Heiniger, O. & Hunziker, G. (1997) Murganggefahr und Klimaänderung—ein GIS-basierter Ansatz (Debris flow hazard and climate change—a GIS based approach). Schlussbericht NFP 31, Zurich, Switzerland (in German with English summary).

Identification of sediment sources in a small rural drainage basin

JEAN PAOLO GOMES MINELLA, GUSTAVO HENRIQUE MERTEN & ROBIN THOMAS CLARKE

Institute of Hydraulic Research (IPH), Federal University of Rio Grande do Sul (UFRGS), PO Box 15029, 91501-970 Porto Alegre, RS, Brazil
merten@iph.ufrgs.br

Abstract In this study, the fingerprinting method, which uses geochemical variables to distinguish sediment sources and classify suspended sediments through multivariate statistical analysis, was applied to a small rural drainage basin in southern Brazil. The drainage basin represents a typical agricultural ecosystem, in which peasants cultivate tobacco in fragile areas (headwaters), liable to severe erosion. By classifying suspended sediments, it was possible to identify the relative contribution from different sources and thus to suggest actions that would limit soil loss. Results showed that the predominant contribution was from fields and roads, contributing 64% and 36%, respectively, of the sediments sampled in the drainage basin control section, with the relative proportion of contribution from each source varying over time and according to management and soil cover in the fields and maintenance-work performed on the roads.

Key words Brazil; composite fingerprinting technique; sediment sources; small drainage basin monitoring

INTRODUCTION

In southern Brazil, peasants cultivate tobacco (*Nicotiana tabacum*) on steep hillsides with shallow soils. In addition to the poor agricultural potential of the land, typical management practices mean that the soil is intensively worked during the period when rainfall is most erosive. This combination of factors favours soil degradation and elevated sediment yield. To solve these problems and to reduce the associated rural poverty, publicly and internationally-financed conservation programmes are being implemented, with the goal of reducing land degradation by controlling soil erosion. The effects of implanting conservation practices on the natural resource base have been evaluated through rural drainage basin monitoring programmes, studying aspects related to hydrology, water quality and sediment yield.

Since sediments originate from a mixture of different sources (roads, fields, pastures and channel banks), it is useful, as part of the monitoring process, to be able to quantify not only the change over time in sediment yield, but also to identify the sources of the sediment, enabling more conclusive results to be drawn about the effect of conservation practices in reducing sediment yield. With this goal in mind, the fingerprinting technique was evaluated for its potential in identifying the origins of eroded sediments in a rural drainage basin.

BACKGROUND

Suspended sediments originate from different sources, with the relative contribution from each source varying over time and space as a consequence of the different erosion processes

Table 1 Chemical properties of sediment source areas.

Element	Pasture	Fields	Unpaved roads	New fields	Channel bank
Fe-total (%)	6.96	7.57	5.01	5.52	2.27
Fe-dithionite (%)	4.20	3.93	2.94	3.33	nd[1]
Fe-oxalate (%)	0.75	0.50	0.29	0.49	nd[1]
Al-total (%)	4.78	4.57	3.16	3.55	3.61
Al-dithionite (%)	0.40	0.34	0.29	0.23	nd[1]
Al-oxalate (%)	0.19	0.18	0.12	0.13	nd[1]
Mn-total	0.09	0.13	0.06	0.09	0.07
Mn-dithionite (%)	0.05	0.07	0.03	0.05	nd[1]
N-total (%)	0.19	0.17	0.08	0.17	0.11
COT (%)	1.94	1.84	0.67	2.21	1.43
Ca (%)	0.05	0.09	0.18	0.13	0.11
P-total (mg kg^{-1})	551.83	770.83	328.33	541.33	479.00
Na (mg kg^{-1})	67.00	76.00	128.33	52.67	867.32
Co (mg kg^{-1})	17.17	26.00	16.33	22.33	11.67
Cr (mg kg^{-1})	10.22	12.95	9.20	7.65	4.48
Zn (mg kg^{-1})	177.17	222.67	211.67	229.17	48.28

nd[1] not determined.

at work in the drainage basin. Much research currently focuses on developing methods that enable these sediment sources to be identified (Symader & Strunk, 1992; Coleman & Scatena, 1986; Horowitz, 1991; Salomons & Stigliani, 1995; Collins & Walling, 2002). One of the most promising of these methods, notable for its efficiency and ease of use, is the fingerprinting technique (Peart & Walling, 1986; Peters *et al.*, 1993; Peart, 1995; Collins *et al.*, 1997; Walling *et al.*, 1999; Walling & Collins, 2000; Collins & Walling, 2002). This method is based on the principle that sediments in suspension maintain some of the geochemical properties of their source material, and that these properties can thus be used as tracers (Table 1).

METHODS

The study area in southern Brazil (Fig. 1) consists of a small rural drainage basin in the headwaters of the Lajeado Ferreira stream, a tributary of the Guaporé River. The area of the drainage basin is 133 ha; however the sediment source study took place in a sub-basin of 57 ha, designated SB1, as identified in Fig. 1. In the larger drainage basin, topography is predominantly rolling (average gradient 7%) with poorly developed soils (average depth 0.50 m), primarily Inceptisoils and Entisoils. The drainage area of SB1 has an average slope of 8%, with a total concentration time of about 30 min.

The land in SB1 is used for tobacco cultivation (31%), pasture (29%), fallow (37%), unpaved roads and paths (3%). Tobacco is grown under traditional management practices, in which draft animals are used to plough the soil during the months of August and September, which correspond with the periods of highest rainfall erosivity. These conditions, in conjunction with the removal of riparian vegetation, favour the process of soil erosion and sediment yield from the fields. In 2002, approximately 57 t km^{-2} $year^{-1}$ of sediment was produced (Merten & Minella, 2003).

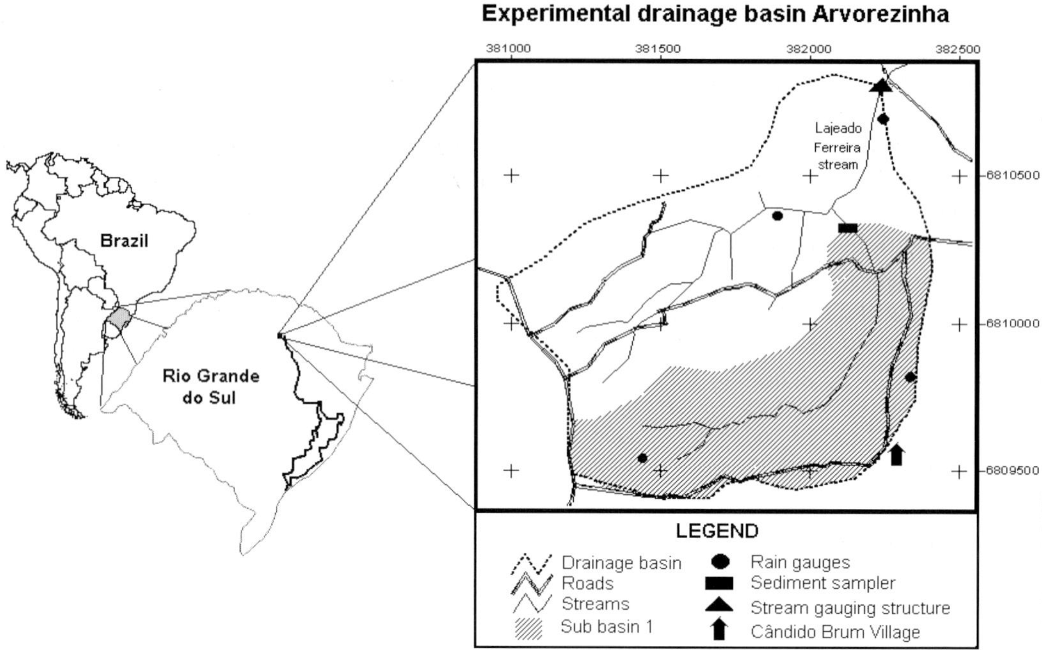

Fig. 1 Location of the study basin on the upper Lajeado Ferreira River.

Flow measurements and suspended sediment and water quality samples have been obtained for a site on the Lajeado Ferreira stream where a Parshall flume has been installed (Fig. 1). Suspended sediment samples were obtained at the location in SB1 shown in Fig. 1 using a sampler similar to the single-stage sampler for suspended sediment (FISP, 1961) in which runoff was stored in a 20-l plastic drum. Modifications made to this equipment enabled sediment samples to be collected during the rising and falling limbs of flood flows (Minella, 2003).

Potential sediment sources were selected after a survey of SB1 to identify the erosion processes operating in the study area. Sources included old and new fields (recently fallow ploughed), pasture, fallow areas, unpaved roads and channel banks. Soil samples from these sources were collected using a trowel, after cutting a small trench in the uppermost layer of the soil (0–0.05 m) which was visibly affected by soil erosion. Six samples were collected to represent each source, with each sample being composed of 10 sub-samples. The amount of material collected for each sample was about 0.5 kg. The samples were dried in the shade and sieved through a 2000 μm sieve before being sent for physical-chemical analysis. Suspended sediment samples were collected between the months of April and October 2002. Ten events were monitored during the course of 1 year. Samples were collected throughout each event (rise, peak and fall). To separate sediment from water, samples were decanted and then dried at 40°C. Samples from ten floods were collected during a period that included road works and changes in soil usage.

Granulometric analysis of the source material and suspended sediment samples was performed using different methods, including laser diffractometry, sieving and pippetting (Minella *et al.*, 2002).

Total levels of N, P, K, Ca, Na, Mg, Cu, Pb, Cr, Co, Zn, Ni, Fe, Mn and Al were measured by extraction after digestion with sulphuric acid (Tedesco, 1995); total organic C was measured by the Walkley-Black method (Tedesco, 1995); Al^{3+} by extraction with KCl (1 M) and titration with NaOH (Tedesco, 1995); Fe and Mn oxides by extraction with dithionite-citrate sodium bicarbonate (Inda Junior, 2002); and Fe oxide by extraction with ammonium acid oxalate (Inda Junior, 2002).

A correction factor $Z_{si} = C_o[100/(100 - F_{ars})]$ was used to correct the concentrations of the chemical elements determined as a function of the granulometric analysis, in which Z_{si} is the corrected concentration of the source (s) and of the chemical element analysed (i); C_o is the average original concentration of the element, and F_{ars} is the percentage of the fraction of the source samples greater than 63 μm (Horowitz, 1991). To assess the potential for discriminating the sources, a number of statistical tests were used, while source ascription for the suspended sediment samples was undertaken using an "optimized sediment mixing model", both of which are described by Walling & Collins (2000).

RESULTS

Assessment of source discrimination potential

The discriminant analysis evaluated five different possible sediment sources: fields (*L*), unpaved roads (*E*), new fields (*LR*), pasture (*P*) and material from the channel banks (*B*). Fallow areas, when analysed by the Kruskall-Wallis test, could not be distinguished from pasture and for this reason were combined with pasture sources. New fields and channel banks did not provide enough data for use in the identification analysis.

For the other sediment sources, the variables that presented discriminatory potential as measured by the Kruskall-Wallis test were: organic carbon (C_{org}), total phosphorus (P), total nitrogen (N), total manganese (Mn_{tot}), total iron (Fe_{tot}), iron dithionite (Fe_{dit}), iron oxalate (Fe_{oxa}), manganese dithionite (Mn_{dit}), total (Al_{tot}), aluminium dithionite (Al_{dit}), aluminium oxalate (Al_{oxa}), total sodium (Na) and total cobalt (Co).

The results of a multivariable analysis of variance, determined using Wilks' lambda (Λ^*), which can be tested using Snedecor's F and the χ^2 distribution as approximations, determined the group of variables that optimized source discrimination and reduced dimensionality. All the possible variables were tested in the variance analysis and those that presented the lowest Λ^* value and significance for the F and χ^2 tests were chosen for use in the classification model. Considering three sediment sources (field, unpaved road and pasture) the best composition was Fe_{tot}, Fe_{oxal}, Al_{oxal}, Mn_{tot}, Ca and P. Considering only two sediment sources, field and unpaved road, there were two optimal compositions: (a) Fe_{tot}, Fe_{oxal}, Fe_{dit}, Al_{oxal}, P and Co; and (b) Fe_{tot}, Fe_{oxal}, Al_{oxal}, Mn_{dit}, P and Co (Table 2).

Besides finding the set that minimized the Λ^* value, another objective of this evaluation is to use the set to explore the different processes associated with soil erosion and sediment transport. The variables should reproduce different mechanisms that occur in their respective sources, such as weathering, redox conditions and soil management. This goal was achieved, as the proposed set presents variables associated with soil management (P and Ca), weathering processes and redox conditions (iron, manganese and aluminium oxides) and the presence of trace metals (Fe_{tot}, Mn_{tot} and Co).

The variables total organic carbon (COT) and total nitrogen (N) were not used as trace variables, despite being selected by the discriminant analysis. The concentrations of these

Table 2 Best sets of tracer variables determined by MANOVA.

	Sources (g = 3): *L, P, E*	sources (g = 2): *L, E*	
	Variables (p = 6): Fe_{tot}, Fe_{oxal}, Al_{oxal}, Mn_{tot}, Ca, P	Variables (p = 6): Fe_{tot}, Fe_{oxal}, Al_{oxal}, Fe_{dit}, P, Co	Variables (p = 6) Fe_{tot}, Fe_{oxal}, Al_{oxal}, Mn_{dit}, P, Co
$\Lambda^* = \lvert W \rvert / \lvert B+W \rvert$	0.0057	0.0563	0.0428
test F			
Λ^* distribution	$[(\Sigma n - p - 2)/p][(1 - \sqrt{\Lambda^*})/\sqrt{\Lambda^*}] \sim F_{2p,\ 2(\Sigma n - p - 2)}$	$[(\Sigma n - p - 1)/p][(1 - \Lambda^*)/\Lambda^*] \sim F_{p,\ (\Sigma n - p - 1)}$	
Statistical test	20.34	13.97	18.63
$F_{v1,v2}$ (α)	$F_{12,\ 20}(0.05) = 2.54$	$F_{6,\ 5}(0.05) = 4.95$	
H_o	rejected	rejected	rejected
Chi-square χ^2			
Chi-square distribution	$-[n - 1 - (p + g)/2]\ln(\Lambda^*) > \chi^2_{p(g-1)}(\alpha)$	$-[n - 1 - (p + g)/2]\ln(\Lambda^*) > \chi^2_{p(g-1)}(\alpha)$	
Statistical test	64.51	20.14	22.06
$\chi^2_{p(g-1)}$ (α)	χ^2_{12} (0.05) = 19.7	χ^2_6 (0.05) = 12.6	
H_0	rejected	rejected	rejected

H_0, no-treatment effects;
B, treatment sum of squares and cross-products;
W, residual sum of squares and cross-products;
n, sample size.

elements in suspended sediment exceeded the capacity of the correction factors. The most probable hypothesis is that, along with the variability of biomass production over time, the load of soluble fertilizers and animal waste released into the drainage system increased the adsorption of N and COT by the sediment. Animal waste (from confined pork production) is released directly into the streams. Thus, N and COT are much higher in the sediment samples than in the soil (Merten & Minella, 2003).

Discriminant analysis determined the existence of variable sets capable of discriminating sediment sources. Classification analysis, performed with an optimization algorithm, was applied to these options, testing the hypothesis that there are a maximum number of possible sources that can be identified by the model, and the influence of the different variables on the model results is evaluated. The efficiency of the model and the applicability of the results were evaluated by relative average error and by comparing results with field observations.

Source ascription for suspended sediment samples

The results of classifying suspended sediment samples in terms of three possible sources (field, unpaved road and pasture) were inadequate. Relative average errors were greater than 54% for all events. The estimated proportions (regression coefficients) were: fields (0.8%), roads (21.3%) and pasture (77.9%). The use of three adjustable regression coefficients significantly increased the margin of error. According to Walling & Collins (2000), estimates with errors greater than 15% should not be used. Field observations confirmed the inconsistency of the results, as it was quite clear that the contribution from fields and unpaved roads was greater than estimated, and that the contribution from pastures was less. However, the model did not present a viable solution within the limits imposed by the errors and with results that conformed to field observations.

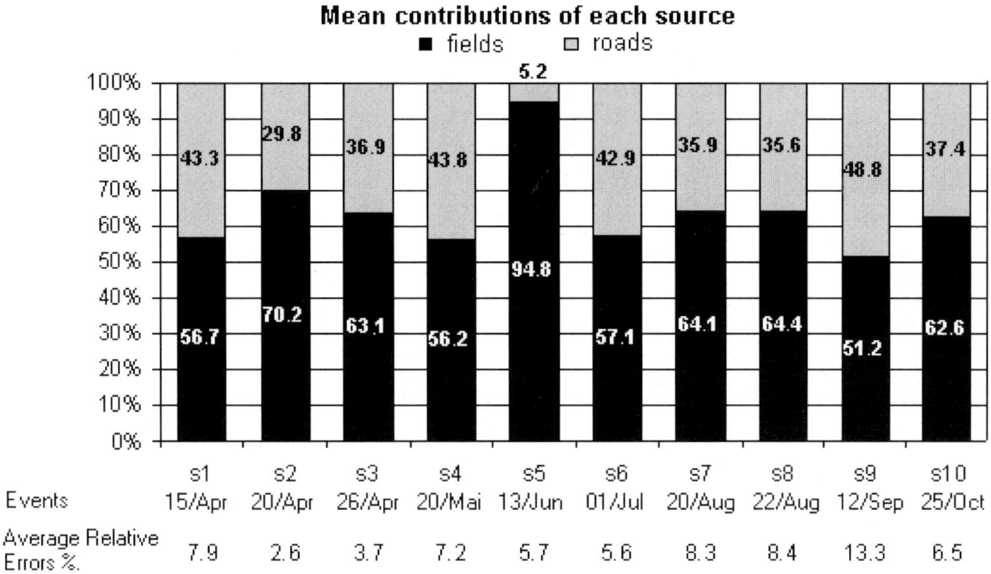

Fig. 2 The relative contributions per event from field and unpaved road sources, showing average relative errors per event.

Based on field evidence and observations of erosion and deposition processes, the hypothesis that unpaved roads and fields are the preponderant sources of sediment was tested. The optimization model found a solution within an acceptable margin of error, 6.92% on average. The relative contributions from each source are presented in Fig. 2. The average contributions were 64% and 36% from fields and from unpaved roads, respectively.

The average relative error is the mean of the sum of the square of the residuals of each variable. This measure is important as it enables one to evaluate individually the behaviour of each of the variables within the linear optimization model.

A wide variation in the relative contribution from each source was observed over time. The average and the coefficient of variation (CV) for the field source were 64% and 19%, respectively. For the unpaved roads, the average was 36% and the CV was 33%. This variability confirms the hypothesis that sediment yield fluctuates depending on factors controlling sediment production and transfer throughout the year.

Considering Fig. 2 and excluding event s5 from the analysis, we see a clear fluctuation in the relative contribution of sediment from the two sources. After April, the soil cover increased, gradually, reducing the contribution from the field source during events s2, s3 and s4. Event s1 also followed this tendency, but there was road-straightening work in early April which increased the contribution from the unpaved roads and, as a result, decreased the relative contribution from the fields.

Between the months of August and September, the fields were ploughed and the results showed a gradual increase in the relative contribution from the fields (s6, s7, s8). Nonetheless, with road-straightening in early September, there was once again an increase in the relative contribution from the unpaved roads.

The results show that the application of the fingerprinting technique for identification of sediment sources was reasonably sensitive to the factors that influenced the proportions of sediment contributed in the drainage basin studied. Future studies using this technique will emphasize the identification of conservative geochemical properties sensitive to changes in soil management, especially by different soil tillage systems. For the purpose of this study, however, the identification of sediment sources showed that conservation practices, as part of the natural resources management programme, should also include strategies to reduce sediment mobilization from unpaved roads.

CONCLUSIONS

The use of the fingerprinting method for identifying sediment sources proved adequate and practical for the purpose proposed in this study, in which fields and unpaved roads were shown to be the most significant sources of sediment. It was also shown that the relative contribution from each source varied over time, depending on plant cover in the drainage basin and unpaved road maintenance work performed during the study period.

Acknowledgements The authors would like to thank technician Adair Giongo for his invaluable field support, the farmers of the drainage basin Arroio Lajeado Ferreira, RS-RURAL, EMATER (Rural Extension Agency of Rio Grande do Sul), FEPAGRO (Foundation for Agricultural Research of Rio Grande do Sul), and FAPERGS (Foundation for Research of Rio Grande do Sul).

REFERENCES

Coleman, D. J. & Scatena, F. N. (1986) Identification and evaluation of sediment sources. In: *Drainage Basin Sediment Delivery* (ed. by R. F. Hadley) (Proc. Albuquerque Symp., August 1986), 3–18. IAHS Publ. 159. IAHS Press, Wallingford, UK.

Collins, A. L., Walling, D. E. & Leeks, G. J. L. (1997) Source type ascription for fluvial suspended sediment based on a quantitative composite fingerprinting technique. *Catena* **29**, 1–27.

Collins, A. L. & Walling, D. E. (2002) Selecting fingerprint properties for discriminating potential suspended sediment sources in river basins. *J. Hydrol.* **261**, 218–244.

Federal Interagency Sedimentation Project (1961) *The Single-stage Sampler for Suspended Sediment*. US Geol. Survey. Report 13, USA.

Horowitz, A. J. (1991) *A Primer on Sediment Trace Element Chemistry*, 2nd edn. Lewis Publishers, Chelsea, UK.

Inda Junior, A. V. (2002) *Caracterização da Goethita e Hematita em Solos Poligênicos* (Characterization of Goethite and Hematite in Poligenetic Soils) Tese de Doutorado (Thesis), Programa de Pós-Graduação em Ciência do Solo. Faculdade de Agronomia. Universidade Federal do Rio Grande do Sul, Brazil (in Portuguese).

Merten, G. H. & Minella, J. P. G. (2003) Relatório interno projeto de monitoramento ambiental de microbacias hidrográficas–RS-RURAL-sub-projeto 7 monitoramento hidrossedimentológico e da qualidade das águas. (Internal report of the environmental monitoring of drainage basins for RS-RURAL sub-project 7: sediment and water quality). Instituto de Pesquisas Hidráulicas. Universidade Federal do Rio Grande do Sul, Brazil (in Portuguese).

Minella, J. P. G., Merten, G. H. & Semelmann, F. R. (2002) Avaliação de dois métodos de dispersão de agregados e de análise granulométrica para o estudo de solos erodidos (Evaluation of two methods of dispersion of aggregates and granulometric analysis for the study of eroded soils) In: *Encontro Nacional de Engenharia de Sedimentos* **5**, 111–119. Anais, São Paulo: ABRH, Brazil (in Portuguese).

Minella, J. P. G. (2003) Identificação de fontes de produção de sedimentos em uma pequena bacia rural (Identification of sediment sources in a small rural drainage basin). Dissertação de Mestrado – Programa de Pós-Graduação em Recursos Hídricos e Saneamento Ambiental. Instituto de Pesquisas Hidráulicas. Universidade Federal do Rio Grande do Sul, Brazil (www.iph.ufrgs.br/dissertacoes) (in Portuguese).

Peart, M. R. (1995) Fingerprint sediment sources: an example from Hong Kong. In: *Sediment and Water Quality in River Catchments* (ed. by I. D. L. Foster, A. M. Gurnell & B. W. Webb), 179–186. John Wiley & Sons, Chichester, UK.

Peart, M. R. & Walling, D. E. (1986) Fingerprinting sediment sources: the example of a drainage basin in Devon, UK. In: *Drainage Basin Sediment Delivery* (ed. by R. F. Hadley) (Proc. Albuquerque Symp., August 1986), 41–55. IAHS Publ. 159. IAHS Press, Wallingford, UK.

Peters, N. E., Hoehn, E. H., Tase, N., Leibungut C. & Walling, D. E. (eds) (1993). *Tracers in Hydrology* (Proc. International Association of Meteorology and Atmospheric Physics and IAHS General Assembly, Yokohama, July 1993). IAHS. Publ. 215. IAHS Press, Wallingford, UK.

Salomons, W. & Stigliani, W. M. (1995) *Biogeodynamics of Pollutants in Soils and Sediments.* Springer, Berlin, Germany.

Symader, W. & Strunk, N. (1992) Determining the source of suspended particulate material. In: *Erosion, Debris Flow and Environment in Mountain Regions* (ed. by D. E. Walling, T. R. Davies & B. Hasholt) (Proc. Chengdu Symp., July 1992), 177–185. IAHS Publ. 209. IAHS Press, Wallingford, UK.

Tedesco, M. J. (1995) *Análises de Solo, Plantas e Outros Materiais* (Analysis of Soil, Plants and Others Materials). Porto Alegre: Departamento de solos da UFRGS, Brazil (in Portuguese).

Walling, D. E. & Collins, A. L. (2000) *Integrated Assessment of Catchment Sediment Budgets: A Technical Manual.* Department for International Development, Exeter, UK.

Walling, D. E., Owens, P. N. & Leeks G. J. L. (1999) Fingerprint suspended sediment sources in the catchment of the River Ouse, Yorkshire, UK. *Hydrol. Processes* **13**, 955–975.

Sediment Transfer through the Fluvial System (Proceedings of a symposium held in Moscow, August 2004).
IAHS Publ. 288, 2004

Sediment redistribution following wildfire in the Sydney region, Australia: a mineral magnetic tracing approach

WILLIAM H. BLAKE[1], PETER J. WALLBRINK[2], STEFAN H. DOERR[3], RICHARD A. SHAKESBY[3] & GEOFFREY S. HUMPHREYS[4]

1 *School of Geography, University of Plymouth, Drake Circus, Plymouth PL4 8AA, UK*
william.blake@plymouth.ac.uk

2 *CSIRO Land & Water, PO Box 1666, ACT 2601, Australia*

3 *Department of Geography, University of Wales Swansea, Singleton Park, Swansea SA2 8PP, UK*

4 *Department of Physical Geography, Macquarie University, North Ryde, Sydney, NSW 2109, Australia*

Abstract Increased sediment and nutrient fluxes arising after wildfires may affect downstream water quality. We use mineral magnetic tracers to elucidate linkage between different slope units, river channel and a reservoir sediment column in the gorge-dissected landscape of a burnt water supply basin. Comparison of magnetic properties of source areas with downstream (sub-aerially stored) channel deposits suggests predominantly ridge-top origin with significant storage of sediment within footslopes. Magnetic properties of the sediment column provide insight into the nature of sediment accumulation on the reservoir floor and the role of immediate post-fire rainfall events. Comparison of source signatures with sub-aqueously stored sediment is complicated by the apparent fragility of some fine pyrogenic mineral grains; these may have to be accounted for in sediment column interpretation. Tracing tools provide river basin managers with important process-based evidence of post-fire sediment redistribution, useful for more effective mitigation of these infrequent, but significant, sediment redistribution events.

Key words Australia; downstream impacts; erosion; mineral magnetics; rainstorm; reservoir; sediment; tracer; wildfire

INTRODUCTION

The dynamics of sediment redistribution following wildfire events, particularly with respect to the timing and intensity of post-fire rainfall, are not well understood. These require attention since post-fire sediment and nutrient fluxes affect downstream water quality. Sediment tracing approaches provide information on river basin sediment dynamics unavailable through direct monitoring approaches (e.g. Walling *et al.*, 1979; Blake *et al.*, 2003). In the context of fire-related sediment redistribution, mineral magnetic tracers have received attention since large quantities of pyrogenic iron minerals, with magnetic properties that can be used to distinguish burnt from unburnt sediment sources, are produced at high temperatures, in the presence of organic matter (Rummery *et al.*, 1979; Gedye *et al.*, 2000; Blake *et al.*, 2004).

The aim of this paper is to explore the potential for using mineral magnetic tracers to elucidate slope–channel–reservoir linkages in water supply basins, near Sydney Australia, which were affected by recent wildfire events (Christmas 2001–2002). Clarification of the link between slope and sediment column provides an opportunity, as a secondary aim, to interpret recent reservoir deposits in the context of the erosional response of the landscape to fire events of variable severity. This is important in the context of both changing hydrological (water repellency) properties of soil with variable fire severity (Doerr *et al.*, 2000) and the increased frequency and severity of ENSO-driven drought. The immediate basin area of the reservoir, Lake Burragorang, is dominated by dry sclerophyllous eucalypt forests which have a history of recurring wildfire. Consequently, we also aim to establish the influence that any sediment eroded after such fires would have on Burragorang water quality over the longer term. We assess this by determining the contribution of fire-eroded soils to the sediment archive deposited on the reservoir floor.

STUDY AREA

Research was undertaken in the Nattai River drainage basin, which flows into the Warragamba River, which in turn forms part of Lake Burragorang (Fig. 1). Lake Burragorang was formed in 1960 by the impoundment of Warragamba Dam and is Sydney's principal water supply reservoir. The basin is underlain by Hawksbury Sandstone and is characterized as a plateau-gorge type landscape. Surface soil materials are typically sandy to sandy loam. The basin is largely forested with a variety of native eucalypt species (Shakesby *et al.*, 2003).

This investigation was prompted by extreme wildfire events of Christmas 2001–2002. Fire events were followed by moderate intensity rainstorm events from January to May 2002 which triggered significant sediment redistribution within the basin (Shakesby *et al.*, 2003).

METHODS

Sampling locations are shown in Fig. 1 and reflect the need to (a) establish the link between slope–channel–reservoir, and (b) interpret the magnetic properties of the sedimentary archive in terms of any temporal trends in fire-driven sediment and nutrient flux to the reservoir.

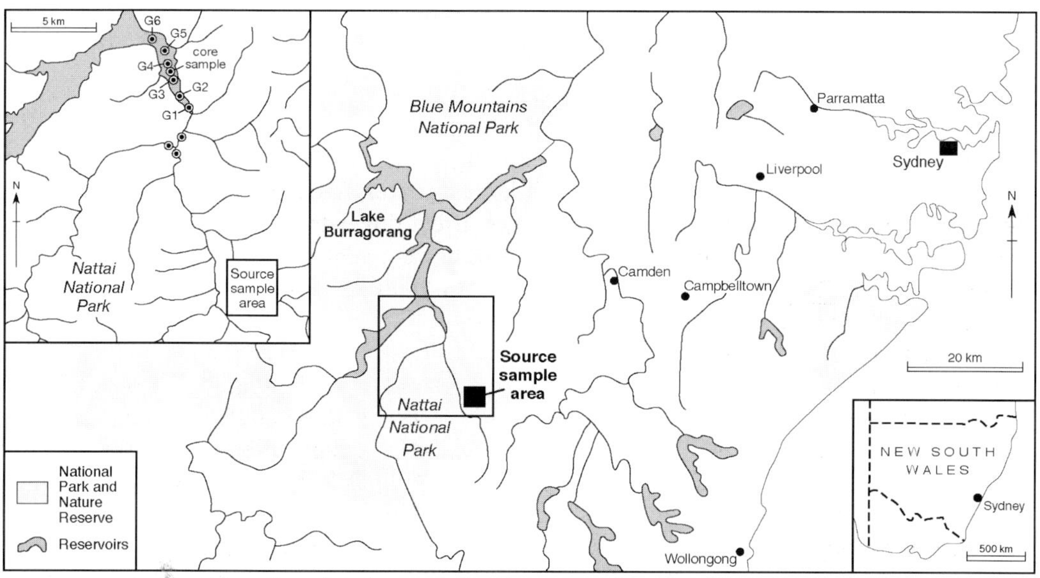

Fig. 1 Location of study area and sediment sampling sites (inset).

Samples of sediment source material were collected within the gorge-dissected landscape, from ridge-top and footslope units representative of areas experiencing contrasting fire severity across the catchment, as described in Shakesby *et al.* (2003) and Blake *et al.* (2004). Each sample population was combined. Samples of recent sediment deposits were collected from the downstream channel network (Fig. 1). A series of grab samples were collected from the mouth of the Nattai River through the Nattai arm of Lake Burragorang (Fig. 1). The <10 μm fraction was extracted from each sample.

A sediment core was taken from the reservoir floor of the Nattai arm (Fig. 2) (the former River Nattai floodplain) using a percussion coring device. The core was divided into stratigraphic units (A–G, Fig. 2) and the <10 μm fraction extracted from each. Units A, C and E displayed characteristics of immediate post-fire-derived material similar to those observed in the recent channel deposits, i.e. black, massive silty clays. Units B and D comprised slightly coarser material with laminations of dark sediment. Units F and G were similar to B and D but with less distinct laminae of finer sediment.

A full suite of mineral magnetic analyses was undertaken following standard procedure. In brief, low-frequency mass-specific magnetic susceptibility (χ_{lf}) is an indicator of the total magnetic concentration of the sample. Frequency-dependent susceptibility ($\chi_{fd\%}$) indicates the proportion of superparamagnetic (SP) (<0.05 μm) minerals present in the assemblage with values ranging from 0 to 14%. Susceptibility of anhysteretic remanent magnetization (χARM) is an indicator of the concentration of magnetic minerals of the 0.02–0.4 μm fraction. Saturation isothermal remanence magnetization (SIRM) indicates the ability of the mineral assemblage to hold a magnetic remanence which in turn can be linked to magnetite concentrations, mineralogy and magnetic grain size (Walden *et al.*, 1999).

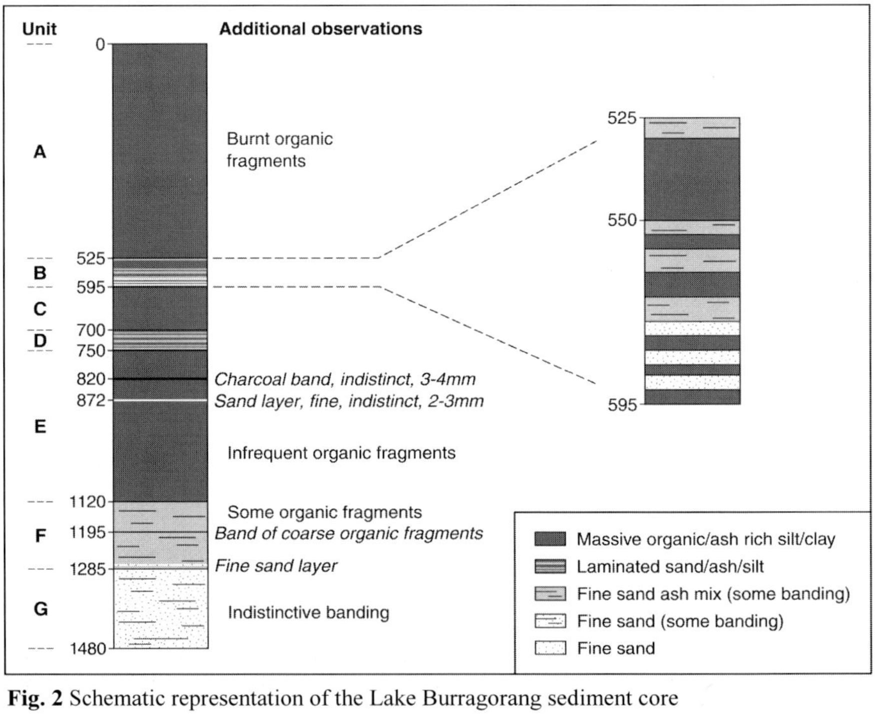

Fig. 2 Schematic representation of the Lake Burragorang sediment core

RESULTS AND DISCUSSION

Linking slope to channel

The main mineral magnetic properties of each sample set (i.e. source, channel, grab sediment and sediment core) are shown in Table 1 (a)–(d).

Comparison of slope material (Table 1(a)) and channel sediment (Table 1(b)) from the Little River channel (using multivariate analysis of the mineral magnetic properties discussed by Blake *et al.* (2004), summarized in Fig. 3) suggests (qualitatively) that the burnt ridge-top area (irrespective of fire severity) dominates as the source of material deposited in the channel of the near-burnt-slope environment (i.e. Little River). This implies that sediment is delivered to the channels via concentrated overland flows directed through existing ephemeral gully features. Conversely, the lower slope locations tend to act as areas of sediment storage rather than source, forming an unusual soil profile where subsurface material carries a fire-enhanced signature as a relic of material burnt and eroded from upslope locations following previous fires. The effect of secondary burn events on this pre-enhanced, stored material appears to be one of further enhancement (compare magnetic concentration severely burnt ridge-top to severely burnt footslope in Table 1(a)) where magnetic grain-size indicators suggest a coarsening of the mineral population (compare χARM and χARM/SIRM in Table 1(a)). However, it is unknown whether the "coarsening" of slope-stored magnetic mineral assemblages reflects "storage diagenesis" (loss of finer grains through dissolution processes; cf. Oldfield *et al.*, 1992) or a mineral formation process linked to multiple burning.

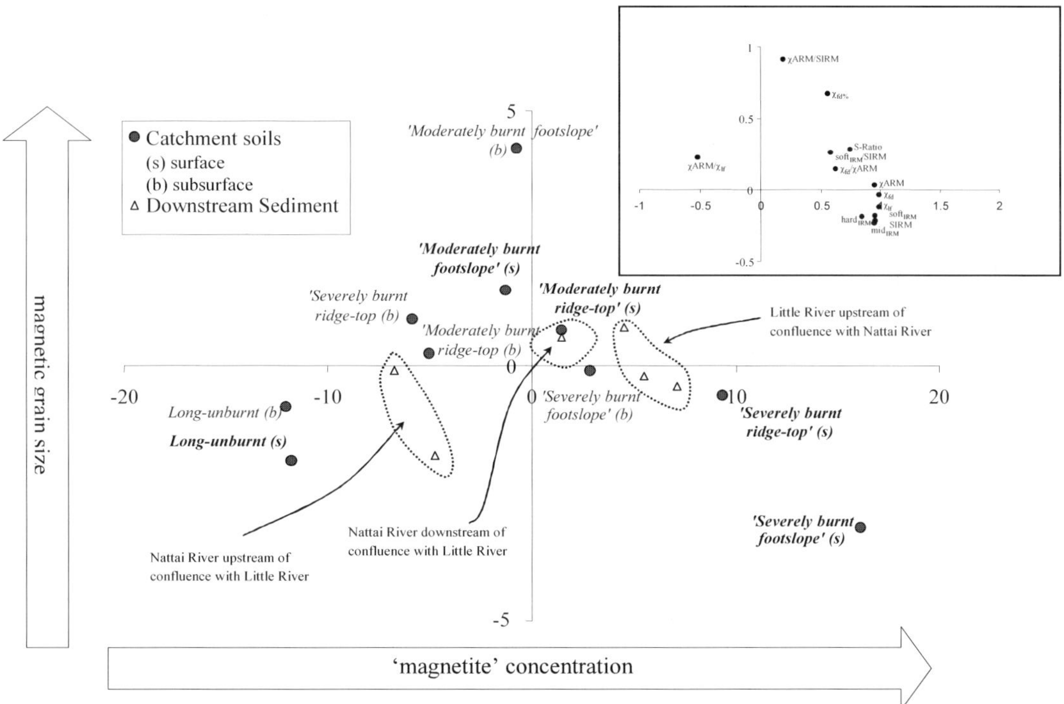

Fig. 3 Multivariate plot (principal components analysis) of sediment source area and downstream sediment properties where proximity within plot space indicates similarity in magnetic signatures after Blake *et al.* (2004).

Table 1 Mineral magnetic characteristics of: (a) source materials on the burnt slopes, (b) channel deposits around the Little River/Nattai River confluence, (c) grab samples of recent lake sediment and (d) stratigraphic units of the lake sediment core.

Table 1(a) *Landscape unit soil*	χ_{lf} (10^{-6} m^3 kg^{-1})	χ_{fd} %	χARM (10^{-5} m^3 kg^{-1})	soft$_{IRM}$ (10^{-5} Am^2 kg^{-1})	hard$_{IRM}$ (10^{-5} Am^2 kg^{-1})	SIRM (10^{-5} Am^2 kg^{-1})	χARM/SIRM (10^{-3} Am)	χ_{fd}/χARM	S-ratio
Unburnt	0.77	8.84	0.35	177	112	829	0.42	0.19	0.79
Mod. burnt footslope	4.06	10.65	1.82	610	378	2368	0.77	0.24	0.96
Mod. burnt ridge-top	4.59	11.18	1.21	676	180	2104	0.58	0.42	0.94
Sev. burnt footslope	10.39	10.68	3.36	1868	746	6700	0.50	0.33	0.97
Sev. burnt ridge-top	8.03	11.24	2.92	1386	399	4751	0.61	0.31	0.99
Table 1(b) *Downstream channel sediment*									
Nattai R. u/s Little R.	2.25	8.54	1.10	562	99	1831	0.60	0.17	0.91
Nattai R. u/s Little R.	3.42	7.32	1.28	666	150	2292	0.56	0.20	0.92
Little R. u/s Nattai R.	6.84	10.82	2.68	1418	317	4478	0.60	0.28	0.93
Little R. u/s Nattai R.	6.10	10.86	2.57	1223	213	3799	0.68	0.26	0.95
Little R. u/s Nattai R.	6.42	10.89	2.41	1235	282	3991	0.60	0.29	0.95
Nattai R. d/s Little R. (*reservoir inlet*)	4.78	10.44	2.08	1015	232	3190	0.65	0.24	0.92
Table 1(c) *Reservoir sediment (grab samples)*									
Grab 1	0.72	9.42	0.64	179.49	23.36	721.75	0.89	0.105	0.89
Grab 2	0.51	8.75	0.57	144.00	17.01	590.35	0.96	0.079	0.89
Grab 3	0.82	7.66	0.62	228.41	18.34	992.79	0.63	0.101	0.93
Grab 4	0.54	7.55	0.70	163.12	23.03	803.27	0.87	0.059	0.92
Grab 5	0.95	8.16	0.92	267.33	44.60	1138.49	0.80	0.085	0.91
Grab 6	0.72	9.42	0.64	179.49	23.36	721.75	0.89	0.105	0.89
Table 1(d) *Reservoir sediment (core samples)*									
Unit A	0.92	6.25	0.70	292	32.5	1347	0.52	0.081	0.93
Unit B	0.67	6.08	0.56	177	30.3	987	0.57	0.071	0.91
Unit C	0.49	5.81	0.40	123	7.8	610	0.65	0.072	0.89
Unit D	0.49	4.75	0.41	119	28.5	611	0.68	0.056	0.86
Unit E	0.34	5.42	0.25	70	25.9	380	0.65	0.075	0.79
Unit F	0.56	7.81	0.35	111	27.5	491	0.71	0.126	0.82
Unit G	1.06	9.25	0.67	251	26.6	903	0.74	0.146	0.90

In contrast, comparison of slope (Table 1(a)) and channel sediment (Table 1(b)) from the Nattai River suggests dilution of the fire-enhanced signal by upstream Nattai River sediment that generally records lower values for each magnetic parameter (Table 1(b)). Possibly, the less-enhanced Nattai River signal results from dilution by unburnt sediment sources, most likely agricultural and urban areas that, in contrast to the Little River basin, form a substantial part of the Nattai basin land use. Fire-related magnetic properties of sediment delivered to the lake have been slightly diluted prior to deposition but carry a significant fire-related signal.

Linking channel to lake

Comparison of the Nattai River channel sediment with surface grab samples of sub-aqueously stored recent sediment suggests a further reduction of the fire-enhanced mineral magnetic signature. However, in this case it is more difficult to reconcile these changes in terms of physical mixing or dilution. Magnetic susceptibility values fall from 4.78 (at the inlet (Table 1(b)) to 0.51–0.95 10^{-6} m^3 kg^{-1} in the surface lake sediment (Table 1(c)). Values of χARM similarly fall from 2.08 to 0.57–0.92 10^{-5} m^3 kg^{-1} and SIRM values fall from 3190 to 590–1138 10^{-5} Am^2 kg^{-1}, but to a lesser degree. These reductions in measures of magnetic concentration coincide with changes in magnetic grain size indicators. Comparison of lake and the Nattai River input sediment shows lower χ_{fd}% values (cf. Table 1(b) and (c)) and lower χ_{fd}/χARM values in the lake sediment, which indicate a "coarser" magnetic assemblage.

This coarser magnetic assemblage could be linked to some loss of the finer grained (SP) fraction of the mineral population to dissolution processes during subaqueous storage (cf. Oldfield *et al.*, 1992). Alternatively, comminution or sorting processes could modify signatures although the effect of the latter should be accounted for with comparison of <10 μm fractions. SP grains will contribute to χ values and also χ_{fd} both of which are significantly reduced. The (less dramatic) reduction in the value of χARM and SIRM indicates that dissolution or distortion might not be restricted to SP grains alone. A loss of pyrogenic grains has important implications for tracing applications both in terms of signal preservation and interpretation of the sediment column. It also contributes to understanding the lack of a "magnetic memory" in soils from surfaces experiencing a fire-event in 1994 but remaining unburnt during the recent event (see unburnt soil in Table 1(a)). Dissolution processes potentially contribute to a resetting of the source signal, together with denudation and bioturbation processes, albeit at a slower pace than in sub-aqueously stored fine sediment.

Interpreting the sedimentary record

The bulk mineral magnetic properties of the reservoir sediment column are shown in Fig. 4. The high values for all properties of units F and G are consistent with results reported for slope-stored eroded burnt material that had potentially been subjected to multiple burnings (Blake *et al.*, 2004). These differences imply that these units represent the former flood plain surface comprising subaerially stored burnt and reburnt material mobilized from upstream. If this assumption is correct, sediment from 0 through to 1020 mm depth can be considered to be lacustrine sediment deposits spanning from the present back to 1960. Dating evidence is required to support this.

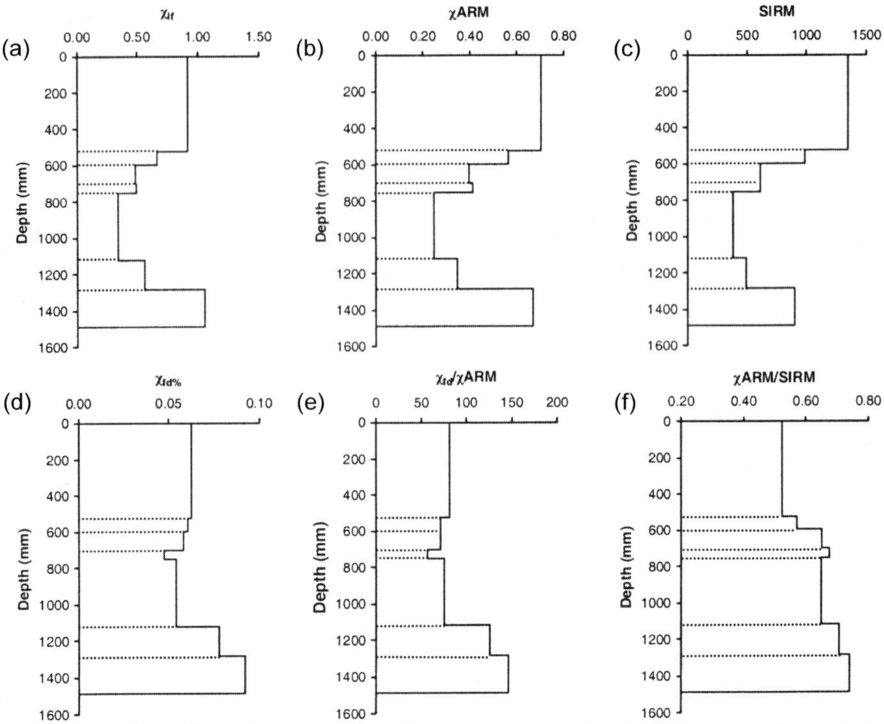

Fig. 4 Depth profiles of average mineral magnetic properties for lake core units A–E (see Table 1 for measurement units).

Stratigraphic evidence (Fig. 2) suggests that during this period there have been at least two additional fire-related, major deposition events (units C and E) of similar character to that of the 2001 erosion events (unit A). Domination of the record by these units, assuming they represent similar conditions to those observed following the study events, demonstrates the importance of immediate post-fire rainstorms in terms of sediment flux in this system.

The magnetic properties of units A, C and E are, however, not similar. A progressive and consistent decline in the magnetic concentration properties (χ_{lf}, χARM and SIRM, Fig. 4 (a)–(c)) is seen with depth. This could imply that either: (a) the dissolution of fine-grained magnetite in sub-aqueously stored sediment, as suggested by comparison of channel and surface lake sediment data above, continues with time, or (b) the fire-events that triggered major redistribution of sediment within the river basin were of different severity and hence led to different levels of enhancement in the source areas (cf. range of values in Table 1(a)). The decline in the proportion of fine magnetic grains in these units with depth suggests the former is more likely (see $\chi_{fd\%}$, χ_{fd}/χARM and χARM/SIRM, Fig. 4(d)–(f)). Evidence from Table 1(a) ($\chi_{fd\%}$) supports this showing that although different severities of fire produce different concentration-related signatures, all are represented by a high proportion of fine SP grains. Spatial variability in the magnetic properties of recent deposits (grab samples in Table 1 (c)), however, should also be noted.

The consistent decline in magnetic properties of the intermediate units (B and D) with depth, which we regard as representing non-immediate post-fire-rainfall and non-fire driven sediment mobilization, supports the above assertion. Interpretation of sediment column mineral magnetic evidence must, therefore, account for possible sub-aqueous distortion of the tracer signal.

CONCLUSION

In the Nattai basin, mineral magnetic properties are valuable tools for tracing sediment redistribution in the sub-aerial post-fire environment, offering information on both sediment source linkages (e.g. ridge-top to channel) and intra-landscape sediment storage. However, complication arises with extension of this tracing tool to the sub-aqueous environment where the pyrogenic magnetic signature of stored burnt sediment appears distorted by loss of some fine-grained pyrogenic magnetite, probably through dissolution.

Our data suggest that substantial post-fire rainstorm events account for 89% of total reservoir sedimentation in the Nattai arm of Lake Burragorang and that, unexpectedly, a significant proportion of that material is sourced from burnt ridge-tops and transported directly, via ephemeral gully flows, to the river channel. Mineral magnetic tracing tools provide river basin managers with important process-based evidence of post-fire sediment redistribution, useful for more effective mitigation of these infrequent, but significant, sediment redistribution events.

Acknowledgements NERC Grant NER/A/S/2002/00143 and NERC advanced fellowship NERC/J/S2002/00662 (SHD); Macquarie University for financial contribution (GSH); Sydney Catchment Authority for logistical support (Glen Capararo) and data (Chris Chafer); Danny Hunt (CSIRO L & W) and Vicky Farwig (UWS) for assistance with field sampling; Gareth James (UWS) for sample analysis; Wirrimbirra Nature Reserve for laboratory space; Nicola Jones (Fig. 1); Jamie Quinn (Fig. 2).

REFERENCES

Blake, W. H., Walling, D. E. & He, Q. (2003) The use of ^{7}Be as a tracer in sediment budget investigations. *Catena* **82A**(2), 89–102.

Blake, W. H., Wallbrink, P. J., Doerr, S. H., Shakesby, R. A. & Humphreys, G. S. (2004) Magnetic enhancement in fire-affect soil and its potential for sediment source-ascription. *Earth Surf. Processes Landf.* (submitted).

Doerr, S. H., Shakesby, R. A. & Walsh R. P. D. (2000) Soil hydrophobicity: its causes, characteristics and hydro-geomorphological significance. *Earth Sci. Rev.* **51**, 33–65.

Gedye, S. J., Jones, R. T., Tinner, W., Ammann, B. & Oldfield, F. (2000) Use of mineral magnetism in the reconstruction of fire history. *Paleogeo. Paleoclim. Paleoeco.* **164**, 101–110.

Oldfield, F., Darnley, I., Yates, G., France, D. E. & Hilton, J. (1992) Storage diagenesis *versus* sulphide authigenesis: implications in environmental magnetism. *J. Palaeolim.* **7**, 179–189

Rummery, T. A., Bloemendal, J., Dearing, J. A., Oldfield, F. & Thompson, R. (1979) The persistence of fire-induced magnetic oxides in soil and lake sediments. *Ann. Géophys.* **35**, 103–107.

Shakesby, R. A., Chafer, C. J., Doerr, S. H., Blake, W. H., Wallbrink, P. J., Humphreys, G. S. & Harrington, B. A. (2003) Fire severity, water repellency characteristics and hydrogeomorphological changes following the Christmas 2001 Sydney forest fires. *Austral. Geog.* **34**, 147–175.

Walden, J., Oldfield, F. & Smith, J. (1999) *Environmental Magnetism a Practical Guide.* Quaternary Research Association Tech. Guide 6, UK.

Walling, D. E., Peart, M. R., Oldfield, F. & Thompson, R. (1979) Suspended sediment sources identified by magnetic measurements. *Nature* **281**, 110–113.

Sediment Transfer through the Fluvial System (Proceedings of a symposium held in Moscow, August 2004).
IAHS Publ. 288, 2004

Erosion in basin geosystems of the Middle Volga (from a landscape analysis perspective)

OLEG YERMOLAEV

Faculty of Ecology, Kazan State University, Department of Landscape Ecology, Kremlevskaya, 18, Kazan, 420008, Russia
oleg.yermolaev@ksu.ru

Abstract Human-induced erosion and accumulation processes on agricultural hillslopes have a number of principal differences from natural erosion. Together with substantially higher rates of the processes, anthropogenic erosion creates a new and higher-level spatial organization of hillslope fluvial geosystems. From the focus-areal type of spatial pattern those transform into belts comprising anisotropic groups of vectorial-organized structures acquiring emergent properties. As a consequence, quantitative evaluation of this kind erosion should be conducted separately from its natural analogue.

Key words basin erosion; landscape; Middle Volga; neural networks; regionalization; regressive model

INTRODUCTION

Multi-factor controls of erosion processes determine the complexity of erosion-affected hillslope geosystem functioning. Relationships between erosion and its major controls change in geographical space, so that slope erosion regimes vary regionally, being determined by landscape conditions. Problems with the quantitative assessment of all types of soil erosion from raindrop to gullies still require a satisfactory solution. In this paper we propose to consider the entire complexity of hillslope erosion processes as a unit termed "basin erosion". The focus of this paper is on the methodological aspects of revealing the roles different landscape conditions play in causing basin erosion.

The Middle Volga basin was chosen as the investigation region due to the wide distribution of a spectrum of erosion processes in the region. It is also where the so-called "Erosion Pole" of European Russia is situated (Fig. 1).

Different combinations of natural and anthropogenic conditions create geocomplexes of different taxonomic levels known as "landscapes". Depending on the degree to which erosion processes are generalized in an investigation, it is necessary to use different geosystem taxon as the basic unit. These may be "landscape zones" for the global analysis of spatial regularities of erosion, or "landscape facies" for the development of empirical models of slope-wash at local or more detailed levels. The significance of the landscape approach is in the complexity of control parameters and the decrease in subjectivity of their choice for multiple classification analysis of erosion processes. This fact is well-understood in the scientific community. The number of published examples of the landscape approach in studying basin erosion is extremely small. The main reason for this are the extremely small number of published large- and medium-scale maps of soil erosion and its factors, the difficulty of their composition and a lack of unified legends.

To evaluate the role different landscape factors play in the development of human-accelerated basin erosion, a landscape map of the Middle Volga region was created. It uses a

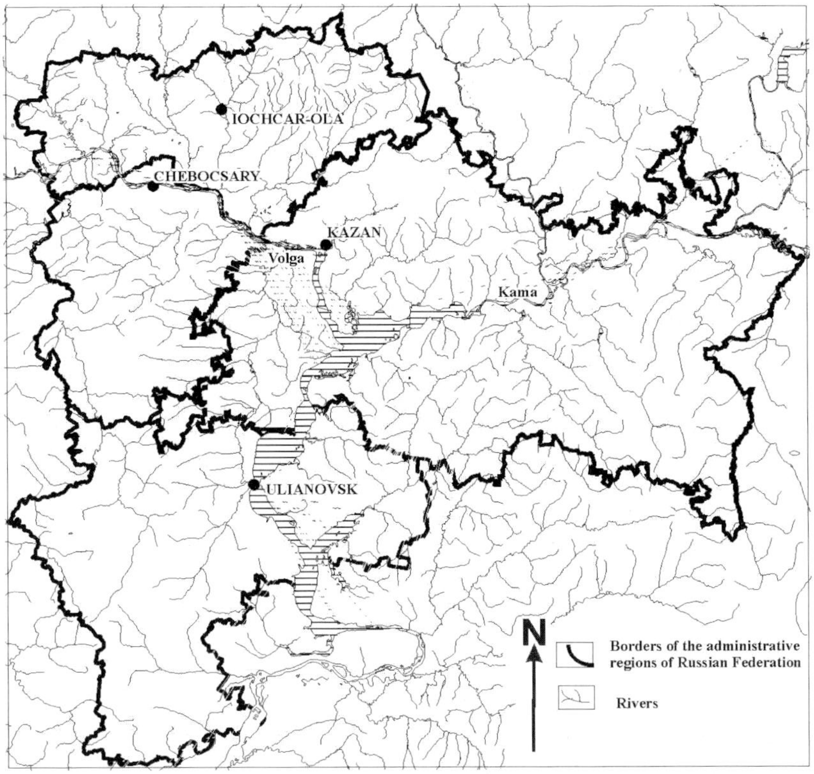

Fig. 1 River network and administrative regions of the Middle Volga territory.

regional level of generalization. Taxons of the landscape region level are presented on the map. The following basic principles were adhered to during the regionalization procedure. Firstly, regions were distinguished based on the most important landscape conditions, their spatial variability being dependent on interactions of both zonal and azonal natural factors. Secondly, for territorial landscape complexes, which must belong to the same taxonomic level, the proportionality requirement was kept at each stage of regionalization. Thirdly, the regionalization was carried out in the "upward" direction, or from individual to general. We chose the most distinctive landscape characteristics because we wanted them to embrace the required range of typical geosystem spaces (Colomitz, 1998), to have sufficient variety of values and to be accessible for measurements aimed at obtaining a massive database. Both zonal and regional factors of landscape differentiation were taken into account with the aim of studying the spatial organization of regional landscape conditions. River basins of the 2–3 order (by A. N. Strahler) have been taken as the basic operational unit of the spatial analysis.

DATA

The total number of parameters used for landscape regionalization comprised more than 50 (including: hydro-climatic, geomorphological, anthropogenic, lithological and landscape-geophysical); 3331 river basins were examined with an average catchment area of 40 km^2.

Examples of the parameters employed are: (a) annual sum of the solar radiation; (b) runoff coefficient; (c) rainfall erosivity index; (d) bedrock lithology; (e) soil texture; (f) slope length and gradient; (g) topographical altitude spectra; (h) landscape primary productivity; (i) soil humus content; (j) annual sum of the positive temperatures; and (k) area of cultivated land, etc.

The study territory is located within the forest, forest-steppe and northern part of the steppe landscape zone of the Russian Plain and comprises more than 130 000 km^2. Information on soil erosion of arable land was obtained for the territory of the Middle Volga, including the basins of the rivers Volga, Vyatka, Cama and Sviyaga. During the last 200 years arable land cover has increased by 40–60% and now comprises about 80–85% of the basins' area. The period of most intensive agriculture in the region began about 200 years ago. According to archive maps, the landscape zones which have experienced the greatest deforestation and subsequent greatest increase in arable areas are the southern forest-steppe and broad-leaved forest landscape zones.

METHODS

As most natural characteristics have continuous spatial distribution, regions with relatively uniform values of parameters were chosen for regionalization by traditionally formal indicators. This formal division leads to creation of artificial borders, loss of information about gradual change of values and, eventually, to incorrect territory regionalization. This is especially true when combined processing of a large number of heterogeneous parameters is undertaken. Modern computer databases storing large volumes of heterogeneous information on spatially distributed landscape characteristics require new methods for their combined processing.

Presentation of all the variety of heterogeneous components in the form of a small set of standard elements should be the best decision in this direction. Application of traditional statistics methods, cluster analysis and regionalization do not allow gradual transitions and connections, common in a real geographical space, to be revealed and visually presented on the map.

In the middle 1990s a new class of non-parametric adaptive algorithms appeared for the analyses of structure distributions—"self-organizing maps of Kohonen" (neural networks methods). The method of "self-organizing maps of Kohonen" (1997) was used as the main approach for automatic regionalization. Effectively it includes a few main stages described below.

In the first stage, 121 classes of units were determined according to a set of landscape characteristics (Yermolaev & Saveliev, 2000; Yermolaev, 2002). Such a number of classes helps to group regions in homogeneous groups presentable for cartographic spatial analysis, by using methods of hierarchical classification. This allows the presentation of topological properties (interarrangement) of distinguished classes in space of characteristics, and therefore a structure of the space of all characteristics for a given territory. This method mainly differs from the others by a large number of degrees of freedom, making it possible to construct any number of precise models. It gives an ability of "self-organizing" to the studied system or an ability to connect and modify its own structure and behaviour taking account of new input data. A valuable property of neural networks is the possibility for generalization, which is expressed in construction of satisfactory models by incomplete or strongly distorted data.

A second stage involved construction of regional mathematical statistics models of natural-anthropogenic modification of basin erosion, the landscape taxons on the last step of hierarchical classification were successfully combined initially into 24, and then into 11 landscape regions. The combination of landscape regions to 11 was pursuing two main targets. Firstly, it was necessary to provide statistically representative samples. In some regions a number of river basins did not satisfy this requirement. Secondly, united regions are close, not only by landscape conditions, but by intensity of basin erosion processes. A further decrease of the number regions is undesirable as it brings to notably distorted accuracy of predictive modelling of erosion results at the accepted generalization level. Supposing that each landscape region has its own specific conditions of basin erosion processes, a general linear regressive model of soil erosion index was constructed for every region. So, equations were obtained for landscape regions of the Middle Volga 11 and for the total territory a piecewise-linear model was made up (in cooperation with S. S. Mukcharamova).

Additional models were constructed during the third stage, aimed at determining the role of factors influencing the intensity and spatial pattern of human-accelerated erosion. Factors were combined in thematic groups: hydro-climatic, geomorphologic, anthropogenic, lithological, landscape-geophysical, and a whole array of factors used for erosion analysis. Separate regionalization of these factor groups by using artificial neural networks with 121 classes was also conducted. It has been found that all these cases have statistically significant (with 99% of confidence probability) correlation between arguments of soil and gully erosion and other variable models. The significance level for each of the factors included in the model is not less than 5%. Therefore none of them can be neglected. However, it does not mean that all the considered regressive models are adequate or possess good forecasting qualities. Characteristics used for evaluation of such qualities: R, coefficient of multiple determination; σ, standard error of estimation; *MAE*, mean absolute error of samples, are given in Tables 1 and 2. The higher the coefficients R^2, R and the lower the values of errors σ and *MAE*, the better and more adequately the model describes processes of basin erosion.

Table 1 Estimation of prognosis accuracy of soil erosion development.

Regionalization by conditions of soil erosion processes' development	"Piecewise-constant" model on the basis of classification of neural networks				"Piecewise-linear" model			
	R	R^2 (%)	σ	*MAE*	R	R^2	σ	*MAE*
Hydro-climatic	0.64	41	0.354	0.279	0.59	0.35	0.373	0.300
Geomorphologic	0.57	32	0.379	0.302	0.50	0.25	0.415	0.334
Anthropogenic	0.54	29	0.387	0.317	0.43	0.18	0.419	0.348
Lithologic	0.61	37	0.365	0.287	0.52	0.27	0.394	0.322
Landscape-geophysic	0.69	48	0.334	0.256	0.68	0.46	0.338	0.264
All indices	0.70	49	0.328	0.250	0.69	0.48	0.330	0.256

Table 2 Characteristics of gully erosion prognosis model quality for the total Middle Volga.

Types of models	σ	*MAE*	R^2 (%)	R
"Piecewise-constant" model on the basis of regionalization by landscape regions	0.689	0.539	52	0.72
"Piecewise-linear" model on the basis of regionalization by landscape regions	0.566	0.42	67	0.82

The fourth stage involves the choice of independent variables (regressors) to be included in a regional model. A number of variables exceeded 50. It should be noted that there is no common statistical procedure for choice of the best subset of regressors. Therefore, a certain subjective decision is always needed. Two opposing criteria exist. On the one side, to provide reliable forecasts the model must include the largest number of variables. On the other side, taking into account efforts for data collection, the equation should include the least number of regressors (Seber, 1980). The choice of "the best subset" is the compromise between these two extremes. With this aim the procedure of regression construction was developed by sorting all possible regressors and accounting expert opinion. At every procedure run, the measure for comparing the forecasting usefulness of different models was calculated.

Earlier, the determination of the coefficient R^2 was used as a measure of agreement between regression model and given data. However, difficulties appeared at the comparison of regressions with different regressors number, because introduction of additional regressor leads to an increase in R^2. We used a different, more suitable measures of model quality called the corrected determination coefficient R^2_c. $R^2_c = 1 - [1 - R^2][N/(N - p)]$, where p is the number of regressors plus one. Regressors making the largest contribution to soil erosion pattern of the given landscape region were distinguished.

RESULTS

The erosion models developed witness the presence of considerable regional differences in control factors. Differences not only concern values of the factors themselves, but also direction and degree of their influence on erosion. Variable complex combinations of landscape conditions and spatial patterns of sheet and rill erosion and gullies on slopes of river basins result in advancing one or other factors to the leading role. For example, for gully erosion in some cases these are geological and geomorphological characteristics (bedrock lithology, soil texture, slope length)—regions 1, 4, 9, 10; in others—climate characteristics (rainfall erosivity index, runoff coefficient)—regions 2, 8; anthropogenic impact (cultivated land area, settlement area)—regions 7, 11 etc. (Fig. 2).

The group of anthropogenic characteristics plays a special role in modelling gully erosion development. Most important of those are interrelated arable land area and forest-covered area. The most widespread reason of important index absence is its weak spatial differentiation in the region or dominance of topographic factors of gully erosion.

The analysis conducted shows that relief parameters, anthropogenic and hydro-climatic factors are the most essential contributors to soil and gully erosion development. Electronic vector maps of typological regionalization, modern and predicted, were created on the basis of developed models of soil and gully erosion processes in the Middle Volga region. They demonstrated very good agreement of results. The models developed will add to a better understanding of regional patterns of human-accelerated erosion processes and may become a basis for land-use plan development and design of soil protection measures.

The landscape approach used to investigate accelerated erosion processes at different taxonomic levels made it possible to address this phenomenon in detail accounting principles of territorial unity, genetic and relative homogeneity. Landscape regions (regional level) reflect zonal and longitudinal-sectoral patterns of basin erosion. Spatial analyses of soil erosion and gullying intensity in the study region based on available information allows

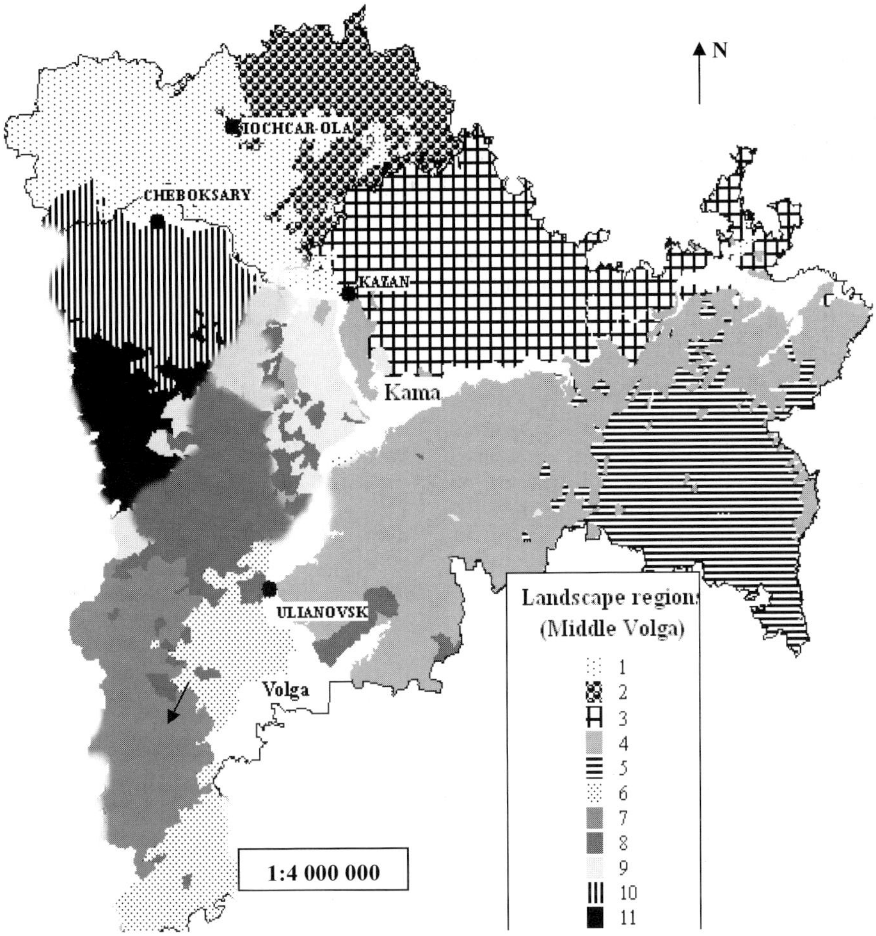

Fig. 2 Landscape regions distinguished by the method discussed of the Middle Volga territory.

us to conclude that: (a) maximum basin erosion is characteristic for upland landscapes of broad-leaved forest zones (Sub-boreal) and the southern part of mixed forest zone; (b) its intensity decreases in both western and eastern directions; (c) to the north and to the south from upland landscapes of broad-leaved forest zones we can also observe lowering of basin erosion intensity; (d) in the north it happens because of lower agricultural activity, and to the south it is due to development of chernozem soils more stable to erosion; and (e) zonality is typical for soil erosion processes. The gully erosion process, on the contrary, is an azonal phenomenon, as it is dependent on the influence of geologic and geomorphologic factors.

The average annual soil losses during the time of agricultural development of territory for different types of soils are in a range of 0.11–0.27 mm $year^{-1}$ (average 0.18 mm $year^{-1}$). Total denudation losses from gully erosion are lower at least by an order of magnitude.

Acknowledgement The author thanks Ms Svetlana Mukcharamova for team-work in the construction of mathematical-statistical models of hillslope erosion for landscape regions.

REFERENCES

Kohonen ,T. (1997) *Self-Organising Maps*, second edn. Springer-Verlag, Heidelberg, Germany.

Kolomitz, F. G. (1998) *Polymorphism of Landscape-Zonal Systems.* Inst. Pustchino, Russia (in Russian).

Seber, J. (1980) *Linear Regression Analysis.* Publ. – "Peace" (in Russian).

Yermolaev, O. P. (2002) *Erosion in Drainage Basin Geosystems.* Kazan, Russia (in Russian).

Yermolaev, O. P. & Saveliev, A. A. (2000) New possibilities of soil erosion factors' analysis using GIS-technologies. *Geomorphologiya* **4**, 46–56 (in Russian).

Using Cs-137 measurements and reservoir deposits to investigate the effects of ceasing cultivation on sediment yields and sediment sources in a small catchment on the Loess Plateau of China

XINBAO ZHANG[1,2], MINGYI FENG[1] & ANBANG WEN[1]

1 *Chengdu Institute of Mountain Hazards and Environment, CAS, Chengdu 610041, China*
zxbao@imde.ac.cn

2 *State Key Laboratory of Soil Erosion and Dry Land Farming on the Loess Plateau, Institute of Soil and Water Conservation, CAS & MWR, Yangling 712100, China*

Abstract The impact of ceasing cultivation on specific sediment yields and relative sediment contributions from the gully and inter-gully areas within a small catchment in the Rolling Loess Plateau of China has been examined by comparing the volume and ^{137}Cs content of sediments deposited in a reservoir before and after 1993. It is suggested that compaction of the formerly ploughed soils caused their erosion resistance to increase but reduced infiltration rates. Rates of soil loss from the inter gully areas therefore decreased but runoff increased. Increased runoff caused increased erosional activity within the receiving gully areas and an increase in the total sediment yield from the catchment.

Key words ceasing cultivation; China; ^{137}Cs measurements; reservoir deposits; Rolling Loess Plateau; sediment sources; specific sediment yield

INTRODUCTION

The Rolling Loess Plateau in Northern Shaanxi and Western Shanxi is one of the most eroded regions of the Loess Plateau of China. Specific sediment yields commonly exceed 1.0×10^4 t km^{-2} year^{-1}. The rolling plateau can be divided into two main geomorphic units, namely, the gully areas and the inter-gully areas, which represent the rolling plateau surface. Sheet and rill erosion are widespread on the inter-gully areas, where the land has been extensively cultivated, while gully erosion and mass movements predominate in the gully areas. Relative sediment contributions from the inter-gully and gully areas are about 30% and 70%, respectively (Jiang *et al.*, 1990; Jing *et al.*, 1997). Since the "Closing cultivation for land rehabilitation" policy was introduced in the 1990s, cultivation has been stopped on much of the sloping cultivated land to promote large scale revegetation of the Loess Plateau. The impact of ceasing cultivation of the sloping lands needs to be assessed, in order to establish its effects on both the geomorphic evolution of the area and the reduction in sediment yield.

Within a previous study undertaken in 1993, the ^{137}Cs technique was used to assess erosion rates on the sloping cultivated land and the relative sediment contributions from the inter-gully and gully areas in the Zhaojia Gully catchment (Zhang *et al.*, 1997). Cultivation has gradually ceased on most of the cultivated land in the catchment since the 1990s. To examine the impact of this ceasing of cultivation, the catchment was visited in 2001 to collect samples of the sediment deposited in the reservoir between 1993 to 1996 for ^{137}Cs measurements. By comparing both the volume of the deposited sediment and its ^{137}Cs content before and after 1993, the changes in erosion rates and sediment yield have been assessed.

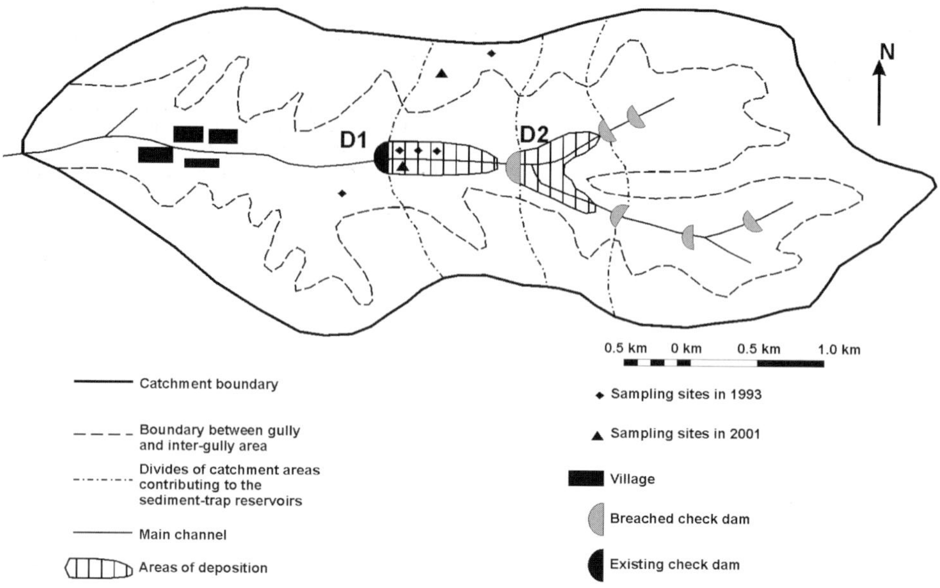

Fig. 1 A sketch map of the Zhaojia Gully catchment and the sampling locations.

THE STUDY CATCHMENT AND THE LAND-USE CHANGES

The Zhaojia Gully catchment is located 3 km east of Zichang Town, and constitutes a small tributary of the middle reaches of the Qingjian River which drains into the Yellow River. The catchment is underlain by loess deposits >100m thick and has a drainage area of 3.86 km^2, an average altitude of 1050 m and a relative relief of 300 m. The inter-gully area and gully areas account for 53 and 47% of the catchment area, respectively (Fig. 1). Sheet and rill erosion were previously widespread on the cultivated land of the relatively gentle inter-gully area with gradients <25°. The gully area consists mostly of steep gully slopes with gradients >30°. Gully erosion and mass movements are the major sediment sources in the gully area. The average annual precipitation is 513 mm, of which approximately 65% occurs during the flood season from July to September.

Land use conditions in the catchment in 1993 are shown in Table 1 (Zhang *et al.*, 1997). The land in the inter-gully area was almost entirely cultivated, and there were a few areas of steep cultivated land on the lower parts of gully slopes near Zhaojia Village in the lower reaches of the catchment. There is little terraced land in the catchment, although a few hectares of such land were constructed in the 1970s.

Several check dams have been constructed in the catchment since the 1970s to trap sediment as well as to create flat fertile farmland (Fig. 1). Dam 2, which is 12 m in height, was constructed in the middle reaches of the catchment in 1971 and had no overflow provision. It controlled an upstream drainage area of 2.03 km^2. Before the 1975 flood season, five small dams, 5–7 m in height and without spillways, were constructed upstream of Dam 2 and the upstream drainage area controlled by Dam 2 was reduced to 0.46 km^2. Unfortunately, those five small dams failed during the big floods of 4–5 July 1977 and Dam 2 was almost filled with sediment and the floodwater spilled over the dam. Subsequently, the crest of Dam 2 was gradually eroded by floodwaters and a narrow straight channel was incised into

Table 1 Event rainfall, mean ^{137}Cs content of deposited sediment, specific sediment yield, and relative sediment contributions from the inter-gully area and from the gully area for the 13 flood couplets deposited during the period 1993–1996.

Date	Rainfall (mm)	Mean ^{137}Cs content (Bq kg^{-1})	Relative sediment contribution from the gully area (%)	Relative sediment contribution from the inter-gully area (%)	Specific sediment yield (t km^{-2})
8–10 August 1996	44.6	0.48	85	15	6537
1 August 1996	53.9	0.18	94	6	10866
7–16 July 1996	52.2	0.72	78	22	3782
26–28 June 1996	29.1	0.00	100	0	1205
15–18 June 1996	94.7	0.00	100	0	21437
29–31 August 1995	56.2	0.79	76	24	10645
25–28 August 1995	58.3	0.00	100	0	8627
5–6 August 1995	58.6	0.73	78	22	6973
17–18 July 1995	61.0	0.42	87	13	8796
31 August 1994	78.7	0.81	75	25	37816
10–12 August 1994	47.1	0.64	80	20	5902
7–8 July 1994	41.1	0.20	94	6	2861
19–23 August 1993	48.8	0.73	78	22	2720

the sediment deposited behind the dam, until it reached the bed of the former gully. Dam 1 was constructed 450 m downstream of Dam 2 before the 1988 flood season and it controls an upstream drainage of 2.63 km^2. It is an earth dam 23 m in height, which incorporates both a side spillway and a bottom culvert with several intakes. The total storage capacity of the reservoir was 0.5×10^6 m^3 and the impounded floodwater was used for downstream irrigation. In order to maintain the storage capacity for large floods, floodwater has been released through the intakes and bottom culvert during the floods that occurred since 1997. During the five dry years between 1997 and 2001, no major floods occurred and all the incoming sediment load was discharged through the open intakes and the bottom culvert and very little sediment was deposited in the reservoir. The cultivated land on the rolling plateau surface area has been gradually abandoned since the early 1990s. A survey undertaken in 2001 showed that nearly all the sloping cultivated land in the catchment had been abandoned by 1995, except for a few areas of sloping cultivated land close to the villages. Overgrazing was widespread in the catchment prior to cessation of cultivation. Grazing has now also been prohibited and the vegetation cover has shown significant recovery, although tree planting has proved unsuccessful.

FIELD SAMPLING AND LABORATORY MEASUREMENTS

In order to assess erosion rates on the sloping cultivated land and the relative sediment contributions from the inter-gully area and the gully area, the sediment deposited within the sediment-trap reservoirs and the soils of the catchment were sampled in October 1993 (Fig. 1) (Zhang *et al.*, 1997). A ^{137}Cs reference profile was collected from the grassland surrounding an old tomb on the rolling plateau surface using a special corer. The sampling depth was 35 cm. The sampling method for cultivated soil profiles was the same as for the reference profile. Samples of surface soil (<5 cm) were also collected. A profile of the sediment deposited in the reservoir behind Dam 1 by the 1993 floods was collected using a corer. A profile of the sediment deposited by the floods occurring from 1973 to 1977 was collected from the wall of the channel incised into the sediment deposited behind Dam 2. All sediment deposit profiles were carefully sectioned to reflect flood deposit cycles and each couplet was usually subdivided into three samples. A few thin couplets were divided into two samples.

In order to establish the impact of ceasing cultivation of the sloping land in the catchment on soil erosion and sediment yield, a pit 9.2 m in depth with a diameter of 1 m was dug 50 m upstream of Dam 1 in November 2001, to collect a profile of the sediment deposited in the reservoir from 1993 to 1996. Depth incremental samples of the profile were collected from the side of the pit, taking account of the flood cycles. Each couplet was subdivided into at least three samples. In addition, five soil profiles for measurement of the ^{137}Cs reference inventory were collected from the 1993 reference sampling site, using the same procedure as in 1993.

All samples were air-dried, disaggregated and passed through a 2 mm sieve prior to analysis of their ^{137}Cs activity by gamma spectrometry using an HPGe detector. The counting times, which were typically about 55 000 s, provided results with an analytical precision of approximately ±6% at the 95% confidence level.

SEDIMENT YIELDS AND RELATIVE CONTRIBUTIONS FROM THE GULLY AND INTER-GULLY AND INTER-GULLY AREAS BEFORE 1993

The main findings of the 1993 investigation of soil erosion and sediment sources in Zhaojia Gully reported by Zhang *et al.* (1997) are as follows:

(a) In the inter-gully area, cultivated land on the gentle hill summits and the steeper hillslopes account for 14% and 36% of the catchment area, respectively, while grass land only accounts for 3%. In the gully area, grass slopes, bare slopes and steep sloping cultivated land account for 21%, 19% and 2% of the catchment area, respectively, while other types of land, such as channels, terrace land and villages account for 5%.

(b) The local ^{137}Cs reference inventory was 2500 Bq m^{-2}. The average ^{137}Cs content of the cultivated soils in the inter-gully area was 3.9 Bq kg^{-1}, while the surface soils on bare slopes and areas affected by mass movements contain little ^{137}Cs (0.02 Bq kg^{-1}).

(c) 1993 was a dry year with an annual precipitation of 429.8 mm and the catchment had a specific sediment yield of 4627 t km^{-2} year^{-1}. The average ^{137}Cs content of the sediment deposited by two floods was 0.74 Bq kg^{-1} and 1.06 Bq kg^{-1}, respectively. Based on these values and the values listed above for the ^{137}Cs content of the potential sources, the mean relative sediment contributions from the inter-gully area and the gully area in 1993 were estimated to be 26.9% and 73.1%, respectively.

(d) The years from 1973 to 1977 were a wet period and the annual precipitation varied between 513.0 mm and 670.3 mm, with a mean value of 521.8 mm. Fifteen floods occurred during this period and the mean specific sediment yield was 16 291 t km^{-2} year^{-1}. The sediment deposits associated with four flood events occurring in the early part of the wet season had a ^{137}Cs content of <1.0 Bq kg^{-1}, indicating that nearly all the sediment mobilized by those events was derived from the gully area and that the relative sediment contribution from the gully area was >85%, with the maximum contribution reaching 98.6%. For the other 11 flood events, the ^{137}Cs content of deposited sediment varied between 1.12 and 1.15 Bq kg^{-1}, and the relative contributions from the gully area varied between 67.0% and 82.5%. The mean specific sediment yield from the catchment for this period was 13 413 t km^{-2} year^{-1} and the equivalent yields for the inter-gully area and the gully area were 6580 t km^{-2} year^{-1}, and 21 118 t km^{-2} year^{-1}, respectively. The relative sediment contributions from the inter-gully area and the gully area were estimated to be 26% and 74%, respectively.

CHANGES IN SEDIMENT YIELD AND THE RELATIVE CONTRIBUTION OF SEDIMENT FROM THE INTER-GULLY AND GULLY AREAS SINCE 1993

Based on evidence provided by down-profile variations in grain size composition and the ^{137}Cs content of the sediment, the rainfall records data and information obtained from local farmers, the sediment profile collected from the pit dug in the reservoir of Dam 1, comprised 13 flood couplets (Fig. 2). The rainfall totals corresponding to these couplets are shown in Table 1.

A field survey undertaken in November 2001 indicated that the surface area of the sediment deposited in the reservoir basin of Dam 1 was 29 310 m^2, 6131 m^2 greater than the area of 1993. The total volume of sediment deposited between 1994 and 1996 was 167 101 m^3, of which the volume associated with the four floods associated with couplets 4, 7, 8, and 9 accounted for 62.6% of the total volume. The major flood of 31 August 1994 (Couplet 4) accounted for 30% of the total volume. Based on the volume of the sediment deposits and assuming a bulk density of 1.4 t m^{-3}, the specific sediment yields for 1994, 1995 and 1996

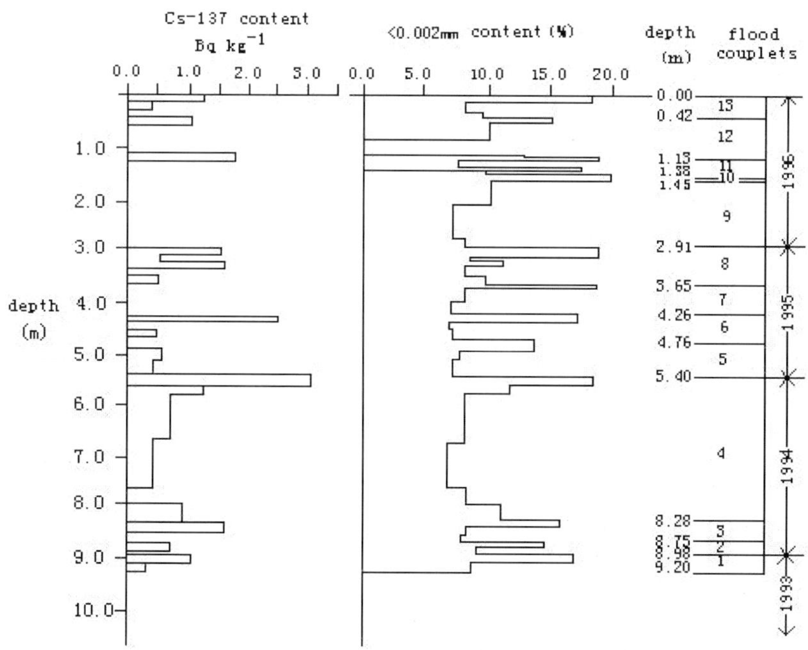

Fig. 2 The ^{137}Cs and clay contents of the sediment associated with the 13 flood couplets deposited between 1993 and 1996.

were calculated to be 33 034, 24 851 and 31 086 t km^{-2} $year^{-1}$, respectively, and the mean value for the three years was 29 650 t km^{-2} $year^{-1}$, which was 2.2 times greater than the average specific sediment yield of the catchment. It appears that cessation of cultivation on the sloping land of the rolling plateau surface did not result in a reduction of sediment yield from the catchment.

The mean ^{137}Cs inventory of the five reference soil profiles collected in 2001 was 1966 Bq m^{-2}, a value very close to the value of 2076 Bq m^{-2}, equivalent to the reference inventory of 2500 Bq m^{-2} measured in 1993 (Zhang *et al.*, 2003). The average annual ^{137}Cs content of the sediment deposits from 1994 to 1996 ranged between 0.18 Bq kg^{-1} and 0.75 Bq kg^{-1} with a mean of 0.45 Bq kg^{-1} (Table 2). The average ^{137}Cs content of the 1993 sediment deposits, sampled in 1993, was 0.90 Bq kg^{-1} (Zhang, 1997). Taking account of decay, this value would be 0.75 Bq kg^{-1} in 2001, a value very close to 0.73 Bq kg^{-1} which represents the average ^{137}Cs content of the 1993 sediment deposits sampled in 2001. Since 1994, the average annual ^{137}Cs content of the deposited sediment has declined markedly from 0.75 Bq kg^{-1} in 1994 to 0.49 Bq kg^{-1} in 1995, and 0.18 Bq kg^{-1} in 1996. This sharp decline in the ^{137}Cs content of the sediment deposited from 1994 to 1996 contrasts with the findings of the 1993 investigation which indicated that: "The limited variation in the mean ^{137}Cs content of the sediment deposits for each year during the period of 1973–1977 indicated that the overall relative contributions of sediment from the rolling plateau and from the gully area remained fairly constant from year to year". It is clear that the relative sediment contribution from the gully area has increased, while that from the inter-gully area has decreased, since 1994 because sediment mobilized from the former area contains little ^{137}Cs.

As in the previous study, the relative sediment contributions from the inter–gully area and the gully area have been estimated using a simple mixing model:

$$Cd = Cm \times fm + Cg \times fg \tag{1}$$

$$fm + fg = 1 \tag{2}$$

where: Cd is the ^{137}Cs content of the deposited sediment (Bq kg^{-1}); Cm is the ^{137}Cs content of the sediment derived from the inter-gully area (Bq kg^{-1}); Cg is the ^{137}Cs content of sediment derived from the gully areas (Bq kg^{-1}); fm is the relative sediment contribution from the inter-gully area (%); and fg is the relative contribution from the gully areas (%).

The average ^{137}Cs content of sediment mobilized from the inter-gully area and from the gully area were 3.9 Bq kg^{-1} and 0.02 Bq kg^{-1}, respectively, in 1993. Allowing for radioactive decay the corresponding values for 2001 were 3.24 Bq kg^{-1}, and 0.017 Bq kg^{-1}, respectively. Using these values for the ^{137}Cs content of sediment delivered from the inter-gully and gully areas, the relative contributions from the two sources during the years 1994–1996 were estimated using Equation (1). Values of 23%, 15% and 6% were obtained for the inter-gully area, whereas the equivalent values for the gully areas were 77%, 85% and 94%, respectively. The associated estimates of specific sediment yield from the inter-gully and gully areas for the years 1994–1996 are listed in Table 2.

During the period 1973–1977, annual precipitation totals ranged from 353.6 mm in 1974 to 670.3 mm in 1977, and the magnitude of the annual sediment yield was closely related to the annual precipitation. However, the relative contributions from the inter-gully and gully areas remained fairly constant and ranged between 17.0% and 24.2% for the inter-gully area and 75.8% and 83.0 % from the gully area. The annual sediment yields during the period1994–1996 were significantly greater than those during the period from 1973 to 1977.

Table 2 A comparison of sediment yields before and after cessation of cultivation in the Zhaojia Gully Catchment.

Year	Annual precipitation (mm)	Flood events	Specific sediment yield of the catchment (t km^{-2} year^{-1})	Mean ^{137}Cs content of deposits (Bq kg^{-1})	Relative sediment contribution %		Specific sediment yield (t km^{-2} year^{-1})	
					Rolling plateau	Gully	Rolling plateau	Gully
1973	553.2	2	9153	1.40	19.5	80.5	4093	14859
1974	353.6	3	9955	1.14	17.0	83.0	3625	17093
1975	513.0	3	7424	1.50	23.2	76.8	3572	11768
1976	518.7	4	22455	1.12	17.2	82.8	8008	38747
1977	670.3	3	32472	1.55	24.2	75.8	16113	50919
1993	429.8	2	4627	0.75	23	77	2008	7580
1994	536.9	3	33034	0.75	23	77	14335	54119
1995	508.1	4	24851	0.49	15	85	7034	44944
1996	540.3	5	31068	0.18	6	94	3517	62136

The specific sediment yields of 1994 and 1996 were 33 034 t km^{-2} year^{-1} and 31 068 t km^{-2} year^{-1}, respectively. These values are close to the specific sediment yield of 32 472 t km^{-2} year^{-1} for 1977, but the annual precipitation totals for 1994 (536.9 mm) and 1996 (540.3 mm) were considerably less than the value of 670.3 mm recorded in 1977 (Fig. 3). The relative sediment contribution from the inter-gully area has also sharply decreased from 23% in 1994, to 15% in 1995 and 6% in 1996, while that from the gully area increased from 77% in 1994, to 85% in 1995 and 94% in 1996.

Cessation of cultivation of the sloping land in the Zhaojia Gully has been shown to have resulted in an increase in the total sediment yield from the catchment, a decrease in the specific sediment yield and relative contribution from the inter-gully area, and an increase in

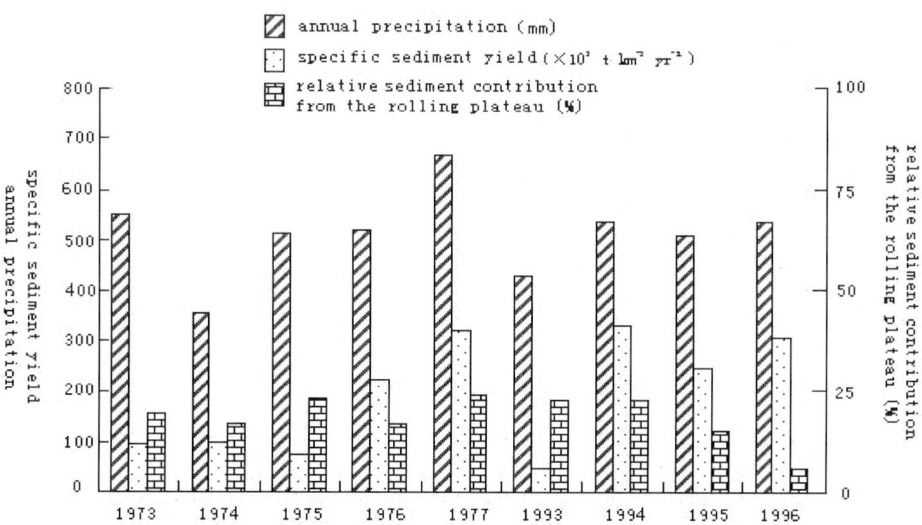

Fig. 3 A comparison of annual precipitation totals, specific sediment yields, and the relative sediment contribution from the inter-gully area before and after the cessation of cultivation in the Zhaojia Gully catchment.

the specific sediment yield and relative sediment contribution from the gully area. It is suggested that compaction of the formerly ploughed soils resulted in an increased resistance to erosion, but decreased infiltration. Thus, although rates of soil loss from the inter-gully areas decreased, runoff amounts increased. An increase of runoff input to the gully areas could be expected to result in an increase in erosional activity and mass movements, causing an increase in the specific sediment yield and relative contribution from the gully areas and in the total sediment yield from the catchment

Loess landscapes are very sensitive to changes in the runoff regime. Changes in the runoff regime caused by cessation of cultivation in the inter-gully area can be expected to result in increased erosion until the terrain adjusts to the new runoff regime. This adjustment may take an extended period and soil erosion is likely to become more severe in the gully area. However, rates of erosion within the gully area and the sediment yield from the catchment can be expected to reduce in the longer-term as the vegetation cover in the catchment fully recovers.

CONCLUSION

The mean specific sediment yield of 29 650 t km^{-2} $year^{-1}$ for the period 1994–1996 after the cessation of cultivation is 2.2 times the long-term mean annual specific sediment yield of 13 413 t km^{-2} $year^{-1}$ from the Zhaojia Gully catchment. This increase in sediment yield associated with cessation of cultivation was coupled with a decrease in both the absolute and relative contribution from the rolling plateau area and an increased absolute and relative contribution from the gully area. It is suggested that compaction of the formerly ploughed soils resulted in increased erosion resistance, but decreased infiltration, so that erosion rates decreased but runoff amounts increased within the inter-gully area. The increased runoff entering the gully area caused an increase in erosional activity and mass movements, resulting in an increased sediment yield and relative sediment contribution from this area of the catchment, as well as an increase in the specific sediment yield from the overall catchment.

Acknowledgements This work was supported by the National Natural Science Foundation of China (Grant nos. 40271015 and 900102017), by Academia Sinica through Knowledge Innovation Projects (KZCX3-SW-330 and KZCX3-SW-422), by IAEA through contract 12322/RO and by the State Key Laboratory of Soil Erosion and Dry Land Farming on the Loess Plateau (contract 10501-109). The assistance with radiometric analysis provided by Professor D. E. Walling and the Environmental Radionuclide Laboratory in the Department of Geography of the University of Exeter is gratefully acknowledged.

REFERENCES

Jiang, Z., Liu, Z. & Jia, Z. (1990) *Research on the Relationship between Topographic Factors and Loss of Soil and Water on Sloping Land.* Memoirs of the Northwest Institute of Soil and Water Conservation, Academia Sinica and Ministry of Water Conservancy no. **12**, 1–8 (in Chinese).

Jing, K., Lu, J. F. & Liang, J. Y. (1997) *Characteristics and Changes of Soil Erosion and its Environment in the Middle Reaches of Yellow River.* Huanghe Water Conservancy Publication House, Zhenzhou, China (in Chinese).

Zhang, X., Walling, D. E. & Quine, T. A (1997) Use of reservoir deposits and caesium-137 measurements to investigate the erosion response of a small drainage basin in the rolling Loess Plateau region of China. *Land Degradation and Development* **8**, 1–16.

Validating the use of caesium-137 measurements to estimate erosion rates in three small catchments in Southern Italy

PAOLO PORTO[1,2], DES E. WALLING[1] & GIOVANNI CALLEGARI[3]

1 *Department of Geography, University of Exeter, Exeter EX4 4RJ, UK*
p.porto@exeter.ac.uk

2 *Dipartimento di Scienze e Tecnologie Agro-Forestali e Ambientali, Università degli Studi "Mediterranea" di Reggio Calabria, Italy*

3 *CNR–Istituto per i Sistemi Agrari e Forestali per il Mediterraneo, Sezione Ecologia e Idrologia Forestale, Rende (Cs), Italy*

Abstract The fallout radionuclide caesium-137 (^{137}Cs) has been increasingly used in recent years to assess soil erosion and deposition at the catchment scale. However, the successful application of the ^{137}Cs approach depends heavily on the availability of reliable conversion models for converting measurements of ^{137}Cs redistribution to estimates of soil redistribution rates. This paper reports the results of a study aimed at validating the use of a theoretical conversion model to convert measurements of ^{137}Cs inventories on uncultivated soils to estimates of soil erosion rates. It is based on three small catchments located in Calabria, Southern Italy, for which measurements of sediment output are available. By comparing the estimates of net soil loss from the catchments derived from ^{137}Cs measurements with the measured sediment output, it is possible to assess the accuracy of the former estimates. The general correspondence between the measured sediment yields and the estimates of net soil loss based on ^{137}Cs measurements, confirms the validity of the theoretical model used for converting ^{137}Cs measurements into estimates of soil redistribution rates.

Key words caesium-137; conversion models; erosion rates; Italy; sediment yield; soil erosion

INTRODUCTION

Increasing concern for both the on-site and off-site impacts of soil erosion in many areas of the world has focused attention on the need to assemble reliable information on rates of erosion and soil redistribution under different environmental conditions. Traditional measurement techniques possess many limitations (Loughran, 1989) and in recent years the potential for using environmental radionuclides, and more particularly caesium-137 (^{137}Cs), has attracted increasing attention (Walling, 1998). However, the successful application of the ^{137}Cs technique depends on the reliability of the conversion model used to convert radiocaesium measurements to estimates of soil redistribution rates. In the absence of empirical conversion models, most conversion models have a theoretical basis and make use of existing understanding of the fate and behaviour of ^{137}Cs in eroding soils to derive a relationship between the erosion or deposition rate and the reduction of the ^{137}Cs inventory relative to the local reference inventory. Several of the assumptions associated with these theoretical conversion models are essentially untested and the models remain largely unvalidated. This paper reports the results of an ongoing study aimed at validating the use of ^{137}Cs measurements to estimate soil redistribution rates on three small catchments with forest and rangeland land use located in Calabria, Southern Italy. Long-term measurements of sediment output are available for the catchments, and by comparing these measurements with estimates of net soil loss from the catchments derived from ^{137}Cs measurements, the accuracy of the latter estimates can be assessed. Some results for two of these catchments have been published previously (Porto *et al.*, 2001, 2003), but this paper introduces data from a third contrasting catchment and provides an overall assessment of the validity of the estimates of erosion rates derived for the catchments using ^{137}Cs measurements coupled with a theoretical conversion model, based on the results from all three catchments.

THE STUDY CATCHMENTS

The three small catchments investigated in this study are located near Crotone in Calabria, Southern Italy (35 m a.s.l., 39°09′02″N, 17°08′10″E). The three catchments, W1, W2, and W3, have drainage areas of 1.47, 1.38 and 1.65 ha respectively and are located in the ephemeral headwaters of the larger Crepacuore basin (Fig. 1), which is incised into the local Upper Pliocene and Quaternary clays, sandy clays and sands. The catchments have never been cultivated. Catchment W1 supports a rangeland vegetation cover, and catchments W2 and W3 were planted with *Eucalyptus occidentalis Engl.* in 1968. These trees have been cut twice (in 1978 and 1990) in catchment W2, and once (in 1986) in catchment W3. The tree cover within catchment W2 is not uniform, and about 20% of the area, located on south facing slopes, is characterized by discontinuous tree and grass cover. In catchment W3, the tree cover is almost continuous and only about 2–3% of the area supports grass. Further details of the topography of the study catchments and the texture of their soils are provided in Table 1. The climate of the area is typically Mediterranean with a mean annual precipitation of ~670 mm, most of which falls during the period extending from October to March.

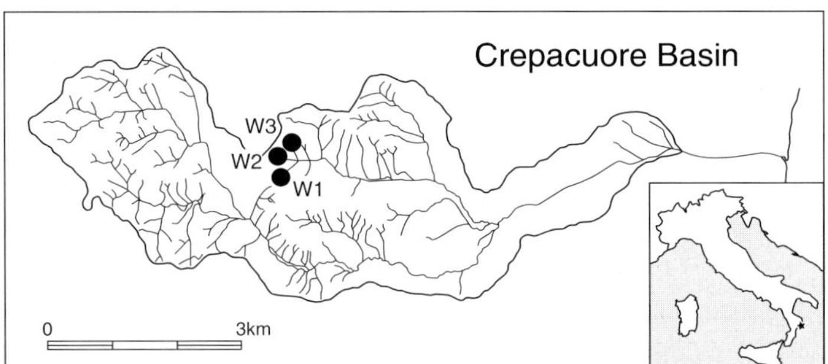

Fig. 1 The location of the study catchments.

Table 1 The characteristics of the study catchments.

Catchment	Drainage area (ha)	Mean altitude (m a.s.l.)	Mean slope (%)	Soil texture Sand (%)	Silt (%)	Clay (%)	Sediment yield (t ha^{-1} year^{-1}): Min	Max	Mean
W1	1.473	122	53	14	44.5	41.5	5.1	38.4	11.6
W2	1.375	103	35	14.6	49.2	36.2	1.7	98.5	20.8
W3	1.654	98	24	20.7	45.5	33.8	2.9	25.7	7.6

In 1978, these catchments were instrumented for measuring rainfall, runoff and sediment yield (Cantore *et al.*, 1980). Precipitation has been recorded using a tipping bucket raingauge and runoff is measured at the outlet of each catchment using an H-flume structure equipped with a mechanical stage recorder. The sediment load passing the gauging structure is measured using a Coshocton wheel sampler installed below the H-flume (Porto *et al.*, 2003).

SEDIMENT YIELDS AT THE CATCHMENT OUTLETS

Sediment yield data available for the catchments for the period 1978 to 1994 (Cantore *et al.*, 1994) have been used in this study. Due to malfunctioning of the sediment sampling equipment in the catchments during some events, values of annual sediment yield are not available for all years. Where the number of unmonitored events was limited, the sediment loads associated with these events have been estimated using the Sediment Delivery Distributed (SEDD) model (Ferro & Porto, 2000) and incorporated into the record. The annual sediment yields for the catchments are shown in Fig. 2 and the range of annual sediment yields documented for each catchment is listed in Table 1.

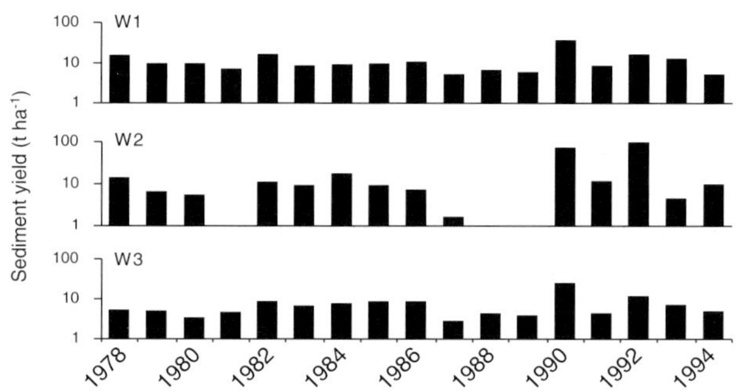

Fig. 2 Measured annual suspended sediment yields from the study catchments during the period 1978–1994.

In order to provide values of sediment yield for the individual catchments relating to the period represented by the ^{137}Cs measurements, the measured data have been extrapolated to cover the period 1956–1998. For catchment W1, the grass cover was assumed to have remained essentially constant over the period 1956–1998 and a simple logarithmic regression between event sediment yield and event rainfall, based on 46 discrete events, was used. Since trees were absent from catchments W2 and W3 during the first 12 years of the period in question and the cover density varied during the remainder of the period under consideration, in response to tree growth and forest harvesting, a multiple regression analysis, involving both event rainfall and an estimate of the time variant cover density based on the age of the trees, was used to develop the equations for predicting event sediment yields from these two catchments (Porto *et al.*, 2003). In the case of catchment W3, both variables were included in the final prediction equation as significant ($p > 0.01$), but for catchment W2 event sediment yield was only significantly related ($p > 0.01$) to event rainfall. The estimates of mean annual sediment yield for the period 1956–1998 for the three study catchments are presented in Table 2, along with the prediction equations used.

Table 2 Relationships for sediment yield calculation.

Catchment	Equation type	Parameters: a	b	c	r	Y (t ha^{-1})
W1	$Y = aX_1^{\,b}$	0.0011	1.72	–	0.64	12.4
W2	$Y = aX_1^{\,b}$	0.0027	1.69	–	0.86	19.2
W3	$Y = aX_1^{\,b}X_2^{\,c}$	0.0033	1.49	–0.17	0.84	7.8

X_1 = rainfall amount (mm); X_2 = canopy cover (%); r = correlation coefficient; Y = mean annual sediment yield 1956–1998.

SOIL SAMPLING AND ^{137}Cs ANALYSIS

Soil sampling within the study catchments involved several campaigns. The first campaigns, which focused on catchments W2 and W3 and the reference sites, were undertaken in 1998, 1999 and 2001 and are described in a previous paper (Porto *et al.*, 2003). The final campaign, undertaken within catchment W1, involved the collection of 68 bulk cores. The cores were collected at the intersections of an approximate 20 m × 10 m grid, with additional cores to take account of topographic variability. The samples were collected using a steel core tube (6.9 cm diameter) driven into the ground by a motorized percussion corer and subsequently extracted using a hand-operated winch.

The samples used to establish the reference inventory were collected from three different areas and following Porto *et al.* (2003) they are referred to as:

- Reference site 1: located in catchment W1 within an area characterized by permanent grassland with minimal slope;
- Reference site 2: located in catchment W3 within a small clearing between the trees;
- Reference site 3: located within adjacent undisturbed rangeland with some scattered oaks (*Quercus pubescens*) at a similar altitude to the study catchments. This third site was selected because of the difficulties in finding an undisturbed area within catchment W2.

At the reference sites, a scraper plate (Campbell *et al.*, 1988) was used to collect samples at depth increments ranging from 1 to 4 cm, to a depth of 50 cm. Six additional 8.6-cm diameter bulk soil cores were collected from each reference site, in order to take account of the local variability of ^{137}Cs inventories. Additional scraper plate samples were obtained from several representative eroding and depositional locations within the study catchments, in order to document the ^{137}Cs depth distribution.

All bulk core and depth incremental scraper plate samples were oven dried at 105°C for 48 h, mechanically disaggregated and dry sieved to recover the <2 mm fraction. A representative sub-sample of this fraction was packed into a plastic Marinelli beaker for determination of ^{137}Cs activity by gamma spectrometry. The measurements were made using high resolution HPGe detectors in the laboratories of the Department of Nuclear Engineering at the University of Palermo, Italy and the Department of Geography at the University of Exeter, UK. Count times were typically approx. 30000s, providing a precision of about ±10% at the 95% level of confidence. All ^{137}Cs measurements were standardized to a fixed date at the end of 2001.

^{137}Cs inventories and depth distributions at the reference sites

The ^{137}Cs inventories obtained for the three reference sites ranged from 2430 to 2492 Bq m^{-2} with a mean value of 2456 Bq m^{-2} (Porto *et al.*, 2003). The ^{137}Cs depth distributions

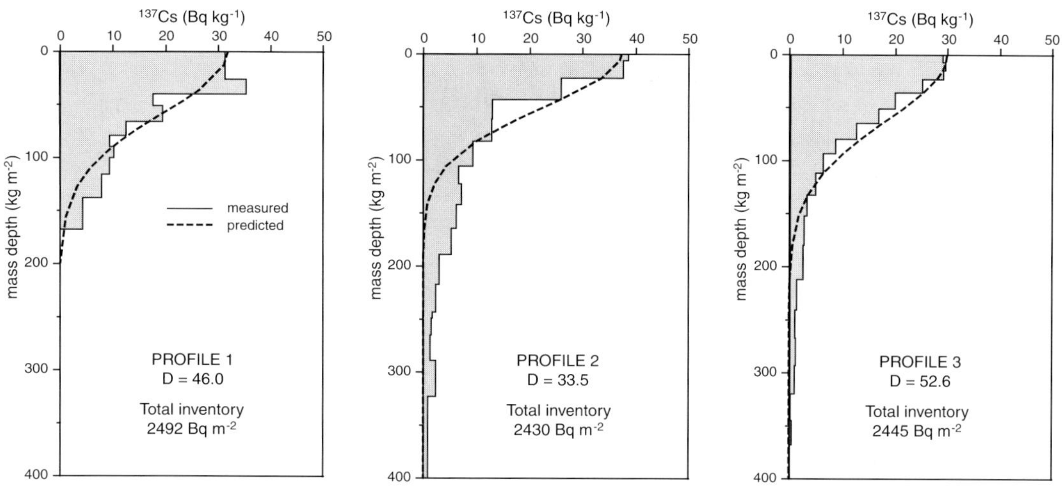

Fig. 3 The ^{137}Cs depth distributions documented for reference sites 1, 2 and 3 and the results of fitting the diffusion and migration model (equation 1) to these profiles.

associated with these profiles are depicted in Fig. 3. The shape of these profiles, showing a steady (exponential) decline with depth from a maximum at, or immediately below, the soil surface, conforms to that expected from undisturbed locations (see Walling & Quine, 1992).

^{137}Cs inventories within the study catchments

The range of the values of ^{137}Cs inventory obtained for the 68 bulk cores collected from catchment W1, and standardized to the year 2001, are listed in Table 3 together with the equivalent values for the samples collected previously from catchments W2 and W3. All values of ^{137}Cs inventory for the cores collected from catchment W2 and W3 are less than the local reference inventory (2456 Bq m^{-2}), suggesting that the surfaces of both catchments have been dominated by net soil loss over the period covered by the ^{137}Cs measurements. In the case of catchment W1, however, the inventories for some cores exceeded the reference inventory, pointing to the existence of depositional areas. Figures 4(a) and 4(b) provide two typical examples of the depth distribution of ^{137}Cs for sampling points within the study catchments. That shown in Fig. 4(a) is representative of an eroding site in catchment W2, whereas that shown in Fig. 4(b), which was collected from catchment W1, is characterized by an inventory greater than the local reference value and is therefore indicative of a depositional site.

Table 3 ^{137}Cs inventories within the study catchments.

Catchment	Min (Bq m^{-2})	Max (Bq m^{-2})	Mean (Bq m^{-2})
W1	4.1	4053	1395
W2	18	2429	867
W3	81	2430	1070

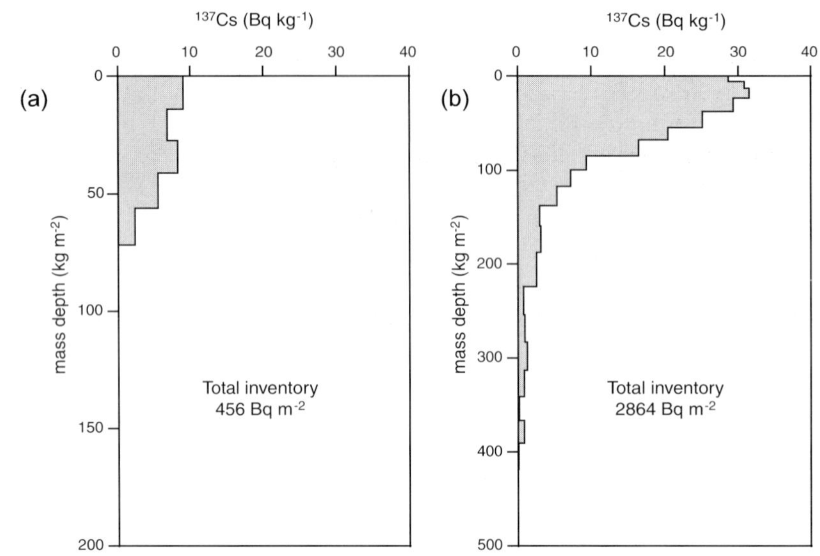

Fig. 4 Representative ^{137}Cs depth distributions for an eroding site in catchment W2 (a) and a depositional site in catchment W1 (b).

USING ^{137}Cs MEASUREMENTS TO ESTIMATE EROSION RATES

Estimation of erosion and deposition rates from ^{137}Cs measurements is commonly based on a comparison of the inventory measured at a specific point with the reference inventory and on the degree of reduction or increase of that inventory. The conversion models, used for uncultivated land to convert the magnitude of the reduction or increase in the ^{137}Cs inventory to an estimate of the rate of soil loss or deposition, employ different theoretical representations of the vertical distribution of ^{137}Cs within the soil and its evolution through time (see Walling & He, 1999). The simple profile distribution model assumes a standard time-invariant exponential depth distribution, which is truncated by erosion, whereas the diffusion and migration model proposed by Walling & He (1999) attempts to model the interaction of surface lowering by erosion with the progressive development of the depth distribution, in response to continuing fallout inputs to the surface and downward diffusion and migration of the radiocaesium.

In this study attention has focused on the diffusion and migration model, which has been shown by Porto *et al.* (2003) to produce more consistent and reliable results than the profile distribution model. Further details of this model can be found in Porto *et al.* (2003), but its main features are briefly summarized below.

The vertical distribution of ^{137}Cs within a soil will vary through time in response to the time-dependent fallout input and the post-depositional redistribution of ^{137}Cs within the profile. For soils with high erosion rates, assuming a constant soil lowering E (kg m^{-2}) and diffusional transport, the vertical distribution of ^{137}Cs within a soil, for any cumulative mass depth and time t', can be expressed (Porto *et al.*, 2003) as:

$$C_e(x,t,t') = \mathrm{e}^{-\lambda(t-t')} \int_0^\infty \frac{I(t')}{H} e^{-\frac{y}{H}} \left\{ \left[e^{-\frac{[(x+E)+y]^2}{4D(t-t')}} + e^{-\frac{[(x+E)-y]^2}{4D(t-t')}} \right] \frac{1}{\sqrt{4\pi D(t-t')}} \right\} \mathrm{d}y \qquad (1)$$

where:
y = integration variable which indicates the depth of soil involved in ^{137}Cs diffusion;
$C_e(x,t,t')$ = the concentration of ^{137}Cs for any cumulative mass depth x and time t' (Bq kg^{-1});
D = the effective diffusion coefficient (kg^2 m^{-4} year^{-1});
H = the relaxation depth expressed as a mass depth (kg m^{-2});
λ = the decay constant for ^{137}Cs (0.023 year^{-1});
x = the mass depth from the soil surface downwards (kg m^{-2});
t = the time since the first deposition of ^{137}Cs (year);
$I(t')$ = the input (Bq m^{-2} year^{-1}) at time t'.

The ^{137}Cs concentration distribution $C_e(x,t)$ (Bq kg^{-1}) in the soil profile at time t can be obtained by integrating $C_e(x,t,t')$ over time t':

$$C_e(x,t) = \int_0^t C_e(x,t,t')\mathrm{d}t \qquad (2)$$

Integration of $C_e(x,t)$ over mass depth x gives the total ^{137}Cs inventory A_u (Bq m^{-2}) for an erosion site at time t:

$$A_u(t) = \int_0^\infty C_e(x,t)\ \mathrm{d}x \qquad (3)$$

Assuming a constant value of H (5 kg m^{-2}), as suggested by Walling & He (1999), equations (1) and (3) can be solved simultaneously for E (kg m^{-2}), with A_u (Bq m^{-2}) representing the measured inventory at an eroding point. The erosion rate R (kg m^{-2} year^{-1}) may then be estimated by dividing the quantity E by the time $t - t_0$ (year) since the commencement of ^{137}Cs fallout. Equation 1 can be fitted to the ^{137}Cs depth distribution at reference locations where E = 0. This was achieved by calculating the parameter D for each profile using least squares. The results of this fitting are depicted in Fig. 3.

For a depositional site, the deposition rate D_R can be estimated from the ^{137}Cs concentration in deposited sediment $C_d(t')$ and the excess ^{137}Cs inventory (defined as the total measured ^{137}Cs inventory A_u less the local reference inventory A_{ref}) using the following relationship (Walling & He, 1999):

$$D_R = \frac{A_u - A_{ref}}{\int_{t_0}^{t} C_d(t')\, e^{-\lambda(t-t')}\mathrm{d}t'} \qquad (4)$$

Assuming that the ^{137}Cs concentration $C_d(t')$ of deposited sediment can be represented by the weighted mean of the ^{137}Cs concentration of the sediment mobilized from the upslope contributing area, $C_d(t')$ can be calculated as:

$$C_d(t') = \frac{1}{\int_S R\, dS} \int_S C_e(t')\, R\, \mathrm{d}S \qquad (5)$$

where S (m^2) is the upslope contributing area and $C_e(t')$ (Bq kg^{-1}) is the ^{137}Cs concentration of sediment mobilized from an eroding point, which can be calculated from equation (1), for $x = 0$.

COMPARISON OF THE ESTIMATES OF SOIL LOSS DERIVED FROM THE ^{137}Cs MEASUREMENTS WITH THE SEDIMENT OUTPUT

The diffusion and migration model provided point estimates of soil redistribution rates (t ha^{-1} year^{-1}), based upon the ^{137}Cs inventories measured for individual soil cores. These point values were interpolated using a kriging procedure to derive a spatially representative estimate of net soil loss for each catchment.

Table 4 provides a comparison between the measured sediment output from the three study catchments and the equivalent estimates of total or net soil loss obtained from the ^{137}Cs measurements. The conversion models were parameterized using the three reference profiles depicted in Fig. 4 and three estimates of soil loss are therefore provided. As reported in previous papers (Porto *et al.*, 2001, 2003), there was no evidence of significant deposition within catchments W2 and W3. In this case, a direct comparison between the estimates of total soil loss derived from the ^{137}Cs measurements and the measured sediment output from the catchments is possible. In contrast, for catchment W1, where evidence of deposition was found, Table 4 lists the estimates of gross soil loss, total deposition, and net soil loss.

Table 4 Comparison between measured and calculated erosion rates using the proposed model (t ha^{-1} year^{-1}).

Profile	Total soil erosion W1	Total deposition W1	Net soil erosion: W1	W2	W3
	Equation (1) and (3)	Equation (4)	Equation (1) and (3)		
1	8.84	0.19	8.6	14.0	11.3
2	7.34	0.18	7.2	11.8	9.4
3	9.16	0.23	8.9	14.7	11.7
Mean annual sediment yield			12.4	19.2	7.8

Table 4 indicates that the estimates of soil loss from the three study catchments derived from the ^{137}Cs measurements using the diffusion and migration conversion model are reasonably consistent with the measured sediment outputs from the catchments. Very close agreement would not necessarily be expected, since both the measured sediment yields and the estimates of net soil loss derived from the ^{137}Cs measurements involve a number of potential errors and uncertainties. For example, the estimates of net soil loss for the individual catchments derived from the ^{137}Cs measurements are clearly sensitive to the number and precise location of the sampling points and uncertainties due to sampling and analytical precision and the spatial interpolation procedure employed. Similarly, the estimates of sediment output from the catchments, based on the measured sediment fluxes, are subject to various uncertainties relating to the measurement procedures, as well as the procedure used to extrapolate the measurements to the longer period covered by the ^{137}Cs measurements. In this context, it can be seen that the relative magnitudes of the estimates of soil loss derived from the ^{137}Cs measurements are not entirely consistent with those of the measured sediment yields. Although both values are highest for catchment W2, the lowest values of sediment yield are associated with catchment W3, whereas the lowest values of estimated soil loss are associated with catchment W1. Furthermore, whereas the measured sediment yields exceed the estimates of soil loss for catchments W2 and W1, the reverse situation exists for catchment W3.

CONCLUSIONS

The broad agreement between the magnitude of the estimates of sediment output from the three study catchments and the estimates of soil loss from the catchments derived from ^{137}Cs measurements, provides a clear confirmation of the validity of the ^{137}Cs technique for estimating erosion and deposition rates from uncultivated areas in general, and of the diffusion and migration conversion model in particular. The differences between the measured sediment yields and the estimates of soil loss derived from the ^{137}Cs measurements may reflect the uncertainties associated with the two approaches. Equally, they could reflect minor limitations in the results provided by the ^{137}Cs measurements. In this context, it is possible that the contrast between the forested (W2 and W3) and grassland (W1) catchments is not adequately represented by the conversion model or the ^{137}Cs depth profiles documented for the reference sites. Further work is clearly required to investigate these minor inconsistencies.

Acknowledgements The study reported in this paper was supported by an EC Marie Curie Fellowship (MCF1-2000-00096) awarded to the first author and IAEA Technical Contract UK-12094. Thanks are also extended to Mr G. Guarnaccia and Mr D. Porcino for assistance with fieldwork and to Sue Rouillard for producing the diagrams.

REFERENCES

Campbell, B. L., Loughran, R. J. & Elliott, G. L. (1988) A method for determining sediment budget using caesium-137. In: *Sediment Budgets* (ed. by M. P. Bordas & D. E. Walling) (Proc. Porto Alegre Symposium), 171–179. IAHS Publ. 174. IAHS Press, Wallingford, UK.

Cantore, V., Iovino, F. & Puglisi, S. (1994) Influenza della forma di governo sui deflussi liquidi e solidi in piantagioni di eucalitti (in Italian). *L'Italia Forestale e Montana* **5**, 463–477.

Ferro, V. & Porto, P. (2000) SEdiment Delivery Distributed (SEDD) Model: Theoretical analysis and experimental validation. *J. Hydrologic Engng* **5**, 411–422.

Loughran, R. J. (1989) The measurements of soil erosion. *Prog. Phys. Geog.* **13**, 216–233.

Porto, P., Walling, D. E. & Ferro, V. (2001) Validating the use of caesium-137 measurements to estimate soil erosion rates in a small drainage basin in Calabria, southern Italy. *J. Hydrol.* **248**, 93–108.

Porto, P., Walling, D. E., Ferro, V. & Di Stefano, C. (2003) Validating erosion rate estimates provided by caesium-137 measurements for two small forested catchments in Calabria, Southern Italy. *Land Degrad. Develop.* **14**, 389–408.

Walling, D. E. (1998) Use of ^{137}Cs and other fallout radionuclides in soil erosion investigations: progress, problems and prospects. In: *Use of ^{137}Cs in the Study of Soil Erosion and Sedimentation*. IAEA-TECDOC-102839–62. IAEA, Vienna, Austria

Walling, D. E. & He, Q. (1999) Improved models for estimating soil erosion rates from cesium-137 measurements. *J. Environ. Qual.* **28**, 611–622.

Walling, D. E. & Quine, T. A. (1992) The use of caesium-137 measurements in soil erosion surveys. In: *Erosion and Sediment Transport Monitoring Programmes in River Basins* (ed. by J. Bogen, T. Fergus & D. E. Walling) (Proc. Oslo Workshop, June 1992), 143–152. IAHS Publ. 210. IAHS Press, Wallingford, UK.

Process interaction and sediment delivery in the Pleiser Hügelland, Germany

N. J. PRESTON[1] **& R. DIKAU**[2]

1 *Department of Physical Geography, Macquarie University, North Ryde, New South Wales 2109, Australia*
npreston@els.mq.edu.au

2 *Geographisches Institut, Universität Bonn, Meckenheimer Allee 166, D-53115 Bonn, Germany*

Abstract Sediment redistribution rates in a small agricultural catchment in the loess-covered hill country of western Germany have been modelled using ^{137}Cs. The study site comprises three parcels, two of which are cultivated and the third is pasture. The pattern of erosion and deposition is broadly controlled by water erosion, but the effect of tillage translocation is important. Tillage appears to enhance rates of water erosion within parcels, but it retards sediment export from the parcel, emphasizing the importance of land-use boundaries.

Key words configuration; ^{137}Cs; Germany; land-use; process interaction; sediment delivery; tillage; water erosion

INTRODUCTION

Tillage translocation has been recognized as an important geomorphic process (e.g. Gov94ers *et al.*, 1994). However, the broader implications of tillage translocation for sediment redistribution dynamics within catchments remain relatively unexplored. Many catchment-scale models of sediment redistribution treat geomorphic processes individually, although it is likely that interaction between processes may play an important part in determining patterns of sediment redistribution.

One of the most powerful tools available for exploring soil and sediment redistribution is ^{137}Cs (e.g. Zapata, 2002). The movement of ^{137}Cs through the landscape represents an integral signal of all sediment redistributive processes that have been active over a period of decades. For the agricultural area of the Pleiser Hügelland, east of Bonn, these principally include water erosion and tillage translocation. There are important differences between these two types of process, relating to distance of transport, size selectivity of particle entrainment and deposition, extent of occurrence and frequency/magnitude spectra. Thus land-use, insofar as it determines the processes by which ^{137}Cs in the soil may potentially be eroded and transported, is an important factor in modelling 9 sediment redistribution.

In this paper we use ^{137}Cs to explore the interaction of water erosion and tillage translocation in the Auf dem Scheid catchment. We have used ^{137}Cs to reconstruct patterns of sediment movement associated with each of these two processes. In addition, the use of a ^{137}Cs-derived mass balance shows variation in sediment delivery with land-use.

STUDY AREA AND METHODS

The Pleiser Hügelland is a predominantly agricultural area to the east of Bonn (Fig. 1). Average annual rainfall is 700 mm, with highest intensities generally occurring in the early

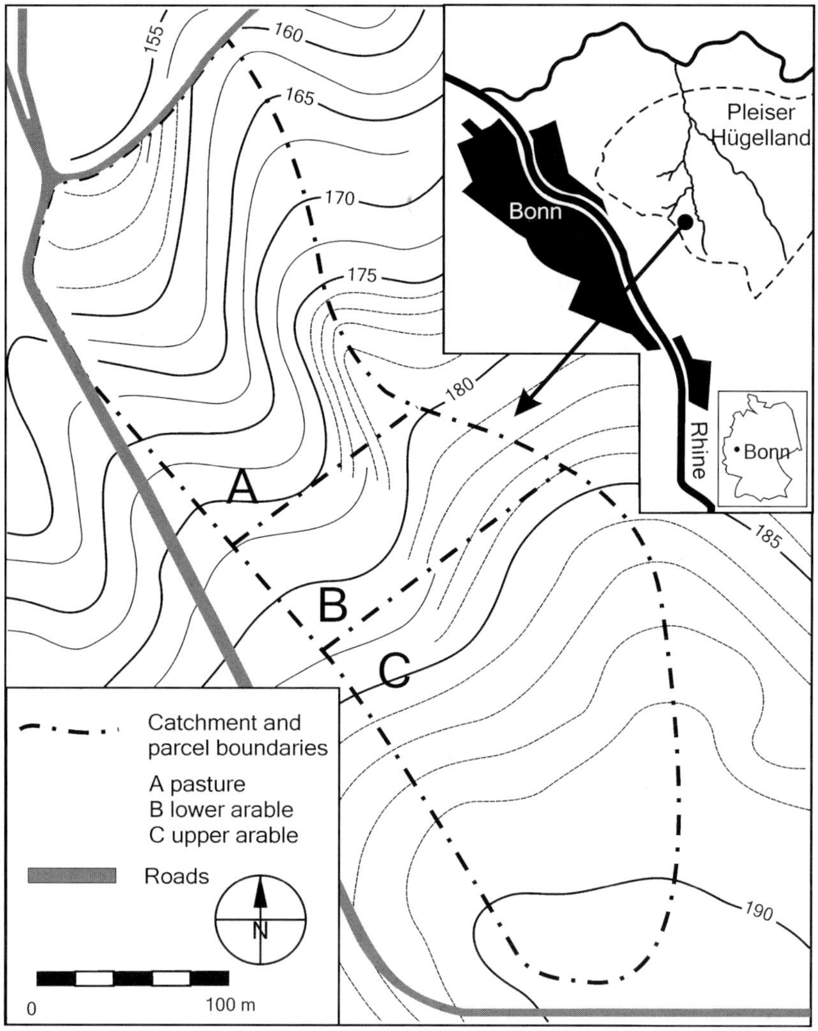

Fig. 1 Location of the study site. Note the three parcels: A, pasture; B and C, arable land.

summer months when fields are least protected by a crop cover. The study site—Auf dem Scheid—is a small (approx. 5.4 ha) zero-order basin. Soils are formed in Pleistocene loess, and most have been extensively eroded as a result of agricultural use. The lower part of the Auf dem Scheid catchment is under pasture, while the upper part is ploughed with a three-way crop rotation. The boundary between these two land-uses is marked by an abrupt break in slope, representing the accumulation of approx. 1 m of material at its greatest. A farm road dating back to at least 1940 defines the lower boundary of the catchment. This road and artificial changes to drainage lines that were undertaken in the 1950s mean that hydrological coupling between the Auf dem Scheid basin and the higher order drainage system is buffered, and under normal conditions Auf dem Scheid is a closed sediment redistribution system.

In an initial survey, ^{137}Cs was found to a depth of 0.3 m at a nearby depositional site. It was therefore considered that sampling to a depth of 0.5 m would enable recovery of all

Table 1 Values used in parameterization of sediment redistribution models.

Factor	Mass balance model	Diffusion and migration model
Proportionality factor (γ)	0.05	n/a
Relaxation mass depth (H)	3.8 kg m^{-2}	5.2 kg m^{-2}
Mass depth of the plough layer	300 kg m^{-2}	n/a
Tillage constant	720 kg m^{-1} year^{-1}	n/a
Diffusion coefficient (D)	n/a	63.5 kg m^{-4} year^{-1}
Migration rate coefficient (V)	n/a	1.13 kg m^{-2} year^{-1}
Reference inventory	5530 Bq m^{-2}	5530 Bq m^{-2}

^{137}Cs present in the soil column. Bulk samples were taken in an open cylinder with a diameter of 6.8 cm. Concentrations of ^{137}Cs per unit weight (Bq kg^{-1}) were determined by high-resolution γ-spectrometry and converted to areal activities (Bq m^{-2}).

Sediment redistribution rates were estimated using numerical models based on the behaviour of ^{137}Cs and the physical processes of sediment redistribution. A mass balance model that incorporates the tillage process was applied to the arable parcels and a migration-diffusion model to the pasture parcel. Full descriptions of these models are given by Walling & He (1999, 2001) and Walling *et al.* (2002); their parameterization is described here and summarized in Table 1.

As input the mass balance model requires:

(a) A proportional factor (γ), representing the proportion of ^{137}Cs fallout that is removed in runoff before incorporation in the plough layer. The value of γ is dependent on the timing of tillage relative to the occurrence of erosive rainfall, and has a maximum value of 1.0 in the hypothetical case of all erosive rainfall occurring immediately prior to tillage. While rainfall in the study area is relatively low, potentially erosive rainfall does typically occur when arable surfaces are at their most exposed. A low proportionality factor of 0.05 was applied to reflect the possibility that loss of ^{137}Cs in runoff may occasionally occur before tillage incorporation.

(b) A relaxation mass depth factor (H, kg m^{-2}) for the depth to which ^{137}Cs initially infiltrates when first delivered to the soil's surface. He & Walling (1997) have published empirical values of H: 3.8 kg m^{-2} for cultivated soil and 5.2 kg m^{-2} for undisturbed surfaces. The former has been adopted for use within the mass balance model for the arable zone.

(c) The mass depth of the plough layer. For Auf dem Scheid, the tillage depth is 20 cm and the average density of soils in the plough layer is 1499 kg m^{-3}. The mass depth of the plough layer is thus estimated as 299.8 kg m^{-2}, approximated to 300 kg m^{-2}.

(d) A tillage constant. This varies with type of tillage machinery and direction, timing and pattern of tillage. The tillage-induced sediment flux is generally higher for contour tillage. Lindstrom *et al.* (1992) report a value of 363 kg m^{-1} for each contour tillage operation. With two contour tillage operations per year in Auf dem Scheid, an approximate value of 720 kg m^{-1} year^{-1} has been adopted for the tillage constant.

The vertical distribution of ^{137}Cs within uncultivated soils is significantly different from that of tilled soils in which ^{137}Cs is mixed throughout the plough layer. In many cases, the vertical distribution of ^{137}Cs in the soil exhibits an exponential decline with depth. This does not remain constant, however, as ^{137}Cs slowly migrates through the soil profile. The diffusion and migration model allows for this. As input the model requires:

(a) A relaxation mass depth (*H*), representing the depth to which ^{137}Cs initially infiltrates. As with the mass balance model, a value published by He & Walling (1997) was used, i.e. 5.2 kg m^{-2} for undisturbed surfaces.

(b) A diffusion coefficient (*D*) and a migration rate coefficient (*V*) based on the depth of maximum concentration of ^{137}Cs and the depth at which concentration decreases to 1/*e* of the maximum. Detailed information on the vertical distribution of ^{137}Cs in Auf dem Scheid soils is not available, and these depths are assumed to be 5 and 11 cm respectively, based on representative profiles for silty soils published by Walling & Quine (1992). *D* is estimated to be 63.5 kg m^{-4} year^{-1} and *V* to be 1.13 kg m^{-2} year^{-1}.

Input data describing measured ^{137}Cs inventories for sample points consists simply of the point inventories. No correction was made for particle size because of the restricted range of soil textures in the Auf dem Scheid catchment, and the empirical observation that *in situ* soils and colluvium have essentially the same textures. The mass balance model specifically applies to slope transects, and requires that input data be arranged sequentially in a downslope direction. The input file thus contains, for each point: the measured ^{137}Cs inventory (Bq m^{-2}), the length of slope segment incorporating the point and the input and output angles to and from that slope segment. Both models require input data describing the temporal distribution of ^{137}Cs fallout. This has been derived from a dataset (Cambray *et al.*, 1989) supplied with the software used to run the model (Walling & He, 2001). Chernobyl fallout in 1986 was not included in this dataset, and a value of 700 Bq m^{-2}, based on values reported by Dörr & Münnich (1987) was added to the fallout dataset.

Undisturbed sites suitable for establishment of a reference inventory were not located within the extensively agriculturally used study area. Records of fallout are not available locally, and the spatial variability of fallout receipts published for other locations is such that values cannot be assumed to apply to this study area. Schimmack *et al.* (2001) have suggested that a mean value from sampled points can give a good approximation of the total ^{137}Cs fallout receipt, if it is assumed that all ^{137}Cs remains within the study area. Given the closed nature of the Auf dem Scheid catchment over the period of ^{137}Cs deposition, this assumption is likely to be satisfied, and the sample mean of 5530 Bq m^{-2} was used as a reference inventory.

RESULTS

The distribution of ^{137}Cs activity is illustrated in Fig. 2 (n = 120, mean = 5530 Bq m^{-2}, standard deviation = 1650.35 Bq m^{-2}, standard error of the mean = 150.06 Bq m^{-2}). Variability of measured activity is high, and the 95% confidence interval for the mean is 5232–5681 Bq m^{-2}.) Of the 120 samples, 20 have activities within this range, and may have influenced the location of boundaries between erosional and depositional areas to a small extent. The pattern is based on a kriging interpolation between points with known values and is thus only indicative. Within the pasture zone there are few points with activities greatly lower than reference. More marked in this zone are the high activities at the base of slopes and especially within the thalweg. The arable zone on the other hand is characterized more by the predominance of values lower than reference. Two patterns can be recognized. In the pasture (A) and lower arable (B) parcels topography appears to exert some control over ^{137}Cs

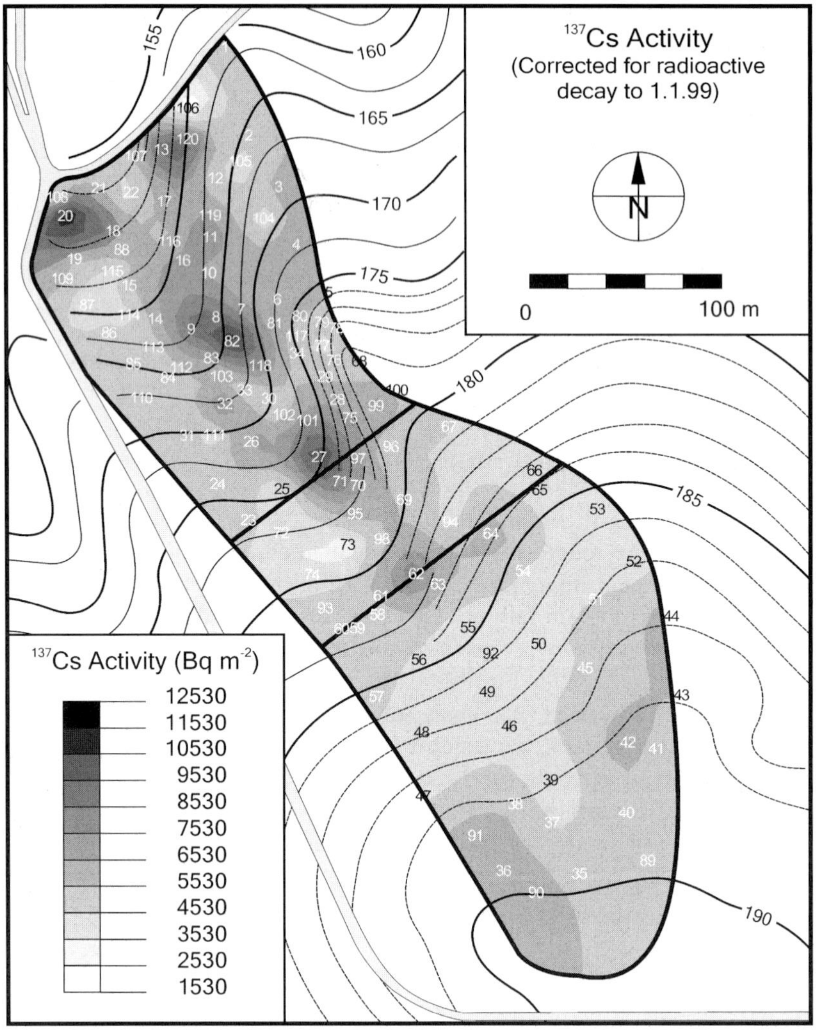

Fig. 2 Distribution of ^{137}Cs activity in Auf dem Scheid. Sample points are numbered. Intervals were chosen to indicate areas with values greater and smaller than the reference inventory of 5530 Bq m^{-2}.

activity. There is a broad relationship between topography and ^{137}Cs activity in the pasture zone, although this is by no means simple. This zone has much steeper slopes, but these do not exhibit especially low ^{137}Cs activities, and indeed in some cases show values greater than reference. Similarly, highest values in the lower arable field are in the thalweg while adjoining slopes have lower values—extremely low in the case of Points 72–74. The relationship of ^{137}Cs activity to topography appears to be reversed in the upper arable parcel (C) with values slightly higher than reference on its higher boundaries, but lower concentrations in the gently sloping middle section. A second pattern can be detected within the arable zone, reflecting its division into two fields with different cropping histories. A zone of higher activity is evident at the boundary between the two arable parcels. Again, there is a concentration of higher values at the lowest point of the boundary separating the lower arable parcel from the downslope pasture zone.

The spatial distribution of modelled net erosion and deposition rates is illustrated in Fig. 3. Clearly, there has been redistribution of sediment associated with water erosion in the pasture zone (Fig. 3(a)), but this has not been extreme, and the arable zone shows greater extremes of both erosion and deposition (Fig. 3(d)). In the arable zone the mass balance model distinguishes between rates of sediment redistribution due to both tillage translocation (Fig. 3(b)) and water erosion (Fig. 3(c)). For the greatest part of the arable zone, tillage produces relatively low rates of erosion. Deposition due to tillage, however, is restricted to a very limited area at the base of the arable zone, representing the development of a colluviation terrace at the arable field boundary. Water erosion produces a more diverse pattern of sediment redistribution and a greater range of values, especially for erosion. Different patterns can be discerned for the two parcels. The lower parcel is characterized by eroding slopes and an accumulating thalweg zone, while the upper parcel has experienced thalweg erosion and some accumulation on its almost flat flanks. The boundary between arable units represents a clear barrier to sediment transport in runoff, as best indicated by the accumulation at Points 62 and 63. This is also true of the land-use boundary dividing the catchment across its middle.

Patterns of ^{137}Cs redistribution also permit inferences to be made regarding sediment delivery. A grid of 25 m^2 cells has been overlaid on the net sediment redistribution map, and volumes of erosion and deposition within each of the three land-use units have been estimated by multiplying median values of erosion or deposition by numbers of cells within each class. Summing these provides estimates of volumes of material eroded from and accumulated within each zone and thus also of sediment delivery ratios (Table 2). The upper arable parcel is estimated to have lost to erosion an average 30.4 t year^{-1} over the period 1954-1999. Only 3.7 t year^{-1} was retained within this unit, which is thus a net sediment exporting zone with a high sediment delivery ratio (87.9%). The lower arable parcel is also a net sediment exporting unit, with erosional and depositional volumes of 10.9 t year^{-1} and 2.5 t year^{-1} respectively, and a sediment delivery ratio of 77.5%. The pasture parcel, by contrast, has accumulated material at a greater rate (6.3 t year^{-1}) than it has been eroded from within its boundaries (5.1 t year^{-1}), resulting in a negative sediment delivery ratio (–22.7%). It is difficult, however, to draw firm conclusions regarding extra-parcel sediment delivery, as there is a discrepancy between volumes of redistribution in absolute terms, apparently invalidating the closed system assumption. This assumption is considered to be justified, however, and the discrepancy may be attributable to the use of different approaches to estimate rates of sediment redistribution in each zone. Further, this sediment budget approach is highly dependent on sampling density and especially on the nature of spatial interpolation between points with known values. Given the costs and measurement time required for ^{137}Cs, 120 samples from an area of < 6 ha, with an error of mean estimation of ~5%, is considered reasonable. Inevitably, however, considerable resolution is missing, which influences the accuracy of interpolation. Comparison between the different land-use types in terms of absolute values is not valid, and it is emphasised that the rates and volumes estimated here are indicative only. Nevertheless, the patterns and results within each land-use unit, within which systematic modelling errors can be considered to be uniform, can be evaluated. These confirm that the arable units are indeed net sediment exporting units, and that the pasture zone has experienced deposition greater than can be accounted for by the sediments generated within its own boundaries.

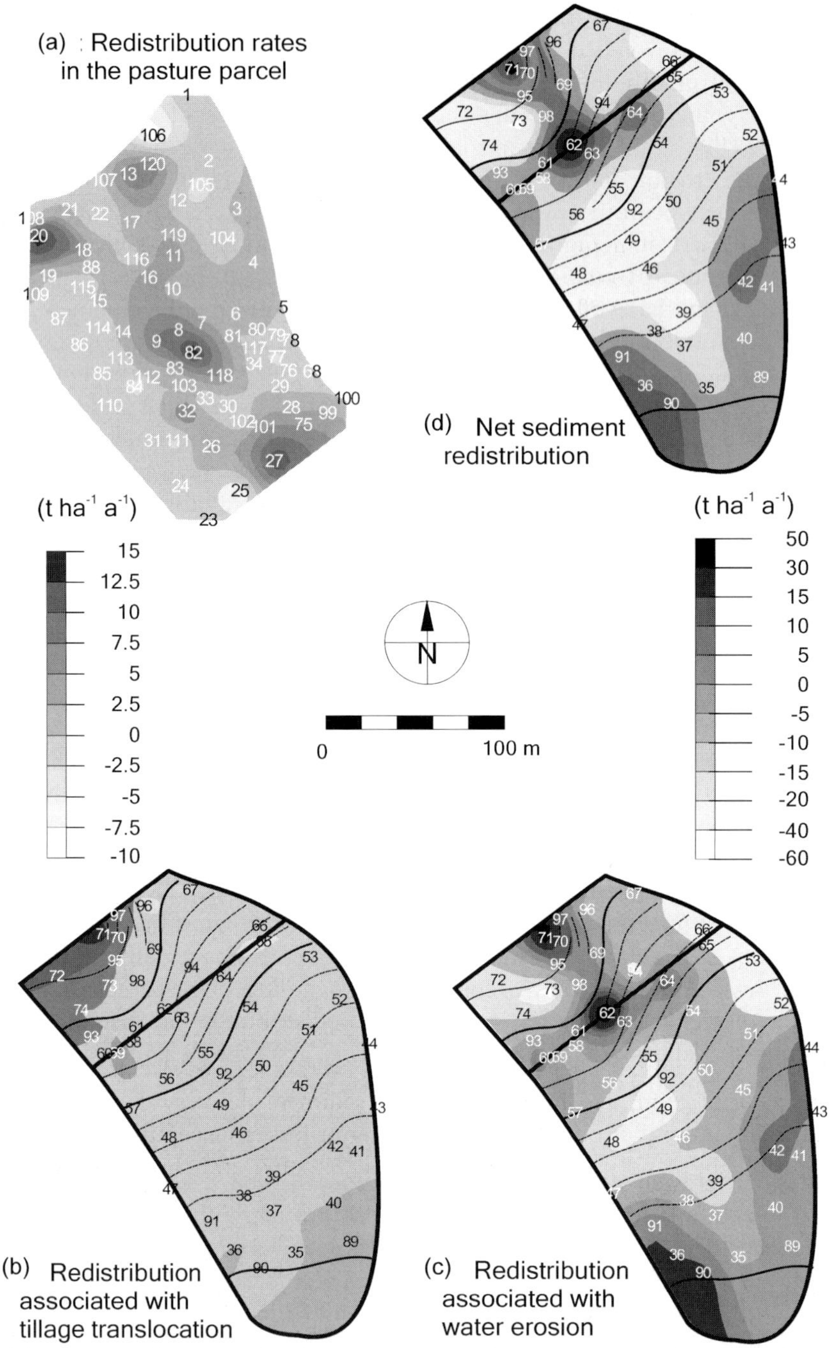

Fig. 3 Net rates of soil erosion and sediment deposition in Auf dem Scheid, modelled on the basis of ^{137}Cs activities. Sample points are numbered. Rates for the pasture zone (a) are modelled with a diffusion/migration model incorporating water erosion only, while those for the arable zone (b–d) are derived from a mass balance model parameterized for both water erosion and tillage translocation. The boundary between arable parcels is marked. Use the scale on the left for (a) and (b) and the right for (c) and (d).

Table 2 ^{137}Cs-derived sediment redistribution for the three land-use units of Auf dem Scheid.

	Class (t ha^{-1} year^{-1})	Median (t ha^{-1} year^{-1})	No. of cells	Area (ha)	Redistribution (t year^{-1})	Net (t year^{-1})	Delivery ratio
Pasture erosion	0–10	5	*410*	*1.025*	*5.125*		
Pasture deposition	0–10	5	431	1.0775	5.3875		
	10–20	15	19	0.0475	0.7125		
	20–30	25	3	0.0075	0.1875		
			453	*1.1325*	*6.2875*	1.16	–22.68
Lower arable erosion	0–10	5	59	0.1475	0.7375		
	10–20	15	108	0.27	4.05		
	20–30	25	54	0.135	3.375		
	30–40	35	13	0.0325	1.1375		
	40–50	45	9	0.0225	1.0125		
	50–60	55	4	0.01	0.55		
			247	*0.6175*	*10.8625*		
Lower arable deposition	0–10	5	37	0.0925	0.4625		
	10–20	15	20	0.05	0.75		
	20–30	25	6	0.015	0.375		
	30–40	35	6	0.015	0.525		
	40–50	45	3	0.0075	0.3375		
			72	*0.18*	*2.45*	–8.41	77.45
Upper arable erosion	0–10	5	306	0.765	3.825		
	10–20	15	280	0.7	10.5		
	20–30	25	185	0.4625	11.5625		
	30–40	35	52	0.13	4.55		
			823	*2.0575*	*30.4375*		
Upper arable deposition	0-10	5	111	0.2775	1.3875		
	10–20	15	55	0.1375	2.0625		
	20–30	25	1	0.0025	0.0625		
	30–40	35	2	0.005	0.175		
			169	*0.4225*	*3.6875*	–26.75	87.89

DISCUSSION

These results show that, in addition to its importance as a process for generating sediment and moving it over small distances, with local morphological effects, tillage may also affect rates of water erosion, and in turn influence sediment delivery dynamics. Tillage can shift soil and sediment into or out of zones of greater or lesser susceptibility to water erosion, thus impacting on the supply of material for water erosion and fluvial transport. Whether this process interaction amplifies or dampens longer-term sediment delivery is likely to depend on the relative frequency/magnitude of each process, and on the specific configuration of tillage operation relative to topography. Modelled values suggest that water erosion is the dominant of the two processes occurring on the arable part of Auf dem Scheid (Fig. 3). Tillage appears to deliver readily erodible sediments to an area in the northwestern corner of the arable zone, where it is further reworked by water erosion, with a net effect of concentrating accumulation in the thalweg outlet to the pasture zone. This suggests that in this particular case the process interaction is acting to enhance sediment redistribution.

While the pattern of net sediment redistribution within the arable zone suggests enhanced sediment redistribution through water erosion, the interaction between water erosion and tillage translocation also has implications for the whole of Auf dem Scheid. There is fluvially-derived deposition at the hydrological outlets of both arable parcels, in addition to the tillage-derived accumulation of sediment along these edges. Clearly, not all of the mobilized material is exported from the parcel. Land-use boundaries thus appear to act as barriers to transport between parcels and to have an effect on sediment delivery. An implication of this is that the interaction between processes may enhance water erosion within the parcel, but that the effect of tillage is to retard sediment export. While this has been the effect in the Auf dem Scheid catchment, it will not necessarily be the case at other locations, or with a different suite of processes or land-uses. The extent to which this is true will partly depend on the relative predominance of each process, and especially on their frequency/magnitude spectra.

Because land-use boundaries can play an important role as barriers and buffers (see also van Oost *et al.*, 2000), it is important that they be properly incorporated within catchment-scale models of sediment redistribution. This will not be trivial, as the effect is likely to be complex and contingent on subtle variations in both land-use practice and the spatial configuration of land-uses. Characterizing the nature and location of tillage is likely to be one of the more important tasks required in the use of such models.

Acknowledgements This study was conducted at Universität Bonn, within the Sonderforschungsbereich 350 project funded by the German Research Foundation. ^{137}Cs activity was measured at the Max-Planck-Institut für Kernphysik in Heidelberg.

REFERENCES

Cambray, R. S., Playford, K. & Carpenter, R. C. (1989) Radioactive fallout in air and rain: results to the end of 1988. *UK Atomic Energy Authority Report 10155.*

Dörr, H. & Münnich, K.O. (1987) Spatial distribution of soil ^{137}Cs and ^{134}Cs in West Germany after Chernobyl. *Naturwissenschaften* **74**, 249–251.

Govers, G., Vandaele, K., Desmet, P. J. J., Poesen, J. W. A. & Bunte, K. (1994) The role of tillage in soil redistribution on hillslopes. *Europ. J. Soil Sci.* **45**, 469–478.

He, Q. & Walling, D. E. (1997) The distribution of fallout ^{137}Cs and ^{210}Pb in undisturbed and cultivated soils. *Appl. Radiation Isotopes* **48**, 677–690.

Lindstrom, M. J., Nelson, W. W. & Schumacher, T. E. (1992) Quantifying tillage erosion rates due to moldboard ploughing. *Soil Till. Res.* **24**, 243–255.

Schimmack, W., Auerswald, K. & Bunzl, K. (2001) Can $^{239+240}Pu$ replace ^{137}Cs as an erosion tracer in agricultural landscapes contaminated with Chernobyl fallout? *J. Environ. Radioactiv.* **53**, 41–57.

van Oost, K., Govers, G. & Desmet, P. (2000) Evaluating the effects of changes in landscape structure on soil erosion by water and tillage. *Landscape Ecol.* **15**, 577–589.

Walling, D. E. & He, Q. (1999) Improved models for estimating soil erosion rates from Caesium-137 measurements. *J. Environ. Qual.* **28**, 611–622.

Walling, D.E. & He, Q. (2001) Models for converting ^{137}Cs measurements to estimates of soil redistribution rates on cultivated and uncultivated soils, and estimating bomb-derived ^{137}Cs reference inventories (including software for model implementation). *A contribution to the IAEA Coordinated Research Programmes on Soil Erosion (D1.50.05) and Sedimentation (F3.10.01).* University of Exeter, UK.

Walling, D. E. & Quine, T. A. (1992) The use of caesium-137 measurements in soil erosion surveys. In: *Erosion and Sediment Transport Monitoring Programmes in River Basins* (ed. by J. Bogen, D. E. Walling & T. J. Day) (Proc. Oslo Symp., August 1992), 143–152. IAHS Publ. 210. IAHS Press, Wallingford, UK.

Walling, D.E., He, Q. & Appleby, P.G. (2002) Conversion models for use in soil-erosion, soil-redistribution and sedimentation investigations. In: *Handbook for the Assessment of Soil Erosion and Sedimentation using Environmental Radionuclides* (ed. by F. Zapata), 111–164. Kluwer Academic Publishers, Dordrecht, The Netherlands.

Zapata, F. (2002) *Handbook for the Assessment of Soil Erosion and Sedimentation using Environmental Radionuclides.* Kluwer Academic Publishers, Dordrecht, The Netherlands.

Applying various methods for assessing soil and sediment redistribution within an intensively cultivated dry valley subcatchment

VLADIMIR R. BELYAEV[1], MAXIM V. MARKELOV[1], VALENTIN N. GOLOSOV[1], YURI R. BELYAEV[2] & ELVIRA V. TISHKINA[1]

1 *The Laboratory of Soil Erosion and Channel Processes, Faculty of Geography, Moscow State University, GSP-2, 119992, Vorob'evy Gory, Moscow, Russia*
river@river.geogr.msu.su

2 *Department of Geomorphology and Palaeogeography, Faculty of Geography, Moscow State University, GSP-2, 119992, Vorob'evy Gory, Moscow, Russia*

Abstract Four independent methods have been employed to evaluate rates of soil and sediment redistribution during a period of intensive agriculture (about 300 years) for a small catchment (area 0.81 km^2) located within the forest zone of the northwestern Russian Plain. These included direct soil survey, radionuclide tracers (^{137}Cs and $^{210}Pb_{ex}$), and USLE-based modelling. Estimates for cultivated slopes vary from 6.4 to 24.2 t ha^{-1} $year^{-1}$. Up to 40% of mobilized soil has been redeposited on arable slopes within the catchment. Aggradation of the dry valley bottom has been estimated at 0.3–0.7 m. The intensity of soil and sediment redistribution is low in comparison with forest-steppe and steppe zones of the Russian plain.

Key words deposition; modelling; radionuclide tracers; small catchment; soil erosion

INTRODUCTION

Agricultural development has significantly influenced contemporary geomorphic processes on plains of the temperate climatic belt. Cultivated land expansion resulted in dramatically increased rates of soil erosion, gully formation and other exogenic processes. This general situation, however, significantly varies regionally, depending on local conditions.

The forested area of the Russian Plain, located to the north of Moscow, has been given much less attention in terms of human-induced accelerated soil erosion and sedimentation studies compared to forest-steppe and steppe zones to the south. We have aimed to begin filling this gap by studying soil redistribution within the small dry valley catchment located in the northern part of the Tver region, about 15 km to the southwest of Torzhok city (Fig. 1). Relief is characterized by hills and hilly ridges of glacial origin, with mainly convex slopes. Slope length is, in most cases, insufficient to promote gully formation under local surface runoff conditions. Therefore, sheet and rill erosion, together with mechanical translocation by cultivation, play a major role in soil and sediment redistribution.

CASE STUDY SITE CHARACTERISTICS

The case study area is located in the northwest of the Russian Plain, within the mixed forest zone. Mean annual precipitation is about 600 mm, relatively uniformly distributed within the year. Erosion events can be caused by both intensive snowmelt runoff and occasional

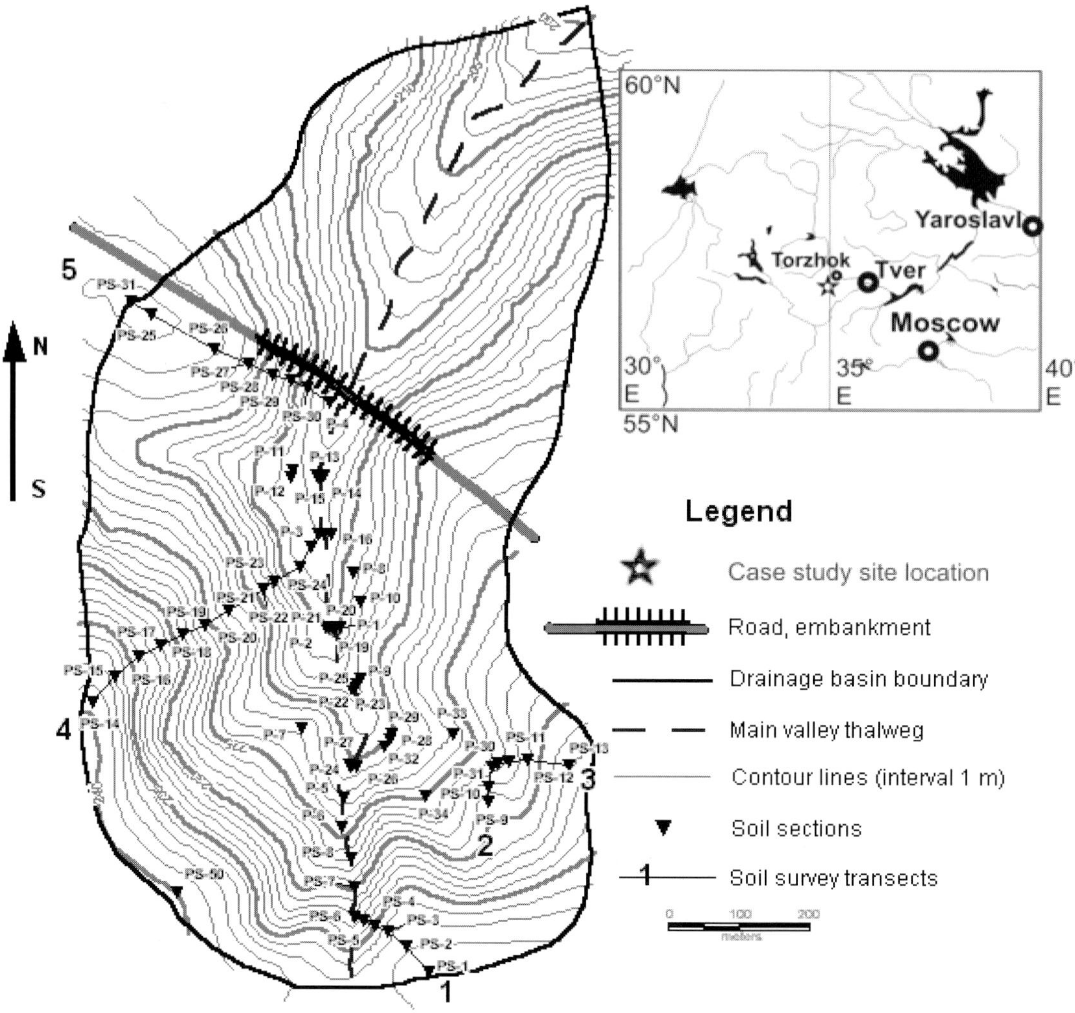

Fig. 1 Map of the northwest of the Russian Plain showing the case study site location and the general study subcatchment scheme with locations of slope transects and soil sections.

summer rainstorms. Rainfall erosivity coefficient of the USLE model is 6.8, estimated for the Torzhok city meteorological station. Soil cover of the region is dominated by podzolic soils formed on glacial or glaciofluvial deposits.

The catchment chosen as a key site for this case study has a drainage area of about 81 ha (Fig. 1). The main valley has a length of about 1.5 km, generally straight long profile (Fig. 2(a)) and average bottom width of 30–40 m. Catchment slopes have different profile shapes, but convex slopes dominate. Slope gradients are moderate, varying from 0.035 to 0.1. A number of slope depressions separate catchment slopes into sections (Fig. 3).

The catchment has been intensively cultivated for at least 300 years. Summer rye, oats, barley and wheat were the main crops before 1917; row crops were not cultivated (Pokrovskiy, 1879). During the Soviet times winter cereals were predominant. After 1991,

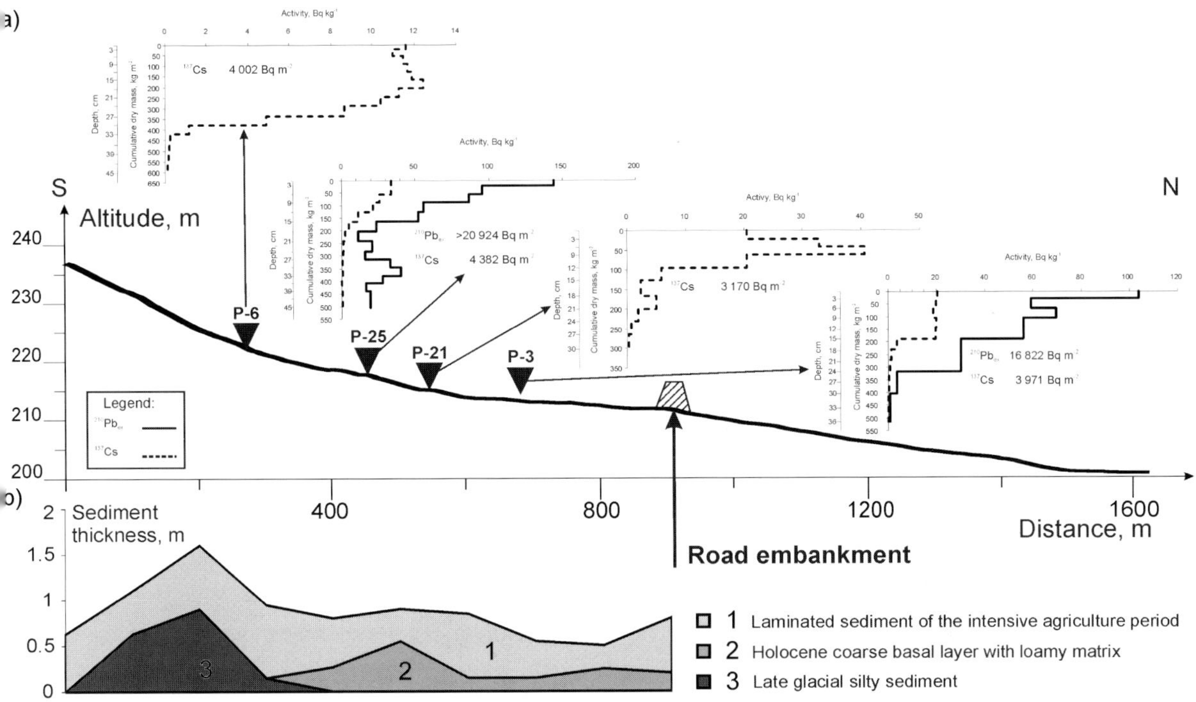

Fig. 2 Studied valley long profile with ^{137}Cs and $^{210}Pb_{ex}$ depth distributions at depositional locations within the bottom (a); and distribution of various sediment layers along part of the valley (b).

arable area slightly decreased and crop rotation changed from cereal-dominated to perennial grass-dominated, which has substantially lowered land vulnerability to water erosion. Today about 80% of the total catchment area is cultivated.

FIELD AND LABORATORY METHODS

The upper part of the study catchment (area about 0.54 km^2) separated by a road embankment, was chosen for the detailed study (Fig. 1). Within this area all major slope types characterizing the entire catchment are represented. The first stage of investigations included topographic and geomorphic mapping. (Figs 1, 3). Soil survey sections along transects within each of the slope types and on the main valley bottom cross sections were excavated and described in detail on the second stage. From some of those, radionuclide tracer soil sampling was carried out. Integral samples were taken from transect 1 (^{137}Cs) and transect 4 (^{137}Cs and $^{210}Pb_{ex}$). Depth-incremental samples were taken from the reference site section PS-31, sections P-3, P-25 (^{137}Cs and $^{210}Pb_{ex}$), P-6 and P-21 (^{137}Cs) (Figs 1, 2(a), 4). Subsequently samples have undergone preparation and counting in the laboratory with counting times not less than 12 h.

Methods employed for estimating soil redistribution rates included soil-morphological method (further—SMM) (Larionov *et al.*, 1973), radionuclide tracers ^{137}Cs (Walling & He, 1999a) and $^{210}Pb_{ex}$ (Walling & He, 1999b), and USLE-based modelling (further—USLE)

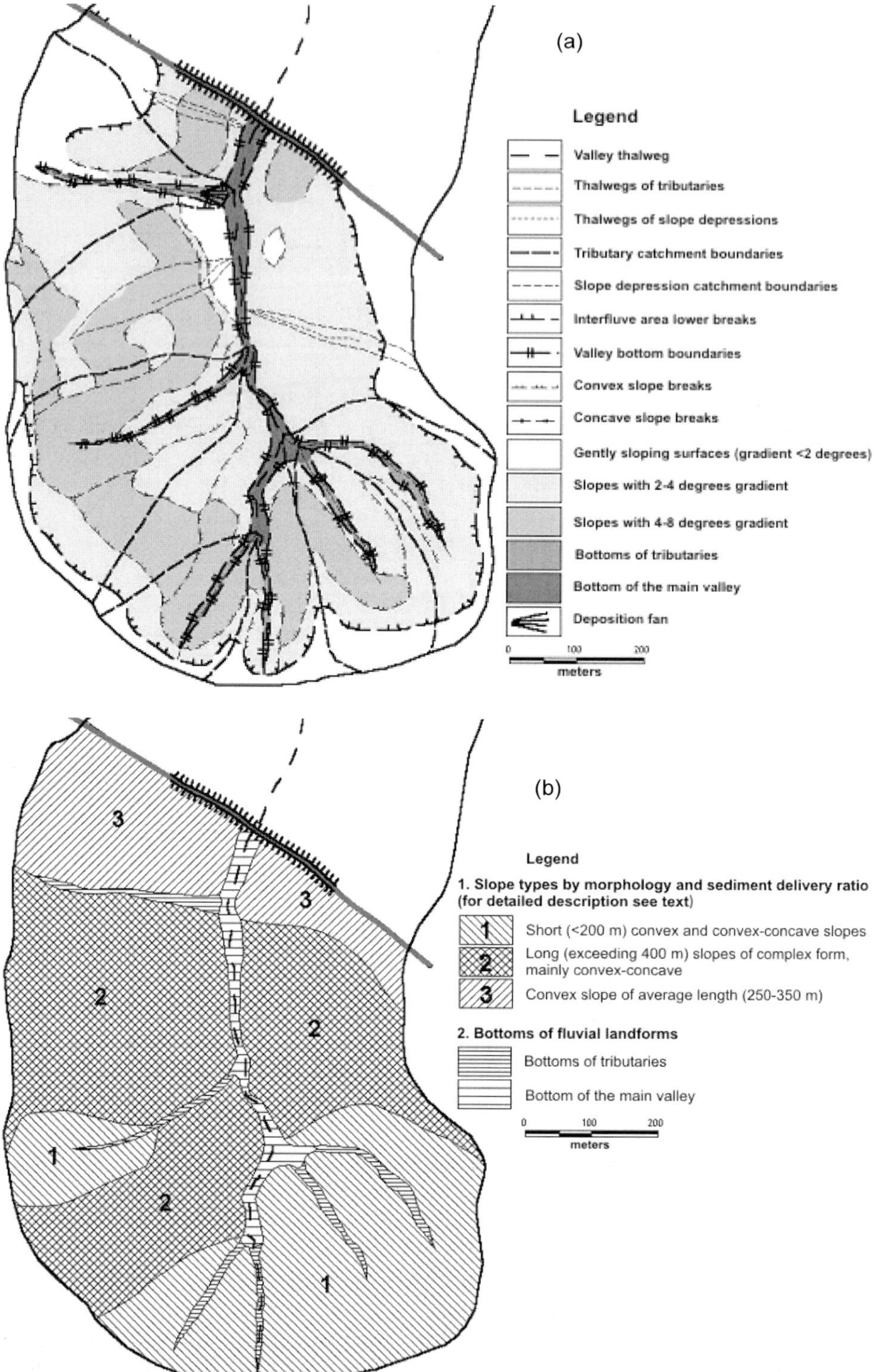

Fig. 3 Slope gradient map (a), and slope morphological types (b) in the study catchment.

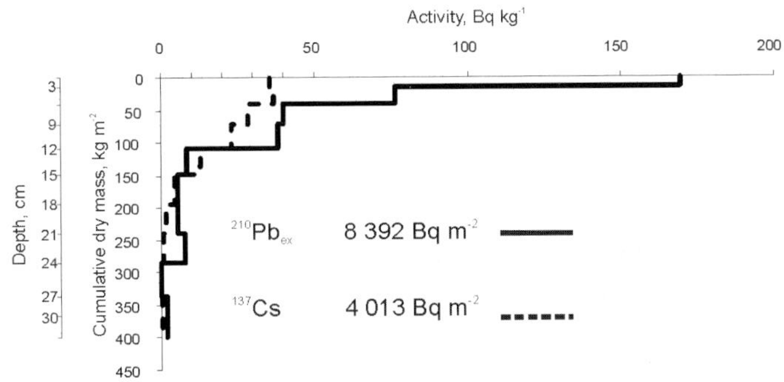

Fig. 4 Depth distribution of ^{137}Cs and ^{210}Pb$_{ex}$ in soil at the reference site (section PS-31).

(Larionov *et al.*, 1998; Krasnov *et al.*, 2001). In addition, the total thickness of the valley bottom sediment layer associated with a period of intensive agriculture has been determined (Fig. 2(b)) from abundant charcoal pieces commonly found at its base; the charcoal originates from forest burning during arable land clearing.

RESULTS AND DISCUSSION

Radionuclide tracers reference inventories

Depth distribution profiles of both ^{137}Cs and ^{210}Pb$_{ex}$ in soil (Fig. 4) at the reference site (section PS-31, Fig. 1) together with detailed description of soil structure led us to conclude that this site has not been cultivated since at least 1954. Obtained values of the ^{137}Cs and ^{210}Pb$_{ex}$ inventories can therefore be used as characteristic for a baseline fallout input in soil redistribution calibration models. About 93% of the ^{137}Cs inventory is found within the upper 15 cm of the section (Fig. 4). Profile shape differs from exponential in the upper part and is characteristic for territories with significant Chernobyl fallout. Modelling of the isotope vertical migration and diffusion (He & Walling, 1997) allowed us to determine its amount as being about 20% of the total inventory. For ^{210}Pb$_{ex}$ about 84% of the total inventory is contained within the upper 12 cm of soil. As ^{210}Pb$_{ex}$ fallout can be treated as a constant process, annual flux of the isotope from the atmosphere can be calculated from the reference inventory and radioactive decay constant. The value obtained is 260.9 Bq m^{-2} year^{-1}, which is in agreement with direct observations of ^{210}Pb$_{ex}$ atmospheric flux (Walling & He, 1999b).

Soil redistribution on the catchment slopes

Three main slope types have been distinguished according to profile shapes, length and slope break locations (Fig. 3(b)), basing on the morphological map of the studied part of the catchment (Fig. 3(a)). These included: (a) short (<200 m) mainly convex or convex-concave slopes (further referred to as Type I slopes); (b) long (≥400 m) slopes of complex form with slope breaks and terrace-like surfaces, but generally convex-concave (Type II); (c) intermediate length (250–350 m) convex slopes (Type III). All three types were characterized by soil survey transects along flow lines (Figs 1, 5, 6, 7).

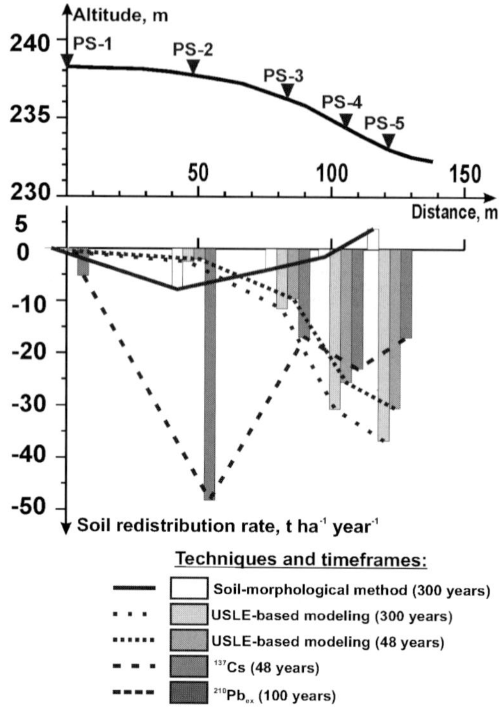

Fig. 5 Slope long profile along transect 1 and soil redistribution rates estimated by different methods.

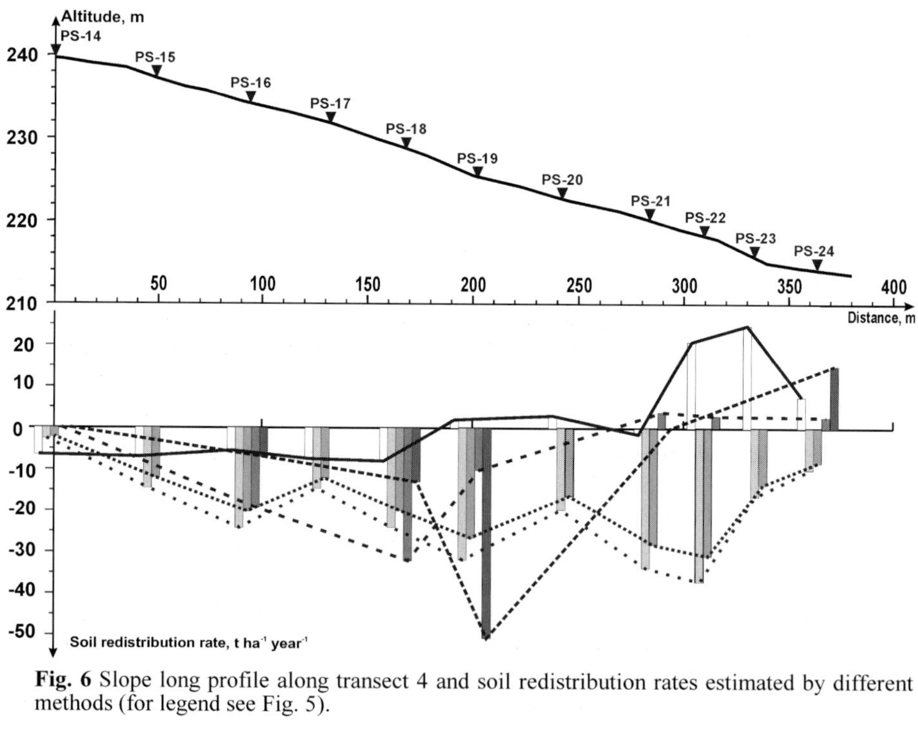

Fig. 6 Slope long profile along transect 4 and soil redistribution rates estimated by different methods (for legend see Fig. 5).

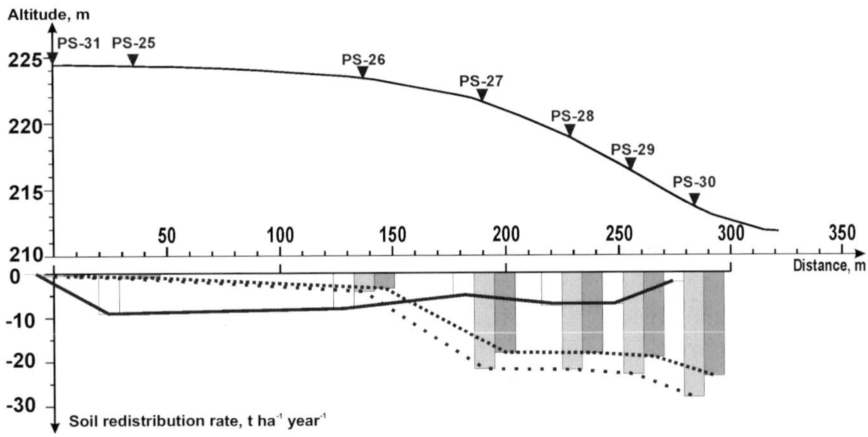

Fig. 7 Slope long profile along transect 5 and soil redistribution rates estimated by different methods (for legend see Fig. 5).

Slopes of Type I are located in the upper part of the studied subcatchment (Fig. 3(b)). These slopes can in turn be subdivided into three subtypes. Shortest (<100 m) convex diverging slopes are located on divides between the main valley and its tributaries. Convex slopes with a flat upper interfluve zone and length of 100 to 200 m characterize the main valley sides. The rest is represented by convex-concave slopes. To characterize all three subtypes, three soil transects were surveyed (Fig. 1, transects 1–3). Data obtained were averaged to characterize all Type I slopes. For transect 1 SMM, USLE and ^{137}Cs methods have been used for soil redistribution rate evaluation (Fig. 5). The first two methods yielded relatively similar values of average soil redistribution rates for the entire cultivation period (6.8 t ha^{-1} year^{-1} from SMM and 9.4 t ha^{-1} year^{-1} from USLE). For the 48 year period of ^{137}Cs presence in the environment (since 1954), this tracer gives a value of 22.6 t ha^{-1} year^{-1}, which is almost three times greater than the value (7.8 t ha^{-1} year^{-1}) obtained from modelling.

In general, all three methods employed gave an adequate evaluation of the dominance of erosion on Type I slopes. However, SMM is the only method to detect deposition. Its average rate is estimated to be 3.3 t ha^{-1} year^{-1}, but its spatial extent is very limited. Zones of erosion occupy about 75% of the Type I slope area, whereas detectable within-slope redeposition occurs only on 14%. Sediment delivery ratios for these slopes vary from 68 to 100% depending on the slope subtype. The only important sediment sinks are slope toes.

The Type II slopes are mainly found in the middle part of the studied subcatchment (Fig. 3b). These are characterized by a complex profile shape and a very gently sloping gradual transition zone to the main valley bottom. For the Type II slopes, all four methods of soil redistribution assessment described above have been employed (Fig. 6). For the 300-year period SMM estimates an average erosion rate of 6.0 t ha^{-1} year^{-1}, whereas USLE estimates 18.5 t ha^{-1} year^{-1}. Estimation for the 100-year period by ^{210}Pb$_{ex}$ (10.0 t ha^{-1} year^{-1}) is essentially close to that from SMM. For the 48-year period, USLE gives an average erosion rate of 15.4 t ha^{-1} year^{-1} and the ^{137}Cs method yields 25.8 t ha^{-1} year^{-1}.

Sediment redeposition within the Type II slopes is again adequately reflected only by SMM. Its average rate is estimated as 8.1 t ha^{-1} year^{-1}. Area of redeposition zones (9.3 ha or 37.6% of the entire Type II slope area) is substantially smaller than that of erosion zones (15.4 ha or 62.4%). The ^{137}Cs and ^{210}Pb$_{ex}$ methods most likely underestimate deposition rates. For sections PS-21, PS-22 and PS-23 (Figs 1, 6) presence of accumulation is only confirmed by high activity of both isotopes below the plough layer (comparable or even exceeding that of the plough layer) and an increase in their concentrations in the topsoil downslope, probably reflecting particle size selectivity. Despite this, good agreement is observed in relative area of erosion- and deposition-dominated zones between radionuclide tracers data and SMM. Sediment delivery ratio estimated from SMM is 19%, highlighting the important role of within-slope sediment sinks for the Type II slopes.

The Type III slopes are located only in the lower part of the studied subcatchment nearby the road embankment (Fig. 3(b)). These are characterized by a relatively long (up to 30% of total slope length) flat interfluve section and simple profile shape below with gradient increasing downward and a sharp boundary between slope toe and valley bottom. For the Type III slopes only SMM and USLE have been applied (Fig. 7). Results obtained are between those for the Type I and Type II slopes. Average erosion rates are 6.4 t ha^{-1} year^{-1} according to SMM and 13.3 (300-year period)—11.0 (48-year period) t ha^{-1} year^{-1} from USLE modelling. Simple morphology determines location of the erosion zone (about 80% of the slope length, except the flat interfluve part) and complete long term sediment delivery to the main valley bottom (ratio 100%).

Deposition in the main valley bottom

The geological structure of the main valley bottom is essentially uniform throughout the studied part of the catchment. Total sediment thickness (including pre-anthropogenic) does not exceed 1.5 m, with mean values of 0.8–1.0 m (Fig. 2(b)). Almost everywhere sediment is underlain by Late Pleistocene glacial boulder clays. The sediment layer associated with a period of intensive agriculture is found throughout the studied part of the main valley bottom. Its recent origin is supported by lateral connections with depositional sheets on adjacent slope toes. Within this layer we have observed a number of secondary erosional boundaries and depositional bodies which confirm repeating episodes of local incision and aggradation. Thickness of this layer varies from 0.3 to 0.7 m, the range of aggradation rates calculated from these data, assuming the 300-year cultivation period, are 1.0–2.5 mm year^{-1}.

Additional time marks in the main valley bottom sediment sequence have been acquired by analysis of four depth-incremental soil sections for radionuclide content (Fig. 2(a)). Radionuclide depth profiles obtained show that deposition rates within the main valley bottom vary significantly from point to point and in time. The ^{137}Cs depth profiles suggest that some parts of the main valley bottom were recently cultivated, as in its upper reach (section P-6) and low left bank terrace in the middle reach (section P-3). In addition, local bottom incision and infill episodes (sections P-21, P-25) influenced radionuclide content making estimations of average aggradation rates problematic (Figs 1, 2(a)). Nevertheless, approximate values of 1.0 mm year^{-1} (^{137}Cs, 48-year period) and 1.6 mm year^{-1} ($^{210}Pb_{ex}$, 100-year period) have been obtained.

General characteristics of sediment redistribution within the study subcatchment

Results of soil and sediment redistribution investigations by different methods for the studied part of the small valley catchment are summarized in Table 1. Average soil erosion rates on

Table 1 Summary of extrapolations of soil and sediment redistribution characteristics obtained from different methods for the entire studied subcatchment area.

Method	Soil-morphological		USLE-based model		^{137}Cs*	^{210}Pb$_{ex}$†
Period (year)	300	48‡	300	48	48	100
Erosion (t ha^{-1} year^{-1})	6.4		14.5	12.0	24.2	10.0
Area of erosion zones (ha/%§)	35.2/65.3		–	–	42.2/78.3	31.9/59.1
Erosion over the period (t)	67350.0	10776.0	220770.3	29323.4	49019.5	31914.0
Within-slope redeposition (t ha^{-1} year^{-1})	6.7		–	–	6.4	5.1
Area of within-slope redeposition zones (ha/%)	13.0/24.1		–	–	8.6/16.0	16.1/29.8
Within-slope redeposition over the period (t)	26190.0	4190.4	–	–	2640.0	8211.0
Sediment yield from slopes over the period (t)	41160.0	6585.6	220770.3	29323.4	46379.5	23703.0
Sediment delivery ratio (t)	61		–	–	95	74.3
Deposition in the valley bottom (t ha^{-1} year^{-1})	21.2**	15.7§	21.2**	15.7††		22.5‡‡
Area of the valley bottom (ha/%)	3.1/5.8					
Deposition in the valley bottom over the period (t)	19690.0**	2332.5§	19690.0**	2332.5††		6975‡‡

*From slope transects 1 and 4 only.
†From slope transect 4 only.
‡Extrapolated from longer-term estimates assuming similar average soil redistribution rates.
§Percentage of the studied subcatchment (total area of approximately 54 ha).
**Estimated from sediment section descriptions.
††Estimated from ^{137}Cs data.
‡‡Estimated from ^{210}Pb$_{ex}$ data.

the catchment slopes are moderate—estimations from different techniques vary from 6.4 to 24.2 t ha^{-1} year^{-1}. In most cases two zones of intensive erosion can be distinguished: upper slope convexity and the lower third of the slope. The former can be attributed to significant contribution of soil translocation by tillage, whereas the latter is associated with water erosion. It is believed that erosion rates have been decreasing since 1991, when a shift to less erosion-prone crop rotations dominated by perennial grasses began. This decrease is confirmed by the USLE calculations, but has no support from ^{137}Cs data because in our case it has overestimated erosion and underestimated deposition rates. Most likely explanations for this problem are: (a) insufficient sampling depth; (b) high spatial variability of the initial fallout has not been adequately statistically assessed; and (c) the calibration model does not account for the influence of extreme erosion events on ^{137}Cs content in soil. The ^{210}Pb$_{ex}$ method is less vulnerable to extremes as its annual fallout can be assumed to be constant (Robbins, 1978).

Within-slope accumulation of sediment is believed to be adequately reflected by SMM only. Deposition zones are most often found at slope toes and along boundaries or transition zones between slope and valley bottom. Additional within-slope sediment sinks exist on longer slopes with complex profile. Those can intercept up to 40% of the material eroded from upslope. The version of the USLE-based model employed is unable to account for within-slope redeposition, therefore its applicability for morphologically complex slopes (Type II) is limited.

Valley bottom aggradation in the studied catchment is much slower than that observed in similar catchments of forest-steppe and steppe zones of the Russian Plain (Golosov, 1998). During the last 48 years, as ^{137}Cs data from depositional locations suggest, the average rate of valley bottom aggradation has significantly decreased. This fact is in agreement with a recent decrease of soil erosion rates on catchment slopes. Local processes of bottom incision and infill are associated with increased erosion and deposition rates, but essentially result in sediment redistribution within the main valley bottom and do not influence the studied catchment sediment budget. In the longer term, as shown by SMM and $^{210}Pb_{ex}$ data, from half to two thirds of mobilized sediment has remained within the catchment, although this value requires further validation. The main results have been the smoothing of geomorphic boundaries, a slight decrease in local topography range and substantial transformation of the soil cover microvariability as various types of degraded, aggraded and degraded-aggraded soils have come into existence.

CONCLUSIONS

Modern rates of soil degradation on the studied catchment slopes (6.4–24.2 t ha^{-1} $year^{-1}$ by different methods) are moderate, but significant in terms of soil degradation as soils in the region are shallow, have low fertility and recover slowly. Sediment delivery from slopes and rates of valley bottom aggradation (1.0–2.5 mm $year^{-1}$) observed are low compared to those reported for similar catchments in forest-steppe and steppe zones of the Russian Plain.

General coincidence of spatial pattern of erosion and deposition zones on the most complex Type II arable slopes determined by soil-morphological and radionuclide tracer methods suggests the main tendency of soil and sediment redistribution has remained similar throughout the period of intensive agriculture. Maximum erosion occurs on upper convexities (associated with tillage translocation) and the steepest middle parts of slopes (associated with water erosion), with maximum deposition occurring at the toe of slopes.

Acknowledgements This research has been made possible thanks to financial support from the Russian Foundation for Basic Research (project 01-05-64503) and the leading scientific schools programme (project NS 1443.2003.5). The authors are grateful to Elena Kvasnikova, Valery Soloviev and Valery Averkov from the Institute of Global Climate and Ecology, Russian Academy of Sciences, for providing facilities and helping with radionuclide sample preparation and analysis.

REFERENCES

Golosov, V. N. (1998) Redistribution of sediments within small river catchments in the agricultural zone of Russia. *Geomorphol.: Relief, Processes, Environ.* **1**, 54–65.

He, Q. & Walling, D. E. (1997) The distribution of fallout ^{137}Cs and ^{210}Pb in undisturbed and cultivated soils. *Appl. Radiat. Isotopes* **48**, 677–690.

Krasnov, S. F., Dobrovolskaya, N. G. & Litvin, L. F. (2001) Prostranstvennye i vremennye aspekty ocenki erozionnogo indeksa osadkov (Spatial and temporal aspects of the rainfall erosivity evaluation) (in Russian). *Soil Erosion Channel Processes* **13**, 8–17.

Larionov, G. A., Kiryukhina, Z. P. & Samodurova, L. S. (1973) Opredelenie tempov ploskostnogo smyva metodom opisaniya parnyh pochvennyh razrezov (Determination of slope wash rates by the method of paired soil pits descriptions) (in Russian). *Soil Erosion Channel Processes* **3**, 162–167.

Larionov, G. A., Dobrovolskaya, N. G., Krasnov, S. F., Liu, B. Y. & Nearing, M. A. (1998) Teoretiko-empiricheskoe uravnenie faktora reliefa dlya statisticheskoy modeli vodnoy erozii pochv (Theoretical-empirical equation of topography factor for a statistical model of soil erosion by water). *Soil Erosion Channel Processes* **11**, 25–44 (in Russian).

Pokrovskiy, V. (1879) *Istoriko-statisticheskoe opisanie Tverskoy Gubernii* (Historic-statistical description of the Tver Region). Tver, Russia (in Russian).

Robbins R. A. (1978) Geochemical and geophysical application of radioactive lead. In: *The Biogeochemistry of Lead in the Environment* (ed. by J. O. Nriagu), 286–383. Elsevier, Amsterdam, The Netherlands.

Sidorchuk, A. Yu & Golosov, V. N. (1995) Calibration of soil erosion models using atmospheric fallout radionuclides. *Eurasian Soil.* **7**, 56–65.

Tsvetkov, M. A. (1957) *Izmenenie lesistosti Evropeiskoy Rossii s kontsa XVII v. po 1914 g* (Dynamics of forested areas in the European Russia from the late XVII century until 1914). RAS Publ., Moscow, Russia (in Russian).

Walling, D. E. & He, Q. (1999a) Improved models for estimating soil erosion rates from cesium-137 measurements. *J. Environ. Qual.* **28**(2), 611–622.

Walling, D. E. & He, Q. (1999b) Using fallout lead-210 measurements to estimate soil erosion on cultivated land. *Soil Sci. Soc. Am. J.* **63**(5), 1404–1412.

2 Interaction of Rivers and River Catchments in Undisturbed and Intensively Cultivated Basins

Predicting slope–channel connectivity: a national-scale approach

D. E. WALLING & Y. ZHANG
Department of Geography, University of Exeter, Exeter EX4 4RJ, UK
d.e.walling@exeter.ac.uk

Abstract Concern for problems associated with increased fine sediment loads in British rivers has focussed attention on the source of the sediment. Available evidence suggests that much of the sediment is derived from erosion of agricultural land and this has in turn directed attention to both rates of soil loss from agricultural land and the connectivity between the catchment surface and the channel network. In order to provide a basis for identifying problem areas and to underpin the development of effective sediment control strategies, the need to provide a national-scale assessment of slope–channel connectivity has been identified. The basis for developing a preliminary national-scale assessment of slope–channel connectivity, based on readily available national data sets and a 1 km × 1 km grid is described.

Key words connectivity; connectivity index; connectivity ratio; sediment delivery; slope-channel connectivity

INTRODUCTION

Recent years have seen a growing awareness of problems linked to a perceived increase in the transport and storage of fine sediment in British rivers. These problems include the siltation of salmonid spawning gravels (e.g. Sear, 1993; Theurer *et al.*, 1998; Walling *et al.*, 2003), the degradation of aquatic habitats and river water quality, and the transfer and storage of sediment-associated nutrients and contaminants in river systems (e.g. Owens *et al.*, 2001). In examining these problems and the potential for their control, attention has been directed to the source of the sediment (Russell *et al.*, 2001; Walling *et al.*, 2003). Available evidence suggests that much of the sediment is derived from erosion of agricultural land and that many of the problems should be seen as reflecting the off-site impact of increasing rates of soil loss from agricultural land.

A considerable body of knowledge exists regarding rates of soil loss within England and Wales (cf. Morgan, 1985; Evans, 1990; Harrod, 1998; McHugh *et al.*, 2002) and this provides a useful basis for identifying areas at risk from accelerated soil loss and for assessing national patterns of sediment mobilization by soil erosion. However, it is well known that estimates of rates of on-site soil loss do not provide a direct indication of the suspended sediment yields from local catchments (Walling, 1983). Much of the sediment mobilized by soil erosion may be deposited prior to reaching a watercourse or stream and the specific sediment yield (t km^{-2} $year^{-1}$) of a catchment may be substantially less than the equivalent local rates of soil loss. Any attempt to identify catchments or areas at risk from increased inputs of fine sediment to the stream system should therefore take account of both the erosion risk and the potential for transfer of the mobilized sediment to the stream network. Where local rates of soil loss are high, but only a small proportion of the mobilized sediment reaches the stream network, the overall risk in terms of increased inputs of fine sediment to the stream system could be relatively low. Equally, areas where the local rates of soil loss are relatively low could still represent a risk, in terms of sediment inputs to the local stream systems, if a substantial proportion of the mobilized sediment reaches the stream system. Information on the efficiency of slope–channel transfer is therefore needed to "convert" estimates of soil loss to estimates of sediment input to the stream system and to identify areas where the local streams and rivers are most at risk from potential increases in rates of soil loss from agricultural land and which should therefore be targeted for implementation of erosion and sediment control measures.

The relationship between on-site rates of soil loss and the sediment yield at a catchment outlet is frequently expressed in terms of the sediment delivery ratio, which represents the ratio of the sediment output from a catchment to the gross erosion within the catchment (Walling, 1983). The sediment delivery ratio will, however, be influenced by both the slope-channel connectivity and deposition and storage of sediment within the channel and flood plain system. In addition, the gross erosion will include mobilization of sediment by erosion from both the slopes of a catchment and from the channel network. In order to characterize the efficiency of slope–channel transfer, the authors have coined the term *connectivity ratio*, to provide a measure which is directly analogous to the sediment delivery ratio, but which refers only to slope–channel transfers.

A NATIONAL SCALE ASSESSMENT OF SLOPE–CHANNEL CONNECTIVITY

The basis

As part of a wider national study of sediment delivery to watercourses (McHugh *et al.*, 2002), a preliminary attempt has been made to identify the broad countrywide pattern of slope–channel connectivity and to produce a raster-based map of England and Wales depicting values for the connectivity ratio on a 1 km × 1 km grid. In view of the preliminary nature of the study, emphasis was placed on using readily available national spatial data sets for potential controlling variables, in order to derive estimates of the connectivity ratio for individual grid cells.

Controlling factors

Existing understanding of the controls on sediment transfer from the slopes of a catchment to the adjacent watercourses, and thus the connectivity ratio, can be seen as highlighting three primary controls, namely, the transport capacity of surface runoff, the spatial distribution and density of the receiving watercourses, and the characteristics of the mobilized sediment. As a first approximation, these three primary controls can, in turn, be represented by six key influencing factors, which need to be incorporated into any algorithm for estimating the connectivity ratio. The six factors are:

(a) a runoff potential factor;
(b) a slope steepness factor;
(c) a slope shape factor;
(d) a drainage pattern factor;
(e) a land-use factor;
(f) a sediment characteristics factor.

The runoff potential factor provides a measure of the amount of surface runoff available to transport sediment and the land-use factor will primarily reflect the surface roughness.

Spatial data sets

The following five data sets were available for use as base data layers within a GIS, in order to derive estimates of the six factors:

(a) A 50 m spatial resolution DEM (Ordnance Survey).
(b) A 25 m resolution land cover data-set for 1990, with land-use classified into 25 categories (CEH, Monks Wood).
(c) Hydrological effective rainfall (HER) data at a spatial resolution of 1 km (derived from the ADAS Magpie database).
(d) A 1:50 000 river network coverage (CEH, Wallingford).
(e) 1-km grid soil data based on the 1: 250 000 map for England and Wales (NSRI), including information on texture and HOST classification from which storm runoff potential (the standard percentage runoff, SPR) can be derived.

The cell-based modelling functions from the GRID module of Arc/Info GIS were used to generate secondary data layers and to scale data to the required spatial resolution (1-km grid cells).

Factor derivation

The parameterization of the controlling factors was necessarily constrained by the availability of relevant national data sets, and was based on existing theoretical functions, empirical relationships and expert judgement. Further details of the derivation of the individual factors are provided below.

The runoff potential factor Surface runoff is the ultimate driver of sediment transfer. Its magnitude will depend primarily on rainfall characteristics, surface condition and soil properties. Two existing spatial data sets were combined to represent the potential for surface runoff generation, namely, the hydrologically effective rainfall (*HER*) and the HOST soil classification (Boorman *et al*., 1995). As part of the HOST classification there is an empirically-derived base flow index (BFI) data set from which the standard percentage runoff (*SPR*) can be estimated using established relationships (Boorman *et al.*, 1995). The runoff potential factor for each cell was estimated as:

$$\text{Runoff potential factor} = (HER \times SPR / 100)^{1.4} \qquad (1)$$

The exponent 1.4 conforms to the generally accepted relationship between sediment transport by runoff and runoff amount.

The slope steepness factor The slope steepness factor is represented as $S^{1.4}$, where S is the sine of the slope angle. The exponent 1.4 is based on the widely accepted relationship between overland flow transport capacity and slope gradient (e.g. Prosser & Rustomji, 2000). To take account of the variability of slope steepness within a 1-km grid cell, derivation of the slope steepness factor was based on a 50-m sub-grid. Values of slope angle for the individual

50-m grid cells were calculated from the DEM using the "slope" function available within the GRID module of Arc/Info GIS. The slope angle for the overall 1-km grid cell was derived using the focal functions within the GRID module.

The slope shape factor It is generally accepted that a surface with a convex profile is more efficient in terms of sediment transfer, because transport capacity increases downslope. In contrast, concave surfaces are more commonly associated with deposition. To derive this factor, a 50-m grid was superimposed onto the DEM and the slope profile shape within each 50-m grid cell was characterized as either dominantly convex or dominantly concave. The proportion of convex elements (cells) within a 1-km grid cell was used to represent the effect of slope shape on sediment transfer and to derive the slope shape factor. Again, several functions available within the GRID module were used for this purpose.

The drainage pattern factor The drainage pattern will influence both the spatial distribution and density of the watercourses within a local area and will therefore influence the slope–channel transfer distance. Although drainage density (km km^{-2}) is widely used to represent the drainage pattern, it only takes account of the total length of the drainage network within a cell and not its spatial configuration. In this study an improved index is proposed, based on the integration of the higher resolution sub-grid (50 m) and the river network coverage (1:50 000). The average (median) distance from all 50 m land cells within a 1-km grid cell to the nearest river channel (D) is used to represent the effects of drainage pattern on sediment transfer. An exponential decay relationship is used to derive the drainage pattern factor:

$$\text{Drainage pattern factor} = e^{-1 \times k \times D} \qquad (2)$$

where k is a scaling constant depending on the units used.

The sediment characteristics factor Sediment transfer is a size-selective process. Fine particles are likely to be more easily transported and are therefore less prone to deposition. The grain size composition of an eroding soil can therefore be expected to exert a significant control on the efficiency of slope–channel transfer. Since most sediment is mobilized by rill and interill erosion processes, the texture of the surface soil is clearly most important. The d_{50} or median grain size is frequently used in sediment transport equations for overland flow, to characterize the grain size distribution of the sediment (Meyer & Monke, 1965; Everaert, 1991). In the absence of detailed information on the texture of surface soils at a national scale, an estimate of the d_{50} of the surface horizons of the soil within each 1-km grid cell was derived from the available data on the percentage sand and silt content. For use in estimating the connectivity ratio, the sediment characteristics factor was expressed as $d_{50}^{-0.5}$. The exponent –0.5 conforms to existing work on the relationship between overland flow transport capacity and particle size (Meyer & Monke, 1965; Everaert, 1991).

The land-use factor Land-use has many effects on environmental systems, but its influence on surface roughness is likely to be most important in influencing the efficiency of slope–channel transfer. Surface roughness can be characterized by Manning's roughness parameter (n) and the parameter $n^{-0.6}$ has been used as the land-use factor. The exponent –0.6 is based on reported relationships with overland flow transport capacity (e.g. Moore & Burch, 1986). Land-use information has been derived from the CEH land cover data set, based on the dominant land-use in the 1-km grid cells, and Manning's n values have been allocated to individual categories, based on existing practice documented for several physically-based soil erosion and water quality models.

Estimating the connectivity ratio

In order to estimate the connectivity ratio for each individual 1-km grid cell, it is necessary to integrate the values obtained for the six controlling factors. In the absence of a detailed physically-based understanding of the influence of the various factors on sediment transfer and their relative importance, and acknowledging the high degree of spatial and temporal lumping involved, a conceptual formulation using a structure analogous to that employed by the Revised Universal Soil Loss Equation (Renard *et al.*, 1991) has been employed. In this approach, a connectivity index (*CI*) is initially derived. This index is seen to be a function of the sediment transport capacity modified by the effects of slope shape and drainage pattern, with the effects of slope shape and drainage pattern being treated as multiplicative:

$$CI = f\,(TC \times f_{sp} \times f_{dp}) \qquad (3)$$

where TC is the sediment transport capacity, f_{sp} is the slope shape factor and f_{dp} is the drainage pattern factor. In the absence of an empirical calibration of equation (3), a simple logarithmic function has been assumed:

$$CI = \log\,(TC \times f_{sp} \times f_{dp}) \qquad (4)$$

where the logarithmic transformation has been used to avoid possible distortion associated with the non-normal distribution of the spatial data sets. To take account of the different magnitudes of the values obtained for the slope shape and drainage pattern factors, these factors were scaled to values between 0 and 1, using a simple linear stretch routine.

The sediment transport capacity parameter TC incorporates the influence of surface runoff magnitude, slope gradient, surface roughness and sediment size. These variables can be represented by the runoff potential factor, the slope gradient factor, the land-use factor and the sediment characteristics factor, respectively. The TC factor can therefore be calculated as:

$$TC = q^{1.4}\, S^{1.4}\, n^{-0.6}\, d_{50}^{-0.5} \qquad (5)$$

Equation 4 was used to derive estimates of *CI* for each 1-km grid cell and these values were scaled to the range 0–1, using a simple linear stretch routine.

In order to derive values for the connectivity ratio, which represents the proportion of the soil mobilized by erosion from the land surface within a grid cell that will reach the stream channels within or adjacent to that grid cell, it is necessary to scale or calibrate the values obtained for the connectivity index (0–1). In the absence of a clear empirical basis for calibrating the values of the connectivity index, these values were scaled to the range 0.2–0.7, based on existing knowledge of the sediment budgets of small agricultural catchments (e.g. Walling *et al.*, 2002). The spatial distribution of the resulting connectivity ratio values for England and Wales is shown in Fig. 1.

PERSPECTIVE

The spatial distribution of the connectivity ratio shown in Fig. 1 conforms to existing understanding of the expected pattern. The broad distinction between higher values in the west and lower values in the east reflects both the increased runoff in western areas and the more impervious nature of the underlying rocks in these areas, which are in turn reflected by

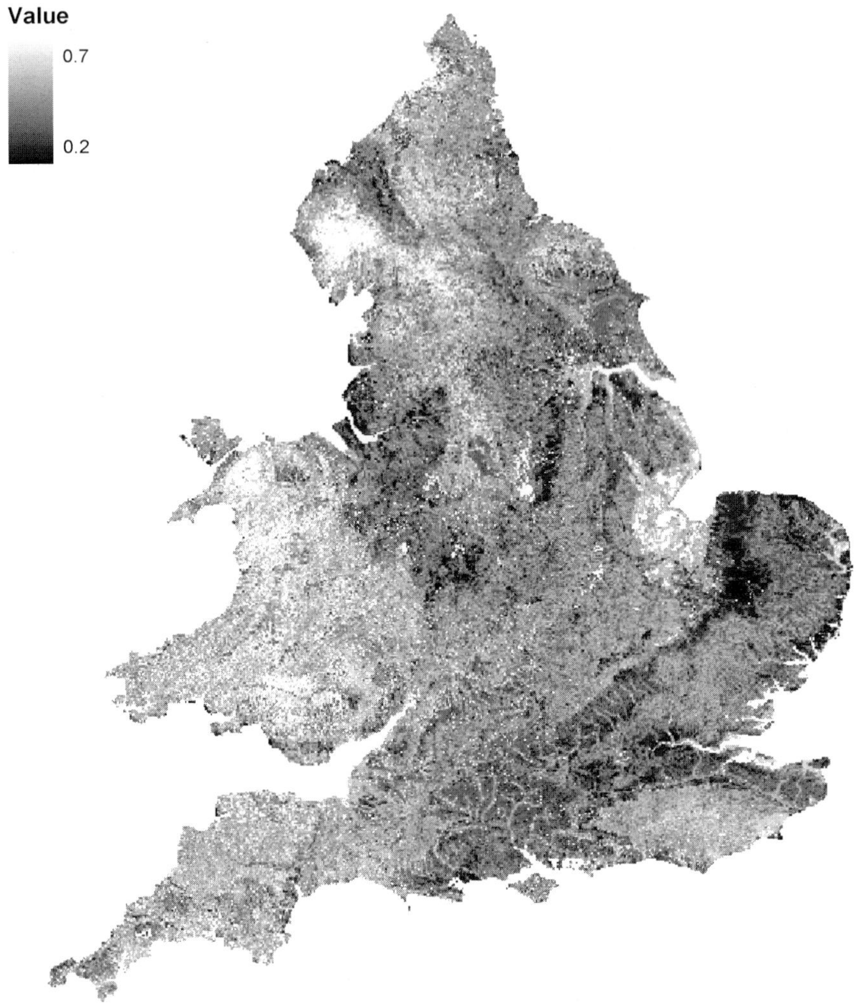

Fig. 1 The spatial distribution of the connectivity ratio.

higher drainage densities and increased amounts of surface runoff. The reduced drainage densities and reduced incidence of surface runoff associated with areas underlain by chalk and limestone are clearly reflected in the low values of connectivity ratio associated with areas underlain by chalk and limestone.

The results presented in Fig. 1 must, nevertheless, be seen as preliminary. Further work is clearly required to refine the approach outlined. The basis of the GIS-based procedure used to derive values of the connectivity ratio for individual grid cells is summarized in Fig. 2, and this should be seen as providing a framework which requires further improvement and refinement. This should include incorporation of additional controlling factors, refinement of the parameterization of existing factors, and more rigorous calibration of the relationship between the connectivity index and the connectivity ratio. In terms of incorporation of additional factors, there is a clear need to take account of the more adventitious effects on

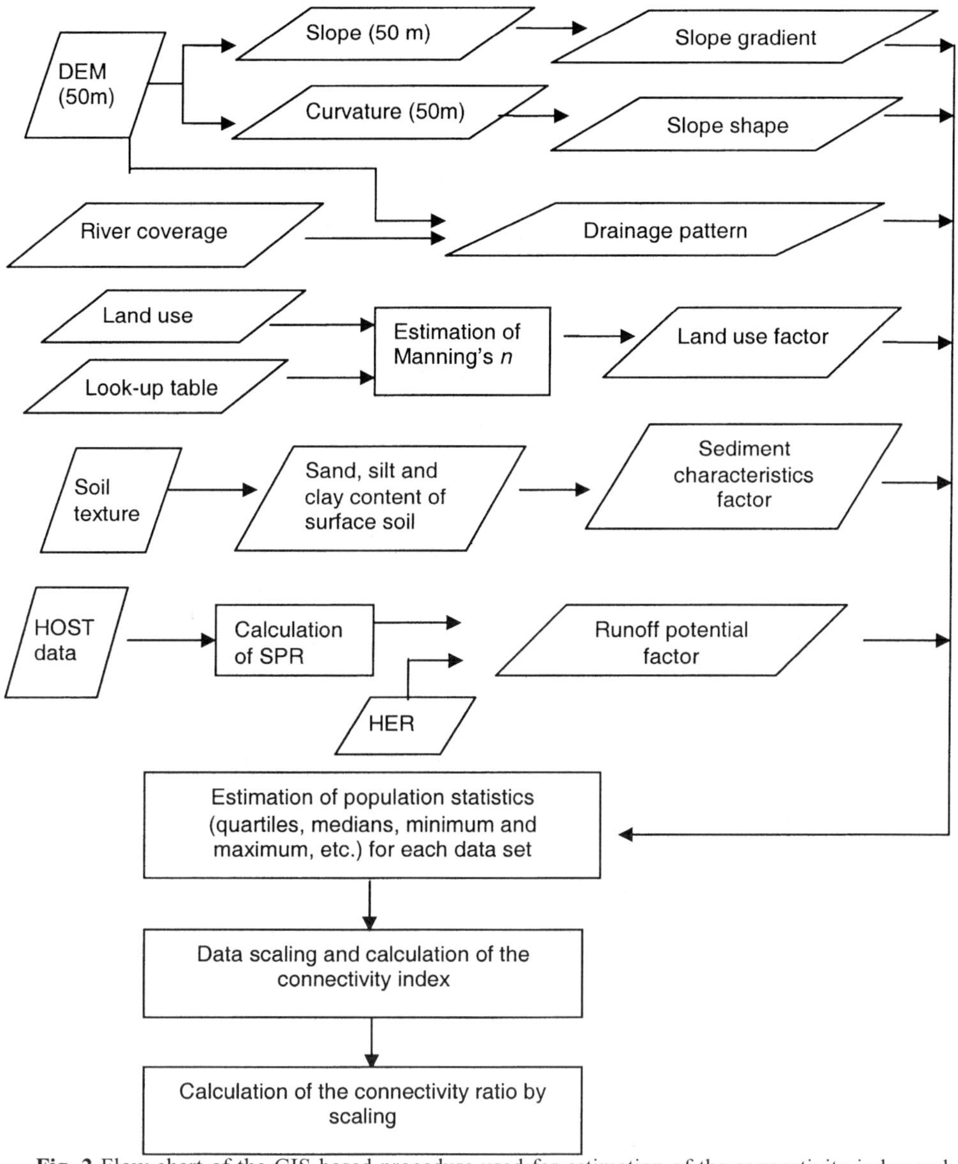

Fig. 2 Flow chart of the GIS-based procedure used for estimation of the connectivity index and ratio.

connectivity associated with features such as field boundaries and associated gateways, as well as the important role of tracks and roads in routing water and sediment towards stream channels.

Acknowledgements The work reported in this paper was undertaken as part of a wider study aimed at predicting sediment delivery to watercourses from land in England and Wales, funded by the Environment Agency (R & D Project no. P2-209). This financial support and the collaboration and help of Antony Williamson, the EA Project Manager, and Marianne McHugh, Gavin Wood and Roy Morgan from the National Soil Resources Institute, Cranfield University, who were responsible for other aspects of the study, are gratefully acknowledged.

REFERENCES

Boorman, D. B., Hollis, J. M. & Lilly, A (1995) *Hydrology of Soil Types: A Hydrologically-based Classification of the Soils of the United Kingdom*. Report no. 126. Institute of Hydrology, Wallingford, UK.

Evans, R. (1990) Soils at risk of accelerated erosion in England and Wales. *Soil Use Manage.* **6**, 125–131.

Everaert, W. (1991) Empirical relations for the sediment transport capacity of interrill flow. *Earth Surf. Processes Landf.* **16**, 513–532.

Harrod, T. R. (1998) A systematic approach to national budgets of phosphorus loss through soil erosion and runoff at National Soil Inventory (NSI) nodes. Final Report to MAFF. MAFF Project NT 1044.

McHugh, M., Harrod, T. R. & Morgan, R. P. C. (2002) The extent of soil erosion in upland England and Wales. *Earth Surf. Processes Landf.* **27**, 99–107.

McHugh, M., Wood, G., Walling, D. E., Morgan, R. P. C., Zhang, Y., Anthony, S. & Hutchins, M. (2002) *Prediction of Sediment Delivery to Watercourses from Land, Phase II*. Environment Agency R&D Tech. Report no. P2-209. Environment Agency, Bristol, UK.

Meyer, L. D. & Monke, E. J. (1965) Mechanics of soil erosion by rainfall and overland flow. *Trans. Am. Soc. Agric. Engrs* **8**, 572–580.

Moore, I. D. & Burch, G. J. (1986) Modelling erosion and deposition: topographic effects. *Trans. Am. Soc. Agric. Engrs* **29**, 1624–1630.

Morgan, R. P. C. (1985) Assessment of soil erosion risk in England and Wales. *Soil Use Manage.* **1**, 127–131.

Owens, P. N., Walling, D. E. Carton, J. Meharg, A. A., Wright, J. & Leeks, G. J. L (2001) Downstream changes in the transport and storage of sediment-associated contaminants (P, Cr and PCBs) in agricultural and industrialized drainage basins. *Sci. Total Environ.* **266**, 177–186.

Prosser, I. P. & Rustomji, P. (2000) Sediment transport capacity relations for overland flow. *Progr Physical Geogr.* **24**, 179–193.

Renard, K. J., Foster, G. R., Weesies, G. A. & Porter, J. P. (1991) RUSLE: Revised Universal Soil Loss Equation. *J. Soil Water Conserv.* **40**, 30–33.

Russell, M. A., Walling, D. E. & Hodgkinson, R. (2001) Suspended sediment sources in two small lowland agricultural catchments in the UK. *J. Hydrol.* **252**, 1–24.

Sear, D. A. (1993) Fine sediment infiltration into gravel spawning beds within a regulated river experiencing floods, ecological implications for salmonids. *Regulated Rivers Research Manage.* **8**, 373–390.

Theurer, F. D., Harrod, T. R. & Theurer, M. (1998) *Sedimentation and Salmonids in England and Wales*. Environment Agency Technical Report P194.

Walling, D. E. (1983) The sediment delivery problem. *J. Hydrol.* **65**, 209–237.

Walling, D. E., Russell, M. A., Hodgkinson, R. & Zhang, Y. (2002) Establishing sediment budgets for two small lowland agricultural catchments in the UK. *Catena* **47**, 323–353.

Walling, D. E., Collins, A. L. & McMellin, G. K. (2003) A reconnaissance survey of the source of interstitial fine sediment recovered from salmonid spawning gravels in England and Wales. *Hydrobiologia* **497**, 91–108.

Land–ocean sediment transfer in palaeotimes, and implications for present-day natural fluvial fluxes

ANDREY PANIN
Department of Geography, Moscow State University, Vorobiovy Gory, 119992 GSP-2 Moscow, Russia
panin@morpho.geogr.msu.su

Abstract Detrital fluxes to the ocean during the period from the late Jurassic to the Pliocene ranged between 2.7–5.2 Gt year^{-1}, but increased up to 9.6–15.5 Gt year^{-1} in the Pliocene–Holocene. Of the Holocene flux, 11–12 Gt year^{-1} was carried by rivers. This estimate is suggested as the natural component of present-day fluvial sediment flux to the ocean. The contemporary flux is 1.1- to 2-fold higher, due to human disturbance of the land. Small basins make a major contribution to this increase, while many large rivers demonstrate stable fluxes. This is due to sediment trapping within the fluvial system, with sediment residence time increasing with increasing basin size. Over long time spans sediment delivery ratios (*SDR*) are less dependent on basin size. At a global land-ocean scale, the *SDR* is estimated to be in the range 5–20% for timescales ranging from decades to centuries but its value increases to 77% for the last 150 million years.

Key words anthropogenic; Cenozoic; denudation; global sediment flux; Holocene; Pleistocene; sediment delivery ratio

INTRODUCTION

Interest in fluvial sediment fluxes to the world's oceans began in the early 1950s, and peaked in the 1970s (Fig. 1). Initial estimates of the magnitude of the fluvial sediment flux were based on limited data sets, and as a result, the various flux estimates differed by as much as an order of magnitude. However, current estimates, developed over the last two decades, tend to fall within the range 13.5×10^9 to 22×10^9 t year^{-1}. Despite increasing reservoir trapping, the contemporary sediment flux to the oceans is still believed to have increased relative to pre-human conditions. The term natural, or pristine has been used to refer to the sediment discharge that would have occurred in the absence of anthropogenic disturbance. It is supposed relatively stable due to the stability of the natural global water balance. The observed global sediment discharge to the oceans may be presented as the sum of an essentially constant natural component and a changing anthropogenic component. To distinguish between the natural and anthropogenic components researchers have employed different approaches including, for example, assessment of the land area subject to accelerated erosion (Bondarev, 1974), using an index of chemical alteration of suspended sediment (McLennan, 1993), making comparisons of sediment yields from basins subjected to different degrees of anthropogenic disturbance (Dedkov & Mozzherin, 2000), etc. The natural flux is thus estimated to lie within the range 6.0×10^9 to 13.5×10^9 t year^{-1} , i.e. with more than a 2-fold range (Table 1). According to different authors, anthropogenic disturbance has raised the global sediment flux by factors of 1.7 (McLennan, 1993), 1.9 (Bondarev, 1974), 2.0 (Milliman & Syvitski, 1992), 2.6 (Dedkov & Mozzherin, 2000), and 3.0 (Harrison, 1994).

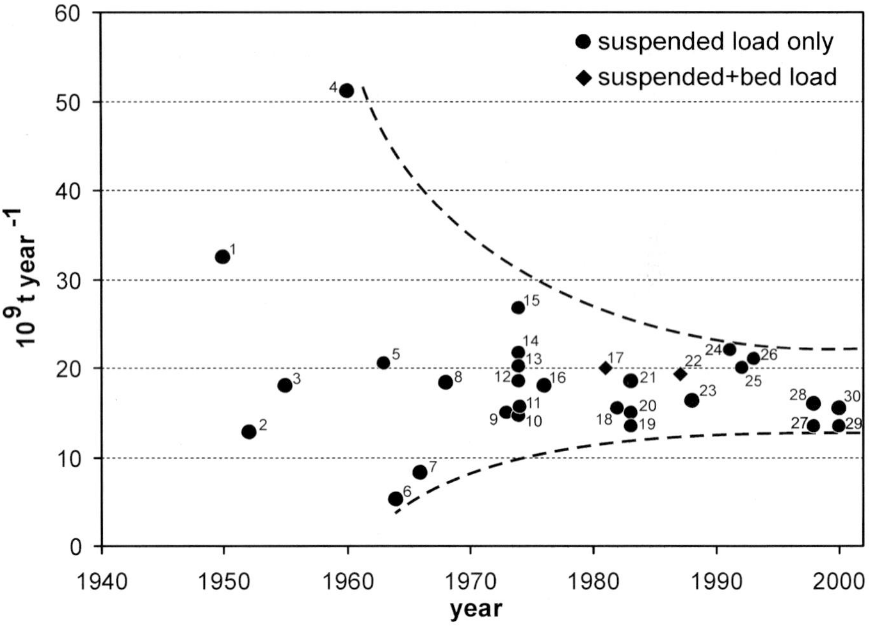

Sources of data: 1 – Kuenen (1950), 2 – Lopatin (1952), 3 – Gilluli (1955), 4 – Fournier (1960)*, 5 – Schumm (1963)*, 6 – Corbel (1964), 7 - Mackenzie & Garrels (1966)*, 8 – Holeman (1968)*, 9 – Sundborg (1973)*, 10 – Bondarev (1974), 11 – Alexeev & Lisitzina (1974), 12 – Lisitzin (1974), 13 – Gudzon (1974 – cit. in Safyanov, 1978), 14 – L'vovich (1974), 15 – Jansen & Painter (1974)*, 16 – Goldberg (1976)*, 17 – Holland (1981), 18 – Meybeck (1982 – cit. in Meybeck, 1988), 19 – Milliman & Meade (1983), 20 – Walling & Webb (1983), 21 – Gordeev (1983), 22 – Vassil'ev (1987), 23 – Meybeck (1988), 24 – Lisitzin (1991), 25 – Milliman & Syvitski (1992), 26 – McLennan (1993), 27 – Stallard (1998), 28 – Ludwig & Probst (1998), 29 – Harrison (2000), 30 – Dedkov & Mozzherin (2000).

*cit. in Walling & Webb (1996).

Fig. 1 Estimates of the global land–ocean river sediment flux produced since the middle of the 20th century.

Table 1 Natural and anthropogenic proportions of the global sediment flux.

Reference	Natural (10^9 t year^{-1})	(%)	Anthropogenic (10^9 t year^{-1})	(%)
Harrison, 1994	6	33	12	67
Dedkov & Mozzherin, 2000	6	38	9.5	62
Bondarev, 1974	7.9	54	6.7	46
Milliman & Syvitski, 1992	<10	50	>10	50
McLennan, 1993	12.6	60	8.4	40

The object of this paper is to evaluate river sediment discharges over geological time on the basis of the sedimentary mass stored in the oceans. Data from the remote past can help to determine typical rates of sediment removal from the land over the Earth's history. Data from the near past could serve as an approximation for the present pristine sediment flux. Within the context of the sediment delivery concept (Walling, 1983), it is also interesting to consider what proportion of the denudation products remained on the land and how much was transferred to the oceans over the geological past. The background of geological history is seen as widening the scope of current understanding of contemporary sediment fluxes.

GLOBAL SEDIMENT FLUX SINCE THE LATE JURASSIC

To calculate sediment fluxes, the volumes of sedimentary rocks on unglaciated land (i.e. Antarctica and Greenland excluded) and oceanic bottom given in Ronov (1993) were used. The mass of clastic marine rocks from a given epoch divided by the duration of this epoch gives the rate at which the products of mechanical denudation were being removed from the continents to the ocean. Glacial marine rocks are excluded to make the results closer to the value of river sediment discharge. Rock masses were corrected for subsequent destruction using an exponential function for observed (remaining, or survival) sedimentation rates given in (Tardy *et al.*, 1989). The obtained rates (Fig. 2) represent mechanical denudation of the continents (glacial denudation excluded), with the major part of it provided by fluvial action. Denudation rates were minimal at the boundary between the Cretaceous and the Cenozoic, which is known as a time of continental planation. In the Cenozoic, erosion rates have increased 3.5-fold, up to 9.6×10^9 t year^{-1} in the Pliocene, as a response to increasing continental relief during the Alpine orogeny.

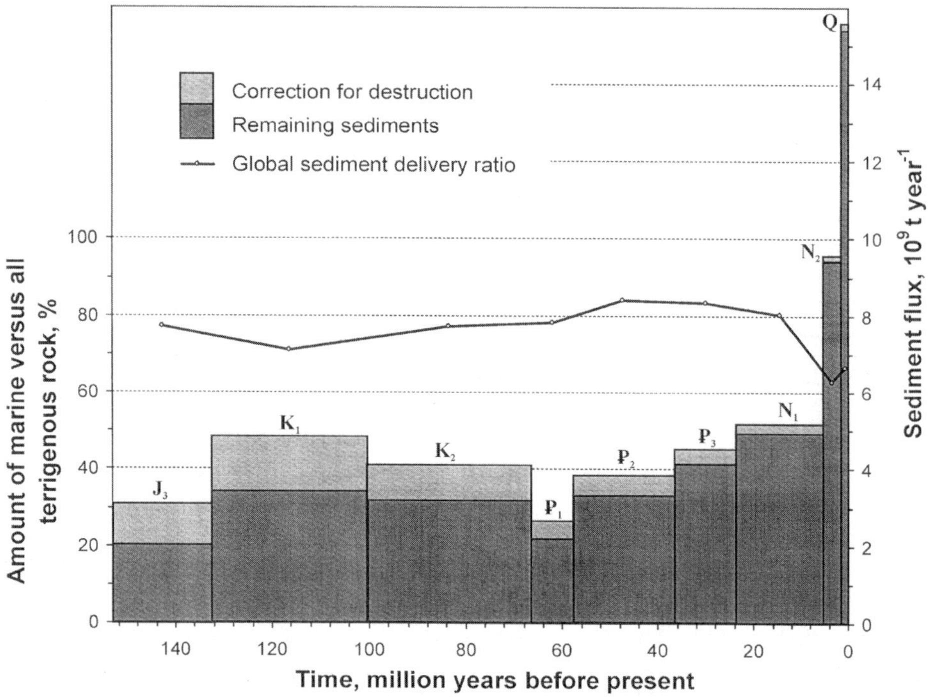

Fig. 2 Rates of non-glacial detrital sediment input to the ocean and the global sediment delivery ratio since the late Jurassic (calculated from data in Ronov (1993) and Hay (1994)).

Table 2 Change of sediment flux during the Late Cenozoic.

Region	Ratio of sediment flux*			Source
	Q to N_2	N_2 to N_1	Q to N_1	
European Alps	1.5	1.7	2.6	Kuhlemann *et al.* (2002)
European Alps	3	6	18	Hay *et al.* (1992—in Kuhleman *et al.*, 2002)
The Meuse River basin (western Europe)	1.8	–	–	Van Balen *et al.* (2000)
The Mississippi River basin	3.9	1.5	5.7	Hay *et al.* (1989)

*Symbols for geological periods (absolute age in millions of years before present given in brackets): N_1, Miocene (23.7–5.3); N_2, Pliocene (5.3–1.6); Q, Pleistocene (1.6–0).

For the Pleistocene (the last 1.6 million years) the only data on global accumulation were reported by Hay (1994). His 24.7×10^{21} g of noncarbonate marine deposits accumulated during the Quaternary represent an average land–ocean sediment flux of 15.4×10^9 t year^{-1}. Evidence from different regions confirms that in the Pleistocene sediment production from the continents continued to rise (Table 2). The lowest of the documented ratios (Table 2) gives a 1.5-fold increase in Pleistocene rates over those of the Pliocene and thus a lower estimate of the average sediment output in the Pleistocene of 14.5×10^9 t year^{-1} which is still close to that calculated from Hay (1994).

There is less information on the Holocene sediment flux than that of the Pleistocene. In middle and high latitudes, sediment yield increased during cold epochs due to glacier action, although the area subjected to river action diminished considerably. In southeastern Asia the summer monsoon was weak during cold epochs (Colin *et al.*, 1999), thus both water and sediment discharges could be expected to decrease. For instance, the Yellow River at that time was periodically dry (Xia *et al.*, 1993). Taking into account the contemporary sediment contribution of the Asian monsoon region, the global sediment flux in the Holocene may be estimated to be close to that in the Pleistocene. This is confirmed by marine sedimentation rates reported by Hay (1994). According to Hay's data (his Tables 1.4 and 1.8 corrected for arithmetical errors), accumulation rates of noncarbonate materials in the deep sea decreased 4- to 5-fold in the Holocene compared to the Pleistocene average (2.8×10^9 *vs* 12.9×10^9 t year^{-1}) because of high sea level after deglaciation. On the other hand, the same reason is responsible for the nearly two-orders of magnitude increase in accumulation on the shelves. The shelf accumulation comprises noncarbonate Holocene sediments on so-called detrital and carbonate shelves—10.3×10^9 and 0.3×10^9 t year^{-1}, respectively. Holocene accumulation rates in marginal seas are highly variable but Hay (1994) suggests that typical rates were 20–25% of those in the Pleistocene, i.e. $0.5–0.6 \times 10^9$ t year^{-1}. Overall, Holocene marine accumulation of noncarbonate sediments, which can be regarded as detrital flux from the land, represents some 14×10^9 t year^{-1}—almost the same as the Pleistocene average given above.

This approximation comprises contributions from different agents of mechanical denudation. Non-fluvial sources may be taken into account by considering their contemporary levels. The aeolian dust deposition over the ocean area has been assessed as (in 10^9 t year^{-1}): 0.06 (Garrels & Mackenzie, 1971), 0.07–0.35 (Goldberg, 1971), 0.5–0.85 (Prospero, 1981), 0.8 (Vassil'ev & Pomortsev, 1988), 0.9 (Duce *et al.*, 1991), 1.6 (Lisitzin, 1974), 2.0 (Bondarev, 1974). Direct glacial input has been estimated at (10^9 t year^{-1}): 0.3–0.4 (Losev, 1989), 1.5 (Lisitzin, 1974), 1.9 (Makkaveev, 1982). Abrasion of the ocean coast may add (10^9 t year^{-1}): 0.15 (Gilluly, 1955), 0.2 (from Kuenen's (1950) 0.12 km^3 year^{-1}), 0.3–0.4 (Losev, 1989), 0.7–1.0 (Bondarev, 1974), 1.0 (Safyanov, 1978). Taking the averages in each case (i.e. 0.9×10^9, 1.2×10^9 and 0.5×10^9 t year^{-1}, respectively) and subtracting them from the total Holocene flux of 14×10^9 t year^{-1}, one may estimate the Holocene fluvial sediment flux to the oceans at (11–12) $\times 10^9$ t year^{-1}. This range may represent the most probable estimate of the natural fluvial flux, i.e. the global river sediment discharge to the ocean that would occur under undisturbed conditions. It is close to the value of natural sediment flux suggested by McLennan (1993), but higher than that suggested by other workers (Table 1).

CHANGES OF RIVER SEDIMENT FLUXES DUE TO HUMAN ACTIVITY

Since the above estimate of the natural sediment flux lies towards the bottom of the range of estimates of contemporary sediment flux proposed during the last two decades (Fig. 1), human activity is likely to have increased global sediment input to the oceans to some extent. Data on sedimentation mass balances provide clearer estimates for some large river basins. The two major contributors of sediment to the world's oceans are of primary interest, namely the Yellow (Huang He) River and the Ganges-Brahmaputra system.

The sediment discharge of the Yellow River is suggested to exhibit the same behaviour as sedimentation rates in the lower Yellow River region, according to Xu (1998): between 12 000 (beginning of the Holocene) and 6000 years ago the sediment flux is taken as constant (S_0), and during the next 4000 years it doubled by 2.2 ka BP (2 S_0). After that sediment discharge is supposed to be proportional to population in the middle Yellow River region. Three models have been explored, namely a linear, exponential and logarithmic-law increase of sediment discharge as a function of population increase since 200 BC, the two former giving almost similar results. The change of sediment discharge during the Holocene is calculated by means of optimization of the variable S_0, assuming a contemporary sediment discharge $S_m = 1.2 \times 10^9$ t year^{-1}, to derive the total Yellow River-derived Holocene marine sediment mass of 3000×10^9 t given in Milliman *et al.* (1987).

During the Holocene the Yellow River changed its course several times, flowing for some 30% of the time into the Gulf of Bohai, and 70% to the south of the Shandong Peninsula. However, some 1330×10^9 t (45%) of the river-derived marine deposits may be attributed to the northern position of the river mouth and only 1680×10^9 t (55%) belongs to the southern position (based on Table 2 in Milliman *et al.* (1987)). The closest distribution of marine sediments is provided by the linear erosion–population model. According to this scenario 2200 years ago at the beginning of sensible land disturbance, sediment discharge was as much as 0.35×10^9 t year^{-1}. This is only two times less than the mean sediment discharge since 1855, which may be estimated at between 0.6 to 0.75×10^9 t year^{-1} on the basis of data presented by Shi *et al.* (2003).

Less evident is the increase in sediment load at the Ganges-Brahmaputra mouth caused by human activity. Kuehl *et al.* (1997) report that the subaqueous part of the Ganges-Brahmaputra delta accumulated some 1.97×10^{12} t of sediment during the last 7000 ^{14}C years (8000 calendar years). The sediment amount leaving the shelf have been assessed from the core data stratigraphy in the Deep Bengal Fan presented by Colin *et al.* (1999), the dimensions of the fan measured on the map from Einsele *et al.* (1996). The total sediment volume over the subaqueous delta and Deep Bengal Fan provides an estimate of the mean sediment delivery to the sea of 440×10^6 t year^{-1} during the last 8000 years. This seems to be an underestimate, since the cores of Colin *et al.* (1999) were located at the eastern edge of the Bengal Fan while a considerable portion of the sediment is transported via the so called Swatch of No Ground Canyon to the west. Michels *et al.* (1998) found that at a decadal and century scale the subaqueous part of the delta accumulates 20% of total load and 35–50% is absorbed by the canyon, so the ratio is 1 : 1.75–2.5 in contrast to the ratio of 2 : 1, which follows from the above estimate. If the ratio found by Michels *et al.* (1998) is spread over the Holocene, the sediment amount in the Bengal Fan would be 3.5–4 times as much as suggested above, and the river sediment input to the Bengal Bay would be calculated as (0.60–0.75) $\times$ 10^9 t $year^{-1}$.

Recent estimates of the modern suspended load of the joint Ganges-Brahmaputra system are 1.06 $\times$ 10^9 (Milliman & Syvitski, 1992) or 1.14 $\times$ 10^9 t $year^{-1}$ (Delft Hydraulics/DHI, 1994—cit. in Michels *et al.*, 1998), with an average of 1.1 $\times$ 10^9 t $year^{-1}$. Flood plain accumulation over the subaerial delta is estimated from 30–45% (Michels *et al.*, 1998) to 40–80% (Milliman & Syvitski, 1992) of the total Ganges-Brahmaputra suspended load. To make it comparable to the ocean sedimentation rate one must also account for sediment losses. If the whole range of loss estimates is taken, the sediment delivery to the sea ranges from 0.22 $\times$ 10^9 to 0.77 $\times$ 10^9 t $year^{-1}$. When this range is compared with the lower and higher estimates derived from the Holocene marine sedimentation rates of 0.44 $\times$ 10^9 and 0.75 $\times$ 10^9 t $year^{-1}$, respectively, the existing sedimentation data provide the possibility of both an increase or decrease of the contemporary sediment contribution to the sea relative to the average Holocene rate.

Other large Asian rivers give less evidence of human-induced increases in sediment flux, for example, the Yangtze (Changjiang) River. An analysis of delta development undertaken by Chen (1996) reveals that rapid deltaic progradation during the last 2000 years is not due to anthropogenic activity, as had been previously thought, but results from river long profile adjustment after deep incision during the last glaciation. Chen (1996) also affirms that there is no evidence of any systematic increase of sediment discharge during the last 2000 years, or a significant difference in sediment discharge between the two periods before and after 2000 years before the present. The Mekong River, the tenth-ranked river for sediment flux (Milliman & Meade, 1983) also shows no evidence of a dramatic increase in sediment discharge due to human activity (Ta *et al.*, 2002). On the basis of sediment volume analysis they estimated the average sediment discharge of the Mekong River for the last 3000 years to be (144 $\pm$ 36) $\times$ 10^6 t $year^{-1}$ which is only 10% less than at present. In the case of the seventh-ranked Mississippi River the present-day sediment load of 210 $\times$ 10^6 t $year^{-1}$ (Milliman & Meade, 1983) is almost half that for the second half of the 19th century reconstructed by Kesel *et al.* (1992), due to sediment trapping in reservoirs.

Data from big rivers reveal that, with the exception of the Yellow River, large basins seem to contribute little to the anthropogenic sediment flux increase. In contrast, sediment discharge from rather small basins is found to respond strongly to land clearing and other human impact. In addition to numerous examples from inland areas (cf. Dearing & Jones, 2003), strong human impacts on near-shore sedimentation have long been reported from many coastal regions supplied with sediment by small and medium-sized rivers draining mountainous or hilly terrain. Wellington Harbour, New Zealand (35 km^2) may serve as an example. Data from several cores taken within 2 km of the coast reveal a significant increase in sedimentation rates since the end of the 19th century (Goff, 1997). Marine sedimentation before European colonization in the middle of the 19th century occurred at rates of 0.3–3.0

mm year^{-1} at a millennial to century scale, while the average rates for the 20th century are 10–50 mm year^{-1}, with increases in individual cores being in the range of 10- to 100-fold.

THE LAND–OCEAN SEDIMENT DELIVERY RATIO

The amounts of marine and continental detrital rock accumulated during a certain epoch not only reflect the rate of mechanical denudation but also the efficiency of transportation processes, namely the proportion of denudation products that reached the ocean and the proportion retained on the land. The marine component of all detrital rock may be regarded as a global sediment delivery ratio (*SDR*) from the land to the ocean during the given epoch, the majority being accomplished by fluvial action. According to Ronov (1993), during the last 150 million years some 67 220 × 10^3 km^3 of continental and 220 430 × 10^3 km^3 of marine clastic sedimentary rocks have been stored all over the Earth (to be exact, these are the volumes that have survived of what had ever accumulated). So the global *SDR* for this timescale is 77%. In the same way the global *SDR* values may be calculated for each geological epoch (Fig. 2). As no age-dependent trend is detected, information losses probably exercise little influence on the *SDR* values and their changes in time reflect the conditions for sediment transport from the continents to the oceans. Sediment delivery was about or higher than average when the sediment flux was low corresponding to planation of continental relief. In contrast, increased relief leads to increasing buffering capacity (term suggested by Walling, 1999): mountain building is accompanied by foredeep subsidence, which creates potential storage, and accounts for much sediment loss.

In the Pliocene the global *SDR* value decreased to its minimal value of 63%, and a slightly higher value is proposed for the Pleistocene (Table 3). The global *SDR* decreased in the Plio-Pleistocene, probably due to the increasing extent of inland areas with no discharge to the ocean. In any case this is much higher than modern *SDR*s obtained at a decadal or century scale. In small basins they are highly variable but still show a trend for rapid decrease with increasing basin area (Fig. 3). Both globally and over vast regions, values of

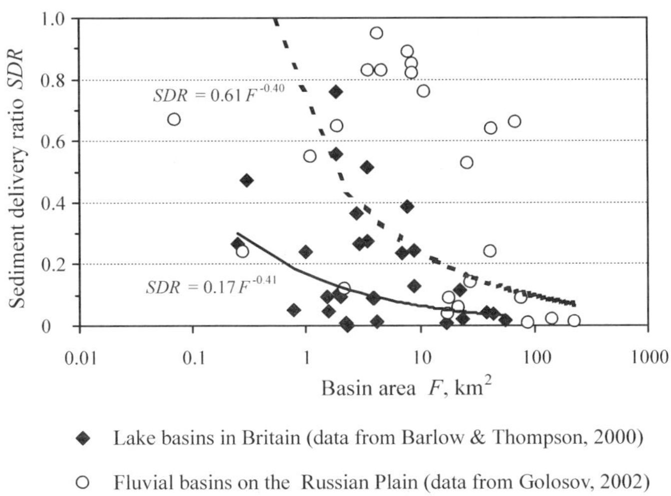

Fig. 3 Sediment delivery ratios in small basins at a decadal to century time scale.

Table 3 The land–ocean sediment delivery ratio (*SDR*) at different temporal scales (calculated from original data of indicated authors).

Region	Period, years	*SDR*	Source
Global land*	last 153 × 10^6	77%	This paper
Himalayas	last 20 × 10^6	74%	Eincele *et al.*, 1996
European Alps	last 5.5 × 10^6	90–95%	Kuhleman *et al.*, 2002
European Alps	last 5.5 × 10^6	90%	Hay *et al.*, 1992 (in Kuhleman *et al.*, 2002)
Global land*	(5.3 ÷ 1.6) × 10^6 (the Pliocene)	63%	This paper
Global land*	last 1.6 × 10^6 (the Pleistocene)	65–69%	Hay, 1994
Meuse River Basin (Western Europe)	last 0.65 × 10^6	83%	Van Balen *et al.*, 2000
United States	n × (10 ÷ 100)	4–5%	Holeman, 1980
Global land*	n × (10 ÷ 100)	13–20%	Walling, 1999, Golosov, 2001
Global land*	n × (10 ÷ 100)	5–10%	Dearing & Jones, 2003

* Antarctica and Greenland not included.

modern *SDR* range, according to different authors, between 5 and 20%, i.e. about an order of magnitude less than that for geological epochs (Table 3). Essential differences between the modern and geological-scale *SDR*s result from repeated redeposition and remobilization of each sediment particle on its way from the initial source to the basin outlet. At any given time the river sediment load comprises only partly new erosion products, the remainder being supplied from different sediment sinks within the basin, namely fans, floodplains, etc. The ratio between these sediment sources is still unknown in large rivers and scarcely known in small rivers.

The delivery ratio seems to be a function of time, with the residence time of different sinks playing a key role. This subject is also poorly investigated, although it appears that the longer the residence times, the slower the sediment movement through the fluvial system and the lower the *SDR* value will be for small time spans. At time scales greater than the basin residence times, the *SDR* value only reflects long-term storage such as active tectonic depressions. This is seen from the fact that modern *SDR*s have essentially higher values in mountain areas than on plains (Walling, 1983) while over geological time scales both mountain and plain basins evidence high delivery ratios (Table 3, Himalayas and Meuse River basin).

Acknowledgements The idea for this research was prompted by Vladimir Myslivets. His indirect participation is greatly appreciated. The author is also grateful to Arthur Horowitz, Nick Preston and Des Walling for improving the English and for valuable comments on the draft paper. This paper is a contribution to the Russian Fund of Basic Research (RFBR) Project no. 03-05-64021.

REFERENCES

Alexeev, V. V. & Lisitzina, K. N. (1974) *Stok vzveshennykh nanosov* (Suspended Sediment Flux). Gidrometeoizdat Publ., Leningrad, Russia (in Russian).

Barlow D. N. & Thompson R. (2000) Holocene sediment erosion in Britain as calculated from lake-basin studies. In: *Tracers in Geomorphology* (ed. by I. D. L. Foster), 455–472. Wiley, Chichester, UK.

Bondarev, L. G. (1974) *Vechnoe Dvizhenie: Planetarnoye Peremeschenie Veschestva i Chelovek* (*Perpetual Motion: Planetary Matter Moving and the Man*). Mysl' Publ., Moscow, Russia (in Russian).

Chen, X. (1996) An integrated study of sediment discharge from the Changjiang River, China, and the delta development since the mid-Holocene. *J. Coast. Res.* **12**(1), 26–37.

Colin, C., Turpin, L., Bertaux, J., Desprairies, A. & Kissel, C. (1999) Erosional history of the Himalayan and Burman ranges during the last two glacial–interglacial cycles. *Earth and Planetary Science Letters* **171**, 647–660.

Corbel, J. (1964) L'erosion terrestre, etude quantitative (methods, techniques, resultats). *Ann. Geogr.* **73**(398), 385–412.

Dearing, J. D. & Jones, R. T. (2003) Coupling temporal and spatial dimensions of global sediment flux through lake and marine sediment records. *Global and Planetary Change* **39**, 147–168.

Dedkov, A. P. & Mozzherin,V. I. (2000) Globalniy stok nanosov v okean: prirodnaya i antropogennaya sostavlyayuschie (Global river sediment discharge into the ocean: Natural and anthropogenic components), 15–23. Erosionnye i Ruslovye Processy. Vypusk 3. Moscow University Publ., Moscow, Russia (in Russian).

Duce, R. A., Liss, P. A., Merrill, J. T. *et al.* (1991) The atmospheric input of trace species to the world ocean. *Global Biogeochem. Cycles* **5**, 193–259.

Einsele, G., Ratschbacher, L. & Wetzel, A. (1996) The Himalaya-Bengal fan denudation-accumulation system during the past 20 Ma. *J. Geol.* **104**, 163–184.

Garrels, R. M. & Mackenzie, F. T. (1971) *Evolution of Sedimentary Rocks*. Norton, New York, USA.

Gilluli, J. (1955) Geologic contrasts between continents and ocean basins. *Geol. Soc. Am. Spec. Pap.* **62**, 7–18.

Goff, J. R. (1997) A chronology of natural and anthropogenic influences on coastal sedimentation, New Zealand. *Mar. Geol.* **138**, 105–117.

Goldberg, E. R. (1971) *Atmospheric Transport. Impingement of Man on the Oceans*. Wiley, New York, USA.

Golosov, V. N. (2001) Pereraspredelenie nanosov v verkhnikh zvenyakh fluvial'noy seti zemledel'cheskykh regionov: teoria voprosa i opyt regional'nogo analysa (Sediment redistribution in upper courses of the fluvial net in agricultural regions: Theory and regional analysis on the plains of the temperate belt), 94–119. Erosia Pochv i Ruslovye Processy. Vypusk **13**. Moscow University Publ., Moscow, Russia, (in Russian).

Golosov, V. N. (2002) Temporal–spatial variations in the sediment delivery ratio of small drainage basins: the Russian Plain example. In: *The Structure, Function and Management Implications of Fluvial Sedimentary Systems* (ed. by F. J. Dyer, M. C. Thoms & J. M. Olley), (Proc. Alice-Spring Sypm. August 2002), 345–353. IAHS Publ. 276. IAHS Press, Wallingford, UK.

Gordeev, V. V. (1983) *Rechnye Vynosy v Mirovoy Ocean i ikh Khimicheskiye Characteristiki* (*River Inputs to the Oceans and their Chemical Characteristics*). Nauka Publ., Moscow, Russia (in Russian).

Harrison, C. G. A. (1994) Rates of continental erosion and mountain building. *Geol Rundsch* **83**, 431–447.

Harrison, C. G. A. (2000) What factors control mechanical erosion rates? *Int. J. Earth Sci.* **88**, 752–763.

Hay, W. W., Shaw, C. A. & Wold, C. N. (1989). Mass-balanced paleogeographic reconstructions. *Geol. Rundschau* **78**, 207–242.

Hay, W. W. (1994) Pleistocene—Holocene fluxes are not the Earth's norm. In: *Material Fluxes on the Surface of the Earth* (ed. by W. W. Hay) 15–27. National Academy Press, Washington, DC, USA.

Holeman, J. N. (1980) Erosion rates in the U.S. estimated by the Soil Conservation Service's inventory. *Am. Geophys. Union Trans.* **61**, 954.

Holland, H. D. (1981). River transport to the oceans. The Sea. In: *The Ocean Lithosphere*, vol. 7 (ed. by C. Emiliani), 763–800. Wiley, New York, USA.

Kesel, R. H., Yodis, E. G. & McCraw, D. J. (1992) An approximation of the sediment budget of the lower Mississippi river prior to major human modification. *Earth Surf. Processes Landf.* **17**, 711–722.

Kuehl, S. A., Levy, B. M., Moore, W. S. & Allison, M. A. (1997) Subaqueous delta of the Ganges-Brahmaputra river system. *Mar. Geol.* **144**, 81–96.

Kuhlemann, J., Frisch, W., Székely, B., Dunkl, I. & Kázmér, M. (2002) Post-collisional sediment budget history of the Alps: tectonic versus climatic control. *Int. J. Earth Sci.* (*Geol. Rundsch.*) **91**, 818–837.

Kuenen, P. H. (1950) *Marine Geology*. Wiley, Chichester, UK.

Lisitzin, A. P. (1974) *Osadkoobrazovanie v Okeanakh* (*Oceanic Sedimentation*). Nauka Publ., Moscow, Russia (in Russian).

Lisitzin, A. P. (1991) *Processy Terrigennoy Sedimentacii v Moryakh I Oceanakh* (Terrigenous Sedimentation in Seas and Oceans). Nauka Publ., Moscow, Russia (in Russian).

Lopatin, G. V. (1952) *Nanosy Rek SSSR: Obrazovanie i Perenos* (Sediments in the USSR Rivers: Origin and Transport). Geografgiz Publ., Moscow, USSR (in Russian)

Losev, K. S. (1989) *Voda* (The Water). Gidrometeoizdat Publ., Leningrad, Russia (in Russian).

L'vovich, M. I. (1974) *Mirovye Vodnye Resursy i Ikh Buduschee* (World Water Resources and Their Future). Mysl' Publ., Moscow, Russia (in Russian).

Ludwig, W. & Probst, J. -L. (1998) River sediment discharge to the oceans: present-day controls and global budgets. *Am. J. Sci.* **298**, 265–295.

Makkaveev, N. I. (1982) Denudacionnaya sostavlyauschaya balansa veschestva v sisyeme ocean—susha I yeye rol' v formirovanii peneplenov (Denudation as a component in the land-ocean mass balance and its role in peneplain formation). *Vodnye resursy* **3**, 147–155 (in Russian).

McLennan, S. M. (1993) Weathering and global denudation. *J. Geol.* **101**, 295–303.

Meybeck, M. (1988) How to establish and use world budgets of riverine materials. In: *Physical and Chemical Weathering in Geochemical Cycles* (ed. by A. Lerman), 247–272. Kluwer, Dordrecht, The Netherlands.

Michels, K. H., Kudrass H. R., Hübscher, C., Suckow A. & Wiedicke, M. (1998) The submarine delta of the Ganges–Brahmaputra: cyclone-dominated sedimentation patterns. *Mar. Geol.* **149**, 133–154.

Milliman, J. D. & Meade, R. H. (1983) World-wide delivery of river sediment to the oceans. *J. Geol.* **91**, 11–21.

Milliman, J. D. & Syvitski, J. P. M. (1992) Geomorphic/tectonic control of sediment discharge to the ocean: the importance of small mountainous rivers. *J. Geol.* **100**, 525–544.

Milliman, J. D., Qin, Y. S. & Ren, M. E. Y. (1987) Man's influence on the erosion and transport of sediment by Asian rivers: the Yellow River (Huanghe) example. *J. Geol.* **95**, 751–762.

Prospero, J. M. (1981) Eolian transport to the world ocean. In: *The Sea. The Ocean Lithosphere*, vol. 7 (ed. by C. Emiliani), 801–874. Wiley, New York, USA.

Ronov, A. B. (1993) *Stratisfera ili Osadochnaya Obolochke Zemli: Kolichestvennoye Issledovanie* (*Stratisphere, or the Sedimentary Sphere of the Earth: Quantitative Research*). Nauka Publ., Moscow, Russia (in Russian).

Safyanov, G. A. (1978) *Beregovaya Zona Oceana v 20-m veke* (The Oceanic Coastal Zone in the 20th Century). Mysl' Publ., Moscow, Russia (in Russian).

Stallard, R. (1998) Terrestrial sedimentation and the carbon cycle: coupling weathering and erosion to carbon burial. *Global Biogeochemical Cycles* **12**, 231–257.

Ta, T. K. O., Nguyen, V. L., Tateishi, M., Kobayashi, I., Tanabe, S. & Saito, Y. (2002) Holocene delta evolution and sediment discharge of the Mekong River, southern Vietnam. *Quatern. Sci. Rev.* **21**, 1807–1819.

Tardy, Y., N'Kounkou, R. & Probst, J.-L. (1989) The global water cycle and continental erosion during Phanerozoic time. *Am. J. Sci.* **289**, 455–483.

Van Balen, R. T., Houtgast, R. F., Van der Wateren, F. M., Vandenberghe, J. & Bogaart, P. W. (2000) Sediment budget and tectonic evolution of the Meuse catchment in the Ardennes and the Roer Valley Rift System. *Global Planetary Change* **27**, 113–129.

Vassil'ev, V. P. (1987) Tverdiy stok v Mirovoy ocean (Solid flux into the World Ocean). *Litologia i Poleznye Iskopaemye* **6**, 19–28 (in Russian).

Vassil'ev, V. P. & Pomortsev, O. A. (1988) Aridnye zony v oceane: ikh granitsy I vklad eolovogo materiala v oceanskuyu sedimentaciyu (Arid zones in the ocean: Their borders and the aeolian contribution to the oceanic sedimentation). *Litologia I Poleznye Iskopaemye* **6**, 3–16 (in Russian).

Walling, D. E. (1983) The sediment delivery problem. *J. Hydrol.* **65**, 209–237.

Walling, D. E. (1999) Linking land use, erosion and sediment yields in river basins. *Hydrobiologia* **410**, 223–240.

Walling, D. E. & Webb, B. W. (1983) Patterns of sediment load. In: *Background to Palaeohydrology* (ed. by K. J. Gregory), 69–100. Wiley, Chichester, UK.

Walling, D. E. & Webb, B. W. (1996) Erosion and sediment yield: a global overview. In: *Erosion and Sediment Yield: Global and Regional Perspectives* (ed. by D. Walling & B. Webb) (Proc. Exeter Symp., July 1996), 3–19. IAHS Publ. 236. IAHS Press, Wallingford, UK.

Xia, D. X., Wu, S. Y. & Yu, Z. (1993) Changes of the Yellow River since the last glacial age. *Mar. Geol. Quatern. Geol.* **13**, 83–88.

Xu, J. (1998) Naturally and anthropogenically accelerated sedimentation in the lower Yellow River, China, over the past 13 000 years. *Geogr. Ann.* **80A**, 67–78.

Recent trends in turbidity and suspended sediment loads in the Murrumbidgee River, NSW, Australia

JON OLLEY & PETER WALLBRINK

Co-operative Research Centre for Catchment Hydrology, and CSIRO, Land and Water, PO Box 1666, ACT 2601, Australia

Jon.olley@csiro.au

Abstract Turbidity and flow data from the Murrumbidgee River, NSW Australia were used to assess if changes in land management have altered suspended sediment loads. The data indicate that over the last 20 years the turbidity and sediment load per unit volume of flow has decreased significantly. The decrease in the sediment load is related to a number of factors, including: (a) a decrease in flow from the tributary catchments; (b) an increase in the proportion of water derived from upstream water storages; (c) a decrease in the total volume of flow in the main channel; (d) stabilization of >50% of the gully network; (e) an increase in the extent of in-stream wetlands; and (f) a massive increase in the number of farm dams. The work illustrates the difficulty in relating changes in sediment loads in large rivers to causal factors.

Key words Australia; channel erosion; land management; Murrumbidgee River; sediment loads

INTRODUCTION

European settlement 180 years ago brought massive changes in land use across the southeast of Australia (e.g. Scott, 2001). The clearing of woodlands, the introduction of grazing stock and rabbits, the drainage of valley bottoms, and the clearing of riparian vegetation caused a large increase in erosion (Prosser *el al.*, 2001; Olley & Wasson, 2003). In < 100 years nearly every valley in southeastern Australia was affected. Massive volumes of sediment were delivered to the rivers, and the form of the rivers and the surrounding landscape were changed dramatically (Olley & Scott, 2002; Olley & Wasson, 2003). Vegetated valley floors were incised; clear flowing waters became turbid; deep pools were filled with sand; and rivers, which abounded with fish and other life, became dominated by algal growth.

Over the last 50 years there have been significant improvements in the way the land is managed. The formation of the Soil Conservation Services, beginning in 1938, started a programme of educating land managers in ways of minimizing soil erosion. The introduction of myxomatosis in the 1950s greatly reduced rabbit numbers, improving the vegetation cover. Many of the gullies are now vegetated and have stabilized. In many tributary catchments in-stream wetlands have formed. The construction of water storage reservoirs and farm dams has also decreased the supply of sediment.

In this paper we used turbidity and flow data to assess temporal changes in the sediment load in the main channel of the Murrumbidgee River, NSW, Australia. An attempt is made to relate variations in the sediment load to changes in the catchment.

BASIN DESCRIPTION AND PREVIOUS WORK

The Murrumbidgee River is one of Australia's largest rivers, draining approximately 84 000 km^2 of the Murray-Darling Basin (Fig. 1). Its catchment can be divided into three distinct

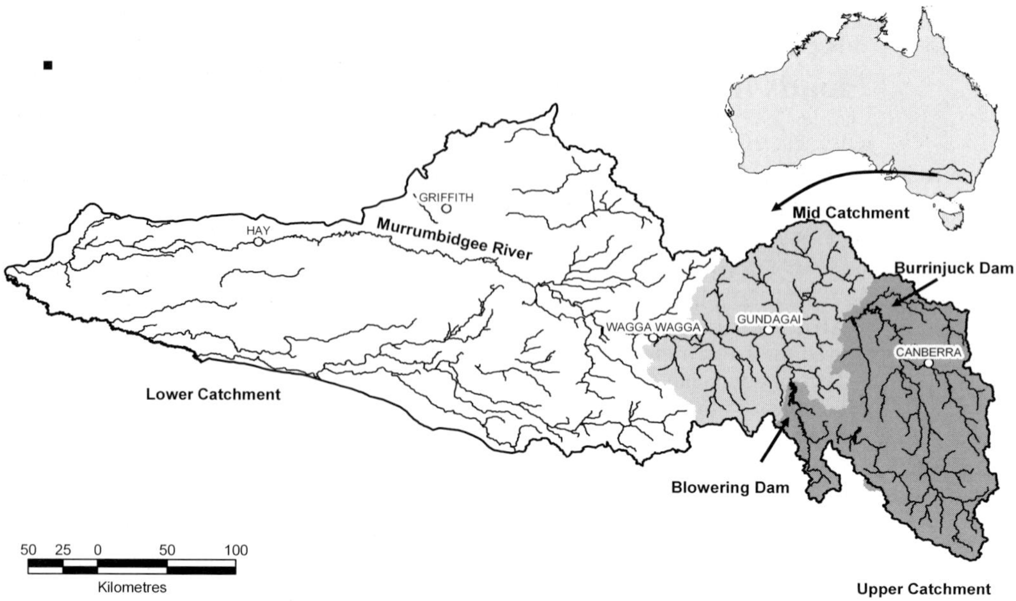

Fig. 1 Map of the basin showing key locations.

geomorphic regions: the upper, mid, and lower Murrumbidgee (Page, 1994). The upper region is mountainous and hilly, and is separated from the mid-region by two large storage reservoirs, Burrinjuck and Blowering (Fig. 1). These dams trap most of the sediment delivered from the upper basin (Wasson *et al.*, 1987). The mid-basin area is characterized by undulating terrain dissected by numerous stream and gully networks, which have been shown to be the primary source of sediment delivered to the lower Murrumbidgee River (Olive *et al.*, 1996; Wallbrink *et al.*, 1998). Most of the major tributaries of the mid to lower Murrumbidgee join the river in this region; their average catchment size is ~1000 km^2. Over the last 20 years a large number of farm dams have been build in these tributary catchments and between 10–20% of the tributary area is now separated from the main channel by farm dams. Many of the gullies have now stabilized with more than 50% of the gully networks in the Wagga Wagga region now well vegetated (Caitcheon, personal communication). In many tributary catchments in-stream wetlands have formed and now trap significant amounts of sediment derived from upstream (Zierholz *et al.*, 2001).

Grazing is the main land use in the upper basin. The region around Wagga Wagga (mid-basin) is a major grain producing area. There is extensive grazing on the riverine and hay plains that make up the lower basin, and intensive cropping and horticulture in the irrigation areas. A strong rainfall gradient exists across the basin; average annual totals vary from 1600 mm in the Snowy Mountains to 300 mm at Balranald. Prior to regulation the river experienced extreme variations in annual flow, with most of the flow concentrated in winter and spring (Ebsary, 1994). The river is now a major source of irrigation water and flow is heavily regulated, both by the two major dams located in the upper catchment, and by a series of weirs below Wagga Wagga (Fig. 1). The combination of regulation and extraction for irrigation has resulted in flow now being spring and summer dominated when water demand by irrigators is highest.

Using the fallout tracers ^{137}Cs and $^{210}Pb_{ex}$, Wallbrink *et al.*, (1998) demonstrated that the dominant source (~90%) of the suspended sediment in the Murrumbidgee River is channel banks and gully walls. Fallout ^{210}Pb data were also used to show that the mean residence times of fine grain sediments within the mid-Murrumbidgee system is unlikely to be more than a few years. Additional evidence from: (a) changes in the mean ^{137}Cs concentrations measured in sediments from flood and low flow waters; and (b) the presence of the short-lived surface-soil tracer ^{7}Be in flood water sediments, showed that the actual transport time of fines in this river is probably weeks to months, rather than years to decades. The low ^{137}Cs concentrations from the tributaries also indicate that most of the subsoil originates from the middle reach tributaries, and not from channel bank erosion along the Murrumbidgee River channel.

THE DATA

Daily turbidity measurements have been made at the Wagga Wagga water treatment plant since 1949. Measurements are made by visual comparison with a set of standards. While the same technique has been used for measurement of turbidity for the entire period the authors recognize that there is no guarantee that the measurements are consistent through time. However, analysis of the record indicates that the range of values has remained relatively constant with no apparent drift in trends. Turbidity has been measured at Gundagai since 1990 using a Hach turbidity meter.

The discharge record used in this study comes from the NSW Department of Water Resources (DWR) continuous gauging stations at the dams, at Wagga Wagga and at Gundagai. These stations are part of a network of gauging stations maintained by DWR throughout NSW. The rainfall data discussed comes from the Australian Bureau of Meteorology.

RESULTS AND DISCUSSION

Turbidity at both Gundagai and Wagga Wagga is linearly correlated with sediment concentration (Olive *et al.*, 1996). However, conversion of the turbidity data to sediment concentration introduces another variable into the analysis of trends in the records. Consequently we have used monthly-flow-weighted-turbidity. This is equivalent to the average turbidity observed per unit volume of water in the month, and equates to the mean monthly sediment concentration. The two records have been cross-calibrated and are shown together in Fig. 2(a). The turbidity data shows an overall decline in sediment supply over the last 50 years, particularly in the last 20 years (indicated by the 2.5 years running average). The 2.5-year running average monthly flow weighted turbidity and Wagga Wagga rainfall data (Fig. 2(b)) tend to co-vary with increases and falls in the turbidity associated with increases and falls in the local rainfall.

The decline in monthly flow weighted turbidity over the last 20 years is correlated with a decrease in flows at Wagga Wagga and an increase in the proportion of water derived from the upstream water storages (Fig. 3(c)). The high turbidities occur at the same time as flow events in the tributary catchments (compare Fig. 3(a) and (b)) supporting the hypothesis

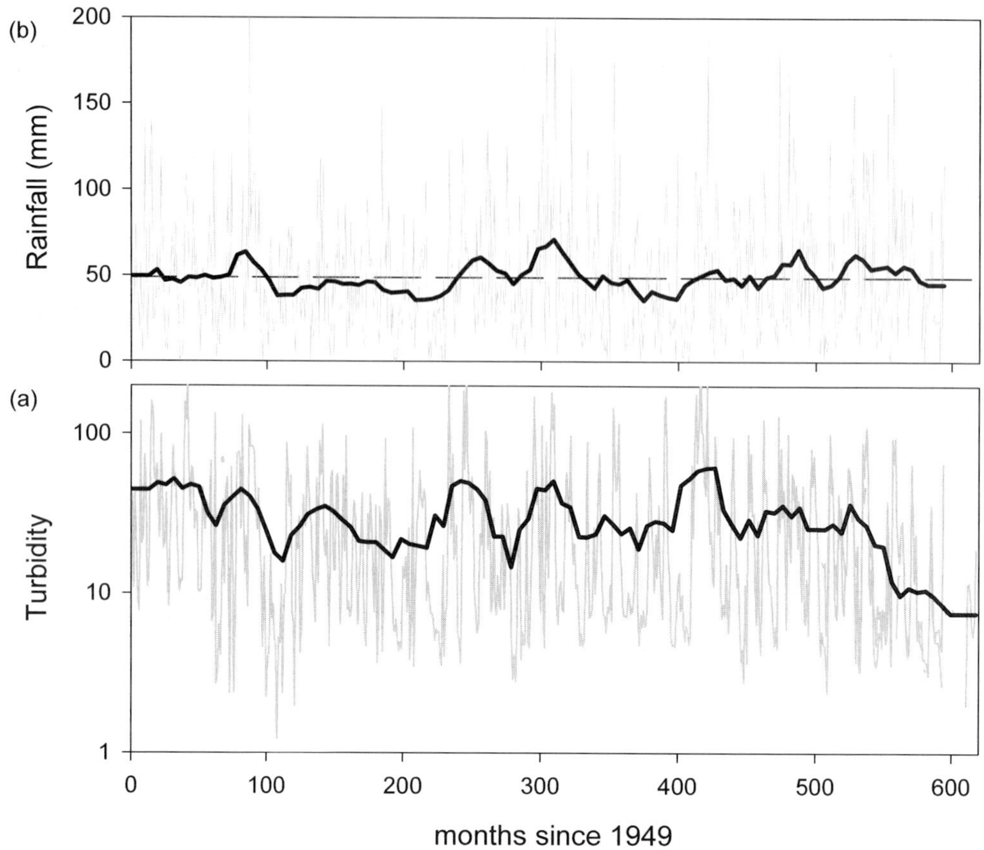

Fig. 2 (a) Monthly flow weighted turbidity for Wagga Wagga (grey line). The thick black line shows the 2.5-year running average on the combined turbidity data. (b) Total monthly rainfall at Wagga Wagga for the period since December 1948 (grey line). The mean monthly rainfall for this period is shown as the horizontal dashed line. The thick black line represents a 2.5-year running average.

proposed by Olive *et al.* (1996) that the tributaries are the primary source of sediment causing the higher turbidity. The decrease in flows from the tributary catchments over the last 20 years may be related to a decrease in regional rainfall. Or it may be a result of the construction of more than 20 000 farm dams in the catchment.

The data presented indicates that over the last 20 years the flow weighted turbidity and hence the sediment load per unit volume of water has decreased significantly at the Wagga Wagga. This improvement in water quality in one of Australia's largest rivers is correlated with a decrease in flows derived from the tributary catchments, a related increase in the proportion of water derived from upstream water storage, and a decrease in the total volume of flow in the main channel. Over the same period there has also been a marked increase in the vegetation cover in the gully networks, an increase in the extent of in-stream wetlands and an increase in the number of farm dams. The synchronicity of these events, all of which may be expected to alter the supply of sediment, make it difficult to discern the primary cause or causes of the improvements in water quality. This is the subject of a continuing investigation.

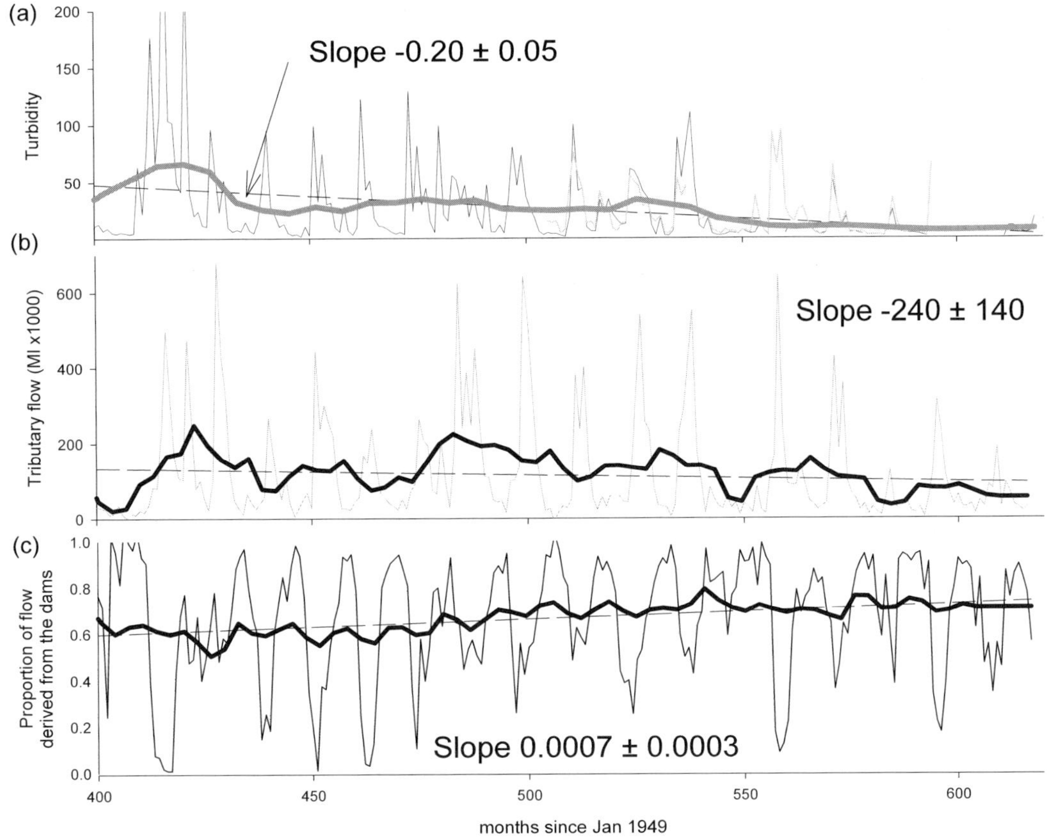

Fig. 3 (a) Monthly flow weighted turbidity for Wagga Wagga (dark grey) and Gundagai (light grey) for the period since 1981. (b) Flows derived from the tributary catchments below the dams. (c) The relative proportion of the water passing Wagga Wagga derived from the dams. In each figure the thick black line shows the 2.5-year running average on the data; the long dashed line shows the linear trends; the slope on the trend line and relative uncertainty are reported in each figure.

REFERENCES

Ebsary, R. (1994) Regulation of the Murrumbidgee River. In: *The Murrumbidgee Past and Present* (ed. by J. Roberts & R. Oliver), 49–59. CSIRO Div. Water Resources, Griffith, Australia.

Olive, L. J., Olley, J. M., Wallbrink, P. J. & Murray, A. S. (1996) Downstream patterns of sediment transport during floods in the Murrumbidgee River, NSW, Australia. *Z. Geomorph.* N.F. Suppl.-Bd. **105**, 129–140.

Olley J. M. & Scott, A. (2002) *Sediment Supply and Transport in the Murrumbidgee, and Namoi Rivers since European Settlement*. Tech. Report 9/02, CSIRO Land and Water, Canberra, Australia.

Olley, J. M. & Wasson R. J. (2003) Changes in the flux of sediment in the Upper Murrumbidgee catchment, SE Australia, since European settlement. *Hydrol. Processes* **17**, 3307–3320.

Scott, A. (2001) *Water Erosion in the Murray-Darling Basin: Learning from the Past*. Tech Report 43/01, CSIRO Land and Water, Canberra, Australia.

Wallbrink, P. J. Murray, A. S., Olley, J. M. & Olive, L. (1998) Determining sediment sources and transit times of suspended sediments in the Murrumbidgee River, NSW, Australia using fallout Cs-137 and Pb-210. *Water Resour. Res.* **34**(4), 879–887.

Zierholz, C, Prosser, I. P., Fogarty, P. J. & Rustomji, P. (2001) In stream wetlands and their significance for channel filling and the catchment sediment budget, Jugiong Creek, NSW. *Geomorphol.* **38**, 221–235.

Sediment Transfer through the Fluvial System (Proceedings of a symposium held in Moscow, August 2004).
IAHS Publ. 288, 2004

Investigation of sediment yield of lowland rivers in Ukraine

I. P. KOVALCHUK[1] & V. I. VISHNEVSKIY[2]

1 *Ivan Franko National University of Lviv, 79000 Lviv, Ukraine*
geomorph@franko.lviv.ua

2 *Central Geophysical Observatory, Kyiv, Ukraine*

Abstract The new map of turbidity in the rivers of Ukraine has been created on the basis of a 50-year data set of sediment yield observed by the Hydrometeorological committee of Ukraine. The trends of sediment discharge have been established. The main causes of increasing or decreasing sediment discharge in rivers of the different regions of Ukraine (Ukrainian Polissia, Podilska and Prydniprovska highlands, Prychornomorska lowland, Donbas, Carpathian Mountains) are explained. Quantitative values of sediment erosion, transit and redeposition in different sections of the catchment system and its flood plain–riverbed complexes have been determined on the basis of a field experimental study of sediment redistribution within the cultivated slope—gully catchment—small river catchment system. The impact of erosion–accumulation processes on small river aggradation has been characterized.

Key words factors; rivers of Ukraine; sediment redistribution; sediments; turbidity

INTRODUCTION

Sediment yield is one of the most important features of the hydrological regime of rivers. It impacts the use of rivers and their hydroecological state. The intensive sheet and gully erosion on the river basin slopes causes the accumulation of sediments in the riverbeds and reduction of the length of the river network. The suspended sediments are a very important factor in the transportation of pollutants and water quality.

These problems have been investigated by many researchers, including Ukrainian scientists. Most of them are members of the "geographical direction-school" of investigation of the river's regime. The representatives of this direction-school consider the spatial and temporal peculiarities of the sediment yield and its dependence upon different factors: the water discharges, the forestry of the river's basins, the slope of the riverbed, etc. Drozd was among the first investigators of turbidity and sediment yield in Ukrainian rivers in the 1960s. His work resulted in the creation of the first map of water turbidity. Goretskaja later followed him (Goretskaja *et al.*, 1988). During the last two decades the problems of sediment transportation and channel processes have been studied by Shvebs & Antonova (1988), Bobrovitskaja (1988), Obodovskiy (2001), Grebin (2000, 2002), Kovalchuk (1992, 1997) and Vishnevskiy (1992, 1994, 1997, 2000, 2001).

There are not many representatives of the "hydraulic direction school" in the Ukraine. It is important to distinguish the investigations of Nikitin (1980) and Bazilevich (Bazilevich *et al.*, 1988; Bazilevich & Vishnevskiy, 1994), who worked in the Institute of Hydromechanics National Academy of Science of Ukraine.

The other scientists' studies focus on the transportation of polluted substances on suspended sediments (Linnik, 1999; Voitsekhovich *et al.*, 1997). The documentation of much greater contamination of fine particles for both various chemical substances and radioactive pollution, when compared to larger particles, was the primary result of these works. In particular, a detailed study in a "near zone" of Chernobyl NPP has shown that most pollutants are associated with particles smaller than 0.05 mm (Voitsekhovich *et al.*, 1997; Voitsekhovich, 2001). This fact significantly affected the distribution of pollution by the Prypiat' River across large distances from Chernobyl NPP.

In spite of many investigations completed during the last decades, some questions were not solved. Among them is the distribution and quantification of natural (not changed by the human impact) turbidity of water.

DATA

The basic source of the sediment yield data, including the size of sediments (both bed and suspended) are the results of monitoring by the Hydrometeorological Committee of Ukraine. The duration of the direct observations on many rivers achieved 40–50 years. The number of gauging stations, on which they are carried out, for 1 January 2002 was equal to 119.

These data allow estimation of the spatial and temporal peculiarities of suspended sediments in rivers. However, it is necessary to consider the impact of ponds and reservoirs. There are 63 000 rivers, 28 800 ponds and 1160 reservoirs in the territory of Ukraine. The total volume of water in ponds and reservoirs exceeds the average water runoff of the Dnepr River or the entire water runoff from the territory of Ukraine.

RESULTS AND DISCUSSION

The highest water turbidity (up to 1000 g m^3 and higher) is observed in the rivers of the Crimean Mountains, which contain steep slopes and sparse vegetation. The rivers of the Carpathian Mountains and the Podillia region have lower turbidities (200–500 g m^3). The smallest turbidity (at a level 10 g m^3) is typical for the rivers of Polissia, in the most northern part of Ukraine (Fig. 1).

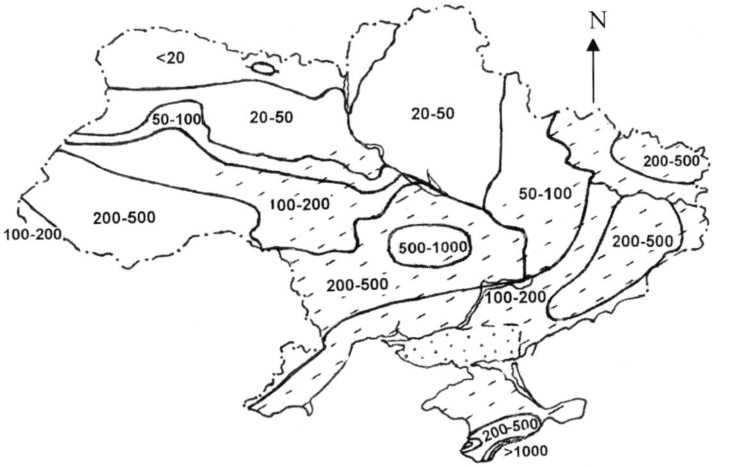

Fig. 1 The map of turbidity of water in Ukrainian rivers, g m^3 (the shaded region represents areas of the territory containing many ponds and reservoirs).

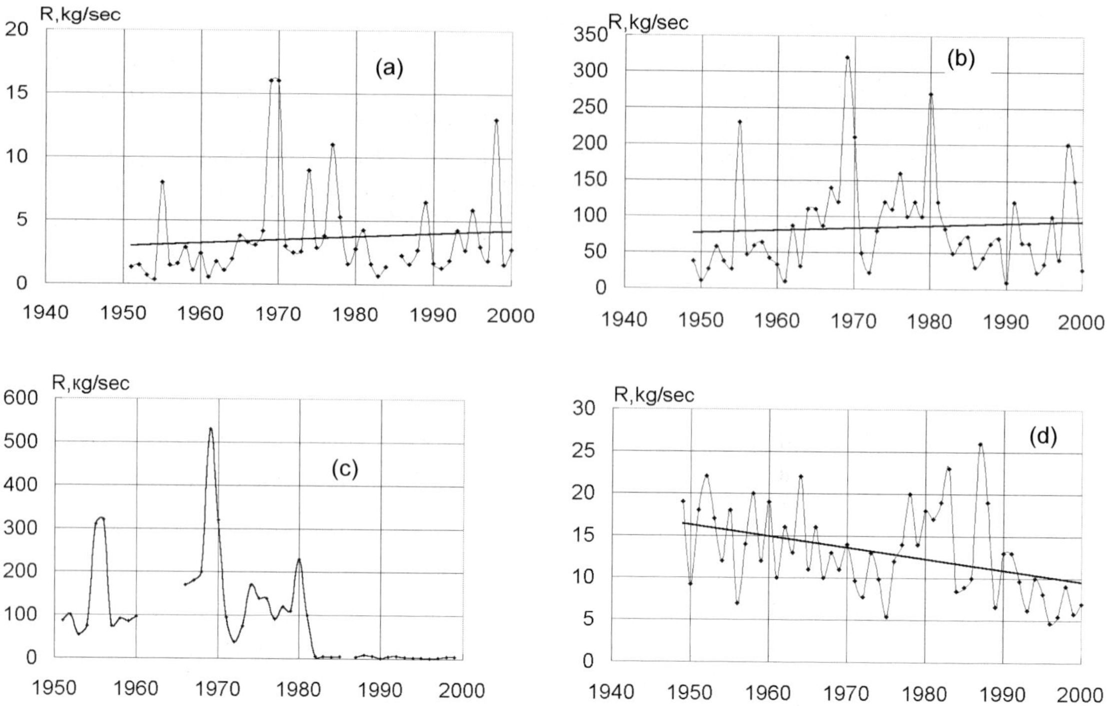

Fig. 2 Long-term changes of suspended sediment yield: (a) Tysa, Rakhiv; (b) Dnister, Zalishchyky; (c) Dnister, Mohyliv-Podilskyi; (d) Desna–Chernihiv.

In the main territory of Ukraine, the sediment yield in rivers is decreasing. The main cause is the creation of ponds and reservoirs. After the creation of the Dnister Reservoir (a total volume of 3.0 km^3) located between gauging stations Zalishchyky and Mohylic-Podilskyi, the sediment yield at the downstream station has been diminished by 30 times.

It is possible to assume that some impact is associated with the reduction of economic activity, including land use during the 1990s. During this decade the cultivated area decreased by 10%. At the same time, the areas cultivated for sugar beets, which has the highest erosion, experienced a 3-fold increase.

Some impact can be explained by climatic changes. During the 1990s, regional snow cover was minimal. The only exception is the increasing sediment yield in the Carpathian rivers, due to timber harvesting. Another factor is the mining of the channel bottom sediments, including the entire largest particle surface layer (Fig. 2).

The significant part of the suspended sediment yield passes during spring flooding. Sediment yield during floods is responsible for 60–70% of the annual quantity in the northern part of Ukraine and 20–30% in the southern part. During the last decades this trend has diminished. For example, the total suspended sediment yield during spring flooding during 1935–1985 was equal to 76%, and for 1991–2000 it was 64% for the Desna River (hydrological station Chernihiv). This reduction can be explained by the reduced amount of surface water runoff during spring snow-melting. In turn, the reduced runoff is explained by an increase of air temperature in the winter period and the decrease of snow storage on the basin slopes (Grebin', 2000; Kovalchuk & Shtoiko, 1992).

It is obvious that the presence of suspended sediments in a river flow is determined not only by their supply from the basin area, but also by the conditions of their transportation. The decreased ability to transport sediment in rivers is the basic reason for sediment deposition and reduction of the length of river network (Vishnevs'kyi, 2000a,b).

It is possible to use the results of some previous investigations to describe sediment erosion and deposition, and to characterize the conditions of sediment transport. The erosion conditions are characterized by the well-known dependence of non-eroding velocity from the size of particles. The conditions of sedimentation are detailed in an investigation by Nikitin (1980). According to these results, the degree of removal of suspended particles by deposition (S) is determined by the equation:

$$S = 1.66 + 0.68 \lg (\omega / V_{ave}) \tag{1}$$

where ω = hydraulic size of particles, V_{ave} = average flow velocity.

Bazilevich *et al.* (1988) used the results of these investigations to obtain analytical dependencies that show the boundary conditions of suspended sediments presence in the river flow. These dependencies are:

$$S = 0 \qquad \omega / V_{ave} = 0.0036 \tag{2}$$

$$S = 1 \qquad \omega / V_{ave} = 0.107 \tag{3}$$

Equation (2) characterizes conditions where sediments do not accumulate ($S = 0$). Hence, when the hydraulic size of sediments is smaller than 0.0036 V_{ave}, the particles are constantly supported in a flow. In these conditions, particle exchange with the riverbed does not occur. Equation (3) characterizes conditions where all sediments contained in a flow accumulate ($S = 1$). In intermediate conditions, there is an exchange of sediments.

It is possible to illustrate the lines of equations (2) and (3) and also the graphic relationship between non-erosive velocity (for the depth 1.0 m) for different sizes of sediments on one figure. This enables the delineation of six zones with various conditions of sediments transportation (Fig. 3).

Zone I. This zone is situated beneath the line of non-erosive velocity. It portrays the absence of movement of the bottom sediments. The flow transports all sediments that are present in the flow.

Zone II. The erosion of the bottom sediments does not occur. Only partial sedimentation of suspended sediments occurs.

Zone III. The velocity of flow is insufficient for erosion of the bottom sediments. Complete deposition of sediments that are carried in flow is observed. Such conditions are observed when the rivers enter lakes, reservoirs and seas.

Zone IV. There is partial erosion of bottom sediment and whole transportation of suspended sediment.

Zone V. This zone, as well as previous zones, is characterized by the erosion of bottom sediments. Simultaneously, a portion of the suspended sediment falls on the bottom. These sediments are re-suspended by velocity pulsations.

Zone VI. This zone is characterized by the erosion of bottom sediments. Simultaneously the majority of suspended sediments falls on the bottom. An intensive exchange of sediments is observed (Vyshnevs'kyi, 2000b).

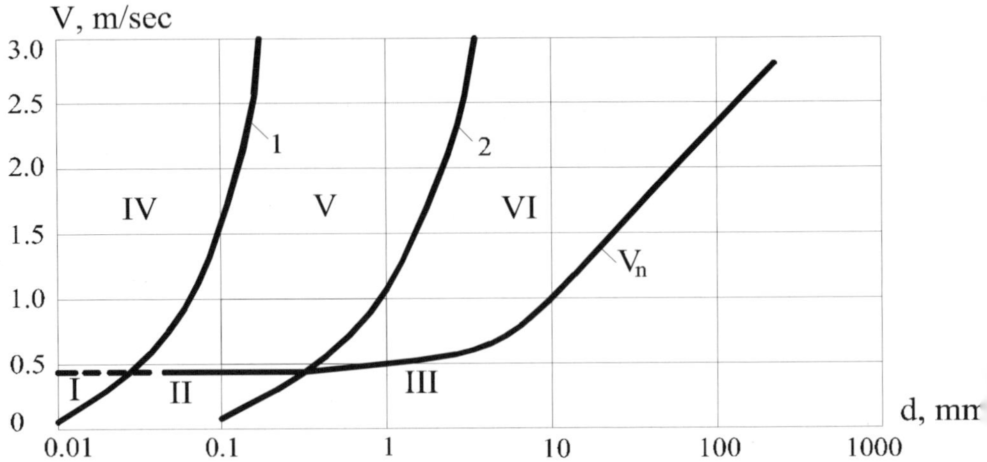

Fig. 3 Distribution of conditions of transportation and deposition of sediments, depending on its size and average velocity of flow: Curves 1 and 2 are plotted from formulas (2) and (3).

These results are an addition to the well-known diagrams (particularly Shields criterion) and show the conditions of sediment transport. These results explain the observed sedimentation of riverbeds of small rivers throughout Ukraine. The flow velocity is much less than 0.4 m s^{-1} in these rivers during the main part of the year.

When the flow velocity is small, and the supply of sediment from the slopes to the river channel is large, the majority of sediment is deposited in the river-bottom (zone II and III of Fig. 3). The spreading of water vegetation in rivers is an additional factor of diminution of the flow velocity. In the rivers of southern Ukraine, most sediment is transported during the warm period of the year. When the hydraulic resistance is great and the flow velocity is small, even fine sediment cannot be transported in suspended form.

The flow's reduced ability to transport sediment can be explained by the alignment of the interior-seasonal distribution of the river runoff. This is the consequence of climatic changes. Increased winter temperatures observed during previous decades has caused a decrease of the water storage in snow. As a result, the maximal discharges during a spring high water have decreased. At the same time, the runoff during dry periods (both winter and summer) became larger.

Field experimental investigation of sediment transportation was carried out on a small catchment of the Dnister River with the following parameters: area is 17.5 ha, a ratio of width and length of 0.56, relative heights are up to 80 m, hillslope angles are 3–10°, thalweg angles are 2.5–4°, exposure of slopes is western and eastern, cultivated area is 100%, and soils are typically comprised of dark grey podsol eroded to various degrees (Kovalchuk, 1997).

This investigation identified the ratio between erosion, transit, and deposition of sediments in a system: slope–primary watershed–thalweg–flood plain of the small river (I or II order)–channel of the small river–transit part (transportation of sediments to the river of higher rank) for the period of snow-melting and fall of rains. The results are shown schematically in Fig. 4.

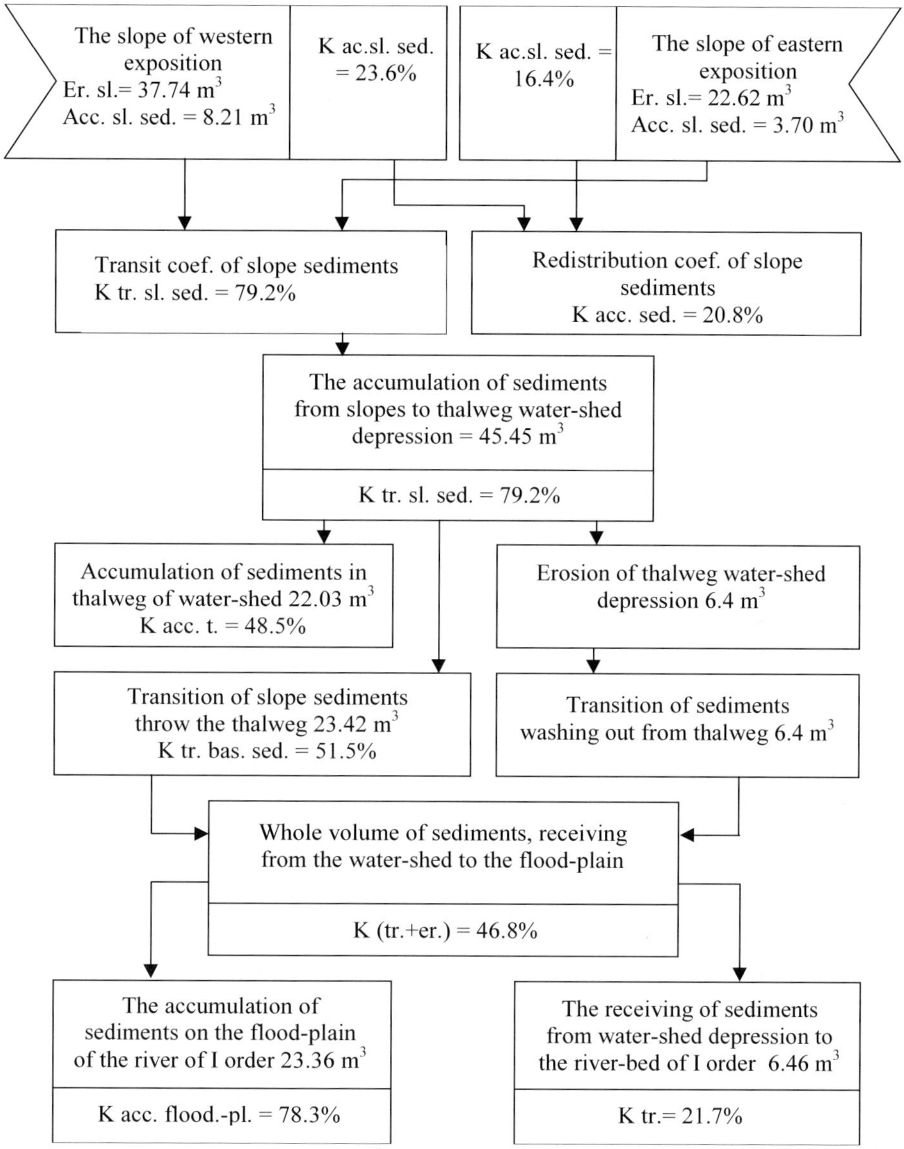

Where: Er sl. = gross soil losses, m^3; Acc.sl.sed. = sediment deposition within slope, m^3; K ac.sl. sed = part of sediment, deposited within slope; K tr. sl. sed.= transition ratio coefficient;

Fig. 4 The ratio of erosion, transition and redistribution of sediments on the different components of a river-basin system (Podillia Highland, Ukraine).

Of sediment eroded from the completely ploughed slope, 16–24% is accumulated (Fig. 4). That corresponds to a sediment ratio coefficient (the ratio of the quantity of sediment delivered to the rivers to the quantity of sediment eroded from the slopes) of approximately 79%. This coefficient is higher when the slopes are steeper. The further fate of slope sediments is different. Part of them (31–48%) accumulates in the bottom of watershed depressions. Here, an additional amount of sediment (from 14% up to 22%) is formed due to erosion of the depression thalweg. The sediment transition coefficient for watershed depressions is 51.5%, or 46.8% if taking into account the sediment of eroded from the thalweg. The main part of this sediment (78.3%) accumulates on the floodplain of small rivers (width of floodplain is 120–184 m) and only 21.7% is transported to first order rivers.

The sediment transition coefficient for the small rivers varies from 17.9 up to 28.1%. It depends on the characteristics of the watershed depressions on the slope and floodplains of rivers. The generalization, provided by Golosov (2003), shows that the above mentioned results correspond to the results of other authors.

Our calculations, based on the results of radiocarbon dating floodplain and riverbed sediments of small rivers of the Podillia highland (the Bilka River, Gnyla Lypa catchment – (left tributary of the Dnister River), show that the average rate of sediment deposition ranges between 0.83–1.04 mm $year^{-1}$ for the last 6.5–7.0 thousand years (according to dating by Marburg Philips University in 1997 (Kovalchuk, 1997). The rate of contemporary erosion from ploughed slopes is 0.3–0.8 mm per season with maximum rates up to 2.9–4.8 mm per season. These data are received as a result of 15 years monitoring of erosion processes on ploughed slopes with inclination $3 – 9^{o}$ at the Bilka catchment.

The change of structure in the river network was investigated using historical and current maps. It was established, that 36–65% of I–III order rivers of the Podillia highland have disappeared during the last 150 years. At the same time, the river length was reduced by 7–31%. Similar results were also obtained by other researchers (Golosov, 2003).

CONCLUSIONS

The following results were obtained from this study:

(a) the new map of turbidity of water in rivers of Ukraine was created;
(b) the anthropogenic impact on water turbidity in the rivers was evaluated;
(c) the modern scheme of sediment transportation in the river flow was obtained;
(d) the characteristics of sediments redistribution in the system of the slope—the watershed depression—the flood plain of first order rivers and riverbed of first order rivers was established.

REFERENCES

Bazilevich, V. O., Gribovskaja, N. K. & Kozitskiy, V. V. (1988) Opriedielieniye raskhoda vzvieshennykh nanosov v pieschanykh ruslakh riek (The evaluation of suspended sediments in sandy channels of rivers). *Gidravlika i Gidrotiekhnika* **46**, 32–38 (in Russian).

Bazilevich, V. O. & Vishnevskiy, V. I. (1994) O raskhodie vzvieshennykh nanosov v malykh ravninnykh riekakh (About the discharge of suspended sediments in small plain rivers). In: *Dinamika i tiermika riek, vodokhranilishch i okrainnykh moriey* (Proc. International Conf., 1994), vol. 1, 189–191. Moscow, Russia (in Russian).

Golosov, V. N. (2003) Erozionno-akkumuliativnye processy v verkhnikh zvieniyakh fliuvialnoy seti osvoennykh ravnin umieriennogo poyasa (Erosion-accumulative processes at the upper parts of fluvial network of moderate zone plains). Doctor of Science Thesis, Moscow State University, Moscow, Russia (in Russian).

Goretskaja, Z. A., Manukalo, V. A. & Sokolov S. B. (1988) Opriedielieniye sridniego godovogo stoka nanosov malykh riek stiepnoy zony pravobieriezhnoy Ukrainy po dannym o zailienii prudov (The estimation of the average annual sediment yield of small rivers of a steppe zone of Ukraine based on the database of the ponds' sedimentation). *Trudy UkrNIGMI* **288**, 49–60. (in Russian).

Grebin', V. V. (2000) Otsinka rezhymu nanosiv richok baseynu Dnipra ,v mezhackh Ukrainy (The assessment of sediment yield of rivers in the Dnipro catchment, on the territory of Ukraine). *Hidrologiya, Hidrokhimiya i Hidroekologiya* **1**, 135–137 (in Ukrainian).

Grebin', V. V. (2002) Vykorystannia geografo-hidrologichnoho metodu dlia vyvchennia stokunanosiv (The use of geographic-hydrological method for the investigation of suspended sediment yield). *Hidrologiya, Hidrokhimiya i Hidroekologiya* **3**, 43–48 (in Ukrainian).

Kovalchuk, I. P. (1997) *Rehionalnyi Ekologo-Geomorfologichnyi Analiz* (*Regional Ecological-Geomorphologic Analysis*). Institute of Ukrainian studies, Lviv, Ukraine (in Ukrainian with English summary).

Kovalchuk, I. P. & Shtoiko, P. I. (1992) Izmienieniya riechnykh sistiem Zapadnogo Podolia v XVIII–XIX v. (Changes of river systems of the Western Podillia in XVIII-XIX centuries). *Geomorphology* **2**, 55–72 (in Russian).

Kovalchuk, I. & Zakharchuk, O. (1997) Analiz stoku nanosiv yak integralnoho pokaznyka skhylovoyi ta ruslovoyi eroziyi (The analysis of sediment yield as the index of slope and riverbed erosion). In: *Ukrains'ka Geomorfologiya: Stan i Perspektyvy* (Proc. International Conf., 1997), 117–119. Merkator, Lviv, Ukraine (in Ukrainian).

Linnik, P. M. (1999) Tiazholyie mietally v povierkhnostnykh vodakh Ukrainy: sodierzhaniye i formy migratsii (Heavy metals in surface waters of Ukraine: the contents and forms of migration). *Gidrobiologicheskiy Zhurnal*, **35**(1), 22–41 (in Russian).

Nikitin, I. K. (1980) *Slozhnyie Turbulientnyie Tiechieniya i Processy Tieplomassopierienosa* (Complicated turbulent currents and processes of heat and mass transportation). Naukova dumka, Kyiv, Ukraine (in Russian).

Obodovs'kyi, O. G. (2001) *Hidrologo-Ekologichna Otsinka Ruslovykh Protsesiv* (Na Prykladi Richok Ukrainy) (Hydrological and ecological estimation of channel process (on example of Ukrainian rivers)). Nika-Tsentr, Kyiv, Ukraine (in Ukrainian).

Shvebs, G. I. (1974) *Formirovaniye Vodnoy Erozii, Stoka Nanosov i ich Otsenka* (Na Primierie Ukrainy i Moldavii) (Formation of water erosion, sediment yield and their estimation (on an example of Ukraine and Moldova)). Gidromietieoizdat, Leningrad, Russia (in Russian).

Shvebs, G. I. & Antonova, S. A. (1988) Sovriemiennyye izmienieniya riezhyma nanosov riek Ukrainy (Modern changes of sediment yield of the rivers of Ukraine). In: *Trudy Piatogo Vsiesoyuznogo Gidrologichieskogo Syezda*, vol. 10, part 2, 155–160. Gidromietieoizdat, Leningrad, Russia (in Russian).

Shvebs, G. I. & Bobrovitskaya, N. N. (1988) Eroziya i stok nanosov s vodosbora (Erosion and sediment transportation from the watershed). In: *Trudy Piatogo Vsiesoyuznogo Gidrologichieskogo Syezda*, vol. 10, part 2, 102–125. Gidromietieoizdat, Leningrad, Russia (in Russian).

Vyshnevs'kyi, V. I. (2000) *Richky i Vodoimy Ukrainy. Stan i Vykorystannia* (The rivers and lakes of Ukraine. The state and the use). Vipol, Kyiv, Ukraine (in Ukrainian).

Vyshnevs'kyi, V. I. (2000) Transportuvannia i osadzhuvannia nanosiv u rivnynnykh richkakh (The transportation and the sedimentation of suspended sediments in plain rivers). In: *Naukovi Pratsi UkrNDHMI* **248**, 91–99. Kyiv, Ukraine (in Ukrainian).

Vyshnevs'kyi, V. I. (2001) Zminy klimatu i richkovoho stoku na terytorii Ukrainy i Bilorusi (The change of climate of and river runoff on the territory of Ukraine and Belarus). In: *Naukovi Pratsi UkrNDHMI* **249**, 89–105 (in Ukrainian).

Vyshnevs'kyi, V. I. & Kulachins'ka, L. N. (1999) Tviordyi stok malykh riek Ukrainy (A sediment yield of small rivers of Ukraine). In: *Melioratsiya i Vodnoye Khoziaystvo* **78**, 78–88 (in Russian).

Voitsekhovich, O. V. (2001) *Upravlieniye Kachiestvom Povierkhnostnykh Vod v Zonie Vliyaniya Avarii na Chernobylskoy AES* (The management of quality of the surface water in zone of influence of Chernobyl NPP accident). UkrNDHMI, Kyiv, Ukraine (in Russian).

Voitsekhovich, O. V., Kanivets, V. V. & Vishnevskiy, V. I. (1997) Sediment associated transport of Chernobyl radionuclides in the Pripyat River, Ukraine. In: *Study of Erosion, River Bed Deformation and Sediment Transport in River Basins as Related to Natural and Man-made Changes*. IHP–V. Technical Documents in Hydrology, vol. 10, 155–164. UNESCO, Paris.

Voloshchuk, V. M., Bojchenko, S. G. Stepanenko, S. M. *et al.* (2002) *Globalne Poteplinnia i Klimat Ukrainy: Regionalni Ekologichni ta Sotsialno-Ekonomichni Aspekty* (The global increasing of air temperature and the climate changes of Ukraine: the regional and social aspects). VPC "Kyivs'kyi Universytet", Kyiv, Ukraine (in Ukrainian).

Zubkova, K. M., Petukhova, G. A. & Tkachova, L. G. (1988) Izmienchivost' stoka nanosov i vliyaniye khoziaystviennoi dieyatielnosti na yego kharaktieristiki (Variability of suspended sediment yield and influence of economic activity on its characteristics). In: *Trudy Piatogo Vsiesoyuznogo Gidrologichieskogo Syezda*, vol. 10, part 2, 149–154. Gidromietieoizdat, Lieningrad, Russia (in Russian).

Soil erosion, suspended sediment sources and deposition in the Maw-Ki-Syiem drainage basin, Cherrapunji, northeastern India

WOJCIECH FROEHLICH
HOMERKA Laboratory of Fluvial Processes, Institute of Geography and Spatial Organization, Polish Academy of Sciences, Frycowa 113, 33-335 Nawojowa, Poland
wfroehlich@pro.onet.pl

Abstract $^{210}Pb_{ex}$ and ^{137}Cs activities in soil and sediments were used to investigate soil erosion, suspended sediment source and deposition in the Maw-Ki-Syiem drainage basin on the southern slopes of the Meghalaya plateau in northeastern India. This area is known for it high rainfall ranging from 8000 to 24 000 mm year^{-1}. It is shown that in the grassland areas hillslope erosion rates are low (0.21 kg m^{2} year^{-1}) with most of the mobilized sediment being trapped on the foot-slopes. Stream and recent flood plain sediments are derived primarily from gully and channel bank erosion. The estimated rates of sedimentation on the flood plain are 0.83 g cm^{-2} year^{-1} using ^{137}Cs and 1.72 g cm^{-2} year^{-1} using $^{210}Pb_{ex}$.

Key words caesium-137; Cherrapunji; flood plain sedimentation; lead-210; overbank deposits; sediment tracing; sediment transfer; soil erosion; suspended sediment source

INTRODUCTION

The Cherrapunji area in northeastern India is known for its high rainfall (8000 to 24 000 mm year^{-1}); with daily falls exceeding 500 mm (Singh & Syiemlieh, 2001). The extreme rainfalls have the potential to cause extreme soil erosion and sediment transfer, and provide an unusual opportunity to study the influence of high magnitude short recurrence interval rainfall events on sediment production and transfer (Froehlich, 2004c).

Much of the Cherrapunji region has been degraded through forest clearing for shifting cultivation, and many previously forested slopes are now grasslands. The cultivation practices, overgrazing, and high rates of sheet flow, produced by heavy rainfalls, accelerate soil erosion, which has caused a substantial degradation of the upper soil profile. If ecologically sustainable land management is to be achieved we need to ensure that the downstream impacts of land uses are minimized. An essential part of this is to reduce the delivery of pollutants to streams. Rectification actions aimed at reducing pollutant delivery need to target the major sources of sediment. In this paper $^{210}Pb_{ex}$ and ^{137}Cs activities in soil and sediments are used to identify the major sources, transport pathways and storage areas in the drainage basin.

THE STUDY DRAINAGE BASIN

The Maw-Ki-Syiem drainage basin is located in the Cherrapunji area on the southern slopes of Meghalaya plateau in Northeastern India. It is 0.21 km^{2} in area and ranges in altitude from 1308 to 1390 m a.s.l. (Fig. 1). The underlying rock-type is thick-horizontally-bedded

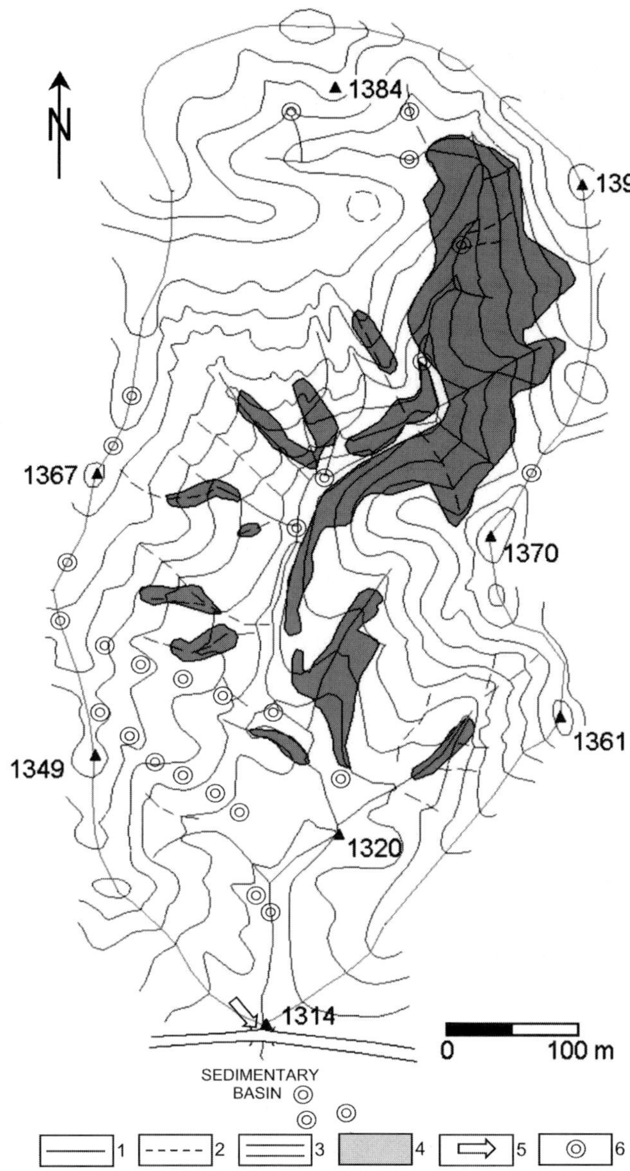

Fig. 1 The Maw-Ki-Syiem study drainage basin (source Starkel *et. al.*, 2002). 1, perennial streams; 2, seasonal streams; 3, main road; 4, woodland and scrub; 5, gauging board; 6, sampling sites (contours every 5 m).

limestone and sandstone. The soils range from Alfisols, Inceptisols to Ultisols (cf. Starkel & Singh, 2004). The slope soils are largely degraded. Long-term erosion has removed much of the fine sediment and produced skeletal debris and a very resistant, compact and impermeable layer on the soil surface, which can be up to 30 cm in depth. Sands and gravels constitute approx. 85% of the upper soil profile, and in some areas exposed bedrock bears testament to intensive soil erosion in the past (Froehlich, 2004a,b,c).

The soil and waste cover is shallowest in the headwater areas and deepest (up to 95 cm) in the eastern part, which was cultivated recently. The variable thickness of soil and waste cover depends on the rate of soil erosion and slope gradients. The foot slopes have a very narrow zone of thin sandy-silty deluvial deposits with a depth of up to 90 cm deep. The dense grass covered foot slope and swampy valley bottom act as a trap retaining eroded sediments from upslope. The valley floors have moist shallow sandy-silty alluvial soils, which cover impermeable bedrock.

The Maw-Ki-Syiem drainage basin can be divided into four geomorphological units: (a) the gullies in headwater areas; (b) steep grassland slopes in middle; (c) swampy valley bottom with step-pool sequence; and (d) bedrock waterfall stream channel and sedimentary basin in the lower reach. Natural grassland is the predominant land use. Small areas of forest and scrub occupy approximately 16% of the drainage basin (Fig. 1).

FIELD AND LABORATORY METHODS

The extreme rainfall events in the study area make it difficult to monitor sediment processes using traditional methods. The use of the fallout radionuclides as a sediment tracer offers a potential tool to investigate soil erosion, sediment source, sediment transfer and sedimentation in areas with limited sediment monitoring programmes (e.g. Walling & Collins, 2000; Froehlich & Walling, 2002; He *et al.*, 2002; Walling, 2002). It is generally assumed that total ^{137}Cs fallout can be treated as homogeneous at the scale of the field or small basin (e.g. Walling & Quine, 1993; Walling & Collins, 2000; Loughran *et al.*, 2002).

To assess the hillslope erosion sampling strategies in the study, drainage basins were based on a transect approach. Samples were collected from two downslope transects with 12 sampling points each. Reference sites for ^{210}Pb$_{ex}$ and ^{137}Cs were sampled on top of grassed hillcrests. The grass cover was similar at each of the sites. In addition cores of footslope waste cover and overbank deposits were collected in the lower reach of the drainage basin (Fig. 1). Soil cores were obtained by hammering a 75-mm diameter steel corer into the ground. The samples were collected at sufficient depth in order to include all ^{210}Pb$_{ex}$ and ^{137}Cs present in the soil and colluvial/alluvial deposits profile.

Information on suspended sediment sources was assembled using the "fingerprinting" approach (Peart & Walling, 1988; Walling & Woodward, 1992; Collins *et al.*, 1997). Samples of surface material from potential sources (grassland slopes, forest slopes, paths, gullies and channel bank) were collected from an area of 1 m^2. It is not possible to directly compare the ^{137}Cs content of these source materials with that of the suspended sediment transported by the stream, because of contrasts in the grain size composition of source materials and suspended sediment and the known enrichment of the finer fractions in ^{137}Cs. The <0.063 mm fraction of the source materials was therefore separated for gamma spectrometry analysis and the resultant values of ^{137}Cs content were used for comparisons with those associated with the fine fresh bottom channel deposits as suspended sediment.

The use of ^{210}Pb$_{ex}$ and ^{137}Cs to estimate soil erosion and deposition rates is based on a comparison of ^{210}Pb$_{ex}$ and ^{137}Cs activity at a sample site with concentration of ^{210}Pb$_{ex}$ and ^{137}Cs measured at the reference site (input site). Sample sites with ^{210}Pb$_{ex}$ and ^{137}Cs concentrations less that the reference sites are eroding, while sample sites with concentrations greater than the reference sites are sites of soil deposition. Measurements of

$^{210}Pb_{ex}$, ^{226}Ra and ^{137}Cs activities in soil and sediment cores collected from the drainage basin were used to estimate soil redistribution rates at representative locations within the study drainage basin. These data were extrapolated to the entire drainage basin, in order to estimate rates of soil loss from the grassland slopes. A comparison of the ^{137}Cs content of the fine fresh deposits of the suspended sediment and material from the potential suspended sediment sources has been used to establish major potential sources. Cores collected from the flood plain areas bordering the stream in the lower reach of the drainage basin and $^{210}Pb_{ex}$, ^{226}Ra and ^{137}Cs measurements on these cores were used to estimate rates of overbank deposition on the flood plains.

Soil and sediment samples were dried at 90°C for 48 h, weighed and passed through a 2-mm screen. Particles >2 mm were discarded. The 100 g of <2 mm fraction was packed into 500 dm^3 Marinelli beakers for gamma-ray analyses. The analyses are made at the Homerka Laboratory of Fluvial Processes Institute of Geography and Spatial Organization Polish Academy of Sciences at Frycowa using the Perkin Elmer (EG &G ORTEC) Digital Gamma Ray Spectrometer (DSP^{EC}) working in 16.384 regime channel spectra with a High-Purity Germanium coaxial detector (45% efficiency and 1.4 keV resolution). The system was calibrated with standard reference materials and the International Atomic Energy Agency Standard IAEA-6 Soil and IAEA-327 Soil. Estimates of $^{210}Pb_{ex}$, ^{226}Ra and ^{137}Cs activity of the soil samples are made using the GammaVision A66-B32, version 5.13 software (1999). The $^{210}Pb_{ex}$, ^{226}Ra and ^{137}Cs, was detected at the 46.52, 185.99 and 661.62 keV photopeak, respectively, and count time for each sample was 86 400 s, providing a measurement precision of ±4 to 6%. Sub-samples of each of the cores collected at each sampling point were combined and analysed for particle size characteristics using a Malvern Instruments MasterSizer S, after pre-treatment with H_2O_2 and disaggregation, and for organic matter content by loss-on-ignition at 600° for 3 h.

SOIL EROSION AND SEDIMENT TRANSFER

Precise estimation of the reference value is fundamental for calculating losses and gains of $^{210}Pb_{ex}$ and ^{137}Cs to determine erosion and deposition rates. The depth distributions of the $^{210}Pb_{ex}$ and ^{137}Cs activities in five reference sites were examined to estimate the reference inventories (Fig. 1). The reference sites were located in areas considered to have similar history of precipitation and fallout to those of the transect sites.

The profiles collected from undisturbed grassland locations show that the majority of $^{210}Pb_{ex}$ and ^{137}Cs fallout had been retained near the soil surface. The profiles exhibit an exponential decline in radioisotopes concentration with depth. Figures 2 and 3 show typical depth distributions of $^{210}Pb_{ex}$ and ^{137}Cs in reference soil profiles, which is typical of an undisturbed soil profile, with highest concentration in the surface soil layer (cf. Walling & Quine, 1993; Walling & Collins, 2000). Below a peak in activity concentration at about 2–3 cm, the profile shows an exponential decline in $^{210}Pb_{ex}$ and ^{137}Cs activity with depth. In each case >75% of the total inventory occurs in the top 15–18 cm, indicating that downward transfer is minimal. Much of the vertical translocation of ^{137}Cs in these profiles can be related to organic matter production at the grassland surface and chemical diffusion connected with heavy rainfalls and also downward transport by macropores in skeletal topsoil profile.

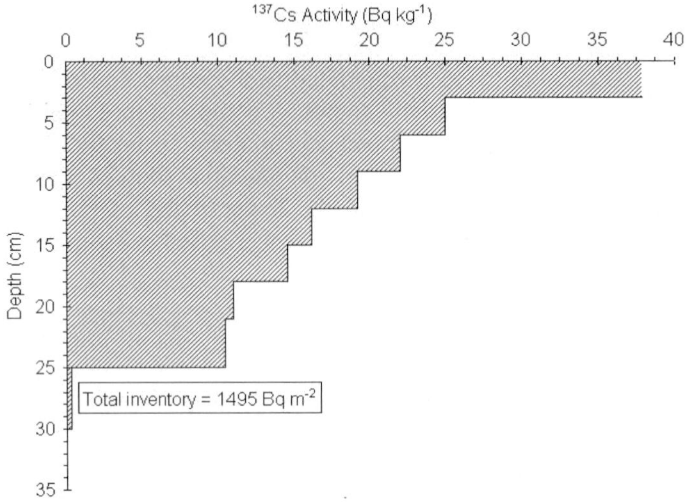

Fig. 2 Distribution of ^{137}Cs in the soil profile collected on watershed (reference site) of the Maw-Ki-Syiem drainage basin.

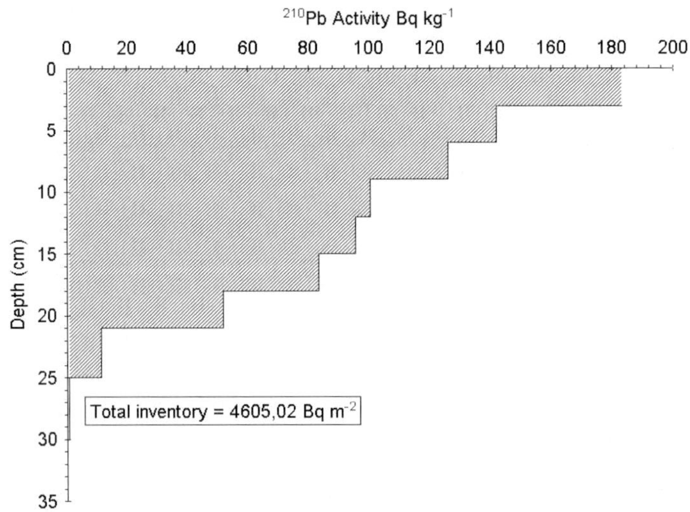

Fig. 3 Distribution of ^{210}Pb in the soil profile collected on watershed (reference site) of the Maw-Ki-Syiem drainage basin.

The reference inventories were estimated from the five sites to be 1425 ± 38.7 Bq m^{-2} for ^{137}Cs and 4605 ± 55.4 Bq m^{-2} for ^{210}Pb. Amounts of fallout generally increase with increasing rainfall (e.g. Longmore, 1982). Further sampling is required to investigate both the local and regional variability of fallout input in the Meghalaya plateau and to assess its influence on the erosion rate estimates.

Comparison of the transect site inventories with those of the reference sites shows that both erosion and deposition are evident along each of the transect (Fig. 4). Based on the ^{137}Cs data, soil redistribution rates were calculated using the Proportional Model (Walling &

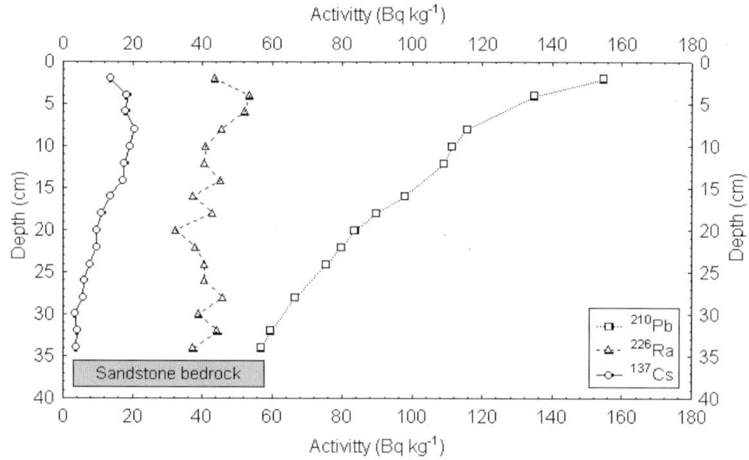

Fig. 4 The $^{210}Pb_{ex}$, ^{226}Ra and ^{137}Cs depth distributions associated with a soil core collected from the grassland footslope in the Maw-Ki-Syiem drainage basin.

He, 1999). Most part (<96%) of the transects represents eroding sites, with a average erosion rate of 0.21 kg m^2 year^{-1}. The deposition area is limited to only 4% of the total length with a mean deposition rate 3.5 kg m^2 year^{-1}. The distribution of ^{137}Cs inventories on the grass covered steep slopes evidences rather minimal soil redistribution. However, the absence of consistent relationships between erosion rates and topographic parameters for the Cherrapunji region, which are commonly used in calibration erosion estimation procedures, raises important questions concerning accuracy of soil erosion rate estimate.

The magnitude of ^{137}Cs transfer from the slopes to valley bottom and its vertical distribution in footslope profiles suggest an accumulation rate of about 0.85 mm year^{-1}. Dense grass covering the footslope and valley bottom is acting as a filter to retain most of the eroding soil particles in the upslope. This indicates that the present day sediment delivery ratio for the study drainage basin is relatively low.

SEDIMENT TRANSFER AND SEDIMENTATION

Sediment source fingerprinting techniques were used to provide information on the main sources of the suspended sediment exported from the drainage basin. The potential sources identified were the surface soil of grassland areas and woodlands drained by dispersed overland flow and sheet flow and linear erosion of paths, channel banks and gullies (Fig. 5).

Low concentration of ^{137}Cs in fine fresh sediment in bottom channel and overbank deposits indicates that main sediment sources of suspended sediment are active gullies, channel bank and path surface material. This indicates that linear erosion plays an important role in sediment delivery.

Cores collected from the flood plain area were used to quantify sedimentation rate on the flood plain in the lower reaches of the study drainage basin (Figs 6 and 7). The cumulative mass depth of the 1963 ^{137}Cs peak was used to estimate the rate of sedimentation and the CFCS model for $^{210}Pb_{ex}$. The $^{210}Pb_{ex}$ provide estimates of sedimentation rate extending back over 100 years (He & Walling, 1996).

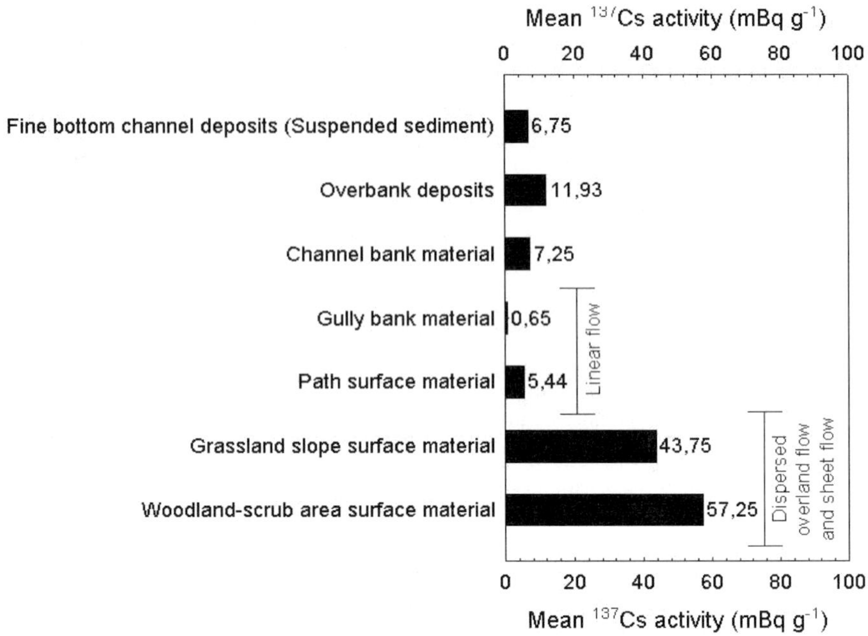

Fig. 5 The ^{137}Cs concentrations associated with suspended sediment and potential source materials within the Maw-Ki-Syiem drainage basin.

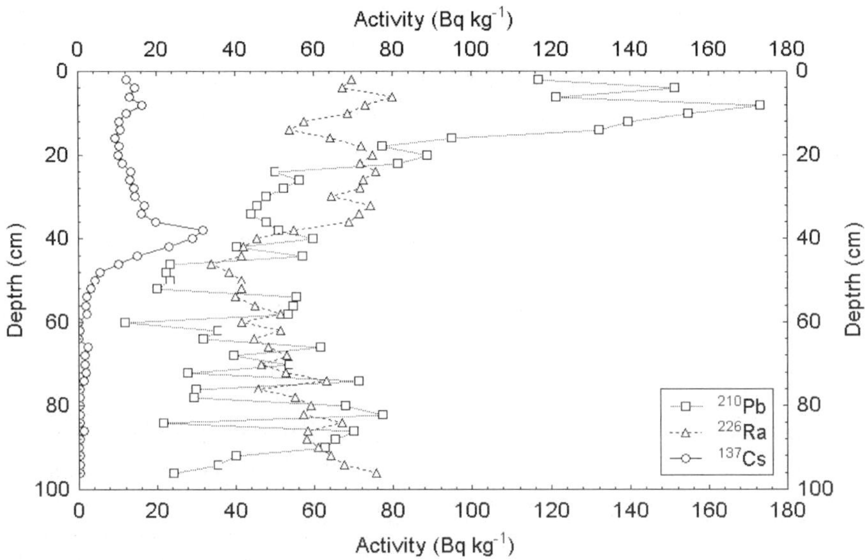

Fig. 6 The $^{210}Pb_{ex}$, ^{226}Ra and ^{137}Cs depth distributions for a sediment core collected from the flood plain of the Maw-Ki-Syiem stream.

The ^{137}Cs sedimentation rates estimated for the individual cores ranged from 0.74 g cm^{-2} year^{-1} to 1.80 g cm^{-2} year^{-1} and 0.91 to 1.65 g cm^{-2} year^{-1} for ^{210}Pb, respectively.

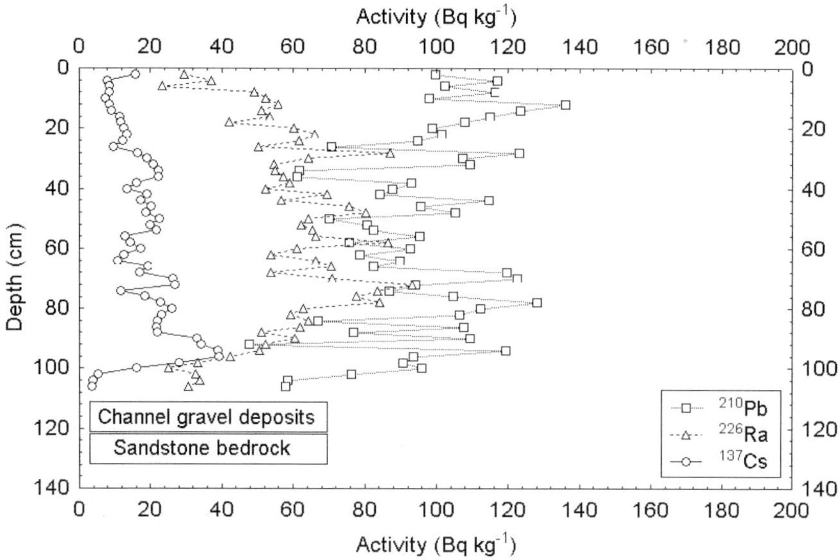

Fig. 7 The $^{210}Pb_{ex}$, ^{226}Ra and ^{137}Cs depth distributions for a sediment core collected from the flood plain of the Maw-Ki-Syiem stream.

CONCLUSIONS

The results show that a very small amount of soil is eroded from the grassland slopes of the Maw-Ki-Syiem drainage basin, and that the eroded soil is deposited at the foot of the slopes. The hillslope erosion rate is estimated to be 0.21 kg m^2 year^{-1} but little hillslope sediment is delivered to the streams. Gullies, channel banks and path surface material were identified as the primary sources of stream sediments. The flood plain sedimentation rates were estimated to be between 0.74 and 1.65 g cm^{-2} year^{-1}. Any strategy aimed at reducing soil erosion and sediment yields from this area should clearly focus on reducing erosion and sediment delivery from the gullies, channel banks and path surfaces. The approach applied in the Maw-Ki-Syiem drainage basin should provide a viable means of investigating soil erosion, sediment transfer and sedimentation in other study areas of extreme rainfalls with limited sediment monitoring programmes.

Acknowledgements This research was undertaken as part of the project entitled "Runoff and soil erosion in the globally extreme humid area in the Cherrapunji region" realized in the framework of the Inter-governmental Programme of Cooperation between the Department of Science and Technology, Government of India, and the Polish Committee for Scientific Research, Government of Poland and scientific cooperation between the Polish Academy of Sciences and Indian National Science Academy. The author wishes to acknowledge Professor Surrendra Singh and Dr Hiambok J. Syiemlieh from North Eastern Hill University, Shillong, India, for help in sending samples to Poland.

REFERENCES

Collins, A. L., Walling, D. E. & Leeks, G. J. L. (1997) Source type ascription for fluvial suspended sediment based on a quantitative composite fingerprinting technique. *Catena* **29**, 1–27.

Froehlich, W. (2004a) Runoff generation. In: *Rainfall, Runoff and Soil Erosion in the Globally Extreme Humid Area, Cherrapunji Region, India* (ed. by L. Starkel & S. Singh), Prace Geograficzne IGiPZ PAN 191 (in print).

Froehlich, W. (2004b) Soil. In: *Rainfall, Runoff and Soil Erosion in the Globally Extreme Humid Area, Cherrapunji Region, India* (ed. by L. Starkel & S. Singh). *Prace Geograficzne IGiPZ PAN 191* (in print).

Froehlich, W. (2004c) Soil erosion. In: *Rainfall, Runoff and Soil Erosion in the Globally Extreme Humid Area, Cherrapunji Region, India* (ed. by L. Starkel & S. Singh). *Prace Geograficzne IGiPZ PAN 191* (in print).

Froehlich W. & Walling D. E. (2002) The use of fallout radionuclides in investigations of sediment sources and overbank sedimentation within the Teesta river basin, Sikkim Himalaya and Ganga-Brahmaputra Plain, India. In: *Communications Kick-off Workshop of the IAHS Decade on Prediction in Ungauged Basins—Hydrological Sciences on Mission* (ed. by P. Hubert, D. Schertzer, K. Takeuchi & S. Koide), 1–11. International Association of Hydrological Sciences, University of Brasilia, Brazil.

GammaVision (1999) A66-B32, version 5.13 software. Perkin Elmer Instruments, USA.

He, Q. & Walling, D. E. (1996) Use of fallout Pb-210 measurements to investigate longer-term rates and patterns of overbank sediment deposition on the flood plains of lowland rivers. *Earth Surf. Processes Landf.* **21**, 141–154.

He, Q., Walling, D. E. & Wallbrink, P. J. (2002) Alternative methods and radionuclides for use in soil-erosion and sedimentation investigations. In: *Handbook for the Assessment of Soil Erosion and Sedimentation Using Environmental Radionuclides* (ed. by F. Zapata), 185–215. Kluwer Academic Publishers, Dordrecht, The Netherlands.

Longmore, M. E. (1982) The caesium-137 dating technique and associated applications in Australia—a review. In: *Archaeometry: An Australian perspective* (ed. by W. Ambrose & P. Duerden), 310–321. Australian National Univ. Press, Canberra, Australia.

Loughran, R. J., Pennock, D. J. & Walling, D. E. (2002) Spatial distribution of Caesium-137. In: *Handbook for the Assessment of Soil Erosion and Sedimentation Using Environmental Radionuclides* (ed. by F. Zapata), 96–109. Kluwer Academic Publishers, Dordrecht, The Netherlands.

Peart, M. R. & Walling, D. E. (1988) Techniques for establishing suspended sediment sources in two drainage basins in Devon, UK: a comparative assessment. In: *Sediment Budgets* (ed. by M. P. Bordas & D. E. Walling) (Proc. Porto Alegre Symp., December 1988), 269–279. IAHS Publ. 174. IAHS Press, Wallingford, UK.

Singh, S. & Syiemlieh, H. J. (2001) Rainstorm characteristics of extremely humid area of the world—Cherrapunji. In: *Forecasting Monsoon From Days to Years and WMO Workshop on Forecasting Monsoon* (Proc. Int. Conf.), 1–23. Delhi, India.

Starkel, L. & Singh, S. (eds) (2004) Rainfall, runoff and soil erosion in the globally extreme humid area, Cherrapunji region, India. *Prace Geograficzne IGiPZ PAN* 191 (in print).

Starkel, L., Singh, S., Soja, R., Froehlich, W., Syiemliech, H. & Prokop, P. (2002) Rainfall, runoff and soil erosion in the extremely humid area around Cherrapunji, India (preliminary observations). *Geographia Polonica* **75**(1), 43–65.

Walling, D. E. (2002) Using environmental radionuclides as tracers in sediment budget investigations. In: *Erosion and Sediment Transport Measurement in Rivers: Technological and Methodological Advances* (ed. by J. Bogen, T. Fergus & D. E. Walling) (Proc. Oslo Workshop, June 2002), 57–78. IAHS Publ. 279. IAHS Press, Wallingford, UK.

Walling, D. E. & Collins, A. L. (2000) *Integrated Assessment of Catchment Sediment Budgets: a Technical Manual.* Produced within the framework of UK Department for International Development Research Project R6868. School of Geography and Archaeology, University of Exeter, Exeter, UK.

Walling, D. E. & He, Q. (1999) Improved models for estimating soil erosion rates from Caesium-137 measurements. *J. Environ. Qual.* **19**, 611–622.

Walling, D. E. & Quine, T. A. (1993) *Use of Caesium-137 as a Tracer of Erosion and Sedimentation: Handbook for the Application of the Caesium-137 Technique.* UK Overseas Development Administration Research Scheme R4579. Department of Geography, University of Exeter, Exeter, UK.

Walling, D. E. & Woodward, J. C. (1992) Use the radiometric fingerprints to derive information on suspended sediment source. In: *Erosion and Sediment Transport Monitoring Programmes in River Basins* (ed. by J. Bogen, D. E. Walling & T. Day) (Proc. Oslo Symp., August 1992), 153–164. IAHS Publ. 210. IAHS Press, Wallingford, UK.

A sediment budget for the Herbert River catchment, North Queensland, Australia

REBECCA BARTLEY[1], JON OLLEY[2] & ANNE HENDERSON[3]

1 *CSIRO Land and Water, PO Box 780, Atherton, Queensland 4883, Australia,* rebecca.bartley@csiro.au

2 *CSIRO Land and Water, GPO Box 1666, Canberra ACT 2601, Australia*

3 *CSIRO Land and Water, Davies Laboratory, Townsville, Queensland 4814, Australia*

Abstract A combination of spatial modelling, sediment tracing techniques and water-quality data were used to determine the major source of fine sediments in the Herbert River catchment, Queensland, Australia. Using modelling, hillslope erosion was predicted to be the dominant source of sediment, contributing 52% of the total sediment load at the estuary. Gully and stream bank erosion contributed equal loads to the estuary (~24%). The ^{137}Cs concentrations measured in this study support the modelled predictions for contributions from different land uses. Results from modelling and sediment tracing also predicted similar ratios of hillslope to channel erosion. The total suspended sediment loads predicted for the downstream freshwater limit of the catchment are within 10% of longer term measured values. These results suggest that the modelling approach used in this study is useful for determining sediment budgets for large tropical catchments.

Key words Australia; Queensland; sediment budget; sediment tracing; SedNet

INTRODUCTION

Sediment supply to the coastal waters adjacent to the Great Barrier Reef World Heritage Area (GBRWHA) in Queensland, Australia, is of increasing concern (e.g. Baker, 2003; McCulloch *et al.*, 2003). The Herbert River is the largest of the wet tropical rivers along the Great Barrier Reef Coast, and is considered to be a "high risk" catchment in terms of its sediment export (Brodie *et al.*, 2001). In this paper, a combination of spatial modelling, sediment tracing techniques, and analysis of water-quality and flow data are used to determine the major source of fine sediments to the Herbert River. This work aims to enable catchment managers to effectively target remedial action to decrease sediment delivery to the GBRWHA.

Herbert River catchment: location and characteristics

The Herbert River catchment covers an area of approximately 10000 km^2, and can be broken up into four distinct physiographic sections (Fig. 1). The upper catchment (~4735 km^2) is dominated by grazing, but has also been subject to various forms of mining, and has other mixed agriculture. The middle section of the catchment (~1825 km^2) is predominantly National Park, State Forests and timber reserves. The lower flood plain section (~3560 km^2) can be divided into two areas; the flood plain area around the Herbert River channel and the southern coastal section which contains streams that flow directly to the coast. Both these areas are dominated by sugarcane cultivation. The geology, geomorphology, soils, rainfall

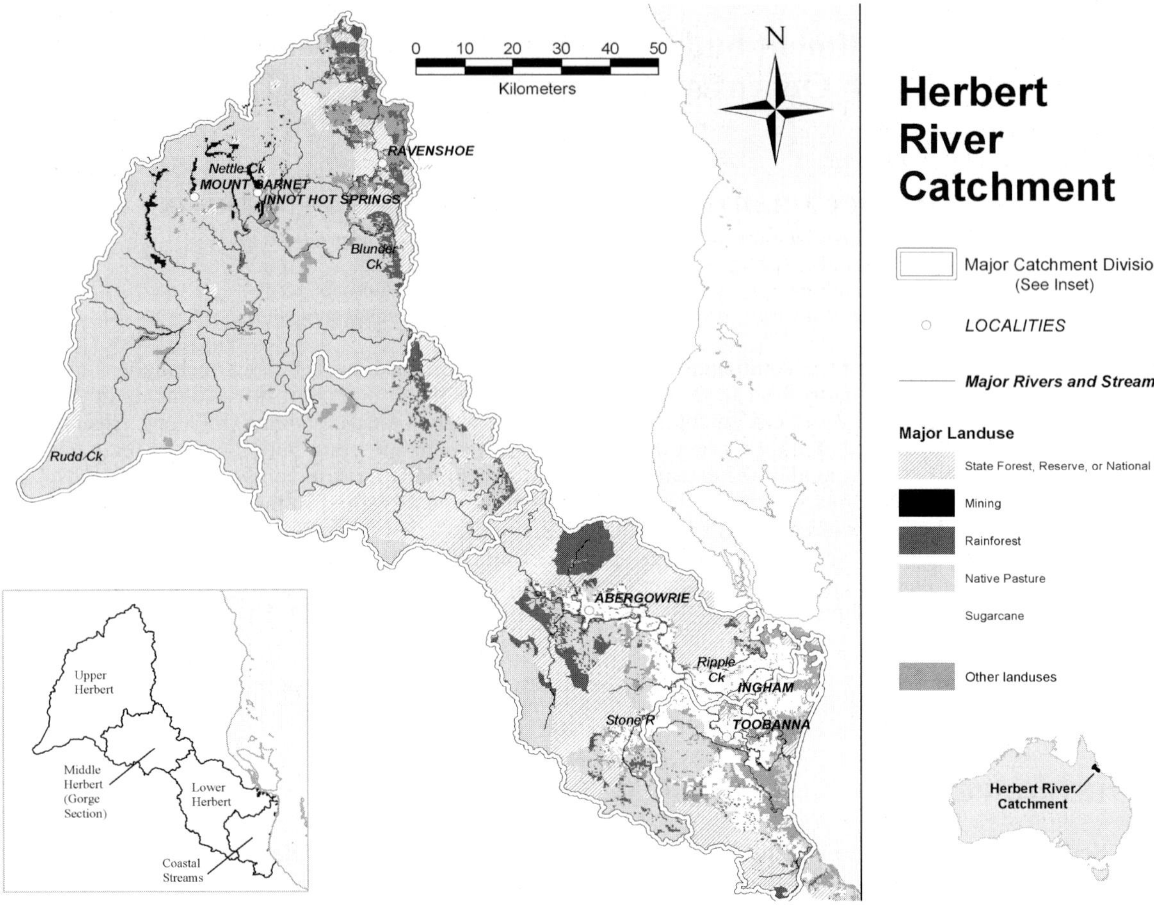

Fig. 1 Major land use units and physiographic areas in the Herbert River catchment.

and land use characteristics are different for the four regions, resulting in differences in the history of erosion and sedimentation for each area. Further bio-geographic information on the Herbert River catchment can be found in Johnson & Murray (1997) and Bartley *et al.* (2003).

METHODS FOR DETERMINING SEDIMENT SOURCES IN THE HERBERT CATCHMENT

Sediments originate from erosion of hillslopes, and from gully and stream bank erosion (channel erosion). The methods used to assess each of these sources and to propagate the sediment through the river network are described below and in more detail in Bartley *et al.* (2003).

Hillslope erosion

Hillslope erosion from sheet and rill erosion was assessed using the Revised Universal Soil Loss Equation (RUSLE; Renard *et al.*, 1997) as applied in the National Land and Water

Resources Audit (NLWRA; Lu *et al.*, 2001). Soils mapping from various sources was collated into a "best available" map from which an erodibility map was derived. The slope length factor was set as a constant value of 1. The slope factor across the Herbert River catchment was derived directly from the 100 m digital elevation model. Land cover mapping at 1:50 000 scale was available over the whole catchment current to 1996. This mapping was used to delineate rainforest and eucalypt dominated areas in the middle and upper Herbert (with some improvements based on Landsat interpretation). Cover factors were estimated for each land cover type using tables from Rosewell (1993), except for sugarcane which was based on results published in Visser (2003).

Gully erosion

The spatial pattern of gullies in the Herbert catchment was determined through aerial photograph interpretation. A gully was defined as a non-permanent watercourse with steep, actively eroding banks. Gullies were mapped using 50 stereo aerial photograph pairs spanning different geographic regions, geologic classes and land uses. Cubist data mining software (Rulequest Research, 2001) was used to predict gully densities throughout the unmapped portions of the catchment (Hughes & Prosser, 2003). Total sediment supplied from gully erosion was calculated as the product of gully density, watershed area, average gully cross-sectional area (~10 m^2) and average dry bulk density of soil (1.5 t m^{-3}), divided by the time over which gullies have developed (100 years). It was assumed that 50% of the gully erosion produced fine suspended sediment.

River bank erosion hazard

A global review of river bank migration data (Rutherfurd, 2000) suggested that the best predictor of bank erosion rate was bankfull discharge (Q_{bf}) (equivalent to a 1.58 year recurrence interval). Rutherfurd (2000) also found a significant relationship between bank erosion and stream power ($\rho g Q_{bf} S_x$, where ρ is the density of water, g is the acceleration due to gravity, Q_{bf} is bankfull discharge in $m^3 s^{-1}$ and S_x is the energy slope approximated to the channel gradient). The linear relationship with stream power was modified using two factors: (a) it is assumed that the bank erosion rate decreases as the proportion of remnant riparian vegetation (PR_x) along the river link increases; and (b) the erosion rate is reduced in narrow valleys, and an exponential relationship between rock exposure and flood plain width (F_x) applied (Hughes & Prosser, 2003). The resultant equation used in this study to predict bank erosion (BE in m $year^{-1}$) is:

$$BE = 0.00002 \times \rho \times g \times Q_{bf} \times S_x (1 - PR)(1 - e^{-0.008 F_x}) \quad (1)$$

The average proportion of riparian vegetation within each link was determined from Moller (1996). The predicted bank erosion rate is converted into sediment supply (kt $year^{-1}$) by multiplying BE by channel length (m), bank height (m), average particle density of bank materials (1.5 t m^{-3}) and dividing by a conversion factor of 1000. It was assumed that 50% of the sediment generated from bank erosion contributed to the suspended sediment supply.

Mine site sources

Mining has long been considered one of the major sources of sediment in the Herbert River catchment. Various forms of mining, particularly tin mining, have been extensive in the upper catchment. There are no published data on the rates of soil loss from mined areas in the upper Herbert catchment. As a result, both the hillslope erosion and gully erosion factors were derived from knowledge of erosion rates from other mine sites around Australia (Carroll *et al.*, 2000). A total erosion rate of 25 t ha^{-1} $year^{-1}$ (as a combination of hillslope and gully sources) was applied to the mine sites.

Sediment transport

To calculate the supply of sediment, deposition and delivery downstream, we used SedNet: the Sediment River Network model. SedNet was developed for the Australian National Land and Water Resources Audit (NLWRA, 2001). The methods used in the construction of input data and implementation of the SedNet program itself are described in detail in a number of CSIRO technical reports (e.g. Prosser *et al.*, 2001a,b, DeRose *et al.*, 2003; Wilkinson & Young, 2004).

The model divides the river into a series of links; a link being the stretch of river between any two stream junctions. Each link has an internal watershed, from which sediment is delivered to the stream network by hillslope, riverbank and gully erosion. Sediment is processed sequentially through the river network, beginning with first order links and terminating at the catchment outlet. The sediment load output from each link is calculated from the supply of sediment from tributary links and the local watershed, less losses though flood plain, reservoir and bed deposition. Flood plain deposition is calculated as the proportion of fine sediment that goes over-bank and settles out during a typical flood. It is calculated as the ratio of the median over-bank flow multiplied by the proportion of fine sediment that would be expected to settle out during over-bank flow. Deposition of coarse sediments on the bed of the river are predicted to occur when there is an excess of sediment supply to a river link beyond the capacity of the link to transport sand sized sediment (Bartley *et al.*, 2003). The sediment yield at the terminating link constitutes the total yield (inputs–storage) of the river network. The model also predicts sediment load and the relative contribution of hillslope and channel (gully and riverbank) erosion throughout the network.

Testing the model predictions

The SedNet sediment load predictions for the Herbert River catchment were tested using two different techniques: (a) ^{137}Cs concentrations; and (b) comparison with measured sediment load data.

The predicted relative contribution of hillslope and channel (combined gully and riverbank) erosion to stream sediments was assessed using concentrations of ^{137}Cs. ^{137}Cs is a product of atmospheric nuclear weapons testing and is concentrated in surface soil. Sediments derived from hillslope erosion have high concentrations of this radionuclide. While those eroded from gullies or channels have little or none. By measuring the concentration in suspended sediments moving down the river, and comparing them with concentrations in sediments derived from hillslope and channel erosion, we determined the relative contributions of each of these processes.

Recently deposited sediment samples were collected from 20 locations along the river during low flow conditions. Source area concentrations were assessed using data from the adjacent Johnstone River catchment (data supplied by Dr Wallbrink, CSIRO). ^{137}Cs concentrations were measured by high resolution gamma spectrometry (Murray *et al.*, 1987). Measurements were made on the <10 μm size fraction (clay and fine silt).

The Australian Institute of Marine Science (AIMS) conducted sediment sampling in the Herbert River between 1995 and 2000. The results indicate that for this period the sediment loads for the Herbert River range between 143 000 t year^{-1} to 900 000 t year^{-1}, with a mean value of ~540 000 t year^{-1} (Furnas, 2003).

RESULTS

Hillslope erosion is predicted to be the dominant source of sediment, contributing 490 000 t year^{-1} (or 52%) of the total sediment load at the river mouth (Table 1). Gully and stream bank erosion contribute equally with approximately 220 000 t year^{-1} (~24%). The highest overall losses from hillslope erosion were from grazing lands and forested areas, however, losses per hectare from hillslopes are greatest from the mine sites, forested areas and agricultural cropping land (Table 2). It is important to note that the areas covered by "forest" (30% of the catchment area) contribute almost the same amount of sediment as grazing (60% of the catchment area). This is because most of the forested sites are on extremely steep slopes with much higher rainfall. The main forest type that contributed to the high loads is the open eucalypt dominated woodland areas that occupy the Herbert River gorge. The erosive potential in the forested areas is much greater than the flat depositional landscapes of the upper catchment grazing lands.

Gully erosion also varies significantly across the catchment with the worst areas being around the mine sites in the Upper Herbert and the steeper grazing areas upstream of the Herbert River gorge. Most of the remainder of the catchment does not suffer from significant gully erosion. Riverbank erosion contributes the same amount of sediment to the stream network as gully erosion, yet there is no obvious pattern to bank erosion in the catchment. The main areas that appear to undergo higher than average rates of bank erosion are in the upper catchment and various sections of the main Herbert River channel. Bank erosion is generally at its worst in river reaches lacking a good riparian zone and where stream power is greatest. The model predicts that roughly 40 000 t year^{-1} of fine sediment is stored on the flood plain.

Table 1 Summary of sediment budget for the Herbert River catchment.

Sediment budget item	Predicted mean annual rate (t year^{-1})
Hillslope Delivery	490 000
Gully erosion rate	220 000
Riverbank erosion rate	220 000
Total sediment supply	930 000
Total suspended sediment stored	40 000
Total bed sediment stored	210 000
Sediment delivery to the estuary	680 000
Total losses	930 000

Table 2 Contribution of hillslope erosion from each of the major land use areas with the Herbert River catchment.

Land use	Area (km^2)	Proportion of total area (%)	Predicted total soil loss (t $year^{-1}$)	Average erosion rate (t ha^{-1} $year^{-1}$)
Native pastures	5840	60	200 000	0.34
Improved pastures	48	0.5	1000	0.10
Forest and other reserves[(1)]	2881	30	200 000	0.69
Sugar cane	690	7	35 000	0.51
Other agriculture	79	0.8	5000	0.64
Residential or industrial	22	0.2	0	0.01
Mining	52	0.5	40 000	7.76
Other (e.g. quarries)	135	1.4	10 000	0.74

[(1)]The forest and other reserves category contains a range of forest types from open eucalypt woodland to rainforest and melaleuca species.

Concentrations of ^{137}Cs measured in the <10 μm size fraction of sediment samples collected from along the Herbert River range from 0.0 ± 0.3 to 4.5 ± 0.8 Bq kg^{-1}, with a mean of 2.4 ± 0.3 Bq kg^{-1}. They generally increase towards the coast indicating a greater contribution from hillslope sources along the river (Fig. 2). Concentrations in sediments derived from hillslope erosion on cultivated and uncultivated land in the adjacent Johnston River catchment were 3.6 ± 0.2 (n = 12) and 8.4 ± 0.6 (n = 9) Bq kg^{-1}, respectively. Those on sediments derived from channel erosion were estimated to be 0.8 ± 0.1 Bq kg^{-1}. The SedNet model predicted that 52% (490 000 t $year^{-1}$) of the sediment at the mouth of the river was derived from hillslope erosion; with 401 000 t $year^{-1}$ of that coming from uncultivated areas (i.e. 42% of the total load). If the model predictions were correct then the ^{137}Cs concentrations in the lower river should be 4.25 ± 0.25 Bq kg^{-1}. This is consistent with measured concentration in the sample from the river mouth (4.5 ± 0.8 Bq kg^{-1}) and with the average concentration in the samples collected in the lower 30 km (3.7 ± 0.5 Bq kg^{-1}) (Fig. 2). Therefore the measured ^{137}Cs concentrations support the predicted relative contributions from different land-uses, and the predicted ratio of hillslope to channel erosion.

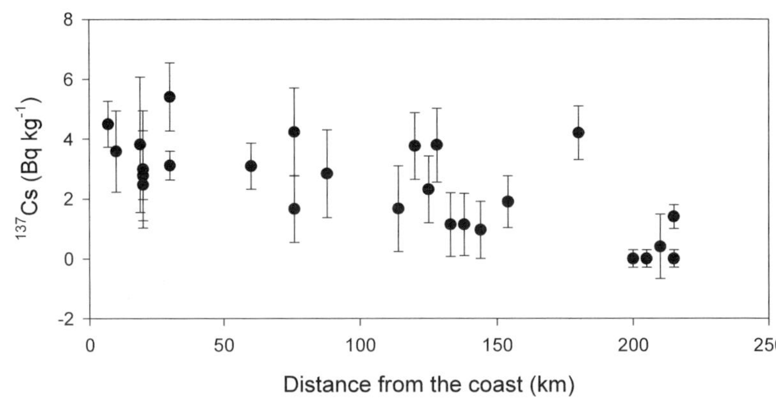

Fig. 2 Concentrations of ^{137}Cs in the <10 μm size fraction of sediment samples collected from along the main channel of the Herbert River.

The model predicted the sediment load at Gairloch Bridge (Ingham) to be ~600 000 t $year^{-1}$. This compares with the 143 000 t $year^{-1}$ to 900 000 t $year^{-1}$, with a mean value of ~540 000 t $year^{-1}$ estimated from water quality and flow data (Furnas, 2003).

DISCUSSION AND CONCLUSION

This study has used a set of GIS-based models (SedNet) to predict the major sediment sources and sinks within a large tropical catchment. The results suggest that hillslope erosion is the dominant source of sediment, however, both gully and bank erosion present significant sources. These results were supported by sediment tracing analysis. The modelled sediment loads estimated at downstream freshwater limit of the catchment were also within 10% of longer term measured values. These results will help better target catchment management practices aimed at reducing soil erosion.

Acknowledgements Ian Prosser, Andrew Hughes, Lucy McKergow, Hua Lu and Christian Roth (formerly from the CSIRO), Jon Brodie and Zoe Bainbridge (James Cook University), Caroline Coppo (Herbert River catchment Coordinator) and members of the Herbert River Catchment Group are gratefully acknowledged for their contributions to this work. The Herbert River Catchment Group, CSIRO and the Reef and Sugar CRC's are also acknowledged for funding various components of this study.

REFERENCES

Baker, J. (ed.) (2003) *A Report on the Study of Land-Sourced Pollutants and their Impacts on Water Quality in and Adjacent to the Great Barrier Reef.* Brisbane, Queensland, Department of Premier and Cabinet, Australia. http://www.premiers.qld.gov.au/about/reefwater.pdf

Bartley, R., Henderson, A., Prosser, I. P, Hughes, A. O., McKergow, L. Lu, H., Brodie, J., Bainbridge, Z. & Roth, C. H. (2003) *Patterns of Erosion and Sediment and Nutrient Transport in the Herbert River Catchment, Queensland.* CSIRO Land and Water Tech Report, Australia. http://www.clw.csiro.au/publications/consultancy/2003/Herbert_River_final_%20report_Aug-03.pdf

Brodie, J., Furnas, M., Ghonim, S., Haynes, D., Mitchell, A., Morris, S., Waterhouse, J., Yorkston, J., Andas, D., Lowe, D & Ryan, M. (2001) *Great Barrier Reef Catchment Water Quality Action Plan.* Great Barrier Reef Marine Park Authority, Townsville, Australia. http://www.gbrmpa.gov.au/corp_site/key_issues/water_quality/action_plan/.

Carroll, C., Merton, L. & Burger, P. (2000) Impact of vegetative cover and slope on runoff, erosion, and water quality for field plots of soil and spoil materials on central Queensland coal mines. *Aust. J. Soil Res.* **38**, 313–327.

DeRose, R. C., Prosser, I. P, Wilkinson, L. J., Hughes, A.O. & Young, W. J. (2003) *Regional Patterns of Erosion and Sediment and Nutrient Transport in the Goulburn and Broken River Catchments, Victoria.* CSIRO Land and Water Tech. Report 11/03, Canberra, Australia (http://www.clw.csiro.au/publications/technical2003/tr11-03.pdf)

Furnas, M. (2003) *Catchments and Corals: Terrestrial Runoff to the Great Barrier Reef.* CRC Reef Research, Townsville, Australia.

Hughes, A. O. & Prosser, I. P. (2003). *Gully and Riverbank Erosion Mapping for the Murray-Darling Basin.* Tech. Reports G and I for the MDBC Basin-wide Sediment Mapping Project, CSIRO Land and Water, Canberra, Australia.

Johnson, A. K. L. & Murray, A. E. (1997) *Herbert River Catchment Atlas.* CSIRO, Davies Laboratory, Townsville, Australia.

Lu, H., Gallant, J., Prosser, I., Moran, C., Moran, C., & Priestly, G. (2001) *Prediction of Sheet and Rill Erosion Over the Australian Continent, Incorporating Monthly Soil Loss Distribution.* Canberra, CSIRO Land and Water, Australia. http://www.clw.csiro.au/publications/technical2001/tr13-01.pdf

McCulloch, M., Fallon, S., Wyndham, T., Hendy, E., Lough, J. & Barnes, D. (2003) Coral record of increased sediment flux to the inner Great Barrier Reef since European Settlement. *Nature* **421**, 727–730.

Moller, G. (1996) *An Ecological and Physical Assessment of the Condition of Streams in the Herbert River Catchment.* Department of Natural Resources, Brisbane, Australia.

Murray, A.S., Marten, R., Johnston, A., Martin, P. (1987). Analysis for naturally occurring radionuclides at environmental concentrations by gamma spectrometry. *J. Radioanalytical Nuclear Chemistry Articles* **115**, 263–288.

NLWRA (2001) *Australian Agricultural Assessment*. National Land and Water Resources Audit, Canberra, Australia. http://www.nlwra.gov.au/

Prosser, I., Hughes, A., Rustomji, P. & Moran, C. (2001a) Predictions of the sediment regime of Australian rivers. In: *Australian Stream Management* (ed. by I. Rutherfurd, G. Brierley, S. Bunn, F. Sheldon & C. Kenyon) (Proc. Third Conf. Cooperative Research Centre for Catchment Hydrology, Melbourne, Australia), 529–533.

Prosser, I., Young, B., Rustomji, P., Moran, C. & Hughes, A. (2001b) *Constructing River Basin Sediment Budgets for the National Land and Water Resources Audit*. CSIRO Land and Water Tech Report 15/01, Canberra, Australia. (http://www.clw.csiro.au/publications/technical2001/tr15-01.pdf)

Rulequest Research (2001) *Rulequest Research Data Mining Tools*. http://rulequest.com.

Renard, K. G., Foster G. A., Weesies D. K., McCool, D. K. & Yoder, D. C. (1997) *Predicting Soil Erosion by Water: A Guide to Conservation Planning with the Revised Universal Soil Loss Equation*. Agriculture Handbook 703, United States Department of Agriculture, Washington DC, USA.

Rosewell, C. J. (1993) *SOILOSS—a program to assist in the selection of management practices to reduce erosion*. Tech. Handbook no. 11 (second edition). Soil Conservation Service of NSW, Sydney, Australia.

Rutherfurd, I. (2000) *Some Human Impacts on Australian Stream Channel Morphology*. John Wiley and Sons, Chichester, UK.

Visser, F. (2003) Sediment budget for cane land on the Lower Herbert River flood plain, North Queensland, Australia. PhD Thesis, Australian National University, Canberra, Australia.

Wilkinson, S. & Young, W. (2004) Estimating mean annual flow and daily flow variability in sparsely gauged river networks. *J. Hydrol*. (in preparation).

The impact of erosion protection work on sediment transport in the River Gråelva, Norway

JIM BOGEN & TRULS E. BØNSNES
Norwegian Water Resources and Energy Directorate, PO Box 5091, Maj. 0301 Oslo, Norway
jbo@nve.no

Abstract The impact of erosion protection works on downstream sediment delivery was evaluated in a study of the River Gråelva in central Norway. Channel degradation and the associated undercutting of adjacent slopes have formerly uncovered pockets of quick clay and triggered major quick clay slides. To prevent further slides the river bed and banks have been reinforced with a layer of rock armouring. The stabilization work started in 1992. When conditions were still close to natural in 1992 and 1993, maximum sediment concentrations were in the range of 15 000–25 000 mg l^{-1} and the calculated annual suspended sediment transport was 163 000 and 99 000 t $year^{-1}$, corresponding to sediment yields of 8150 and 4950 t km^2 $year^{-1}$. Maximum concentrations in 2000 and 2001 did not exceed 6000 mg l^{-1} and the annual sediment transport had decreased enormously, to 11 800 and 18 500 t $year^{-1}$, giving sediment yields of only 590 and 925 t km^2 $year^{-1}$. However, a year-to-year variability controlled by climatic variables was still present.

Key words erosion protection work; quick clay slides; Norway; suspended sediment

INTRODUCTION

It is sometimes desirable to reduce erosion activity within catchments. High sediment loads may reduce water quality and have a negative impact on ecology. It is thus of interest to study the effect of erosion protection work on erosion activity and sediment transport. In the marine clay areas in Norway, erosion activity may trigger large magnitude quick clay slides. The clays were deposited in the sea in front of the retreating ice sheet during the last deglaciation about 10 000 years ago, when the land was still considerably depressed by glacio-isostatic loading. Subsequent emergence of the shallower areas through isostatic recovery gradually exposed large tracts of clay as dry land. These can extend almost to the marine limit (ML), defined as the highest level that was once covered by the sea at any location after the ice sheet withdrew. In the lower-lying parts, large thicknesses of clay had been deposited.

When clays are deposited in seawater, the individual particles (which are flat plates) are bonded edge-to-centre by the salt, producing a strong card-house structure (Fig. 1). However, once uplifted above sea level, groundwater leaching of the salt commences and the clays are transformed into quick-clays (Rosenqvist, 1960). This is a slow process, millennia rather than centuries, and is followed by the development of a drying crust. As the surface soil dries and hardens, the quick zone is not usually visible close to the surface. Quick-clay has reduced mechanical strength and is highly susceptible to liquefaction if subjected to shock waves or excess loading. Slope failure caused by the removal of downslope support, as in river undercutting, has a similar effect. The hollow card-house structure then collapses, releasing the enclosed pore water to form a clay slurry. A body of quick-clay may be kept in

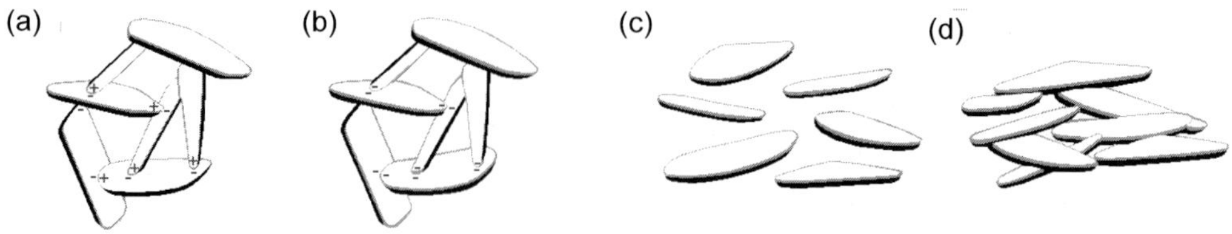

Fig. 1 Illustration of the formation of quick-clay: (a) Deposition of clay particles in card–house structure in saltwater. Particles are edge-to-face bonded by the salt; (b) Above sea level: salt is washed away. As bonding is destroyed the structure becomes unstable; (c) Quick-clay subject to liquefaction. Structure collapses, excess water produces a low viscosity slurry; and (d) After the slide: Flat-lying grain structure contains less water and becomes stable.

place if the surrounding soil is stable. Fluvial erosion in areas with marine clay deposits in Norway has triggered numerous large quick-clay slides, Bjerrum (1971). The clay areas are often densely incised by V-shaped gullies and thus highly vulnerable to erosion. In Scandinavian literature these landforms are termed "ravines", a usage employed here.

The River Gråelva drains an area of 47.6 km^2, of which about 20 km^2 is covered by marine clay, situated below 150 m a.s.l. (Fig. 2). Over two thirds of the trunk stream channel is situated in the clay area. A survey of the soils in the clay areas has revealed that there are large pockets of quick clay in the catchment (Gregersen, 1991). Over the years a number of large magnitude slides have occurred. A list of major slides occurring between 1676 and 1995 is given in Table 1. Seven large volume slides were recorded during this period. To remove the slide hazard, it was decided to carry out erosion protection work along the

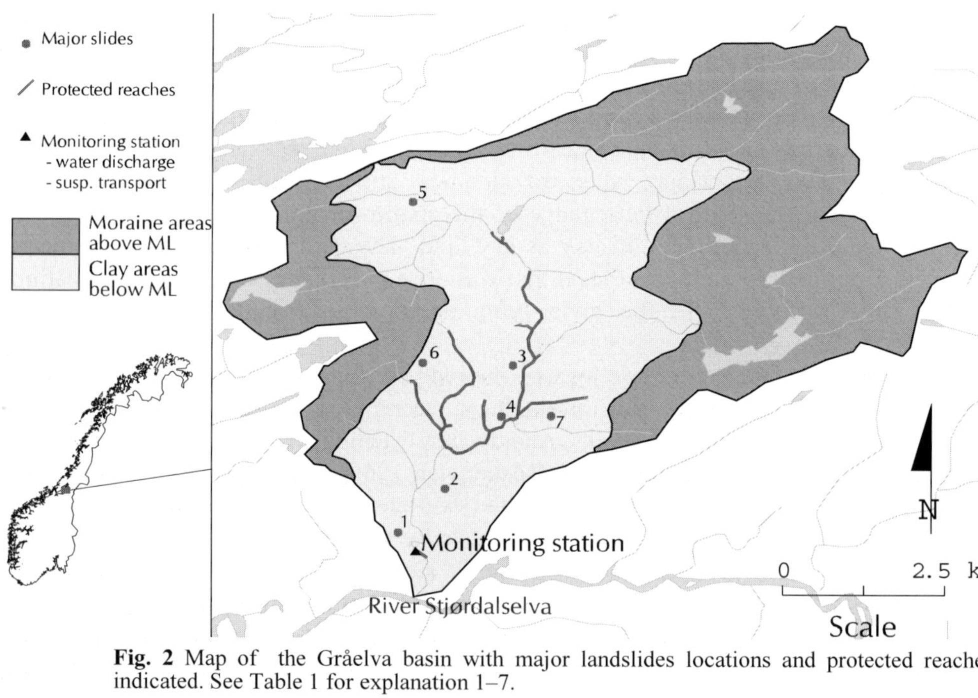

Fig. 2 Map of the Gråelva basin with major landslides locations and protected reaches indicated. See Table 1 for explanation 1–7.

Table 1 Major slides in the Gråelva river catchment.

No	Year	Location	Incidence
1	1676	Børstadfallet	The farm Børstad destroyed
2	1686	Kylloraset	7 people killed
3	1893	Mørsetfallet	Landslide, 800 000 m^3 clay
4	1921	Kvålsvedjan	Landslide, 1 ha land
5	1962	Hovfallet	Landslide 20–30 hectare land, 1 killed
6	1975	Bangmarka	Landslide 1 ha woodland
7	1995	Skjelstadmarka	Landslide 10^{-1} ha land

degrading reaches of the river. A sediment transport monitoring programme was initiated at the same time as the construction work.

SEDIMENT SOURCES AND EROSION PROTECTION WORKS

In areas above ML the overburden is thin and discontinuous, providing negligible sediment yield compared with the clay area. The most important sediment source in the Gråelva catchment is supplied by mass movement caused by river channel degradation. A number of river reaches have been found to degrade. A river at bankful discharge will tend to degrade until a stable equilibrium profile or a stable armouring layer is established. Bjerrum (1971) studied the longitudinal profiles of rivers in clay areas in Norway and found an empirical relation describing the equilibrium profile:

$$S_{cr} = 0.0116A^{-0.55} \qquad (1)$$

where S_{cr} is the critical gradient and A is the catchment area of the reach. Bogen & Sandersen (1991) investigated a similar clay area in southern Norway and found that the large floods were of decisive importance in this connection. During these events, high gradient reaches tended to degrade until a new equilibrium was obtained. Sandersen (1993) used equation (1) to compute the stable equilibrium profile of the River Gråelva. The gradient of the present longitudinal profile is in some reaches 2–4 times that of the calculated stable gradient (Fig. 3(a)). Unless an armouring layer slows down the erosion, the main channel will continue to degrade and in turn destabilize the tributaries. This may cause a further instability of the slopes adjacent to channels and trigger minor landslides and accelerate the soil creep as illustrated by the cross sections in Fig. 3(b). The soil creep is caused by repeated freeze/thaw and wetting/drying cycles. Creep velocity increases with slope gradients. In steep ravines in the clay areas there is a continuous supply of sediment to the river channels. Sandersen estimated the sediment delivery to the Gråelva from creep to be of the order of 10 kg year^{-1} m.

Landslides occur during periods of high precipitation and snowmelt causing high pore water pressure on high-gradient slopes. Vegetation has an important influence too, as it has a binding effect and reduces the soil water content by evaporation and consumption of water. The volumes of the landslides along the river have been observed to vary mostly from a few to several thousand m^3. However, slides of 100 000 m^3 have been triggered by fluvial erosion in the Gråelva catchment. On some reaches the bed has been protected naturally by boulders and stones eroded out from the soil but lateral erosion on these reaches may move the channel, initiating a new phase of degradation.

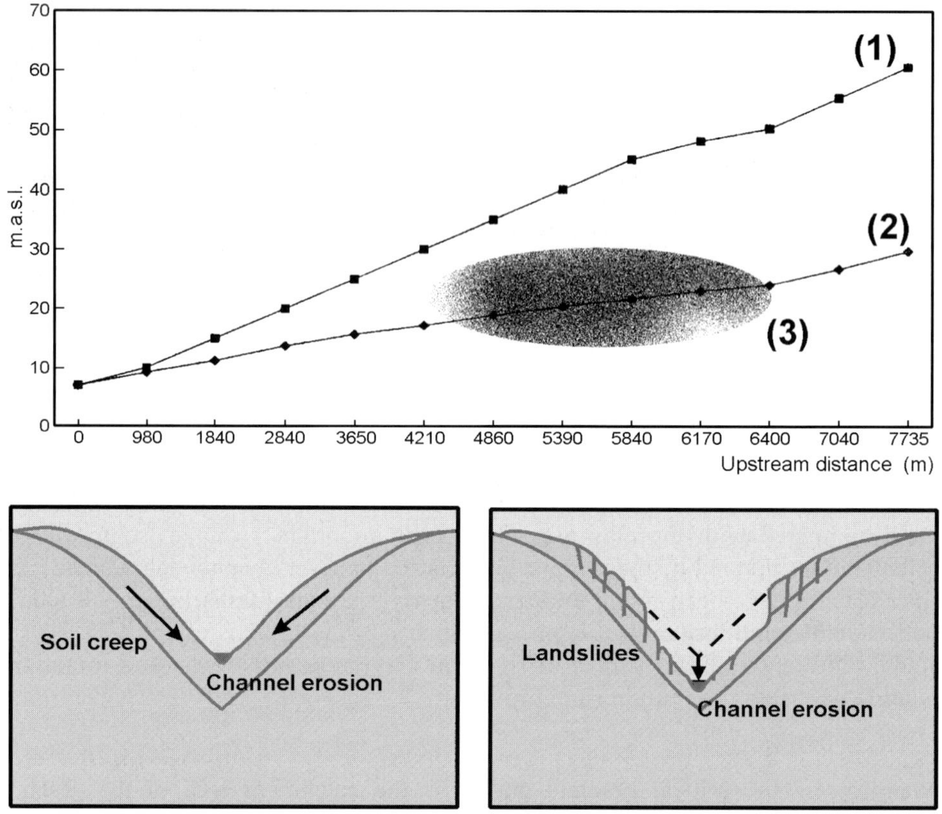

Fig. 3 (a) (1) Channel gradient of River Gråelva; (2), Calculated theoretical equilibrium slope; (3), Schematic illustrating channel incising pocket of quick clay. (b) Cross sections illustrating the processes of erosion associated with channel degradation.

The erosion protection procedure is schematically illustrated in Fig. 4. The bed of the main river was raised and covered by an armouring layer of cobbles and boulders. To prevent further degradation of tributaries, their erosion bases were also artificially raised. The level of the channel floor was thus in some places built up about 1.5–2 m above the natural bed level. The work was initiated in July 1992 but progress varied much from year to year. In 1999 a 4 km long reach was secured, whereas in some years like 1995 and 1998 there was little or no construction activity. By 2001 a total of 10 km had been secured along the main channel and tributaries.

SUSPENDED SEDIMENT TRANSPORT

Sediment concentration and water discharge were measured at Børstad (Fig. 2). The sampling programme involved an ISCO automatic sampler programmed for sampling at a rate of 2–4 samples a day throughout the season. Water samples were filtered through Whatman GF/C filters and the concentration of organic and inorganic particulate matter was

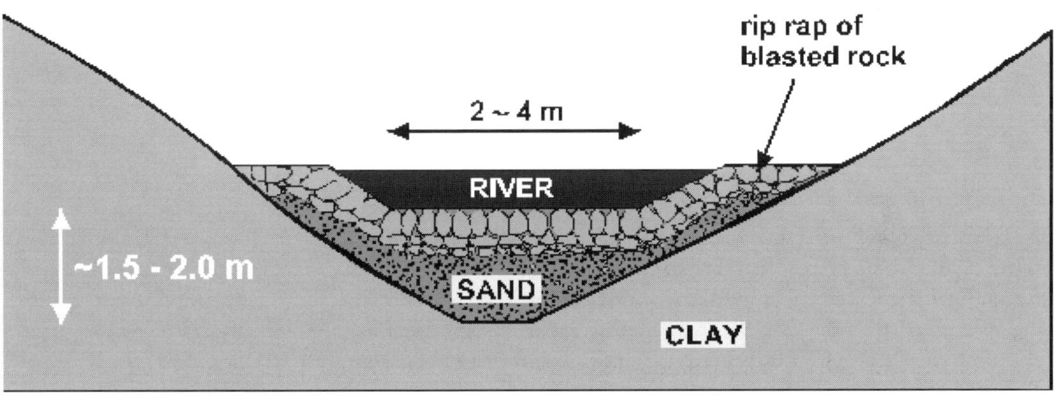

Fig. 4 The bed of the main river was raised and covered by an armouring layer of cobbles and boulders.

determined by weight loss on ignition at 500°C for 2 h. For further description of the sampling methods, see Bogen (1988, 1992). The hydrological regime in the River Gråelva is characterized by low self regulation. The flood events are peak floods of short duration alternating with long periods of very low water discharge. Sometimes floods also occur during winter, when runoff may take place on frozen topsoil. As infiltration is limited under such conditions, the upper part of the soil may become oversaturated and easily eroded.

The seasonal variation in suspended sediment concentration in 1993 is selected as an example of conditions close to natural (Fig. 5(a)). In January and February there were several winter flood events caused by rain in temperatures above 0°C. Flood events due to combined snowmelt and rain occurred in April and May. The summer months were dry but from late July until October there were frequent rainfloods. The maximum recorded concentration of suspended sediments in 1993 was >15 000 mg l^{-1} and on three other occasions that year the concentrations exceeded 10 000 mg l^{-1}. In 2002 the pattern of flood events and the magnitude of water discharges were comparable to those in 1993. The concentrations nevertheless remained low, with a maximum of 3500 mg l^{-1}; the concentration exceeded 1000 mg l^{-1} only on one other occasion that year (Fig. 5(b)).

The calculated annual sediment transport rates show a distinct overall decrease during the period of measurement, from 163 000 t year^{-1} in 1992 to 5800 t in 2002 (Fig. 6). The

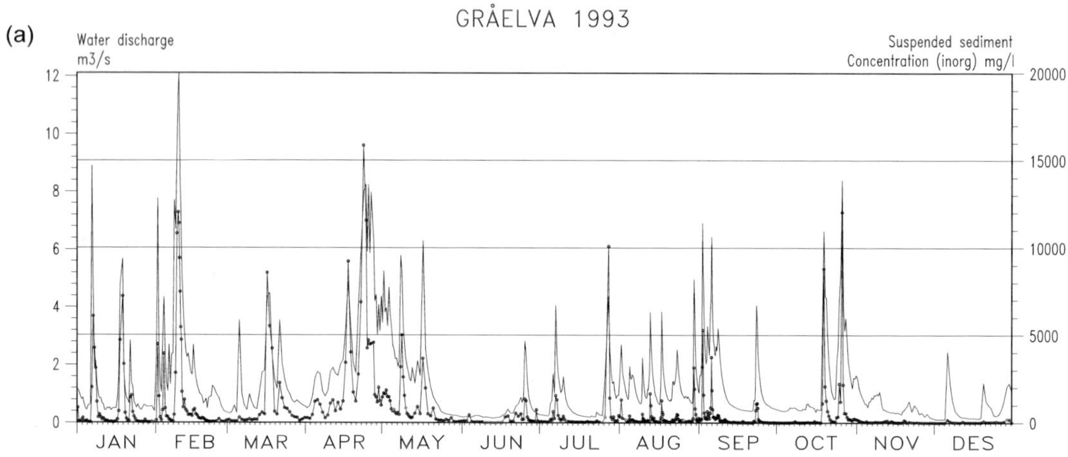

Fig. 5 The seasonal variation in suspended sediment concentration in 1993 and 2003. Water discharge (line) and suspended sediment concentrations (dots). (a) 1993 (at early stage of protection work) (b) 2002 (after 10 km of channel protected).

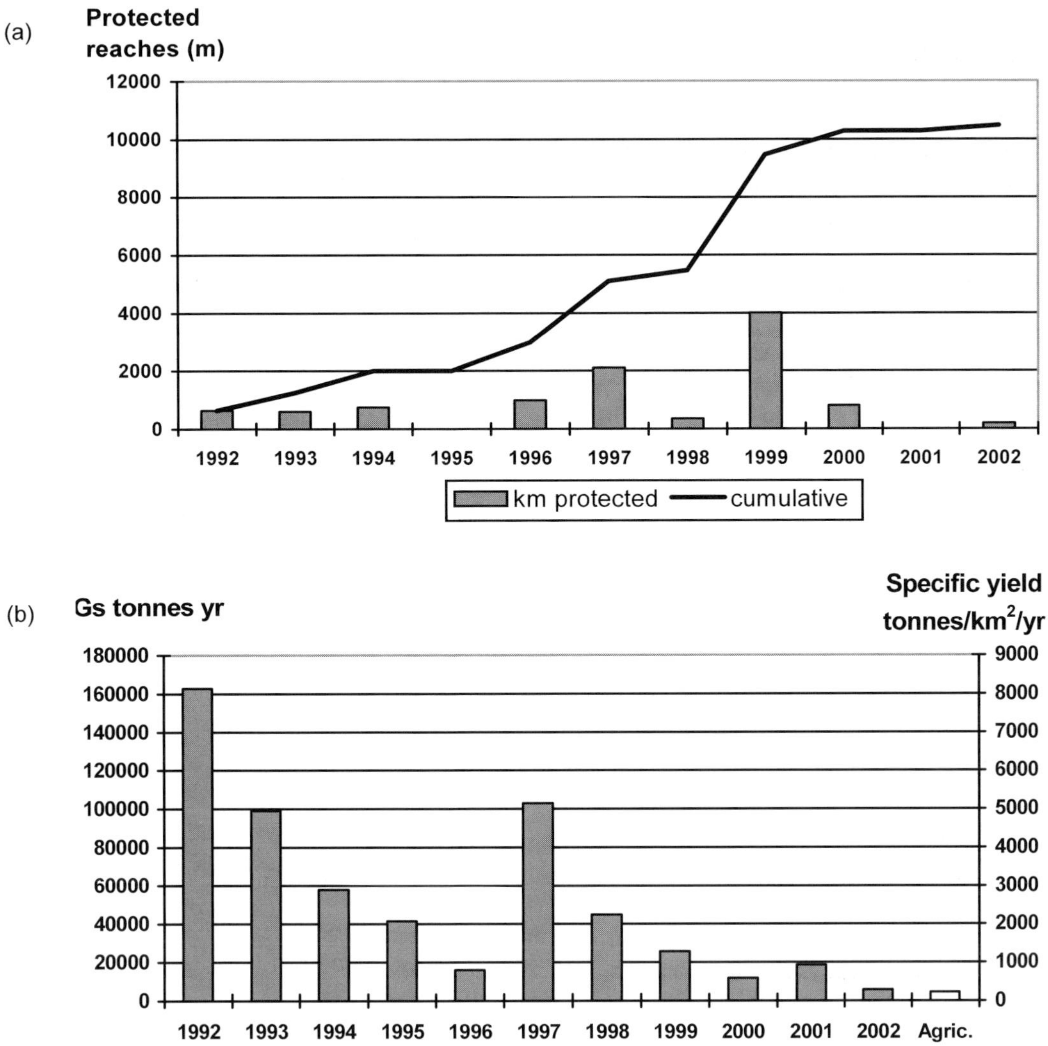

Fig. 6 Annual suspended sediment transport (Gs) of the River Gråelva (b) plotted against erosion protection work progress (a). Agric, indicates estimate of mean annual sediment yield from cultivated land.

decrease coincided with the progress in the erosion intervention programme, see Fig. 6 (upper). The primary cause of the reduction in particle concentration and sediment yield was the decrease in landslide activity and lateral erosion. Bønsnes *et al.* (2000) found a significant correlation between suspended sediment concentration and water discharge. This probably reflects the fact that a major part of the sediment sources are situated close to the river channel and are thus affected by a rise in water discharge. The regression curves did not change very much throughout the period of measurements. It is therefore likely that the dominating erosion processes remain the same but in total have much less effect than previously.

A comparison between sediment yields in other Norwegian rivers (Bogen, 1996) and those measured in the River Gråelva when conditions were still close to natural shows that the latter are the highest on record. Despite the large reduction in sediment transport, the sediment yield is still high when compared to other areas. The sediment yield of agricultural land within the River Gråelva catchment (225 t km^{-2} $year^{-1}$) is included in Fig. 6 for comparison. When the erosion protection works have been completed, erosion by surface runoff (including the contribution from agricultural land) will become the most important sediment source.

Some of the year-to-year variation in sediment transport is not explained by the influence of the erosion protection work alone. The decreasing trend was interrupted by a sharp increase from 16 000 t in 1996 to the 100 000 t measured in 1997. In 1998 the transport decreased once more, but only to 45 000 t, which is still higher than the annual yields in 1995 and 1996. However, in 2002 the transport was down to 5800 t.

The overall reduction in sediment transport totals as sediment sources were cut off may be due to the natural variability controlled by weather conditions. The number of flood events and the pattern of precipitation are subject to large seasonal and annual variations. To study the impact of such climatic variables an analysis was carried out involving a simple classification of the number of water discharge events above a certain threshold level.

To avoid including interrelated floods, the water discharge had to fall below a certain prestated level before the next flood was regarded as a separate one. The result of an analysis of the total number of flood events within each year is shown in Fig. 7. A peak over threshold technique was applied, involving three different threshold levels: 3.7 m^3 s^{-1}, 4.6 m^3 s^{-1} and 5.6 m^3 s^{-1}. These levels correspond to 2–3 times the standard deviation of the mean water discharge for the period of observation (Shaw, 1983).

The results of the analyses indicate that the low transport total in 1996 was largely due to the lack of floods that year, while the high total in 1997 may be explained by the occurrence of a number of flood events. The very large transport total in 1992 was also associated with the number of floods, but especially with the fact that one of these floods occurred during winter conditions. The large flood in January that year accounted for 80 000 t transport measured during just 1 week. During the period 1992–1996 a decrease in the number of floods augmented the effect of erosion protection work. During the period 1997–2002 there was again a marked decrease in sediment transport, despite the number of flood events

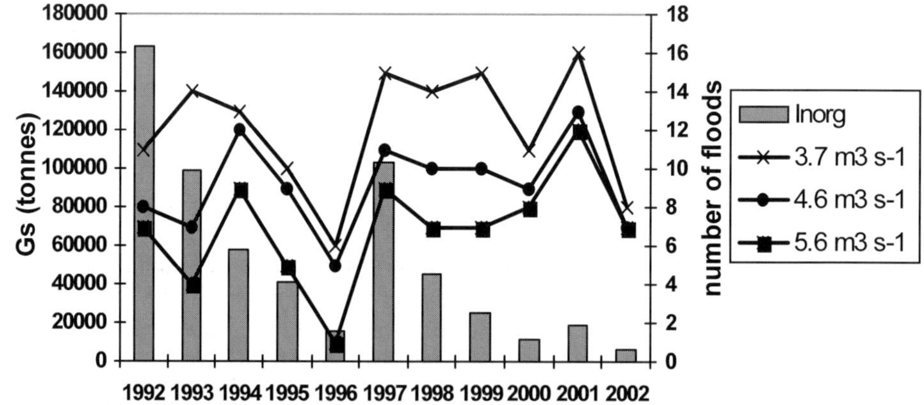

Fig. 7 Annual inorganic suspended sediment transport in the River Gråelva plotted together with the annual number of flood events exceeding the specified magnitudes.

increasing or remaining constant. Most of the decrease in sediment transport during this period can therefore be attributed to the effectiveness of the erosion protection work.

Berger *et al.* (1997) investigated the ecological impacts on the freshwater ecosystem after the stabilisation of the river bed and concluded that the erosion protection work seemed to have had a positive effect. There have been increases in the density of both bottom living organisms and younger fish within the river reaches affected by construction. This was attributed to an increase in algae production because of improved light conditions, and to improved fish habitats.

CONCLUSIONS

The aim of this paper is to report from investigations carried out to study the impact of erosion protection work on sediment transport in a river, the Gråelva, flowing in one of the marine clay areas of Norway.

River channel degradation and the associated undercutting of adjacent slopes followed by frequent small scale mass movement events are the main erosion processes. In the past this erosion activity has uncovered areas of quick clay and triggered several major slides. To reduce the slide hazard, erosion protection work has been carried out along the degrading reaches of the river. At the start of the construction work a sediment transport monitoring programme was initiated.

During 1992 and 1993, when the river conditions were still close to natural, there were several flood events with maximum concentrations in the range of 15 000–25 000 mg l^{-1}. In 2002 the maximum concentration was only 3500 mg l^{-1} and otherwise exceeded 1000 mg l^{-1} on only two occasions that year. The annual sediment transport rate had thus decreased from 163 000 t $year^{-1}$ in 1992 to 5800 t $year^{-1}$ in 2002. This great reduction in annual sediment yield, resulting from the cutting off of major sediment sources, did not result in a continuous decrease in sediment transport totals. A year to year variability controlled by climatic variables was still present, manifesting itself in the number and character of flood events.

During the period 1992–1996 a decrease in the number of floods thus strengthened the reducing effect of the erosion protection work. The marked decrease in sediment transport during the period 1997–2003 can be attributed to the impact of the erosion intervention programme as the number of flood events was increasing or constant during this period. The erosion protection work also had a positive impact on the freshwater ecosystem. There has been an increase both in the density of bottom living organisms and younger fish within the river reaches affected by construction.

REFERENCES

Berger, H. M., Breistein, J. B., Larsen, B. M. & Nøst, T. H. (1997) Gråelva–Mindre leirslam gir mer bunndyear og fisk. Sluttrapport 1991–1995. River Gråelva (Less suspended clay gives more invertebrates and fish) (in Norwegian). *NINA oppdragsmelding* **468**, 1–42.

Bjerrum, L. (1971) Kvikkleireskred. Et studium av årsaksforhold og forbygningsmuligheter. (Quick clay slides. A study of causes and controls). *NGI publ.* **89**, 1–14 (in Norwegian).

Bogen, J. (1986) Transport of suspended sediments in streams. In: *Partikulært bundet stofftransport i vann og jorderosjon* (*Particle bound matter in water and soil erosion*), (ed. by B. Hasholt), 9-21. *NHP-rep.* **14**, KOHYNO (in Norwegian).

Bogen, J. (1992) Monitoring grain size of suspended sediments in rivers In: *Erosion and Sediment Transport Programmes in River Basins* (ed. by J. Bogen, D. E. Walling & T. Day), 183–190. IAHS Publ. 210. IAHS Press, Wallingford, UK.

Bogen, J. 1996 Erosion and sediment yield in Norwegian rivers. In: *Erosion and Sediment Yield: Global and Regional Perspectives* (ed. by D. E. Walling & B. W. Webb), 73–84. IAHS Publ. 236. IAHS Press, Wallingford, UK.

Bogen, J. & Sandersen, F. (1991) Sedimentkilder, erosjonsprosesser og sedimenttransport i Leira-vassdraget på Romerike. (Sediment sources, erosion processes and sediment transport in river Leira, Romerike. Report within the research project," Pollution resulting from clay erosion and the significance of erosion protection measures). *NVE–publ.* **20**/91, 1–126 (in Norwegian).

Bogen, J., Berg, H. & Sandersen, F. (1993) Forurensning som følge av leirerosjon og betydningen av erosjonsforebyggende tiltak. (Sediment pollution caused by clay erosion and the significance of erosion protection measures). *Sluttrapport. NVE–Publ.* **21**, 1–86 (in Norwegian).

Bogen, J., Sandersen, F. & Berg, H. (1994) The contribution of gully erosion to the sediment budget of the river Leira. In: *Variability in Stream Erosion and Sediment Transport* (ed. by L. J. Olive, R. J. Loughran & J. A. Kesby), 307–315. IAHS Publ. 224. IAHS Press, Wallingford, UK.

Bønsnes, T. E., Bogen, J. & Husebye, S. (2000) Effekten av erosjonssikring på sedimenttransporten i Gråelva, Nord Trøndelag. (The effect of erosion protection measures on the sediment transport in Gråelva, Northern Trøndelag). *Norges Vassdrag og Energidirektorat-rep* **12**, 1–88 (in Norwegian).

Gregersen, O. (1991) Grunnundersøkelser og forslag til sikringstiltak i Gråelva, Stjørdal. (Ground investigations and proposals for protection work along the river Gråelva, Stjørdal). *NGI-rep 900053–2.* Norwegian Geotechnical Institute, Oslo, Norway (in Norwegian).

Rosenqvist, I. T. (1960) Marine clays and quick clays slides. In: Geology of Norway (ed. by O. Holtedahl). *Norwegian Geological Survey* no. 208, 463–471.

Sandersen, F. (1993) Gråelva, Stjørdal. Kartlegging av sedimentkilder og estimat av sedimentbudsjett (Survey of sediment sources and sediment budget estimate). *NGI-rep 924037–1.* Norwegian Geotechnical Institute, Oslo, Norway (In Norwegian).

Shaw, E. M. (1983) *Hydrology in Practice.* Van Nostrand Reinhold Co. Ltd., UK.

Sources of fine grained sediment in incised and un-incised channels, Jugiong Creek, NSW, Australia

PETER WALLBRINK[1,2] & JON OLLEY[1,2]

1 *Co-operative Research Centre for Catchment Hydrology, PO Box, 1666, ACT 2601, Australia*
2 *CSIRO, Land and Water, PO Box 1666, ACT 2601, Australia*
peter.wallbrink@csiro.au

Abstract In southeastern Australia land-use changes due to European settlement (~180 years ago) triggered widespread channel incision, and erosion from these features delivered massive volumes of sediment to rivers. It is proposed that the channels remain a major source of sediment. We use measurements of ^{137}Cs concentrations in soils and sediments to examine sources of fine sediment in incised and un-incised channels. In un-incised channels the ^{137}Cs concentrations decreased to around 40–60% of surface hillslope values by stream order 3, whereas the ^{137}Cs concentrations in incised channels decreased to 10–15% of surface hillslope values in the 1st order streams. In both cases we attribute the decreases in ^{137}Cs concentrations to input of sediment from the channel walls; the input is four times greater in the incised channels. It is concluded that erosion from the walls of incised channels remains a significant source of sediment to rivers in southeastern Australia.

Key words catchment management; channel bank erosion; Cs-137; deposited sediments; gully erosion; land-use change

INTRODUCTION

Subsoil derived from stream incision has been identified as the major contributor to the suspended sediment loads in many large Australian rivers (Prosser *et al*., 2001a). The rates of release of this material are believed to be much higher than naturally occurred prior to European settlement 180 years ago (Olley & Wasson, 2003). This sediment is recognized as a significant river contaminant causing decreases in stream water quality and alterations to physical habitats (Baldwin *et al*., 2002). Catchment managers are now required to plan river rehabilitation strategies that achieve greatest benefits in sediment reduction for the least cost. This task is best undertaken by identifying the most active erosion sources. While subsoil sources have been clearly shown to be a significant contributor of sediment to rivers in southeastern Australia (Olley *et al*., 1993; Wallbrink *et al*., 1998), the question remains as to whether this material is derived from recent erosion of channel banks or from reworking of sediment stored in the river network. In this paper we use measurements of ^{137}Cs on soils and sediments to compare the sources of sediment in incised and un-incised channels, and determine the contribution of channel bank erosion in these two systems. We focus our attention on material <10 μm, and make the assumption that this represents recent erosional activity given the short residence time of fine sediment in these systems (Wallbrink *et al*., 1998).

Caesium-137 is a product of atmospheric nuclear weapons testing and is concentrated in surface soil (Walling & He, 1997). It has a half life of 30 years and deposition from testing effectively ceased in Australia by 1970 (Longmore *et al*., 1983). Sediments derived from hillslope erosion have high concentrations of this radionuclide, while those eroded from gullies or channels have little or none (Wallbrink & Murray, 1993). In this study we compare concentrations of ^{137}Cs on hillslope soils to those on sediments collected from along both un-incised and incised streams of increasing stream order.

STUDY AREA

The Jugiong catchment (34050′S 148040′E) is 2170 km^2 in area and located on the south-western slopes of New South Wales (NSW), Australia (Fig. 1). It was previously identified as the major contributor of fine grained sediment to the Murrumbidgee River, one of Australia's largest rivers (Olive *et al*., 1996). Agriculture for cereal crops (wheat and canola) occurs on ~20% of the area. Cropping is rotated with pasture for sheep and cattle grazing which occupies the majority of the catchment area. The catchment is dominated by undulating terrain in the northern areas on Silurian granodiorite and is steeper in the south on Silurian dacite, where the density of channel incision is higher (Zierholz *et al*., 2001).

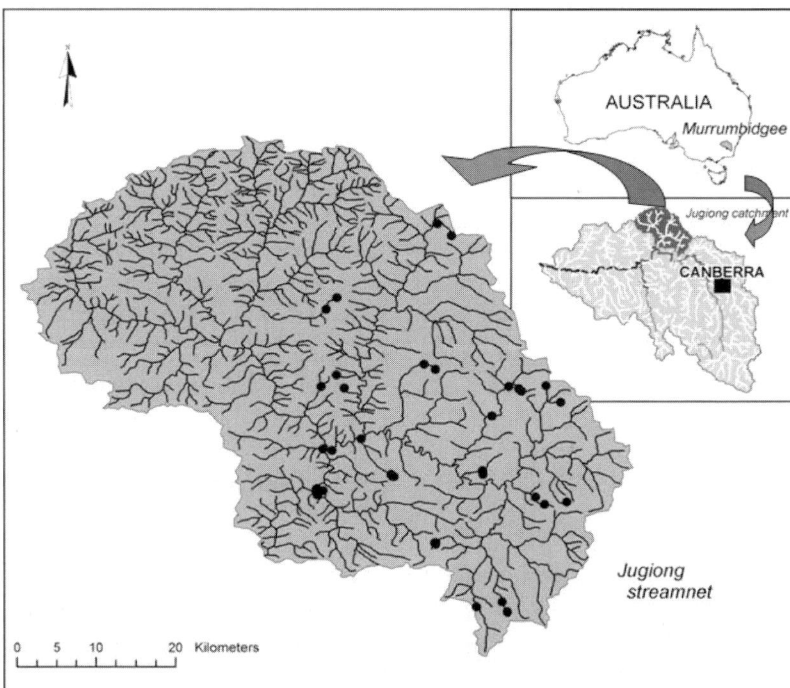

Fig. 1 Site location diagram, Jugiong Creek, NSW, Australia.

SAMPLE COLLECTION AND TREATMENT

The incised and un-incised channel networks in the catchment have been mapped and incorporated into a GIS database by Zierholz *et al*. (2001). We use this database in our study to identify stream sections from order 1 to 6, according to Strahler (1969), in both un-incised and incised sections of the network. A summary of the number of replicates of each order

Table 1 Number of locations for which ^{137}Cs concentrations have been analysed on the <10 μm fraction of a representative mixed sediment sample.

Channel type	Hillslope positions	Stream order (according to Strahler, 1969) 1st	2nd	3rd	4th	5th	6th
Un-incised	8	16	5	2	1	1	0
Incised		6	4	0	1	1	3

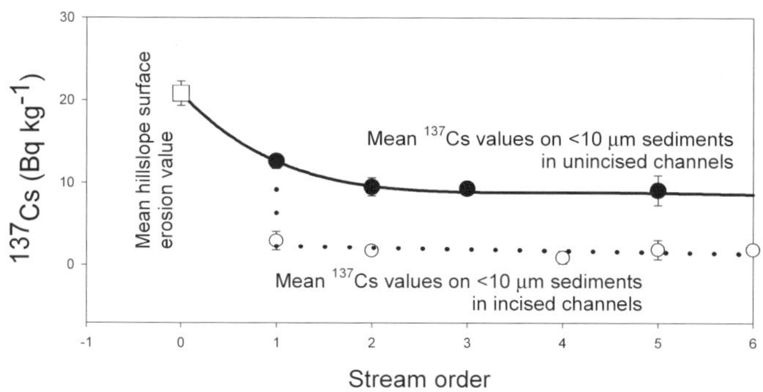

Fig. 2 Concentrations of ^{137}Cs (Bq kg^{-1}) on <10 μm fraction of soils (open squares) and sediments in un-incised (closed circles) and incised (open circles) channels, Jugiong Creek, NSW, Australia.

stream for each channel type is given in Table 1. As many reaches of the same stream order were sampled as practicably possible; though the number of high order streams that were un-incised in their lower order tributaries was limited, thus constraining the number of sites available. The locations of the reaches are given as circles in Fig. 2. In each of the selected reaches 20–40 grab samples of recently deposited sediment was collected. These samples were then combined to give one sample from each reach. Each of these reach samples was then analysed for its ^{137}Cs concentration (see below).

To characterize the ^{137}Cs concentration on sediments derived from hillslope erosion a series of samples were taken from uncultivated slopes around the Jugiong catchment incorporating a range of different slope positions and slope angles. At each location a number of surface scrapes (20–30) to 10 mm depth were then taken over a hillslope surface area of ~400 m^{-2}. This material was then mixed together to provide a single "averaged" ^{137}Cs concentration value for that hillslope position and location. The mixing procedure was used to average out the variability in soil ^{137}Cs concentrations that arise from initial fallout (Wallbrink *et al.*, 1994). The number of hillslope sampling locations is also summarized in Table 1.

Measurements were made on samples fractionated by a combination of wet-sieving and settling to recover the <10μm fraction (clay and fine silt). This is the fraction that corresponds best to sediment transported in the suspended form. Furthermore, most of the nutrients and contaminants associated with sediment are in this fraction. All discussion hereafter pertains to our analyses of this fine sediment fraction. The ^{137}Cs concentrations were determined by routine high resolution gamma spectrometry, using the methods described by Murray *et al.* (1987). The detectors used in this study are n-type closed ended co-axials. Detection limits were about ±0.3 Bq kg^{-1} for ^{137}Cs. Count times were typically one day. The data was time corrected to June 2000.

RESULTS

The ^{137}Cs concentrations (Bq kg^{-1}) on the <10 μm fraction from the Jugiong catchment hillslopes were in the range 15–24 Bq kg^{-1} with a weighted mean of 20.6 ± 1.5 Bq kg^{-1} (uncertainties given as standard errors: Fig. 2). These values are consistent with those from surface soils in the general region (see Wallbrink & Murray, 1993; Wallbrink *et al.*, 1998). The mean ^{137}Cs concentrations on deposited sediments in the un-incised channels are also given in Fig. 2 (closed circles). The mean concentrations decreased from ~12.6 ± 1.0 Bq kg^{-1} in the first-order streams to 9.1 ± 1.8, and 9.3 ± 1.8 Bq kg^{-1} respectively in the 3rd and 5th order streams. Given the uncertainties, this indicates a varying hillslope contribution to these stream orders of between 60–40%, assuming a two component mix between surface soils and channel wall material (Wallbrink *et al.*, 1996).

The ^{137}Cs concentrations from the incised channels are given as the open circles in Fig. 2 and are much lower (by a factor of 4) than those in the un-incised channels. The mean concentration in the first order streams of 3.0 ± 1.1 Bq kg^{-1} indicates a hillslope contribution of only ~15%. The mean value is 1.8 ± 0.5 Bq kg^{-1}in 2nd order streams and then remains low and constant in higher order streams thereafter. Using the procedure above, these data indicate a varying hillslope contribution of only 10–15% for incised stream orders 1–6, and that the major contribution from channel walls has probably already occurred by 3rd order streams. These ^{137}Cs sediment concentrations are also similar to those observed in the main channel of the Murrumbidgee River (reported by Wallbrink *et al.*, 1998) and are a marked contrast to those in the un-incised channels.

DISCUSSION

A USLE based erosion map of Lu *et al.* (2001) shows that erosion is in the range 0–10 t ha year^{-1} for hillslopes of both incised and un-incised streams in the Jugiong catchment. Thus, if we assume that the hillslope contribution to the un-incised and the incised streams are similar, then the ^{137}Cs data indicate that channel erosion in incised streams contributes ~4 times more sediment than that in un-incised channels. It is unlikely that these differences are due to lithology as the incised and un-incised channels were selected across the two major rock types in the catchment. The largest decrease in ^{137}Cs concentrations occurs in the first order streams, with concentrations in the un-incised stream being half those of the hillslope sediments, and those in the incised stream being 1/7th of values from hillslope sediments. A similar large decrease in ^{137}Cs concentrations was shown to arise from sediment input from 1st order channel walls in rainfall simulator experiments in the Goulburn region, NSW. In these experiments the ^{137}Cs concentrations on material eroding directly from gully walls was extremely low and sediment production from them substantially diluted the high concentrations on the material eroded from the surrounding surface areas (Wallbrink & Murray, 1993).

The ^{137}Cs concentrations in the 3rd to 6th order incised channels 1.9 ± 0.6 Bq kg^{-1} are also similar to those observed in the main channel of the downstream Murrumbidgee River 2.1 ± 0.3 Bq kg^{-1} (time corrected from data in Wallbrink *et al.*, 1998). This result is consistent with sediment in the main channel being primarily derived from erosion of channel walls in 1st to 3rd order incised streams, and that the relative proportions of hillslope and channel erosion remain approximately constant thereafter.

CONCLUSIONS

We have used the reduction in ^{137}Cs concentrations on deposited stream sediments relative to the initially high concentrations of hillslope eroding material, to infer that channel walls dominate sediment supply in incised channels. This work indicates that management action to reduce sediment delivery to the main channel of the Murrumbidgee River needs to focus on channel erosion, particularly in the 1st to 3rd order incised streams. The management effort should concentrate on improving and protecting riparian vegetation along channels. It should be noted that most sediment will be delivered during infrequent floods (Olive *et al.*, 1996) so management strategies must be robust enough to cope with erosion and sediment delivery occurring during these events.

Acknowledgments The authors would like to thank Mr Danny Hunt (field sampling and processing of samples), Mr Haralds Alksnis (radionuclide analyses) and Mr Ron De Rose (assistance in Fig. 1) from CSIRO Land and Water.

REFERENCES

Baldwin, D. S., Mitchell, A. M. & Olley, J. M. (2002) Pollutant–sediment interactions: sorption, reactivity and transport of phosphorus. In: *Agriculture, Hydrology and Water quality* (ed. by P. Haygarth & S. Jarvis), 265–280. CABI Publishing, Wallingford, UK.

Longmore, M. E., O'Leary, B. M., Rose, C. W., & Chandica, A. L. (1983) Mapping soil erosion and accumulation with the fallout isotope Caesium-137. *Aust. J. Soil Res.* **21**, 373–385.

Lu, H., Gallant, J., Prosser, I., Moran, C. & Priestley, G. (2001) Prediction of sheet and rill erosion over the Australian Continent, incorporating monthly soil loss distribution. *CSIRO Land and Water Technical Report 13/01*. CSIRO Land and Water, Canberra, Australia.

Murray, A. S., Marten, R., Johnston,. A. & Martin, P. (1987) Analysis for naturally occurring radionuclides at environmental concentrations by gamma spectrometry. *J Radio. Nucl. Chem.* **115**, 263–288.

Olive, L. J., Olley, J. M., Wallbrink, P. J. & Murray, A. S. (1996) Downstream patterns of sediment transport during floods in the Murrumbidgee River, NSW, Australia. *Z. Geomorph. NF, Suppl.Bd* **105**, 129–140.

Olley, J. M., Murray, A. S., Mackenzie, D. H. & Edwards, K. (1993) Identifying sediment sources in a gullied catchment using natural and anthropogenic radioactivity. *Water Resour. Res.* **29**(4),1037–1043.

Olley, J. M. & Wasson, R. J. (2003) Changes in the flux of sediment in the Upper Murrumbidgee catchment, SE Australia, since European settlement. *Hydrol. Processes* **17**, 3307–3320.

Prosser, I. P., Rutherfurd, I. D., Olley, J. M., Young, W. J., Wallbrink, P. J. & Moran, C. J. (2001a) Large-scale patterns of erosion and sediment transport in rivers networks, with examples from Australia. *Mar. Freshwater Res.* **52**, 8–99.

Prosser, I. P, Rustomji, P., Young W., Moran, C. J. & Hughes, A. O. (2001b) Assessment of river sediment budgets for the national land and water resources audit. *CSIRO Land and Water Tech. Report 15/01*. CSIRO Land and Water, Canberra, Australia.

Strahler, A. N. (1969) Quantitative analysis of erosional landforms. In: *Physical Geography*, 481–500. John Wiley and Sons, New York, USA.

Wallbrink, P. J. & Murray, A. S. (1993) The use of fallout nuclides as indicators of erosion processes. *Hydrol. Processes* **7**, 297–304.

Wallbrink, P. J., Olley, J. M. & Murray, A. S. (1994) Measuring soil loss using ^{137}Cs: implications of reference site variability. In: *Variability in Stream Erosion and Sediment Transport* (ed. by L. J. Olive, R. J. Loughran & J. Kesby), 95–103. IAHS Publ. 224. IAHS Press, Wallingford, UK.

Wallbrink, P. J., Olley, J. M., Murray, A. S. & Olive, L. J. (1996) The contribution of subsoils to sediment yield in the Murrumbidgee River basin, NSW, Australia, In: *Erosion and Sediment Yield: Global and Regional Perspectives* (ed by D. E. Walling & B. Webb), 347–355. IAHS Publ. 236. IAHS Press, Wallingford, UK.

Wallbrink, P. J. Murray, A. S., Olley, J. M. & Olive, L. (1998) Determining sediment sources and transit times of suspended sediments in the Murrumbidgee River, NSW, Australia using fallout Cs-137 and Pb-210. *Water Resour. Res.* **34**(4), 879–887.

Walling, D. E. & He, Q. (1997) Use of fallout Cs-137 in investigations of overbank sediment deposition on river floodplains. *Catena* **29**, 263–282.

Zierholz, C., Prosser, I. P., Fogarty, P. J. & Rustomji, P. (2001) In stream wetlands and their significance for channel filling and the catchment sediment budget, Jugiong Creek, NSW. *Geomorphology* **38**, 221–235.

Sediment Transfer through the Fluvial System (Proceedings of a symposium held in Moscow, August 2004).
IAHS Publ. 288, 2004

Relative effects of fluvial processes and historical land use on channel morphology in three sub-basins, Napa River basin, California, USA

SARAH A. PEARCE & ROBIN M. GROSSINGER
San Francisco Estuary Institute, 7770 Pardee Lane, Oakland, California 94621, USA
sarahp@sfei.org

Abstract Fluvial geomorphology and historical geomorphology studies were conducted on three sub-basins in the Napa River basin. Despite proximity of the sub-basins, differing physical and climatic settings, land-use histories, and channel modifications have resulted in substantially different sediment sourcing, storage, and transport to the Napa River. We present three examples in which the partnering of field-based fluvial geomorphic data and archival data has fostered a greater understanding of the observed channel morphology and fluvial processes, and how the channel has responded to anthropogenic modifications. Comparatively, sediment sourcing and transport to the Napa River is greatest in Sulphur Creek, followed by Carneros and Soda Creeks, respectively.

Key words California; channel modification; channel morphology; historical ecology; land use; Napa River; sediment supply

INTRODUCTION

It is generally recognized that channel conditions observed in a given watershed are the product of a combination of physical, climatic, and anthropogenic processes. Identifying which factors are most responsible for observed conditions is often a challenge, typically necessitating the involvement of other disciplines in addition to applied fluvial geomorphology. Through a series of related studies in the Napa River basin, we had the opportunity to conduct synchronized field geomorphology and historic geomorphology studies in three sub-basins. Due to excess sediment in the Napa River and associated habitat degradation of key native fish species (Stillwater Sciences, 2002), and the need to determine the contribution of local basins to the overall sediment budget of the San Francisco Bay (McKee *et al.*, 2002), there is substantial current interest in determining the natural variation and anthropogenic influence on channel morphology, sediment supply, and sediment transport through small tributaries.

Despite close geographic proximity (within 30 km of each other), and generally similar, semi-rural settings, each sub-basin has distinctly different underlying lithology, annual precipitation totals, historic and current land use, and extent of anthropogenic modification to the fluvial system. From these studies, we quantify and describe the three distinct sub-basin scale channel morphologies, dominant geomorphic processes, primary sediment sources, styles of sediment storage, and relative volumes of sediment supply to the Napa River. We present evidence from example locations that illustrate how historic land use and channel modification have affected modern channel morphology, process and management priorities.

STUDY AREA

The Napa River basin comprises an 1103 km^2 area that drains into the San Francisco Bay, in northern California, USA. The basin has a Mediterranean climate, receiving an average of 90% of annual precipitation between November and April. Precipitation depth varies generally with elevation from about 500 mm at the river mouth (sea level) to about 1500 mm in the headwaters (maximum elevation 1325 m). Headwater first order drainages exhibit gradients of up to 30%. Large variability exists in the amount of water and sediment delivered to the main stem by the 46 named tributaries. This paper details studies that were completed on three tributary sub-basins: Soda, Sulphur, and Carneros Creeks during 2001 and 2002 (Fig. 1). The primary goal of these studies was to understand the relative influences of natural and human activities on sediment processes in a subset of physical settings within the larger Napa River basin.

METHODS

Fluvial geomorphology study

Process-based field studies focused on collecting detailed data for each channel, including: identifying sediment sources, storage volumes and locations, and transport processes and relative volumes; measuring in-channel large woody debris (LWD) densities; and measuring

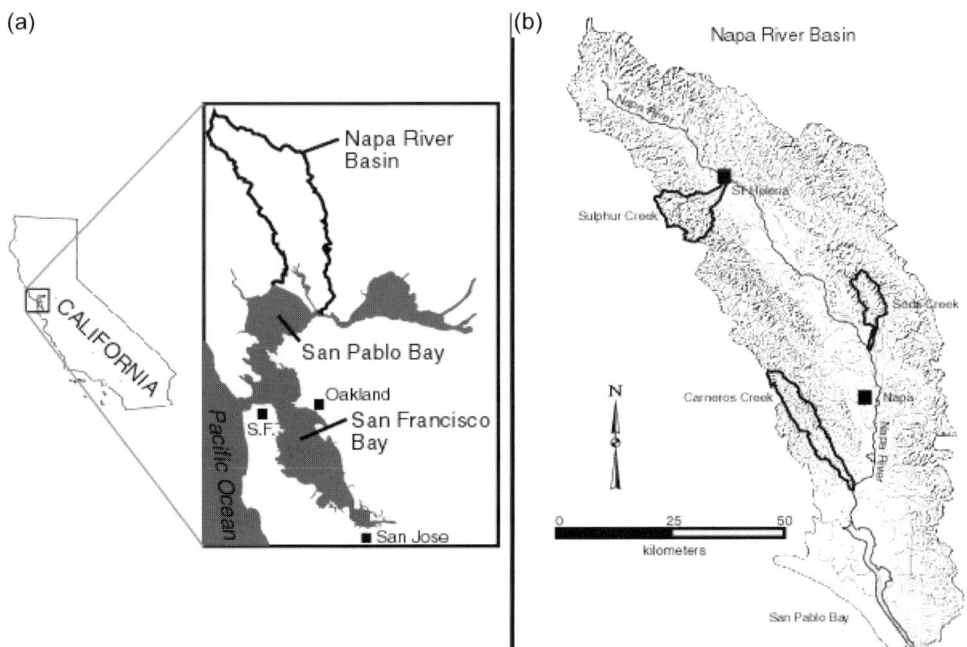

Fig. 1 (a) Map of San Francisco Bay and the Napa River basin area. (b) Map of the Napa River Basin showing the location of the Soda Creek, Carneros Creek, and Sulphur Creek sub-basins, and the towns of Napa and St Helena.

pool depths, spacing, and sediment in-filling. A stratified random sampling plan was developed to collect detailed data in 10 sample reaches (each reach is 25 bankfull widths in length) of each creek. These sample reaches were selected to represent the natural variability found within each sub-basin, allowing interpretation of the form and function of the entire channel length from a smaller but more detailed sample reach dataset.

Historical geomorphology study

Archival research techniques were used to build a reliable historical dataset for analysing questions about historical land use and channel change in the three sub-basins. Many types of historical data were utilized, including: aerial photographs, maps, land grant surveys and court documentation, historical documents, long-time resident interviews, field documentation of residual features, and historical climatic records. The historical dataset was calibrated and synthesized using techniques described in Grossinger (2001). Data from both study components were collected iteratively and analysed together, in order to determine the physical processes responsible for sediment supply and transport through each basin.

RESULTS

Current channel morphology in each sub-basin reflects a unique adjustment to physical, climatic and anthropogenic processes. The following details major findings in each sub-basin (Table 1).

Table 1 Geomorphic and historic characteristics of three Napa River sub-basins.

	Soda Creek	Carneros Creek	Sulphur Creek
Primary source of sediment	Hillslopes, bank erosion	Bank erosion, land use, landslides	Landslides
Total volume of measured sediment in storage (m^3)	1447	3053	9848
Volume of sediment storage per unit channel length (m^3 m^{-1})	0.2–2.3	0.5–3.2	0.2–14.1
Primary storage location	Large in-channel bars	In-channel bars and terraces	Alluvial fan, in-channel bars, terraces
Current relative supply to the Napa River	Low	Moderate	Moderate to High
Spanish era grazing?	No	~25 years intensive grazing	No
First railroad crossing the creek	None	1905	1866
Extensive historical logging/clearing?	No (?)	Yes	Yes
Substantial channel reaches rerouted?	Yes, 600 m	No	Yes, 600 m
Mouth of channel relocated?	Yes	No	No
Flood control channel?	No	No	Yes

Soda Creek

Soda Creek is a 12.2 km^2 sub-basin underlain by Plio-Pleistocene pyroclastic Sonoma Volcanics, with 610 (mouth) to 914 mm (headwaters) mean annual precipitation and comparatively low sediment production and transport to the Napa River. The channel transitions from a narrow headwater stream with little sediment storage, to a broad channel with many bedrock outcrops in the middle reach, to an entrenched alluvial channel in the lowest reach. Generally, the channel bed has a coarse grain size distribution, largely dominated by cobble and boulders, with very little fines. The current primary source of sediment is from hillslope interactions with small zero-order channels as well as from localized bank erosion. The highest volume of bank erosion was measured in the modified lowest reach (1.5 m^3 m^{-1}). However, this reach also stores the most sediment, in the form of a few large bars. Throughout the sub-basin, stored sediment volume ranged from 0.2 to 2.4 m^3 m^{-1} of channel. Shear stress analyses suggest that only two of 10 sampled reaches have ratios of total to threshold shear stress low enough to encourage channel aggradation.

Soda Creek has experienced relatively non-intensive land use since the time of European contact (*c.* 1810). The lowest reach of Soda Creek, however, has experienced significant channel modification that had not been previously recognized. Historical maps and photographs show that Soda Creek was disconnected from approximately 5 km of its historical drainage system by the filling of a 600 m section of stream channel, and the relocation of the channel mouth (Fig. 2). Prior to this diversion, the creek joined the neighbouring sub-basin along with two other creeks, before joining with the Napa River farther downstream than its current confluence. This tributary system had low channel gradient, stored large volumes of sediment as large bars and flood plain deposits, and likely provided high quality riparian habitat for animals and anadromous fish such as a local endangered trout species (*Oncorhynchus mykiss*). Between 1938 and 1942, the channel was redirected from its previous tributary system, to a more direct connection with the Napa River, presumably to allow faster drainage of Soda Creek while increasing the area available for agriculture.

Today, the redirected reach of Soda Creek is entrenched, sinuous, and highly dynamic, contains deep pools and large bars, and has the highest rates of bank erosion measured in the entire sub-basin. In addition, this reach now experiences backwater flood hazards when the Napa River is in flood. Evidence for flooding over the current Soda Creek banks towards its former route is provided by local recollections and bank stabilization efforts at that location. The likely effects of changing the course of Soda Creek include: an adjustment in grade to the new local base level and associated increased sinuosity, a greater connectivity for sediment delivery to the main stem Napa River, decreased length and quality of fish habitat, increased bank erosion associated with channel incision, decreased access to the flood plain, and increased backwater flood hazards during high Napa River stages. These findings provide context for any future channel restoration of this reach.

Sulphur Creek

Sulphur Creek is a 24.2 km^2 sub-basin underlain by Cretaceous and Jurassic Franciscan Formation (accretionary wedge prism) notorious for its erosion potential, with 886 mm mean

Fig. 2 Comparison of the lower reaches of Soda Creek. Aerial photographs from 1942 (COF) and 1999 (WAC Corp.). (a) Young riparian vegetation suggesting recent channel relocation. (b) Previous route of Soda Creek. Notice the mature vegetation. (c) Agricultural reservoir occupying the previous channel location. Third panel highlights the changes in channel route visible in the photographs.

annual precipitation, high sediment production, and comparatively moderate to high transport of sediment to the Napa River. The channel transitions from a narrow, boulder dominated headwater stream to a wider alluvial channel, to a broad, braided alluvial channel on the alluvial fan, finally to a highly entrenched and modified flood control channel in the lowest reach. The primary source of sediment is from landslides and other hillslope mass-movements that are directly connected with the fluvial system, including a single large slide named Devil's Slide that has been documented back to 1869. Sediment contribution from bank erosion attains a maximum of 17 m^3 m^{-1} in reaches where landslides impinge on the channel. Sulphur Creek's large alluvial fan is the most significant location for sediment storage; however other areas such as terraces and bars downstream of landslides are locally significant. Throughout the sub-basin, stored sediment volume ranges from 0.3 to 14.1 m^3 m^{-1} of channel, with the highest volumes measured in the alluvial fan reaches. The abundant sediment supply maintains a gravel bed texture, despite high bed shear stresses that would otherwise be expected to produce channels armoured with cobbles and boulders. Not surprisingly, the lowest shear stress values were calculated for the alluvial fan reaches.

In comparison to Soda Creek, Sulphur Creek has experienced a relatively dynamic and intensive land-use history. At the time of European contact, the basin was managed by native California peoples, who used controlled burns to shape vegetation patterns. After Europeans arrived, portions of the upper watershed were grazed and logged. Beginning in the late 1800s, gravel mining operations took advantage of the gravel storage in the alluvial fan and its natural replenishment during wet season floods. The mine was in operation until 1999, and provided much of the material for railroad and roadbeds throughout the Napa Valley. Over the past half century, the growing town of St Helena gradually encroached upon the creek, eventually necessitating channel modifications, including channel redirection and a flood control channel. Recent changes in land use, including increased road densities and hillside viticulture, are increasing the overall sub-basin sediment supply. However, our findings suggest that the contribution from landslides still far outweighs contributions from other sources.

Many major, but differing, channel modifications were also identified in this sub-basin, especially in the alluvial fan reach. The high natural sediment supply (documented by early descriptions of large landslides) in this watershed helped to develop the alluvial fan, with its large bars and terraces. However, for over a century, this sediment storage was counter-balanced by anthropogenic extraction of gravel from the channel. Interviews with gravel miners confirmed the methods and volumes of historic gravel withdrawal, and the seasonal natural replenishment of sediment in the extraction reach. Nearly 5 years have elapsed since the cessation of mining, and interviews with locals suggest that substantial channel aggradation has occurred. If this trend continues, the alluvial fan reach of Sulphur Creek may cause flood hazards for the community.

Carneros Creek

Carneros Creek is a 23.0 km^2 sub-basin underlain by Cretaceous and Miocene sandstones, siltstones and mudstones, with 710 mm mean annual precipitation and comparatively moderate sediment production and transport to the Napa River. The channel transitions from a narrow, boulder dominated headwater reach, to a channel with bedrock outcrop and high rates of bank erosion in the middle reaches, into a highly entrenched channel with some large bars and fine sediment deposits in the lowest reaches. The primary source of sediment is localized bank erosion, with smaller contributions from hillslope landslides and slumps, and land uses such as grazing and viticulture. The highest volumes of bank erosion were measured in the middle reaches, averaging 3.7 m^3 m^{-1}. Sediment storage occurs as active channel deposits (annually mobile sediment in riffles, glides and runs) in the upper middle reaches, large bars in the lower-middle reaches, and pool deposits in the lowest reaches, with volumes ranging from 0.5 to 3.2 m^3 m^{-1} of channel. Shear stress analyses suggest the lowest reaches have a very low threshold shear stress, indicating that the stored volumes of fine sediment in the long, slow-velocity pools can be easily re-entrained.

Carneros Creek has experienced a complex agriculturally-oriented land use-history, yet with few major direct modifications to the channel. Due to the proximity of the Sonoma Mission (one of 21 Spanish missions in California), the Carneros lands were used as pasture for cattle and sheep as early as the 1820s. Later, the basin was developed for a series of agricultural crops, and continued grazing. In recent decades, viticulture has mostly replaced orchards and grains. Land use since European contact has likely increased the volume of sediment supplied to the channel. The primary sediment supply, bank erosion in the middle reaches, is likely caused by historic and recent intense cattle grazing. Interviews with landowners confirm that bank trampling and nearly complete removal of vegetation on the banks by grazing occurred, providing a likely cause for the current measured high rates of erosion. Grazing operations have ceased in this location, allowing re-vegetation of the banks, and the potential for future bank stability. Upstream, some grazing still occurs, but efforts to fully fence cattle out of the channel are contributing to increased bank stability.

Despite the intensive and prolonged land use, there is only evidence of minor changes to channel banks and little evidence of major channel modifications on the order of those documented in Soda and Sulphur Creeks. Analysis of historic maps indicates that Carneros Creek has experienced little redirection, straightening, meander loss, or incision since European contact (Fig. 3). Although the highly entrenched lower reach is providing a source of sediment to the channel, conclusive evidence of recent channel incision was not found. However, evidence of channel widening, or bank lay-back was documented. Aerial photographic comparisons show that the total width of the riparian canopy in this reach has been expanding since the first available images (1942). Currently, the banks are lined with a single row of mature bay laurel trees (approximately 50 to 100 years old), which are severely undercut. This suggests that the banks have been dynamic over this time period, possibly due to natural adjustments in channel morphology, or possibly due to a period of increased incision caused by early grazing. This process likely will continue to occur, contributing sediment to the channel, until the banks reach an angle of equilibrium.

CONCLUSIONS

Using both field and archival geomorphic methods, we studied three tributary sub-basins of the Napa River. Despite the proximity of these basins, each has a unique suite of dominant physical processes, channel morphologies, land use and channel modification histories. We have found that studies in historical ecology and geomorphology are natural partners to traditional field geomorphology studies, often providing critical information to understanding

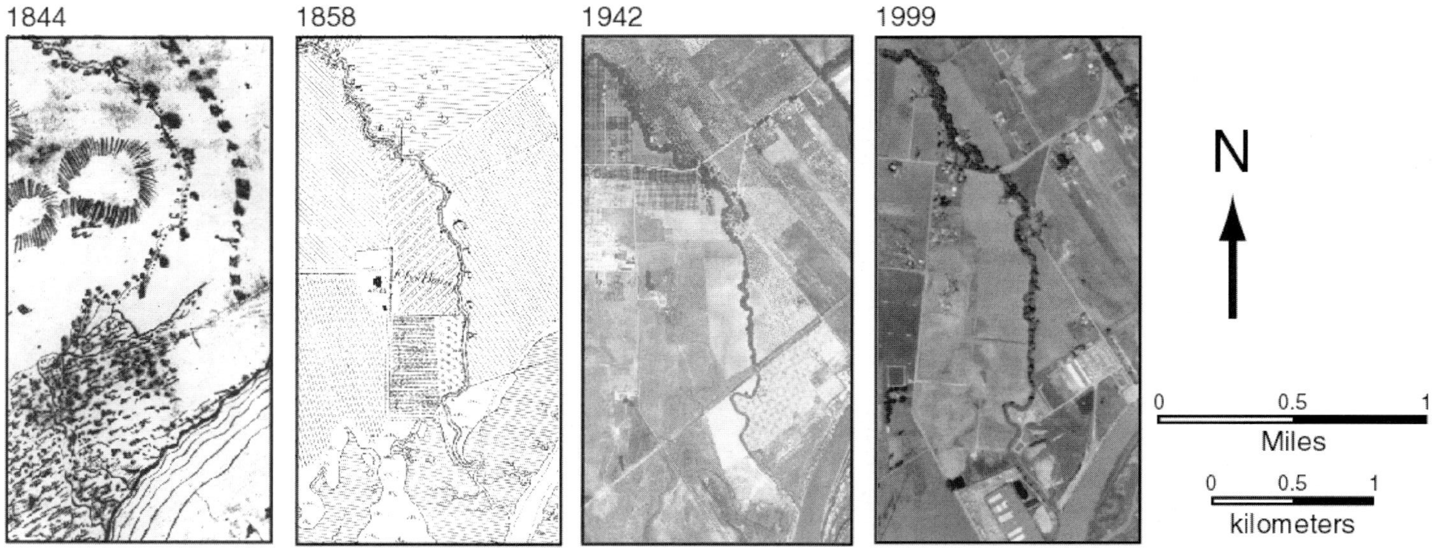

Fig. 3 Comparison of the lower reaches of Carneros Creek. Panels include: 1844 Huichica land grant (O'Farrell, 1844), 1858 US Coast Survey (USCS), 1942 (COF) and 1999 (WAC Corp.) aerial photography. Note the similarity of channel location and pattern through time.

the observed processes and morphologies, and recommending appropriate management practices. Relative contributions of sediment to the Napa River are highly variable throughout the Napa River basin, as well as between the three example sub-basins. Comparatively, Sulphur Creek contributes more sediment than Carneros Creek, while Carneros contributes more than Soda Creek. The lowest reach of Soda Creek is still responding to a 1940s era redirection of the channel, causing high bank erosion, backwater flood hazards, and increased sediment transport to the Napa River. In light of high natural sediment loads, Sulphur Creek residents must address the ramifications of increased channel bed aggradation in the alluvial fan since the cessation of gravel mining. Carneros Creek residents must address the supply of sediment to the fluvial system from high rates of bank erosion, likely caused by historic and recent grazing, in both the middle and lower reaches. Due to changing land uses and channel modifications (mainly channel simplification), the relative contribution of sediment supply and transport to the Napa River appears to have increased from pre-European contact in all three sub-basins.

Acknowledgements The authors thank the following individuals and groups for support during these studies: Napa County Resource Conservation District, USDA Natural Resources Conservation Service, the Carneros Creek Stewardship Group, the Sulphur Creek Stewardship Group, landowners of Soda Creek, Matt O'Connor and Brenda Rosser of O'Connor Environmental Inc., Lester McKee and Chuck Striplen of San Francisco Estuary Institute, and Elise Brewster of Brewster Design Arts. Funding was provided by CALFED and a Clean Water Act 205(j) grant.

REFERENCES

COF (1942) Aerial photography of the Napa Valley. 1:20 000, black and white, stereo overlap. Source: Napa County Resource Conservation District, California, USA.

Grossinger, R. (2001) Documenting local landscape change: the bay area historical ecology project. In: *The Historical Ecology Handbook: A Restorationist's Guide to Reference Ecosystems* (ed. by D. Egan & E. Howell). Island Press, Washington, DC, USA.

McKee, L., Leatherbarrow, J., Pearce, S. & Davis, J. (2002) A review of urban runoff processes in the Bay Area: Existing knowledge, conceptual models, and monitoring recommendations. A report prepared for the Sources, Pathways and Loading Workgroup of the Regional Monitoring Program for Trace Substances. SFEI Contribution 66. San Francisco Estuary Institute, Oakland, California, USA.

O'Farrell, J. (1844) Diseño Rancho de Huichica. Map D-11 in the Land Case "Huichica, Jacob P. Leese, Claimant, 50 ND." [literal attribution on map: "Jaspar Farral Agremensor y Arquitecto San Rafael"; Obtained from The Bancroft Library].

Stillwater Sciences (2002) Napa River basin limiting factors analysis. A report prepared for the San Francisco Bay Water Quality Control Board, and the California State Coastal Conservancy. Stillwater Sciences, Berkeley, California, USA.

United States Coast Survey (USCS) (1858) Napa Creek and Napa City California, Plane Table Sheet XXXII, Register no. T777.

WAC Corporation (1999) Aerial photography, 1:24 000, natural color, stereo overlap.

Erosion and fluvial sediment supply in undisturbed and cultivated basins: the case of the Desix and Maury river basins (western Mediterranean area)

PIERRE SERRAT[1] & WOLFGANG LUDWIG[2]

1 *UMR 6042–CNRS, Géodynamique des Milieux Naturels et Anthropisés, Université Blaise Pascal, Maison de la Recherche, 29 boulevard Gergovia, F-63037 Clermont-Ferrand, France*
pierre.serrat@univ-perp.fr

2 *UMR 5110–CNRS, CEFREM Université de Perpignan, 52 avenue de Villeneuve, F-66860 Perpignan, France*

Abstract Two small neighbouring rivers basins (western Mediterranean area) were compared: the Maury and the Desix rivers which belong to the Agly River basin (1045 km^2). In this area, the climate is characterized by a marked contrast between dry and humid conditions, with the occurrence of torrential downpours during only a few days of the year. As a result, the bedload and nearly all the suspended sediment load are transported during given events. The main difference between these two catchments is the land-use patterns: viticulture and garrigue (a typical Mediterranean vegetation type) are mainly growing at the Maury basin, while forest occupies the main part of the Desix basin. As a consequence, erosion marks are clearly visible on the bare slopes of the Maury basin. This river provides an important suspended sediment load and a great part of bedload to the Agly River.

Key words bedload; erosion; fluvial dynamics; land-use; Maury and Desix rivers; Mediterranean area

INTRODUCTION

During recent years, river basins were studied using an approach which requires integrated studies based on multipurpose investigations (Snishchenko, 1992; Ludwig & Probst, 1996; Bobrovitskaya & Zubkova, 1997). It consists of understanding and linking different factors, from the local to the entire basin scale, including the fluvial dynamics (Hadley *et al.*, 1985; Ludwig & Probst, 1998; Serrat, 2003). In this paper we suggest linking different physical characteristics of two catchments, with their sediment supply at the outlet. For that, a mid spatial scale was selected, and we studied the Desix and the Maury basins, located in a Mediterranean environment. The main features of the physical patterns of these catchments have been compared (altitude, slope, drainage density, etc.). One of the main differences between both sub-basins is the land cover. The ground surface conditions of the basins may therefore affect the river sediment supply. The purpose of this study is: (a) to evaluate the influence of the viticulture on slope erosion; and (b) to identify the relationships between erosion and sediment supply. For that, the composition of bedload at the mouths of the Desix River and the Maury River has been studied, as well as their respective supply to the main river.

STUDY AREA

This comparative study is focused on the Desix and the Maury basins, which belong to the Agly basin (1045 km^2), located in the northeastern boundary of the French Pyrenees (latitude

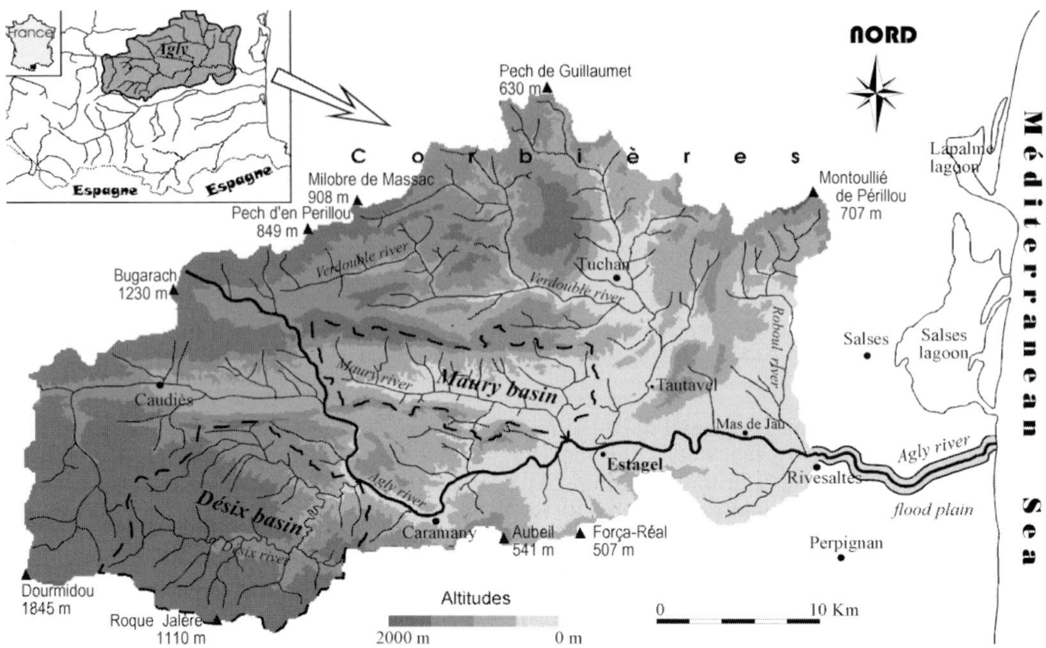

Fig. 1 Location map of the Agly basin and its sub-basins.

N: 42°40′/42°59′; longitude E: 2°16′/3°02′). Structural and tectonic features have a great impact on the Agly drainage network, stretched in an east–west direction (Fig. 1).

The average elevation of the Agly drainage basin is 397 m and the average slope 18.5°. The main part of this basin is underlain by compact limestone from the Jurassic and Cretaceous periods. The annual rainfall is about 700 mm for the whole basin compared to only 579 mm at Perpignan (1946–1996), which is the main regional city near the Mediterranean coast line. A major part of precipitation can occur during storm events (for example 304 mm in 6 h on 13 October 1986 at Torreilles), which cause heavy floods (MISE, 1993; Rigo & Llasat, 2003). These flash floods that only last a few hours are terrible for the population and are well-known in this region (Pardé, 1941; Serrat, 2001). While the mean annual flow of the Agly River is 6.1 $m^3 s^{-1}$ (1969–1997), floods can reach over 2000 $m^3 s^{-1}$, as occurred on 12–13 November 1999. Thus, the Mediterranean region is often affected by strong erosion processes, and fluvial dynamics depend on several basic conditions, such as basin morphology, land-use and vegetation cover.

CHARACTERISTICS OF THE MAURY AND DESIX BASINS

Morphology and hydrology

The morphology of the Agly River basin has been obtained from a Digital Elevation Model (10 m grid). The main morphological characteristics of the Agly basin and sub-basins including the Maury and Desix rivers basins are presented in Table 1. Drainage network is derived from the Digital Elevation Model. The Maury and Desix basins are characterized by

Table 1 Morphological characteristics of local river basins.

	Agly	Verdouble	Boulzane	Desix	Maury	Roboul
Basin area (km^2) - (A)	1045	317	156	143	87	66
Basin perimeter (km) - (P)	196	91	63	51	49	43
Compactness index (Ki)	1.71	1.44	1.42	1.20	1.48	1.49
Maximal altitude (m)	1845	964	1845	1542	964	707
Mean altitude (m)	397	370	739	697	241	245
Mean slope (degrees)	10.50	12.20	18.50	13.60	9.70	6.70
Relief aeration index (I_A)	4.60	2.60	2.50	2.20	4.00	2.90
Main drain lenght (km)	73.20	42.80	33.80	31.60	19.80	21.2
Drainage density	1.77	1.76	1.68	1.79	1.98	1.67

Compactness index : $Ki = P\,(2\sqrt{\pi A})^{-1}$ Relief aeration index $I_A = Alt_{max} / Alt_{mean}$

high drainage densities. The slope gradient is more important for the Desix (13.6°) than for the Maury basin (9.7°). The torrential hydrology of some rivers (the Boulzane River, the Desix River, the Maury River) can be explained by their morphological characteristics.

According to recent observations (1995–2000) mean annual water discharge at the outlet the Desix River is 1.7 $m^3\ s^{-1}$. However, maximum observed water discharges were about 250 $m^3 \cdot s^{-1}$ (e.g. on 26 September 1992 and 16 December 1996). The flood levels arise very quickly because of steep slopes and the circular configuration of the basin (Serrat & Depraetere, 1997).

Unfortunately, there is no gauging station on the Maury River. According to data (period observation 1968–1997) from neighbouring gauging stations located within basins with similar morphological characteristics, the mean annual water discharge of the Maury River may be about 1.1 $m^3\ s^{-1}$. Similar to the Desix River, flash-floods occur in the Maury basin, with maximum estimated water discharge up to 500 $m^3\ s^{-1}$ on the 12 November 1999, when a rail track near the river was completely destroyed.

Vegetation and land use

In many parts of the world, it has been shown that vegetation, land cover and agriculture practice have severe consequence on erosion and sediment delivery to the fluvial system (e.g. Johnson, 1988). In this study, in order to characterize the land cover of the two basins, remote sensing data were used (CNES Spot-Image). Spectral and reflectance analysis allow the quantification of the biomass which covers and protects the soil from erosion. The digital data result from the computation of the NDVI (Normalized Difference Vegetation Index) given by the equation:

$$NDVI = (N_{IR} - R) / (N_{IR} + R) \qquad (1)$$

where N_{IR} is near infra red (0.79–0.89 μm) and R is red (0.61–0.68 μm).

These data have been completed by a supervised classification. This involved identifying a set of sample locations that have been visited in the field. The land cover mapped in the field was then compared with the one mapped in the image at the same location. To simplify, we considered that the signature analysis produced three groups of vegetation: a set of conifer forest and beech grove; the olm oak forest; and the garrigue and the vineyards.

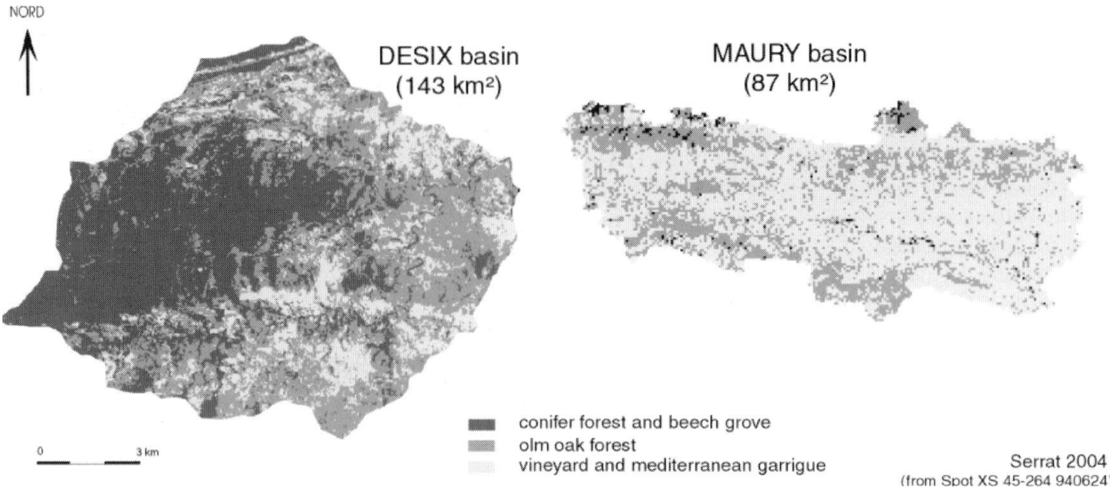

Fig. 2 Land cover maps of the Desix and the Maury basins from remote sensing data (CNES-Spot-image).

Table 2 Land cover of the studied basins (Desix and Maury).

Basin	Conifers, beech grove	Olm oak forest	Garrigue, vineyard
Desix	58.9 km^2 (41.2%)	60.8 km^2 (42.5%)	23.2 km^2 (16.3%)
Maury	1.6 km^2 (1.8%)	27.9 km^2 (32.1%)	57.5 km^2 (66.1%)

These three groups make up three different heights of vegetation and as many types of protection against the intensive rainfalls which occur between September and February at Mediterranean region. Height and density of vegetation play an important role in the processes of interception of the drops and their high fall velocity.

The Maury basin area is mainly covered by vineyards and garrigue. A main part of the Desix basin is covered by different forests (Fig. 2, Table 2). A striking feature is the opposition for the third group (garrigue, vineyards) between the Maury and Desix basins. So, what are the consequences on erosion and sediment supply?

IMPACTS OF LAND USE: EROSION, SUSPENDED SEDIMENT AND BEDLOAD

We investigated erosion, suspended sediment and bedload, using different approaches in order to study their relations with the land cover.

Slope erosion

In the Desix basin, forests occupy a greater area and ensure an efficient protection of the ground against heavy rainfall. Except for vineyards in the bottom valley sides, there are only few erosion marks.

Fig. 3 Gully erosion in the vineyard of the Maury basin.

On the other hand, overland flow and rill erosion were frequently observed in the Maury basin. Gullies of 0.5 m depth have been often observed on many parcels of vineyard (Fig. 3) This has been observed, in relation to specific conditions, in other areas of the world (Mura *et al*., 1988). Studies of these processes allow an understanding of why the suspended matter is so important at the outlet of the Maury River.

Suspended sediments

Sediment concentrations of the Desix River are 0.3–0.9 g l^{-1}. Up to now, sediment fluxes are not available because of an insufficient number of samples and too short a period of flow observation.

During floods events of the Maury River, high concentrations were measured (2.1–4.8 g·l^{-1}). At the confluence with the Agly River, the colour of the Maury River water was very brown and this high density of the flow did not allow mixing with the main river water (Fig. 4). For the entire Agly basin the mean total suspended sediment is 107 t $year^{-1}$ for the period 1968–1997 (Serrat, 2003). The X-ray analysis of clay minerals reveals for the Maury River >10% illite and 52% of quartz, which means a strong degree of erosion. However, these values are low for the Desix and Agly rivers downstream (Table 3).

Bedload

A petrographic study of bedload was initiated for evaluating the respective sediment input of each tributary to the main river. Nine sites were investigated in order to study the

Fig. 4 Outlet of the Maury channel: high density of the streamflow and fluvial gravel bar (bed transport) carried during floods in December 1996.

Table 3 X-ray analysis of suspended sediments (clay minerals in %).

Basin	Smectite	Illite	Kaolinite	Chlorite	Quartz	Feldspar	Carbonates	Total
Desix	6.3	3.8	2.4	1.2	35.2	21.5	29.6	100
Maury	0.2	10.4	0.2	0.3	52.1	0.7	36.1	100
Agly downstream	1.5	9.9	0.7	0.8	26.6	2.0	58.5	100

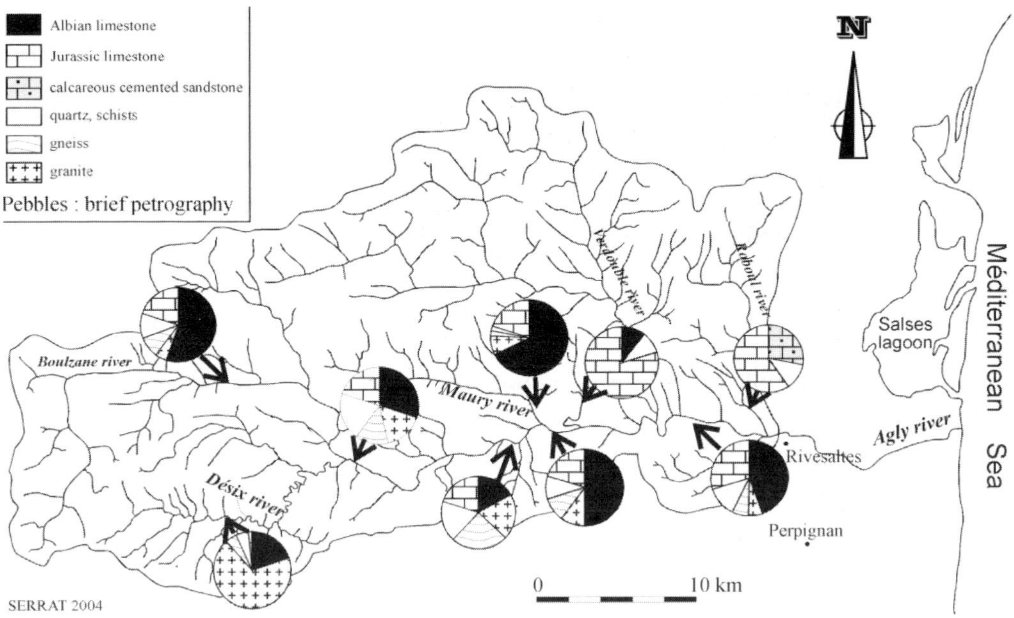

Fig. 5 Petrography of gravel and pebbles along the Agly River and its tributaries.

percentages of rock types. The best tracer is Albian limestone, pebbles of which are well represented in the Maury riverbed and downstream in the Agly River. It is possible to conclude that an important part of bedload enters from the Maury basin (Fig. 5). For the entire Agly basin, bedload is 1000 t year^{-1}, obtained by bathymetric measurements from 1991 to 1999 in a small retention dam near Rivesaltes (Serrat, 1999). By taking this result and using numerical values of repartition of pebble petrography shown in Fig. 5, we solved a system of four equations. The bedload of the Maury River was estimated: 300 t year^{-1}, including 210 t year^{-1} of Albian limestone. As a confirmation of this result, after the high flood of December 1996, we find a new bar of coarse sand and gravel at the outlet of this basin, with a volume of 1100 m^3 (Fig. 4). If we considered the results of Reid & Dunne (1996) and also observed suspended sediments downstream in the Agly River (Serrat, 1999), the expected percentage of bedload in total load is only 2%. The Maury basin provides about 15 000 t year^{-1} (172 t km^2 year^{-1}), which represents about 15% of the sediment supply of the entire Agly.

CONCLUSION

Although further studies should be done, this work is a first step towards understanding the erosion processes involved in the Agly basin. Erosion is significantly greater on unprotected soils of the Maury basin than on the forested soils of the Desix basin. In the Maury basin, thin soils contribute to low infiltration capacities, and during periods of heavy rain, overland flow has been frequently observed. Mineral tracers such as carbonates for suspended sediment, and Albian rocks for bedload, confirm the importance of sediment supply provided by the small Maury basin (87 km^2). The anthropogenic factors are dominant. The Desix basin, which has a greater area (143 km^2), higher mean altitude and more steep slopes, provides less sediment load. Further studies should be undertaken: small runoff-erosion plots will be used to assess soil loss in undisturbed forest conditions and in vineyards. As a general rule, viticulture leads to important soil loss (e.g. Loughran *et al.*, 2000). In the south of France, intensive viticulture, which originates from the mid 19th century, is perhaps unsustainable under long-term clean-till management practices.

REFERENCES

Bobrovitskaya, N. N. & Zubkova, K. M. (1997) Improved methodology for the computation of normal annual yield of suspended sediments from rivers In : *Study of Erosion, River Bed Deformation and Sediment Transport in River Basins as Related to Natural and Man-made Changes*, 92–103. UNESCO-IHP-V, no. 10. Paris, France.

Hadley, R. F., Lal, R., Onstad, C. A., Walling, D. E. & Yair, A. (1985) *Recent Developments in Erosion and Sediment Yield Studies*. UNESCO-IHP, Paris, France.

Johnson, R. C. (1988) Changes in the sediment output of two upland drainage basins during forestry land use changes. In: *Sediment Budgets* (ed. by M. P. Bordas & D. E. Walling), 463–471. IAHS Publ. 174. IAHS Press, Wallingford, UK.

Loughran, R. J., Elliott, G. L., Maliszewski L. T. & Campbell, B. L. (2000) Soil loss and viticulture at Pokolbin, New South Wales, Australia. In: *The Hydrology–Geomorphology Interface: Rainfall, Floods, Sedimentation, Land Use* (ed. by M. A. Hassan, O. Slaymaker & S. M. Berkowicz) (Proc. Symp., Porto Alegre, December 1988), 141–152. IAHS Publ. 261. IAHS Press, Wallingford, UK.

Ludwig, W. & Probst, J. -L. (1996) A global modelling of the climatic, morphological, and lithological control of river sediment discharges to the ocean. In: *Erosion and Sediment Yield: Global and Regional Perspectives* (ed. by D. E Walling & B. W. Webb) (Proc. Symp., Exeter, July 1996), 21–28. IAHS Publ. 236. IAHS Press, Wallingford, UK.

Ludwig, W. & Probst, J. -L. (1998) River sediment discharge to the oceans: Present-day controls and global budgets. *Am. J. Sci.* **296**, 265–295.

Mission Inter-Service de l'Eau (1993) *La crue du 26 Septembre 1992 dans les Pyrénées Orientales. La plus forte depuis l'aiguat d'octobre 1940.* Préfecture du département des PO, Perpignan, France.

Mura, R., Cambon, J. -P., Combes, F., Meunier, M. & Oliver J., (1988) La gestion du bassin versant expérimental de Draix pour la mesure de l'érosion In: *Sediment Budgets* (ed. by M. P. Bordas & D. E. Walling), 251–258. IAHS Publ. 174. IAHS Press, Wallingford, UK.

Pardé, M. (1941) La formidable crue d'octobre 1940. *Rev. Géogr. Pyr. et Sud-Ouest* **XII**(fasc. 3), 237–279.

Reid, L. M. & Dunne, T. (1996) *Rapid Evaluation of Sediment Budgets.* Catena Verlag, Reiskirchen, Germany.

Rigo, T. & Llasat, M. C. (2003) Flash floods and heavy rain events in Catalonia: analysis of the 1996–2000 period. In: *Hydrology of the Mediterranean a Semiarid Regions* (ed. by E. Servat, W. Najem, C. Leduc & A. Shakeel) (Proc. Montpellier Symp. April 2003), 269–275. IAHS Publ. 278. IAHS Press, Wallingford, UK.

Serrat, P. (1999) Dynamique sédimentaire actuelle d'un système fluvial méditerranéen: l'Agly (France). *C.R. Acad. Sci. Paris* **329**(série II), 189–196.

Serrat, P. (2001) Du risque naturel à la catastrophe: la crue-éclair de la Grave à Estagel, le 12 novembre 1999. In: *Au chevet d'une catastrophe: les inondations des 12 et 13 novembre 1999 dans le Sud de la France* (ed. by Medi-Terra), 95–110. Presses University Perpignan, Perpignan, France.

Serrat, P. (2003) L'Agly, fleuve côtier des Pyrénées orientales: évolution sur le long terme et dynamique actuelle. *La Houille Blanche* **5**(Paris), 137–146.

Serrat, P. & Depraetere, C. (1997) Incidences de facteurs géomorphologiques dans le fonctionnement d'un bassin en situation de crue: l'exemple de l'Agly (Pyrénées orientales). *Géomorphologie* **1**, *SEDES*, Paris, 73–88.

Snishchenko, B. F. (1992) Multipurpose studies of erosion and sedimentation in the Upper Ob basin. In: *Erosion and Sediment Transport Monitoring Programmes in River Basins* (ed. by J. Bogen, D. E. Walling & T. Day), 405–411. IAHS Publ. 210. IAHS Press, Wallingford, UK.

Drainage basin controls on geochemical heterogeneity of modern stream sediments in the Guadalhorce basin (Spain)

E. N. ASEYEVA[1], N. S. KASIMOV[1], S. B. KROONENBERG[2] & G. J. WELTJE[2]

1 *Faculty of Geography, Moscow State University, Vorob'evi Gory, Moscow 119892, Russia*
aseyeva@mail.ru

2 *Department of Applied Earth Sciences, Delft University of Technology, Mijnbouwstraat 120, PO Box 5028, NL-2600GA, Delft, The Netherlands*

Abstract Geochemical data on modern stream sediments in the Guadalhorce basin (Province of Malaga, Spain) were used to explore the role of parent lithology and drainage basin size as well as grain-size differentiation. Sand-sized sediments (189 samples) with median grain sizes ranging from 2 to 0.125 mm were collected at 136 locations along different river segments. These segments represent streams of various order draining areas with different rock types. For the individual sampling site the area of influence, i.e. the sample catchment, was defined. The catchments were classified on the basis of quantitative parameters: catchment size and the areal proportions of parent rock sources. Geochemical data were described in terms of the relative abundances of major elements (SiO_2, TiO_2, Fe_2O_3, Al_2O_3, MnO, MgO, CaO, Na_2O, K_2O, P_2O_5) in a bulk sample measured with X-ray fluorescence spectroscopy after pre-treatment of the samples, including loss on ignition (LOI) determination. The compositional data were used as the input for end-member modelling which allowed an objective unravelling of the distinct sediment populations.

Key words catchments; chemical composition; classification; end-member modelling; Guadalhorce basin; stream sediments

INTRODUCTION

Chemical composition of stream sediments within a single river basin usually shows a high degree of spatial variations. These variations reflect several factors, such as: (a) provenance; (b) fluvial transport and deposition; (c) post-depositional processes (Kroonenberg, 1992; Johnsson, 1993). There is a long history of research that uses stream sediments as the principle sampling medium to characterize the geology of catchments, to isolate areas with atypical geochemistry and to relate these anomalies either to natural sources or to man's environmental activities (Rose *et al.*, 1979; Förstner, 1989). For better interpretation of sediment geochemical survey results the quantitative consideration of all the factors which influence the spatial variability of sediment is required (Bonham-Carter *et al.*, 1987). This study was undertaken to investigate the relative importance of such factors as basin lithology, basin size and grain-size differentiation in determining sediment geochemistry in a heterolithic river basin. The study area is the Guadalhorce basin, located in the south of the Iberian peninsula (Province of Malaga, Spain). Recent alluvial sands of different granulometry were used as the main sampling medium. Relative abundances of 11 macroelements (analysed with XRFS) in bulk samples were regarded as geochemical signals from drainage area and fluvial processes and treated as multivariate geochemical measurements. The sources as well as spatial patterns of geochemical variability were detected and described using end-member modelling. The modelling results were integrated with the bedrock geology and basin-size data to observe the relationship between source area and composition of the sediments.

STUDY AREA AND SOURCE ROCKS

The Guadalhorce River is a Mediterranean fluvial system of the 6th order. Its drainage basin covers approximately 2600 km^2 and is located in the western part of the Betic Cordilleras, a mountain chain belonging to the actively uplifting Alpine-Mediterranean Fold Belt. The Betic Cordilleras is traditionally subdivided into the Internal zone, most closely to the coast, and the External zone (Weijermars, 1991). The Guadalhorce basin intersects both the structural units of the Betic orogen (Fig.1). The upper Guadalhorce basin (891 km^2) and the catchment area of its right tributary, the Turon River (224 km^2), are situated in the Subbeticum, which is the allochtonous subdivision of the External zone. It is composed by non-metamorphic carbonate rocks of essentially Mesozoic and Tertiary age (limestones and marls partly gypsiferous). The lower Guadalhorce basin is located in the Betic zone where the bedrock is represented mainly by Triassic and older low-grade metamorphic and metasedimentary rocks (phyllite, crystalline schist and shale), Lower Liassic dolomites and limestones, as well as plutonic (ultramafic) masses. The average differences in major element concentrations can be grouped into five broad units (Table 1) according to the lithologies occurring in the Guadalhorce basin.

SAMPLING AND CATCHMENT ANALYSIS

During the low-water period a total 186 sediment samples with median grain size ranging from 2 to 0.125 mm were collected from active channels and the lower flood plain in the

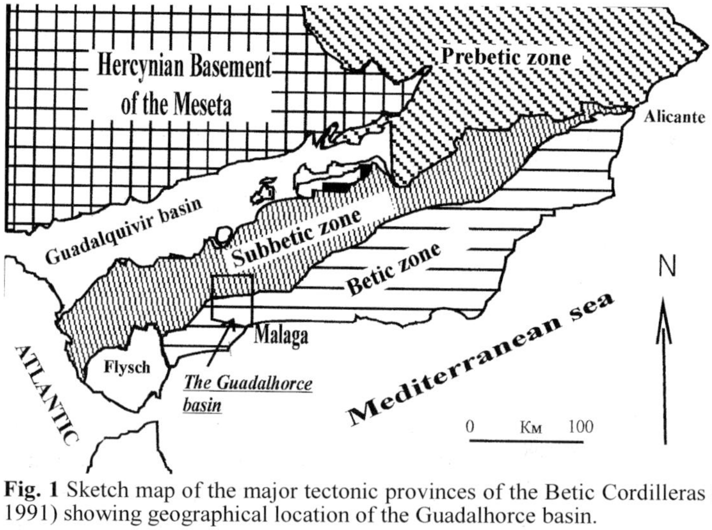

Fig. 1 Sketch map of the major tectonic provinces of the Betic Cordilleras (after Weijermars, 1991) showing geographical location of the Guadalhorce basin.

Table 1 Lithogeochemical types of rock sources in the Guadalhorce basin.

Lithogeochemical units	Typical macroelements	Rocks in the Guadalhorce basin	% total area
Calcareous units	Ca (Mg)	Limestones, dolomites, argillaceous limestones, marbles, limestones and dolomites	13.4
Calcareous and argillaceous	Ca, Al (Mg)	Marls and clays of flysch formations argillaceous limestones	17.8
Gypsiferous and marly	Ca, S, Al, Na, K	Transitional and shallow water facies (Triassic), dominated by speckled clays in association with evaporites and dolomites	12.1
Arenaceous siliciclastic	Si, K, Na	Sands, conglomerates, sandstones dominating in terrigenous clastic formations. Quaternary alluvial, colluvial and alluvial fan deposits.	32.9
Schistose	Al, Fe, Ti	Gneiss, phyllite, crystalline schist and shale	20.8
Ultramafic	Mg, Fe, (Ti, Ca)	Peridotite, pyroxenite and serpentinite	3.0

Guadalhorce basin. The sampling was performed at 136 locations. The choice of the locations was related to the scale of spatial variations in rock types and the size of the corresponding drainage areas. For each sample point the area of influence, i.e. the sample catchment, was defined and characterized in terms of its size and parent lithology. Stream order indices were used as a proxy of drainage basin size and fluvial transport distance. The indices were given according to Rzhanitsin (1960). The rock parentage was characterized using the rock groupings described in Table 1. According to the areal proportion of these units all subcatchments were assigned to three broad categories of terrain: monolithic catchments, differentiated catchments and compound heterolithic catchments. The monolithic catchments are areas composed of one type of lithology. The geochemically differentiated catchments are represented by areas with two or more parent rock types, one of which occupies more than 50% and less than 90% of the total area. Compound heterolithic catchments are terrains with several lithogeochemical units, each occupying less than 50% of the total area. Chemical analysis of the sediments representing all the studied sub-catchments was performed with a Philips PW1410 X-ray fluorescence spectrometer in the laboratory of the Department of Soil Science and Geology at Wageningen Agricultural University (The Netherlands). The analysis was carried out after the pre-treatment of the samples and loss on ignition (LOI) determination. The XRFS suit of elements used in this study included SiO_2, TiO_2, Fe_2O_3, Al_2O_3, MnO, MgO, CaO, Na_2O, K_2O and P_2O_5.

END-MEMBER MODELLING

Context

The present study was built upon the assumption that compositional variation among recent alluvial sands in relatively small river basins of mountainous regions originates mainly from the physical mixing of heterogeneous erosional material, derived from different rock sources. In this case spatial heterogeneity of sediments can be conveniently described in terms of a linear mixing model. Data sets which conform to this model can be expressed as mixtures of a limited set of end members (EMs). In natural systems the exact nature of the mixing process is often unknown. For such cases the mixing parameters are estimated by means of inverse modelling techniques. In this study we used the inversion "unmixing" algorithm proposed by Weltje (Weltje, 1994; Prins & Weltje, 1999) to estimate the number of EMs (the first modelling stage), the EM compositions and the EM proportional contributions to each observation (the second modelling stage). The choice of q, the number of linearly independent EMs in the mixing model, is a crucial step in the unmixing procedure. Different measures can be used as guidelines for the best choice of q. In this approach a minimum number of EMs which allows one to "explain" the observed compositional variations is chosen on the basis of the goodness-of-fit statistics by calculating the mean determination coefficient across the variables. After the "best" solution has been chosen, the bilinear problem of estimating of fully non-negative mixing proportion matrix and "conservative" end-member compositions is solved. The algorithm has been tested on two different types of compositional data: on petrographic data of modern beach sands (Weltje, 1994) and on the grain-size distribution data of terrigenious deep-sea sediments (Prins & Weltje, 1999). In this study the modelling is performed on geochemical data.

Interpretation of the modelled end-members in terms of their chemical composition

The results of EM modelling revealed that the compositional variations within three grain-size groups of sediments in the Guadalhorce and Turon basins are adequately described as mixing of four EMs. Silica and aluminium, the major constituents of aluminosilicates, were two dominant components in the first EM of all grain-size groups (Table 2). The concentrations of Fe, Ti, K, Na as well as of Mg were in agreement with the abundance of these elements in sedimentary rocks that are dominated by aluminosilicate minerals—feldspars, micas and clays. Therefore the first EM was interpreted as an *aluminosilicate component* of alluvial sands. The major compositional variations of the aluminosilicate EM across the grain-size spectrum are related to SiO_2 abundance, which is enhanced in the finer sand fractions. The increase of SiO_2, which apparently can be induced by the admixture of quartz, causes the "dilution" effect—the relative decrease of concentrations of other elements except Na, also suggest the minor feldspar enrichment in sediments of coarse and medium grain size and in the fine fraction.

The major chemical constituents of the second EMs for all grain-size groups of sandy sediments are CaO and the LOI. Other elements, including Mg, are not abundant. These EMs can be interpreted as the *calcareous component* of alluvial sands. The calcareous phase is obviously dominated by carbonate minerals. Across the grain-size spectrum, the calcareous EMs of the finer sediments have higher concentrations of the elements associated with aluminosilicate minerals: Si, Al, and Fe. The increase of SiO_2 up to 10% for the group of coarse and medium alluvial sands suggests that the calcareous EM includes not only a minor amount of aluminosilicate minerals, but also quartz (Table 2).

The third EM in all groups of the sediments is characterized by a very high amount of Mg, relatively high concentrations of Si and Fe, and a high LOI. These EMs were interpreted as the *magnesian component* of alluvial sands. Low Al concentrations, especially in the group of very coarse sands, indicate that the principle carriers of MgO are silicates and/or

Table 2 Modelled end-members of recent alluvial sands in the Guadalhorce and Turon basins.

Grain-size groups of alluvium	Chemical composition, %	End member 1 (aluminosilicate)	End member 2 (calcareous)	End member 3 (magnesian)	End member 4 (quartzose)
Very coarse sand (2–1 mm)	SiO_2	55.92	0.00	41.91	86.23
	TiO_2	1.09	0.28	0.24	0.13
	Al_2O_3	21.26	0.75	3.61	3.13
	Fe_2O_3	8.34	0.15	8.10	1.75
	MnO	0.33	0.33	0.33	0.33
	MgO	2.01	1.41	33.64	1.31
	CaO	0.00	48.69	0.19	4.71
	Na_2O	0.78	0.15	0.32	0.56
	K_2O	3.44	0.65	0.24	0.59
	P_2O_5	0.33	0.33	0.33	0.33
	LOI	6.49	47.26	11.08	0.91
Coarse and medium sand (1–0.250 mm)	SiO_2	62.76	10.43	44.84	86.07
	TiO_2	0.98	0.19	0.12	0.13
	Al_2O_3	17.94	2.71	5.17	1.39
	Fe_2O_3	7.04	1.75	7.95	0.81
	MnO	0.16	0.19	0.16	0.18
	MgO	1.69	0.77	30.41	1.65
	CaO	0.00	44.00	1.12	5.24
	Na_2O	0.95	0.00	0.31	0.22
	K_2O	3.39	0.64	0.39	0.00
	P_2O_5	0.17	0.17	0.17	0.17
	LOI	4.92	39.14	9.37	4.13
Fine sand (0.250–0.125 mm)	SiO_2	66.27	3.11	52.79	83.64
	TiO_2	1.00	0.14	0.35	0.22
	Al_2O_3	18.29	1.85	8.66	2.58
	Fe_2O_3	7.00	1.11	6.79	1.58
	MnO	0.20	0.20	0.20	0.20
	MgO	0.61	1.30	20.28	1.93
	CaO	0.00	48.77	0.43	3.89
	Na_2O	0.84	0.02	0.78	0.15
	K_2O	2.72	0.28	1.52	0.72
	P_2O_5	0.20	0.20	0.20	0.20
	LOI	2.86	43.01	8.00	4.89

Mg-rich carbonates. Across the grain-size spectrum of the magnesian EMs we observe the decrease of Mg for the sediments with finer granulometry. In parallel we found the decrease of Fe and the LOI and higher concentrations of elements of aluminosilicate association—Si, Al, K and Na—in the fine sediments. This trend can be explained both by instability of magnesium silicates in the supergene zone and/or the enriching of the finer EM sediments with aluminosilicate material.

The major chemical parameter of the fourth EM for all grain size groups is silica. In natural environments very high SiO_2 concentrations, coupled with very low contribution of other elements, are the main geochemical feature of sediments derived from quartzose rock sources. Since the concentrations of SiO_2 in the fourth EMs agrees well with the chemical data on average quartz arenites, the fourth EMs were interpreted as the *quartzose component* of the alluvium. The relative increase of Al and Fe may correspond to the presence of aluminosilicate minerals. The minor enrichment with Mg, Ca, and an increase of LOI in the finer EM sediments can be caused by the interference of fine calcareous material.

SPATIAL ANALYSIS OF GEOCHEMICAL SUBPOPULATION OF ALLUVIAL SEDIMENTS

Spatial changes in geochemistry of sediments are expressed as changes in relative abundances of the EMs. The analysis of the relative EM contributions to the sediments in the studied basins combined with the quantitative consideration of the corresponding source terrains (catchment lithology and size) revealed that the aluminosilicate, calcareous, and magnesian EMs have mainly detrital origin and can be attributed to the erosion of specific rock sources. The abundance of quartzose EM is dominantly controlled by the distance of fluvial transport (stream order), which is accompanied by maturation of sands and mixing of sediments with material from older alluvial sequences (Fig. 2).

The relative abundance of the aluminosilicate EM is dominantly influenced by the erosion of schistose lithology (schists, gneiss, slates, phillites, shales). The calcareous EM is matched with carbonate-rich sources such as limestones. The magnesian EM is mainly derived from ultramafic lithology. The sands produced by these rock types in monolithic catchments represent non-mixtures and form the distinct sediment subpopulations within the Guadalhorce basin. The proportion of the quartzose component in non-mixtures reflects the mineralogical composition of the eroding rocks: the lowest levels are typical for sediments in calcareous or ultramafic monolithic catchments.

In differentiated and compound catchments the sediment geochemistry responds differently to the catchment lithology depending on the catchment scale. Relatively small catchments might generate monocomponent sediments. The main controlling factor in such cases is the local supply from the nearest lithology through bank and channel erosion. In the 4th

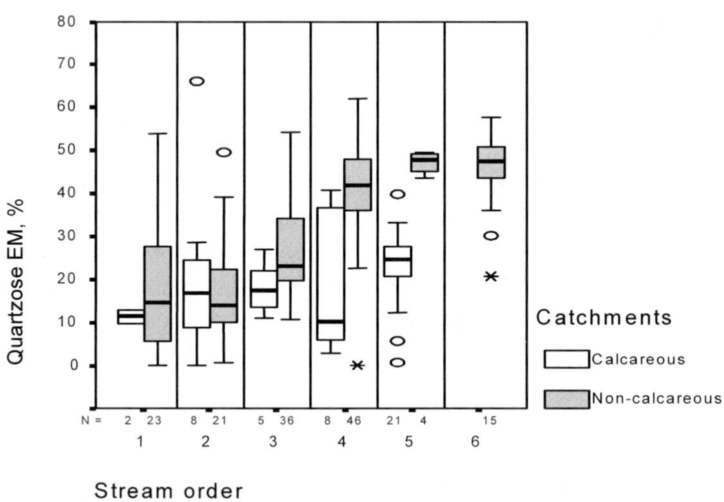

Fig. 2 Box plots of quartzose EM abundances in alluvial sands subdivided by dominant catchment lithology. Median, 75% and 25% quartiles, maximum and minimum values as well as outliers (open circles) and extremes (stars) are indicated.

order streams and larger rivers mixing and deposition processes dominate over erosion. The sands produced in such catchments show a wide spectrum of EM abundances. Major control on the sediment geochemical subpopulation appears to be rock assemblage and areas of different bedrock types and also the capacity of a given rock source to produce sand with respect to the others. To understand the relationship between the three EMs considered (aluminosilicate, magnesian and calcareous) and the proportions of the corresponding source lithologies, data from both sources and sands were analysed together. The analysis revealed that calcareous EM is poorly represented compared with the source lithology. In contrast, the aluminosilicate component is always over-represented in sediments. Schistose lithology seems to have the highest rates in generating sand. The sediments of the lower course of the Guadalhorce trunk river, intersecting the Internal zone, are largely ternary mixtures with a dominant influence of aluminosilicate EM. In the northern reaches of the Guadalhorce trunk river with sedimentary rock assemblages in the source terrains, the alluvial sands are dominantly composed of a mixture of aluminosilicate and calcareous EMs.

CONCLUSIONS

Major proportions of the geochemical variation in stream sediments (sands) in the studied basin can be explained on the basis of mixing of four EMs. The compositional resemblance of the modelled EMs to the geochemistry of some sedimentary rocks allowed interpretation of their dominant mineralogical assemblages as aluminosilicate, calcareous, magnesian and quartzose components of the alluvium. Spatial analysis matched the EMs either to the erosion of specific lithologies (aluminosilicate, calcareous, magnesian EMs) or to fluvial transport distance (quartzose EM).

Due to the strong dependence on lithological and fluvial factors the sands produced in different sub-catchment areas of the studied heterolithic basin show a wide spectrum of EM abundances. The relationships between source lithologies and sand composition found for the Guadalhorce basin confirm major inferences made in petrological studies by Palomares & Arribas (1993) and Arribas *et al.* (2000).

The EM modelling also revealed that grain-size differentiation seems to be a less important factor for geochemical heterogeneity of alluvial sands in the Guadalhorce basin: the EMs across different grain-size groups are geochemically similar. The minor variations shown by similar EMs might indicate the systematic grain-size changes in mineralogy that occur during weathering and fluvial transport, or suggest the difference in lithological parentage for the sediments of different granulometry.

The results presented above show that EM modelling can be a valuable tool both for unravelling complex patterns of variations in chemical composition of sediments and for evaluating the controls on sediment geochemistry.

REFERENCES

Arribas, J.,Critelli, S., Le Pera, E. & Tortosa, A. (2000) Composition of modern sand derived from a mixture of sedimentary and metamorphic source rocks (Henares River, Central Spain). *Sedim. Geol.* **133**, 27–48.

Bonham-Carter, G. F., Rogers, P. J. & Ellwood, D. J. (1987) Catchment basin analysis applied to surficial geochemical data, Cobequid Highlands, Nova Scotia. *J. Goechem. Explor.* **29**, 259–278.

Förstner, U. (1989) *Contaminated Sediments*. Lecture notes in Earth Sciences 21. Springer Verlag, New York, USA.

Johnsson, M. J. (1993) The system controlling the composition of clastic sediments. In: *Processes Controlling the Composition of Clastic Sediments* (ed. by M. J. Johnsson & A. Basu), 1–19. Geological Society of America Special Paper, 284. Boulder, Colorado, USA.

Kroonenberg, S. B. (1994) Effect of source rock, sorting and weathering on the geochemistry of fluvial sands from different tectonic and climatic environments. In: *Proc. 29th Int. Geol. Congr, Part A* (ed. by F. Kumon & K. M. Yu), 69–81. VHS Publishers, Utrecht, The Netherlands.

Palomares, M. & Arribas, J. (1993) Modern stream sands from compound crystalline sources: Composition and sand generation index. In: *Processes Controlling the Composition of Clastic Sediments* (ed. by M. J. Johnsson & A. Basu), 313–322. Geological Society of America Special Paper, 284. Boulder, Colorado, USA.

Prins, M. A. & Weltje G. J. (1999) End-member modelling of siliciclastic grain-size distributions. The Late Quaternary record of eolian and fluvial sediment supply to the Arabian Sea and its paleoclimatic significance. In: *Numerical Experiments in Stratigraphy: Recent Advances in Stratigraphic and Sedimentologic Computer Simulations.* SEPM Special Publications no. **63**, 91–111.

Rzhanitsin, N. A. (1960) *Morphologicheskije i hydrologicheskije zakonomernosti strojenija rechnoi seti (Morphometrical and hydrological regularities river network structure).* Hydrometeoizdat, Leningrad, USSR (in Russian).

Rose, A. W., Hawekes, H .E. & Webb, J. S. (1979) *Geochemistry in Mineral Exploration.* Academic Press, London, UK.

Weijermars, R. (1991) Geology and tectonics of the Betic zone, SE Spain. *Earth-Sci. Rev.* **31**, 153–236.

Weltje, G. J. (1994) Provenance and dispersal of sand-sized sediments: Reconstruction of dispersal patterns and sources of sand-sized sediments by means of inverse modelling techniques. *Geologica Ultraiectina* **121**, 47–120.

3 Large River System Functioning

The relationship between sediment yield and drainage basin area

ALEXEI DEDKOV

Department of Geography and Geoecology, Kazan State University, Kreml'evskaya Str. 18, 420008 Kazan, Russia
dedkov@mail.ru

Abstract Hydrological data for small and intermediate sized rivers in the plains and mountains of the temperate belts of Eurasia were analysed. Total suspended sediment yield increases downstream faster than the basin area for rivers with undisturbed or slightly disturbed basins. Rivers with intensively cultivated basins are also characterized by an increase in total suspended sediment yield downstream, but this increase is slower than the increase in basin area and the specific suspended sediment yield therefore decreases downstream. Deposition very often prevails over erosion in the lower reaches of these rivers. Both relationships are not distinctly expressed for river basins with intermediate proportions of cultivated area and for large rivers.

Key words agricultural activity; basin area; downstream trend; human activity river; suspended sediment yield

INTRODUCTION

Suspended sediment yield data are widely used for evaluating the effects of various natural and anthropogenic factors on the intensity of erosion processes. However, sediment yield is also influenced by basin area. Makkaveev (1955) suggested that this dependence is the main reason why accurate estimates of erosion rates cannot be derived from information on specific sediment yield. As a result, researchers frequently use sediment yield data for only a restricted range of basin sizes. However, this reduces the size of the database. Researchers may therefore find it necessary to deal with comparatively large ranges of basin area within which the effects mentioned above cannot be completely excluded.

The question arises as to whether there is a general relationship between total and specific sediment yield and basin area and what this relationship is. An inverse relationship between specific suspended sediment yield and drainage basin area has been generally reported in the literature (Karaushev, 1977; Schumm, 1977). This trend indicates that the total suspended sediment yield increases downstream, but not as fast as the basin area increases. As a result, specific suspended sediment yield decreases as basin area increases. A positive relationship between specific sediment yield and drainage basin area observed for some rivers was usually considered to be a deviation from the general trend (Karaushev, 1977; Owens & Slaymaker, 1992). It is clear that analysis of the above mentioned trends could provide an improved understanding of the characteristic features of erosion and deposition in river basins.

DATA

A database of suspended sediment yield and controlling factors for 4200 hydrologic stations located on rivers all over the world was used. Attention focused on the data from rivers in the plains and low mountain areas of the Eurasian temperate belt, with basin areas up to 250 000 km^2. (This provided data from monitoring stations in 352 river basins). Some of the results of the analysis have been published previously (Dedkov & Mozzherin, 1984, 1992, 1996.)

The river basins were split into three categories according to the area of arable land, viz:

(a) uncultivated basins or basins with limited cultivation (tillage area <30%, forest area >70%);
(b) basins with intermediate amounts of cultivation (tillage and forest 30–70%);
(c) intensively cultivated basins (tillage area >70%, forest area <30%).

Data on basin areas (S, km^2), water discharge (Q, l s^{-1}), total suspended sediment yield (R, t year^{-1}), specific suspended sediment yield (r, t km^{-2} year^{-1}) were used for the analysis.

Some new information was also obtained specifically for the analysis. Firstly, *differential specific suspended sediment yield* (DSSSY), characterizing the basin area between two monitoring stations on the same river, was determined using the following equation:

$$r_d = (R_1 - R_2) / (S_1 - S_2) \text{ t km}^{-2} \text{ year}^{-1} \quad (1)$$

where index 1 and index 2 refer to the two monitoring stations, with 1 downstream from 2 and r_d is DSSSY.

It is clear that DSSSY reacts to changes in suspended sediment yield, and, consequently, changes in the relationship between erosion and accumulation, faster than the specific suspended sediment yield itself (*the integral specific suspended sediment yield*, characterizes erosion on the whole basin area upstream from a given hydrologic station) (Fig. 1). Increasing values of DSSSY indicate that erosion increases, and reducing values show that sedimentation prevails.

In addition, several coefficients, which characterize the increase downstream of the following parameters were also calculated: basin area (S_n / S_m), water discharge (Q_n / Q_m), a total suspended sediment yield (R_n / R_m), specific suspended sediment yield ($r_n - r_m$), where index n is a hydrologic station located downstream from the other hydrologic station (index m).

RESULTS

The results of the analysis demonstrate that the relationship between specific suspended sediment yield and drainage basin area in the forest and forest-steppe zones of the temperate belt of Eurasia is different for basins with a relatively undisturbed landscape and basins which are part-cultivated.

Rivers with uncultivated basins or basins with limited cultivation (category a)

These rivers are characterized by low suspended sediment yields. The specific yield in plain areas seldom exceeds 20–30 t km^{-2} year^{-1}, and in low mountains it increases to 50–70 t km^{-2} year^{-1}. Where there is a dense vegetation cover, surface (sheet and gully) erosion is limited, and the main sediment supply is provided by erosion of river channel and banks.

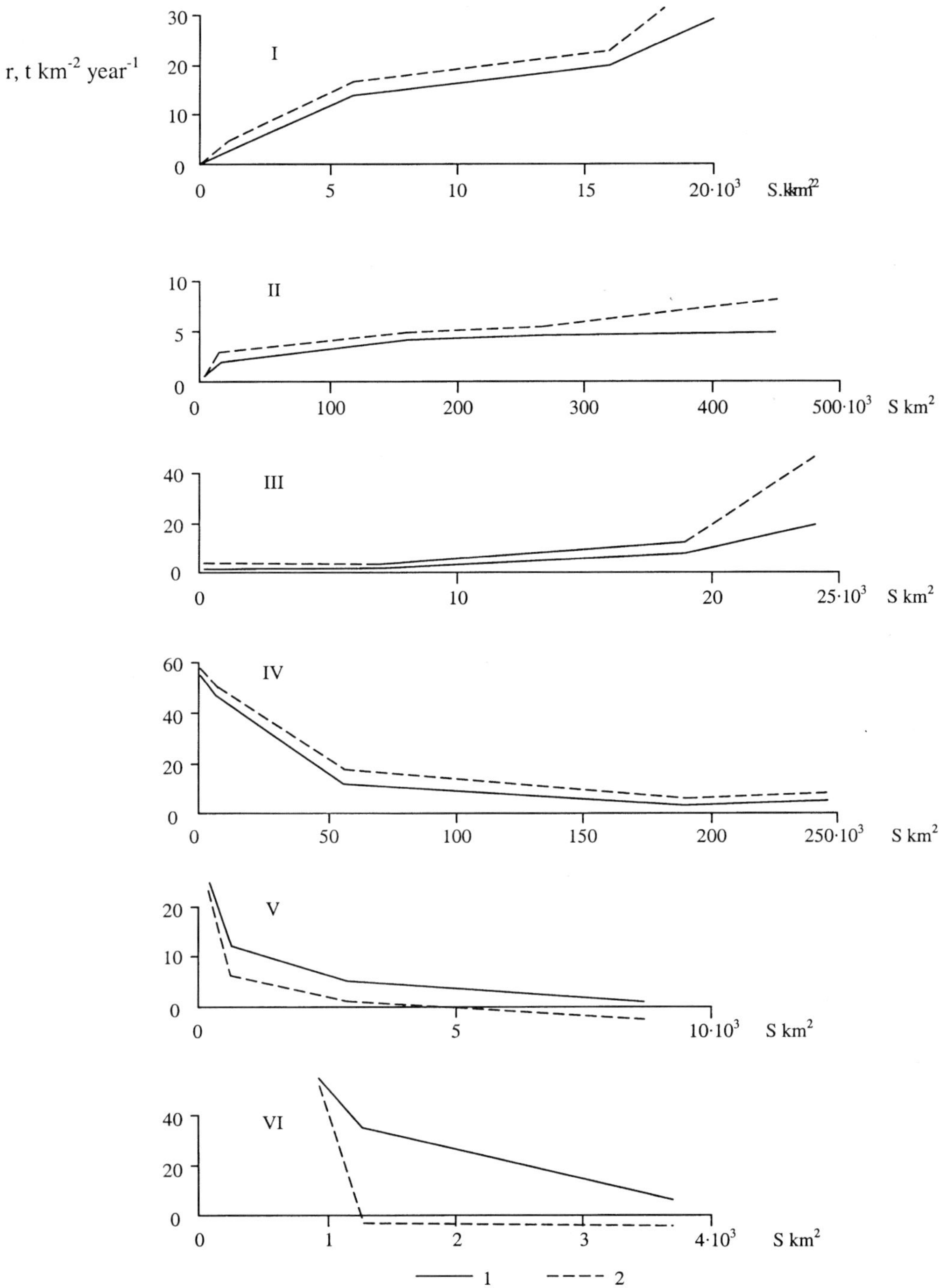

Fig. 1 The relationship between specific suspended sediment yield (*r*) and drainage basin area (*S*, 10^3 km^2) for rivers with basins with a limited cultivation area (I, II, III) and intensively cultivated basins (IV, V, VI). (1), integral specific sediment yield; (2), differential specific sediment yield (DSSY). Rivers: I, Chusovaya; II, Nizhnyaya Tunguska; III, Vym'; IV, Oka; V, Tersa; VI, Kalmius.

Water discharge commonly increases downstream approximately proportionally to the increase in catchment area, though more often the rate of increase in the former is less than the increase in the latter ($Q_n / Q_m \leq S_n / S_m$). Total sediment yield increases even more rapidly downstream, due to the approximately cubic relationship between sediment discharge and water discharge ($Q_s = Q_w^n$, where $n = 2.6$ for lowland rivers and $n = 3$ for mountain rivers) (Makkaveev, 1955). The increase in total sediment yield for a given river always exceeds the increase in basin area ($R_n / R_m > S_n / S_m$) (Fig. 2). Therefore the values of specific suspended sediment yield (both integral and to an even greater extent, differential) increase regularly downstream (Fig. 1). This positive relationship between specific suspended sediment yield and drainage basin area (Fig. 3(a)) is typical for uncultivated basins or basins with limited cultivation.

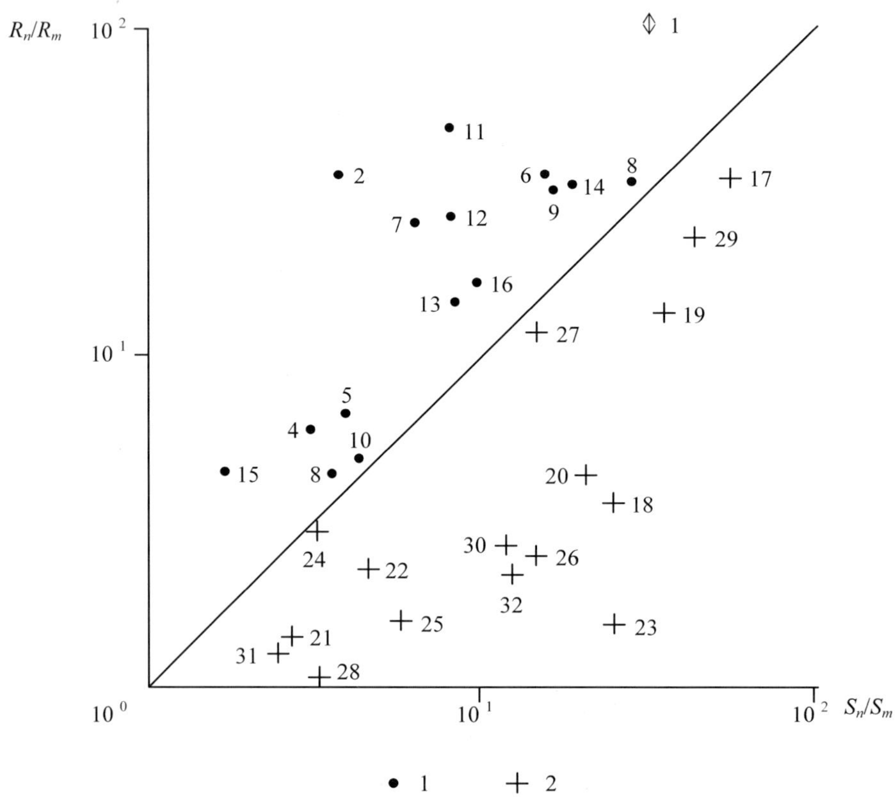

Fig. 2. The relationship between sediment yield increase (R_n / R_m) and basin area increase (S_n / S_m) for sparsely cultivated (1) and intensively cultivated (2) river basins. Rivers: Category a: 1, Chusovaya; 2, Wym'; 3, Soswa; 4, Nizhnyaya Tunguska; 5, Ussuri; 6, Dirysa; 7, Om'; 8, Belaya; 9, Malka; 10, Bureja; 11, Wasygan; 12, Kan; 13, Witim; 14, Aldan; 15, Zeja; 16, Ai; 17, Upper Lena; 18, Indigirka; 19, Kamchatka; 20, Tom'; Category c: 21, Hoper; 22, Ilovlja; 23, Svijaga; 24, Samara; 25, Sal; 26, Eruslan; 27, Tersa; 28, Dema; 29, Kalaus; 30, Sheshma; 31, Tobol; 32, Ishim; 33, Mius; 34, Kalmius; 35, Medvediza; 36, Seversky Donez; 37, Oka; 38, Argun; 39, Osym; 40, Bolshoy Irgiz.

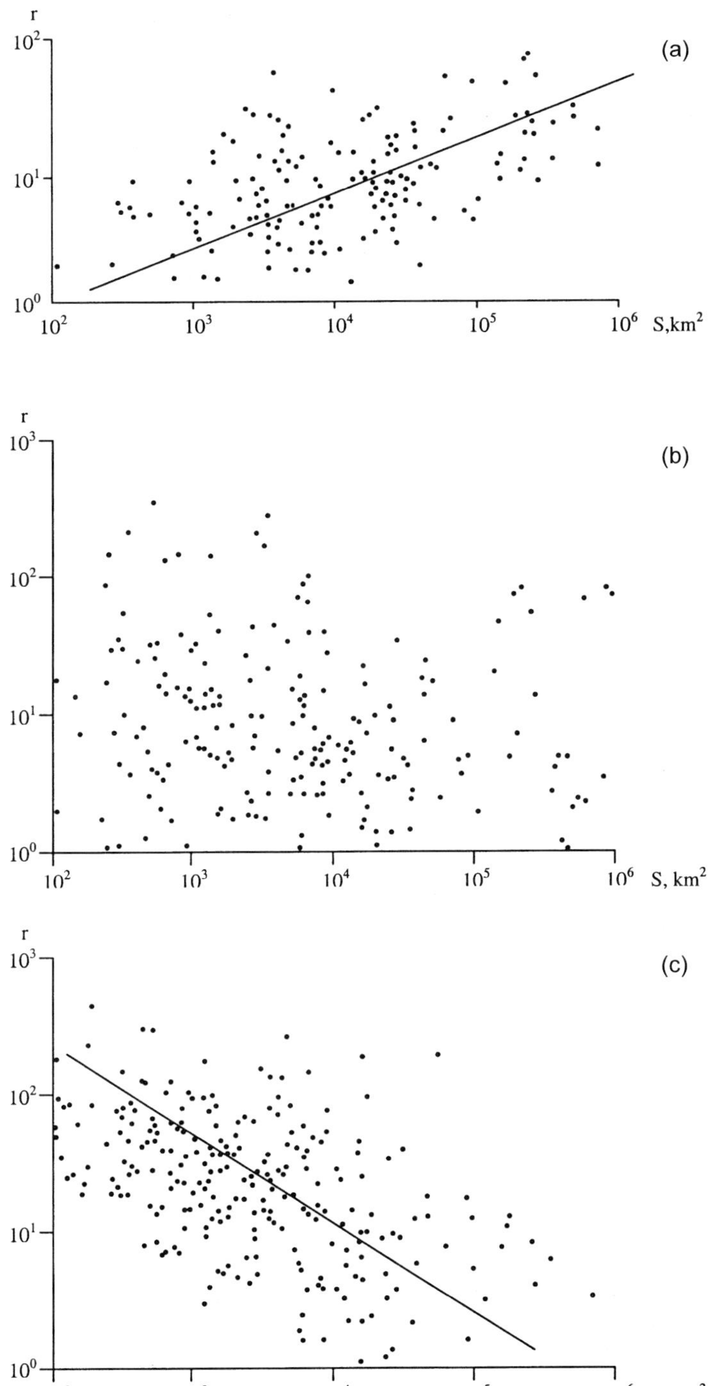

Fig. 3 Relationships between specific suspended sediment yield (r, t km^{-2} year^{-1}) and drainage basin area (S, km^2). (a) uncultivated or sparsely cultivated basins; (b) intermediate cultivated basins; (c) intensively cultivated basins.

Rivers with basins with intermediate proportions of cultivated land

Figure 3(b) shows the relationship between specific suspended sediment yield (integral) and drainage basin area for small- and medium-sized river basins in category b for the temperate belt of Eurasia. The significant dispersion of points and the low correlation coefficient can be explained by the high variability of the catchments with respect to their area, relief, geological structure, and the area of arable land *vs* forest land.

Because of the dominance of channel erosion in basins of category a, there is a relatively high proportion of bed load sediment, which constitutes about 20% of the suspended sediment load (Mozzherin, 1994).

Rivers with intensively cultivated basins (category c)

Rivers of this group are characterized by the largest specific suspended sediment yields for the plain rivers. The averages values of specific sediment yield are 60–100 t km^{-2} year^{-1}, and the maximum is almost an order of magnitude larger (Fig. 3(c)). Such high values of suspended sediment yield reflect intensive surface erosion on arable lands. Surface erosion is a key component of the suspended sediment yield. The proportion of bed load sediment reduces to 5–10% of the suspended sediment load, because sheet erosion primarily supplies the rivers with fine material.

Rivers with intensively cultivated basins are also characterized by an increase in total suspended sediment yield downstream. However, this increase occurs more slowly than the increase in basin area (R_n / R_m < S_n / S_m) (Fig. 2). The rivers are overloaded by sediment derived from the catchment surface and some of this sediment is depositional, causing the increase in sediment yield to slow downstream. Correspondingly, specific suspended sediment yield, both integral and even to a greater extent DSSSY, decrease downstream. Moreover, in the lower reaches of some rivers the total sediment yield reduces due to its deposition. In these regions, the DSSSY has negative values (Fig. 1,V and VI).

A negative relationship between specific sediment yield and drainage basin area is observed on the rivers of category c in periods with intensive snow melting and after high intensity rains, which erode the surface of the catchment. During low water periods, the input of sediment from the catchment surface is close to zero, and a positive relationship between specific sediment yield and drainage basin area exists.

Contemporary sedimentation on flood plains with deposits up to 2–2.5 m thick is widespread in the agricultural belt of Eurasia (Dedkov & Mozzherin, 1984; Owens & Slaymaker, 2000) and is the consequence of intensification of erosion of the catchment surface after slope cultivation. Radiocarbon dating of these sediments on the flood plains confirms that they are synchronous with periods of cultivation.

If arable lands are for some reason abandoned and become covered by forest, then the reverse process takes place. The rivers transport less sediment derived from the catchment surface and erosion of the river bed and banks increases, markedly promoting the formation of a new lower level flood plain. A similar phenomenon was described by Polish researchers in the upper reaches of the Visloka River (the right-hand tributary of the Vistula River) where between 1938 and 1990 the forest area increased from 30% to 67% on abandoned lands and, correspondingly, the area of arable land reduced (Lach & Wyzga, 2000).

It should be noted that almost for a quarter (23%) of the river basins of category c an inverse relationship between specific sediment yield and drainage basin area is not found, or is only faintly expressed. This can be explained by the existence of various hydrotechnical structures, and sharp changes in hydrological, geological and other conditions. However, the predominance of the negative relationship between specific sediment yield and drainage basin area is typical for cultivated regions of the temperate belt of Eurasia (Fig. 3(c)).

CONCLUSION

The negative relationship between specific sediment yield and drainage basin area which was, until recently, considered by many researchers to be a general global relationship for the temperate belt (without arid regions), is only characteristic of intensively cultivated basins. It was shown that this relationship was most clearly defined for temperate belt specific sediment yields. A positive relationship between specific sediment yield and drainage basin area is observed for uncultivated basins or basins with limited cultivation, which are characterized by the lowest suspended sediment yields.

The conclusions presented are gradually gaining recognition. For example, Walling & Webb (1996) wrote that the questioning of the traditional view, provided by Dedkov & Mozzherin, shows that the inverse relationship between specific sediment yield and basin area is to a large extent a reflection of human impact on the fluvial system, rather than a basic precept (Walling & Webb, 1996). They also point out that "the basic precept" probably results from the dominance of studies undertaken in the USA, which is characterized by strong anthropogenic impact. It can be suggested that in Russia also, the conclusion about the universal character of inverse relationships was also primarily based on studies undertaken in cultivated regions.

In this connection the study of erosion and sediment yield, carried out by Renwick (1996) is of interest. He showed that specific sediment yields increase, and the inverse dependence on basin area becomes more strongly expressed, the more intensively the region is cultivated. Thus, for the USA as well, changes in specific sediment yield downstream show different trends, depending on the relative areas of arable lands and forest land within the river basin.

REFERENCES

Dedkov, A. P. & Mozzherin, V. I. (1984) *Erosia I stok nanosov na Zemle (Erosion and Sediment Yield on the Earth)*. Izdatelstvo Kazanskogo Universiteta, Kazan, Russia (in Russian with English summary).

Dedkov, A. P. & Mozzherin, V. I. (1992) Erosion and sediment yield in mountain regions. In: *Erosion, Debris Flows and Environment in Mountain Regions* (ed. by D. E. Walling, T. R. Davies & B. Hasholt) (Proc. Chendy Symp., July 1992), 29–36. IAHS Publ. 209. IAHS Press, Wallingford, UK.

Dedkov, A. P. & Mozzherin, V. I. (1996) Erosion and sediment yield on the Earth. In: *Erosion and Sediment Yield: Global and Regional Pespectives* (ed. by D. E. Walling & B. W. Webb) (Proc. Exeter Symp., July 1996), 29–33. IAHS Publ. 236. IAHS Press, Wallingford, UK.

Lach, I. & Wizga, B. (2000) Projection of climate changes, variations in land use and urban development of the drainage basin. In: *Geomorphic Response to Land Use Changes* (Int. Symp.) Abstracts, 37. Bratislava, Slovakia.

Makkaveev, N. I. (1955) *Ruslo reki i erosia v eye basseine* (Channel of the River and Erosion in its Basin). Izdatelstvo Akademii Nauk SSSR, Moskva, Russia (in Russian).

Mozzherin, V. I. (1994) *Geomorphologichesky Analiz Tverdogo Rechnogo Stoka Gumidnikh Ravnin Umerennogo Poyasa* (Geomorphic Analysis of Sediment Yield of Humid Plains of Temperate Belt). St Petersburg, Russia (in Russian).

Owens, P. N. & Slaymaker, H. O. (1992) Late Holocene sediment yield in small alpine and subalpine drainage basins, British Columbia. In: *Erosion, Debris Flows and Environment in Mountain Regions* (ed. by D. E. Walling, T. R. Davies & B. Hasholt) (Proc. Chendy Symp., July 1992), 147–154. IAHS Publ. 209. IAHS Press, Wallingford, UK.

Renwick, W. H. (1996) Continental scale reservoir sedimentation patterns in the United States. In: *Erosion and Sediment Yield: Global and Regional Perspectives* (ed. by D. E. Walling & B. W. Webb) (Proc. Exeter Symp., July 1996), 513–522. IAHS Publ. 236. IAHS Press, Wallingford, UK.

Schumm, S. A. (1977) *The Fluvial System*. John Wiley and Sons, New York, USA.

Karaushev, A. V. (ed.) (1977) *Stok Nanosov, ego Izucheniye i Geographicheskoe Rasprostranenie (*Sediment Yields, their Study and Geographical Distribution). Hydrometeoizdat, Leningrad, Russia (in Russian).

Walling, D. E. & Webb, B. W. (1996) Erosion and sediment yield: a global overview. In: *Erosion and Sediment yield: Global and Regional Perspectives* (ed. by D. E. Walling & B. W. Webb) (Proc. Exeter Symp., July 1996), 3–19. IAHS Publ. 236. IAHS Press, Wallingford, UK.

Morphological expressions of river sediment transport and their role in channel processes

ROMAN S. CHALOV
The Laboratory of Soil Erosion and Fluvial Processes, Moscow State University, GSP-2, Vorob'evy Gory, Moscow 119992, Russia
river@river.geogr.msu.su

Abstract Relationships between erosion, sediment transport and deposition processes in river channels under different conditions of sediment yield formation and variable contribution of suspended (basin-derived) and bed (channel-originated) sediment are discussed. The role of the latter in total sediment yield and as a factor of morphological channel type differentiation is shown.

Key words channel deformations; channel processes; morphological type; runoff; sediment transport

INTRODUCTION

Sediment yield is the major driver of channel processes. The latter can also be considered as an expression of sediment yield (Makkaveev, 1955; Kondratiev *et al.* 1959). Fluvial erosion (mobilization of sediment particles from deposits or lithogenic substrate through their interaction with water flow to form the sediment load), sediment transport (movement of particles through the fluvial system) and sediment deposition (reverse transformation from one category to another) make up the complex of channel processes (Makkaveev & Chalov, 1986; Kondratiev *et al.*, 1982; Chalov, 1997). Sediment can originate from the channel itself if bed erosion occurs under the condition of virtual velocity V_{vir} excess over the non-eroding velocity V_n, i.e. $V_{vir} > V_n$, (Fig. 1). Sediment can also be basin-derived, if it is delivered to the channel from drainage basin sources. The latter mainly comprises suspended load, whereas the sediment of channel origin comprises bed load, although it may become suspended load if fine flood plain deposits, deposits in backwaters etc. are eroded. Furthermore, gullies, talus on mountain slopes, debris flows etc. provide a source of bed load in rivers.

Unlike water runoff, values of total sediment yield and contribution of its suspended and bed load components remain difficult to assess. Main reasons for that are insufficient number of monitoring points, short periods of observations, laboriousness of measurements, lack of reliable information on bed sediment transport and some others. The main aim of this paper is to consider relationships between sediment yield and forms of transport in river channels on one side and their expression in channel morphology and dynamics on the other, and to give geographical evaluation of these differences.

GEOGRAPHICAL VARIATION OF SEDIMENT YIELD AND RIVER CHANNEL MORPHOLOGY

Bed load and part of the suspended sediment load comprise the channel-forming sediment. Cessation of its transport results in the "prevailing content in the channel bed deposit" (Karasseff, 1975). However, the proportion of suspended load comprising the channel-

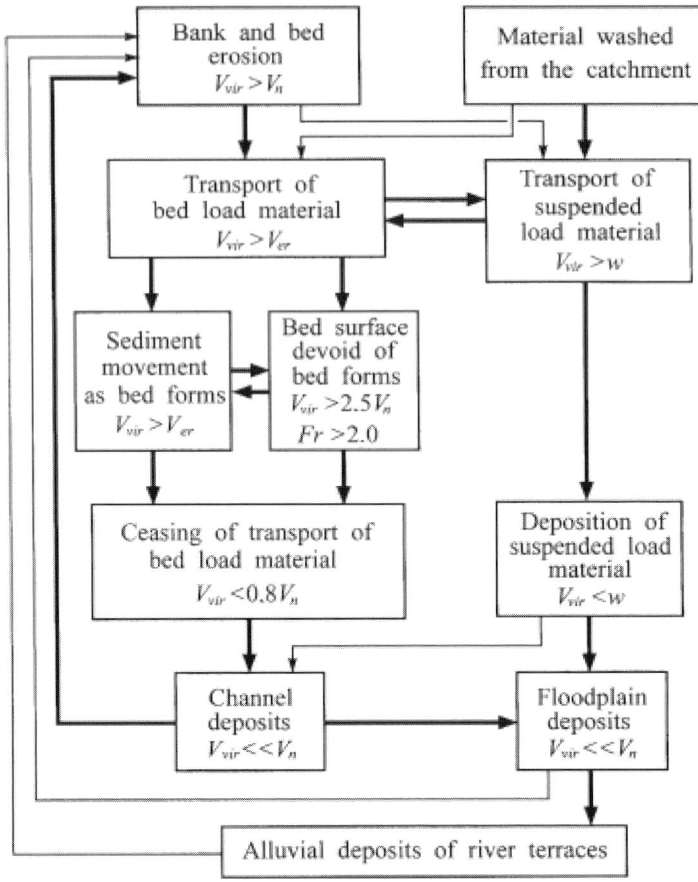

Fig. 1 Scheme of channel erosion interaction, transport and sediment accumulation in channel processes (arrow thickness defines primary or secondary role in material transport). V_{vir}, average virtual flow velocity; V_n, non-eroding velocity; V_e, eroding velocity; w, hydraulic size; Fr, Froude number.

forming sediment, as well as the ratio of suspended load to bed load yield W_S/W_B in the total sediment yield vary over a wide range of values. This range is controlled by the geographical laws of sediment yield formation, transport capacity R_{tr} conditions, and spatial–temporal changes of channel flow hydraulic characteristics (from axis to the peripheral channel parts, from high-water to low-water period etc.). The role of suspended sediment increases under the domination of finer material in the overall basin sediment yield. The coinciding tendency of increase is observed for total sediment yield W_{S+B} (Table 1), which increases from north to south and reaches its maximum in the rivers of southeastern Asia.

Suspended sediment-dominated rivers

The main sediment sources of the southern rivers are plateaus composed of loess or weathering crusts. As a result, basin-derived material dominates the sediment load, which is

Table 1 Annual sediment yield of several large rivers of Eurasia (based on Chalov *et al.*, 2000; Shtankova, 2002).

River-gauging station	Annual discharge Q, ($m^3 s^{-1}$)	Sediment load Suspended load W_R, (t × 10^6)	Bed load W_G, (t × 10^6)	Total W_{R+G}, (t × 10^6)	$\frac{W_G}{W_{R+G}}$
Vyatka–Kirov		0.50	0.54	1.04	52.0
Lena–Kusur	7150	22.66	17.45	40.11	43.5
Sev. Dvina–Abramkovo	1930	1.3	0.7	2.03	35.0
Oka–Murom		1.54	0.40	1.94	20.5
Belaya–Birsk		2.62	0.54	3.18	12.1
Ob–Kolpashevo	4260	16.65	2.89	19.54	14.8
Ob–Barnaul	1470	8.10	0.57	8.67	6.6
Yangtze–Yichang	14300	512	13.08	525.1	2.48
Yellow River–Mijin	1523	1320	19.2	1321.9	0.14
Vichegda–Fedyakovo	1010	1	0.54	1.54	0.35

in turn dominated by the suspended component. It provides the maximum loading of flow by sediments ($R_{tr} \leq W_{S+B}$). Changes of R_{tr} and the hydraulic characteristics of river flow are caused by the irregularity of the temporal pattern of runoff. When river discharge decreases, a portion of the suspended load can be transformed into bed load. With $V_{vir} >> V_n$, bed load transport takes place without dune formation. Under such conditions bed sediment is deposited during the flood recession, forming continuous shoals. During low water periods its transport ceases completely, thus transforming bed sediment into channel deposits. Dried out shoals make high-level channels look like sandy fields extending along the rivers, among which the low-level channel forms meander (sediment bodies—point bars) or braids (sediment bodies—braid bars).

Subsequent re-entrainment of that sediment during the following high-water period causes wandering of channels, classified by the intensity of lateral deformation as absolutely unstable channels (the Yellow River, the Amu-Darya River): Lohtin index $L = d/I < 1.0$, where d is average sediment size and I is channel bed gradient (Chalov *et al.*, 2000). In turn, it causes channels to maintain relatively straight courses, since the numerous dried sediment bodies have no time to be fixed by vegetation and transformed into fragments of flood plain (meanders spurs, islands). This is illustrated by the plot of the changes in the position of the flow dynamic axis in the Yellow River (near Zhengzhou) during the high water period (Fig. 2). It shows that wandering occurs within a strip of more than 2 km wide, with an average rate of lateral channel shift from 25 to 144 m day^{-1}. These inter-annual dynamics are superimposed onto a background of longer-term changes (i.e. 3–5 years) within a strip up to 10.6 km wide (Chalov *et al.*, 2000). On the Amu-Darya River (Berkovich *et al.*, 1973) seasonal flow dynamic axis migration takes place within a strip up to 1.3 km wide with long-term position changes near the left or the right bank occurring every 7–10 years.

Condition $R_{tr} < W_{S+B}$ causes sediment deposition in the channel, i.e. preservation of sediment in the form of channel bed deposits. It results in aggradation of the river bed. For example, the Yellow River in its lower reaches aggrades at a rate of 4.5–7.7 cm $year^{-1}$ (Chalov *et al.*, 2000), the Amu Darya River—3.4 cm $year^{-1}$ (Zvetkova, 1963), and the Brahmaputra River—3.4 cm $year^{-1}$ (Gusarov, 2002). Such river channels are buffered by natural levees, and even low-water level exceeds the elevation of the adjacent land (on the Yellow River up to 7 m).

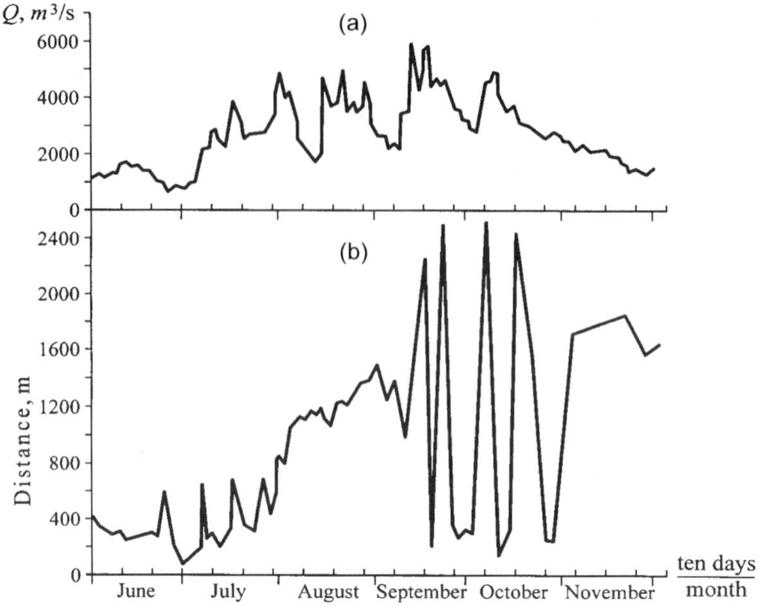

Fig. 2 Changes in the Yellow River flow dynamic axis through the high-water period (based on Chien, N. & Wan, Z., reproduced from Chalov *et al.* (2000)): (a) hydrograph; (b) flow dynamic axis position.

Bed load-dominated rivers

When coarser material (sand, gravel) is delivered into the channel, the basin-derived component of the sediment load decreases and bed load increases. As bed sediment availability increases, a hierarchy of bed forms is produced. Its structure depends on the flow discharge at different phases of the hydrological regime and position in the channel (axis or peripheral parts). During high floods, bed forms of all generations are actively transformed over the entire channel width. At low water macro- and medium-size bed forms (point bars and braid bars) stabilize, their higher parts emerge, and microforms (dunes) move only between the latter. The width of the belt of active sediment transport at low water may decrease up to two–three times.

Each dune is a depositional landform, in which sediment is temporarily stored, forming a stable nucleus. However, erosion on the dune stoss side leads to the entrainment of sediment, which accumulates again on the lee side. Active dunes as depositional channel bed forms are not necessarily evidence of the formation alluvial deposits. They may represent a temporary cessation of transport of sediment particles, which resumes with the process of dune movement as a whole. Nevertheless, each dune possesses a potential ability of transformation, partially or completely, from such seasonal deposits into longer-term alluvial deposits. It occurs through the influence of dunes on the kinematic structure of channel flow, causing erosion of channel banks opposite to drying dunes, vegetation development on dune surfaces and their transformation into flood plain segments without further movement. The stabilization of channel bed forms by vegetation is accompanied by deposition of suspended load material (with hydraulic size $w > V_{vir}$) on their surface and the initiation of thin flood plain deposits. Accumulation of suspended material is the dominant process on the flood plain surface. An increase in high-water suspended load (and, consequently, flow turbidity) causes the intensity of overbank deposition and the depth of the thin flood plain deposits to increase. As a result, the initial bends in the flow dynamic axis between point bars are transformed into channel meanders, and by flow divergence around braid bars into braided channels (point bars form meander spurs, braid bars—islands). Thus, a channel of a certain morphodynamic type is formed, where the further movement of bed forms determines lateral channel deformation (meander development, redistribution of discharge, deepening and shallowing of branches).

The major condition allowing such bed form transformation into channel form is the higher stability of sandy and gravelly alluvial channels (as a rule, Lokhtin index $L > 2$–3) in comparison with silty-sandy channels. Relatively low bed load transport is usually associated with straight channels, since there are no slowly moving macroforms of the channel bed to form initial planform irregularities. First order streams are usually characterized by relatively straight channels, since channel-flow interaction does not supply the stream with sufficient amounts of sediment to generate dunes. Relatively straight channels are also maintained by large rivers with unstable channels characterized by highly mobile bed forms. The latter have no time to be stabilized by vegetation to form new flood plain segments. This is also the case for rivers where suspended load material provides the channel-forming sediment (Chalov *et al.*, 1998). The largest bed forms (riffles) are characterized by long-term and seasonal changes: accumulation at high water (floods), erosion of the riffle crest at low water. However, the opposite situation occurs when a series of riffles/pools develops. Continuing movement of smaller-size bed forms during low water periods leads to an increase in the channel depth range and a decrease of riffle depth, causing shallowing at low water (Turykin, 1997).

Sandy and silt-sandy rivers are usually characterized by relatively balanced time-averaged values of load and transport capacity, i.e. $W_{S+B} \approx R_{tr}$. Usually R_{tr} slightly exceeds W_{S+B}. That is the main reason for the dominant tendency of slow (<1 mm year^{-1}) incision for many rivers. However, this can only be observed over historical or geological intervals of time. Possible reflections of this slow incision are steps in the flood plain topography as well as transformation of upper flood plain levels into unflooded terraces. The inverse situation (W_{S+B} slightly exceeds R_{tr}) is found in lower reaches and river mouths. As the excess of W_{S+B} over R_{tr} is usually low, rates of direct sedimentation are normally low and compensated by overbank accumulation on the flood plain. As a result there are no natural levees on the flood plain, but overlying flood plain segments (where younger deposits overlie older ones) or single-level flood plains can form (Makkaveev & Chalov, 1986).

The formation and movement of dunes in rivers with gravel-cobble bed load is very specific. Suspended load material does not take part in this process. During low-water periods, when $V_{vir} < V_n$, water turbidity decreases to around zero and sediment transport stops. On a flood rise, appreciable sediment transport only takes place after destruction of the armouring, when $V_{vir} > V_n$ (for relatively uniform armour-forming sediment particles). At this moment, the underlying mixed-size alluvium is characterized by $V_n << V_{vir}$. Avalanche-like growth of the bed load discharge takes place due to practically instant entrainment of finer material, as well as because of the decrease of V_n at increased sediment concentration in the flow. During the flood recession, transport of the largest particles ceases first. These produce armouring that overlies and protects the mixed-size deposit of the channel bed. On the other hand, because of the sediment coarseness and the presence of armouring, the hierarchy of bed forms is reduced. Usually it is limited to macro- and partly medium-sized forms. Those are largely more or less stable, and not eroded during flood recession and low-water periods. Therefore most gravel-cobble channels can be classified as stable or completely stable (Lokhtin index $L > 10–50$).

Usually in rivers with gravel-cobble alluvium $W_{S+B} << R_{tr}$. Because of a lack of sediment, bedrock channels (non-alluvial river reaches with channels cut in bedrock) can form (Chalov, 2003). Variations in geological structure then begin to play a leading role in determining channel morphology. For example, a change from volcanic to sedimentary rocks leads to an increased sediment supply, channel widening, and a decrease of R_{tr}. A similar situation is observed when mountain rivers enter piedmont areas or intermontane depressions. Rapid transition from constrained flow in a channel occupying the entire bottom of an incised gorge-like valley to relatively unconstrained flow in a widened braided channel takes place. The R_{tr} reduction here is also associated with a decrease in gradient. As a result, high sediment transport is replaced by the formation of dunes. These form gravel-cobble fields, within which the river is broken up into numerous braids (analogous to the sandy fields of the wandering rivers with complete domination of the suspended load described above).

In mountain rivers extremely high stream velocities determine specific modes of sediment transport and bed form movement. At relatively low channel gradients for Froude number Fr from 0.8 to 2.0, bed sediment is transported forming antidunes (so-called alluvial mountain channels with developed bed forms), whereas for $Fr > 2.0$ massive unstructured (so-called alluvial mountain channels without bed form generation) sediment transport is observed. At higher channel gradients, non-alluvial channels form (so-called channels with rapids and falls), which are able to transport extremely large rock fragments under the combined action of full flow energy (both kinetic and potential) and gravitational force (Makkaveev & Chalov, 1986).

Bank erosion returns the alluvial deposits comprising flood plains and terraces to sediment transport. Hence, flood plains also represent one of the expressions of sediment transport (Makkaveev, 1955). An increase of flood plain width and channel stability is connected with an increase in the amount of sediment preserved in flood plain deposits. At the same time, eroded fragments of unflooded terraces may become an additional source of sediment supply to the river channels. Intensive erosion of sandy terraces leads to the formation of a series of shallow riffles downstream, represented by either chaotic combinations of numerous various-sized dunes or the most complex (in terms of their morphology and deformation regime) braided reaches. Examples are the Severnaya Dvina River below Tolokonnaya Mountain (Chalov *et al.*, 2000) and the Zeya River below Belaya Mountain (Vinogradov, *et al.*, 2003).

CONCLUSIONS

River channel morphology, rates and tendencies of their deformations are determined largely by sediment yield and relative contribution of its suspended (basin-derived) and bedload (channel-originated) components. The most important characteristic appears to be size composition of channel-forming sediment and flow transportation capacity. Domination of different forms of sediment transport for material of various sizes results in morphological and dynamical differentiation of river channels with similar total sediment yield values under different natural conditions. On the other hand, similar channel forms can be produced under different conditions of sediment yield and transport.

Acknowledgements The work was conducted with financial support from the Russian Foundation for Basic Research (RFBR) (Project no. 03-05-64302) and the Program for Support of Leading Scientific Schools (Project no. 1443.2003.5).

REFERENCES

Berkovich, K. M., Lodina, R. V. & Chalov R. S. (1973) Stok nanosov i zakonomernosti ruslovyh deformacii v verhnem techenii Amudar'i (Sediment yield and laws of channel adjustment in headwaters of the Amu-Darya river). In: Soil *Erosion and Channel Processes* (ed. by R. Chalov) Iss. 3, 241–262. MSU Publishing House, Moscow, Russia (in Russian).

Chalov, R. S. (1997) *Obshee i geographicheskoe ruslovedenie* (General and Regional River Channel Studies). MSU Publishing House, Moscow, Russia (in Russian).

Chalov, R. S. (2003) Skal'nye rusla: rasprostranenie, morphologiya, speciphika i usloviya deformacii (Rocky channels: distribution, morphology, specificity and conditions of deformations) (in Russian). In: *Problem of Erosion, Channel and River Mouth Processes* (ed. by R. Chalov) (Proc. 18th Interinstitutional Coordination Meeting), 211–212. Kursk University Publ., Russia.

Chalov, R. S., Alabyan, A. M., Ivanov, V. V., Lodina, R. V. & Panin A. V. (1998) *Morphodimamika rusel ravninnyh rek* (*Channel Morphodynamics of the Plain Rivers*). GEOS, Moscow, Russia (in Russian).

Chalov, R. S., Liu Shuguang & Alekseevskiy, N. I. (2000) *Stok nanosov i ruslovye processy na bol'shih rekah Rossii i Kitaya.* (*Sediment Yield and Channel Processes on the Great Rivers of Russia and China*). MSU Publ. House, Moscow, Russia (in Russian)

Gusarov, A. V. (2002) Tendencii erosii i stoka vzveshennyh nanosov v Azii vo vtoroi polovine XX stoletiya (The tendencies of erosion and suspended load runoff in Asia in second half of XXth century). *Geomorphologiya* **4**, 70–87 (in Russian)

Karaseff, I. F. (1975) *Ruslovye processy pri perebroske stoka* (*Channel Processes under Runoff Diversion*). Gidrometeoizdat, Leningrad, Russia. (in Russian).

Kondratiev, N. E., Lyapin, A. N., Popov, I. V., Pinkovsky, S. I., Fedorov, N. N. & Yakunin I. I. (1959) *Ruslovye processy* (Channel Process). Gidrometeoizdat, Leningrad, Russia. (in Russian).

Kondratiev, N. E., Popov, I. V. & Snischenko, B. F. (1982) Osnovy *hydromorphologicheskoi teorii ruslovogo processa* (Basics of Hydromorphological Theory of the River Channel Process). Gidrometeoizdat, Leningrad, Russia (in Russian).

Makkaveev, N. I. (1955) *Ruslo reki i eroziya v ee basseine* (River Channel and Erosion in it's Basin). USSR Academy of Sciences Press, Moscow, Russia (in Russian).

Makkaveev, N. I. & Chalov, R. S. (1986) *Ruslovye processy* (Channel Processes). MSU Publishing House, Moscow, Russia (in Russian).

Shtankova, N. N. (2002) Stok nanosov na rekah basseina Volgi (Sediment yield on the rivers of Volga basin). In: *Dynamics of Small Fluvial Landforms and Channel Processes* (ed. by R. Chalov), 112–119, MSU Publishing House, Moscow, Russia (in Russian).

Turykin, L. A. (1993) Gryadovyi rel'ef rechnogo rusla i ego vliyanie na sezonnye dephormacii perekatov (Dune bed configuration of river channel and its influence on seasonal riffle adjustment). *Geomorphologiya* **2**, 83–88 (in Russian).

Vinogradov, V. A., Klaven, A. B., Nikitin, V. N. & Turutina, T. V. (2003) Neravnovesnye processy v phormirovanii rusel rek (Nonequilibrium processes in forming of rivers channels). AVN works, Iss. **9**, 55–68 (in Russian).

Zvetkova, N. A. (1963) Rezhim nanosov r. Amu-dar'ya (Sediment regime of the Amu Darya River). *Problems of Hydraulic Engineering*, Iss. 13, Tashkent, Russia (in Russian).

Sediment Transfer through the Fluvial System (Proceedings of a symposium held in Moscow, August 2004).
IAHS Publ. 288, 2004

The relationship between sediment yield and catchment characteristics in the middle Yellow River basin of China

LU JINFA[1] **& HUANG XIUHUA**[2]

1 *Institute of Geographical Sciences and Natural Resources, Chinese Academy of Sciences, Beijing 100101, China*
lujf@igsnrr.ac.cn

2 *Institute of Remote Sensing Application, Chinese Academy of Sciences, Beijing 100101, China*

Abstract Nearly 100 drainage basins with gauging stations were chosen to investigate relationships between sediment yield and catchment characteristics in the Middle Yellow River basin. The size of the catchments ranged from 50 to 25 000 km^2 and they were grouped into seven categories according to their different physiographic characteristics. A database comprising information on sediment yield and catchment characteristics, including vegetation cover, surface material and morphology, was assembled for these basins using remote sensing combined with ground truthing and desk analysis. Based on this information, an index of basin characteristics *VSD* was produced to relate to specific sediment yield and a significant positive relationship was obtained for the study catchments. Multiple regression analysis was also employed to demonstrate that the most important variable influencing specific sediment yield was vegetation coverage, followed by basin morphology and surface material.

Key words catchment characteristics; China; Middle Yellow River basin; sediment yield

INTRODUCTION

The relationship between sediment yield and catchment characteristics has long attracted the interest of geomorphologists and hydrologists (e.g. Gregory & Walling, 1973; Schumm, 1977; Hadley *et al.*, 1985; Chien *et al.*, 1987; Walling & Webb, 1996; Chen *et al.*, 1988). However, uncertainties still remain in our understanding of this important topic. For example, the relative importance of different controlling variables is difficult to establish in an environmentally complex region, such as the Middle Yellow River basin, where rainfall, vegetation, surface material, and morphology exhibit considerable local variability. Against this background, Hadley *et al.* (1985) and Walling & Webb (1996) pointed to the need for more detailed analysis of sediment yield data from different regions of the world and from different environments, in order to explore general relationships between sediment yield and catchment characteristics.

The Middle Yellow River basin of China is well known for the severe soil erosion that occurs over much of its area. From north to south, the basin extends across the arid and desert steppe zone, the semiarid steppe zone, and the warm temperate semihumid forest-steppe zone and its morphology is characterized by hills and mountains alternating with intermountain basins (Fig. 1). The landscape of the region therefore ranges from rocky mountains through rock-loess mountains and hills to gullied loess hills and loess tablelands, resulting in considerable diversity of catchment characteristics and sediment yields in the tributary drainage basins.

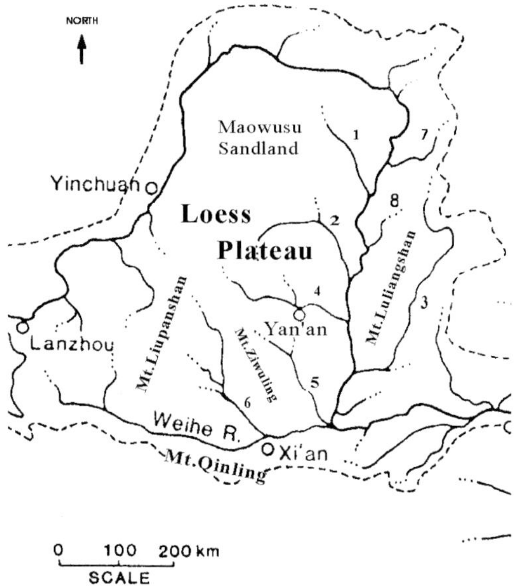

Fig. 1 A sketch map of the Yellow River basin. The dotted line shows the watershed of the river basin. The numbers refer to some of the main tributaries: (1) Kuyehe River; (2) Wudinghe River; (3) Fenhe River; (4) Yanhe River; (5) Beiluohe River; (6) Jinghe River; (7) Zhujiachuan River; (8) Qiushuihe River.

Nearly 100 tributary basins within the main Middle Yellow River basin for which sediment yield data were available were selected for this study. With catchment areas ranging from 50 to 25 000 km^2, these basins were classified into seven categories according to their physiographic characteristics. Using the database of basin characteristics, which was assembled using remote sensing, morphometric analysis and field investigation, an integrated study was undertaken to explore the relationships between sediment production and catchment characteristics.

DATA SOURCES

All the specific sediment yield data used in this study represent mean annual values. The sediment yield records extend back to the establishment of the measuring stations and cover periods ranging from 10 to 32 years. The data were abstracted from the hydrological yearbooks prepared by the Yellow River Conservancy Commission for internal use.

Vegetation cover density, surface material and drainage density were selected as key catchment characteristics for this study. The procedure for assembling these data was as follows. Firstly, each river basin was subdivided into a number of terrain units, and the vegetation cover and surface material were determined for each terrain unit by visual interpretation of Thematic Mapper (TM) images, combined with thematic maps and field investigations. Secondly, morphometric data were obtained from topographic maps (1:50 000) for each terrain unit by sampling. Finally, the value of each index for each basin was obtained by area-weighting all terrain units within the basin.

Table 1 Relative erodibility values for different types of surface material.

Type of surface material	Loess			Easily erodible material					Hard rock
	Sandy loess	Loess	Clayey loess	Siltstone/ mudstone	Red earth1	Red earth2*	Sandstone/ shale	Weathered granite	
Values	6.0–9.0	4.0–6.0	<4.0	9.0–10.0	2.5–5.0	2.5–3.0	2.0–2.7	1.5–2.0	0.5–1.0

* Red Earth 1 and 2 are both Late Tertiary strata. The former is widespread in western part of the Middle Yellow River Basin and the latter in north part of the basin. The latter is rich in illite and montmorillonite and is easily weathered.

Five classes of vegetation cover density were delimited viz. (1) >70%, (2) 50–70%, (3) 30–50%, (4) 10–30%, and (5) <10%, and five types and 11 subtypes of surface material were identified. Information on vegetation cover density was obtained primarily by interpretation of TM images, based on the percentage of red patches and exposed bedrock as well as the clarity of the gully pattern. For the purpose of this investigation, surface material was primarily an expression of geology and thus the incidence of loess and bedrock with associated weathering horizons. The relative erodibility was used as an index to characterize the resistance of surface materials to erosion and was determined using a rating procedure based on both expert knowledge of the erosion features associated with each type of surface material in the field and data on specific sediment yield for small catchments dominated by a particular surface material. Some laboratory measurements, including tests for resistance to erosion and scour, were also used to establish the relative erodibilities of sandy loess, loess, clayey loess and red earth. The relative erodibility rating values for each type of surface material are listed in Table 1.

In order to establish how different sets of physiographic conditions can influence the relationship between sediment yield and catchment characteristics, seven types of river basin with different physiographic conditions were identified as follows: (a) semiarid sandy loess and siltstone and mudstone basins; (b) semiarid hilly gullied loess basins; (c) semiarid and sub-humid hilly loess basins with some rocky mountains; (d) thin loess hilly basins; (e) sub-humid rocky mountain basins with some areas of gullied loess hills; (f) sub-humid loess plateau basins; and (g) sub-humid rocky mountain basins.

RESULTS

The relationship between specific sediment yield and an index of catchment characteristics, *VSD*

Due to the complex physiographic background of the study catchments, the mean annual values of specific sediment yield of the individual tributary basins within the Middle Yellow River basin varied from 25 t km^{-2} year^{-1} to 20000 t km^{-2} year^{-1} (Table 2). Highest values of specific sediment yield are associated with river basins underlain by sandy loess and Pishayan strata (Permian siltstones and mudstones) and with a low vegetation cover density, followed successively by loess and mixed loess-rock basins. The basins underlain primarily by hard rocks and with a good vegetation cover have the lowest values of specific sediment yield.

Table 2 The range of specific sediment yields for different types of river basin.

Basin types	Semiarid sandy loess and siltstone–mudstone basin	Semiarid loess gullied hilly basin	Semiarid and sub-humid loess hilly basin partly with rocky mountain	Thin layered loess hilly basin	Sub-humid loess tableland basin	Sub-humid rocky mountain basin partly with loess gullied hills	Sub-humid rocky mountain basin
Specific sediment yield (t km^{-2} $year^{-1}$)	12500–21000	7500–14000	3100–8300	2000–2800	1500–2000	400–1800	<400

In this paper *VSD* was introduced in an attempt to provide an integrated index of basin characteristics to relate to specific sediment yield. It was defined as follows:

$$VSD = (1 - V_c) + S_m + D_d \qquad (1)$$

in which V_c (%) is the vegetation cover density, S_m is the relative erodibility of the surface material, and D_d is the drainage density. All values of V_c, S_m and D_d were standardized to lie within the range 0–1 before they were summed. Since specific sediment yield was positively related to surface material and drainage density, but negatively related to vegetation cover the vegetation index was expressed as $(1 - V_c)$, in order to maintain a common positive relationship (Lu, 2001, 2002; Lu & Huang, 2003). *VSD* therefore reflects the capacity of a river basin to resist erosion. As *VSD* increases, the resistance of the river basin to erosion reduces and sediment production increases for a given energy condition and *vice versa*.

Figure 2 shows that there is a well-defined positive relationship between specific sediment yield and *VSD*, with the correlation coefficient reaching 0.93. The different types

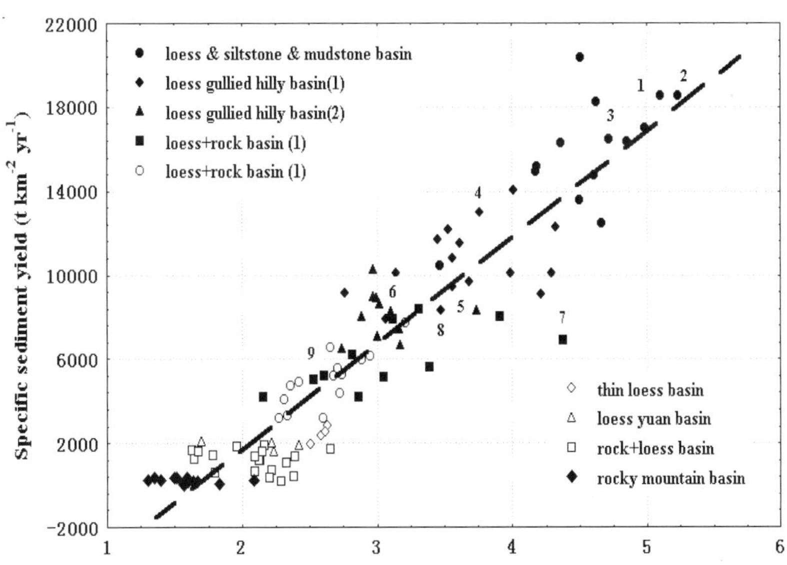

Fig. 2 The relationship between specific sediment yield and the index *VSD* for catchments in the Middle Yellow River basin. The numbers refer to specific rivers: (1) Huangpuchuan; (2) Gushanchuan; (3) Jialuhe; (4) Kuyehe (above Shenmu); (5) Kuyehe (above Wenjiachuan); (6) Wulanmulun; (7) Tuweihe; (8) Pianguanhe; (9) Lanyihe.

of river basin have been distinguished on Fig. 2 using different symbols and it can be seen that the different types of basin fall on different parts of the line, according to the resistance of the river basin to erosion. For example, the Huangpuchuan River, the Jialuhe River and the Gushanchuan River in North Shaanxi Province, which are mostly underlain by sandy loess and Pishayan strata and are poorly vegetated and highly dissected, are all located in the upper part of the plot. These are followed in turn by loess basins and loess-rocky basins, more resistant rocky-loess basins and resistant rocky basins, respectively. This sequence clearly demonstrates the influence of catchment characteristics on sediment production. The Wulamulun River, the Kuyehe River above Shenmu County and the Upper Tuweihe River are found lower in the plot than the above rivers, because they are situated in the transitional zone between the Loess Plateau and the Erdos High Plain and the Maowusu Sandland, and are characterized by an undulating morphology and an intermittent coverage of sand.

Basin morphology can also exert an important influence on specific sediment yield. For example, although they are located at a similar latitude to the above basins, the Pianguanhe and Lanyihe Rivers in Western Shanxi Province are also found in the lower part of the plot, because 60–70% of their catchment area is composed of less dissected rocky mountains or loess-covered rocky mountains with vegetation coverage as high as 50%, and 10–50% of their catchment area lies on the Tertiary peneplain.

Multivariate analysis of the relationship between specific sediment yield and catchment characteristics

It has been shown above that in the Middle Yellow River basin specific sediment yield is inversely related to vegetation cover and positively related to the erodibility of the surface material and the drainage density. In order to explore further the relative importance of these three variables and also catchment area, multiple regression analysis was employed. All variables were standardized to the range 0–1 before the regression analysis was undertaken. The final regression equation (R = 0.96) took the form:

$$Qs = -9119.89 - 0.695 \times \ln V_c + 0.343 \times D_d + 0.10 \times S_m + 0.030 \times S_a \qquad (2)$$

Where: V_c, D_d, S_m and S_a are vegetation cover density, drainage density, the relative erodibility of surface material and catchment area respectively, and Qs is the specific sediment yield. The correlation coefficient for the multiple regressions was 0.96, and the variance analysis for the relationship is presented in Table 3.

Table 3 Variance analyses for the stepwise regression between sediment yield and basin characteristics for catchments in the Middle Yellow River basin.

	SUMS OF	DF	MEAN_SQU	F	P_LEVEL
Regress.	2.77E + 09	4	6.93E + 08	320.2516	0
Residual	2.19E + 08	101	2164867		
Total	2.99E + 09				

SUMS OF, Sums of Squares;
DF, Degree of Freedom Mean;
SQU, Mean Squares;
F, F-test values;
P-Level, Significance level.

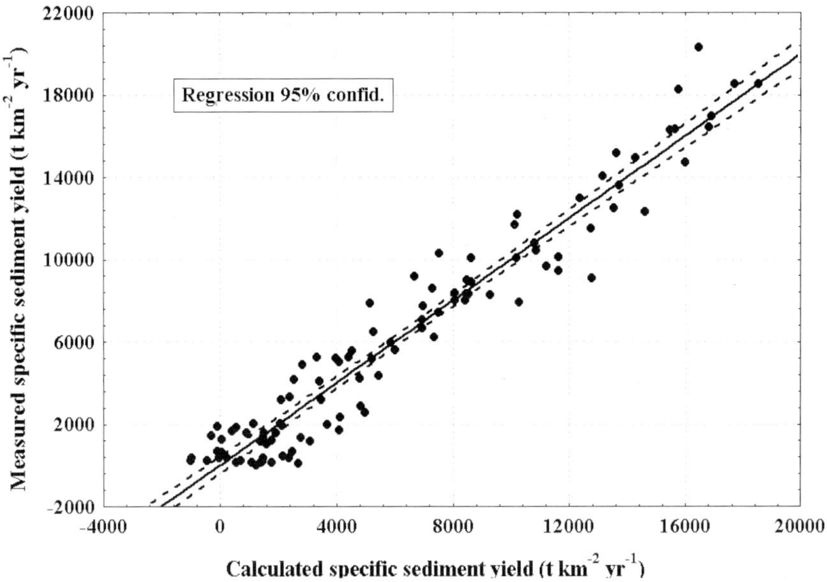

Fig. 3 The relationship between estimated and measured specific sediment yield for catchments in the Middle Yellow River basin.

Table 4 Partial correlations between sediment yield and basin characteristics for catchments in the Middle Yellow River basin.

	BETA_IN	PARTIAL COR.	SEMIPART COR.	TOLERNCE	R_SQUARE	T_103_	P_LEVEL
$\ln V_c$	0.695	0.807	0.370	0.283	0.717	13.741	7.43E-25
D_d	0.343	0.703	0.267	0.606	0.395	9.930	1.28E-16
S_m	0.100	0.171	0.047	0.217	0.783	1.741	0.084771
S_a	0.030	0.107	0.029	0.952	0.048	1.083	0.281231

BETA_IN, The beta in (standard regression coefficient for the respective variable if it were to enter into the regression equation as an independent variable);

PARTIAL COR, The partial correlation (between the respective variable and the dependent variable, after controlling for all other independent variables in the equation);

SEMIPART COR, The semi-partial (part) correlation (the correlation between the unadjusted dependent variable with the respective variable after controllingfor all independent variables in the equation);

TOLERNCE, The tolerance for the respective variable (defined as 1 minus the squared multiple correlation between the respective variable and all independent variables in the regression equation);

R_SQUARE, The minimum tolerance (the smallest tolerance among all independent variables in the equation if the respective variable were to be entered as an additional independent variable);

T_103_, The t-value associated with these statistics for the respective variable;

P_LEVEL, The statistical significance of the t-value.

A comparison between observed sediment yields and those estimated for the same catchments using the multiple regression equation is presented in Fig. 3, on which the 95%confidence limits for the relationship, have also been plotted. The partial correlation coefficients and beta coefficients for the multiple regressions have been calculated and these are listed in Table 4. The partial correlation coefficients and the beta coefficients indicate that vegetation cover density exerts the strongest influence on specific sediment yield, followed by drainage density, the relative erodibility of the surface material and catchment

Table 5 The result for a simple regression between the relative erodibility of surface material and drainage density for catchments in the Middle Yellow River basin.

	BETA	ST. ERR.	B	ST. ERR.	T_103_	P_LEVEL
Intercept			–2.261	0.989	–2.286	0.024
D_d	0.562	0.081	1.743	0.251	6.937	3.51E-10

BETA, the standardized regression coefficients;
B, non-standardized regression coefficients;
ST. ERR., standard error.

area. The relationship between specific sediment yield and basin area is relatively weak and the positive trend differs from the negative relationship between specific sediment yield and catchment area commonly reported in the literature (cf. Walling, 1983).

The above results indicate that in an environmentally complex area, such as the Middle Yellow River basin, the influence of surface material and drainage area on specific sediment yield is secondary to the control on specific sediment yield exerted by the vegetation cover and the morphology of a catchment, and only becomes clear when vegetation cover and morphology show limited variation. In addition to the above, the relatively weak relationship between specific sediment yield and surface material is probably attributed to the fact that relative erodibility of surface material is closely related to drainage density, i.e. surface material does not act as an independent variable in the multiple regression (Table 5).

CONCLUSIONS

In an environmentally complex region such as the Middle Yellow River basin, where rainfall, vegetation, surface material, and morphology exhibit considerable spatial variability, a good relationship was obtained between specific sediment yield and basin characteristics by introducing an index of basin characteristics *VSD*. A multiple regression analysis was also carried out and this indicated that the best correlation occurs between specific sediment yield and vegetation cover, followed in turn by drainage density and surface material. Therefore, within the study area in the Middle Yellow River basin, vegetation cover must be seen as the most important factor affecting sediment production, compared with morphology and surface material.

Acknowledgements This manuscript is based on the results from three projects that were funded by both the National Natural Science Foundation of China and the Innovation Fund of the Institute for Geographical Sciences and Natural Resources, CAS (Grant nos. 40171013, 59890200 and CXIOG-A02-01). This funding is gratefully acknowledged. The authors also wish to extend their sincere thanks to The Yellow River Conservancy Commission for permitting access their unpublished data.

REFERENCES

Chen, Y., Jing, K. & Cai, Q. (1988) *Recent Erosion and its Management in the Loess Plateau*. Science Press, Beijing, China (in Chinese).

Chien, N., Zhang, R. & Zhou, Z. D., (1987) *Channel Changes*. Science Press, Beijing, China (in Chinese).

Gregory, K. J. & Walling, D. E. (1973) *Drainage Basin Form and Process: a Geomorphological Approach*. Edward Arnold, UK.

Hadley, R. F., Lal, R., Onstad, C. A., Walling, D. E. & Yair, A. (1985) *Recent Developments in Erosion and Sediment Yield Studies*. UNESCO, Paris, France.

Lu, J. (2001) Response of sediment yield to climate in the Middle Yellow River Basin. *J. Soil Erosion & Soil and Water Conservation* **15**(3), 27–30 (in Chinese).

Lu, J. (2002) Effect of basin morphology on sediment yield in The Middle Yellow River. *J. Geogr. Res.* **21**(2), 171–178 (in Chinese).

Lu, J. & Huang X. (2003) Effect of land cover on sediment yield of river basin in the Middle Yellow River. *J. Geogr. Res.* **5**, 571–578 (in Chinese).

Schumm, S. A. (1977) *The Fluvial System*. John Wiley & Sons, Chichester, UK.

Walling, D. E. (1983) The sediment delivery problem. *J. Hydrol.* **65**, 209–237.

Walling, D. E. & Webb, B. W. (1996) Erosion and sediment yield: a global overview. In: *Erosion and Sediment Yield: Global and Regional Perspectives* (ed. by D. E. Walling & B. W. Webb) (Proc. Exeter Symp., July 1996), 3–19. IAHS Publ. 236. IAHS Press, Wallingford, UK.

Sediment yields and erosion rates in the Napo River basin: an Ecuadorian Andean Amazon tributary

A. LARAQUE[1], C. CERON[1], E. ARMIJOS[2], R. POMBOSA[2], P. MAGAT[1] & J. L. GUYOT[3]

1 *HYBAM (UR154 LMTG), IRD - BP 64 501, F-34394 Montpellier Cedex 5, France*
alain.laraque@mpl.ird.fr

2 *INAMHI –700 Iñaquito y Correa, Quito, Ecuador*

3 *HYBAM (UR154 LMTG), IRD – Casilla 18 1209, Lima18, Peru*

Abstract This paper presents the first results obtained by the HYBAM project in the Napo River drainage basin in Ecuador during the period 2001–2002. Three gauging stations were installed in the basin to monitor suspended sediment yields, of which two are located in the Andean foothills and the third station on the Ecuador–Peru border in the Amazonian plain. At the confluence of the Coca and Napo rivers, the suspended sediment yield transported from the Andes Mountains is 13.6 10^6 t $year^{-1}$ (766 t km^{-2} $year^{-1}$). At the Nuevo Rocafuerte station 210 km downstream, the suspended sediment yield reaches 24.2 10^6 t $year^{-1}$ for an annual mean discharge of 2000 m^3 s^{-1}. These values indicate intensive erosion processes in the Napo Andean foreland basin between the Andean foothills and the Nuevo Rocafuerte station, estimated to be 900 t km^{-2} $year^{-1}$. These high rates of erosion are the result of the geodynamic uplift of the foreland.

Keywords Amazon Basin; Andes; Ecuador; erosion; hydrology; Napo River; suspended sediment

INTRODUCTION

Most (95%) of the sediment discharged to the Atlantic Ocean by the Amazon River comes from the Andes Mountains although the range only covers 12% of the surface area of the Amazon basin. The eastern half of Ecuador, which comprises the eastern slopes of the Andes and the foreland, covers around 135 000 km^2. This represents less than 3% of the total surface area of the Amazon basin. The foreland corresponds to the edge of the humid, equatorial plain of the Amazon, which continues in Peru and Brazil. This is the region in which the Andes present the most important tectonic activity characterized by intensive volcanic and seismic activity.

The international HYBAM programme (Hydrological and Geodynamic Research Program on the Amazon Basin), in which France, Brazil, Bolivia, Ecuador and Peru are taking part, is carrying out global studies on the climatological, hydrological, sedimentological, geochemical and geodynamic aspects of the Amazon basin. The overall objective of the project is to understand how the Amazon basin functions and to study the link between this and its geodynamics. The studies aim to identify the spatial distribution of the erodable zones, and of the transfer and deposit of sediments. One of the other purposes is to estimate the processes of each, linked with the phenomenon of upward movements in the Andes and subsidence in the Amazonian depression. Starting in January 2001, a network of seven reference stations was launched in Ecuador, supported by the IRD and INAMHI Institutions, to study the processes of the transfer of solid material from the eastern side of Ecuadorian Andes, by means of the study of its principal basin, the Napo River (Fig. 1).

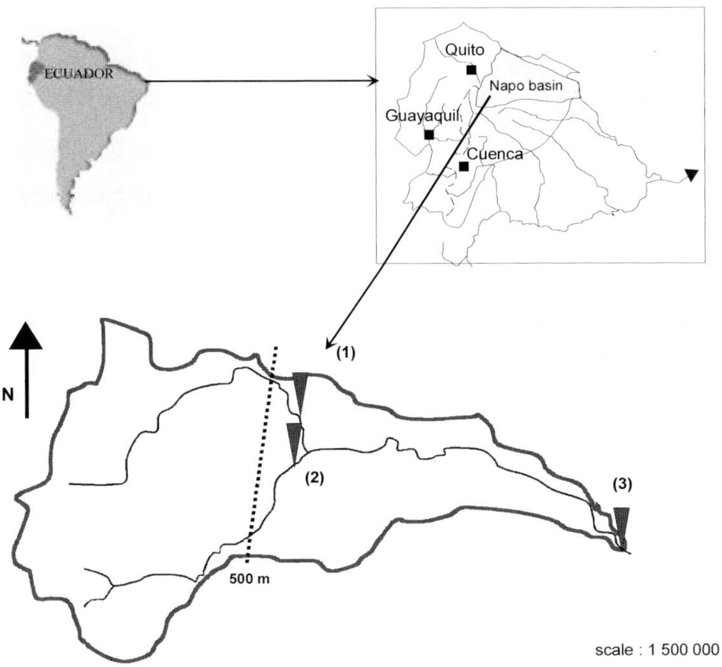

Fig. 1 Location of the Napo basin and its reference stations (see number explanation, Table 2).

THE STUDY AREA

The high precipitation (mean annual rainfall 2900 mm) has a heterogeneous spatial distribution. The 2-fold pluviometric and hydrological regime of the upper basin tends to become one in the lower basin (Moreno & Tapia, 2001). Recorded at daily intervals, these very irregular regimes are characterized by flash floods, reflecting their extreme sensitivity to rain events. The Napo basin covers an area of 26 860 km^2, i.e. equivalent to 20% of the eastern part of Ecuador. It is situated between the southern latitudes of 00°01′ and 01°10′ and the western longitudes of 77°20′ and 78°30′ (Perez Suasvana, 2000; Armijos, 2001).

The Napo River is 1300 km long and starts in the Ecuadorian Andes at an altitude of more than 5000 m. After the first 460 km, the river leaves Ecuador at an altitude of 189 m and enters Peru, where it forms the left tributary of the fluvial axis of the Marañon/Amazon at Iquitos. Figure 1 is a schematic presentation of the basin with the location of the three gauging stations used on the two main rivers. The figure also shows the altitude limit of 500 m, to indicate the boundary between the Andean and Amazonian regions.

DATA AND METHODS

Three reference stations were selected in Ecuador: San Sebastian and Fr. De Orellana, which control the output of the Andes, and Nuevo Rocafuerte which closes the basin at the outlet from Ecuador. One of the three stations was opened at the beginning of 2001 (San Sebastian on the Coca River) and the others (Fr. De Orellana and Nuevo Rocafuerte on the Napo River), have kept limnimetric records since 1980 but needed calibrating. The gauging curves were established over only two hydrological cycles using a 1200 kHz ADCP (Acoustic Doppler Current Profiler) (RDI, 1999). At these stations, samples of water and suspended matter (SM) were also taken every 10 days. The relations between these surface SM concentrations and the average of the cross section were established using several complete solid discharge measurements taken during an entire annual flow season (Laraque *et al.*, 2003) (Figs 2 and 3).

Using a point sampling protocol, in each solid gauging section, *n* samples were taken over all the depth of *p* vertical lines uniformly distributed over the area of the section (with $n > 3$, $p < 5$). The point-to-point concentrations were weighted by the velocities corresponding to the sampling points. The samples were filtered using 0.45 μm cellulose acetate filters to separate the suspended matter. There were thus two years' (2001–2002) data available on both water discharge and SM load from the three stations. These data allowed us to calculate the first annual fluxes of SM in the principal sub-basins of the Napo River and to compare the balance of the transport between the stations.

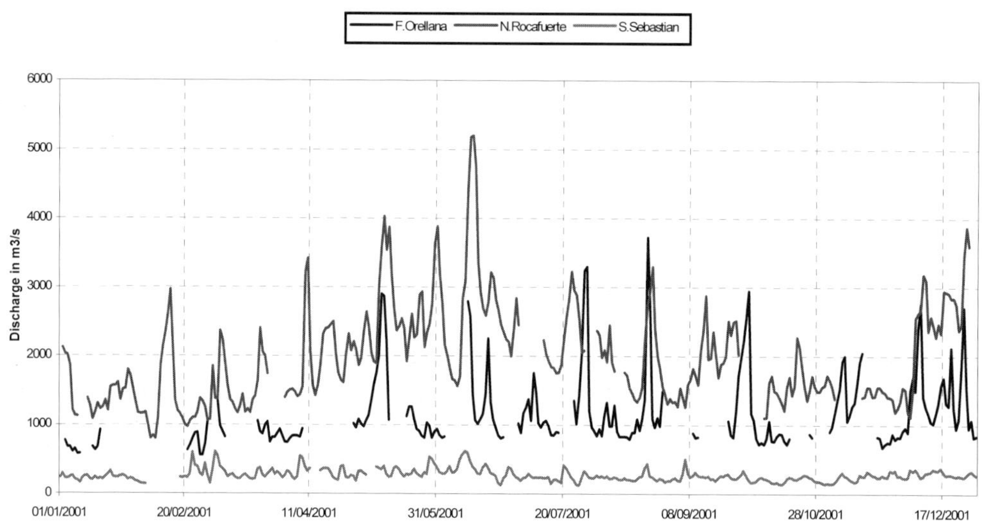

Fig. 2 Daily hydrographs of the Coca River at Sebastian, and of the Napo River at Fr. De Orellana and Nuevo Rocafuerte stations (year 2001).

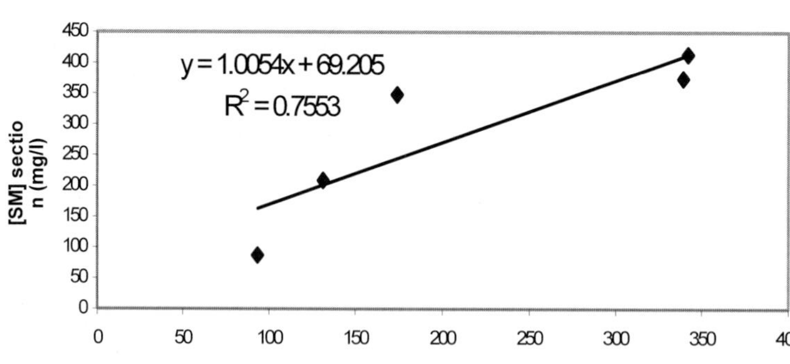

Fig. 3 Relation between surface SM load *vs* mean SM load in the section of the Napo River at Nuevo Rocafuerte station.

RESULTS AND DISCUSSION

Table 1 presents the first annual hydrological balance of the basins under study for the period 2001–2002. The mean annual discharge of the Napo River at Nuevo Rocafuerte was 2000 $m^3 s^{-1}$, which corresponds to an annual volume of 63 km^3. The specific discharges indicate a high degree of spatial irregularity, from 47 $l s^{-1} km^2$ in the basin of the Coca at San Sebastián to 99 $l s^{-1} km^2$ for the Napo River at Fr. De Orellana.

The SM load concentration varied from 10 to 2800 mg l^{-1} from one station to another and with the seasons. The ratio between the highest and lowest daily values decreased downstream: San Sebastian (161), Fr. De Orellana (146) and Nuevo Rocafuerte (31) as a function of altitude, of the increase in the size of the area drained and of the geographical origin of stream flow (only mountains or mountains and plain). These sediment concentration ratios were much higher than the highest ratios for extreme daily discharges: San Sebastian (4.9), Fr. De Orellana (6.7) and Nuevo Rocafuerte (6.6). Table 2 presents the first computations of the specific sediment fluxes for the years 2001 and 2002.

The rates of specific sediment yield (QSs) varied between 557, 1260 and 901 t km^{-2} $year^{-1}$, respectively for the Napo at F. Orellana, Coca at San Sebastian, and Napo at Nuevo Rocafuerte, for basin areas ranging from 5270 to 27 000 km^2. The maximum QSs correspond to the Coca River basin, which is the principal provider of the sediment, although it has the lowest specific stream flow (Qs) of the basins studied. Its annual fluxes of SM are similar to those of the Napo basin at Fr. De Orellana station, whose surface area is two times bigger. The Andean part of this basin produced 13.6 x 10^6 t $year^{-1}$ of SM, but a total of 24.2 × 10^6 t $year^{-1}$ leave the Ecuadorian Napo Basin downstream. The area between the stations of Fr. De Orellana and San Sebastian, both located at the foot of the Andes, and the station at Nuevo Rocafuerte, covers 9146 km^2. Thus, this area provided a net gain of 10.6 × 10^6 t of SM to the outlet of the basin, which corresponds to 45% of the Napo River sediment yield at Nuevo Rocafuerte. The specific sediment yield from the lower part of the basin was about 1160 t km^{-2} $year^{-1}$. This rate seems very high, considering there are almost no uplands in this intermediary area, which is protected by a dense forest cover. The high erosion rate here corroborates the very active uplift tectonics evidenced geomorphically (Bès de Berc *et al.*, 2004).

Table 1 Details of the drainage basins studied with location of gauge sites.

Gauging station	River	Latitude	Longitude	Altitude (m)	Drainage area (km^2)	Annual mean discharge: ($m^3 s^{-1}$)	($mm year^{-1}$)	($l s^{-1} km^{-2}$)
San Sebastian	Coca	00°20'30"S	77°00'21"W	285	5272	247	1478	46.9
F. Orellana	Napo	00°28'15"S	76°58'38"W	262	12443	1234	3127	99.2
Rocafuerte	Napo	00°54'18"S	75°25'48"W	189	26861	1996	2343	74.3

Table 2 Sediment fluxes and specific export in the Napo basin from mean SM concentration in the sections (year 2001–2002).

River and gauging station	Q annual ($m^3 s^{-1}$)	Mean annual SM (mg l^{-1})	Fluxes of SM (10^3 t $year^{-1}$)	(t km^2 $year^{-1}$)	Surface area (km^2)
Coca at San Sébastian = 1	247	399	6644	1260	5272
Napo at Fr. De Orellana = 2	1234	158	6929	557	12443
Napo at Nuevo Rocafuerte = 3	1996	285	24195	901	26861
Intermediate area = 3 – 2 – 1	515		10622	1161	9146

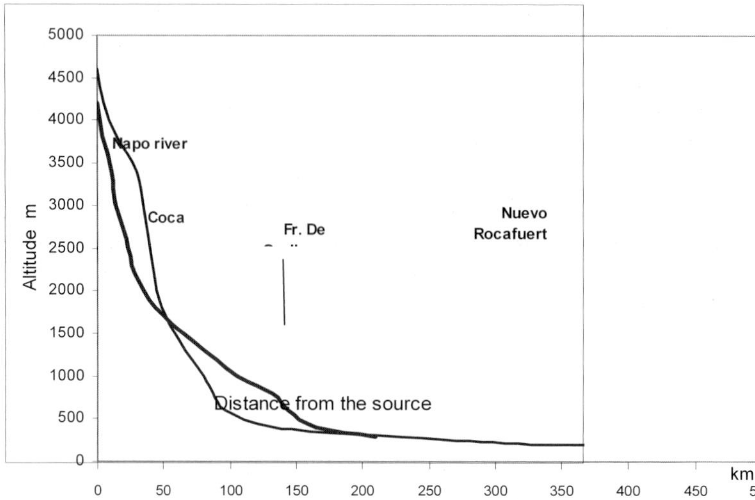

Fig. 4 Longitudinal profiles of the Napo and Coca rivers.

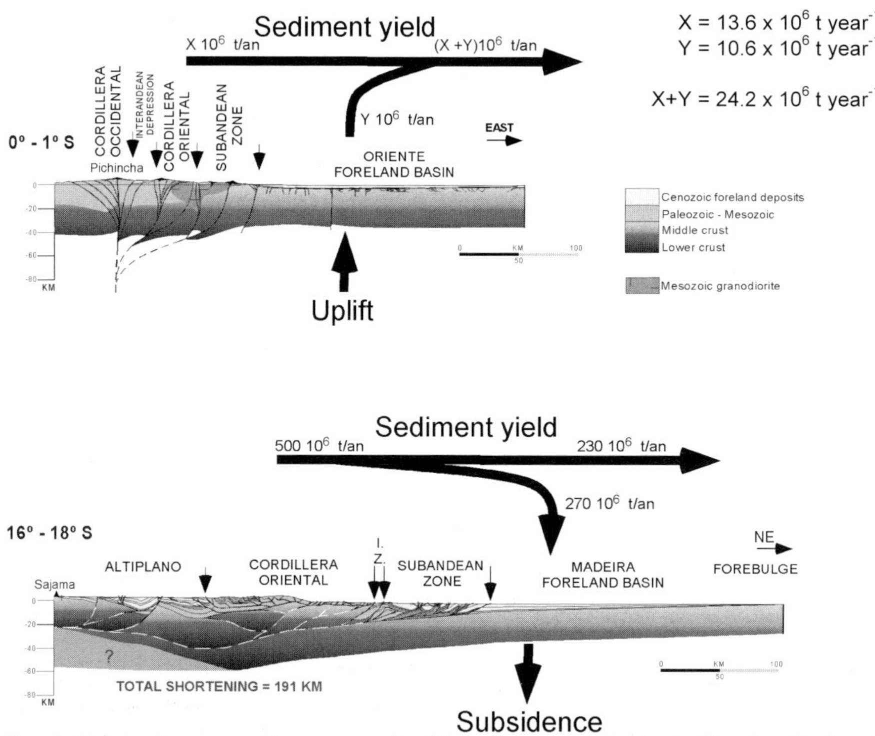

Fig. 5 Structural cross-sections across the High Amazonian basin in Ecuador (top) and Bolivia (bottom), and respective sediment yields (from Baby *et al.*, 1999).

A process of remobilization of fine material occurs along the 200 km section between Fr. De Orellana and Nuevo Rocafuerte. The slope of this section of the river is still high

(35.5 cm km^{-1}, Fig. 4), i.e. more than 10 times higher than in the rest of the Amazon basin, where the slope varies between 3 cm km^{-1} and 0.3 cm km^{-1} in its central depression. The steep slope explains the relatively rectilinear course of the Napo River as it leaves the Andes. The opposite condition has been observed at the foot of the Bolivian Andes (Guyot *et al.*, 1996; Baby *et al.*, 2004), where the streams meander through an area undergoing subsidence (slope < 3 cm km^{-1}) and deposit more that half of their load of suspended matter (Fig. 5). This study enabled us to confirm the observation made by Baby *et al.* (1999), that "in Ecuador, the Oriente basin is uplifting by transpressive tectonics influence and no Pliocene and Quaternary sedimentation has been recorded. It is actually in process of erosion and the sediments are transported towards the east and the southeast".

CONCLUSION

The first results of this study on the Napo basin show that it exports annually a total of 63 × 10^9 m^3 of water and 24 × 10^6 t of SM. Around 45% of the solid fluxes result from the remobilization of fluvial sediments along the 200 km section between the foot of the Andes and the Ecuador exit. This part of the basin presents a slope of 35 cm km^{-1}, which is high for this part of the Amazon plain river. These results confirm those of other studies on geodynamics (Burgos *et al.*, 2004) that were conducted over a longer period, and which show that since the end of the Neogene period and during the Quaternary, under the influence of the Andes uplifts, the foreland basins now present a zone of sediment transfer and erosion of deposits that were deposited at a time when the slopes were less steep.

Acknowledgements We would like to thank participants in HYBAm/Ecuador project from Ecuador and France, and their institutions of origin: INAMHI, INOCAR, IRD.

REFERENCES

Armijos, E. (2002) Estudio hidrofisico de las cuencas de los ríos Napo, Pastaza y Santiago dentro del proyecto Hybam. Hydrological feature of the Napo, the Pastaza and the Santiago basins, Hybam project. MSc Thesis, Central University of Ecuador, Quito, Ecuador.

Baby, P., Guyot, J. L., Deniaud, Y., Zubieta, D., Christophoul, F., Rivadeneira, M. & Jara, F. (1999) The high Amazonian basin: tectonic control and mass balance. In: *Proc. Hydrological and Geochemical Process in Large-scale River Basins, Manaus Symp.* Manaus, November, 1999.

Baby, P., Guyot, J. L. & Herail G. (2004) The high Amazonian basin of Bolivia: tectonic control, erosion and sedimentation. *Hydrol. Processes*. (in press).

Bès de Berc, S., Soula, J. C., Baby, P., Souris, M., Christophoul, F. & Rosero, J. (2004) Geomorphic evidence of active deformation and uplift in a modern continental wedge-top foredeep transition: example of the eastern Ecuadorian Andes. *Tectonophysics* (in press).

Burgos, J. D., Baby, P., Christophoul, F., Soula, J. C. & Rochat, P. (2004) Cuantificación de las erosiones terciaria y plio-cuaternaria en la parte sur de la Cuenca Oriente del Ecuador. Erosion rates during the Tertiary and Plio-Quaternary period in the South of the Ecuador Oriental basin. In: *Geología de la Cuenca Oriental del Ecuador* (*Geology of Oriental Basin of Ecuador*). *IFEA, IRD, Petroproduccion, Lima, Peru.*

Guyot, J. L., Filizola, N., Quintanilla, J. & Cortez, J. (1996) Dissolved solids and suspended sediment yields in the Rio Madeira basin, from the Bolivian Andes to the Amazon. In: *Erosion and Sediment Yield: Global and Regional Perspectives* (ed. by D. E. Walling & B. W. Webb) (Proc. Exeter Symp., July 1996), 55–63. IAHS Publ. 236. IAHS Press, Wallingford, UK.

Laraque, A., Pombosa, R., Cerón, C., Armijos & E. & Magat, P. (2003) La rede de estaciones de referencias del proyecto HYBAM en Ecuador - Primeros resultados (Network of reference station of the Hybam project, First results). Informe Interno IRD-INAMHI. Report of IRD-INAMHI, Quito, Ecuador.

Moreno, F. & Tapia, A. (2001) Regionalización hidrometeorológica de las cuencas amazónicas del Ecuador (Hydrometeorological Regionalization of the Ecuatorian Amazon basin). MSc Thesis, Central University of Ecuador, Quito, Ecuador.

Perez Suasnavas, V. P. (2000) Balance hídrico superficial de la cuenca del Río Napo (Hydrological balance of the Napo Basin). MSc Thesis, Central University of Ecuador, Quito, Ecuador.

RDI (1996) *Acoustic Doppler Current Profilers – Principles of Operation: A practical Primer* (ed. by broadBand ADCPs). San Diego, California, USA.

Sediment Transfer through the Fluvial System (Proceedings of a symposium held in Moscow, August 2004).
IAHS Publ. 288, 2004

The grain size characteristics of overbank deposits on the flood plains of British lowland rivers

D. E. WALLING, D. FANG, A. P. NICHOLAS & R. J. SWEET
Department of Geography, University of Exeter, Exeter EX4 4RJ, UK
d.e.walling@exeter.ac.uk

Abstract River flood plains are attracting increasing attention because of their wide-ranging environmental significance. In the UK, little is currently known regarding the grain size composition of the fine-grained overbank deposits that dominate most lowland river flood plains. This paper reports the findings of a study aimed at investigating the grain size composition of the overbank flood plain deposits associated with the flood plains of six rivers in southern Britain, embracing a range of catchment characteristics and hydrological conditions. More than 900 cores were collected from representative reaches along the flood plains of the six rivers and their grain size composition was determined. Analysis of the results obtained focuses on contrasts between the rivers, on within-reach and longitudinal variability in grain size composition and on the relationship between grain size composition and sedimentation rate.

Key words fine sediment; flood plain; grain size; overbank deposits; sediment cores; sedimentation rate; southern Britain

INTRODUCTION

River flood plains have long attracted the interest of geomorphologists, both in terms of their contemporary development as readily identifiable fluvial landforms and the potential of their sediments to provide information on past catchment response to environmental change (Hudson, 2003). However, in recent years there has also been a growing recognition of their wider environmental significance. This significance relates particularly to the flood plains of lowland rivers, where overbank deposition, rather than channel migration, represents the key mode of flood plain development. Such flood plains represent an important interface between the river channel and the surrounding catchment and a key component of the river corridor, with its distinctive habitats (Dyer *et al.*, 2002). In addition, these flood plains frequently serve as natural flood control reservoirs. Furthermore, when viewed in a wider catchment context, the overbank deposition that constitutes the primary mechanism for their development also represents an important sink in the transfer of both sediment and sediment-associated nutrients and contaminants from a catchment to its outlet (Walling *et al.*, 1999; Walling & Owens, 2003). Although relatively stable at present, future changes in river response associated with climate change could result in remobilization of the large stores of sediment that have accumulated in these river flood plains.

Against this background, there is a need to develop an improved understanding of the physical properties of the fine overbank deposits that are a key feature of lowland river flood plains. Although, such deposits are known to result primarily from deposition of suspended sediment transported by the river during overbank flood events, and are therefore necessarily fine-grained, relatively little detailed information is currently available in the UK regarding their grain size composition, and the spatial variation of this composition, both within the flood plain of a particular river and between rivers. This contribution reports information on the grain size composition of overbank flood plain deposits obtained from a recent investigation of sediment deposition on the flood plains of six rivers in southern Britain.

THE STUDY RIVERS AND THEIR FLOOD PLAINS

The six rivers were selected to be representative of a range of conditions within southern Britain, including both the underlying geology and terrain characteristics of their catchments and their hydrological response. The location of the rivers selected is shown in Fig. 1 and further details of their catchment and hydrological characteristics are provided in Table 1. Key contrasts include those between the Rivers Exe and Usk, which rise in the wetter upland areas of Exmoor and the Brecon Beacons, respectively, and the Rivers Culm and Stour, whose catchments are more lowland in character and receive substantially less rainfall. The Rivers Torridge and Axe can be seen as intermediate between these two extremes, draining areas of moderate relief. Equally, the underlying geology ranges from the indurated Devonian and Carboniferous slates and sandstones occupying much of the catchments of the Exe, Usk and Torridge, to the softer and more permeable Permian, Triassic, Jurassic and Cretaceous sandstones, marls, shales and clays and the chalk found in the catchments of the Culm, Axe and Stour. These contrasts in topography, underlying geology and annual precipitation give rise to further contrasts in soils and land use. Information on the suspended sediment yields of the study catchments is limited, but estimates of the specific suspended sediment yields are also provided in Table 1. These values must be seen as relatively low by world standards. Similarly, only limited information is available on the grain size composition of the suspended sediment loads of the study rivers, but available data indicate that the sediment loads are relatively fine. Median grain size or d_{50} values for the absolute size distributions are likely to be within the range 5–15 μm and the percent sand content is typically about 10% (Walling & Woodward, 2000).

The flood plains of the study rivers are characterized by subdued microtopography, with the local relief rarely exceeding 1 m amplitude. Flood plain widths along the main channel systems of the study rivers range from a minimum of about 160 m to a maximum of 1150 m, with mean values for the individual rivers ranging from ~300 m (River Torridge) to 630 m

Table 1 Some characteristics of the study rivers and their catchments.

River	Catchment area[a] (km^2)	Mean altitude (m)	Mean slope ($m\ km^{-2}$)	Dominant lithology	Mean annual precipitation (mm)	Specific sediment yield[c] ($t\ km^{-2}\ year^{-1}$)
Axe	303	134	90	sandstone, marl, shale	991	ca 50
Culm	276	140	71	sandstone, marl	952	ca 25
Exe[b]	908	208	116	sandstone, slate, marl	1148	ca 30
Stour	1067	85	51	clay, chalk	861	ca 30
Torridge	723	156	83	slate, sandstone	1176	ca 75
Usk	1124	273	149	sandstone	1304	ca 50

[a] The catchment area listed represents the catchment area above the lowest flood plain reach investigated.
[b] The data for the River Exe are based on the combined catchments of the Rivers Exe and Culm.
[c] The values of specific suspended sediment yield listed are estimates based on available data.

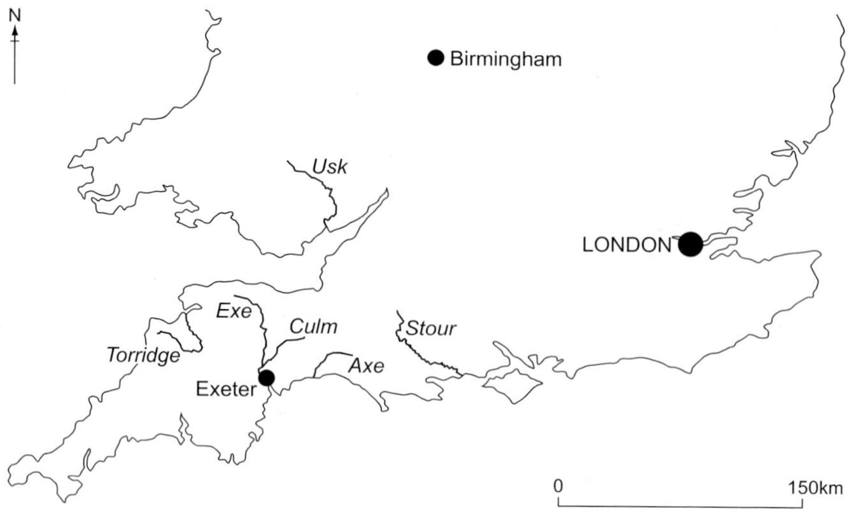

Fig. 1 The six study rivers and their location in southern Britain.

(River Stour). Distinct natural levees are generally absent from the flood plains, probably as a result of the low sand content of the suspended sediment loads transported by the rivers. The flood plains of the study rivers are regularly inundated on several occasions each year, typically during the winter months, although the depth of inundation does not generally exceed 1 m.

DATA COLLECTION

Attention focused on the flood plains bordering the main channels of the six study rivers. In each case, the main channel was subdivided into a number of segments (six for the Exe and eight for the other rivers), and representative reaches of flood plain (~600 m in length) were identified for each of these segments. Within these reaches, 20 bulk sediment cores were collected to a depth of ~0.6 m using a motorized percussion corer equipped with a 6.9 cm diameter core tube. The cores were collected from a representative range of locations, aimed at embracing likely contrasts in sedimentation rate and sediment properties associated with both distance from the channel and different morphological features. Visual inspection of the cores confirmed that, in all cases, they were composed entirely of fine overbank deposits and that there was no clear evidence of major downcore changes in texture.

After drying, disaggregation and sieving (2 mm) to remove coarse organic debris, the grain size composition of a representative subsample of sediment from each core was determined. Grain size analysis was undertaken on chemically dispersed samples, from which organic matter had been removed by treatment with hydrogen peroxide, using a Coulter LS 200 laser granulometer. Estimates of the specific surface area of the samples ($m^2\ g^{-1}$) were derived numerically from the grain size distributions, assuming spherical particles. As part of wider investigations, the cores were also assayed for their Cs-137 inventories by gamma spectrometry, in order to derive estimates of medium-term sedimentation rates at the sampling points (Walling & He, 1997; Walling *et al.*, 1999).

RESULTS

Information regarding the mean values for selected measures of grain size composition obtained for the cores collected from the individual study rivers are presented in Table 2. These measures comprise specific surface area ($m^2 g^{-1}$), median diameter or d_{50} (μm), percent clay, and percent sand. The results presented in Table 2 emphasize the fine-grained nature of the overbank deposits associated with the study rivers. The mean sand content ranges from 20 to 36%, and in all cases, therefore, the overbank deposits are dominated by clay- and silt-sized material, with the mean clay content ranging from 7.5 to 14.0%. Comparisons with the limited information on the grain size composition of the suspended sediment loads of the study rivers referred to above, suggest that the overbank deposits are characterized by both a greater d_{50} and an increased sand content, relative to the suspended sediment load, and are thus appreciably coarser. This is likely to reflect preferential deposition of the coarser fractions, although it must also be recognized that the grain size composition of the transported sediment may be coarser during high magnitude overbank flood events that inundate the flood plains, than during more normal storm-period events. The presence of significant proportions of clay in the overbank deposits (i.e. ~10%) could appear inconsistent with the preferential deposition of the coarser fractions of the suspended load, but can be readily accounted for in terms of the importance of flocculation and the associated occurrence of larger composite particles with increased settling velocities (Walling & Woodward, 2000). Preferential deposition of the larger composite particles will result in the deposition of smaller clay- and fine-silt-sized particles contained within the composite particles.

Table 2 A comparison of the mean, standard deviation and coefficient of variation of selected measures of grain size composition obtained for the sediment samples collected from the flood plains of the study rivers

River	Specific surface area ($m^2 g^{-1}$)			d_{50} (μm)			Percent clay			Percent sand		
	$\bar{x}$	sd	CV	$\bar{x}$	sd	CV	$\bar{x}$	sd	CV	$\bar{x}$	sd	CV
Axe	0.32	0.09	0.28	32	22	0.70	7.8	2.5	0.32	29	14	0.47
Culm	0.33	0.08	0.24	35	33	0.93	8.0	2.2	0.28	36	13	0.38
Exe	0.37	0.06	0.17	20	18	0.89	7.8	1.7	0.22	25	11	0.46
Stour	0.51	0.11	0.22	16	15	0.99	14.0	3.6	0.26	20	11	0.53
Torridge	0.37	0.08	0.22	29	25	0.86	9.2	2.1	0.23	30	13	0.44
Usk	0.31	0.08	0.25	39	27	0.68	7.5	2.3	0.31	34	14	0.42

Despite the clear contrasts between the study rivers in terms of relief, lithology, catchment area and mean annual precipitation (see Table 1), with the exception of the River Stour, there is relatively little difference between the rivers in the grain size composition of their overbank flood plain deposits. The flood plain sediments from the River Stour flood plain stand out as being significantly finer, and therefore as having a significantly higher specific surface area than those from the other five rivers. This reflects the clay lithology of the upper part of the Stour catchment. Differences between the rivers have been further explored using a simple *t*-test to compare the mean values for the four measures of grain size composition obtained for the six study rivers. This analysis shows that both the Exe and Stour have values of d_{50} and percent sand that are significantly different from those found in all the other five rivers, with the Stour and Torridge exhibiting similar contrasts for percent clay and the Stour for specific surface area. There is therefore some evidence that the flood plain deposits of the River Exe are also finer than those of the Rivers Axe, Culm, Torridge and Usk, and that the River Torridge has a relatively high clay content when compared with the Rivers Axe, Culm, Exe and Usk. The increased clay content of the overbank sediments collected from the flood plain of the River Torridge may again reflect a lithological control, in that the Carboniferous Culm Measures, which underlie most of this catchment, weather to produce soils with a high clay content. The somewhat finer sediment found on the flood plains of the River Exe may also reflect the outcrop of Carboniferous Culm Measures within the central part of this catchment. Overall, however, the important differences in lithology between the six study catchments are not reflected by major differences in the grain size composition of the overbank flood plain deposits. Rather, they seem likely to account for more minor and subtle differences between the catchments.

Table 2 also lists standard deviation and coefficient of variation values for the four measures of grain size composition obtained for the flood plain sediments collected from each of the six study rivers. The relatively high standard deviation values, which are commensurate with coefficients of variation in the range 17 to 99%, indicate that there is appreciable spatial variability in the grain size composition of the flood plain sediments. This variability could reflect variability both within the individual reaches and longitudinally. Maximum values of the coefficient of variation are associated with the median diameter (d_{50}) and minimum values with specific surface area, whilst the values for percent sand are consistently greater than those for percent clay. This pattern suggests that the magnitude and composition of the coarser fractions, as reflected by the percent sand and d_{50}, evidence greater spatial variability than for the finer fractions, as reflected by the percent clay and specific surface area. Overall, the River Stour appears to show the greatest variability in the grain size composition of its flood plain sediments. This may reflect the considerable range of rock types outcropping within the catchment, which includes both clays and chalk.

In an attempt to distinguish the within-reach and longitudinal components of the spatial variability in the grain size composition of overbank flood plain sediments evidenced by the six study rivers, Table 3 provides a comparison of the mean coefficient of variation (CV) associated with the individual sampled reaches along a particular river with the coefficient of variation representing all the samples collected from the flood plain of that river. The values listed show a substantial degree of uniformity in the CV values for both the reaches and the entire flood plain and comparison of the two values indicates that most of the variability associated with the grain size composition of the flood plain sediments reflects within-reach variability, rather than longitudinal variability. In the case of within-reach variability, this is likely to reflect a general fining of the deposited sediment with increasing distance from the river (Marriott, 1992), spatial contrasts in the relative importance of diffusive and convective

Table 3 A comparison of the mean coefficient of variation for the individual study reaches with the value for all samples collected from the study river.

River	(a) Mean CV for study reaches	(b) CV for all samples	(a)/(b)
Axe	0.22	0.28	0.77
Culm	0.19	0.24	0.81
Exe	0.15	0.17	0.88
Stour	0.17	0.22	0.80
Torridge	0.17	0.22	0.77
Usk	0.20	0.25	0.81

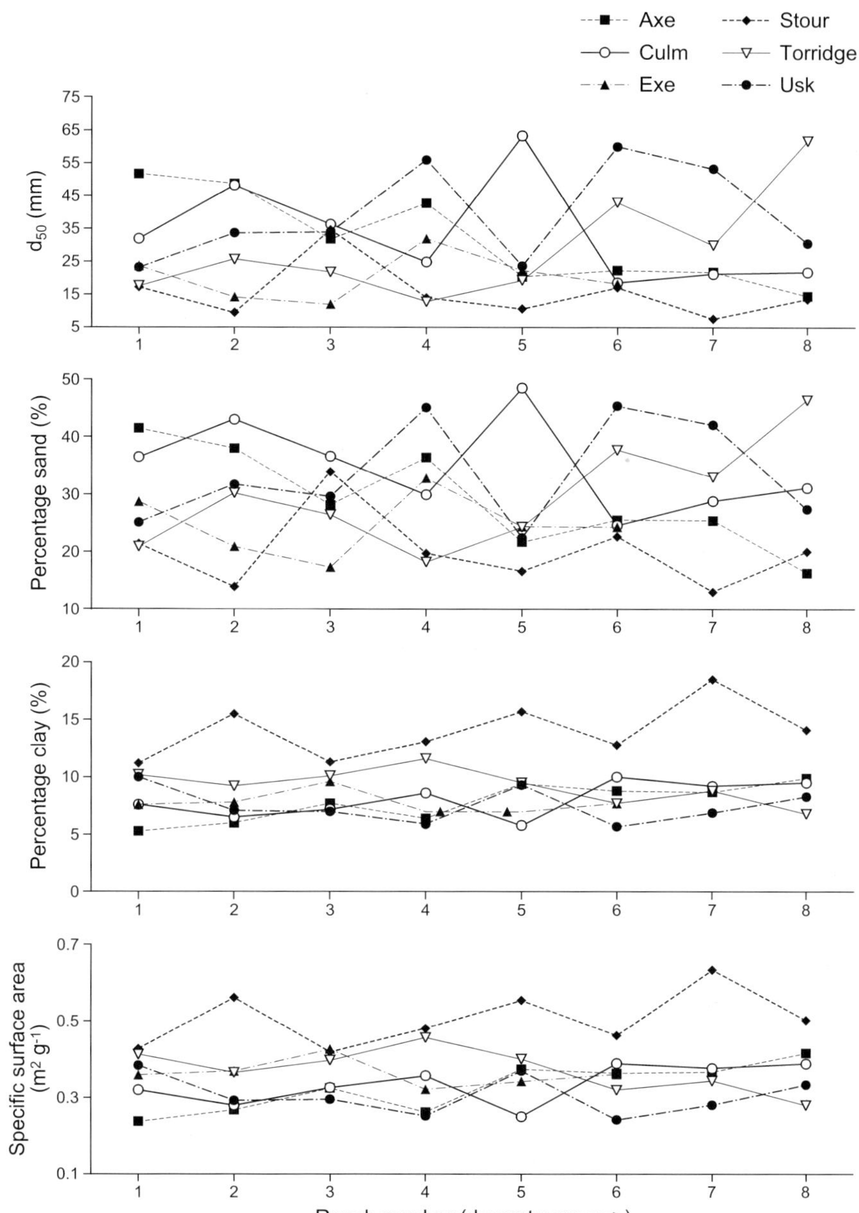

Fig. 2 Downstream changes in the mean values of the four grain size parameters for individual flood plain reaches.

transfer of sediment from the river onto the flood plain (James, 1985; Pizzuto, 1987) and the influence of local morphology on water depth and flow velocity across the flood plain (Walling *et al.*, 1996).

In order to investigate further the downstream or longitudinal variation in the grain size composition of the overbank flood plain deposits, Fig. 2 presents plots of the mean values of

Table 4 Statistically significant relationships between the selected measures of grain size composition for the flood plain sediments and distance downstream.

River	d_{50}	Specific surface area	Percent clay	Percent sand
Axe	–	+	+	–
Culm	0	0	0	0
Exe	0	0	0	0
Stour	0	0	0	0
Torridge	+	–	–	+
Usk	0	0	0	0

\+ Significant positive relationship at the 0.05 level
– Significant negative relationship at the 0.05 level
0 No statistically significant relationship

the four measures of grain size composition obtained for individual reaches *vs* reach number. Although there is no clear consistent pattern evident for all rivers, there is some evidence of a trend for specific surface area and percent clay to increase, and for percent sand to decrease, downstream. Such trends could be accounted for in terms of a progressive downstream fining of the transported sediment, as the coarser sediment is deposited, and thus a downstream fining of the deposited sediment. The existence of statistically significant longitudinal trends was explored by testing the significance of the relationship between the mean values of the four measures of grain size composition obtained for the individual reaches along a river and distance downstream. The results of this analysis presented in Table 4, indicate that only the Rivers Axe and Torridge provide evidence of a statistically significant downstream trend. Interestingly, the trends shown by the Rivers Axe and Usk are opposite. Thus, whereas the River Usk provides evidence of statistically significant downstream decreases in the median diameter (d_{50}) and percent sand content of the flood plain sediment, and increases in the percent clay and specific surface area, which are all consistent with downstream fining, the Torridge provides evidence of downstream coarsening, with the d_{50} and percent sand increasing downstream and the specific surface area and percent clay decreasing. It is difficult to account for the trend evidenced by the River Torridge in terms of hydraulic controls and it is suggested that the downstream coarsening probably reflects local controls associated with inputs of coarser sediment from the River Okement, a southern tributary of the River Torridge, which rises on Dartmoor, an area underlain by granite.

Since estimates of medium-term (i.e. *c.* 40 year) sedimentation rate (g cm^{-2} $year^{-1}$) were available for each of the cores collected from the flood plains of the six study rivers, based on their Cs-137 inventories, the influence of sedimentation rate on the variability of the grain size composition of the overbank sediment deposits associated with the individual rivers was also explored. Low levels of explanation were associated with simple linear relationships between the four measures of grain size composition and sedimentation rate, with r^2 values ranging between 0.001 and 0.11. Only 14 out of the 24 relationships were statistically significant at the 0.05 level. The highest r^2 values and the greatest number of statistically significant relationships (five out of six rivers) were associated with the measures primarily reflecting the finer fractions, namely percent clay and specific surface area. Both these measures showed the expected inverse relationships with sedimentation rate. Figure 3 presents plots of the relationships between specific surface area and sedimentation rate for the six study rivers. In this case a simple exponential function has been fitted to the

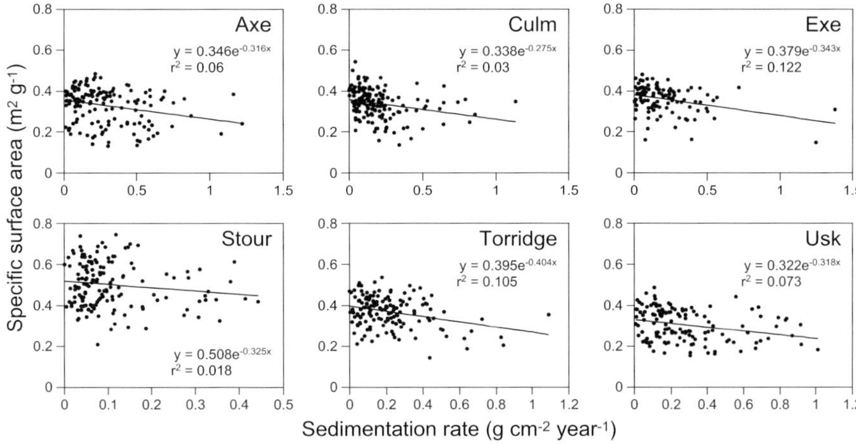

Fig. 3 The relationship between the specific surface area of samples collected from the flood plains of the six study rivers and the sedimentation rate at the sampling points estimated from Cs-137 measurements.

relationship. The lack of a well-defined relationship and the considerable scatter around the fitted lines reflects, at least in part, the influence of composite particles in masking any clear relationship between the two variables. Nevertheless, it also further emphasises that sedimentation rate is not a key control on the spatial variability of the grain size composition of the deposited sediment. Further work is required to identify the key controls on this variability.

PERSPECTIVE

The results presented above provide representative information regarding the grain size composition of overbank deposits on the flood plains of six rivers in southern Britain. Despite appreciable difference between their catchments, in terms of both underlying geology and other terrain characteristics, there is relatively little variation in the grain size composition of flood plain sediments between the six rivers. However, the sediments from the flood plain of the River Stour are shown to be significantly finer than those from the flood plains of the other five rivers. Other relatively minor differences between the rivers can be linked to the influence of catchment lithology. Considering all the samples collected from the flood plains of the individual rivers, there is evidence of appreciable spatial variability in the grain size composition of the overbank flood plain sediments, which reflects both within-reach and longitudinal variability. The former is shown to be the dominant source of variability and the study rivers show only limited evidence of significant downstream trends in the grain size composition of their overbank deposits. Only two out of the six rivers evidence statistically significant downstream trends, but the trends associated with these two rivers are opposite. The Axe shows downstream fining, whereas the Torridge shows evidence of downstream coarsening, which appears to reflect local controls associated with tributary inputs. Examination of the relationships between the measures of grain size composition and sedimentation rate, estimated for the individual sediment cores using Cs-137 measurements, provided only limited evidence of statistically significant relationships. The strongest relationships were associated with the measures primarily reflecting the magnitude of the fine fraction (i.e. percent clay and specific surface area). Overall, however, sedimentation rate was shown not to be a key control of the spatial variability in the grain size composition of the flood plain sediments. Further work is required to identify the key controls on the documented variability.

Acknowledgement The work reported in this paper formed part of a wider study of the role of river flood plains as sediment sinks supported by a NERC Research Grant (GR3/12635). This support, the assistance of local landowners in permitting access to the study reaches and the help of Helen Jones in producing the figures are gratefully acknowledged.

REFERENCES

Dyer, F. J., Thoms, M. C. & Olley, J. M. (eds) (2002) *The Structure, Function and Management Implications of Fluvial Sedimentary Systems* (Proc. Alice Springs Symp., September 2002). IAHS Publ. 276. IAHS Press, Wallingford, UK.

Hudson, P. F. (2003) Flood plains: environment and process. *Geomorphology* **56**, 223–224.

James, C. S. (1985) Sediment transfer to overbank sections. *J. Hydraul. Res.* **23**, 435–452.

Marriott, S. (1992) Textural analysis and modelling of a flood plain deposit: River Severn, UK. *Earth Surf. Processes Landf.* **17**, 687–697.

Pizzuto, J. E. (1987) Sediment diffusion during overbank flows. *Sedimentology* **34**, 301–317.

Walling, D. E. & He, Q. (1997) Use of fallout ^{137}Cs in investigations of overbank sediment deposition on river flood plains. *Catena* **26**, 263–282.

Walling, D. E. & Owens, P. N. (2003) The role of overbank floodplain sedimentation in catchment contaminant budgets. *Hydrobiologia* **494**, 83–91.

Walling, D. E. & Woodward, J. C. (2000) Effective particle size characteristics of fluvial suspended sediment transported by lowland British rivers. In: *The Role of Erosion and Sediment Transport in Nutrient and Contaminant Transfer* (ed. by M. Stone) (Proc. Waterloo Symp., July 2000), 129–139. IAHS Publ. 263. IAHS Press, Wallingford, UK.

Walling, D. E., He, Q. & Nicholas, A. P. (1996) Flood plains as suspended sediment sinks. In: *Flood Plain Processes* (ed. by M. G. Anderson, D. E. Walling & P. D. Bates), 399–440. Wiley, Chichester, UK.

Walling, D. E., Owens, P. N. & Leeks, G. J. L. (1999) Rates of contemporary overbank sedimentation and sediment storage on the flood plains of the main channel systems of the Yorkshire Ouse and River Tweed, UK. *Hydrol. Processes* **13**, 993–1009.

Decreasing sediment yields in northern California: vestiges of hydraulic gold-mining and reservoir trapping

L. ALLAN JAMES
Geography Department, University South Carolina, Columbia, South Carolina 29208, USA
ajames@sc.edu

Abstract Suspended sediment loads in large rivers are increasing globally, but this trend has reversed in some basins due to dam construction, particularly in developed countries. Sediment loads in lower Sacramento Valley basins began decreasing by 1900 (as shown by G. K. Gilbert's classic 1917 study of hydraulic gold-mining debris), preceding most USA reductions by up to 50 years. From 1853 to 1884, hydraulic mining generated >3.7 billion tonnes of tailings with specific sediment productions up to 16 807 t km^{-2} year^{-1}. Double-mass curves of post-1950 suspended sediment and runoff indicate decreasing sediment responses to streamflow in the lower Sacramento and Feather Rivers continued through the late 20th century. Independent evidence indicates that these reductions were caused by dam closures and the reduced availability of historical alluvium stored below the dams.

Key words historical alluvium; Sacramento Valley, California; suspended sediment; trends

INTRODUCTION

This study examines evidence of a decreasing trend in suspended sediment transport in the lower Sacramento River, a large river draining much of northern California, USA, and several of its main tributaries (Fig. 1). Decreasing sediment loads have been observed in many large

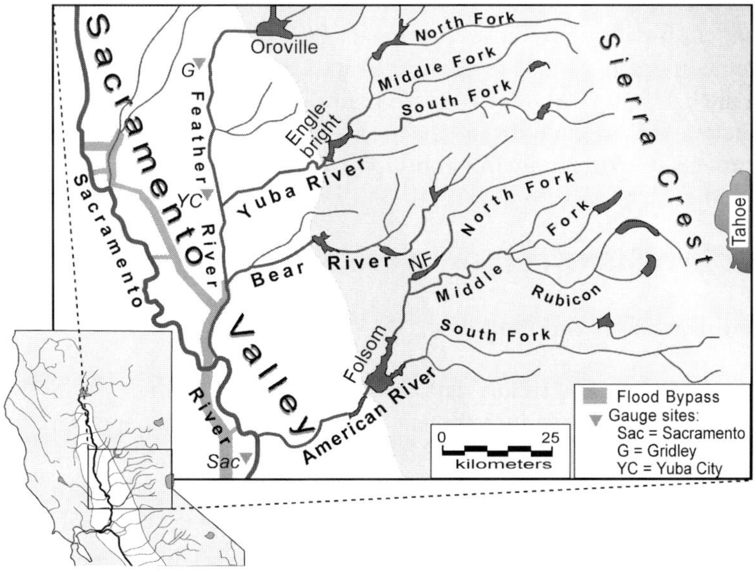

Fig. 1 Lower Sacramento Valley with major tributaries, reservoirs, and gauges.

rivers that were dammed in the mid-20th century, but reductions began early in the lower Sacramento Valley due to relaxation from episodic sedimentation by hydraulic mining from 1853 to 1884. Sediment deliveries were affected by extensive levee construction beginning in the 1860s, abrupt cessation of mining in 1884, and damming by the late 1960s of all large tributaries draining mining districts. The upper Sacramento River received far less historical sediment than the Yuba, Bear, and North Fork American Rivers that have large masses of mining sediment stored where they enter the Sacramento Valley.

The global sediment flux to lowland rivers is of the order of 24 Gt $year^{-1}$ including bedload (Savitsky, 2003) and causes a 1% decrease in reservoir capacity per year at an estimated annual cost of $6 billion (Walling, 1997). Large uncertainties remain in estimates of the global flux of suspended sediment due to the dearth of uniform long-term instrumental records and temporal trends in suspended sediment at individual gauges. Accelerated erosion by human activities has greatly increased sediment production while dams have decreased yields during the period of instrumental records, but these changes have varied greatly in time and space (Walling, 1997). Vörösmarty *et al.* (1997) calculate that the 633 largest reservoirs trap approximately 16% of the global suspended sediment budget, and that all reservoirs may trap more than 25% of global suspended-sediment loads. Systematic temporal changes in large river sediment loadings are so pervasive that Walling & Fang (2003) found statistically significant trends—both positive and negative—in approximately half of the 145 major river long-term sediment records they examined.

Interpretations of extensive sediment-discharge records require an understanding of the magnitude and timing of human impacts on sediment production and retention behind reservoirs (Walling & Fang, 2003; Vörösmarty *et al.*, 2003). Recognition of trends and associated processes is needed if sediment records are to be used for future projections, past reconstructions, or calibration of simulation models. The growing importance of long-established gauges also calls for an understanding of sediment processes in those instrumented basins. Hence, past and present characteristics of basins that have long sediment-concentration records should be carefully evaluated.

Due to the importance of historical sediment in the region, the Sacramento River has a relatively long record of suspended sediment measurements. Interpretation of this record is easily confounded, however, by an exceptional history of sediment production and storage, extensive levee construction, and regulation of all large tributaries by reservoirs. This study examines evidence of temporal trends in sediment fluxes in the lower Sacramento and Feather Rivers and factors influencing those changes. These trends are placed in the context of channel adjustments to extreme hydrogeomorphic changes in the basins.

RECORDS OF SEDIMENT FLUX AND CHANNEL ADJUSTMENTS

Suspended sediment data and methods

Daily suspended sediment and runoff data from US Geological Survey (USGS, 2003) gauging stations are used to show trends in sediment production and transport. Data from five gauges (Table 1) provided instrumental records of suspended sediment concentrations and corresponding water discharges from as far back as 1957. These data were used by the USGS to calculate daily sediment fluxes and mean daily runoff (USGS, 2003). The runoff data have few missing values, but occasional periods of missing sediment concentrations required the USGS to empirically estimate daily loads from discharge. Uniform data-collection

Table 1 Stream gauges cited in this report.

Gauge name	ID	Basin area, km^2	Water years
Deer Creek near Smartsville	11-4185.00	219	1974–1979
Feather River at Gridley	11-4071.50	9521	1965–1992
Feather River at Yuba City	11-4077.00	10 293	1964–1976
Sacramento River at Sacramento	11-4475.00	60 885	1957–1979
Sacramento River at Freeport	11-4476.50	~61 000	1979–1989

methods utilized by the USGS (Edwards & Glysson, 1999), enhance the reproducibility and interpretability of analyses derived from these data. Double-mass curves are used here to isolate long-term trends by removing the seasonal and inter-annual variability in sediment loads associated with discharge fluctuations. Walling (1997) and Walling & Fang (2003) used this method with cumulative mean annual values of specific sediment yield and runoff. Here, cumulative values are calculated from mean daily data and are on the order of 365 times larger than values drawn from mean annual data.

The suspended sediment record for the lower Sacramento River is derived by combining data from two gauge sites. Measurements were made at the Sacramento gauge from water years 1957–1980, after which the gauge moved approximately 17 km downstream to Freeport where another 10 years of data were collected. No large tributaries enter between the two gauge sites.

Suspended sediment double-mass curves

Suspended sediment responses to stream flow on the Feather and Sacramento Rivers steadily declined in the second half of the 20th century as is shown by the progressive lowering of slopes on double-mass curves. At the Sacramento gauge, no change in sediment flux rates with runoff was apparent in 1980, suggesting that the change in gauge site had little effect on sediment responses to discharge (Fig. 2). Sediment responses decrease gradually, however, throughout the period from 1957 to 1989. Events such as the 1965 and 1986 floods and the 1978 high flows following a severe drought, appear as brief sediment-flux events that are minor when compared to the long-term trend at this site. The sediment flux during the record 1986 flood was less than that of the 1965 flood indicating a decreased response in sediment to runoff. Although the mean daily discharge of the1986 flood was 16% greater than the 1965 flood at this gauge, it produced less than half the maximum mean daily sediment flux (3.3 t day^{-1} km^{-2}) as the 1965 flood (7.8 t day^{-1} km^{-2}). In fact, the four largest mean daily sediment fluxes on the Sacramento River for the 32-year period of record at the Sacramento and Freeport gauge sites were in 1965 and the fifth largest flux was in 1963. Singer & Dunne (2001) examined 17 years (1963–1979) of annual sediment discharge data from the Sacramento gauge and found no statistically significant evidence of a temporal trend over that period. While the double-mass analysis presented here does not constitute a significance test, the gradual long-term reduction in sediment responses to runoff shown by double-mass curve analysis together with the limited loads carried by the 1986 flood suggest that this longer record has a trend.

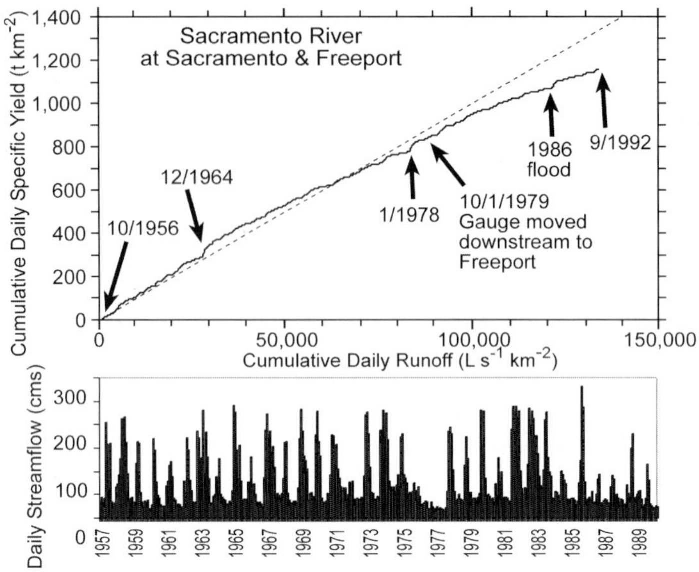

Fig. 2 Double-mass curve showing decreasing suspended sediment responses to discharge. Reference line (dashed) represents sediment flux of one t per 100 l s^{-1}. Hydrograph shows annual floods and floods of record in water years 1965 and 1986.

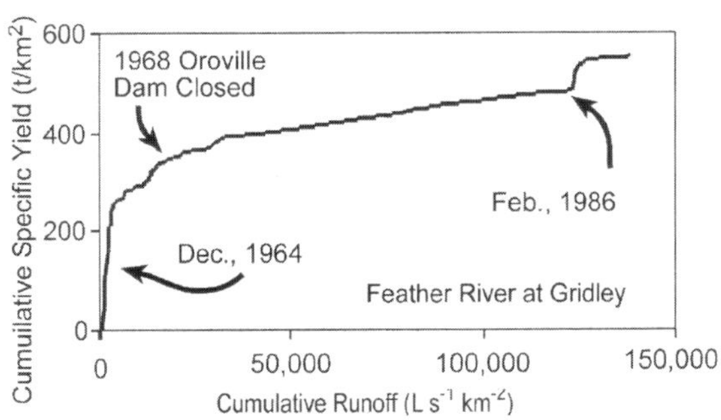

Fig. 3 Double-mass curve of mean daily suspended sediment and runoff for Feather River at Gridley. The December 1964 flood is a major sediment event; subsequent flux rates, including the record 1986 flood, are much lower.

A similar pattern of long-term decreasing sediment flux rates was revealed at the Gridley and Yuba City gauges on the Feather River below the Oroville Dam and above the Yuba River confluence at Marysville (Table 1). The double-mass curve for the Gridley gauge (Fig. 3) shows a pronounced sediment response to the 1965 flood that was equally apparent at the Yuba City gauge approximately 30 km downstream. In a 30-day period ending in January 1965 the flood carried 250 t km^{-2} at the Gridley gauge and almost 150 t km^{-2} at Yuba City. This comprised 45% of the entire suspended load over the 28-year period at the Gridley gauge and almost 20% of the entire load for the 12-year period at Yuba City. Closure of Oroville Dam in 1968 did not immediately lower sediment responses, presumably due to production of sediment from local channel storage during moderate-magnitude floods and lack of large floods until 1986. The 1986 flood, the flood of record at many gauges in this region, carried much less sediment at Gridley than the 1965 flood. At the Gridley gauge, the 1986 flood (4130 $m^3 s^{-1}$) had essentially the same magnitude as the 1965 flood (4220 $m^3 s^{-1}$) due to flood storage in Oroville Reservoir, but the mean daily sediment concentration on the day of peak flow was almost three orders of magnitude less in 1986 (456 mg l^{-1}) than in 1965 (345 000 mg l^{-1}). This subdued sediment response to the 1986 flood was presumably due to trapping of sediment in Oroville Reservoir and reduction of erodible sediment below the dam.

The variability of suspended sediment with discharge after 1968 was also reduced at the Gridley gauge. A power function was fit to the suspended sediment and discharge data and regression residuals from this long-term rating curve were calculated. Positive residuals largely correspond to the rising limbs of hydrographs at this gauge indicating the occurrence of a "first-flush" during floods. When plotted against time, the largest positive residuals cluster prior to 1968 (Fig. 4). Subsequent reduction of the first-flush response shows that supplies of sediment easily eroded during the rising limb were greatly reduced after closure of the Oroville Dam. These supplies may include both mountain sources above the dam and historical alluvium in the lower river, but they suggest an important change in the timing and sources of sediment following dam closure.

Few small tributaries in the region have long records of suspended sediment. Deer Creek, a small Yuba River tributary, is a small basin in the heart of the mining districts for which a short record of daily suspended sediment discharges is available. Deer Creek is unique in that it not only received large volumes of mining sediment, but also remained relatively unregulated by reservoirs below areas of extensive mining. The Deer Creek double-mass curve for suspended sediment and runoff for the six-year period 1974 to 1979 shows a high sensitivity of sediment transport to two runoff events (Fig. 5). High flows in 1974 and 1978 were not large runoff events, but they generated high proportions of the total sediment load for this period. Sensitivity of sediment transport to runoff in Deer Creek is also reflected by the close relationship between daily runoff and daily suspended sediment in the sediment rating curve (Fig. 6). This is characteristic of a transport-limited system with abundant erodible sediment. In sediment-starved basins, sediment-discharge relationships tend to be weak due to dominance of intermittent hillslope processes as sediment production mechanisms.

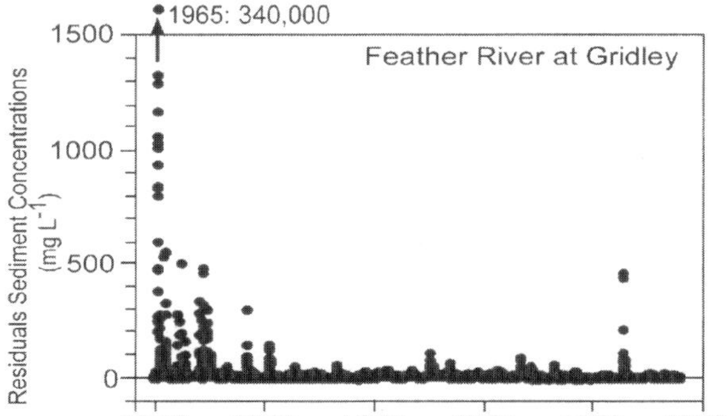

Fig. 4 Residuals from regression of daily mean suspended-sediment concentration on daily mean discharge for Feather River at Gridley, 1965 to 1988. Large positive residuals in 1960s indicate high loads on hydrograph rising limbs prior to Oroville Dam closure in 1968.

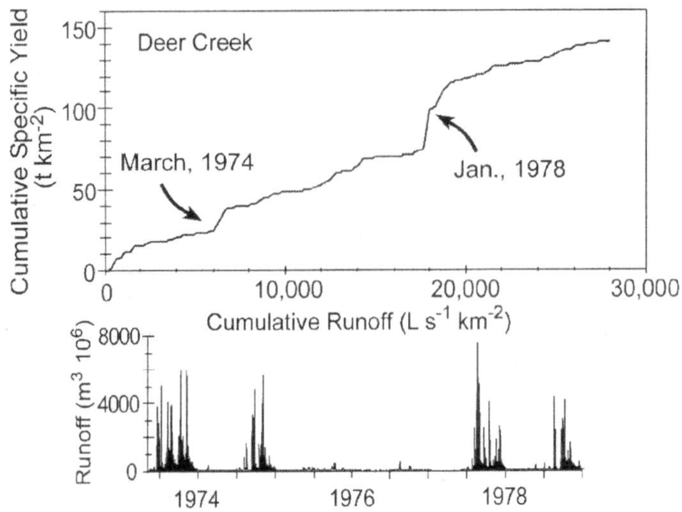

Fig. 5 Double-mass curve for Deer Creek, a small tributary to lower Yuba River that received much hydraulic mining sediment and has only minor regulation by dams.

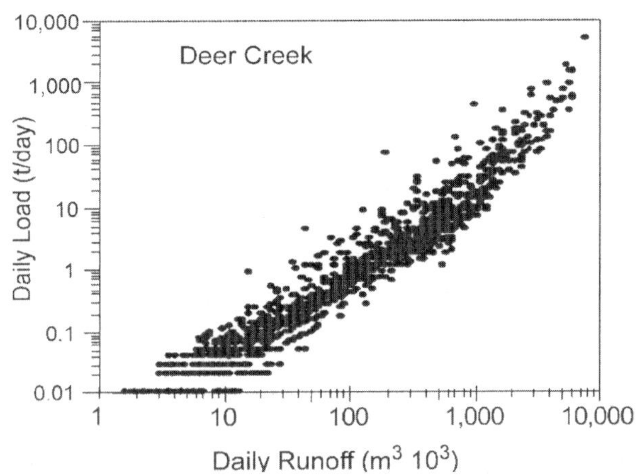

Fig. 6 Sediment rating curve for Deer Creek shows close relationship between mean daily discharge and daily suspended sediment load from 1974 to 1979.

It is difficult to statistically test the available suspended sediment data for the significance of temporal trends caused by reservoir trapping. Extensive records of sediment data are not available for the Yuba, Bear, and American Rivers that received the most mining sediment. Sediment data collected at the Feather River gauges did not begin until three or four years prior to closure of Oroville Dam and include the large 1965 flood in the pre-dam period. The Sacramento River gauge at Sacramento integrates sediment trapping by numerous dams constructed over an extended period, so it is not possible to isolate the effects of one dam. Independent evidence, however, indicates regional channel deepening and enlargement associated with substantial sediment production that supports the interpretations of a decreasing trend in sediment loads.

Stage lowering as evidence of sediment production

Stage–discharge relationships developed from archival USGS hydrographic records provide an independent test for trends, push the observational database back another 50 years, and expand the spatial resolution of observations to locations where suspended sediment data are not available. Changes in channel morphology and sediment production can be inferred from the lowering of flow stages through the examination of channel cross-sections and hydraulic geometry (James, 1999). Stage–discharge statistical regressions presented in previous studies document the timing of flow-stage lowering at several foothill gauges. For example, substantial flood-stage reductions were documented at three streamgauge sites on the lower American River including almost 2 m of flood-stage lowering at a critical bend in Sacramento (NRC, 1995; James, 1997). Stage lowering at these sites was shown to correspond to local channel incision and enlargement and the timing was related to closure of the North Fork Dam upstream. Similarly, at a site on the lower Bear River, flow stages decreased about 2 m from the early 1950s to 1970 (James, 1991). The timing of incision at this site was not related to the enlargement of Camp Far West Reservoir upstream in 1963 as hypothesized, but was delayed because the channel had avulsed onto an armoured layer in the 1870s.

DISCUSSION: SPATIAL AND TEMPORAL PATTERNS OF SEDIMENT LOADS

Sediment from the mountain areas is now prevented from reaching the Sacramento Valley by large dams, but below the dams are large repositories of historical sediment. Clearly, these storage sites were actively producing sediment during the 20th century. Important questions to be addressed are the extent to which these sources continued to supply sediment during the late 20th century, their present importance, and systematic trends. Singer & Dunne (2001) analysed gauge data on the Sacramento River using a time-series analysis of daily suspended sediment records for the period from 1948 to 1979. They constructed a sediment budget using six gauges on the main river and an additional gauge on each of four major tributaries including the Feather River. Their budget identifies zones of increasing and decreasing sediment loads that define a distinct spatial pattern of sediment production in the Sacramento Valley. The greatest erosion occurred in the lower Sacramento River between the mouth of the Feather River and the Sacramento gauge where aggradation by historical mining sediment from the Yuba, Bear, and American Rivers was most severe. Erosion there is consistent with an interpretation of long-term readjustments to hydraulic mining sedimentation. Singer & Dunne (2001) emphasize the importance of failed levee banks and bed erosion as sediment sources and cite evidence of 42 cm of bed incision in the Sacramento River above the Feather River between 1965 and 1979. Additional (perhaps the dominant) sediment sources to the Feather River and lower Sacramento below the Feather confluence are historical alluvium stored in terraces along the Yuba, Bear, lower Feather, and lower American Rivers. More study is needed of sediment production by channels widening and eroding these terraces.

Mining sediment deposits, levee construction, and dams

Sediment budgets in the lower Sacramento Valley cannot be fully appreciated without an understanding of the massive sediment deposits remaining along the east margin of the

Table 2 Nineteenth century mining sediment produced (1853–1884).

Basin	Drainage area (km^2)	Volume produced [A] (m^3 10^6)	Vol. year^{-1} 31 years (m^3 10^6 year^{-1})	Denudation (mm)	(mm year^{-1})	Mass Produced @ ρ =2.2[B] (Mt)	Mass year^{-1} (Mt year^{-1})	Spec. prod. (t km^{-2} year^{-}
Feather River								
at Yuba City	10301	77	2.5	7.4	0.2	168	5.4	527
Yuba River	3499	523	16.9	149.6	4.8	1151	37.1	10616
North Yuba[C]	1351	165	5.3	122.4	5.3	364	11.7	8685
Middle Yuba	536	109	3.5	203.7	6.6	240	7.7	15192
South Yuba	988	165	5.3	167.2	5.4	363	11.7	11794
Deer Creek	233	29	0.9	126.4	4.1	65	2.1	8968
Bear Basin	1143	271	8.7	236.8	7.6	596	19.2	16807
American	5014	197	6.3	39.2	1.3	433	14.0	2783
North Fork	900	164	5.3	181.7	5.9	360	11.6	13069
Middle Fork	1586	33	1.1	20.9	0.7	73	2.4	1486
Totals:	25437	1663	53.7	65.4	2.1	3659	118.0	4640

[A] Production volumes from Benyaurd *et al.* (1891) multiplied by 1.51 as recommended by Gilbert (1917).
[B] ρ = 2.2 g cm^{-3} is a conservative approximation of densities of bedrock materials mined.
[C] North Yuba above South Yuba confluence not including South or MiddleYuba.

Sacramento Valley. Hydraulic mining was associated with rapid sediment production and resulted in dramatic sedimentation along the Yuba, Bear, and North Fork American Rivers (Gilbert, 1917; James, 1999). Volumes and masses of sediment produced by 19th century hydraulic mining are given in Table 2. Volumes were derived from Benyaurd *et al.* (1891) adjusted using a coefficient of 1.51 based on Gilbert's (1917) surveys of mine-pits exhumed in regolith and consolidated rock—cemented conglomerates with a late Tertiary volcanic overburden. Volumes were converted to mass using a density of 2.2 g cm^{-3}. Mining sediment was produced for 31 years at an average rate of 118 Mt year^{-1}, equivalent to 0.6% of the modern global annual suspended sediment yield delivered to lowland rivers (approx. 20 Gt year^{-1}). However, only a fraction of the sediment produced was delivered to the lower Sacramento River and much was stored in the mountains and in low-gradient fans along the margin of the Sacramento Valley. Upland denudation in these large basins during the mining period was as high as 237 and 214 mm at rates of 7.6 and 6.6 mm year^{-1} for the Bear River and Middle Yuba Basins, respectively (Table 2).

Most of the mining sediment remains in terraces in the mountains and deep flood plain deposits along the lower reaches of the rivers where they enter the Sacramento Valley. These historical deposits continue to be reworked, although dams prevent much of the sediment from being delivered to the Sacramento River. Levees isolate large volumes of the historical alluvium, but they also deepen flows and facilitate erosion of historical alluvium stored in terrace and channel-margin deposits (Fig. 7). Deep deposits of stratified sands and silts along these rivers are highly erodible and continue to be reworked (Fig. 8). For example, repeated topographic surveys on the lower Bear River in 1985 and 1989 documented 82 m^2 of net cross-section enlargement by the 1986 flood representing 82 000 m^3 km^{-1} of sediment supplied to the lower Sacramento from this reach below all dams (James, 1993). Scour down to the pre-mining surface was indicated by exhumed tree stumps, and additional historical alluvium remaining on the left channel margin indicates that the channel capacity has not yet been restored at this site.

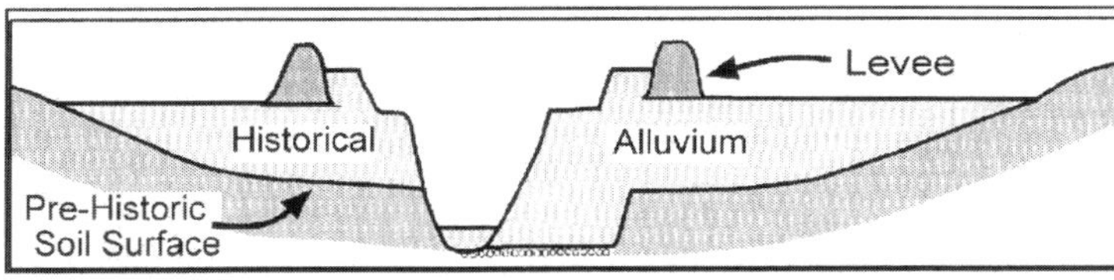

Fig. 7 Typical relationship between pre-mining surface, historical alluvium, and levees in major tributaries from mining districts in Sacramento Valley. Vertically exaggerated.

Fig. 8 Exposure of erodible historical alluvium in right bank of lower Yuba River above railroad bridge near Feather River confluence. (A) Stratified silts in vertical face undercut at base by sand layer. (B) Upper layer of clean white sand wasting down face of exposure.

CONCLUSION

Understanding the long-term sediment dynamics in the lower Sacramento basin requires an appreciation for the extensive deposits of erodible historical alluvium in major tributaries of the Sierra foothills. Channels along the margins of the Sacramento Valley incised and enlarged in the 20th century in response to reduced sediment loads after mining stopped and dams were constructed. This widespread erosion represents a tremendous flux of sediment from storage. Double-mass curves provide evidence of decreasing sediment fluxes for a given discharge from the mid- to late-20th century. Stage-discharge relationships and other evidence of channel incision and enlargement indicate that much of the sediment production came from channel erosion. While field evidence indicates that erosion of historical alluvium in the lower Sacramento Valley continues, the double-mass curves suggest that available supplies are decreasing. The relative importance of sediment production from historical alluvium in the lower Sacramento River system should be systematically quantified.

The on-going reworking of historical alluvium in the lower Sacramento Valley has important implications to the longevity of episodic sedimentation events. Much of the vertical adjustment in these channels occurred rapidly (Gilbert, 1917), but after initially incising, the channels have continued to widen during large floods, except where prevented by levees or other channel-protection measures. Gilbert's (1917) symmetrical wave model implies a relatively short-lived sediment response followed by rapid stabilization and reduction of sediment loads to pre-sedimentation levels. Use of the wave model to infer sediment yields is based on the conflation of low-flow channel-bed elevations with sediment loads, however, and does not stand up to scrutiny in the lower Sacramento Valley where it was initially conceived. If future global sediment fluxes are projected on the basis of this assumed wave symmetry, they will likely underestimate long-term sediment discharges following episodic sedimentation events. When sediment production exceeds channel capacities to the extent that large sediment volumes are stored in floodplain and terrace deposits, elevated sediment yields may persist for centuries.

REFERENCES

Benyaurd, W. H. H., Heuer, W. H. & Handbury, T. (1891) Report of Board of Engineers on Mining-Débris question in State of California. Appendix VV, Report of Major W.H. Heuer. House Exec. Doc. 267. 51st US Congress, 2nd Session.

Edwards, T. K. & Glysson, G. D. (1999) Field methods for measurement of fluvial sediment. Chapter C2. In: *Techniques of Water-Resources Investigations of the US Geol. Survey*; Book 3, *Applications of Hydraulics*. US Dept. Interior, Reston, Virginia, USA.

Gilbert, G. K. (1917) Hydraulic-Mining débris in the Sierra Nevada. *US Geol. Survey Prof. Paper 105*.

James, L. A. (1991) Incision and morphologic evolution of an alluvial channel recovering from hydraulic mining sediment. *Geol. Soc. Am. Bull.* **103**, 723–736.

James, L. A. (1993) Sustained reworking of hydraulic mining sediment in California: G. K. Gilbert's sediment wave model reconsidered. *Z. Geomorphol. N.F; Suppl.-Bd.* **88**, 49–66.

James, L. A. (1997) Channel incision on the lower American River, California, from streamflow gage records. *Water Resour. Res.* **33**, 485–490.

James, L. A. (1999) Time and the persistence of alluvium: river engineering, fluvial geomorphology, and mining sediment in California. *Geomorphol.* **31**, 265–290.

National Research Council (NRC) (1995) *Flood Risk Management and the American River Basin: An Evaluation*. National Academy Press, Washington, DC, USA.

Savitski, J. P. M. (2003) Supply and flux of sediment along hydrological pathways: research for the 21st century. *Global & Planetary Change* **39**, 1–11.

Singer, M. B. & Dunne, T. (2002) Identifying eroding and depositional reaches of valley by analysis of suspended sediment transport in the Sacramento River, California. *Water Resour. Res.* **37**, 3371–3381.

US Geological Survey (2003) Suspended-sediment Database; Daily Values of Suspended Sediment and Ancillary Data. http://webserver.cr.usgs.gov/sediment/.

Vörösmarty, C. J., Meybeck, M., Fekete, B. & Sharma, K. (1997) The potential impact of neo-castorization on sediment transport by the global network of rivers. In: *Human Impact on Erosion and Sedimentation* (ed. by D. E. Walling & J.-L. Probst) (Proc. Rabat, Morocco Symp., April 1997), 261–274. IAHS Publ. 245. IAHS Press, Wallingford, UK.

Vörösmarty, C. J., Meybeck, M., Fekete, B., Sharma, K., Green, P. & Syvitski, J. P. M. (2003) Anthropogenic sediment retention: major global impact from registered river impoundments. *Global & Planetary Change* **39**, 169–190.

Walling, D. E. (1997) The response of sediment yields to environmental change. In: *Human Impact on Erosion and Sedimentation* (ed. by D. E. Walling & J. -L. Probst) (Proc. Rabat, Morocco Symp., April 1997), 77–89. IAHS Publ. 245. IAHS Press, Wallingford, UK.

Walling, D. E. & Fang, D. (2003) Recent trends in the suspended sediment loads of the world's rivers. *Global & Planetary Change* **39**, 111–126.

Patterns of erosion and sediment transport in the Murray-Darling Basin

RONALD DE ROSE[1], IAN PROSSER[2] & MARTIN WEISSE[3]

1 *CSIRO Land and Water, GPO Box 1666, Canberra, ACT 2601, Australia*
ronald.derose@csiro.au

2 *Land & Water Australia, GPO Box 2182, Canberra, ACT 2601, Australia*

3 *Institute of Geography, Schneiderberg 50, D-30167 Hanover, Germany*

Abstract The Murray-Darling Basin in the southeast of Australia covers approx. 1×10^6 km^2, equivalent to 14% of the country's total area. Accelerated erosion, primarily in upland regions, has greatly degraded river habitats over the past century. Here we describe the results from the basin-wide sediment modelling for this region using SedNet: a spatially distributed sediment routing model which predicts sediment loads, together with bed, flood plain and reservoir deposition. Comparisons are made between current (past 100 years) and natural (pre-European) conditions. The results demonstrate that the basin is one of sediment redistribution rather than net sediment export. The basin is estimated to have 18% of rivers with >100 times natural loads. Erosion of gullies and riverbanks has also resulted in 11 000 km (13% of the total) to have historical accumulation of over 0.3 m of sand and gravel averaged over the river length.

Key words erosion; Murray-Darling Basin, Australia; sediment budget; sediment load; spatial model

INTRODUCTION

Extensive modification to the natural environment in many parts of Australia began with the arrival of European farmers in the early part of the 19th century. Modification of the natural environment, principally through clearance of the native Eucalypt forests, initiated a phase of accelerated erosion: the effects of which continue to the present day (Scott, 2001). Many of the ephemeral streams and creeks that previously flowed across swampy meadows and formed "a chain of ponds", began to incise into alluvial valley fills, releasing large volumes of sediment that subsequently caused extensive sand-bed aggradation and increased suspended sediment loads. While there is evidence for declining erosion rates in recent times, erosion rates remain high in areas where poor land management practices continue (Scott, 2001).

The impacts of land degradation on river systems are an important issue for land management authorities. Targeting and prioritizing the source areas of erosion that contribute disproportionately to river loads is an efficient way of maximizing returns from investment into soil, land and river rehabilitation (Lu *et al.*, 2003). To achieve this necessitates the mapping of erosion processes and rates and the construction of a sediment budget which links river loads with upstream sources of sediment. The only practical way to assess the patterns of erosion and sediment transport across large complex areas such as the Murray-Darling Basin (Fig. 1) is a spatial modelling framework. There are often few direct measurements of sediment transport in regional catchments, and it is unrealistic to initiate sampling programmes of the processes now and expect results within decades (Prosser *et al.*, 2001a).

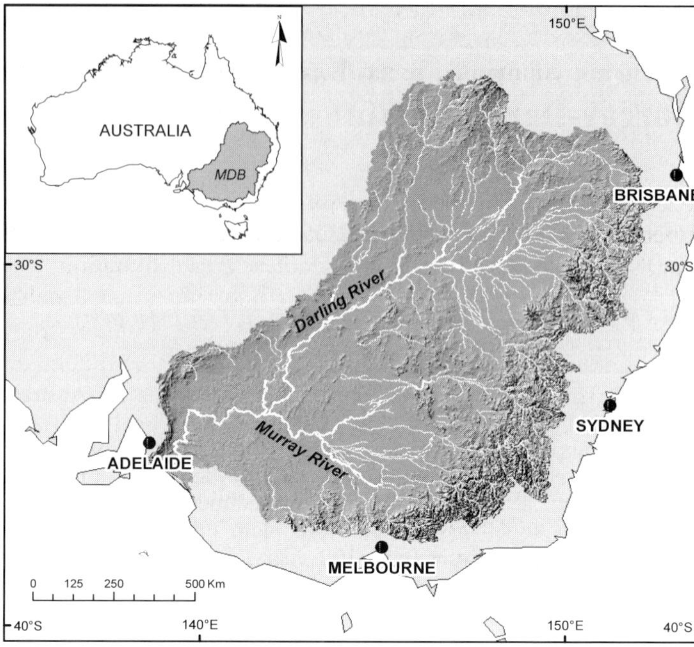

Fig. 1 The Murray-Darling Basin and river network.

This paper summarizes results from application of a spatially distributed sediment model (SedNet) to the Murray-Darling Basin (DeRose *et al.*, 2003). SedNet couples erosion, hydrological, and sediment transport modelling to predict sediment loads and deposition throughout river networks, utilizing digital elevation models (DEM) and regional resource surveys (e.g. land use, vegetation, erosion, soil, flood plain maps). The model was first developed at national scales for the Australia NLWRA (Prosser *et al.*, 2001b), but is now routinely applied in regional scale assessments.

STUDY AREA

The Murray-Darling Basin (MDB) is large, covering approx. 1×10^6 km^2 or about 14% of Australia's land area (Fig. 1). The Basin contains >20 major river systems, including the three longest rivers in Australia, the Darling, Murray, and Murrumbidgee.

The MDB covers a diverse range of environments from the alpine margin of the Great Dividing Range in the east, to tropical Queensland in the north, to the central low-lying riverine plain which dominates the region, and to the sparsely vegetated arid interior in the west. Average annual rainfall shows a strong regional gradient from >1000 mm along the Great Dividing Range in the east, to as little as 180 mm in the arid west. Rainfall is highly variable and can swing from periods of above average rainfall and frequent flooding to protracted periods of drought, such as in recent years, when rainfall was well below average and many streams and rivers ceased to flow.

The natural vegetation at the time of European settlement ranged from dry and wet *sclerophyll* forest communities in the east to more open woodlands including *mallee* forests and native grasslands in the west, all of which were largely dominated by *Eucalyptus* spp. At least half of the pre-European vegetation cover has now been removed and converted for cropping, grazing, and less commonly, horticulture, viticulture or plantation forests.

Most rivers begin in the wetter and more mountainous region along the eastern and northern margins of the Basin where, although drainage networks are initially well defined, they soon flow onto the extensive riverine plain and become much more sinuous and anabranching in form. There are at least 9 major distributaries systems in the MDB where rivers split into between 2 to 5 anabranches, each diverging by up to 10–20 km, before coalescing to form the Murray and Darling Rivers. The greatest continuous length of river in the Basin extends 3750 km from the source of the Condamine, near Brisbane, to the mouth of the Murray, near Adelaide (Fig. 1). Only streams that carry sufficient flow merge with the major rivers and most smaller streams and rivers in the western arid zone dry up well before reaching the Darling River.

All major rivers are now heavily regulated due to demands on fresh water for domestic consumption, agricultural production and industry. There are many dams, reservoirs and river off-takes throughout the MDB that have reduced both mean annual discharge and flow variability throughout the river network. Outflow from the mouth of the Murray River now averages 4.9 Gl year^{-1}. This represents 36% of the long-term flow of 13.7 GL year^{-1} estimated under natural conditions, despite there being significant inflows from coastal rivers as part of the Snowy Mountain Hydroelectric Scheme. Annual diversions have gradually climbed from 3 GL in the 1930s to 11 Gl in the early 1980s and now remains capped at this level (Murray-Darling Basin Commission).

METHODS

The conceptual framework for the SedNet model is described by Prosser *et al.* (2001a) and details of model structure, parameterization, assumptions and limitations are given by Prosser *et al.* (2001b). A brief overview of the SedNet model follows. SedNet is a sequence of Arc-GIS routines coded in ARC Macro Language (AML) script. The model first builds a link representation of the river network from a digital elevation model (DEM), separated by tributary junctions or nodes. For the MDB, the river network was defined from a 9″ resolution (approx. 250 m) DEM producing 9900 river links averaging 12 km in length and with an upslope contributing area >50 km^2. Each link in the river network has an associated link watershed area averaging 100 km^2. The river links and associated watersheds form the basic processing elements for the model.

Each river link, receives a mean annual supply of suspended (silt and clay) and bedload (sand and gravel) sediment from upstream tributaries, and from gully, hillslope, and riverbank erosion in the associated watershed area. Each river link similarly loses sediment by deposition on the flood plain (suspended) and on the bed of the channel (bedload) when the sediment transport capacity of the link is exceeded. Reservoirs and lakes trap all passing bedload while a proportion of the suspended load is trapped depending on mean annual inflow and storage capacity.

Each of the three main erosion sources are separately modelled and calibrated to local measurements where possible. Hillslope (sheetwash and rill) erosion is modelled using the RUSLE model, adjusted for seasonal variation in rainfall and cover factors (Lu *et al.*, 2001), and scaled to 50 km^2 by the sediment delivery ratio (SDR). The SDR represents the probability of eroded soil reaching the river network and is predicted to average 0.044 and vary mostly between 0.001 in arid regions and 0.4 in wetter regions (Lu *et al.*, 2003). As hillslope erosion generates mostly fine sediment it was considered to contribute only to the suspended load. Gully erosion was estimated using rule-based statistical extrapolation of air photo measurements of gully density (Hughes & Prosser, 2003) while river bank erosion is based on an empirical model relating a global review of meander migration rate (Rutherfurd, 2000) to stream power, extent of riparian vegetation, and flood plain width (Hughes & Prosser, 2003). Limited available data and field observations of bank texture suggest a relatively even contribution of sediment (e.g. Dietrich & Dunne, 1978) from riverbank and gully sources to the suspended and bedload of rivers. A suspended to bedload sediment contribution ratio of 40:60 was found to provide the best overall agreement between observed river loads and known extents of bedload accumulation.

A number of hydrological and channel morphological parameters (width, depth) are used to predict sediment transport and deposition. Flood plain deposition is based on the proportion of flow that goes over bank during the median flood event. Bank full discharge was considered equivalent to the 1.58 year recurrence interval flow on the annual maximum time series. The hydrological parameters were calculated from the time series of daily flows measured at hydrometric gauging stations and interpolated across the remainder of the river network by correlation with mean annual flow (Q_a). Q_a was in turn derived by regionalization of the rainfall runoff coefficient. The model further considers regulated flow and diversion of flow along anabranching networks.

The sediment budget for natural (pre-1820) conditions assumed no significant gully erosion, the RUSLE model applied to natural vegetative cover (Lu *et al.*, 2001) and a level of riverbank erosion equivalent to 5% of current levels.

SEDIMENT BUDGET

The sediment budget (Table 1) shows that only a small portion of the 28.7 Mt year^{-1} of sediment entering the river network reaches the mouth of the Murray River. The MDB is therefore one of net sediment redistribution rather than net sediment export. Hillslope erosion is predicted to be the dominant source of suspended sediment whereas gully erosion is the dominant source of bed load. Of the sediment eroding from upland areas of the basin 46% is deposited on flood plains, 20% is trapped in reservoirs and 33% is deposited within

Table 1 Sediment Budget for the Murray-Darling Basin.

Sediment budget item	Predicted mean annual rate (Mt year^{-1})		
	Suspended	Bed load	Natural
Hillslope delivery	8.4		0.3
Gully erosion rate	4.6	7.0	0
Riverbank erosion rate	3.5	5.2	0.4
Total sediment supply	16.5	12.2	0.7
Flood plain deposition	13.1		0.3
Reservoir deposition	3.3	2.6	0
Total bed accumulation		9.6	0.3
Sediment export	0.1	0.0	0.1

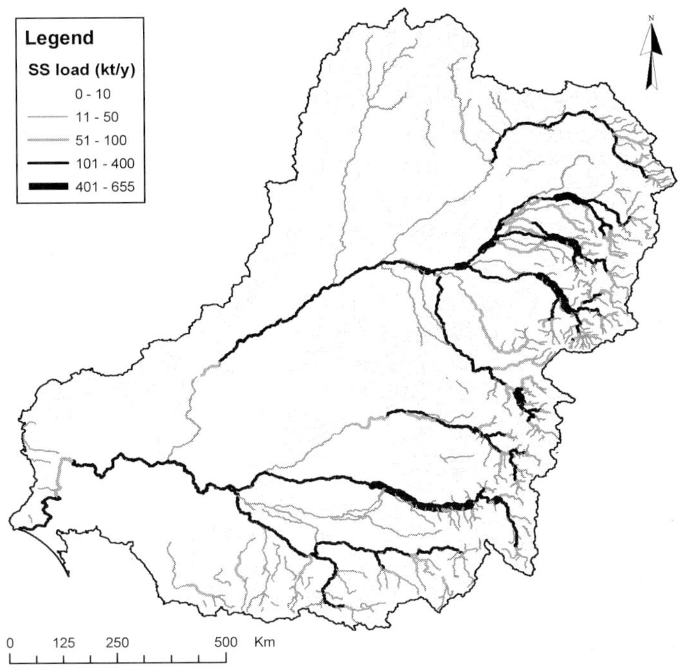

Fig. 2 Suspended sediment loads (SS) across the MDB.

channels as sand slugs. Under natural conditions, only 3% of the present day amount of sediment is predicted to enter the river network, but as no reservoirs and much higher flows existed prior to regulation, then rates of sediment export from the basin as a whole would appear to have been similar to the present day.

River suspended sediment (SS) loads vary considerably across the Basin (Fig. 2). In general they are greatest, typically exceeding 100 and occasionally 400 kt year^{-1}, along the main reaches of rivers draining the eastern and southern basin margins. In these regions specific SS yields are equivalent to 10–50 t km^{-2} year^{-1} and thus low by international standards. Yields decline with distance downstream and are predicted to be below 1 t km^{-2} year^{-1} for much of the lowland river network. Regions of the river network with highest loads, together with upland areas having extensive soil erosion, are those most impacted by increases in SS relative to natural conditions. Overall, 63% and 18% of the river network is predicted to have SS loads above 20 and 100 times natural rates.

The sediment budget model also predicts extensive sand accumulation as a result of accelerated gully and river bank erosion. In all, some 20% of the river network is affected by sand deposition and 13% is predicted to have sand accumulation over 0.3 m in thickness where significant impacts on fresh water habitat can be expected. The river reaches most affected by sand accumulation are those lying below regions of extensive gully and bank erosion and where there is a decrease in sediment transport capacity owing to decrease in channel slope. Consequently most sand slugs are predicted to occur where there is a transition from the higher energy streams of upland areas to the lower energy of the more sinuous lower gradient rivers of the lowlands. This pattern is generally consistent with field observations of the sites of sand slug deposition in regional catchments.

MODEL EVALUATION

Each of the erosion process sub-models contain significant localized error due to limited spatial resolution of data sources. This, together with model assumptions where many of the parameters are treated as global constants (e.g. bank full discharge = 1.58 year recurrence flow, coarse to fine sediment production ratio of 0.6:0.4) means that there is potentially significant spatial error in predicted loads.

The performance of the present model results can be determined by comparison with measured loads from river quality sampling sites and the results of radionuclide ($^{137}Cs/^{210}Pb$) tracer studies which determine the relative proportion of river sediment derived from subsoil (gully and riverbank) and hillslope sources (Wallbrink *et al.*, 1998). Figure 3 shows that despite the uncertainties in source terms there is reasonable agreement between predicted and observed specific SS yields for a number of sites within major southern tributaries of the Murray River. Sites with upslope basin areas of >3000 km^2 have a relative error of 0.30 while those for smaller basin areas have a greater error of 0.82, suggesting a decrease in model accuracy with decreasing scale. Care is therefore needed in the interpretation of predicted loads for the smaller rivers and streams.

At the lower end of the river system, regular measurement of turbidity (Mackay *et al.*, 1998) indicates that the Murray has a lower average turbidity (30 NTU) than the Darling (109 NTU) above the confluence of the two rivers. This contrasts with average annual SS concentrations of 35 and 12 mg l^{-1}, respectively, based on predicted SS yields and discharge at the same locations. Given that river loads appear reasonably well represented for tributaries of the Murray River (Fig. 3) then this difference tends to imply under prediction of loads in the Darling River. Radionuclide studies in the Namoi River, a tributary of the Darling, suggest that >70% of transported sediment is derived from predominantly bank erosion sources (Olley & Scott, 2002) and this contrasts with a much lower model estimate of approx. 30%. This, together with estimates of river loads based on turbidity measurements (Olley & Scott, 2002) which are about four times those predicted by SedNet, point to significant underestimation of the levels of river bank erosion for some basin areas and this has led to under prediction of river loads for the Darling River. Furthermore, Olley & Caitcheon (2000) suggest that, on the basis of major element chemistry, most of the sediment presently transported in the Darling-Barwon River is derived from weathered granite and

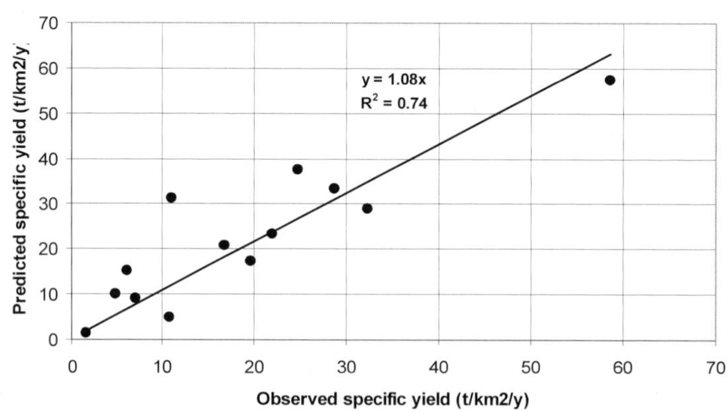

Fig. 3 Comparison of predicted and observed specific SS loads.

sedimentary bedrock in lowland areas rather than originating from contemporary erosion of intensely farmed basalt terrain in upland areas.

MODEL DEVELOPMENT

The present sediment budget and basin-wide estimates of SS loads provide a synoptic overview of erosion and sediment transport throughout the MDB. When compared with measured loads, the present results also provide an examination as to where potential deficiencies in the model occur. Currently there are few measurements of river bank erosion in Australian rivers and this means that significant localized under or overestimation in river loads will occur due to a lack of effective model calibration or parameterization. Similarly, as routing of suspended sediment through river networks depends on channel geometry and the frequency of bank full discharge and overbank flooding, then poor spatial representation of these parameters will also lead to significant localized error in prediction of river loads. There is growing evidence that the statistically derived 1.58-year recurrence interval flow provides poor spatial representation of, and may in fact underestimate, bank full discharge for many Australian basins. Lower than actual bank full discharge may have led to overestimation of the levels of flood plain deposition and is therefore also a likely cause for under prediction of river loads along lower reaches of the Darling River.

The present sediment budget is considered a major improvement over previous studies and is the best that can be achieved using currently available data sources and knowledge. While it provides us with a reasonable picture of the spatial patterns of sediment source areas, deposition and river loads, it has also highlighted where future development work is needed before the model can be applied at much smaller regional scales. Future research and model development of SedNet aims to provide better spatial representation of rates of river bank erosion and to provide a much more rigorous understanding of channel geometry and how this affects bank full discharge and routing of sediment through river networks.

REFERENCES

CSIRO Land and Water Technical Reports. http://www.clw.csiro.au/publications/technical2003/.

De Rose, R. C., Prosser, I. P., Weisse, L. & Hughes, A. O. (2003) Regional patterns of erosion and sediment and nutrient transport in the Murray-Darling Basin. *CSIRO Land & Water Tech. Report no 32/03*. CSIRO Land and Water, Canberra, Australia.

Dietrich, W. E. & Dunne, T. (1978) Sediment budget for a small catchment in mountainous terrain. *Z. Geomorphol.* **29**, 191–206.

Hughes, A. O. & Prosser, I. P. (2003) Gully and riverbank erosion mapping for the Murray-Darling Basin. *CSIRO Land and Water Tech. Report no 3/03*. CSIRO Land and Water, Canberra, Australia.

Lu, H., Gallant, J., Prosser, I. P., Moran, C. & Priestley, G., (2001) Prediction of sheet and rill erosion over the Australian continent, incorporating monthly soil loss distribution. *CSIRO Land and Water Tech. Report no 13/01*. CSIRO Land and Water, Canberra, Australia.

Lu, H., Moran, C. J., Prosser, I. P. & DeRose, R. C. (2003) Spatially distributed investment prioritization for sediment control in the Murray Darling Basin. *CSIRO Tech. Report G to the Murray Darling Basin Commission*. CSIRO Land and Water, Canberra, Australia.

Prosser, I. P., Rutherfurd, I. D., Olley J. M., Young, W. J., Wallbrink, P. J. & Moran, C. J. (2001a) Large-scale patterns of erosion and sediment transport in river networks, with examples from Australia. *Mar. Freshwat. Res.* **52**, 81–89.

Prosser, I. P., Hughes, A. O., Rustomji, P., Young, W. & Moran, C. J. (2001b) Assessment of river sediment budgets for the national land and water resources audit. *CSIRO Land and Water Tech. Report no 15/01*. CSIRO Land and Water, Canberra, Australia.

Mackay, N., Hillman, T. & Rolls, J. (1988) Water quality of the Murray River. *Murray Darling Basin Commission Water Quality Report No 1*. Murray Darling Basin Commission, Canberra, Australia.

Olley, J. & Caitcheon, G. (2000) Major element chemistry of sediments from the Darling-Barwon River and its tributaries: implications for sediment and phosphorous sources. *Hydrol. Processes* **14**, 1159–1175.

Olley, J. & Scott A. (2002) Sediment supply and transport in the Murrumbidgee and Namoi Rivers since European settlement.. *CSIRO Land and Water Tech. Report no 9/02*. CSIRO Land and Water, Canberra, Australia.

Rutherfurd, I. (2000) Some human impacts on Australian stream channel morphology. In: *River Management: The Australasian Experience* (ed. by S. Brizga & B. Finlayson), 11–49. John Wiley & Sons, Chichester, UK.

Scott, A. (2001) Water erosion in the Murray-Darling Basin: learning from the past. *CSIRO Land and Water Tech. Report no. 43/01*. CSIRO Land and Water, Canberra, Australia.

Wallbrink, P. J., Olley, J. M., Murray, A. S. & Olive, L. J. (1998) Determining sediment sources and transit times of suspended sediment in the Murrumbidgee River, NSW, Australia using fallout ^{137}Cs and ^{210}Pb. *Water Resour. Res.* **34**, 879–887.

The assessment of Ukrainian riverbed deformation

ALEXANDR G. OBODOVSKY
Geographical Faculty, Kyiv National Taras Shevchenko University, Vladimirska St., 64 Kyiv-33, 01033, Ukraine
obodovskiy@mail.univ.kiev.ua

Abstract Two types of riverbed deformation (vertical and horizontal) were assessed for Ukrainian rivers. In zones of mixed forests and forest-steppe, vertical deformation was manifested in the form of a decrease in water levels, with associated washing out of the riverbed. For rivers of steppe zone and some rivers of forest-steppe, vertical deformation was manifested in the accumulation of alluvium due to increasing water level. Freely meandering rivers are characterized by intensive horizontal deformation in the zone of mixed forests. The least horizontal deformation is observed for in-cut riverbeds. Criteria are provided which allow for the assessment of both vertical and horizontal deformation types.

Key words activity of rivers; criteria correction; riverbed deformation types; riverbed formation; Ukrainian rivers

INTRODUCTION

Riverbed deformation reflects space and time dynamics of riverbed processes. It is one of the components in the structure of hydroecological analysis (Obodovsky, 2001). River deformation varies in direction (vertical and horizontal) and rate as influenced by various human and natural factors.

Riverbed deformation develops freely, if soil resistance towards washing out is not significant. It was observed that rivers with soil beds and banks are easily erodable. In the Ukraine, such rivers are located in the territory of Polessie, Pridneprovska (Dnipro) and Prichernomorska (Black Sea) lowlands. Development of riverbed deformation is limited if the resistance ability of covering soils is more than the washing ability of the flow. Such regions are located in the territory of Ukrainian crystal shell, at Podolskaya and Priazovskaya (Azov Sea) uplands. Depending on riverbed course development, there are two types of riverbed deformation (Chalov, 1981, 1994):

(a) vertical–attenuation or denudation, which cause transformation of longitudinal river profile and changes of level of river bottom;
(b) horizontal–migration of channel.

Scientists have defined one more type of riverbed deformation: movement of alluvium masses, resulting in spit, rifts and sandbanks (Makaveev & Chalov, 1986).

This work presents a discussion on the first two types of river deformation in the lowland part of the Ukraine.

DATA

In order to define the intensity and direction of vertical deformation for Ukrainian rivers, the author chose 52 hydrological stations (later HS), located at 37 lowland rivers. Curves of the relationship between water discharge and water level were drawn for the hydrological stations for the period 1960–1997. On average, 6–7 curves per station covering 5 to 6 years each were produced. The total number of curves produced was approximately 350.

Besides the above data, in order to find criteria for riverbed deformation assessment, the author has applied the data for the period 1963–1997 to changes in water levels (ΔH) and changes in grade of alluvium ($\Delta\ d_{50}/d_{95}$) (d_{50}, value of average fraction, mm; d_{95}, value of maximum fraction, mm) for those rivers.

Horizontal riverbed deformation was assessed using the data from 53 parts of riverbeds, located at 39 lowland rivers of the Ukraine. Spatial changes of the riverbeds location were defined by comparison of large-scale maps produced between 1923 and 1992. The earliest map was for the Sluch River, Sarny town.

In order to assess more objectively horizontal riverbed transformation, the author used relative indicators of: B/z_{cp} (B, width of riverbed in low water period, m; $z_{m.}$ is average intensity of bank washing out during many years, m) and $Q_{фр}/Q_{ЭДВ}$ ($Q_{фр}$ is riverbed forming discharge, observed within the limits of river course lines, m^3 s^{-1}, $Q_{ЭДВ}$ is minimal water discharge, which does not cause environmental damage in the period of low water, m^3 s^{-1} (Obodovsky, 2001).

All primary information on morphometry of riverbeds, bottom alluvium, water levels and discharges was taken from resources on surface waters (annual and long-term data).

METHOD

Hydrological methods were applied to investigate vertical riverbed deformation. The most reliable method is to analyse curves of relevant water levels and water discharges: $Q = f(H)$ (Makaveev & Chalov, 1986; Tsayts & Obodovsky, 1993). Up or down displacement of the curves in correlation charts correspondingly shows accumulation of alluvium or riverbed erosion. Change in water level, having the same discharge, shows intensity of accumulation or deep erosion at a river section. During the investigation the flood plain component of the curves was eliminated to allow for a more objective assessment of riverbed vertical deformation.

The method above only provides general direction of vertical riverbed deformation. The depth of flow significantly distorts the curves. Therefore, the dimensionless indicator $\Delta H/h_{ЭДВ}$ (ΔH is a decrease or an increase in water level, m; $h_{ЭДВ}$ is average depth of flow in case of minimal water discharge, which does not cause environmental damage, m) was suggested. Use of this criterion allowed for the effective evaluation of water level dynamics with river depth during low water periods. Change in grade of river alluvium can be tracked using $\Delta\ d_{50}/d_{95}$. This formula characterizes changes in relatively large bottom sediments over time. The two criteria can be related as follows:

$$\Delta H/h_{ЭДВ} = f(\Delta\ d_{50}/d_{95}) \qquad (1)$$

Using these criteria in conjuction with regression analysis allows for the assessment of the influence of the above parameters on the unlimited and limited manifestation of riverbed deformation. Hydromorphological analysis (combining of topographical maps of the same scale [1:50 000, 1:25 000], made in different years) was applied during horizontal riverbed deformation investigations. Average annual intensity of deformation (z_{cp}, m year^{-1}) at a given riverbed cross-section is expressed as:

$$z_{cp} = z/N \quad (2)$$

where z is the amplitude of displacement of the bank line, m; N is the number of years between maps.

An additional objective way of assessing horizontal deformation for lowland rivers, is by relating the following ratios; riverbed width (at low water) (B) to average annual intensity of planned deformation (z_{cp}) to relative course-forming discharge ($Q_{фр}$) and minimal water discharge, which does not cause environmental damage ($Q_{ЭДВ}$) (Obodovsky, 2001):

$$B/z_{cp} = f(Q_{фр}/Q_{ЭДВ}) \quad (3)$$

In addition to the above river deformation analysis, river width (B/h [B is riverbed width in low water period, m; h is depth of riverbed in low water period, m]) was assessed to characterize widening of the riverbed (Velikanov, 1958). The value of relative width of river is not the same in conditions of unlimited and limited development of riverbed processes. In conditions of unlimited development, it increases during active manifestation of horizontal riverbed deformation; in the conditions of limited development and of permanent erosion of riverbed, it decreases. Criteria, relating widening of riverbeds with flow related riverbed formational activity, are:

$$B/h = f(Q_{фр}/Q_{ЭДВ}) \quad (4)$$

$$B/h = f(P_{Qфр}) \quad (5)$$

where $P_{Qфр.}$ is probability of riverbed forming discharge, observed within the limits of river course lines, %.

Criterion (Equation 4) enables the analysis of the impact of relative flow related riverbed formational functions on the widening of riverbeds. Criterion (Equation 5) assesses the impact of riverbed forming discharge on the riverbed itself (Obodovsky, 2001). In order to assess these criteria, regression and correlation analyses were used.

RESULTS AND DISCUSSION

Concerning vertical riverbed deformation, analysis of space and time dynamics of criterion $Q = f(H)$ for lowland rivers of Ukraine showed quite clear regularities. The majority of rivers located both in mixed forests (Polessie) and forest-steppe are characterized by a general decrease in water levels for relevant discharges and "subsidence" of riverbeds. For freely meandering rivers, its value differs from 0.20 cm year^{-1} (the Styr River, HS Zhyrovtsy) to 3.05 cm year^{-1} (the Desna River, HS Letki). For riverbeds flowing in conditions of limited development of riverbed deformation, this value is much lower and it is from 0.0 (the Ros River, HS Korsun-Shevchenkovsky) to 1.05 cm year^{-1} (the Southern Bug River, HS Sabarov). This shows vertical deformation associated with erosion processes in riverbeds.

Tendency for water levels to increase was observed for the majority of rivers in the steppe zone (in particular, tributaries of the Seversky Donets River and rivers near the Azov Sea) and for some rivers in the forest-steppe zone (the Psel and the Vorskla Rivers). This is explained by deposition of alluvium in the riverbeds and domination of deposition processes. The biggest increase in water levels was typical for the rivers in conditions of unlimited development of riverbed deformation and it fluctuates from 0.04 cm year^{-1} (the Seversky Donest River, HS Chuguev), to 1.45 cm year^{-1} (the Obitochnaya River, HS Primorsk). For in-cut riverbeds, average values change from 0.02 cm year^{-1} (the Seret River, HS Chortkiv) to 1.29 cm year^{-1} (the Kalchik River, HS Kremenevka).

Criteria assessment of vertical riverbed deformation enabled us to define the following relationship (approximated by level of regression for the rivers in conditions of unlimited riverbed formation):

$$\Delta H/h_{ЭДВ} = -1.89\ (\Delta\ d_{50}/d_{95}) - 0.04 \qquad (6)$$

Error here is $\sigma = 0.92$, correlation coefficient is $r = -0.67$ (Fig. 1(a)).

Equation (6) shows that when the criterion $\Delta\ d_{50}/d_{95}$ increases (namely, when alluvium grade decreases), subsidence of water levels in riverbeds increases. It shows how active processes of erosion influence alluvium in the riverbeds (the Styr, the Goryn, the Uzh, the Desna, the Sula rivers), and how a decrease in $\Delta\ d_{50}/d_{95}$ and an increase in $\Delta H/h_{ЭДВ}$ shows

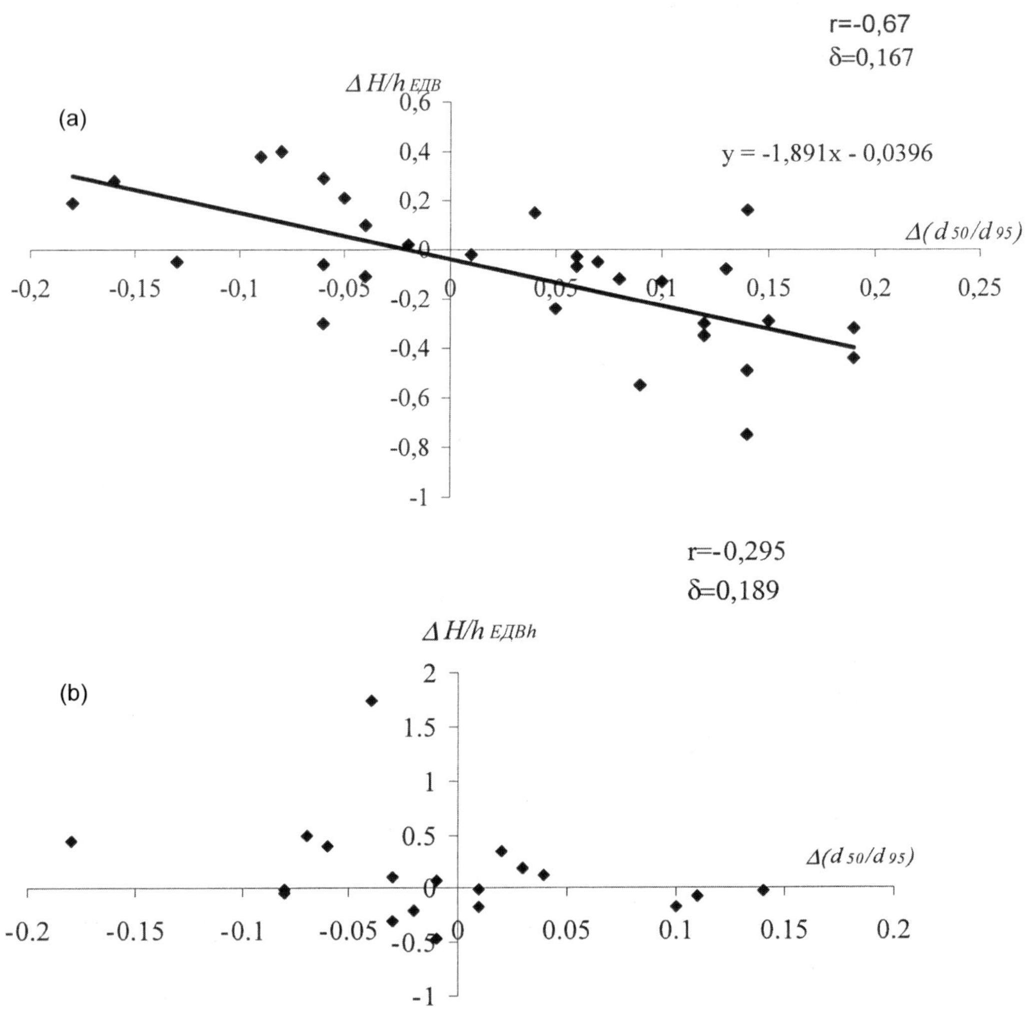

Fig. 1 Charts of Dependencies $\Delta H/h_{ЭДВ} = f\ (\Delta d_{50}/d_{95})$. (a) for rivers of Ukraine, flowing in conditions of free development of riverbed deformations; (b) for rivers of Ukraine, flowing in conditions of limited development of riverbed deformations.

increasing heterogeneity of riverbed alluvium with concomitant "accumulative" direction of vertical deformation (Opel, Samara, Seversky Donets, Molochnaya, Obitochnaya rivers).

For rivers flowing in conditions of limited development of riverbed deformation, there is no correlation between the above parameters ($r = -0.29$) (Fig. 1(b)). Therefore, processes of riverbed formation were mainly dependent on geological and geomorphologic structure of the riverbed with flow playing a less active role.

The highest intensity of horizontal riverbed deformation is typical for rivers in conditions of unlimited riverbed formation. On average, its value varies from 0.20 m year^{-1} (the Vorskla River, HS Chernetchina) to 5.65 m year^{-1} (the Sluch River, HS Sarny). These indicators are significantly smaller and fluctuate from 0 for the riverbeds where limited conditions of riverbed formation dominate (the Southern Bug River, HS Alexandrovka) to 1.97 m year^{-1} (the Ingulets River, HS Mogilevka).

Criteria assessment of this type of deformation showed that conditions of their unlimited development can be expressed by the following regression formula:

$$B/z_{cp} = 3.32(Q_{фр}/Q_{ЭДВ}) + 16.76 \qquad (7)$$

with error $\sigma = 6.72$ and correlation coefficient $r = 0.71$ (Fig. 2(a)).

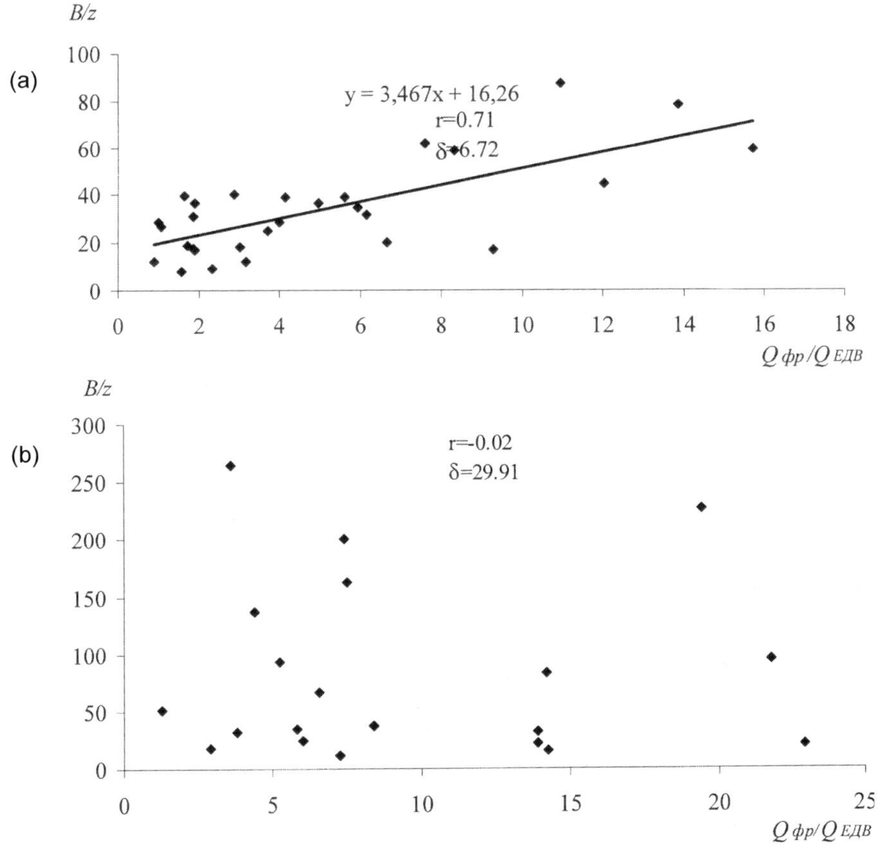

Fig. 2 Charts of dependencies $B/zcp = f(Qфр/QЭДВ)$ (a) for rivers of Ukraine, flowing in conditions of free development of riverbed deformations; (b) for rivers of Ukraine, flowing in conditions of limited development of riverbed deformations.

Equation (7) shows that an increase in relative riverbed forming water discharge leads to an increase in relative widths of the riverbeds and a decrease in the scale of horizontal deformation compared with riverbed morphometry. There is no such correlation ($r = 0.02$) for the rivers located in conditions of limited manifestation of riverbed deformation (Fig. 2(b)). Here flow and bank erosion depend on geology.

Analysis of changes of relative riverbed width (spreading), caused by riverbed formation activity of flow, shows different correlations for rivers in conditions of unlimited and limited riverbed formation. In the first case, there is a clear tendency for relative width of riverbed to increase, if relative riverbed forming discharge increased. This is shown in the following equation;

$$B/h = 2.35\ (Q_{\text{фр}}/Q_{\text{ЭДВ}}) + 23.61 \qquad (8)$$

with error $\sigma = 8.7$ and direct correlation $r = 0.68$ (Fig. 3(a)).

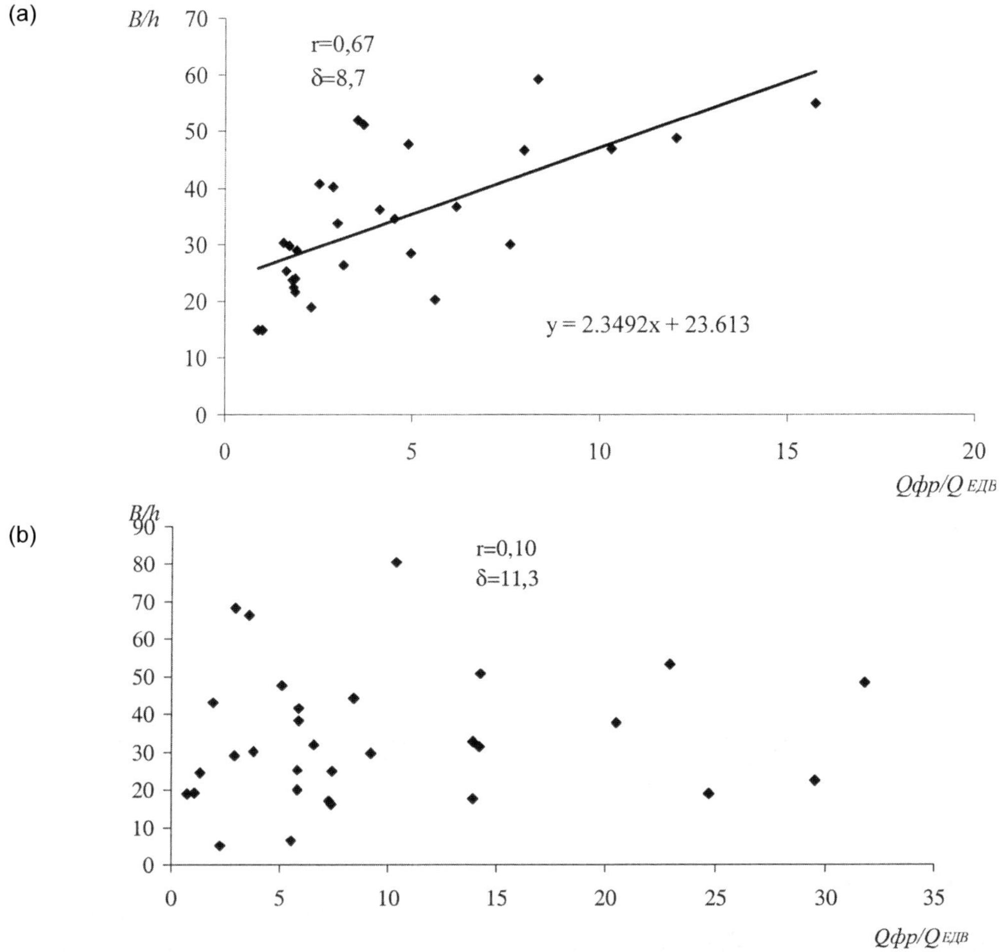

Fig. 3 Charts of dependencies $B/h = f(Q_{фр}/Q_{ЭДВ})$ (a) for rivers of Ukraine, flowing in conditions of free development of riverbed deformations; b) for rivers of Ukraine, flowing in conditions of limited development of riverbed deformations.

At the same time, in conditions of unlimited riverbed formation, when water supply $Q_{фр}$ (namely, absolute values of discharges) increases, relative width of the riverbed also increases. This is expressed by the following equation:

$$B/h = -0.47(P_{Qфр}) + 44.71 \quad (9)$$

with error $\sigma = 22.1$ and reverse correlation $r = -0.57$ (Fig. 4(a)).

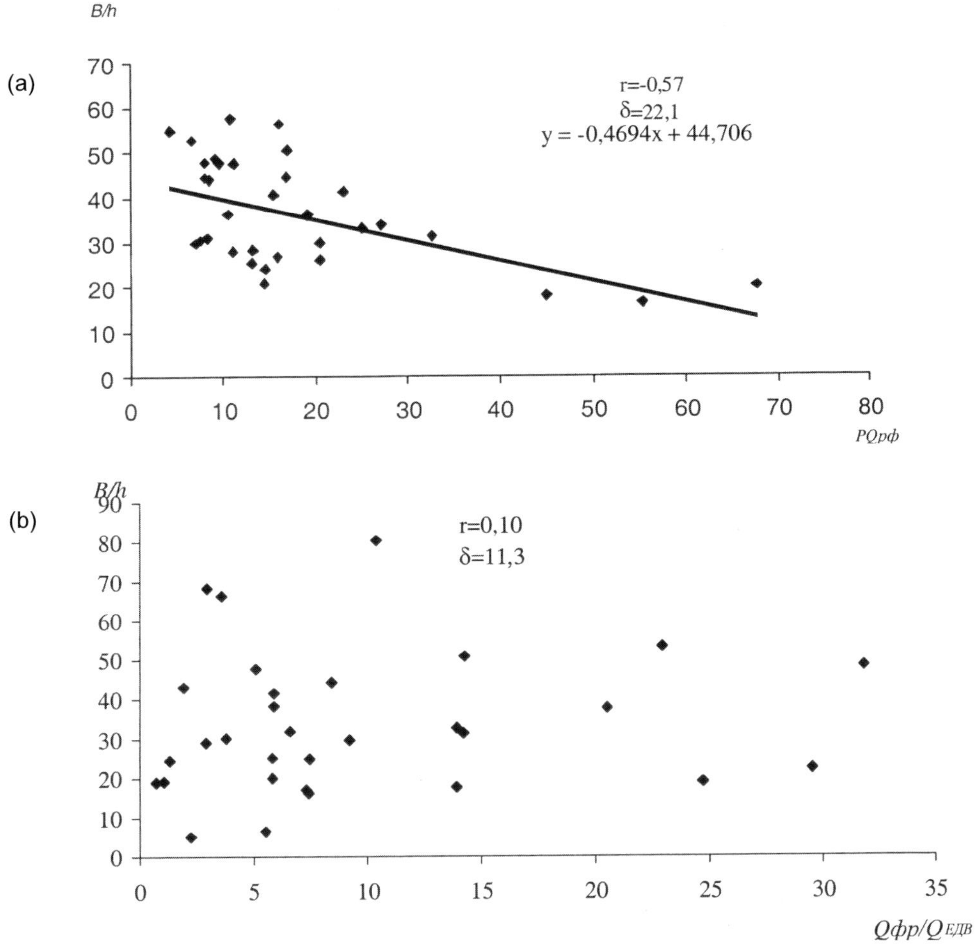

Fig. 4 Charts of dependencies $B/h = f(PQфр.)$ (a) for rivers of Ukraine, flowing in conditions of free development of riverbed deformations; (b) for rivers of Ukraine, flowing in conditions of limited development of riverbed deformations.

The above demonstrates that the greater the discharge, the greater is the potential for meandering (deformation) of the riverbed. For rivers in conditions of limited development of horizontal riverbed deformation, there is no correlation, described by the above equations (Figs 3(b), 4(b)). They show that neither lower relative riverbed forming discharges, nor their absolute values in the given conditions affect the establishment of riverbed morphometry.

CONCLUSION

Quantitative parameters of flow are the main active factor of vertical as well as horizontal manifestations of deformations for freely meandering rivers. Riverbed morphometry is to a large extent defined by space–time dynamics of grade of alluvium. At the same time, there are no such regularities for the rivers in conditions of limited manifestation of riverbed deformation. Geological and geomorphological conditions in the river basins are the main factor for riverbed formation processes.

REFERENCES

Chalov, R. S. (1994) Geomorphologicheskoe otrazhenie gorizontal'nyh ruslovyh deformacii (Geomorphologic manifestation of horizontal riverbed deformation). *Geomorphologiya* **3**, 31–40 (in Russian).

Makkaveev, N. I. & Chalov R. S. (1986) *Ruslovye processy* (*Riverbed Processes*). Moscow, Publishing House of Moscow State University, Moscow, Russia (in Russian).

Obodovsky, A. G. (2001) *Hydrological-Environmental Assessment of Riverbed Processes* (*using Ukrainian Rivers as Example*). Nika-Center, Kyiv, Ukraine (in Ukranian).

Obodovsky, A. G. (2001) Ocenka vertika'nyh ruslovyh deformacii na ravninnyh rekah Ukrainy (Assessment of vertical riverbed deformation at lowland rivers of Ukraine). *Zaschita okruzhayuschei sredy ot antropogennoi nagruzki* **5**(7), 49–57 (in Russian).

Tsayts, E. S. & Obodovsky, A. G. (1993) Opredelenie padeniya urovnei vody pri razrabotke ruslovyh kar'erov (Definition of water level dropping during development of riverbed quarry). *Meteorologiya, Climatologiya i Hydrologiya* **79**, 123–127.

Velikanov, M. A. (1958) *Ruslovye processy* (Riverbed Processes). Gosphysmatizdat, Moscow, Russia (in Russian).

Monitoring of channel processes on the interfluve between the Kama and the Vyatka rivers

IVAN RYSIN & LARISA PETUKHOVA
Department of Geography, Udmurt University, Izhevsk, Russia
geo@uni.udm.ru

Abstract Results of a survey of lateral channel shift from 54 control sections of different rivers that flow across various landscapes of Udmurtia are presented and discussed. The intensity of bank downcutting is controlled by a number of natural factors. Human economic activity in the drainage basin also exerts an important influence on the pattern of channel migration.

Key words bank erosion; channel morphological type; channel shift; factors of channel processes; meandering; monitoring; Udmurt Republic

INTRODUCTION AND STUDY AREA

A network of numerous rivers belonging to the Kama and Vyatka river basins, dissect the plain territory of the Udmurt Respublic (UR). The most widespread type of river channel morphology is meandering. A survey of morphometrical and morphological characteristics of river channels has been carried out using published topographic maps (scale 1:25 000, 1:50 000, 1:100 000) to distinguish the type of meander. Rivers of Hortonian orders 2–4 have been chosen for this survey. The main aim of the work presented is a geographical analysis of morphodynamic river channel type distribution within the UR and determination of channel lateral migration rates (Fig. 1). Influence of natural and environmental factors on river channel dynamics is also discussed.

Two main morphological types of river channels are distinguished within the region according to their pattern and tendency of lateral shift: relatively straight and meandering channels (Chalov *et al.*, 1998). Relatively straight channels dominate the upper reaches of rivers and comprise up to 25% of the total river length in the Vala basin and 38% in the Izh basin. Channel gradients of these river sections have maximum values of 0.009–0.01, coefficient of sinuosity is minimum.

Free meanders (not confined by bedrock slopes) prevail among meandering channels. It is well known that meanders pass certain stages in the process of their development. Each stage of development differs in intensity and direction of lateral deformation of the meander, kinematic flow structure and hydraulic characteristics. There are five types of meandering channels identified on the UR rivers (Table 1): segmented gradual meanders ($1.15 < L/l < 1.40$), segmented developed ($1.40 < L/l < 1.70$), segmented steep ($1.70 < L/l < 2.00$), loop-shaped ($L/l > 2.00$), cut-off ($1.50 < L/l < 2.00$), where L/l is a ratio of channel length to meander step (sinuosity coefficient).

The most widespread are segmented meanders in which downstream shift dominates. On average these occupy about 63% of the total channel length (from 55% in the Kilmez River basin to 71% in the Toyma River basin). Segmented meanders exhibit different stages of development which is evidence for active transformation of river channel planform as a whole. All three main stages in the development of segmented meanders, from gradual to

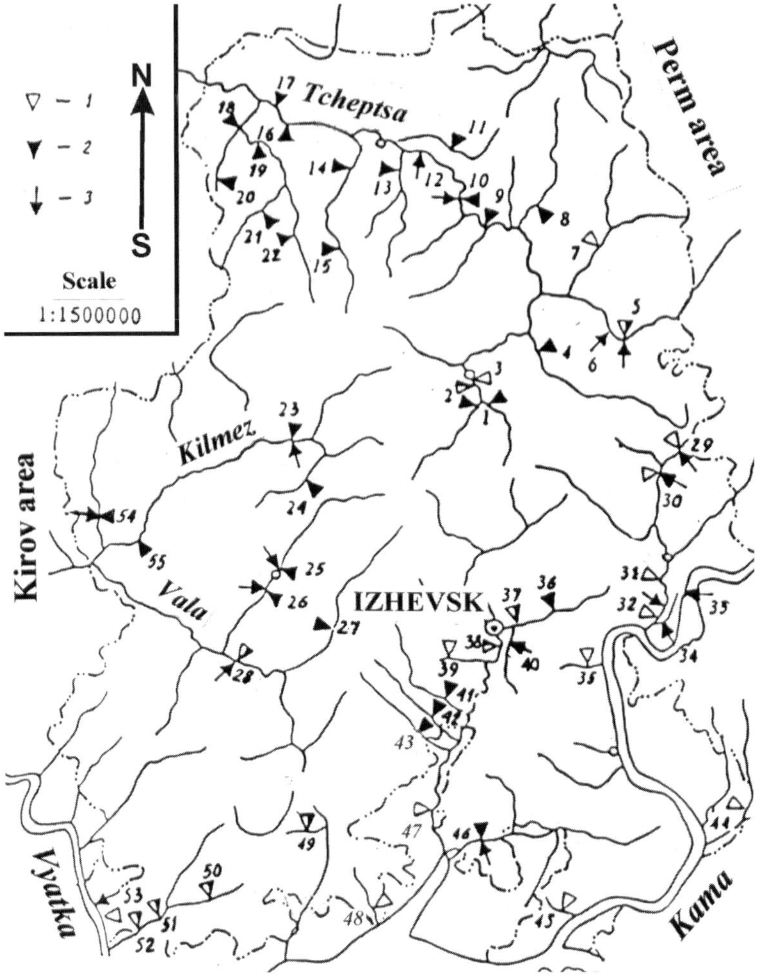

Fig. 1 Schematic of bank erosion monitoring points locations on rivers of the UR. (1) marks and reference points installed in 1999; (2) fixed marks and control points installed in 2000; (3) tacheometric survey of eroded bank sections.

Table 1 Distribution of morphodynamic channel types in different river basins of the UR (%).

Morphodynamic channel type	Tcheptsa	Vala	Kilmez	Siva	Izh	Toyma and the left Vyatka tributary	The right Kama tributary
1. Relatively straight	28	25	32	35	38	28	32
2. Free meandering	72	75	68	65	62	72	68
(a) gradual segmented meanders	29	53	29	46	37	48	60
(b) developed segmented meanders	14	11	23	9	17	19	6
(c) steep segmented meander	13	–	3	4	4	4	2
(d) loop-shaped meander	12	6	10	4	2	1	–
(e) cut-off meanders	4	5	3	2	2	–	–

steep (Chalov *et al.*, 1998), are present at certain river reaches. The subsequent stages of meander development under different conditions are associated with either straightening of the river bed and formation of a cut-off meander, or with further curvature increase and transformation into a loop-shaped meander. For the latter, increase in meander belt dominates over downstream shift. Cut-off meanders occupy about 2–5% of the total river length of the UR, and loop-shaped meanders—10–12% in the basins of medium sized rivers and 1–4% in basins with smaller rivers.

Sinuosity and asymmetry coefficients have been calculated in order to determine more precisely morphodynamic channel type. The sinuosity coefficient (C_s) suggests that the predominant channel type is straight and gradual segmented meandering with the C_s value ranging from 1.1 to 1.4. The asymmetry coefficient for the river valley slopes was estimated as:

$$K = 1 - L_k/L_p$$

where L_k, length of steeper slope; L_p, length of more gradual slope.

The valleys of both large and medium rivers are characterized by sharp asymmetry of slopes. Within the territory of UR two main genetic types of asymmetry (planetary and climatic), and two secondary types (tectonic and hydrodynamical) are present. The spatial distribution of asymmetrical sections of river valleys is uneven. Within the UR, average value of asymmetry coefficient is 0.44. The number of symmetrical valley sections is 7.6% and sharply asymmetrical ones is 58.4%. From the south to the north there is an increase in the proportion of symmetrical valley sections from 5–7% to 19%. The number of sharp asymmetrical sections decreases accordingly from 50–55% to 35%.

THE STUDY METHODS AND RESULTS

Meandering channels actively develop in time and space, changing their planforms. The main causes of this are lateral channel deformations. Comparison of air photos taken in 1934 and 1987 (scale 1:10 000) has allowed us to estimate the rates of lateral channel migration for the Tcheptsa River and its main tributaries (Table 2). For the main river, average channel lateral shift in 53 years has been estimated at 55.3 m (maximum, 105 m; minimum, <10 m), giving an average bank downcutting rate of 1.02 m year^{-1}. On the tributaries of the Tcheptsa River, the channel deformation has been found to vary from 30 to 160 m, and the range of bank retreat rates are 0.30–0.90 m year^{-1}.

Since 1999, regular annual surveys of lateral channel shift have been carried out on some rivers of the UR in summer. For this purpose, about 300 fixed marks and control points

Table 2 Rates of systematic channel shifts in the Tcheptsa River basin during 1934–1987.

River basin	Number of sections	Maximum shift (m)	Minimum shift (m)	Average shift (m)	Average rate of shift (m year^{-1})
Tcheptsa	47	105	9	55.3	1.02
Loza	5	98	35	63.3	1.17
Ita	4	60	30	48.7	0.90
Lekma	14	100	10	22.8	0.42
Ubit	18	30	10	16.1	0.30
Sada	5	30	12	19.0	0.35
Sepitch	50	160	10	40.6	0.75

Table 3 Morphometric characteristics of studied rivers.

Groups of rivers	1	2	3
Length (km)	<25	25–100	>100
Channel width (m)	5–10	15–50	>50
Meander radius (km)	0.1–0.15	0.20–0.25	>0.7
Gradient (m km^{-1})	2–2.5	0.6–1.0	<0.2

have been installed on actively transforming sections (often on high flood plain levels). Repeated theodolit-tacheometrical surveys have been conducted on 30 sections annually. The river sections studied are rather diverse, with the intensity of channel processes differing greatly. This provided for classification of rivers into three groups according to their morphometry: 1, very small; 2, small; 3, medium and large (Table 3).

Values of annual bank retreat vary considerably. During 2003 average bank downcutting rate in the UR was 0.47 m $year^{-1}$. Rates were on average greatest in 2001 (Table 4). The greatest average annual rates of lateral channel shift of 1.2 to 3.0 m $year^{-1}$ took place on rivers of the third group, which are characterized by high discharges and flow velocity. The maximum annual rate of bank retreat occurred along river reaches where the channel is at least 5–10 m wide. For the second group of rivers, average rates of bank downcutting are between 0.4 and 0.7 m $year^{-1}$ with maximum rates of up to 2–3 m $year^{-1}$. For the streams of the first group, estimated values are lower: 0.1–0.3 m $year^{-1}$, though at certain points a shift of 1 m $year^{-1}$ and more has been observed.

The method of fixed marks and control points has its shortcoming: only certain points along actively retreating sections are studied and it is rather difficult to evaluate the whole process. More detailed investigation is possible by making systematic tacheometric surveys of entire channel sections. Conventional theodolits and, later, electronic tachometers have been used for carrying out geodetic surveys. The length of the survey sections ranged from 50–100 to 400–500 m.

Survey data processing has allowed us to create eroded bank plans. From these, not only average and maximum bank retreat rates, but also areas and volumes of erosion can be determined. To compare these, ratios of area and volume of eroded bank to the length of the certain section have been calculated and called the specific area and volumetric erosion rates. The greatest specific areas and volumetric rates of bank downcutting have also been observed on large rivers, varying from 0.4 to 0.8 m^2 m^{-1} and from 2.0 to 17.2 m^3 m^{-1} (Table 5). For smaller rivers specific erosion areas vary from 0.1 to 0.3 m^2 m^{-1} and specific volumes from 0.4 to 0.6 m^3 m^{-1}.

DISCUSSION

The data obtained for the last 4 years has allowed us to consider the dynamics of lateral channel shifts and make some conclusions about the influence of different factors. The major agent in the channel formation is water discharge. This is demonstrated by the higher correlation coefficients for the relationship between channel lateral shift and average annual water discharge (Table 6). Table 6 also shows that along with the discharge, other natural factors exert considerable influence in bank downcutting intensity. These are rock lithology, slope gradients, radius of meander curvature, soils and vegetation cover.

Table 4 Average annual rates of bank retreat of rivers of the UR for the period of 2000–2003.

Number	Rivers	Control section (settlement)	Average annual rates of bank retreat (m $year^{-1}$)			
			2000	2001	2002	2003
1	Loza	Kushya	–[1]	0.15	0.26	0.12
2	Nyas	Sundur	0.0	0.23	0.10	0.63
3	Loza	Loza	0.14	0.09	0.38	0.36
4	Ita	Zura	–	0.27	0.41	0.23
5	Tcheptsa	Debesy	0.48	0.44	0.44	0.52
6	Lyp	Sosnovy bor	0.43	0.19	0.24	0.24
7	Pysep	Bany	–	0.86	0.32	0.68
8	Tcheptsa	Kamenoe sadelye	–	1.80	–	1.2
9	Tcheptsa	Kozhilo	–	1.33	1.25	0.31
10	Varyzh	Keldikovo	–	0.42	0.32	0.29
11	Sepitch	Glazov	–	–	0.82	0.49
12	Ubyt	Tchura	–	0.34	0.13	0
13	Ubyt	Palagay	–	0.23	0.21	0.16
14	Tcheptsa	Yar	–	0.43	0.44	1.44
15	Tcheptsa	Dizmino	–	1.27	1.65	1.05
19	Lekma	Nizhny Ukan	–	0.16	0.07	0
20	Sada	Yur	–	0.17	0.10	0.05
21	Lema	Shamardan	–	0.10	0.05	0.08
22	Lekma	Potchinky	–	0.28	0.10	0.14
23	Kilmez	Goloviznin Yazok	–	2.24	1.43	0.56
24	Arlet	Tchibir-Zyunya	–	0.29	0.23	0.13
26	Uva	Uva-Tuklya	–	0.45	0.51	0.37
27	Nylga	Hilga	–	0.57	063	0.45
28	Vala	Makarovo	0.52	0.98	0.52	0.38
29	Bilibka	Shoner	0.10	0.34	0.10	0.16
30	Sharkan	Titovo	0.51	0.12	0.56	0.15
31	Siva	Gavrilovka	0.27	0.60	0.85	0.23
35	Golyanka	Goliany	0.38	0.20	0.25	0.3
36	Pozim	Kabanikha	0.44	0.05	0.12	0.28
37	Pozim	Pozim	–	0.10	0.10	0.27
38	Izh	Bolshaya Vehya	0.45	0.27	0.28	0.24
39	Ludzinka	Yusky	0.33	0.46	0.27	0.13
41	Agryzka	Bagrash-Bigra	–	–	0.92	3.15
42	Postolka	Postolsky	–	–	0.45	018
43	Bobinka	Abdes-Urdes	–	–	0.09	0.61
44	Piz	Novokreshenka	–	0.48	0.04	–
45	Kobylka	Klestovo	0.35	0.22	0.23	0.13
46	Kirikmas	Tavzyamal	–	0.55	0.18	0.72
47	Izh	Russkaya Sharshada	0.22	0.12	0.14	0.46
48	Varzinka	Yumyashur	–	0.15	0.11	0.2
49	Alnashka	Alnashy	–	0.19	0.21	0.15
50	Adamka	Grakhovo	0.20	0.25	0.40	0.15
51	Umyak	Pussky Kuyuk	0.52	0.45	0.18	0.15
52	Umyak	Bazhenikha	–	0.15	0.13	0.13
53	Vyatka	Krimskaya Sludka	3.25	3.97	4.53	3.0
54	Lumpun	Kharlamovskaya	–	–	–	0.16
55	Kilmez	Maliye Syumsy	–	–	–	0.65
Average total			0.48	0.54	0.50	0.47

[1] – indicate "not measured".

Table 5 Areas and volumes of bank retreat in rivers of the UR for the period of 2001–2002 year.

River	Control section (settlement)	Area (m^2) / specific area (m^2 m^{-1})		Volume (m^3) / specific volume (m^3 m^{-1})	
		2001	2002	2001	2002
Vyatka	Krimskaya Sludka	117.7/0.39	–	5107.7/17.2	–
Kilmez	GoloviznínYazok	86.9/0.75	97.0/0.85	378.8/3.29	423.0/3.7
Tcheptsa	Lnozavod–1	170.0/0.41	43.5/021	884.0/2.2	226.2/1.10
Tcheptsa	Lnozavod–2	30.92/0.12	30.5/0.13	160.78/0.64	159.7/0.65
Tcheptsa	Adam–1	28.4/0.12	67.0/0.26	147.68/0.61	348.4/1.36
Tcheptsa	Adam–2	–	170.0/0.41	–	885.4/4.29
Tcheptsa	Debecy–1	17.84/0.27	24.0/0.18	1340.23/1.0	175.2/1.29
Tcheptsa	Debesy–2	–	44.0/0.21	–	168.5/0.81
Tcheptsa	Varny	–	37.0/0.19	–	208.0/1.08
Siva	Metlyaky–1	–	116.0/0.47	–	464.0/1.9
Siva	Metlyaky–2	16.9/0.16	46.5/0.61	67.5/0.67	186.0/2.5
Kirikmas	Tavzyamal–1	–	38.5/0.16	–	200.2/0.82
Kirikmas	Tavzyamal–2	15.84/0.08	21.5/0.10	82.37/0.41	179.4/0.82
Uva	Uva	11.74/0.13	17.8/0.18	44.5/0.49	68.2/0.71
Kama,res.	Berkuty	25.76/0.42	29.0/0.51	762.49/12.7	858.9/15.3
Kama,res.	Galevo	–	27.0/0.22	–	810.0/5.53
Sharkanka	Titovo–1	–	24.0/0.27	–	68.3/0.78
Sharkanka	Titovo–2	–	16.5/0.24	–	42.02/0.60
Sharkanka	Shoner	–	10.0/0.17	–	32.5/0.54
Vala	B.Volkovo	38.55/0.19	–	115.05/0.58	–
Bidvayka	Zavyalovo	–	12.0/0.08	–	31.9/0.22

Table 6 Coefficients of correlation between rate of bank retreat and different natural factors.

Groups of rivers	1	2	3
Water discharge	0.585	0.658	0.782
Stream gradient	0.577	0.451	0.758
Meander radius	0.687	0.381	0.812
Erosion stability	–0.622	–0.440	–0.690
Forest vegetation (% of basin area)	0.314	0.350	0.589

Intensity of erosion depends on such soil characteristics as resistance to erosion (bank strength) and permeability/infiltration capacity. Soil resistance to erosion has been determined using the approach proposed by Bastrakov (1990). It essentially involves measuring the impact of a single water jet with a known hydraulic power onto an undisturbed soil sample placed in a cylindrical steel container. Experiments were carried out on a specially designed apparatus in a laboratory.

Obtained values of resistance to erosion (R) characterize the ability of soil to withstand erosion. The values of R vary greatly from 3.15 to 15.5 Newton (N). Erosion stability depends mainly on the soil texture. Average index of R for sandy soils is 3.7 N; loamy-sandy soils, 4.3 N; medium-loamy, 6.6 N; and clayey-loamy, 10.0 N.

The relationship between erosion stability and bank retreat rates is separately determined for all three groups of streams by the calculation of the correlation coefficient. As expected, these are negatively correlated, i.e. the higher the erosion resistance, the lower is the lateral channel shift rate. This relationship is especially clear in the first and third groups of rivers: $r = -0.622$ (first group) and $r = -0.690$ (third group). For the most abundant second group of rivers this relationship is weaker ($r = -0.440$) because of the higher variability of both bank retreat rate and soil resistance to erosion (Table 6).

Channel gradient greatly influences the intensity of channel deformations and erosion. When the gradient is high, erosion and transportation capacity of flow increases. Calculated values of correlative ratio confirm this fact (Table 6). This linkage is especially clear for large rivers ($r = 0.758$). The values of the first and the second groups of rivers are lower ($r = 0.577$ and $r = 0.451$, respectively).

The value of the meander curvature radius also influences the rate of lateral erosion. The higher is the channel curvature, the higher is the flow velocity along the eroded concave bank of a meander. Circular currents directed from concave to convex meander bank are formed within the main flow, which intensify bank erosion significantly. This relationship is rather weak for the second group of rivers ($r = 0.381$). The relationship is closer for the first group of streams ($r = 0.687$) and the best for the large rivers of the third group ($r = 0.812$).

Another factor influencing the intensity of riverbank erosion is vegetation cover and in particular forest vegetation percentage (FVP). Values of FVP at control sections change from 10–20% in the southern part of the UR to 70% and more to the north. Calculated values of correlative ratio between the bank retreat rate and the FVP show that some relationship exists between them. However, as the linkage is rather weak then FVP does not play the main role in river channel deformation.

The present condition of river channels is also strongly influenced by the extent of anthropogenic activity in river valleys. Its degree and extent is very diverse, especially for large rivers. Small rivers dominate in the region of the UR, and the extent of human impact on their valleys is relatively low. On the other hand, such rivers react to natural and anthropogenic changes within their drainage basins more intensively and quickly.

Deforestation and intensive cultivation of the drainage basins result in delivering great amount of sediments to small river valleys and channels. It causes intensified deposition and siltation of channels. Earth dams are often constructed in small river valleys, in order to maintain water supply in summer months. During spring floods many such dams can be breached by intensive runoff, flow velocities increase greatly and erosion accelerates. For example, significant bank downcutting has been observed on the Agrizka River (the Izh River basin) in 2003 as a result of earthen dam breach. The maximum value of the bank retreat was more than 8 m in this instance.

Construction of some engineering structures, communication lines etc. on river banks and in channels also result in the process of channel regime changes. One of the oldest forms of local engineering effects to river channels is bridge construction. It usually causes flood capacity decrease, which results in the formation of hydraulic jump above the bridge (head wave is greatest at water conveying outlets of road embankments, crossing small rivers) and hydraulic fall below the bridge during floods. That increases specific water discharge and erosion intensity. Sharp changes of the channel morphology and intensification of lateral deformations can take place below the bridges. Some control sections give clear examples of the differences in channel deformation intensity upstream and downstream from bridges where bank retreat rates upstream of bridges are far lower (Table 7).

Table 7 Rates of bank retreat upstream and downstream from bridges (2003 year data).

River (settlement)	Average annual rate of bank retreat upstream from bridges (m year^{-1})	Average annual rate of bank retreat downstream from bridges (m year^{-1})
Ita (Zura)	0.10–0.15	0.25–0.40
Lyp (Sosnovy Bor)	0.01–0.10	0.50–0.90
Golianka (Golyany)	0.10–0.20	0.30–0.50
Ludzinka (Yusky)	0.05–0.10	0.30–0.40

During melioration activities, rivers serve as water receivers and transmitters, which increases their discharge. Such increase may not be visible at all for larger rivers, but is very a serious factor for smaller rivers and can sometimes cause very serious erosion where active incision of valley floors and active bank collapses take place. An example of such negative anthropogenic activity has been observed at the control section in the area of Nylga settlement (the Nylga River). A network of drainage ditches was constructed there, through which the flood plain water was diverted back into the channel. For this reason active development of meandering process, deformation of segmented meanders, high rate of lateral erosion (average bank retreat rate for 2003 is 0.45 m year^{-1}, maximum 1.15 m year^{-1}) has became characteristic for this river section.

Under the influence of anthropogenic factor, rivers can also negatively affect the life of people. Engineering structures, communications, residences can be damaged as a result of bank erosion. Valuable agricultural fields and forest areas can also be lost. The data obtained in this study will be used in the forecasting of hazardous situations along the rivers and streams of the UR.

REFERENCES

Aleksentseva, A. I. (1981) On symmetry and assymmetry of Prekamie region topography. In: *Physics-Geographical Foundations of Development and Distribution of Productive Forces of Nechernozem Urals*. Perm, Russia.

Bastrakov, G. V. (1990) Theoretical principles and methods of determining soil resistance to erosion and infiltration capacity. In: *Soil Erosion Characteristics of Some Regions of Russia*. Bryansk, Russia.

Chalov, R. S., Alabyan, A. M., Ivanov, V. V., Lodina, R. V. & Panin, A. V. (1998) *Plain River Channel Morphodynamics*. M. Geos, Russia.

Analysis of coarse sediment connectivity in semiarid river channels

J. M. HOOKE
Department of Geography, University of Portsmouth, Buckingham Building, Lion Terrace, Portsmouth PO1 3HE, UK
janet.hooke@port.ac.uk

Abstract Coarse sediment in river channels influences the channel morphology but channel morphology also influences the supply, sediment transfer and sedimentation. Coarse sediment is mainly stored on bars: lack of bars in a reach may reflect lack of supply, lack of storage availability, or high competence. A conceptual model of sediment connectivity has been developed and is applied to two channel systems in the semiarid area of southeast Arizona, USA. Detailed evidence of sediment sources was collected from field mapping and aerial photographs. Sediment size was sampled at intervals down the channel and competence was calculated from surveyed cross-sections and hydraulic data. The spatial patterns of sediment transfer and storage were shown to be related to a complex combination of sediment supply, valley morphology and channel gradient. Deposits and sediment segregation in such systems can reflect recessional flows. The implications of spatial variability in coarse sediment transfer and storage are discussed.

Key words channel bars; channel morphology; connectivity; flow competence; fluvial deposits; sediment

INTRODUCTION

Understanding of coarse sediment transport in fluvial systems is important because of its influence upon morphology through deposition in channels, mainly as bars. Morphology itself influences the transport and deposition of coarse material. There is increased emphasis in the literature on identification of sediment sources, transfers and sinks/stores and on the use of sediment budgets as a framework for analysis. However, it is important to identify whether there is direct connectivity of the coarse sediment. Inferences can be made from the geomorphological and sedimentological evidence.

CONCEPTUAL MODEL

Conventional theory assumes that there is sediment connectivity down through a channel system. i.e. that sediment has the potential to move through the system. If sediment is too coarse to be moved then it is assumed that accumulation takes place such that channel hydraulics are altered to accommodate that movement over time. Situations where this connectivity of transport of particular sizes of sediment do not apply have been recognized, e.g. in studies by Ferguson & Ashworth (1991) and others of Allt Dubhaig in which a fining downstream is seen to be the result of selective transport due to a marked decrease in valley slope. Much debate has taken place on the role of selective transport and abrasion in downstream fining of sediment. Newson (1992) long ago recognized that fluvial systems may function as a series of individual systems in sequence downstream with discontinuities between. Barriers obviously act as discontinuities, but in other cases the connectivity may not be obvious. Coarse sediment moves in steps through systems, during high flows and much research has been undertaken into lengths of steps or distances of travel, e.g. Hassan *et al.* (1991).

The major storage locations in channels are bars. Reaches with channel bars are therefore indicative of a sediment flux, though only from the nearest upstream source (within or out of channel). Reaches which lack bars are not necessarily indicative of a lack of coarse sediment transport but may arise from several different causes or situations (Fig. 1):

(a) A lack of supply of sediment to be transported and stored, (due to lack of sources, deposition upstream or absorption of load by lateral movement);
(b) A lack of competence to transport sediment through the reach;
(c) High competence such that coarse sediment is flushed through the reach and not stored.

Inferences about the status or functioning of such reaches can be made from the morphological and sedimentological evidence in the sequence of reaches upstream and downstream. For example, lack of coarse sediment supply would generally be indicated by erosion and lack of bed sediment upstream. A lack of competence would result in a storage zone of coarse sediment upstream of the incompetent reach. A highly competent reach would result in coarse deposition downstream where competence decreases but could only be demonstrated if there were no new sources in between. The competence of a reach can be calculated using various published formulae and relationships.

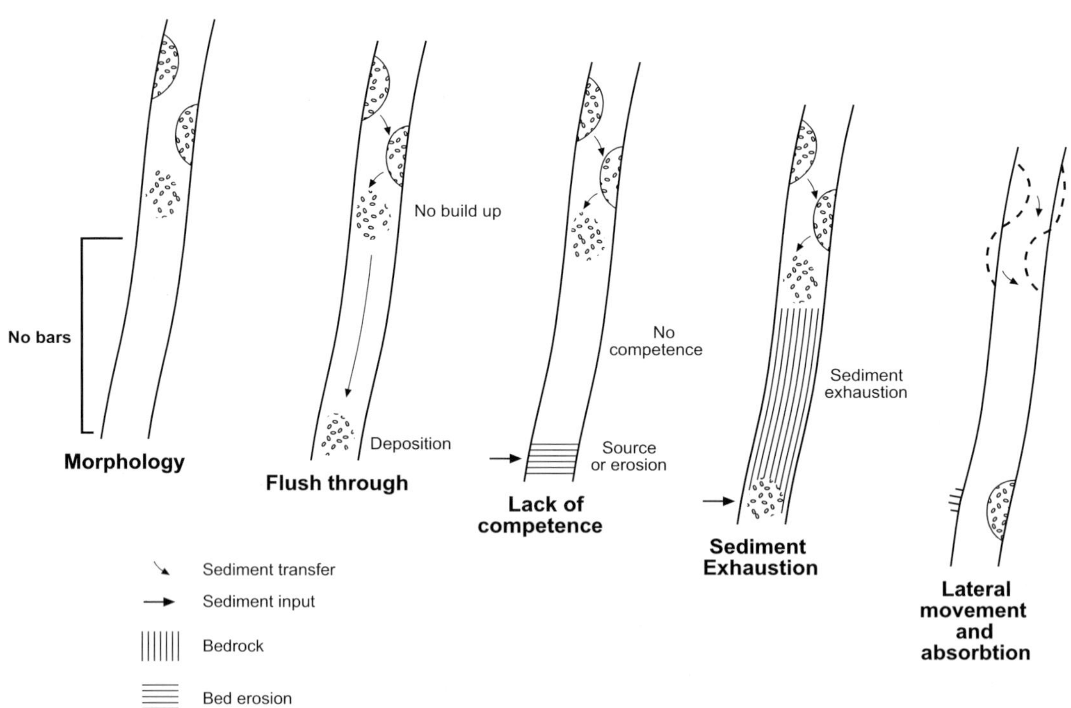

Fig. 1 Situation and explanation of reaches without coarse sediment bars.

METHOD

It is suggested that inferences can be made about the functioning of a reach by careful mapping and measurement of morphological features and sediment characteristics in reaches and analysis of the spatial patterns and sequences. These methods have been applied to several channel systems in contrasting environments. Hooke (2003) showed that an active meandering system, with perennial flow, such as the River Dane in northwest England, may be functioning as a disconnected system, with deposition close to sources of sediment.

In this paper two channel systems from a dryland area, differing in size and hydrological regime, are examined. In each case, long sections of channel incorporating reaches of differing characteristics have been mapped morphologically and sediment has been sampled. Competence has been calculated for representative cross-sections in each reach.

RESULTS

Case Study 1: River Gila in Safford Valley, Arizona

The Gila River in Safford valley in southeast Arizona is a large river, flowing through an arid area, with a highly variable but perennial discharge. Though much of the sediment is very coarse and appears abundant, morphological mapping of the sources from very large-scale aerial photographs and field verification indicate a wide range of sources, including gullies and arroyos, valley wall, channel banks and bars, and high spatial variability in their occurrence (Fig. 2). There is evidence of high mobility of sediment in floods from other published studies (e.g. Huckleberry, 1994) and by direct observation of deposits. Detailed cross sectional and sediment size measurements have been made at 1-km intervals through the valley (Fig. 3) (Hooke, 2000). Competence calculations have been made using various methods and various discharges. The values from Graf's (1983) equation are plotted in Fig. 4

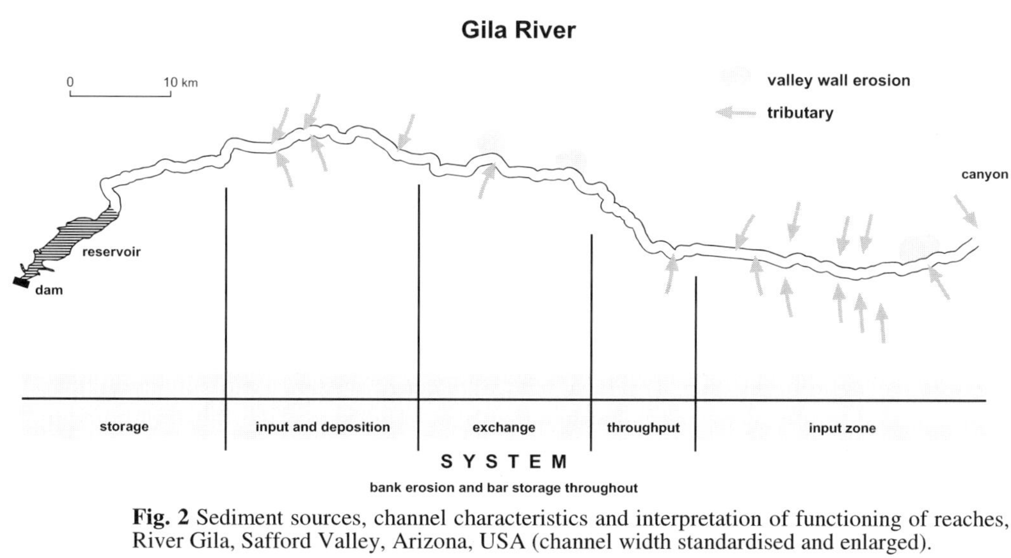

Fig. 2 Sediment sources, channel characteristics and interpretation of functioning of reaches, River Gila, Safford Valley, Arizona, USA (channel width standardised and enlarged).

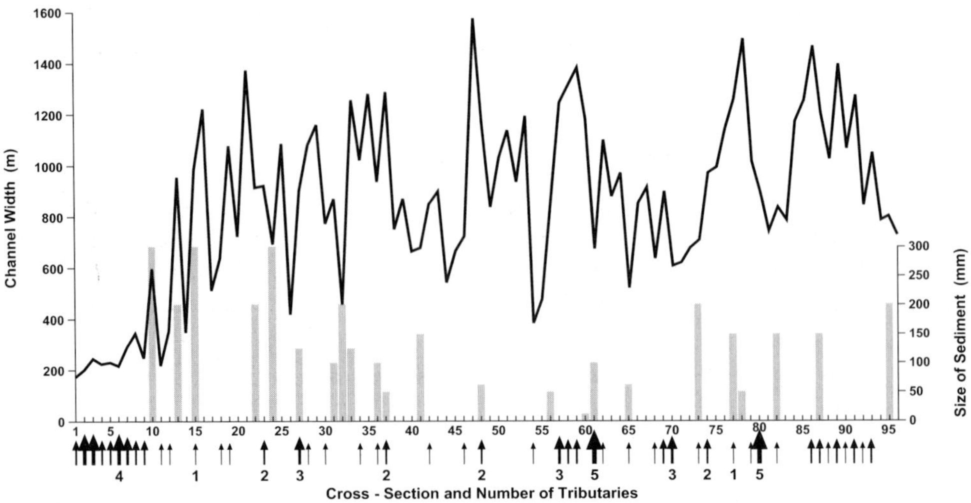

Fig. 3 Spatial pattern of channel width, maximum sediment size and tributaries, River Gila.

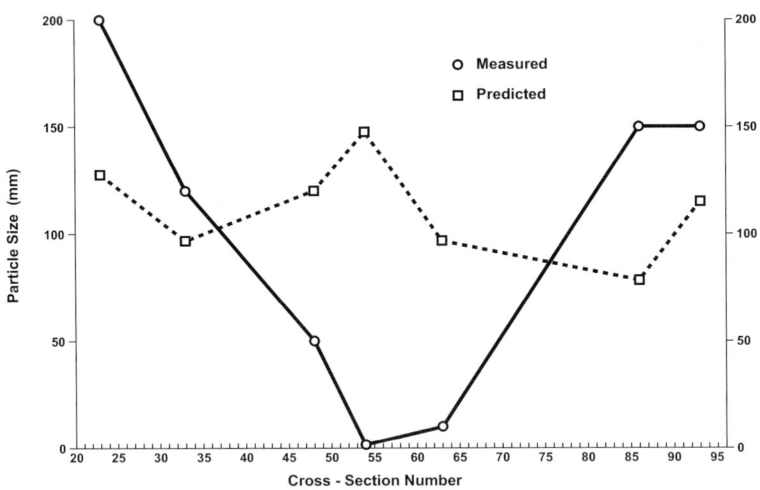

Fig. 4 Competence of flow and variation in sediment size in River Gila, Safford Valley, Arizona, USA.

for sample sections and compared with maximum sediment sizes measured in the channel. The size of material deposited does vary through the valley. In particular, in one reach around cross-section 54 the deposited material is much finer. This is a steep and narrow reach and competence calculations indicate that coarse material could be carried through here in moderate flows (Fig. 4) (Hooke, 2000). There are two possible interpretations of this narrow, fine sediment reach, lacking coarse bars. One is that the decline in sediment size downstream implies an exhaustion or lack of supply of coarse sediment. Alternatively, the coarse sediment may be flushed through on high flows and only fine material is deposited at low flow. It would appear that in highly competent reaches the deposits represent the residual flow rather than the maximum competence. There is some evidence from the upstream values of sediment size to suggest sediment exhaustion in this case.

Case Study 2: Walnut Gulch, Arizona

Walnut Gulch is an ephemerally flowing channel in southeast Arizona, USA. It is an instrumented catchment that has been run by the US Agricultural Research Service since the 1950s. In this study the channel of one sub-catchment has been intensively mapped and cross-sections and sediment sampled at intervals through the catchment. Fig. 5 indicates that there is a range of hillslope and channel sources, though many of the hillslope channels do not connect with the main channel. Overland flow from the hillslopes is unlikely to transport coarse material. Some coarse material is derived from valleyside erosion of older deposits. The channel material is dominantly gravel, with some sand and a few cobbles, but most of the coarse material is found on the bars. The channel material is mostly rather homogenous, d_{50} of channel samples at most sections being in the range 1–4 mm diameter. There are some zones of bars and braiding but their density varies and they are more common in the uppermost and lowermost reaches.

Calculations of the hydraulic competence of flows indicate that even a flood of 5-year recurrence interval could transport very coarse material. This is confirmed by measurements made after such a flood in 1993 when there was evidence of cobbles up to 160 mm diameter having been transported and evidence from bulldozed trenches across the channel that the whole channel bed to a depth of about 20 cm had been mobile. The question is whether the coarse material was actually moved through the reaches lacking bars. Hydraulic calculations indicate that it could have been moved. Some ancillary evidence indicates that in these narrower parts the coarse sediment may be flushed through because a few cobbles were found trapped in vegetation in the middle of the channel. Elsewhere, the coarse sediment is deposited on the tops of bars and edges of the channel, indicating deposition during high flow. Finer material is deposited in the channels, indicating continued transport in deeper

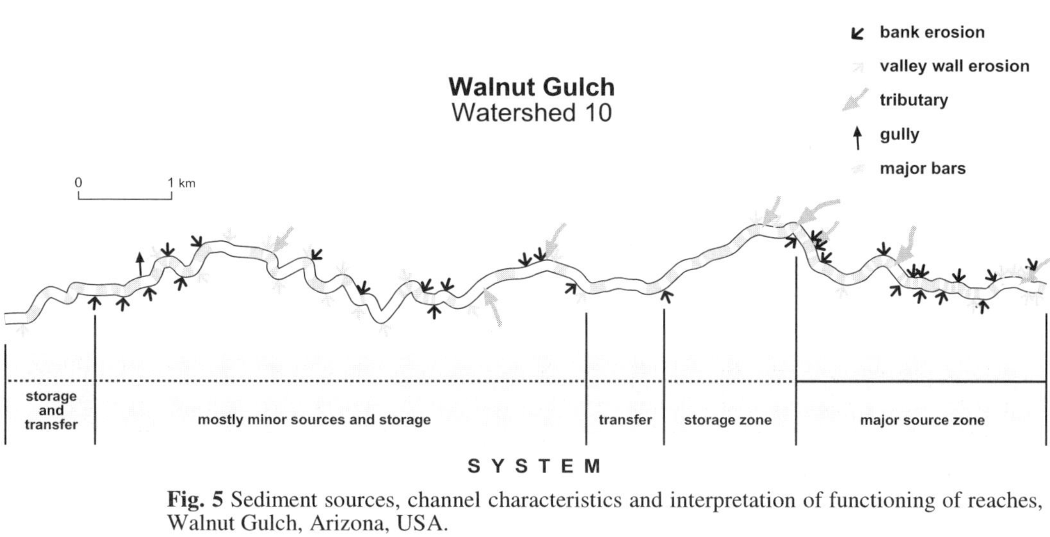

Fig. 5 Sediment sources, channel characteristics and interpretation of functioning of reaches, Walnut Gulch, Arizona, USA.

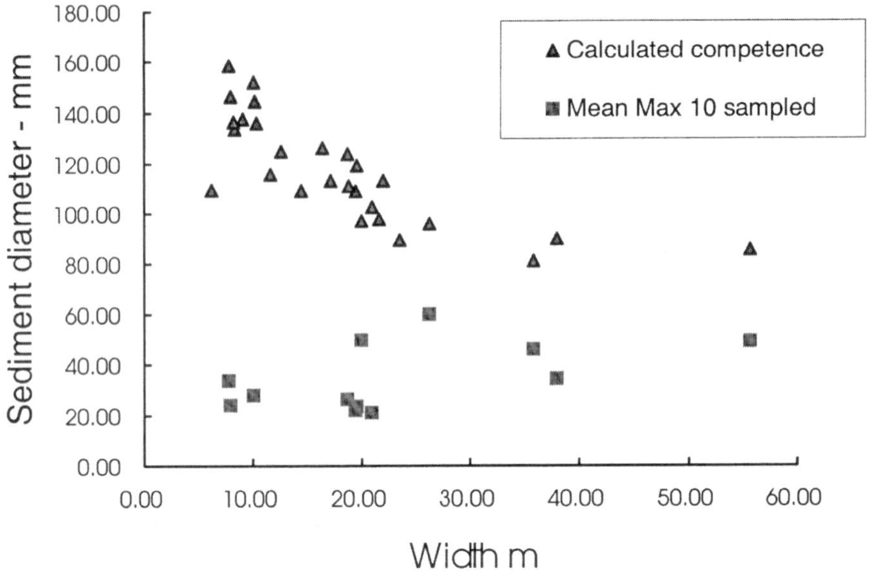

Fig. 6 Flow competence calculated for 5-year recurrence interval flow, and mean sizes of largest 10 particles sampled.

flow until the highly mobile, fine gravel and sand sediment is deposited in the late stages of flow recession. An analysis of the competence of the cross sections in relation to width indicates a non-linear inverse relation. However, measurements of coarsest material deposited show a positive relation to width (Fig. 6).

DISCUSSION AND CONCLUSIONS

Both of these channel systems appear to be highly connected, or potentially highly connected if sediment is available, for coarse sediment, though their morphological and sedimentological characteristics vary down through the system and they have reaches lacking bars that are of greater length than commonly quoted step lengths of coarse particle transport. High flows in dryland systems are generally highly competent and transport large loads. In both these channel systems it would appear that the steep narrow reaches without bars have the potential to flush coarse material through. However, there is some question of the availability of coarse material, particularly in the case of the Gila.

Comparison of competence calculations with actual maximum sizes deposited (with evidence of transport) shows an inverse relation, indicating deposition of coarse material at lowest competences. Under the conditions of transport in such channels, with a wide range of sizes of material, then the material deposited is related to loss of competence and recession of flows so that coarse material is deposited at the upper and outer edges of flow, on bars, and finer material continues to be carried until low flow in the most competent reaches, producing fine channel deposits. In the short term the wide shallow sections, with bars, having low competence, will tend to be perpetuated and the steep narrow sections will tend to flush material through and offer no opportunities for storage.

The interpretation of functioning of reaches is important for understanding the actual and potential for coarse sediment transport through reaches. Direct measurements of flux also need to be carefully interpreted in relation to sources and stores and their connectivity, because measurements in one reach may not be representative of what is moving though the whole channel system or even through adjacent reaches. Geomorphological analysis of the spatial relations and functioning of channel reaches, based on morphological and sedimentological evidence, can provide useful insights into the connectivity of the system and a framework for interpretation of any direct measurements.

REFERENCES

Ferguson, R. & Ashworth P. (1991) Slope-induced changes in channel character along a gravel-bed stream: the Allt Dubhaig, Scotland. *Earth Surf. Processes Landf.* **16**, 65–82.

Graf, W. L. (1983) Variability in sediment removal in a semiarid watershed. *Water Resour. Res.* **19**, 643–652.

Hassan, M. A., Church, M. & Schick, A. P. (1991) Distance of movement of particles in gravel bed streams. *Water Resour. Res.* **27**, 503–511.

Hooke, J. M. (2000) Spatial variation in channel morphology and sediment dynamics: Gila Rivers, Safford Valley, Arizona. In: *The Hydrology–Geomorphology Interface: Rainfall, Floods, Sedimentation, Land Use* (ed. by M. A. Hassan, O. Slaymaker & S. M. Berkowicz) (Proc. Conf., Jerusalem, May 1999), 251–272. IAHS Publ. 261. IAHS Press, Wallingford, UK.

Hooke, J. M. (2003) Coarse sediment connectivity in river channel systems: a conceptual framework and methodology. *Geomorphology* **56**, 79–94.

Huckleberry, G.(1994) Contrasting channel response to floods on the middle Gila River, Arizona. *Geology* **22**(12), 1083–1086.

Newson, M. (1992) Geomorphic thresholds in gravel-bed rivers. In: *Dynamics of Gravel-Bed Rivers* (ed. by P. Billi, R. D. Hey, C. R Thorne, P. Tacconi), 3–15. Wiley, Chichester, UK.

Sediment Transfer through the Fluvial System (Proceedings of a symposium held in Moscow, August 2004).
IAHS Publ. 288, 2004

Channel adjustments in response to human alteration of sediment fluxes: examples from Italian rivers

NICOLA SURIAN[1,2] & MASSIMO RINALDI[3]

1 *Autorità di Bacino dei fiumi dell'Alto Adriatico, Dorsoduro 3593, I-30123 Venice, Italy*
nicola.surian@libero.it

2 *Present address: Dipartimento di Geografia, Università di Padova, via del Santo 26, I-35123 Padova, Italy*

3 *Dipartimento di Ingegneria Civile, Università di Firenze, via S. Marta 3, I-50139 Florence, Italy*

Abstract In response to various types of human impacts, most Italian rivers have experienced considerable channel adjustments during the last two centuries. Human impact includes reforestation, channelization, construction of dams and sediment mining. The most important effect of human impact has been an alteration of sediment fluxes, and specifically a remarkable decrease in sediment supply to river channels. The five rivers selected in northern Italy (Tagliamento, Piave, Brenta, Trebbia and Vara), which have or used to have a braided morphology, have undergone channel narrowing (between 58 and 85%), decrease of braiding intensity and incision (up to 4–5 m). Narrowing and incision have been the dominant processes during the last two centuries, particularly intense from the 1950s to the 1990s; however, recent data suggest that those processes could now be exhausted since other kinds of adjustments, specifically channel widening and (local) aggradation, have occurred during the last 10–15 years.

Key words braided rivers; channel adjustments; channel narrowing; channel widening; human impact; Italian rivers; sediment fluxes

INTRODUCTION

During the last century most of the Italian rivers, like many other rivers throughout the world (e.g. Williams & Wolman, 1984; Wyzga, 1993; Kondolf, 1994; Winterbottom, 2000; Liebault & Piegay, 2001), have experienced considerable morphological changes in response to various types of human interventions. Two recent papers, about the rivers of Tuscany (central Italy) (Rinaldi, 2003) and Italian rivers (Surian & Rinaldi, 2003), have documented that: (a) incision and narrowing have been the dominant processes in river channels; and (b) these processes have been mainly induced by a dramatic reduction in sediment supply. In these papers two classification schemes of channel evolution (for Tuscany and Italian rivers, respectively) were developed.

The aim of this paper, which is based on some recent case studies (Sogni, 2003; Simoncini, 2003; Surian, 2004), is to: (a) improve the existing knowledge about channel adjustments in Italian rivers, and in particular existing temporal trends of width adjustment; (b) analyse the most recent channel changes (the last 10–15 years) for which there is very little known; and (c) try to improve the existing scheme of channel evolution (Surian & Rinaldi, 2003) for rivers with a braided, or initially braided, morphology.

GENERAL SETTING

The five selected rivers are in northern Italy (Fig. 1): three drain from the Alps (Tagliamento, Piave and Brenta) and two from the Apennines (Trebbia and Vara). Their

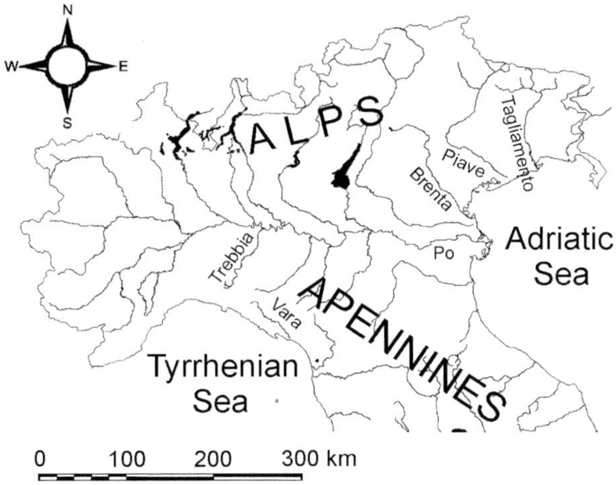

Fig. 1 Location map of the selected rivers.

Table 1 Hydrologic and drainage basin characteristics of the selected rivers.

River	Drainage basin area (km^2)	Length (km)	Length of the study reach (km)	Basin relief (m)	Precipitation (mm $year^{-1}$)	Mean annual discharge ($m^3\ s^{-1}$)	Flood peak discharge ($m^3\ s^{-1}$)
Tagliamento	2580	178	39	2696	2150	109	4650
Piave	3899	222	23	3162	1330	132	5300
Brenta	1567	174	18	3079	1386	71	2400
Trebbia	1070	116	35	1753	1400	24	3430
Vara	572	62	6	1603	1742	23	820

hydrological and drainage basin characteristics are reported in Table 1. The drainage basin areas range from 572 to 3899 km^2 and river lengths from 62 to 222 km. The basin relief is higher in the Alpine rivers than in those draining from the Apennines (about 3000 m and 1600–1700 m, respectively), whereas precipitation is relatively high in all the rivers (it ranges from 1330 to 2150 mm $year^{-1}$). Floods are relatively flashy and high in magnitude: the ratio between flood peak discharge and mean annual discharge ranges between 34 (Brenta) and 143 (Trebbia).

In long reaches of their course, these rivers have (or used to have) a braided morphology and have channel beds mainly composed of gravel. The following historical analysis on channel adjustments is focused on those braided reaches (or a part of those reaches), in which channel morphology is very sensitive to changes in sediment fluxes.

HUMAN IMPACT ON SEDIMENT FLUXES

Several human interventions (reforestation, torrent control works, constructions of levees, dams, sediment mining) have taken place in the selected rivers during the past centuries, and in particular during the last one (Table 2). Such interventions have had different impacts on the fluvial system (e.g. runoff changes, decrease of low flows), but the most important

Table 2 Human interventions in the selected rivers.

River	Drainage area upstream from dams (%)	Time of dam closure	Time of intense sediment mining	Construction of levees and other bank protection structures	Reforestation and torrent control works in the drainage basin
Tagliamento	3	1950s	1970s–1980s	19th–20th century	20th century (?)
Piave	54	1930s–1950s	1960s–1980s	14th–20th century	since the 1920s
Brenta	40	1954	1950s–1980s	19th–20th century	since the 1920s
Trebbia	25	1920s–1950s	1950s–1980s	not considerable	not considerable
Vara	36	1930s	1960s–1970s	20th century	20th century

impact on channel processes is the alteration of sediment fluxes, and specifically a remarkable decrease in sediment supply to river channels.

The chronology of human intervention is quite similar in the selected rivers, although there are some differences (Table 2). Reforestation, which occurred after several centuries of intense logging, and torrent control works generally started in the first decades of the 20th century, whereas construction of dams took place between the 1930s and 1950s. Later, between the 1950s and 1980s, sediment mining from river channels was carried out, frequently with extraction rates largely exceeding replenishment rates. These interventions have altered sediment fluxes in different ways, both in relation to the type of intervention (in particular the volumes of sediment affected by the specific intervention) and the location of the intervention (in the drainage basins and within the river channels). For instance, while sediment mining has had a dramatic impact in terms of the volume of sediment removed from the river channel in the short term, the impact of reforestation and dams probably have a greater impact in the long term.

CHANNEL ADJUSTMENTS

A historical analysis of river morphology was performed using maps, aerial photographs, longitudinal profiles and cross sections. Channel width, which was measured on maps and aerial photographs, has undergone dramatic changes during the last two centuries (Fig. 2). Average channel width has decreased by several hundreds of metres. Referring to the original widths in the early 19th or 20th century, a width reduction varying between 58 and 85% was measured (Table 3). It is worth noting that the selected rivers show very similar temporal trends, and in particular two phases of narrowing followed by a very recent phase of widening. As regards narrowing, data clearly show that the process has not been constant through time, and two different periods can be recognized: a first phase of narrowing took place from the early 19th century (in the case of the Tagliamento, Brenta, Trebbia and Vara Rivers) or early 20th century (for the Piave) to the 1950s, whereas a second one took place from the 1950s to the 1990s (Fig. 2). As for the amount of narrowing, in two out of five rivers, about 70% of the whole narrowing occurred in the second phase, whereas in the other three rivers narrowing was almost the same in the two phases (Table 3). The average rate of narrowing has always been higher (between 2.5 and 7.0 times) in the second phase than in the first one (Table 3) (e.g. 5.7 m year^{-1}, in the second period, and 2.5 m year^{-1}, in the first period, in the Vara River and 4.2 and 0.6 m year^{-1} in the Brenta River).

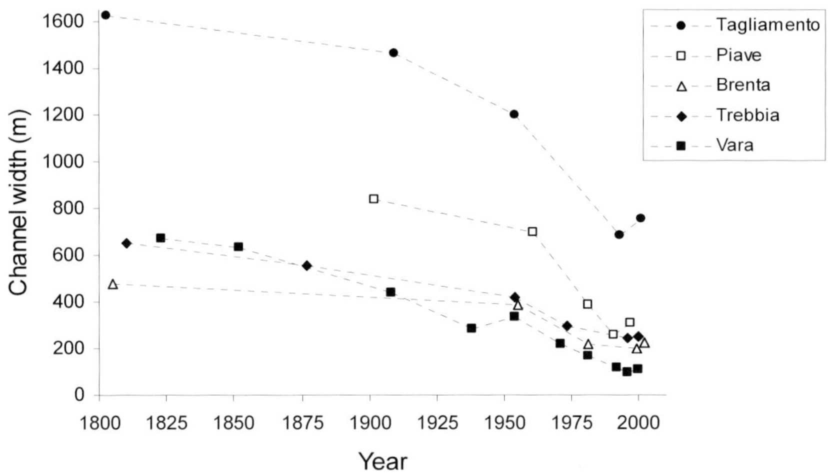

Fig. 2 Changes in channel width during the last 200 years in the selected rivers.

Table 3 Channel narrowing in the selected rivers.

River	Total narrowing (m)	Total narrowing (%)	Narrowing in the 1st phase (% of the total)	Narrowing in the 2nd phase (% of the total)	Rate of total narrowing (m year^{-1})	Rate of narrowing in the 1st phase (m year^{-1})	Rate of narrowing in the 2nd phase (m year^{-1})
Tagliamento	942*	58	46	54	5.0	2.8	13.2
Piave	580**	69	25	75	6.5	2.4	14.6
Brenta	276*	58	33	67	1.4	0.6	4.2
Trebbia	403*	62	56	44	2.2	1.6	4.2
Vara	569*	85	58	42	3.3	2.5	5.7

Total narrowing: refers to the period from early 19th (*) or 20th (**) century to the 1990s;
Narrowing in the 1st phase: refers to the period from early 19th or 20th century to the 1950s;
Narrowing in the 2nd phase: refers to the period from the 1950s to the 1990s;
Rate of narrowing: average rate of narrowing estimated over the different time periods considering initial and final channel widths.

Table 4 Channel widening in the selected river.

River	Widening (m)	Period of documented channel widening	Widening compared to original channel width (%)	Widening compared to previous channel width (%)	Rate of widening (m year^{-1})
Tagliamento	72	1993–2001	4	11	9.0
Piave	49	1991–1997	6	19	8.2
Brenta	21	1999–2002	4	10	7.0
Trebbia	4	1996–2000	0.6	2	1.0
Vara	13	1996–2000	2	13	3.3

Widening compared to original channel width: represents the ratio between the amount of widening and channel width in the early 19th or 20th century;
Widening compared to previous channel width: represents the ratio between the amount of widening and channel width at the beginning of the widening process;
Rate of widening: average rate of widening estimated over the different time periods considering initial and final channel widths.

The most recent data point out that channel narrowing has stopped in all the selected rivers and channel widening has been taking place since the 1990s (Fig. 2). The magnitude of widening is low if the original channel width (width in early 19th or 20th century) is considered, but relatively high if the channel width at the end of narrowing is taken as a reference. In this latter case, in four out of five rivers widening is more than 10% (e.g. 19% in the Piave River), with an average rate ranging between 3.3 and 9.0 m $year^{-1}$ (Table 4).

Braiding intensity has also changed considerably. A decrease of braiding indices has occurred in all these rivers, but, differently from channel width, temporal trends show some variability from one river to another (Fig. 3). Braiding indices used to be around 2.0 in two of the rivers and much higher in the other three (5.7, 3.9 and 3.8, respectively in the Tagliamento, Piave and Vara), whereas nowadays they are equal or less than 1.5 in four out of five rivers. Three rivers (Tagliamento, Trebbia and Brenta) show a remarkable decrease of braiding intensity in the period 1950s–1990s (corresponding to the major phase of narrowing), whereas the other two rivers (Piave and Vara) show a less constant decreasing trend in the same period (Fig. 3).

In addition to channel narrowing and decrease of braiding intensity, channel incision has been clearly documented in some of the selected rivers (Castiglioni & Pellegrini, 1981; Surian, 2004). Incision, which has been up to several metres (e.g. 4 and 5 m in the Piave and Brenta), occurred together with the major phase of narrowing, whereas in the last 10–15 years there is evidence, from cross-sections, that bed-level lowering could be over. As for the Vara River, in some reaches incision has been followed by aggradation during the last 10 years, promoted by the construction of some grade control structures. On the contrary, less information is available as regards bed-level adjustments during the 19th century and the first decades of the 20th century.

Such channel adjustments (narrowing, decrease of braiding intensity, incision) have produced significant changes in the channel configuration of those initially braided rivers. In several reaches there has been a modification from braided to wandering, and in some cases

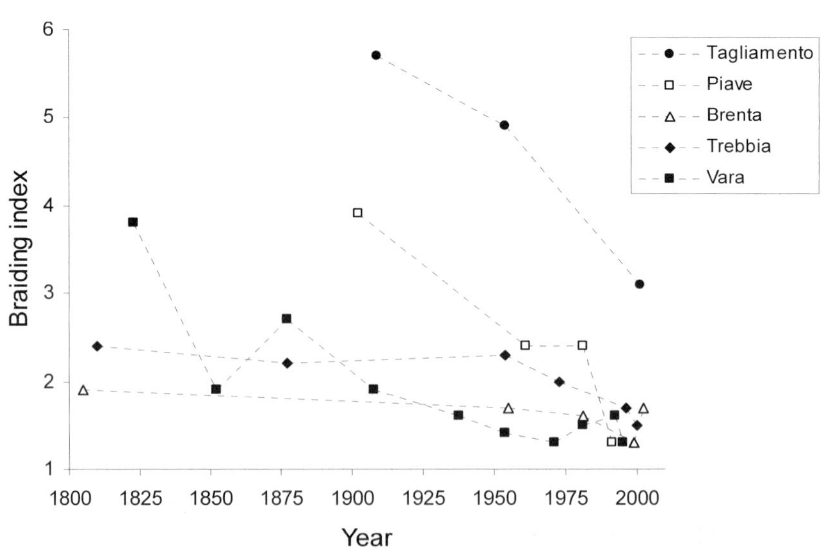

Fig. 3 Changes in braiding index during the last 200 years in the selected rivers.

to a single-thread morphology (e.g. in the Brenta). Only the Tagliamento, notwithstanding significant channel changes, has retained a clear braided morphology along the whole study reach.

DISCUSSION AND CONCLUSIONS

On the basis of the data discussed above, and in particular of the temporal trends of channel width considered in this paper (only the trend of the Piave River was available previously), it is possible to further develop the scheme of channel evolution of Italian rivers originally proposed by Surian & Rinaldi (2003). Four stages of channel evolution can be recognized in the selected braided rivers (Fig. 4). Considering the channel morphology in the early 19th or 20th century as the initial stage (stage I), a second stage of channel evolution can be identified after a first phase of narrowing and incision (in such phase bed-level changes are not well documented, but there is evidence that, at least during the first half of the 20th century, some incision occurred). Stage III represents the channel morphology after the major phase of narrowing and incision, which took place from the 1950s to the 1990s, whereas stage IV represents the present morphology which is the result of some widening processes. The recognition of the very recent channel adjustments occurred in the last 10–15 years (channel evolution from stage III to stage IV), is meaningful since it has pointed out that the relatively long period of narrowing and incision should be exhausted and other processes, in particular channel widening, can become dominant at present.

The causes of these channel adjustments are mainly human interventions, and in particular those interventions, such as sediment mining, dams, reforestation, which have

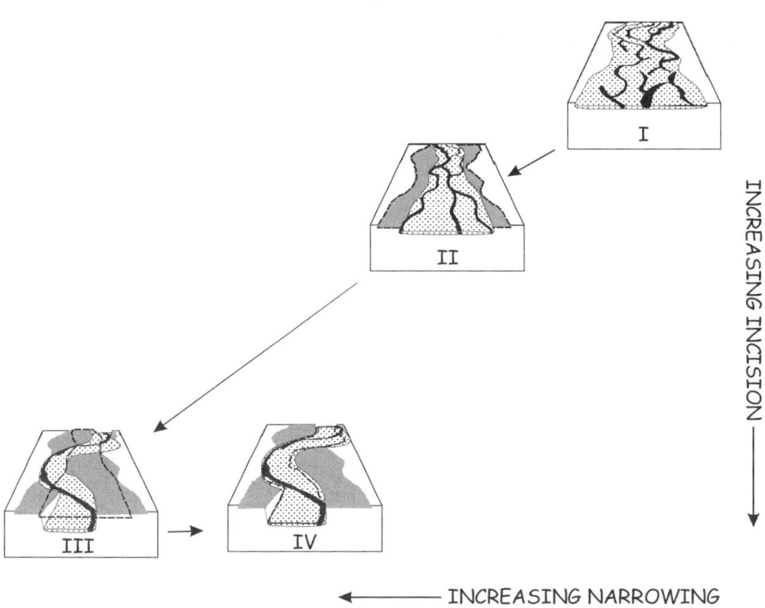

Fig. 4 A channel evolution model for Italian braided rivers; "stage I" represents channel morphology in the early 19th or 20th century, "stage IV" represents the present morphology. Note: active channel is represented by dots and abandoned channel in grey.

induced a remarkable alteration of sediment fluxes. There is a strong temporal relationship between human interventions and channel adjustments as for the main phase of adjustment (from stage II to stage III). Two factors, sediment mining and construction of dams, had a main role in this phase of adjustment, even if some other factors (e.g. reforestation and torrent control works) could have had some effects on sediment fluxes (see also Table 2). On the contrary, there is more uncertainty regarding the first phase of adjustment (from stage I to stage II). Reforestation and torrent control work should have played a significant role during the first half of the 20th century, whereas the construction of bank protection structures (levees, groynes) and climate changes could be significant factors in explaining channel adjustments during the 19th century. The first of the latter factors (bank protection structures) is well documented, whereas at the present stage of the research the second one (climate changes) is only a hypothesis. In fact, the climatic transition occurred during the 19th century, at the end of the Little Ice Age, could have had some effects on sediment fluxes in the selected rivers. Finally, the most recent channel adjustments (from stage III to stage IV) could be due to an increase of sediment availability in river channels (mainly in relation to a remarkable reduction of sediment mining) which promotes the formation of bars and, therefore, widening of the active channel.

Finally, issues that will require further research to improve the present knowledge about channel adjustments in Italian rivers include: (a) bed-level changes: it is necessary to gain more data about these changes during the 19th century and the first half of the 20th century and about the temporal trends of such changes; (b) human interventions: some of them, in particular their development through time, are still not well documented (e.g. reforestation in drainage basins and along river channels, torrent control works); (c) present channel processes and sediment fluxes: since different processes have occurred in the last few years, specifically widening after a long period of channel narrowing, monitoring of present processes is very important as well as information about the present and future sediment fluxes in Italian rivers (for instance in rivers with no dams, or where dam effects are little, an increase of sediment fluxes could take place in the next future).

REFERENCES

Castiglioni, G. B. & Pellegrini, G. B. (1981) Two maps on the dynamics of a river bed. In: *Erosion and Sediment Transport Measurement* (Proc. Symposium, Florence, 22–26 June 1981), 223–228. IAHS Publ. 133. IAHS Press, Wallingford, UK.

Kondolf, G. M. (1997) Hungry water: effects of dams and gravel mining on river channels. *Environ. Manage.* **21**, 533–551.

Liebault, F. & Piegay, H. (2001) Assessment of channel changes due to long-term bedload supply decrease, Roubion River, France. *Geomorphology* **36**, 167–186.

Rinaldi, M. (2003) Recent channel adjustments in alluvial rivers of Tuscany, Central Italy. *Earth Surf. Processes Landf.* **28**, 587–608.

Simoncini, C. (2003) Studio geomorfologico del Fiume Vara finalizzato alla riqualificazione fluviale (Geomorphological study of the Vara River aimed to river restoration). Thesis, University of Florence, Italy (unpublished) (in Italian).

Sogni, D. (2003) Evoluzione morfologica recente dell'alveo del tratto inferiore del F. Trebbia (Recent morphological channel evolution in the lower reach of the Trebbia River). Thesis, University of Milan, Italy (unpublished) (in Italian).

Surian, N. (2004) Effects of human impact on braided river morphology: examples from northern Italy. In: *Braided Rivers 2003* (Proc. Conf., Birmingham, 7–9 April 2003) (submitted).

Surian, N. & Rinaldi, M. (2003) Morphological response to river engineering and management in alluvial channels in Italy. *Geomorphology* **50**, 307–326.

Williams, G. P. & Wolman, M. G. (1984) Downstream effects of dams on alluvial rivers. *US Geol. Survey Prof. Paper 1286*.

Winterbottom, S. J. (2000) Medium and short-term channel planform changes on the Rivers Tay and Tummel, Scotland. *Geomorphology* **34**, 195–208.

Wyżga, B. (1993) River response to channel regulation: case study of the Raba River, Carpathians, Poland. *Earth Surf. Processes Landf.* **18**, 541–556.

The settling behaviour of fine sediment particles: some preliminary results from LISST instruments

N. D. WILLIAMS[1], D. E. WALLING[1] & G. J. L. LEEKS[2]

1 *Department of Geography, University of Exeter, Exeter EX4 4RJ, UK*
n.d.williams@exeter.ac.uk

2 *Centre for Ecology and Hydrology, Wallingford, Oxfordshire OX10 8BB, UK*

Abstract The settling velocity of suspended particles is a dominant factor in controlling the transfer and fate of sediment and sediment-associated substances. The properties of fine particles can vary significantly throughout a catchment, especially in terms of the degree of aggregation/flocculation, but relatively little is known about the consequences this has on settling velocity. This study attempts to explore the significance of the particle size distribution in influencing the settling behaviour of natural particles. Particles were collected from a range of sources across two contrasting catchments, giving natural variability in the grain size composition and degree of aggregation/flocculation of the samples. Particle size and settling velocity were measured using novel LISST-100 and LISST-ST laser diffraction devices. Significant differences in settling velocity were found between samples, notably between aggregated/flocculated and dispersed samples, and between individual size classes. The results emphasize the importance of aggregation/flocculation in the hydraulic behaviour of sediment.

Key words aggregate; fine sediment; floc; LISST-100; LISST-ST; particle size; settling velocity

INTRODUCTION

The transport and fate of fine sediment play a key role in the transfer of nutrients and contaminants in river basins, and in the physical degradation of aquatic habitats in the hyporheic zone. Several recent studies have investigated the spatial variation of the physical and chemical characteristics of fine sediment, including particle size (Walling & Moorehead, 1987, 1989), nutrient and organic matter content (Droppo *et al.*, 1997; Walling *et al.*, 2001; Ankers *et al.*, 2003) and trace elements (Foster & Charlesworth, 1996; Ankers *et al.*, 2003; Krein *et al.*, 2003). Source tracing studies have highlighted the importance of the catchment surface as a source of fine sediment in aquatic systems (Collins & Walling, 2002). However, relatively few studies have specifically considered the transport mechanisms involved in the transport of this material, which are critical to the understanding and modelling of the movement of fine sediment and associated substances. In part, this reflects a limited understanding of the hydraulic significance and effects of aggregation and flocculation, the importance of which has been demonstrated by comparisons of effective size distributions sampled *in situ* with equivalent absolute (dispersed) size distributions measured in the laboratory (Walling & Moorehead, 1987, 1989; Phillips & Walling, 1995, 1999; Droppo *et al.*, 1997, 1998, 2000). Aggregates are densely packed, well rounded composite particles, formed by non-aqueous processes, which retain their structure during transport through the system (Walling & Woodward, 2000). In contrast, flocs are composite particles formed by inter-particle interactions within the water column, which are known to be much more loosely bound, irregular in shape and of relatively low density (Droppo, 2001). While it is assumed that the two classes of composite particle are structurally and functionally different, few studies have been able to explore the implications of these differences for their hydraulic behaviour. The fate of particles during transport can be inferred from their settling velocity; a measure of the potential for transport or deposition. The settling velocity of a suspension is dependent on a range of variables, including particle size, density and shape, the concentration of the suspension and the viscosity of the suspending medium. However, the dominant control can be assumed to be particle size, which in turn is dependent on particle composition and source. This study uses novel laser diffraction devices to investigate the settling behaviour of structurally contrasting particles, representative of sediment from a range of sources. Particular attention is paid to particle size and size distribution, with reference to aggregated/flocculated and dispersed samples.

METHODS

Samples of fine-grained bed sediment (surficial fine-grained laminae (SFGL), Droppo & Stone, 1994), storm suspended sediment, bank sediment and soils from a range of land uses were collected from two contrasting catchments in southern England. The River Dart is a tributary of the River Exe, lying approximately 15 km north of Exeter, Devon. It has a catchment area of 46 km^2. The catchment is developed on Upper Carboniferous sandstones, shales and mudstones, with some alluvial deposits in the steep valley bottoms. Soils are predominantly brown earths and surface-water gleys. The dominant land use is pasture, with some permanent deciduous woodland and some arable agriculture. The Chilfrome catchment forms part of the headwaters of the River Frome in Dorset, located approximately 10 km northwest of Dorchester. It has an area of approximately 36 km^2. The catchment is entirely underlain by chalk, and is characterized by argillic brown earth and brown rendzina soils. Land use is primarily pasture, with areas of arable farming and some deciduous woodland.

Soil samples collected from both catchments were classified as ploughed, pasture, deciduous woodland, or track/bridleway. The soil sampling locations were selected to represent sites that had been observed to generate surface runoff, and were generally gateways from fields on the steeper slopes. Samples were collected using a polyethylene scoop and stored in large, unsealed polyethylene bags. Although these were stored in dark, refrigerated conditions, it should be noted that they can only provide an approximate representation of field conditions, given the period of storage required when running numerous lengthy settling experiments.

Soil aggregates were isolated by placing the soil sample into a plastic tray fixed at a 30° angle and spraying the tray with filtered river water from the appropriate catchment, in order to simulate the generation of sediment-rich saturated overland flow. Runoff from the trays was collected in 500 ml polyethylene bottles, kept well mixed by gentle agitation and analysed as soon as possible. Samples of true bed sediment from the study rivers were collected in 500 ml polyethylene bottles. Fine bed sediment was entrained into the water column by disturbing the SFGL prior to sampling. The samples were stored in a dark refrigerator and resuspended by gentle agitation prior to analysis in the laboratory. Suspended sediment samples were collected from the water column during storm events, stored and resuspended in the same way. Duplicates of all samples were mechanically dispersed by ultrasonication to provide the absolute size distributions. Sodium hexametaphosphate was not used for dispersion because of its potential effects on water buoyancy and viscosity.

The grain size distributions of all samples were measured using a LISST-100 (Type C) laser diffraction particle sizer, which provides results in 32 logarithmically spaced size classes in the range of 2.5 to 500 μm. Settling velocity was measured using a LISST-ST (Type C), which permits the measurement of fall velocity for eight separate size classes in the same overall range. Both instruments employ the principle of small angle forward laser scatter to measure particle size. The sensing area in the LISST-100 is designed to be non-intrusive for *in situ* field deployment, whereas in the LISST-ST it is located at the bottom of a 30-cm settling column. Fall velocity is calculated from the evolution of particle size spectra, with reference to a preliminary measurement of the initially well-mixed sample. Repeat measurements are taken at known time increments, with a size class that was measured at the beginning of the experiment being deemed to have settled out of suspension as soon as it is no longer detected. Since the LISST-ST only resolves settling velocity for eight size classes and does not provide detailed information on size distributions, it was used in conjunction with the LISST-100, which provides a higher resolution size distribution. Volumetric mean particle size was calculated from the LISST-100 size results, but detailed size distributions are not presented here. Since both instruments use the same operating principles, results from the two can be directly compared. Further details of these instruments are provided by Agrawal & Pottsmith (2000).

Prior to laboratory measurements, suspensions of each sample were diluted to produce a concentration of 400 $\mu l\ l^{-1}$, using filtered river water from the appropriate catchment. The value of 400 $\mu l\ l^{-1}$ was selected as close to the instrument's minimum optical transmission of $\tau = 0.3$, when using a 50% optical path reduction module. One hundred discrete measurements of the diluted sample were made at a rate of 4 Hz and the mean 32-class size distribution recorded before the sample was transferred to the LISST-ST and the settling velocity measured over an 11-h duration. Sample preparation and handling, as described above, are assumed to have had no significant impact on the structure of the composite particles. Floc structure is known to evolve in response to changing hydraulic conditions (e.g. Phillips & Walling, 1995), but is thought to be essentially stable for at least 80 s (Phillips & Walling, 1995). The timescale of sample preparation in the adopted methodology was therefore considered acceptable.

It is necessary to dilute samples so that they can be accurately measured using the LISST devices, and because concentration is one of several potential controls on settling behaviour. Water viscosity was assumed to be constant, since temperature was essentially constant in the laboratory. It was not possible to measure particle shape and it was therefore assumed that the only significant variation between samples was the particle size distribution. This encompasses grain size and density, because the experimental set up considers the effects of changes in density due to aggregation and flocculation.

RESULTS

The relationships between particle size and settling velocity for all samples are presented in Fig. 1. Particle size refers to the volumetric mean, calculated from the 32-class size distribution provided by the LISST-100. Mean settling velocity is calculated from the eight class values provided by the LISST-ST. Mean values therefore refer to the 2.5 to 500 μm size range of the instruments. Figure 1 illustrates the broad variation in both mean particle size and mean settling velocity associated with a range of sediments and potential source

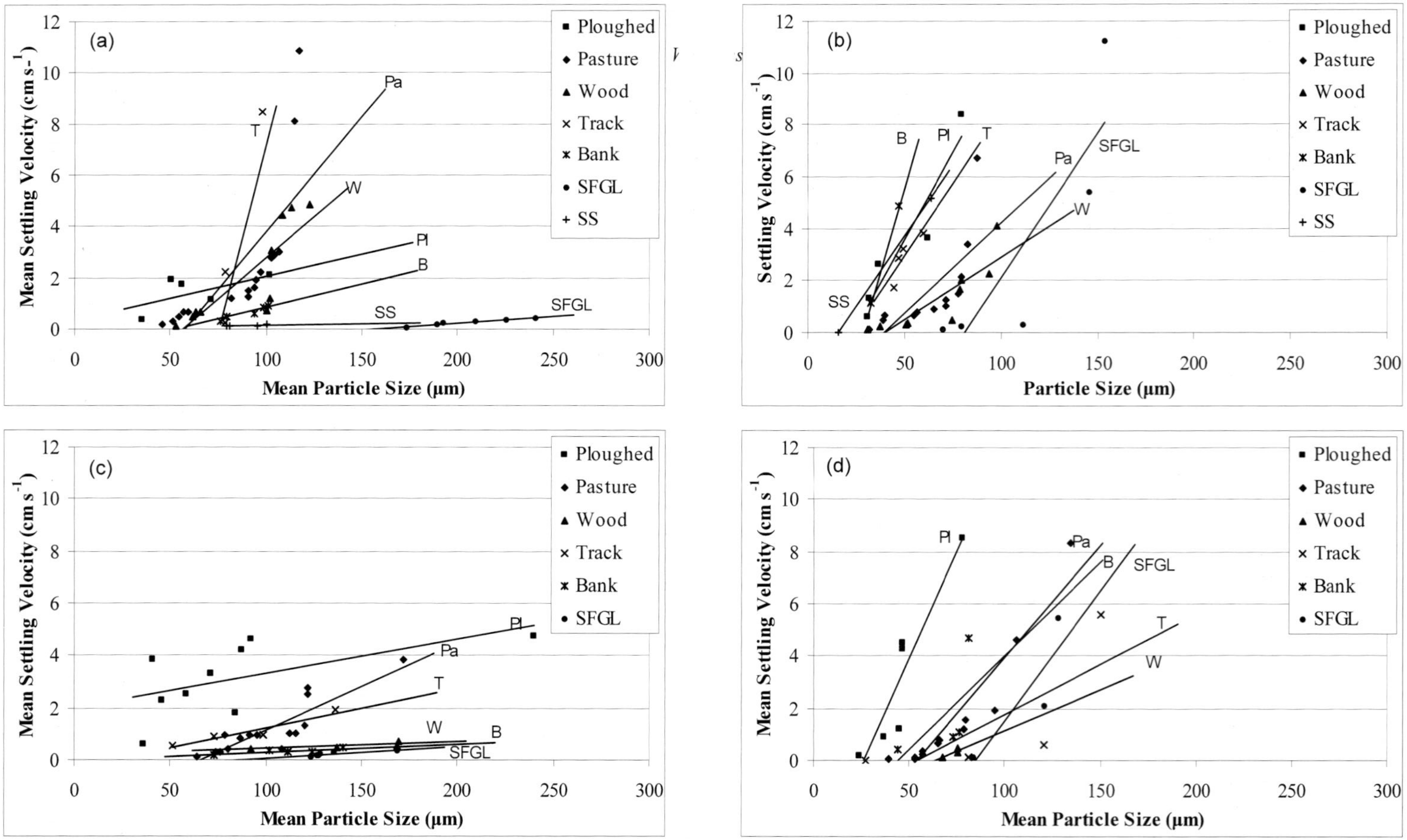

Fig. 1 Regression lines for the relationships between mean settling velocity and mean particle size for sediment from different sources distinguished by catchment and by APSD or EPSD. (a) Dart effective size distributions, (b) Dart absolute, (c) Chilfrome effective, (d) Chilfrome absolute. For all plots Pl = sediment from ploughed areas, Pa = pasture, W = wood, T = track, B = bank, SFGL = fine bed sediment and SS = suspended sediment. Lines have been extrapolated for visual clarity.

materials collected from within relatively small catchments. However, it is also apparent that significant trends exist when samples are grouped according to sediment type. Intra-group variation in both fall velocity and particle size is significantly less than inter-group variability. This is true for both the effective particle size distribution (EPSD) (Figs 1(a) and 1(c)) and absolute particle size distribution (APSD) (Figs 1(b) and 1(d)). Within all sample groups, it is apparent that mean settling velocity increases with mean particle size.

The rate of increase in settling velocity with particle size is generally more rapid for absolute distributions than effective size distributions. This can be attributed to the effects of aggregation/flocculation and the associated decrease in particle density caused by inefficient particle packing, and to the increased porosity of composite particles compared to discrete particles of equivalent size. This assertion is reinforced by the fact that the largest shift in regression gradient between the EPSD and APSD is for suspended sediment and SFGL flocs in both catchments. Flocs are much more loosely bound than aggregates, with relatively high water content within the floc matrix, whereas aggregates are densely packed particles. The dominant effect of dispersing aggregates is therefore to reduce fall velocity as a consequence of lower particle size as opposed to the more radical structural effects of dispersing flocs. It is particularly interesting to note the similarity in overall scatter of points for dispersed samples between the two catchments, when untreated samples show pronounced inter-catchment variation. This highlights the effect of particle structure (a consequence of source, composition and formation mechanisms) on sediment hydraulic behaviour. Non-dispersed (natural) samples from the Chilfrome catchment (chalk) generally exhibit a lower mean settling velocity and slightly larger mean particle size than samples from the Dart catchment (sandstone/shale/mudstone). This can be attributed to material from the Chilfrome drainage basin being of relatively low density, due to the dominance of chalk and a higher organic content of the soil.

The discrepancies between the EPSD and APSD for individual size classes are explored in Fig. 2. This emphasizes the positive trend between particle size and fall velocity. It also shows that the greatest disparity in settling rate between the APSD and EPSD is at the upper end of the size range. This is attributable to flocculation/aggregation and decreasing density with increasing size.

In both catchments, the greatest differences between the EPSD and APSD behaviour are found for the SFGL, where the low density blanket of fine sediment is preferentially entrained from larger bed material during resuspension in the form of large flocs, with relatively low numbers of equivalent sized individual grains. SFGL flocculated particles appear to settle faster than the equivalent size discrete particles up to around 80 μm. This is suggested to be because of a strong bias towards the high end of each size class for the EPSD because of the number of fine particles constituting a single floc, which has the effect of increasing fall velocity until a threshold density is reached. Similar patterns exist for all samples from ploughed land and tracks, and for samples from pasture land, in the Chilfrome catchment. This is attributed to the selection of sampling sites, which were depositional points within areas that generate extensive surface runoff over bare soils. It is likely that extensive sorting and preferential transport of fine particles occurs during overland flow, leading to a bias towards finer particles and an upward-fining depositional profile, the surface of which is more likely to be sampled. If size spectra evolve by *in situ* aggregation between storm events then the same effective/absolute trends as in SFGL will be seen within the settling size classes.

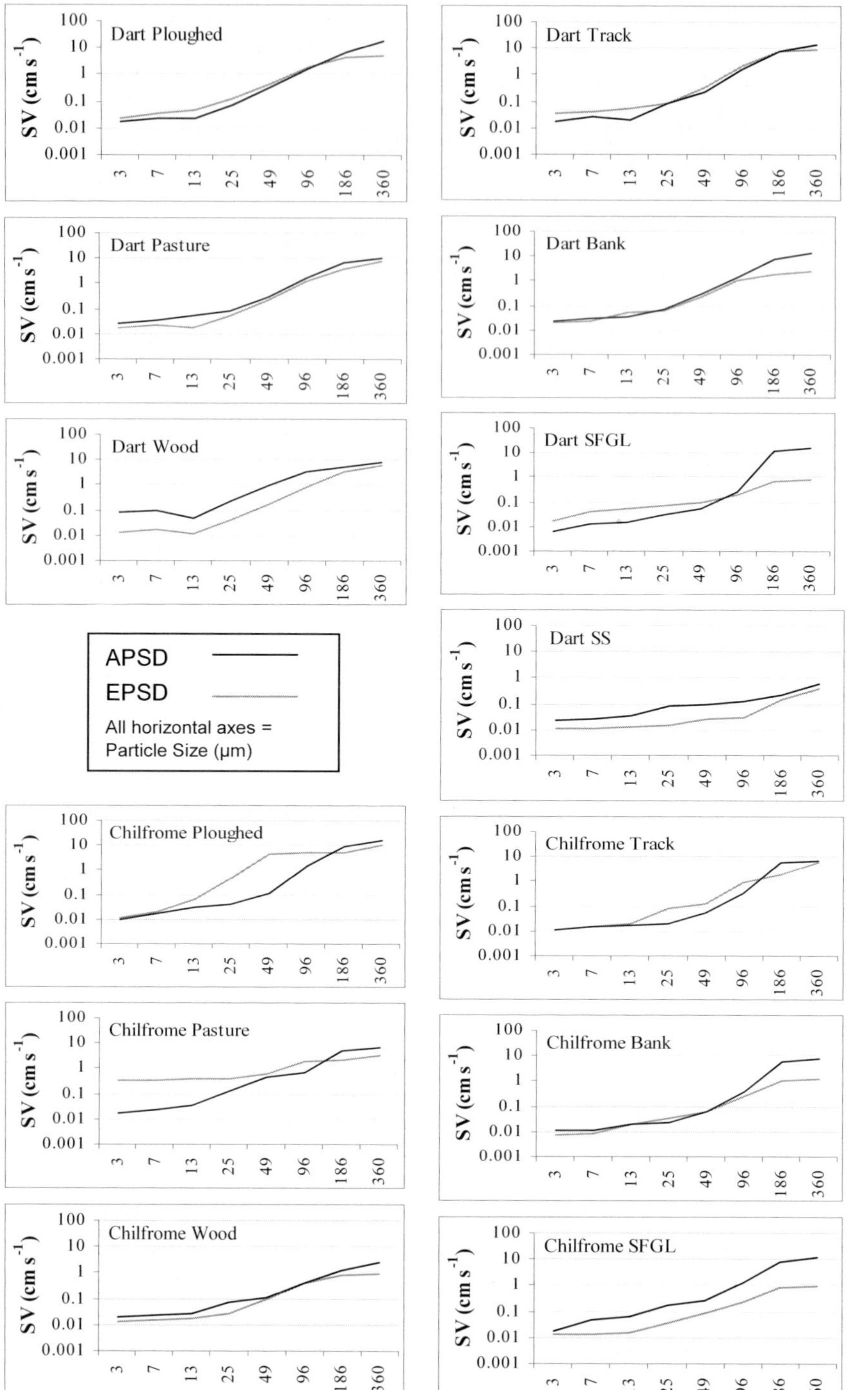

Fig. 2 Comparisons of the relationship between settling velocity (SV) and particle size for effective and absolute size distributions for representative samples of each type. Note that particle size is plotted as the median of each of the eight size classes of the LISST-ST, where the instrument range is 2.5 to 500 μm.

Samples of suspended sediment and from woodland soils in the Dart drainage basin and of SFGL in the Chilfrome catchment show the most significant increases in settling velocity by size class as a result of particle dispersal (Fig. 2). The mean absolute size of Chilfrome SFGL is greater than in the Dart, negating the effects of aggregation-enhanced settling within the finer fractions. Suspended sediment is known to be low density and loosely bound, due to the mechanics of fluvial transport, resulting in large differences between effective and absolute distributions. Sediment from wooded areas can be assumed to contain a much higher proportion of low density organic material than other land use types, giving a low overall effective settling velocity for fine particles bound to larger organic detritus.

In all cases it should be stressed that mean settling velocity is higher for the APSD than for the EPSD. This is due to the effects of particle structure and density changes, brought about by the dispersal of flocs/aggregates.

CONCLUSIONS AND IMPLICATIONS

The LISST devices are useful tools for investigating the complex relationships between settling velocity and particle size, particularly as fall velocity is reported for a range of size classes within a sample. It has been shown that mean particle size is a significant control on mean settling rate, but general empirical relationships cannot be established, due to the complexities induced by particle composition and structure. This is most pronounced in the differences between effective and absolute size distributions, and is also evident in the differences between flocs and aggregates. The upper end of the particle size range is most likely to be significantly affected by flocculation/aggregation, and this likely to represent a large proportion of the sample volume, though not necessarily particle numbers. Observations of settling rate versus particle size for separate fall velocity classes show that aggregation/flocculation may have the effect of increasing or decreasing settling velocity, although for the overall size ranges of natural particles, mean effective settling velocity is always lower than mean absolute settling velocity. This may be a useful aid in interpreting particle formation processes. The discrepancies in particle fall velocity that result from differences between APSD and EPSD have important implications for the understanding of particle settling, and therefore transport characteristics, and for the modelling of such processes, since fluvial particle size research has traditionally focused on the absolute particle size range, with a reliance on derivations of Stokes' law in estimations of fall velocity. The time taken to run a large number of settling experiments prohibits the generation of very large data sets, but the findings herein suggest that the subject clearly warrants further investigation.

Acknowledgements The authors gratefully acknowledge the support of the Natural Environment Research Council in providing a research studentship (N. Williams) and the funding to purchase the LISST equipment. The Centre for Ecology and Hydrology also supports the studentship through CASE funding.

REFERENCES

Agrawal, Y. C. & Pottsmith, H. C. (2000) Instruments for particle size and settling velocity observations in sediment transport. *Marine Geology* **168**, 89114.

Ankers, C., Walling, D. E. & Smith, R. P. (2003) The influence of catchment characteristics on suspended sediment properties. *Hydrobiologia* **494**, 159–167.

Collins, A. L. & Walling, D. E. (2002) Selecting fingerprint properties for discriminating potential suspended sediment sources in river basins. *J. Hydrol.* **261**, 218–244.

Droppo, I. G. (2001) Rethinking what constitutes suspended sediment. *Hydrol. Processes* **15**, 1551–1564.

Droppo, I. G. & Stone, M. (1994) In-channel surficial fine-grained sediment laminae (Part I); physical characteristics and formational processes. *Hydrol. Processes* **8**, 101–111.

Droppo, I. G., Leppard, G. G., Flannigan, D. T. & Liss, S. N. (1997) The freshwater floc: a functional relationship of water and organic and inorganic floc constituents affecting suspended sediment properties. *Water Air and Soil Pollut.* **99**, 43–53.

Droppo I. G., Walling, D. E. & Ongley, E. D. (1998) Suspended sediment structure: implications for sediment and contaminant transport modelling. In: *Modelling Soil Erosion, Sediment Transport and Closely Related Hydrological Processes* (ed. by W. Summer, E. Klaghofer & W. Zhang) (Proc. Vienna Symp., 1998), 437–444. IAHS Publ. 249. IAHS Press, Wallingford, UK.

Droppo I. G., Walling, D. E. & Ongley, E. D. (2000) The influence of floc size, density and porosity on sediment and contaminant transport modelling. In: *The Role of Erosion and Sediment Transport in Nutrient and Contaminant Transfer* (ed. by M. Stone) (Proc. Waterloo Symp., 2000), 141–147. IAHS Publ. 263. IAHS Press, Wallingford, UK.

Foster, I. D. L. & Charlesworth, S. M. (1996) Heavy metals in the hydrological cycle: trends and explanation. *Hydrol. Processes* **10**, 227–261.

Krein, A., Petticrew, E. & Udelhoven, T. (2003) The use of fine sediment fractal dimensions and colour to determine sediment sources in a small watershed. *Catena* **53**, 165–179.

Phillips, J. M. & Walling, D. E. (1995) An assessment of the effects of sample collection, storage and resuspension on the representativeness of measurements on the effective particle size distribution of fluvial suspended sediment. *Water Res.* **29**, 2498–2508.

Phillips, J. M. & Walling, D. E. (1999) The particle size characteristics of fine-grained channel deposits in the River Exe Basin, Devon, UK. *Hydrol. Processes* **13**, 1–19.

Walling, D. E. & Moorehead, P. W. (1987) Spatial and temporal variation of the particle size characteristics of fluvial suspended sediment. *Geografiska Ann.* **69A**, 47–59.

Walling, D. E. & Moorehead, P. W. (1989) The particle size characteristics of fluvial suspended sediment: an overview. *Hydrobiologia* **176/177**, 125–149.

Walling, D. E. & Woodward, J. C. (2000) Effective particle size characteristics of fluvial suspended sediment transported by lowland British rivers. In: *The Role of Erosion and Sediment Transport in Nutrient and Contaminant Transfer* (ed. by M. Stone) (Proc. Waterloo Symp., 2000), 129–139. IAHS Publ. 263. IAHS Press, Wallingford, UK.

Walling, D. E., Russell, M. A. & Webb, B. W. (2001) Controls on the nutrient content of suspended sediment transported by British Rivers. *Sci. Total Environ.* **266**, 113–123.

In-channel storage of fine sediment in rivers of southwest England

A. J. WILSON[1], D. E. WALLING[1] & G. J. L. LEEKS[2]

1 *Department of Geography, University of Exeter, Devon EX4 4RJ, UK*
andrew.j.wilson@exeter.ac.uk

2 *Centre for Ecology and Hydrology, Wallingford, Oxfordshire OX10 8BB, UK*

Abstract The in-channel storage of fine sediment is an important, yet relatively poorly understood, component of sediment transfer through river systems. Previous research has shown it to be a significant factor in controlling the suspended sediment flux through aquatic systems. Additionally, it may also be of significance in the degradation of aquatic ecosystems. This paper presents the results of a comparative investigation of in-channel fine sediment storage and deposition rates for four contrasting rivers in southwest England over a period of 27 months. The results obtained demonstrate significant spatial and temporal variations in the amounts of fine sediment deposited and remobilized from the beds of the study rivers and indicate that the potential role of in-channel fine sediment storage in regulating the suspended sediment flux varies significantly between the study rivers.

Keywords fine sediment; in-channel sediment storage; sediment deposition; suspended sediment loads

INTRODUCTION

The transport of fine sediment in suspension through river systems is commonly an intermittent process, with sediment transfer occurring primarily during flood events and with sediment often being stored on the channel bed between transport episodes. The in-channel storage of fine sediment is thus potentially a significant component of the drainage basin sediment budget, due to its capacity to regulate the transmission of material to the basin outlet. In addition, such storage may also be of ecological significance in the degradation of aquatic ecosystems through the siltation of salmonid spawning gravels (Walling *et al.*, 2003a), clogging of aquatic vegetation and accumulation and release of sediment bound pollutants, such as phosphorus and heavy metals (Walling *et al.*, 2003b) The in-channel storage of fine (<0.063 mm) sediment within UK river systems has been examined by several studies. However, these studies have involved either medium-term investigations of a single drainage basin (e.g. Lambert & Walling, 1988; Walling *et al.*, 1998; Walling & Amos, 1999) or river reach (e.g. Smith *et al.*, 2003), or short-term "snapshot" investigations comparing a number of drainage basins (e.g. Heywood, 2002). There remains a need to undertake medium-term investigations of several drainage basins, in order to assess inter-river variability in the dynamics of in-channel storage of fine sediment. Furthermore, previous investigations have focused primarily on either the role of fine sediment in environmental degradation or the flux of contaminants through river systems. Less attention has been given to the role of in-channel fine sediment storage within the overall drainage basin sediment budget. The study reported here focused on this latter consideration, providing a medium-term study of several catchments with contrasting characteristics and comparing their response.

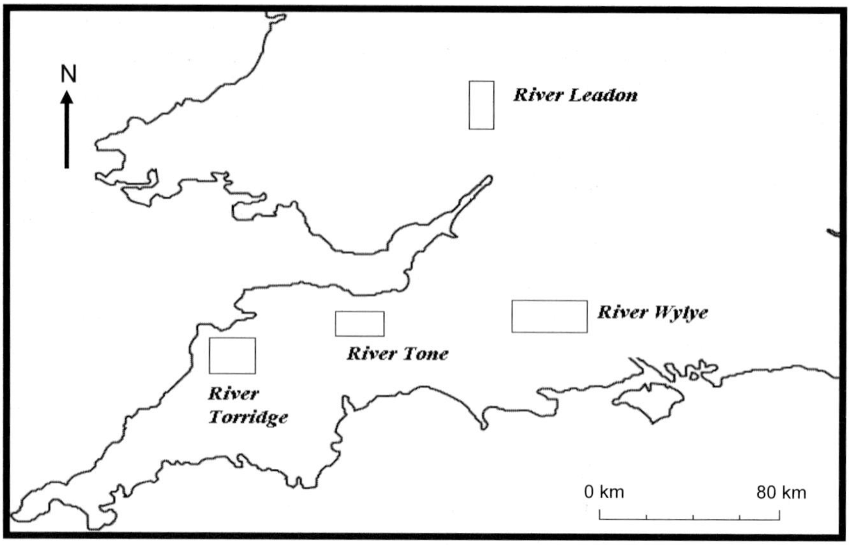

Fig. 1 Location of the study rivers in southwest England.

Table 1 The main characteristics of the study basins.

Drainage basin	Basin size (km^2)	Max. altitude (m AOD)	Basin geology	Basin land use	Mean annual rainfall (mm)	Mean daily flow ($m^3 s^{-1}$)	Max. storm flow ($m^3 s^{-1}$)
Leadon	293	180	Devonian sandstones, Triassic mudstones	Rural, mixed agriculture	685	4.14	63.21
Tone	84	390	Devonian shales and slates, Triassic sandstones	Rural, mixed agriculture	851	1.14	49.84
Torridge	258	220	Carboniferous shales and sandstones	Rural, 80 % pasture	1186	7.54	109.06
Wylye	443	270	Cretaceous Chalk 90%, Cretaceous Upper Greensand 10%	Rural, mixed agriculture, military firing range	830	2.02	24.71

THE STUDY BASINS

The study reported examined the in-channel storage of fine sediment in four representative, but contrasting, drainage basins; their locations are shown in Fig. 1. The selection of the Rivers Leadon, Tone, Torridge and Wylye was primarily based on their contrasting drainage basin characteristics, although logistical considerations were also important. Further details regarding the characteristics of the individual basins are presented in Table 1.

METHODS

Suspended sediment concentrations were continuously monitored in each river between spring 2001 and May 2003, using Hydrosphere™ self-cleaning optical-backscatter turbidity probes coupled to data loggers. Continuous records of suspended sediment concentration were obtained from the turbidity records via calibration relationships developed using manually collected suspended sediment samples. Suspended sediment loads were calculated by combining the suspended sediment records with the continuous discharge records obtained from adjacent Environment Agency gauging stations.

A river reach approximately 100 m long was selected in the lower reaches of each basin, close to the suspended sediment monitoring site, for measuring fine sediment deposition and storage. Each reach encompassed a pool and riffle sequence in order to account for the local variability in river behaviour. These reaches were all of similar gradients (~2 m km^{-1}). The deposition and storage of fine sediment are difficult to measure precisely, because of the problems of replicating natural conditions and the difficulty of documenting a continuous process with periodic measurements. In the absence of generally accepted techniques, two different approaches were adopted.

Firstly, sediment deposition was documented directly, by means of tray traps (0.107 m^2 surface area, 9.0 cm, deep) similar in design to those described by Frostick *et al.* (1984) and Walling & Amos (1999). The trays were installed flush with the bed and subsequently filled with representative bed material cleaned to exclude all sediment of less than 2 mm. One tray was installed at the beginning and end of the pool and riffle in each river reach. Traps were emptied on a monthly basis, although the outbreak of Foot and Mouth disease prevented measurements in three of the catchments during the period March to August 2001. One limitation of this monthly interval is that the estimates of the mass of sediment deposited during the preceding month will represent a *minimum* estimate of fine sediment deposition during the period between trap emplacement and emptying, since some of the sediment deposited during the measurement interval could have been remobilized prior to the emptying of the trap. Nevertheless, the approach is seen as providing an effective means of comparing the individual study rivers.

Secondly, fine sediment storage was quantified using the resuspension technique described by Lambert & Walling (1988). This entailed placing a 1-m high galvanized steel cylinder (area 0.18 m^2) on the river bed. Both the water within the cylinder and the upper 10 cm of the gravel bed were then agitated to resuspend the fine sediment stored on and within the upper part of the channel bed and a sample of the turbid water was taken. The sediment content of this sample was assumed to reflect the remobilization of fine sediment mantling the surface and contained within the bed material matrix. By knowing the area of bed enclosed by the cylinder and the volume of water in the cylinder (derived from a measurement of mean depth) it was possible to calculate the quantity of stored sediment from the values of sediment concentration obtained from the samples. These measurements of the bed storage of fine sediment were made at points close to where the deposition trays were installed, again at monthly intervals. As such they provide periodic instantaneous estimates of the total amount of fine sediment stored on the channel bed. The amount of sediment stored on the channel bed can clearly be expected to vary during the periods between measurements and the estimate obtained could therefore under- or over-estimate the mean value for the period. By calculating the change in fine sediment storage between the individual monthly measurements, it was possible to estimate whether the intervening period had been one of net remobilization or net deposition and to produce corresponding estimates of sediment deposition (or remobilization).

An attempt was also made to estimate the total amount of fine sediment stored on the channel bed of the main channel system of each study basin, by using a modification of the approach described by Lambert (1986). First, the main channel network upstream of the study reach was subdivided into reaches. Second, at a representative point within each upstream reach, fine sediment storage was measured using the resuspension technique at both a pool and riffle site. The channel geometry of each upstream reach was also documented, to assist determination of the total channel bed area in the reach and the relative proportions of this area occupied by pools and riffles. The total fine sediment storage in each upstream reach was then calculated by extrapolating the measured values and the estimates were summed to provide an estimate for the entire main channel system. This value represented an estimate of instantaneous storage at the time of sampling in August 2002. It was adjusted to provide an estimate of the mean storage over the study period by multiplication by the ratio of mean monthly storage to storage in August 2002 derived for the main study reach.

After transfer to the laboratory, the sediment recovered from the deposition trays was wet sieved through a 0.063 mm sieve and the <0.063 mm fraction was freeze dried. The samples provided by the resuspension technique were filtered through 1.2 μm membrane filters, in order to determine the suspended sediment concentration.

RESULTS

Suspended sediment response

Table 2 presents summary information on suspended sediment concentrations and loads derived from the continuous records of suspended sediment concentration for the four study catchments, provided by the recording turbidity meters. The results highlight substantial variation between the study rivers. Maximum concentrations were found in the River Tone, where storm-period concentrations exceeded 2000 mg l^{-1}, more than eight times the maximum concentrations found in the River Wylye. Discharge-weighted mean concentrations ranged between 13 mg l^{-1} for the River Wylye and 77 mg l^{-1} for the River Leadon. Specific annual suspended sediment yields (November 2001–October 2002) varied by more than an order of magnitude, from 4 t km^{-2} $year^{-1}$ for the River Wylye to approximately 90 t km^{-2} $year^{-1}$ for the River Tone. By UK standards, the suspended sediment yield of the River Wylye is low, whilst the sediment yield of the River Tone is high (cf. Walling & Webb, 1981). The sediment yields of the Rivers Leadon and Torridge are more typical of those for UK river systems.

Table 2 Suspended sediment characteristics of the study basins.

River	Mean discharge weighted suspended sediment concentration (mg l^{-1})	Max. suspended sediment concentration (mg l^{-1})	Annual sediment load (t $year^{-1}$)	Annual specific sediment yield (t^{-1} km^2 $year^{-1}$)
Leadon	77	1513	12748.6	43.51
Tone	22	2137	7512.6	89.65
Torridge	37	1501	16929.2	65.67
Wylye	13	230	1756.3	3.96

Sediment deposition

Figure 2 and Table 3 and indicate that the estimates of fine sediment deposition provided by the tray traps demonstrate significant differences between the four study rivers. Table 3 presents

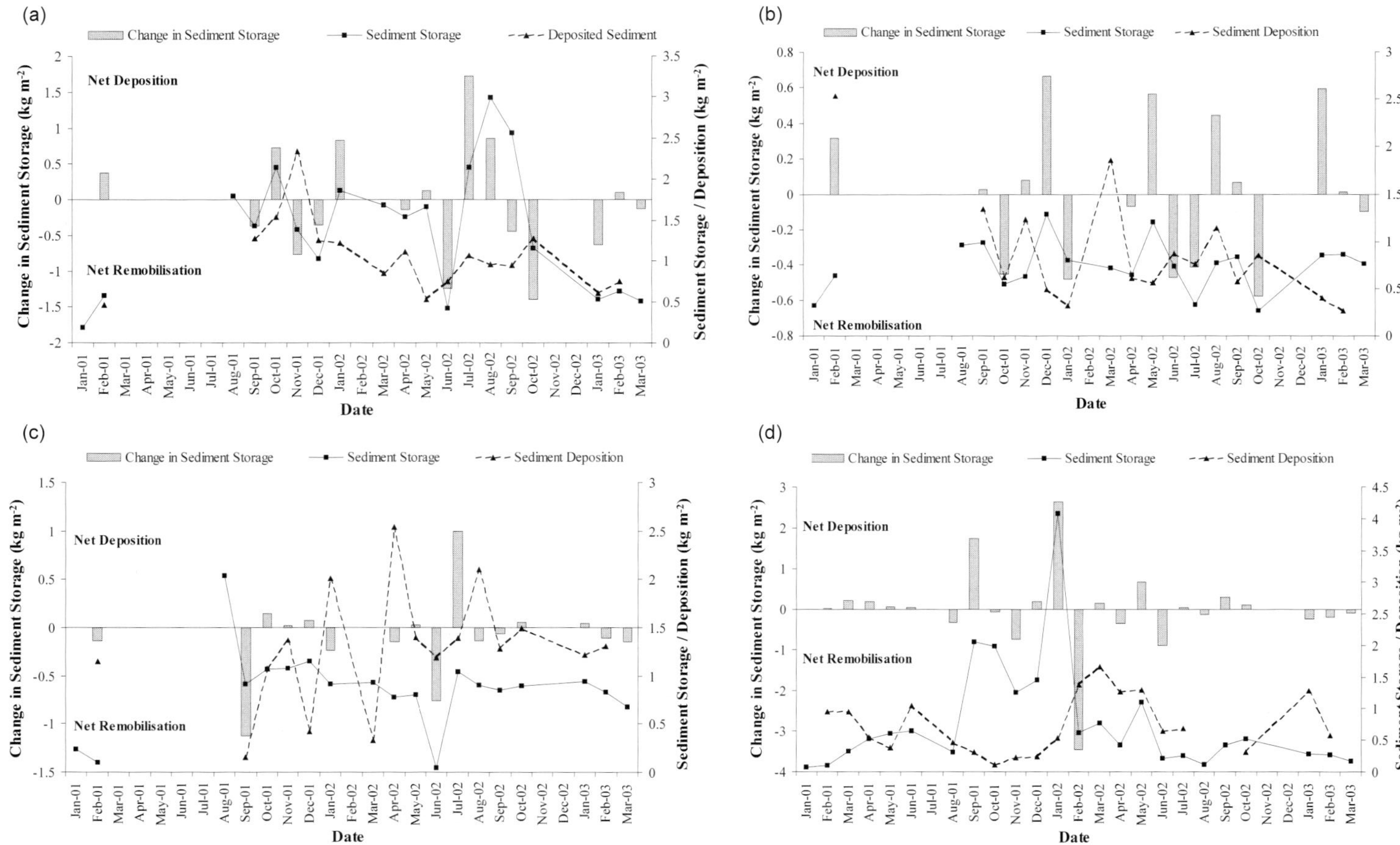

Fig. 2. Monthly measurements of fine sediment deposition and storage and estimates of net deposition and remobilisation derived from the monthly storage measurements, for the study rivers. (a) River Leadon, (b) River Tone, (c) River Torridge, (d) River Wylye.

Table 3 The maximum, minimum and mean monthly values of fine (<0.063 mm) sediment deposition and storage for the study rivers.

River	Fine sediment deposition (kg m^{-2})				Fine sediment storage (kg m^{-2})			
	Max.	Min.	Mean	Standard deviation	Max.	Min.	Mean	Standard deviation
Leadon	3.54	0.52	1.56	0.71	3.00	0.42	1.48	0.78
Tone	2.54	0.26	0.90	0.61	1.28	0.26	0.74	0.29
Torridge	2.54	0.15	1.28	0.63	2.04	0.10	0.87	0.26
Wylye	1.67	0.12	0.74	0.46	2.05	0.07	0.82	1.05

the maximum, minimum and mean of the average deposition amounts measured for all the tray traps within the study reach of a particular river, for each measurement interval, with the standard deviation of the average values for each measurement interval providing a measure of the temporal variability of sediment deposition. The highest maximum and mean values of 3.54 and 1.56 kg m^{-2}, respectively, were recorded for the River Leadon, whereas the lowest maximum and mean values of 1.67 and 0.74 kg m^{-2}, respectively, were recorded for the River Wylye. The broad similarity between the values obtained for the four study rivers is, however, worthy of note. Whereas the values of maximum suspended sediment concentration and specific sediment yield reported in Table 2 vary by around an order of magnitude, the values of maximum and mean deposition listed in Table 3 vary by a factor of only ~2. Equally, although the lowest values of sediment deposition are found in the River Wylye, and therefore coincide with the lowest values of maximum concentration and specific suspended sediment yield, the maximum values of deposition are found in the River Leadon, which is characterized by only intermediate values of maximum suspended sediment concentration and specific suspended sediment yield.

Figure 2 presents estimates of monthly deposition amounts obtained from both the tray traps (measured) and from the monthly measurements of sediment storage (inferred). The measured values suggest that sediment deposition is a continuous process in each river, because some fine sediment was always recovered from the tray traps at each measurement. However, these results are somewhat inconsistent with the estimates of net deposition and remobilization inferred from the monthly measurements of sediment storage, since several periods are shown to be characterized by net remobilization, and no deposition might therefore be expected. Furthermore, in several cases, high values of net remobilization coincide with relatively high measured values of deposition. These apparent inconsistencies are evident for all four rivers and undoubtedly reflect the nature and basis of the measurements employed. In the case of the tray traps, episodes of both deposition and remobilization could be included within the measurement period and, if the period of remobilization occurred at the beginning of the measurement period, this would not be reflected by the amount of sediment collected. Equally, significant periods of both deposition and remobilization could have occurred between the measurements of sediment storage and these would not necessarily be reflected by the storage measurements undertaken at the beginning and end of the period involved. Despite these limitations, and the need for careful interpretation of the results obtained, the results presented in Fig. 2 are seen as providing a useful indication of contrasts in fine sediment deposition and storage between the four study rivers.

Perhaps the two most important features of the results presented are, firstly, the temporal variation in the amounts of sediment recovered from the tray traps and the alternation of periods of net deposition and net remobilization as inferred from the storage measurements, and, secondly, the lack of a common temporal pattern or trend for the four rivers. Periods of increased or reduced deposition and gains or losses from storage occur at different times in the individual rivers, despite the similar hydrometeorological conditions experienced by their catchments. An example of these contrasts is provided by the period of increased measured fine sediment deposition occurring in the Rivers Leadon, Tone and Torridge during the late autumn and early winter of 2001, which coincides with a period of reduced measured deposition in the River Wylye.

Sediment storage

Table 3 presents summary results for the monthly measurements of fine sediment storage on and within the upper part of the river bed, undertaken on the four rivers. Maximum, minimum and mean values of the monthly average values and the associated values of standard deviation are presented. Clear contrasts are apparent between the catchments. As with the measurements of fine sediment deposition provided by the tray traps, the highest maximum and mean values are found in the River Leadon. Interestingly, however, the lowest values of maximum and mean storage are those for the River Tone, the river with the highest specific suspended sediment yield and maximum suspended sediment concentration. Similarly, almost identical intermediate values of maximum and mean storage are listed from the Rivers Torridge and Wylye, despite the marked differences in specific sediment yield and maximum and mean suspended sediment concentration evident between these catchments (cf. Table 2). As with the values of sediment deposition provided by the tray traps, there is no clear relationship between the magnitude of the measurements of sediment storage reported for the individual rivers and the magnitude of the associated values of specific sediment yield and suspended sediment concentration. The standard deviation values indicate that the Wylye and Leadon are characterized by substantial temporal variation in fine sediment storage, whereas the storage values recorded for the Rivers Tone and Torridge evidence much less temporal variability. Figure 2 provides further information on this temporal variability and shows that the increased variability associated with the Rivers Wylye and Leadon reflects the marked increase in sediment storage documented between August 2001 and February 2002 in the River Wylye and between June and October 2002 in the River Leadon. Perhaps more importantly, however, Fig. 2 shows that, although each of the rivers shows evidence of cyclical variations in storage, with periods of accumulation separated by periods of remobilization, there is little evidence of any common temporal (seasonal) pattern of increase and decrease in storage for the four rivers. The only common trend appears to be that more sediment was stored in the river channels during winter 2001 as compared to winter 2002.

Table 4 presents estimates of the mean total fine sediment storage, on and within the upper part of the river bed, for the entire main channel system of each of the four rivers. Values range from 278.7 t in the River Wylye to 66.7 t in the River Tone. The differences between the four rivers can be largely explained in terms of differences in basin size, since storage must be expected to increase with increasing basin size and channel length. Previous studies (e.g. Walling *et al.*, 1999) have examined the total amount of fine sediment stored on the channel bed and related it to the annual suspended sediment load, in an attempt to assess

Table 4 Average in-channel fine sediment storage in the main channel systems of the study rivers expressed as a percentage of the annual suspended sediment load.

River	Average in-channel sediment storage (t)	% of annual suspended sediment load
Leadon	193.2	1.5
Tone	66.7	0.9
Torridge	239.7	1.4
Wylye	278.7	15.9

its role in the basin sediment budget and its potential significance in terms of a conveyance loss. These values are also reported in Table 4 and indicate that, although there is little difference in magnitude between the values reported for the Rivers Leadon, Tone and Torridge, that for the River Wylye is substantially greater. This in turn suggests that channel storage plays a greater role in the sediment budget for the catchment of the River Wylye, since the mean total storage amount is equivalent to ~16% of the total annual sediment output from the basin. Because the total amount of sediment moving into, and out of, storage is likely to be substantially greater than the estimate of mean total storage, it is clear that channel storage can potentially exert an important influence on the transmission of fine sediment through the channel system of this river, through storage and attenuation of the sediment transfer. However, the contrast between the River Wylye and the other catchments primarily reflects the much lower suspended sediment yield of the River Wylye, rather than increased channel storage in this catchment (cf. Table 4).

CONCLUSION

In reviewing the findings presented above, three key findings merit emphasis. First, the four rivers exhibit significant contrasts in both the magnitude and the temporal behaviour of fine sediment deposition and storage. Although their catchments experience similar hydro-meteorological regimes, there is little evidence of common temporal patterns. Second, there is no clear link between the relative magnitude of fine sediment deposition and storage in the four rivers and the relative magnitude of their specific suspended sediment yields and concentrations. Furthermore, the major differences between the four rivers evidenced by their sediment yields and concentrations are not matched by equivalent differences in fine sediment deposition and storage. Factors other than the magnitude of the suspended sediment loads and the ambient concentrations appear to control the magnitude of fine sediment deposition and storage and the contrasts between the catchments noted previously. Further work is clearly required to elucidate these controls. Thirdly, the results suggest that for three of the catchments channel storage is of limited importance in the overall basin sediment budget, whereas such storage is likely to exert an important influence on the sediment response of the River Wylye. However, the significant amounts of fine sediment deposition and storage documented for all four rivers, and particularly in the Rivers Leadon and Wylye, could impact on their aquatic ecology, through siltation of spawning gravels, clogging of vegetation and the accumulation and release of pollutants. Again, further work is required to develop an improved understanding of the dynamics of fine sediment deposition and storage in these and similar rivers.

Acknowledgements The support of the Natural Environment Research Council and CEH Wallingford in providing a postgraduate studentship for A. J. Wilson, the valuable assistance of local landowners in permitting access to field sites and the generosity of the Environment Agency in permitting the siting of turbidity monitoring equipment at the river gauging stations and in providing flow data, are gratefully acknowledged.

REFERENCES

Frostick, L. E., Lucas, P. M. & Reid, I. (1984) The infiltration of fine matrices into coarse-grained alluvial sediments and its implications for stratigraphical interpretation. *J. Geol. Soc. London* **141**, 955–965.

Heywood, M. J. T. (2002) Sedimentation of salmonid spawning gravels: an investigation of associated sediment dynamics in the Hampshire Avon catchment. PhD Thesis, University of Exeter, Exeter, UK.

Lambert, C. P. (1986) The suspended sediment delivery dynamics of river channels in the Exe basin. PhD Thesis, University of Exeter, Exeter, UK.

Lambert, C. P. & Walling, D. E. (1988) Measurement of channel storage of suspended sediment in a gravel bed river. *Catena* **15**, 65–80.

Smith, B. P. G., Naden, P. S., Leeks, G. J. L. & Wass, P. D. (2003) Characterising the fine sediment budget of a reach of the River Swale, Yorkshire, UK during the 1994 to 1995 winter season. *Hydrobiologia* **494**, 135–143.

Walling, D. E. & Amos, C. M. (1999) Source, storage and mobilisation of fine sediment in a chalk stream system. *Hydrol. Processes* **13**, 323–340.

Walling, D. E. & Webb, B. W. (1981) Water quality, In: *British Rivers* (ed. by J. Lewin), 126–129, George Allen & Unwin, London, UK.

Walling, D. E., Collins, A. E. & McMellin, G. K. (2003a) A reconnaissance survey of the source of interstitial fine sediment recovered from salmonid spawning gravels in England and Wales. *Hydrobiologia* **497**, 91–108.

Walling, D. E., Owens, P. N. & Leeks, G. J. L. (1998) The role of channel and floodplain storage in the suspended sediment budget of the River Ouse, Yorkshire, UK. *Geomorphology* **22**, 225–242.

Walling, D. E., Owens, P. N., Carter, J., Leeks, G. J. L., Lewis, S., Meharg, A. A. & Wright, J. (2003b) Storage of sediment-associated nutrients and contaminants in river channel and floodplain systems. *Applied Geochemistry* **18**, 195–220.

Sediment Transfer through the Fluvial System (Proceedings of the Moscow Symposium, August 2004).
IAHS Publ. 288, 2004

Alluvial relief structure and bottom sediments of the lower Volga River

V. N. KOROTAEV, V. V. IVANOV & A. YU. SIDORCHUK
Geographical Faculty, Moscow State University, 119899 Moscow, Russia
sidor@yas.geogr.msu.su

Abstract The alluvial relief of the lower Volga River has a complicated hierarchical structure. This structure includes megaripples, three orders of dunes, two orders of bars, islands and meanders. The main channel of the Volga and its branch the Akhtuba, form a parallel-channel system. This complicated structure is one of the main characteristics of unconfined large rivers with a fine bed load.

Key words alluvial relief; bottom deposits; hierarchical structure; lower Volga River channel; sonar measurements

INTRODUCTION

Alluvial relief is characterized by a well-defined hierarchical structure (Sidorchuk, 1996; Alekseevskiy, 1998). The most complicated hierarchy can be observed at the unconfined reaches of the large rivers having sandy alluvium. The lower Volga River is one of the best examples of the high diversity and dynamics of alluvial features. The lower Volga valley in the Volgograd and Astrakhan' districts is well developed. Most settlements are situated along the river banks and changes in the river channel can cause environmental hazards. Sediment dynamics and river bank erosion lead to instability in hydro-technical constructions and cause sedimentation of water intakes and quays, and erosion around pipelines and bridges. These negative effects of the channel processes need to be considered in the planning and practice of the river valley development. Therefore the hydrological and morphological features of the lower Volga River channel require investigation.

METHODS OF INVESTIGATION

Investigations of the morphology and dynamics of the main lower Volga River channel and its distributaries along the section from Volgograd to Astrakhan' in the period 1995–2003 complemented significantly the existing information about fluvial processes, bottom sediment size and thickness. The bottom relief and alluvium characteristics were measured with a complex of hydro-sonar equipment (side-looking and profiler), designed in the Sonar Laboratory of the Institute of Oceanology, RAS, and the coordination of the survey was performed using a GPS system. Sonar measurements were interpreted using existing coring data (held by the Gidroproyekt and Soyusmorniiproyekt institutes), and analyses of bottom sediment particle size sampled during 1990–2003 (Korotaev & Ivanov, 2000). The observations covered the 520-km long section of the lower Volga River from Volgograd to Astrakhan' (Fig. 1(a)).

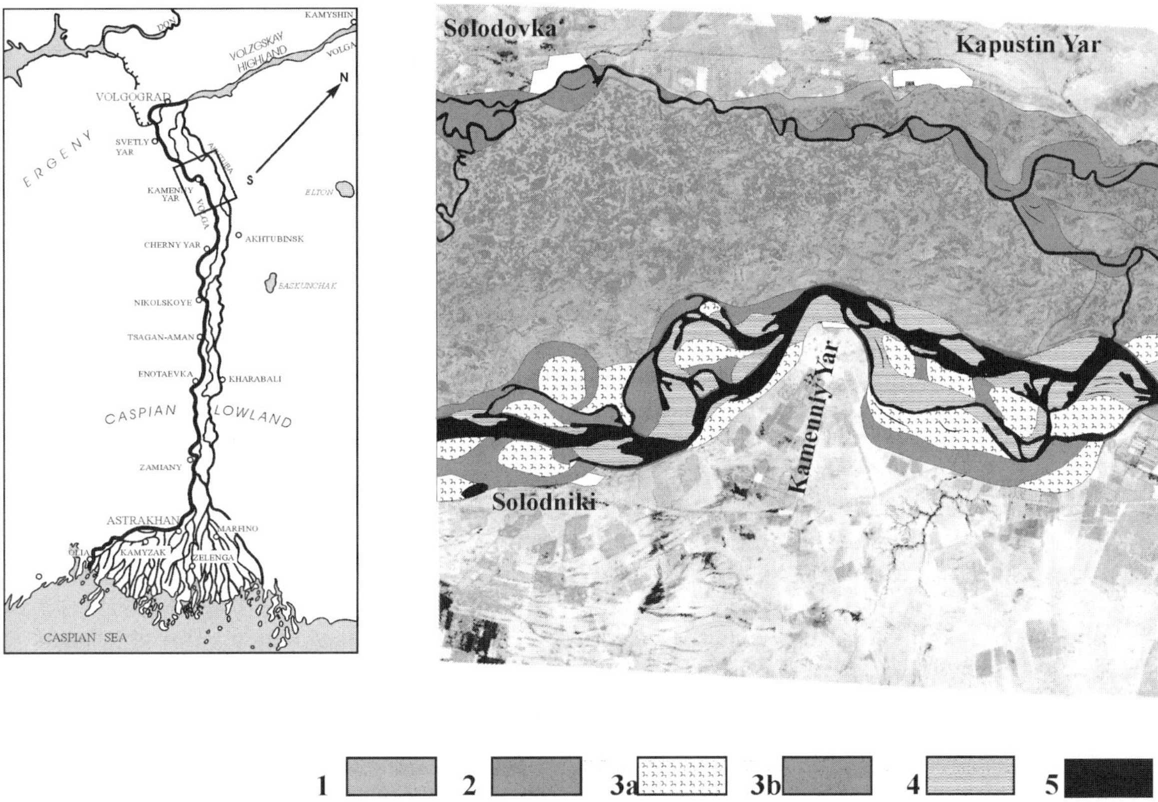

Fig. 1 The lower Volga River valley: (a) general view, and (b) main morphological features of the section near Kamenniy Yar. 1: high Volga-Akhtuba floodplain; 2: meandering belt of Akhtuba; 3: floodplain of the Volga River main channel (a: high surface, b: low channel remnants); 4: second order alternation and braid bars; 5: river channels and branches.

HYDROLOGICAL REGIME OF THE LOWER VOLGA RIVER

The lower Volga River is a typical lowland river with mean slope 0.031 m km^{-1}. The annual flow is 259 km^3 near Volgograd and 253 km^3 at the river outlet to the Caspian Sea. The lower Volga valley is situated in a semi-desert region with high evapotranspiration. The water regime of the Volga River was changed by the construction of a system of hydro-electric dams with large reservoirs. The lower Volga is mainly influenced by the Volgograd reservoir, constructed in 1959. The mean maximum discharge was reduced from 34 500 m^3 s^{-1} before 1959 (with the extreme 51 900 m^3 s^{-1} in 1926) to 26 800 m^3 s^{-1} in 1959–1999 (with the extreme 34 100 m^3s^{-1} in 1979). The water regime regulation led to concentration of the water flow in the main river channel and abandonment of the small flood plain distributaries.

Bed load and suspended load deposition in the reservoirs led to significant decrease of the sediment input to the lower Volga River. In 1938–1953 the annual sediment load near Volgograd was 12×10^6 t, and it was 13×10^6 t at the delta head. Some increase of sediment load was observed along the lower Volga River channel due to local erosion. Sediment transport measurements in the Volgograd reservoir were terminated after reservoir construction, therefore comparative figures are available only for the delta head. Here the annual load decreased nearly two-fold to 7.9×10^6 t, the annual maximum load decreased from 3 900 to 2 100 kg s^{-1}, the mean annual suspended sediment concentration decreased from 56 to 34 g m^{-3}, and the annual maximum suspended sediment concentration decreased from 250 to 170 g m^{-3}.

CHANNEL BOTTOM DEPOSITS

Bottom deposits in the Volga River channel are only partly formed due to accumulation of suspended and wash load. These finer sediments (silty sand and silt) have accumulated on the surface of the Volga-Akhtuba floodplain and in the Volga River delta. The main source of bottom deposits is the transit and deposition of alluvial bed load. In the channel segment below the Volgograd reservoir dam the bed load and bottom deposits are represented by sand with median grain size 0.15–0.50 mm (Fig. 2). Collection and analysis of sediment samples shows two main lithological parts of the channel: (a) from Volgograd to Tsagan-Aman the most frequent deposits (~60% of the bottom area) are medium-sized sands with a median diameter 0.25–0.45 mm; (b) fine sand (0.1–0.25 mm) covers the main part of the channel bed between Tsagan-Aman and Astrakhan'. The coarsest material (median diameter 0.6–0.7 mm) is observed along the eroded bed rock valley sides.

Sonar profiler measurements show a mean thickness of 6–8 m of bottom alluvial deposits in the lower Volga River, and the actual thickness varies from 0 to 15 m. Along the eroded banks (mainly at concave meander loops) the channel bed is composed of marine clay and flood plain loams. This cohesive matter (the marine clay) is quite often exposed in deep pools. Locally, about a third part of the channel bottom is not covered with alluvium.

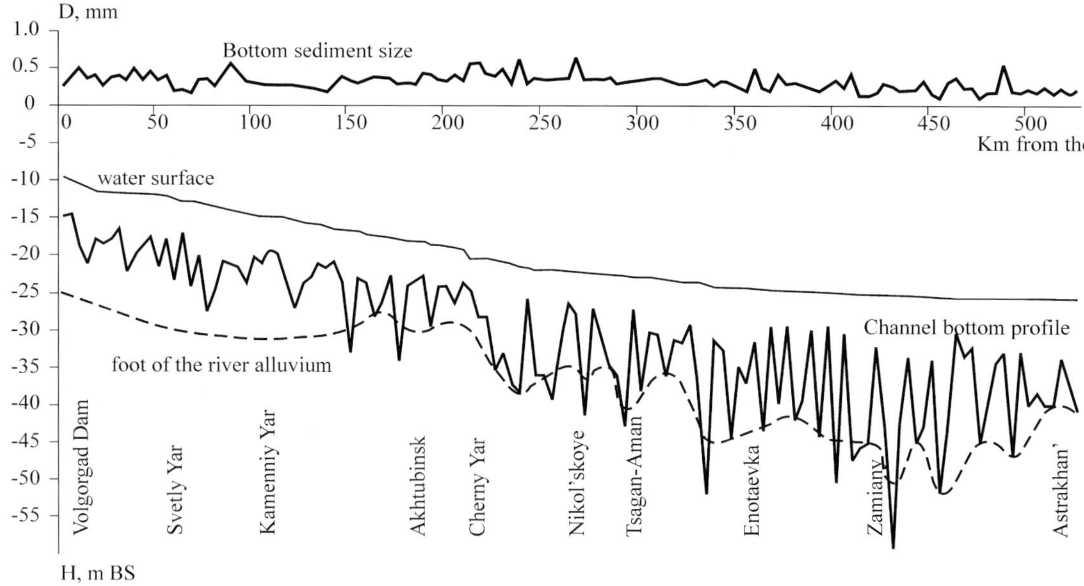

Fig. 2 Changes of the bottom sediment size and channel alluvium thickness along the lower Volga River channel.

RIVER CHANNEL MORPHOLOGY

The lower Volga River is characterized by two main channels: the Volga River main channel and the Akhtuba river branch, located respectively along the western and eastern borders of the valley floor. These two channels are divided by a broad flood plain and are connected by a complicated net of flood plain distributaries of different size.

The main channel has a mean width of 1.5 km between the flood plain banks and has a sinuous pattern with multiple secondary anabranches (local term is "volozhka"), divided by islands, covered by vegetation (Fig. 1(b)). These islands, of mean length 10–13 km, may be solid, and may be composite, consisting of several islands with dividing branches. Anabranches mainly occur in the upper part of the lower Volga River, whereas a single channel with well-shaped meanders is more common in the lowermost segment.

The Akhtuba branch has a mean width of 250 m and has a meandering pattern. The typical meander wavelength is 2 km. The Akhtuba flows parallel to the main channel of the Volga River for a distance of about 520 km, but only twice (near Akhtubinsk and Tsagan-Aman) are these two channels joined to each other. This morphological pattern was formed due to the great width of the lower Volga valley bottom: up to 40 km.

Being separate channels, the main Volga River and Akhtuba are connected by the net of small (30–50 m wide) flood plain branches and distributaries, mostly meandering. The water feeds these branches and numerous flood plain lakes during high floods. Due to flood control by the reservoir system the Volga-Akhtuba flood plain is now flooded only quite rarely and the network of flood plain water bodies has degraded.

STRUCTURE OF THE ALLUVIAL RELIEF IN THE LOWER VOLGA CHANNEL

Analysis of functions of spectral density for channel bottom elevations and of histograms of the length of alluvial features (Sidorchuk, 1996), combined with observations of side-looking sonar images of the channel bottom, shows the complicated structure of the bottom relief in the Volga River channel. The complex of bottom forms is hierarchical and consists of four levels: 1: megaripples of mean wavelength 5 m; 2: first order dunes (40 m); 3: 2nd order dunes (140 m); and 4: 3rd order dunes (580 m). With the alternating and braided first order (3100 m) and second order (5900 m) bars, as well as the islands described above, and meanders (13 000 m), the hierarchical structure of the lower Volga River alluvial relief consists of seven levels. This complicated structure was investigated during conditions of stable low water discharge (7600 $m^3 s^{-1}$) within single channel segments of the river.

Megaripples

Mean length L_R of megaripples is 2–5 m, and their height is <0.1 m. Megaripples are three dimensional bottom forms, easily recognized on the bathymetric profiles and side-looking sonar images.

Dunes

Dunes are most common in the Volga River channel with sandy alluvium. They are marked on the spectrum of channel bottom elevations with a well-defined local maximum.

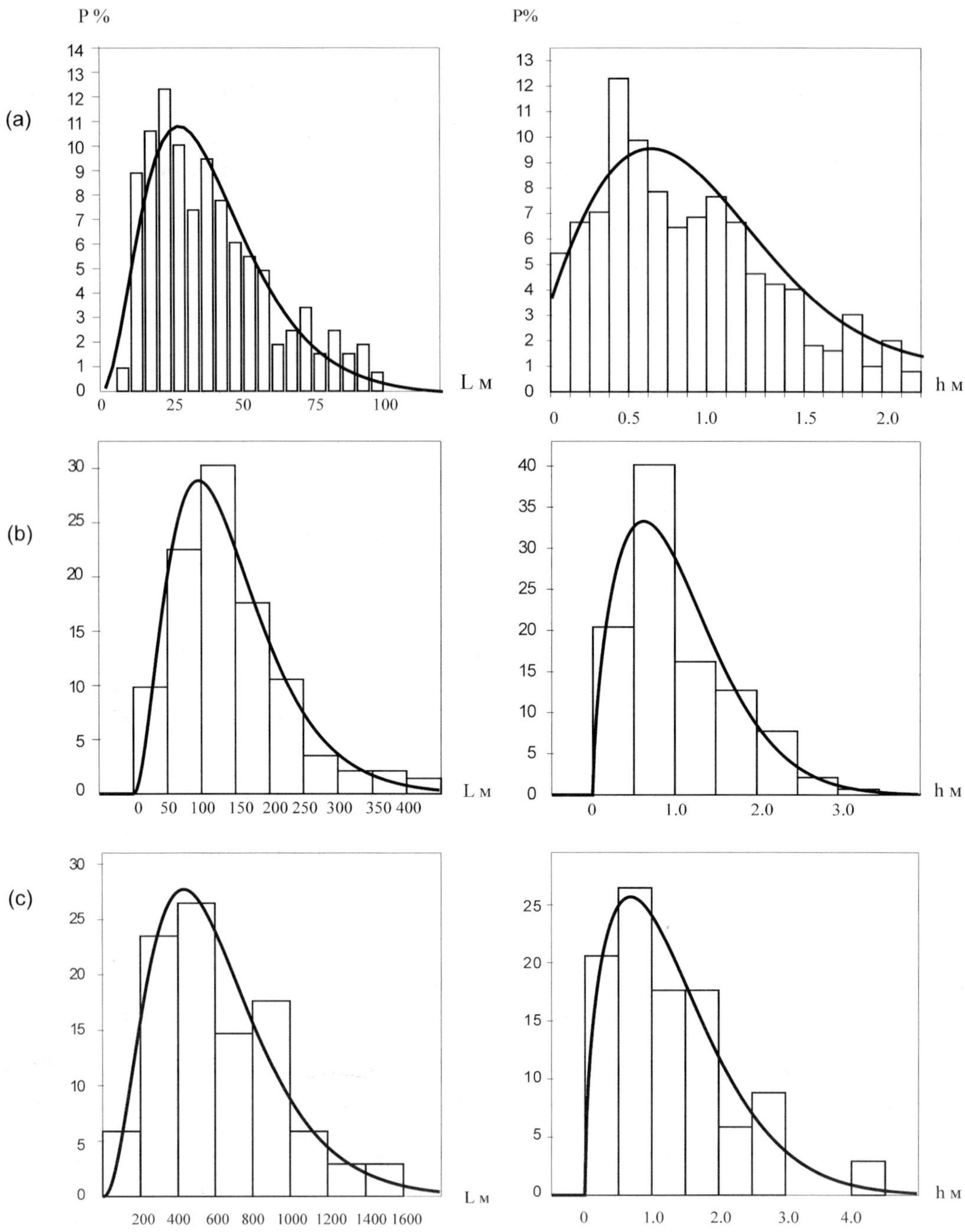

Fig. 3 Histograms and probability density functions for dune length (a), and height (b). For details see the text.

First order dunes have a mean length of 45 m and height 0.8 m. Their distribution fits well to a two-parameter gamma-distribution (Fig. 3(a)):

$$dp = \frac{\eta^{\mu}}{\Gamma(\mu)} L_1^{\mu-1} \exp(-\eta L_1) dL_1 \tag{1}$$

where Γ is the gamma function. Parameters $\mu = 3.51$ and $\eta = 0.089$ are correlated with the first order dunes mean length and standard deviation: $L_1 = \mu/\eta$; $\sigma = \sqrt{\mu/\eta^2}$.

The Weibull distribution fits the height variability of the first-order dunes well (Fig. 3(b)).

$$dp_h = \alpha\lambda h_1^{\alpha-1} \exp\left(-\lambda\, h_1^{\alpha}\right) dh_1 \tag{2}$$

The Parameters of the Weibull distribution can also be estimated using the mean value and standard deviation:

$$h_{1m} = \lambda^{-1/\alpha}\Gamma\left(\frac{1+\alpha}{\alpha}\right);\quad \sigma_h = \lambda^{-1/\alpha}\sqrt{\frac{2}{\alpha}\Gamma\left(\frac{2}{\alpha}\right) - \frac{1}{\alpha^2}\left[\Gamma\left(\frac{1}{\alpha}\right)\right]^2} \tag{3}$$

For the first-order dunes of the lower Volga River during low water periods, $\alpha = 1.89$, $\lambda = 1.11$.

About 30% of the first-order dunes at the channel bottom were nearly isometric in both longitudinal and cross section. First-order dune asymmetry is described by a normal distribution with the mean equal to 1. The steepness of the downward first-order dunes slope is well described by a gamma distribution with $\mu = 1.95$ and $\eta = 37.8$. That corresponds to a mean steepness of 0.052 and standard deviation 0.037.

First-order dunes are well defined on the side-looking sonar images and are marked two-dimensional bottom features at the low water conditions. Their tops form straight parallel lines across the channel, and their typical sizes can be measured using only one longitudinal bathymetric profile.

Second-order dunes, with a mean length 140 m and mean height 1.0 m, are also common features in the lower Volga River. Their morphology is described with the same distribution curves (Fig. 3), as for first-order dunes. The second-order dune gamma-distribution for length has parameters $\mu = 3.09$ and $\eta = 0.0217$. Second-order dunes profile asymmetry is well approximated by the normal distribution. The Weibull distribution for height has parameters $\alpha = 1.60$ and $\lambda = 0.87$. Mean steepness of downward slope is 0.0234 with standard deviation 0.027, gamma distribution with $\mu = 2.05$ and $\eta = 244.5$ fits well to steepness data. The relatively low steepness of second-order dunes the downward slope makes them practically invisible on the side-looking sonar images, because steeper first-order dunes completely predominate in the image.

Dunes of the third order with mean length 580 m and mean height 1.2 m are less frequent features in the lower Volga River. Nevertheless, there are enough empirical data to obtain statistically reliable distribution curves of their geometry (Fig. 3). Gamma-distribution for third-order dunes length has parameters $\mu = 3.41$ and $\eta = 0.0058$. The Weibull distribution for height has parameters $\alpha = 1.49$ and $\lambda = 0.69$. The mean steepness of the downward slope is 0.0067 with a standard deviation 0.0048; a gamma distribution with $\mu = 2.13$ and $\eta = 318.5$ fits well to the steepness data. Third-order dunes are mainly three-dimensional bottom forms, but during low water periods some of them can appear above the water surface. In that case they look like small bars and locally define the configuration of the water flow.

Bars

Alternating bars and braid bars form the internal structure of the channel morphology. These alluvial features have no vegetation or are only partly vegetation covered, and are submerged during the average flood. Alternating bars and braids cause additional sinuosity in the low water flow. There are two orders of bars. Bars of the first order have mean length 3100 m with standard deviation 830 m. A gamma-distribution with $\mu = 14.46$ and $\eta = 0.0046$ can be used for their length description. The height of first-order bars ranges from 2–3 to 11–12 m.

Second-order bars have a mean length of 5900 m with standard deviation 1410 m, their height ranges from 4–5 to 12–15 m. A gamma-distribution with $\mu = 20.5$ and $\eta = 0.0035$ can be used to describe their length. Their movement along the islands causes the alteration of resistance to flow in the main channel and the anabranch, which appears in the quasi-periodic decrease and increase of the discharge in the main channel.

RESULTS AND CONCLUSION

The channel of the lower Volga River was formed in a broad valley bottom where the confining factors of the fluvial processes were weak. The bed load and bottom sediments of significant thickness are mainly fine sand, easily reworked by the river flow. Therefore self-organising processes of the fluvial morphology evolution are well developed here, and complicated hierarchical structure characterizes the channel relief.

The highest level of this hierarchy is the two-parallel-channel pattern of the main Volga River channel and Akhtuba, with an anastomosed net of small flood plain distributaries. The two channels flow parallel to each other for a distance about 520 km, and only twice are connected within short segments near Akhtubinsk and Tsagan-Aman.

The Akhtuba and small distributaries mostly have a meandering pattern. The main Volga River channel only meanders close to the delta head. The major part of the channel is sinuous, and shaped like a ∞ (infinity symbol), with the main branch wider and deeper, and the anabranch narrower and shallower, divided by a large island. The sinuosity of the main channel is formed by the combination of the main branches, so if the main branch is right at the upper part of the ∞, it is left at the lower part of the symbol, and *vice versa*. The sinuosity of the main channel is complicated: during the low water period the water flow forms secondary curves due to the existence of alternating bars of the second-order, third-order smaller curves due to the influence of first-order bars, and even fourth-order, the smallest curves, due to local shallows of associated with third-order dunes.

The bed forms of the lower Volga River channel are organized into a four level hierarchy: megaripples and dunes of three orders. They are partly two-dimensional, and partly three-dimensional features. Their statistical characteristics are similar to those of the bed forms in other rivers and large flumes: the length is described with a gamma distribution and the height with a Weibull distribution. The rather low asymmetry of the bed forms in plan or profile in general is not typical for low water conditions, but can be characteristic for the largest rivers of the lower Volga type.

Acknowledgements The Russian Foundation of Basic Research funded these investigations. Valuable comments by Dr J. Hooke and an unknown referee were incorporated into the text.

REFERENCES

Alekseevskiy, N. I. (1998) Formation and transport of the river sediments (Formirovanie I transport rechnyh nanosov). Moscow University Press, Moscow, Russia (in Russian).

Korotaev, V. N. & Ivanov, V. V. (2000) Transformation of the lower Volga River channel (Ruslovye deformacii na Nizhnei Volge) (in Russian). *Moscow Univ. Bull. Ser. 5* (Geography), **6**, 37–43.

Sidorchuk, A. (1996) The structure of river bed relief. In: *Coherent Flow Structures in Open Channels* (ed. by F. Ashworth, S. Bennett, S. Best & J. McLelland), 397–421. Wiley, Chichester, UK.

Sediment Transfer through the Fluvial System (Proceedings of a symposium held in Moscow, August 2004).
IAHS Publ. 288, 2004

Sediment transport and morphodynamics of the Tanaro River, northwestern Italy

ANNUNZIATO SIVIGLIA[1], BIANCA FEDERICI[1], IGNAZIO BECCHI[2] & MASSIMO RINALDI[2]

1 *Dipartimento di Ingegneria Ambientale, Università di Genova, via Montallegro 1, I-16145 Genova, Italy*
nunzio@diam.unige.it

2 *Dipartimento di Ingegneria Civile, Università di Firenze, via S. Marta 2, I-50139 Firenze, Italy*

Abstract This paper describes a study to determine sediment transport processes and morphodynamics of the Tanaro River in northwestern Italy to support river management strategies. An integrated hydraulic-geomorphic approach was used to: (a) assess geology, land use and climate controls affecting sediment yield at the catchment scale; (b) evaluate changes in channel morphology and sediment transport processes; (c) model river channel change. Numerical simulations were used to evaluate the transient solution for flow and bed profile due to the propagation of the flood wave. It is concluded that Alessandria town is the most critical reach from the flooding point of view and so different design solutions were tested in order to verify whether geometric alteration of the river bed would allow for an increase in flood capacity.

Key words bed equilibrium configuration; channel changes; morphodynamic; sediment transport; Tanaro River, Italy

INTRODUCTION

River management programmes in Italy rarely address sediment transport processes in the design of flood control structures (Autorità di Bacino del Fiume Po, 1997; Autorità di Bacino del Fiume Arno, 2000). More recently, there has been an increasing awareness of regional sediment issues and channel morphodynamics as an integral part of river management. Consequently, there is a need to develop or further refine methodological approaches that include an assessment of sediment transport processes and morphodynamics to ensure their application to widespread river management practice. A multidisciplinary study is required to quantify catchment and reach-scale processes, forms and causes of instability as a basis for quantitative hydraulic modelling and analyses (Environment Agency, 1998; Thorne, 1998).

A large flood event, with an estimated return period of about 100 years, occurred along the Tanaro River in northwestern Italy during November 1994. During the event, hundreds of landslides occurred in the drainage basin and sediment transfer to and by the river resulted in damage to several towns. This event emphasizes the need for developing better flood control strategies and to take into account sediment transport processes and morphodynamic aspects of river management.

The objective of this paper is to examine morphodynamic and sediment transport processes in the Tanaro River to provide appropriate information for river management. Data on channel morphology and sediment transport are used to develop a numerical model to calculate the equilibrium configuration and possible effects of bed changes on flood capacity.

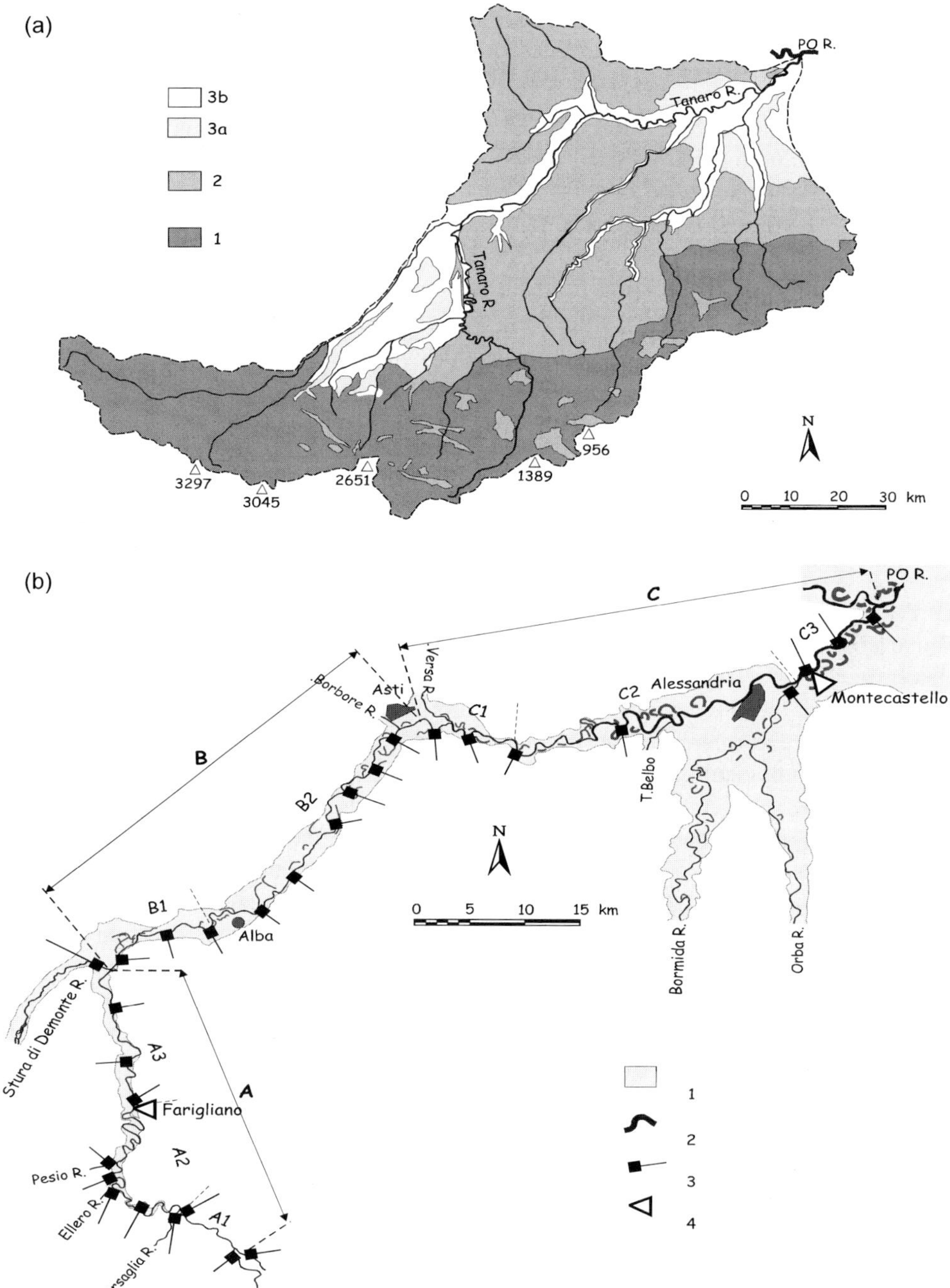

Fig. 1 General characteristics of the drainage basin and of the River Tanaro.
(a) Geological sketch. 1: rocks antecedent to the Tertiary (mainly metamorphic and calcareous rocks); 2: Tertiary sedimentary rocks (mainly marls, sandstones, sands and clays); 3a: Pleistocene (alluvial deposits); 3b: Holocene (alluvial deposits).
(b) Geomorphological classification in river reaches, with location of sediment samples and sediment transport evaluation. 1: alluvial deposits (Holocene); 2: palaeo-meanders; 3: location of sediment samples and geomorphological river reconnaissance; 4: location of gauging stations used for sediment transport evaluation.

STUDY AREA AND METHODS

Study area

The river basin in northwest Italy drains an area of about 8300 km^2 (Fig. 1). Three dominant lithologies including pre-Tertiary rocks, Tertiary Piedmont basin, and Quaternary alluvial deposits, cause variable erosion rates and sediment supply in the river basin. The hilly areas of the Tertiary Piedmont basin constitute the main areas of suspended sediment supply which is generated by soil erosion and earth flows involving the surface soil cover (Biancotti, 1981). The main supply of bed load is from the mountain areas which consist of metamorphic and calcareous rocks that enter the river from rock falls and mass wasting processes.

The basin is located in a temperate climatic zone, but with significant differences between the lower part (Po Plain) and the Alpine ridge. Annual rainfall is extremely variable in relation to relief, ranging from 640 mm on the Po Plain (Alessandria) to about 1500 mm on the Alpine ridge. Mean daily discharge of Tanaro River ranges from about 23 $m^3 s^{-1}$ in the upper course (Farigliano) to 80 $m^3 s^{-1}$ in the lower course (Montecastello).

Geomorphological analysis

A desk study was conducted to provide an understanding of the geomorphology of the Tanaro basin. The geology, soils, topography, land use and geomorphology of the basin were determined to investigate factors influencing sediment yield and to identify the main sources of sediment at the catchment scale. A more detailed assessment of the channel characteristics was based on the interpretation of aerial photographs and river reaches were divided into similar morphological characteristics, based on valley-floor morphology (direction of the valley and degree of confinement of the river) and channel planform.

The alluvial portion of the river was classified in a series of reaches, starting from the boundary between the pre-Tertiary rocks and the sedimentary units of the Tertiary basin. Three main segments (A, B, and C) reflect the major structural controls (direction and confinement of the alluvial valley floor), while a second further division in sub-units is mainly based on channel morphology, resulting in a total of eight sub-reaches (Fig. 1(b)). Reach A is characterized by a sinuous channel (A1 and A3) alternated with a central sub-reach of meanders confined in the bedrock (A2). Reach B is characterized by a sinuous, transitional channel morphology with a significant increase of the alluvial plain and channel width, while reach C exhibits typical meandering morphology.

Sediment survey

A series of sedimentological and geomorphological field surveys were conducted in July and August, 2002. Sediment was collected from channel bars and pebble counts were conducted at a total of 23 locations along the Tanaro River and an additional six locations along the main tributaries (Fig. 1(b)). A river reconnaissance survey was conducted for each site using a stream reconnaissance sheet described by Thorne (1998) and specifically adapted to the scope and resources of the project.

Sediment transport was evaluated for two gauging stations (Garzonotti, 2003). The two stations (Farigliano and Montecastello) are located along reaches A and C (Fig. 1(b)) and are considered to be representative of the upper and the lower course of the river, respectively.

Standard sediment transport formulae were used to calculate mean annual bed load, suspended load, and total sediment load (Table 1). Although errors resulting from the use of bed load transport formulas are widely recognized, the formulas provide results of the same order of magnitude and can be considered a first approximation of the sediment transport of the Tanaro River.

Table 1 Evaluation of sediment transport at Farigliano and Montecastello (location of the two sites is shown in Fig. 1(b)).

Location	Bed load (m^3 $year^{-1}$)				Suspended load (m^3 $year^{-1}$)		Total load (m^3 $year^{-1}$)	
	A&M	E&F	MPM	P90	VR	S&ML	B	E&H
Farigliano	16457	11369	10149	14622	93134	100300	106017	151289
Montecastello	23825	30253	38953	36072	152123	122590	123875	172411

A&M: Ashida & Michiue; E&F: Engelund & Fredsoe; MPM: Meyer-Peter & Muller; P90: Parker (1990); VR: Van Rijn; S&ML: Smith & McLean; B: Brownlie; E&H: Engelund & Hansen.

RESULTS

Recent channel adjustments

Longitudinal profiles of the channel bed from 1973 and 2002 were compared to assess changes in bed elevation. Reach A was affected only in some short reaches by incision of the order of 1.5 m to a maximum of 2.2 m due to the presence of several grade control structures and bedrock outcrops. Reach B had the highest amount of bed erosion. Maximum values of 6 m were observed in sub-reach B1 and there was a slight decrease downstream of 1.5 and 4 m in sub-reach B2. Reach C was incised in the first part (sub-reach C1 and part of C2) from 1 to 2.5 m, while downstream reaches were characterized by erosion incision and deposition of up to 1 m. The incision rates are comparable to those observed in many other Italian rivers (Surian & Rinaldi, 2003), and have been related to various types of human intervention during the last 100 years, mainly sediment extraction, dams and channelization.

NUMERICAL MODELLING OF CHANNEL CHANGES

Model formulation

A one-dimensional mathematical model is used to describe longitudinal bed profiles, longitudinal free surface profiles and sediment transport as a function of time and hydraulic flow conditions. The governing equations adopted for the hydro-morphodynamic problem are the de Saint-Venant equations (1) and (2) for the liquid phase and the Exner equation (3) for the solid phase. Because flow conditions in the Tanaro River are nearly always subcritical ($Fr < 0.8$), meaning that the rate of bed morphological evolution is of a lower order of magnitude than flow changes with adequately low sediment concentration, we adopted a decoupled solution (Ferreira & Leal, 1998; Siviglia, 2003). It is possible to find the stationary solution of the fluid phase over a frozen bottom topography by solving equations (1) and (2), and then updating the bed elevation by solving equation (3). Defining a Cartesian coordinate system (x,y,z) with the x longitudinal axis lying on the bottom, y transversal axis, and the z axis upward normal to them (Fig. 2), the governing equations for hydro-morphodynamics are:

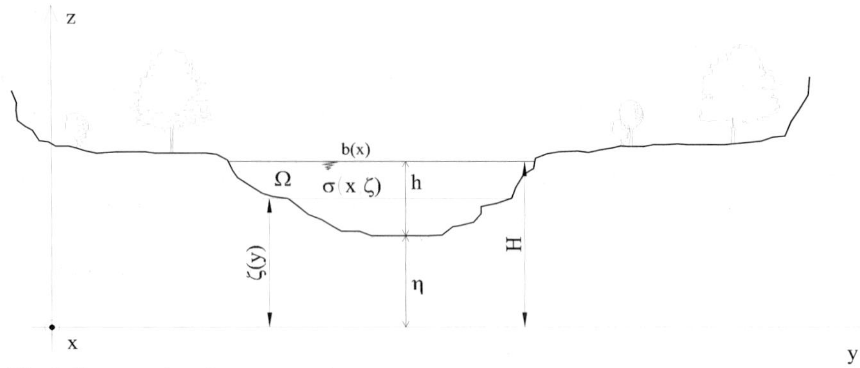

Fig. 2 Cross-sectional geometry.

$$\frac{\partial \Omega}{\partial t}+\frac{\partial Q}{\partial x}=q_l \tag{1}$$

$$\frac{\partial Q}{\partial t}+\frac{\partial}{\partial x}\left(\beta\frac{Q^2}{\Omega}\right)+g\Omega\frac{\partial H}{\partial x}+g\Omega j=0 \tag{2}$$

$$(1-p)b_{eff}\frac{\partial \eta}{\partial t}+\frac{\partial Q_s}{\partial x}=q_{sl} \tag{3}$$

where the energy loss per unit length of the channel is expressed through a friction coefficient:

$$j=\frac{Q^2}{g(\Omega^2C^2R)} \tag{4}$$

The unknowns of the full problem are the wetted cross-sectional area Ω, the volumetric discharge Q, and the minimum bottom elevation η. Moreover, H is the water level, R the hydraulic radius, q_l the discharge per unit length due to lateral inflow/outflow, β is the correction coefficient for the momentum, b_{eff} is the width of the mobile bed. Due to the complexity of natural geometry, the calculation of the quantities Ω, β, $(\Omega\, C^2\, R)$ has been done evaluating the integrals across the sections following the Engelund approach (Engelund, 1964). Application of the above method leads to the following terms:

$$\Omega=\int_b (H-\zeta(y))\mathrm{d}y \tag{5}$$

$$\beta=\frac{\Omega\int_b c^2(y)[H-\zeta(y)]^2\mathrm{d}y}{\left[\int_b c(y)[H-\zeta(y)]^{\frac{3}{2}}\mathrm{d}y\right]^{\frac{3}{2}}} \tag{6}$$

$$\Omega^2C^2R=\left[\int_b c(y)[H-\zeta(y)]^{\frac{3}{2}}\mathrm{d}y\right]^2 \tag{7}$$

where $\zeta(y)$ is the bed elevation, η is the minimum of $\zeta(y)$, $c(y)$ is the local conductivity which is a function of the transversal coordinate y. In equation (3), b_{eff} is the cross-sectional area effective width where the solid transport holds, p is the porosity, q_{sl} the solid discharge per unit length due to lateral inflow, while Q_s is the global solid discharge integrated all over the effective cross-sectional area.

In order to model sediment transport, we have identified in the Tanaro River different consecutive reaches. We have considered reaches short enough to neglect the longitudinal sorting effect. Thus, we have characterized the mobile bed of each reach by two grain sizes representative of bed load and suspended load respectively. The global solid discharge (Seminara *et al.*, 1996) is determined as:

$$Q_s = F_b \int_{b_{eff}} \Phi\sqrt{(s-1)gd_s^3}\,\mathrm{d}y + (1-F_b)\int_{b_{eff}} \psi U(y)(H-\zeta(y))\mathrm{d}y \qquad (8)$$

where s is the ratio of the sediment and the water density, g is the acceleration due to gravity, d_s is the average sediment diameter, F_b is the percentage of sediment transported as bed load, Φ and Ψ are the dimensionless bed load and suspended load discharge respectively, which are evaluated by empirical relations available in the literature (Meyer-Peter & Müller (1948) for bed load and Van Rjin (1984) for suspended load).

Numerical modelling results

Numerical results were obtained for first an unsteady fixed bed simulation along all the Tanaro River; second a stationary mobile bed simulation and eventually an unsteady mobile bed simulation along a short reach. The main purpose of the fixed bed simulation was to tune the local conductivity parameter of each section, reproducing real flood events. Such a parameter is fundamental for the correct evaluation of the global sediment discharge.

The mobile bed simulations examined a 35-km long reach of the Tanaro River near Alessandria, from the confluence with the Belbo River to the confluence with the Po River, because it is the most critical from the flooding point of view. The study reach is characterized by six bridges, including the ancient Cittadella Bridge that is protected by an apron producing a large scour hole downstream. The width of the main channel varies from 60 to 200 m. The maximum safe water discharge flowing below the Cittadella Bridge is about 2600 $m^3 s^{-1}$, while the 100-year discharge is estimated to be about 3500 $m^3 s^{-1}$.

First, we performed stationary mobile bed simulations. Such simulations should be interpreted as the first step to understanding the influence of channel changes on the bed profile and to highlight the critical points. These computations showed that the actual configuration of the bed topography of the reach downstream of Alessandria is very close to the equilibrium one, whereas in the neighbourhood of the town significant erosive processes occur (Fig. 3). It is worth noting that high erosion rates "at equilibrium" in some sections were over predicted in two ways: (a) the vertical sediment distribution in the alluvial deposit, i.e. the sediment coarsening with the depth, is neglected, and (b) we imposed constant water discharge assuming that the peak of the flood lasts to infinity, i.e. neglecting the increasing and decreasing flood phases. No information is derived about the time scale which is required to achieve the equilibrium configuration. This information, which is crucial from an engineering point of view, is given by the unsteady morphodynamic model which

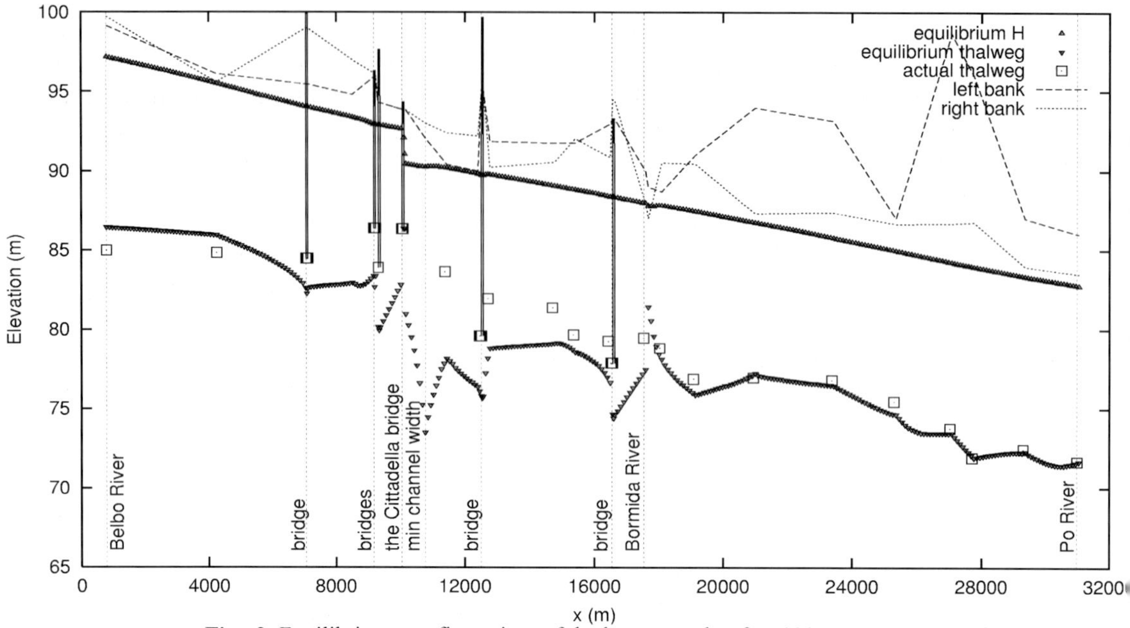

Fig. 3 Equilibrium configuration of bed topography for 100 years water and sediment discharge considering the actual geometry of the cross-sections.

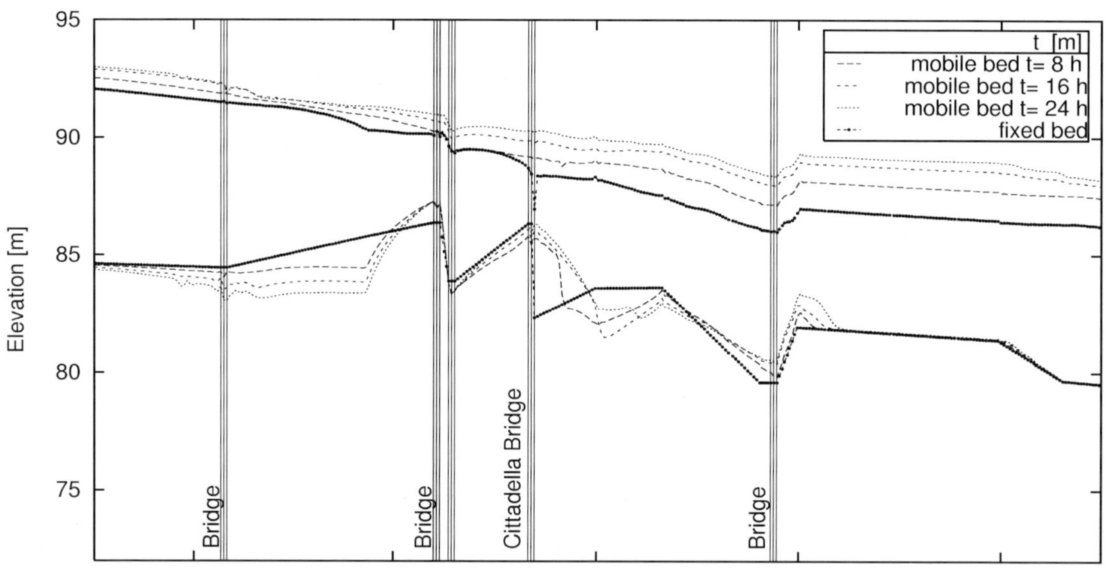

Fig. 4 Unsteady simulation: bed topography evolution during the increasing phase of an intense flood event.

predicts the magnitude of bed variations and the time scale required to occur. In Fig. 4 the evolution of the bed during the increasing phase of an intense flood event is shown. It is seen that, despite the fact that the configuration of the bed profile during a flood event is very far from the one predicted by the stationary model, at some points it slowly moves towards the equilibrium conditions. Unsteady mobile bed calculations also allow evaluation of the maximum scour in correspondence of critical sections, i.e. bridges and narrowing, during a real flood. This allows the civil engineer to verify correctly the stability of structures such as piers and banks.

Finally, we employed such numerical tools to verify whether geometric alterations of the river bed allow for an increase in flood capacity. We found that lowering the elevation of the apron of the Cittadella Bridge by about 2 m, and recalibrating the city reach so that the main channel width is constant, leads to decrease of the water level upstream of the Cittadella Bridge by up to 15% so increasing the safety of the whole city reach. This last application is an outstanding example as to how this tool is very versatile and useful for future river monitoring and management.

Acknowledgement This work has been partially developed within the framework of the National Project co-funded by the Italian Ministry of Universities and Research and the University of Genoa (COFIN 2001): Morphodynamics of fluvial networks, partially funded by Fondazione Cassa di Risparmio di Verona, Vicenza, Belluno e Ancona (Progetto RIMOF).

REFERENCES

Autorità di Bacino del Fiume Arno (2000) *Linee guida per la progettazione delle casse di laminazione* (Guide-lines for designing storage tanks). Quaderni dell'Autorità di Bacino, 9, Italy (in Italian).

Autorità di Bacino del Fiume Po (1997) *Piano stralcio delle fasce fluviali* (Plan for fluvial zones). Relazione, Autorità di Bacino del Fiume Po, Parma, Italy (in Italian).

Biancotti, A. (1981) Geomorfologia dell'Alta Langa (Piemonte meridionale) (Geomorphology of Alta Langa, Southern Piemonte) (in Italian). *Memorie della Società Italiana di Scienze Naturali e del Museo Civico di Storia Naturale di Milano* **XXII**, Fasc.III, 59–104.

Engelund, F. (1964) *Book of Abstracts*. Basic Research Tech. Report no. 6, University of Denmark, Denmark.

Environment Agency (1998) Sediment and gravel transportation in rivers. A procedure for incorporating geomorphology in river maintenance. In: *National Centre for Risk Analysis and Options Appraisal* (prepared by the University of Newcastle Upon Tyne), Guidance Note 23, Executive summary, 19–39. Environment Agency, Newcastle, UK.

Ferreira, R. M. L. & Leal, J. G. A. B. (1998) 1D Mathematical modelling of the instantaneous dam-break flood wave over mobile bed: Application of TVD and flux-splitting schemes. In: *CADAM Proceedings, Munich Meeting*. http://www.hrwallingford.co.uk/projects/CADAM/CADAM/Munich/contents.html

Garzonotti, M. (2003) Studio della dinamica del Fiume Tanaro. Tesi di Laurea in Ingegneria per l'Ambiente ed il Territorio, Facoltà di Ingegneria, Università di Firenze, Italy.

Meyer-Peter, E. & Müller, R. (1948) *Formulas for Bedload Transport* (2nd IAHR Congress). Int. Assoc. Hydraul. Res., Stockholm, Sweden.

Seminara G, Colombini, M. & Parker, G. (1996) Nearly pure sorting waves and formation of bedload sheets. *J. Fluid Mechanics*, **312**, 253–278.

Siviglia, A. (2003) Numerical solutions for hydrodynamic, morphodynamic and mudflow modelling. PhD Thesis, University of Genoa, Italy.

Surian, N. & Rinaldi, M. (2003) Morphological response to river engineering and management in alluvial channels in Italy. *Geomorphology* **50**, 307–326.

Thorne, C. R. (1998) *Stream Reconnaissance Handbook. Geomorphological Investigation and Analysis of River Channels*. John Wiley & Sons, Chichester, UK.

Van Rijn, L. C. (1984) Sediment transport: suspended load transport. *J. Hydraul. Engng ASCE* **110**(11), 1613–1641.

Sediment Transfer through the Fluvial System (Proceedings of a symposium held in Moscow, August 2004).
IAHS Publ. 288, 2004

The stratigraphy, mode of deposition and age of inset flood plains on the Barwon-Darling River, Australia

M. C. THOMS[1] & J. M. OLLEY[2]

1 *CRC for Freshwater Ecology, University of Canberra, Canberra, ACT, Australia*
martin.thoms@canberra.edu.au

2 *CSIRO Land and Water, Canberra, ACT, Australia*

Abstract Inset flood plains are a common feature of dryland river systems. These depositional landforms are attached to the bank between the riverbed and the main flood plain surface. Along the Barwon-Darling River in New South Wales, Australia, seven inset surfaces were identified. We used optical dating techniques and the presence of numerous European artefacts to show that these in-channel features range in age from ~10 to 2200 years. Three main stratigraphic sequences were recorded: a general fining upward sequence; a series of fine laminated sediments; and a distinct cut and fill sequence. The latter of which has not been previously reported for these deposits. Given their age and stratigraphy it is suggested that large quantities of sediment are exchanged between these temporary storage areas and the main channel over a period of 10–2000 years. The implication of these transfers on the ecology of this dryland river ecosystem is discussed.

Key words dryland rivers; Inset flood plains; sediment storage

INTRODUCTION

Inset flood plains are common features along the lowland sections of many Australian rivers. These relatively horizontal depositional landforms are bank attached sediment bodies that occur at intermediate elevations between the riverbed and the main flood plain surface. They have also been termed in-channel benches by Woodyer (1968) and have been recognized on many rivers worldwide (e.g. Kilpatrick & Barnes, 1964; Miller *et al.*, 1971). In-channel benches are important alluvial sediment storages, the character of which is dependent on a number of factors including catchment conditions, sediment supply, and prevailing hydraulic conditions during flood events (Thoms, 2003). Indeed, up to 87% of the sediment budget of some river systems can be in the form of temporary flood plain deposits like benches (Marron, 1992). The importance of in-channel benches for retaining organic material in lowland sections of dryland rivers has been demonstrated (Thoms & Sheldon, 1997). Large amounts of organic matter can accumulate on the surface of benches and the presence of these in-channel features increase the ability of these river systems to retain organic material.

Relatively few studies have detailed the sedimentology of in-channel bench deposits. Erskine & Livingstone (1999) organized the stratigraphy of channel deposits in the Hunter River, New South Wales, into three classes: stratic sediments, massive sediments and cumulic sediments, which together with a series of repeated channel cross sections suggest these benches are unstable. Catastrophic floods in the Hunter River—those with recurrence intervals greater than a 100 years and peak discharges 10 times the mean annual flood—cause the complete destruction of benches with their subsequent construction occurring over longer periods of time by smaller flood events. This cyclic formation contrasts to the long-term stability of benches described by Woodyer *et al.* (1979) along the Barwon-Darling River in western New South Wales, Australia. Benches in this low energy river system accrete both laterally and vertically at rates depending on their elevation and to a lesser extent their situation in the channel. Finely laminated accretionary sedimentary sequences up to 5 m in depth and a lack of erosional contact surfaces attest to the long-term stability of the various sedimentary sequences. However, Thoms & Sheldon (1997) reported significant changes to the cross sectional morphology of the Barwon-Darling over the last 100 years, with notable changes to the morphology of in-channel benches. Hence, the nature of in-channel benches differs between and within rivers. Different modes of bench formation will have implications for both the physical and ecological functioning of riverine ecosystems. Indeed, exchanges of sediment between various components of a river system and at different time scales, are an important ecosystem process and one that is recognized in various riverine ecosystem models such as the River Continuum Concept of Vannote *et al.* (1980), the Flood Pulse Model of Junk *et al.* (1989) and the Riverine Productivity Model of Thorp & Delong (1994).

The objectives of this paper are 3-fold: (a) to describe the stratigraphy of in-channel benches along the Barwon-Darling River; (b) to determine the age of the benches; and, (c) to comment on how they are formed.

STUDY AREA AND METHODS

The Barwon-Darling River drains 650 000 km^2 of the north-westerly portion of the Murray-Darling Basin in southeast Australia (Fig. 1(a)). Most of its tributaries (the Condamine-Balonne, Macintyre, Gwydir, Namoi, Castlereagh and Macquarie Rivers) drain the western margins of the Great Dividing Range in northern New South Wales and southern Queensland. Others, notably the Warrego and the Paroo Rivers, have their headwaters in the more arid west and are intermittent contributors, only providing significant runoff during periods of intense rainfall. The catchment is characterised by extreme climatic variability and runoff. Average annual rainfall and evaporation range from 200–1000 mm and 500–1800 mm, respectively (Thoms & Sheldon, 2000).

The Barwon-Darling is a suspended load river with characteristic high bankfull width to depth ratios (>32) and a highly sinuous channel (sinuosities >2). It also has "complex" bankfull cross-sections (see Woodyer 1968; Woodyer *et al.*, 1979; Thoms & Sheldon, 1997) because of the presence of inset flood plains or in-channel benches. Woodyer *et al.* (1979) identified and described the stratigraphy of four inset flood plain surfaces within the Barwon-Darling channel near Walgett (Fig. 1(b)). The two lower surfaces were considered to be formed by suspended-load deposition; either point, concave, convex and lateral benches and are composed of essentially horizontal laminations (ranging in thickness from 0.1 to 14 cm) of fine inorganic sediments and organic rich mud (Woodyer *et al.*, 1979). The upper surfaces, also termed benches, are relic surfaces and part of the present flood plain being inundated about once in every 15 years (Woodyer, 1968). However, recent research by Thoms & Sheldon (1997) has shown there to be at least seven different bench levels along the Barwon-Darling (Fig. 1 (c)). Regardless of the number and type of feature, each surface in the channel reflects a response to a change in flow regime (Woodyer, 1968; Woodyer *et al.*, 1979; Thoms & Sheldon, 1997). Similar in-channel bench features have been reported along the lower River Murray in South Australia by Thoms & Sheldon (1997) and along the coastal rivers of New South Wales by Erskine & Livingstone (1999).

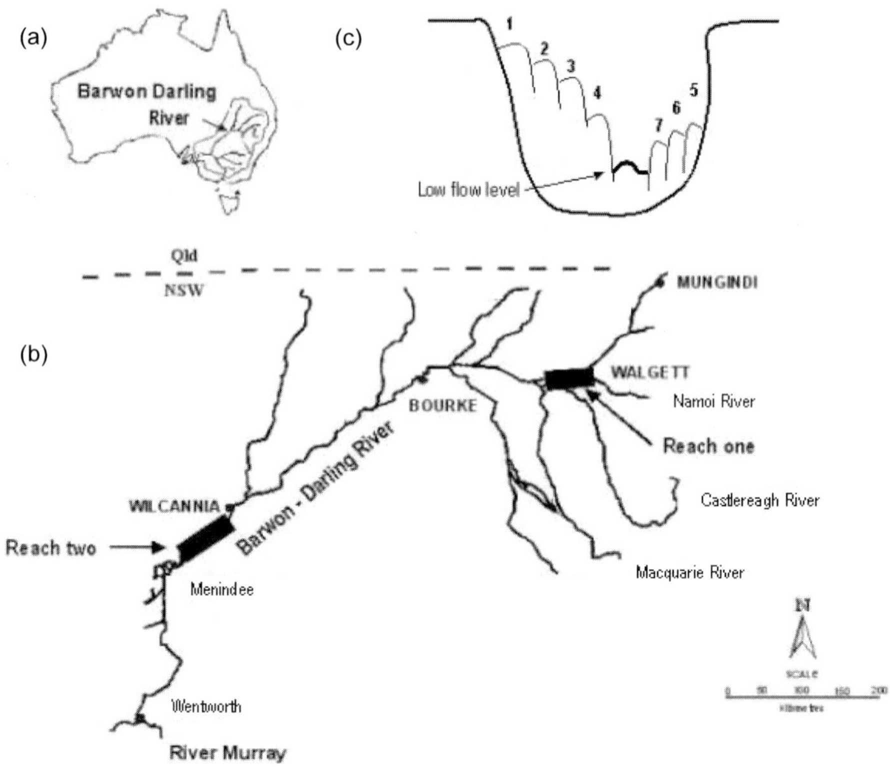

Fig. 1 (a) The Barwon-Darling catchment. (b) The Barwon-Darling River showing location of study reaches. (c) A schematic diagram of the river channel showing location of in-channel benches.

A series of pits, trenches, exposures and auger holes were dug in 98 in-channel benches along two 10 km reaches of the Barwon-Darling River. Reach one is located near Walgett in a section of river studied by Woodyer (1968) and Taylor & Woodyer (1978). Reach two is located just downstream of Wilcannia (Fig. 1(b)). The position of each bench was located using data of Thoms & Sheldon (1997) with benches being numbered sequentially from higher (Bench 1) to lower (Bench 7) elevations within the main channel (Fig 1(c)). In this study, the seven different bench levels were sampled seven times along each reach. The stratigraphy of each bench was recorded; from the surface to the low flow level and sediment samples from seven different bench levels in reach two were collected for textural analysis and optical dating.

Optical dating of sediments

Optical dating can be used to estimate the time elapsed since buried sediment grains were last exposed to sunlight (Aitken, 1998). This method of sediment dating makes use of the fact that daylight releases charge from light-sensitive electron traps in crystal lattice defects in minerals such as quartz and feldspar. The release of trapped charge by light resets the optically stimulated luminescence (OSL) signal; this process is commonly referred to as bleaching. When grains of quartz are buried and hidden from light, they begin to accumulate a trapped-charge population due to the effects of ionising radiation, such as that arising from radionuclides naturally present in the deposit. This trapped-charge population increases with burial time in a measurable and predictable way. As a result, the time elapsed since sediment grains were buried can be determined by measuring the OSL signal (burial-dose) from a sample of sediment and estimating the ionising radiation to which it has been exposed since burial (the dose rate) such that: burial-time = burial-dose/dose rate. Optical dating has been successfully used to date aeolian, freshwater and marine sediments (e.g. Bailey *et al.*, 2001; Murray & Clemmenson, 2001; Radtke *et al.*, 2001; Hilgers *et al.*, 2001; Olley *et al.*, 1999, 2004; Murray & Olley, 2002).

Analytical methods

All OSL measurements were made on two Risø automated TL/OSL readers, each fitted with an EMI 9635QA photomultiplier tube and three U-340 transmission filters. The readers are also equipped with green-plus-blue light sources (420–550 nm), giving an illumination intensity of about 25 mW cm^{-2} on the sample (H. Christiansen, personal communication). Small aliquots (40–60 grains) of quartz were analysed using the regenerative-dose protocol described by Roberts *et al.* (1998), which was modified from those presented by Murray & Roberts (1998). The dose (D_e) for each aliquot was calculated as:

$$D_e = (L_n/L_r) \times (T_2/T_1) \times \text{regenerative dose} \quad (1)$$

where L_n, L_r, T_1 and T_2 are the OSL signals produced by the natural, regenerative, test 1, and test 2 doses. The test dose signals are used to correct for any changes in OSL sensitivity between the natural and regenerative dose cycles. The samples were illuminated for 125 s at 125°C. In each case, the OSL signal was integrated over the first 20 s of illumination, and the OSL signal integrated over the final 20 s was subtracted as background. The reported uncertainties are based on the counting statistics, curve fitting errors and incorporate calibration uncertainties for the beta sources. A preheat temperature of 240°C for 10 s was used for L_n and L_r measurements, and a cut-heat to 160°C was given after each test dose.

To determine the field dose rate a sub-sample was taken from each of the OSL samples. These were analysed by high-resolution gamma spectrometry for ^{238}U, ^{226}Ra, ^{210}Pb, ^{228}Th, ^{228}Ra, and ^{40}K concentrations. Sample masses of about 200 g were cast in resin and counted for 24 h. The intrinsic germanium detectors were calibrated using the Canadian Centre for Mineral and Energy Technology (CANMET) uranium ore BL-5, and thorium nitrate refined in 1906 (Amersham International). Independent checks on calibration were performed using various standards from the USA National Bureau of Standards, and IAEA inter-comparisons.

Textural analysis

Particle size analysis was done on a 5 g subsample which was ultrasonically dispersed in a 5% sodium hexametaphosphate solution before being sized by a Malvern Autosizer, with a 63 mm lens. Results were expressed in phi (ϕ) units, where $\phi = -\log_2$ (mm). Each sediment sample was analysed three times to check instrument precision and to calibrate the instrument. National Bureau Standards of known sphere size were run after every 25 samples.

RESULTS

Particle size distribution and bench stratigraphy

Sediments contained in the various in-channel bench deposits were dominated by medium to fine sand and silt-clay mixtures. Median grain sizes ranged from 1.09 to 3.99 ϕ. Distinct variations in sediment colour and texture occur both between and within the different bench deposits. Lower level bench deposits (bench levels 5–7, Fig. 1(c)) are generally associated with coarser sediments (median grain sizes: 1.09–2.69 ϕ) in comparison to higher-level bench deposits (median grain sizes of bench level 1–4: 2.29–3.99 ϕ). Two distinct stratigraphic sequences were recorded in the higher-level benches (bench levels 1–4) (Fig. 2). The first consists of an intricate series of fine laminated deposits (Fig. 2(a)). Here, lenticular sand layers were common, some reaching a thickness of 35 cm although they did decrease in thickness up profile. In general, sand layers were separated by thin layers of a silt-clay mixture which, contained variable levels of organic matter (loss on ignition: 5.86–39.56%). Graded bedding was common and three different grading configurations were recorded; a simple grading from either sand to silt-clay or silt-clay to sand; a complex grading of sand to silt-clay to sand or silt-clay to sand to silt-clay; and, multiple grading in which there was several sequences of complex grading. The deposits contained flat and wavy parallel laminations as well as cross laminations and all contacts between the individual layers were generally depositional in nature. The second stratigraphic sequence recorded in the higher benches differs to that just described in that several erosional contacts were noted in some benches (Fig. 2(b)). Distinct cut and fill sequences were recorded in a number of level 2, 3 and 4 benches and these were traced along their length. In one level 4 bench, three cut and fill sequences were recorded. The silt-clay layers found in bench levels 2, 3 and 4 contained elevated levels of organic matter compared to that found within the bench level 1 deposits (loss on ignition of bench level 1: 3.24–12.45% and bench levels 2–4: 23.34–45.67%). A general fining upward sequence grading from a coarser basal layer of well-sorted medium sands through to a very fine sand coarse silt mixture at the surface was recorded in the lower level benches: bench levels 5–7 (Fig. 2(c)).

Optical dating

From Reach two, a sediment sample was collected at the boundary of the main channel–bench deposit boundary, thereby providing an age for the main channel of the Barwon-Darling. Further samples were collected from within the higher bench deposits, especially from those positions above and below notable erosional contacts. Dose rates were calculated using the conversion factors of Olley *et al.* (1996) and the computer program listed in Roberts *et al.* (1993). The water content measured in the samples ranged from 7.8 to 28.1 percent of their dry weight. These water concentrations are taken to be representative of the long-term average, and have been assigned relative uncertainties of ±50%. For all samples the dry dose rate, determined by gamma spectrometry, was corrected for these water concentrations, following Aitken (1998).

The cosmic-ray dose rates were calculated from Prescott & Stephan (1982) and Prescott & Hutton (1988). Beta-attenuation factors were taken from Mejdahl (1979) and the effective alpha dose rate contribution has been estimated using an alpha-efficiency "a" value for

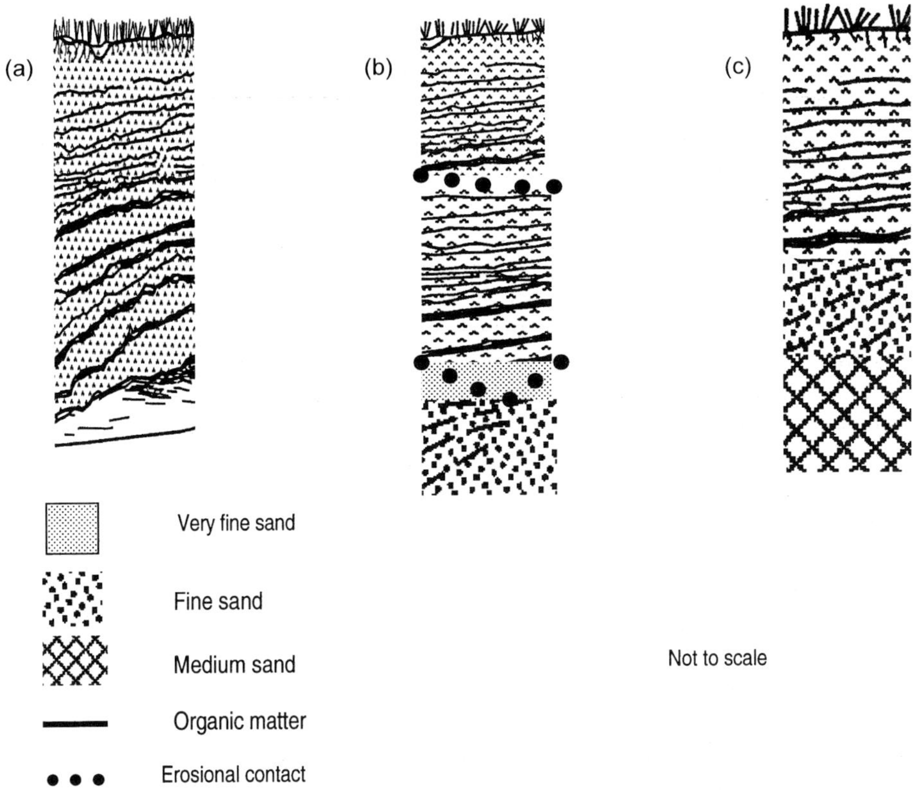

Fig. 2 Stratigraphy of in-channel benches. (a) Series of finely laminated sediments; (b) cut and fill sequence; and, (c) general fining upward sequence

quartz of 0.10 ± 0.02. The alpha dose rate contribution is about 5% of the total dose rate. The calculated total dose rates are presented in Table 1 and range from 1.64 ± 0.16 to 2.52 ± 0.22 mGy year^{-1}.

Doses measured on individual aliquots of each of the samples using the regenerative-dose, single-aliquot OSL protocol are presented in Table 1. There is clearly a wide spread of doses present in all of the samples, indicating that the sediments were not fully bleached at the time of deposition. For example, in sample CS–D3 the measured doses range from 0.00 ± 0.08 Gy to 22.6 ± 1.5 Gy. In such circumstances the best estimate of the burial dose

Table 1 Measured water contents (% dry weight), dose rates (D_r), dose range, burial dose estimates (D_b), and calculated burial ages for fluvial sediment samples CS–D1 to CS–D5.

	Water content %	D_r (mGy year^{-1})	Dose range (Gy)	D_b (Gy)	Age (years)
CS–D1	16.5	1.76 ± 0.17	0.41 ± 0.03 to 60 ± 2	0.43 ± 0.04	240 ± 50
CS–D1a	15.2	1.96 ± 0.18	0.754 ± 0.015 to 86 ± 4	0.76 ± 0.01	390 ± 60
CS–D2	7.8	2.18 ± 0.21	0.11 ± 0.05 to 60 ± 2	0.21 ± 0.01	95 ± 20
CS–D3	28.1	2.05 ± 0.25	0.00 ± 0.08 to 22.6 ± 1.5	0.06 ± 0.07	30 ± 20
CS–D4	17.9	1.64 ± 0.16	3.32 ± 0.17 to 54 ± 4	3.61 ± 0.16	2200 ± 250
CS–D5	8.4	2.52 ± 0.22	26 ± 3 to 140 ± 3	32.6 ± 2.1	13 000 ± 1500

will be provided by the aliquots containing the lowest doses (Olley *et al.*, 1998, 1999, 2004). Consequently, the burial dose for each sample has been calculated using the lowest dose population determined using the minimum age model (Galbraith *et al.*, 1999). The burial dose (D_b) and calculated ages are presented in Table 1.

DISCUSSION

Our understanding of many basic ecosystem processes in large dryland river systems is poor in comparison with those from more humid and temperate regions (Thoms, 2003). Current models of river system function cannot be easily applied to those in a dryland setting (Graf, 1988; Walker *et al.*, 1995) partly because these systems have highly variable and unpredictable flow and sediment regimes, with episodic connections between the main river channel and adjacent flood plain. The development of seven bank attached bench deposits along the Barwon-Darling River may be considered as a morphological adjustment in response to the highly variable flow and sediment regimes of the region (Thoms *et al.*, 2004). As a result of their presence, the Barwon-Darling River has a "nested compound" channel where lower flow channels are contained or "nested" within a series of higher flow channels, with each nested channel being marked by the horizontal surface of each bench. The nested channels are markedly younger than the main channel—the main channel had a buried date of 13 000 years compared to burial dates ranging from <20 to 2200 years for the in-channel deposits. The nested compound channel of the Barwon-Darling River differs from the compound channels in the Gila River, Arizona as described by Graf (1988). The Gila has a well-defined inner low flow channel that meanders within a much larger outer flood channel, which is often braided in planform. This reflects two modes of operation (Graf, 1988): a single low water channel and a wider high water channel represent an adjustment to a particular flow regime that is dominated by near continuous low flows coupled with a few rare high-discharge events. The multiple nested channels of the Barwon-Darling may then reflect multiple modes of operation. Hence, within-system morphological variability and its apparent relationship to hydrological and sediment regimes illustrate the complexity of dryland river systems.

Large quantities of sediment are stored within the main channel of the Barwon-Darling River as evidenced by the presence of seven distinct benches. Collectively, the morphogenesis of in-channel benches along the Barwon-Darling River is highly complex. Level 1 benches located at the highest elevations in the channel are relatively stable, displaying a stratigraphy characterized by sequences of thin interbedded sand and silt-clay layers (Fig. 2(a)). They are also the oldest in-channel deposits with a burial date of ~2200 years (Fig. 3). These benches are similar in nature to those described by Woodyer *et al.*, (1979). Level 2–4 benches, located at more intermediate elevations within the channel are not as stable and are younger than Level 1 benches. Whilst the overall stratigraphy of Level 2–4 benches was similar to Level 1 benches distinct cut and fill sequences were evident in the former and not the latter. Fill sediments were much younger with burial dates of 30–95 years compared to those immediately below erosional contacts—burial dates between 240 and 340 years (Fig. 3). Level 5–7 benches, at the lowest elevations in the channel, were the youngest deposits (all deposits <20 years) and all displayed a general fining upward sequence. These benches are probably formed in a similar manner to those described by Erskine & Livingstone (1999) where larger flood events completely rework these lower

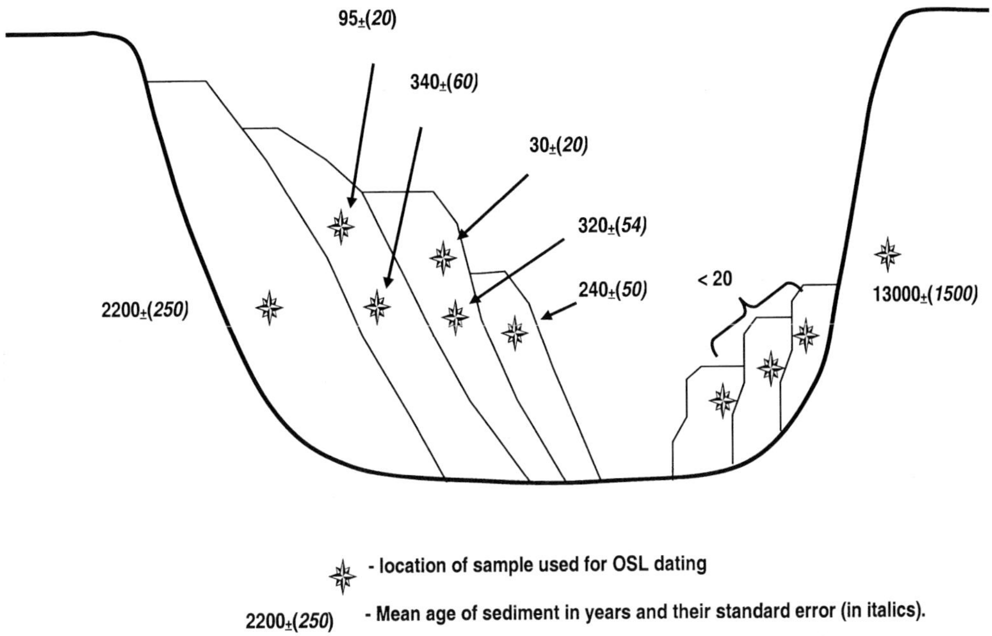

Fig. 3 A schematic diagram of a typical series of in-channel bench deposits and the age of the various in-channel benches.

elevation deposits and smaller floods are responsible for their construction. The variable sedimentology of the in-channel deposits of the Barwon-Darling reflects the variable discharge and sediment regimes and geomorphic scales of operation of this dryland river system.

Three modes of bench formation occur along the Barwon-Darling River:

(a) The long-term vertical and lateral accretion of benches located at higher channel elevations.
(b) Vertical and lateral accretion interrupted by the partial reworking of in-channel deposits at intermediate channel elevations.
(c) The complete reworking of lower level benches during flood events followed by their formation by smaller events.

Modes 1 and 3 have only been previously reported (i.e. Woodyer *et al.*, 1979; Erskine & Livingstone, 1999). Partial reworking and bench deconstruction followed by an accretionary stage similar to that recorded for benches located at higher elevations in the channel reflects the presence of regular cut and fill sequences within Level 2–4 benches of the Barwon-Darling. Olley & Caitcheon (2000) show that the sediment currently in transport in the Barwon-Darling does not originate from contemporary upland erosion, but is derived from lowland areas of the catchment that contain more weathered material. We propose that partial reworking of these in-channel bench deposits at time intervals up to 95 years is an important source of sediment to the river; a finding consistent with this previous observation. Moreover, large amounts of organic matter are present within these bench deposits and this material may represent an important albeit longer-term source of organic carbon to the food web of this dryland river.

Inset flood plains are temporary sediment storage areas. Combining data on the morphology of the different benches, their stratigraphy and age, approximately 1.186 m m^3 of sediment is stored within the in-channel benches of the two study reaches. The residence time of sediment differs between the various bench levels. It is estimated that 55 600 m^3 of sediment is reworked over a 20-year time span from the lower benches (benches 5–7)—5% of the total volume of sediment in the bench deposits. By comparison, 181 440 m^3 (15%) would be made available from the upper sections of the mid level benches (benches 2–4) every 20 to 100 years whilst 214 560 m^3 (18%) is reworked every 100 to 400 years from the lower sections of these benches. Larger volumes of sediment—735 000 m^3 (62%) are made available from the high level benches (bench level 1) over time periods up to 2000 years. Thus sediments contained within the in-channel benches constitute a large secondary and local sediment source in this dryland river system.

There has been a strong trend in recent years to view rivers as ecosystems. This requires a holistic framework that recognize:

(a) interconnections between the physical, chemical and biological components of riverine ecosystems and the different scales of operations of each;
(b) linkages between upstream–downstream and the river channel–flood plain; and;
(c) that different parts of the river system may operate over different time scales.

Large rivers are often considered less retentive than small streams for accumulating organic material, mostly the result of a decrease in retentive structures (Webster *et al.*, 1994). The apparent decrease in the availability of retentive structures in large dryland rivers ignores the role of in-channel benches. These variable geomorphic surfaces along river valleys are known to create complex physical patterns that are reflected in the development of riparian plant communities and the distributions of aquatic biota (Gregory *et al.*, 1991). In the Barwon-Darling River, in-channel benches not only retain large quantities of surface organic material (Thoms & Sheldon, 1997) but bench deposits are also a source and a sink of organic carbon that may be made available to aquatic food webs over time intervals of up to 100 years. Dryland rivers do experience relatively frequent within-channel floods that inundate in-channel "bench" features at one or more levels, depending on the magnitude of flow. Geomorphic in-channel complexity and its ability to retain organic material, therefore, means that although the dominant lateral movements of organic material from the flood plain and riparian zone into the channel will still occur during large overbank flows, smaller "pulse" inputs will also occur with each in-channel rise and fall in water level and during partial reworking of individual bench features. In dryland rivers, where large overbank flows only occur infrequently smaller "pulse" inputs of organic material from both the surface and during erosion events may also` be vital for the integrity of these ecosystems.

Acknowledgements The authors would like to thank Dr Ian Maddock, Oscar Mamalai and Matt Henseliet for their expert and tireless trenching abilities and general entertainment in the field. Discussion with Heather McGinness on sources and sinks of carbon in dryland rivers is gratefully acknowledged.

REFERENCES

Aitken, M. J. (1998) *An Introduction to Optical Dating*. Oxford University Press, Oxford, UK.

Bailey, S. D., Wintle, A. G., Duller, G. A. T.& Bristow, C. S. (2001) Sand deposition during the last millenium at Aberffraw, Anglesey, North Wales as determined by OSL dating of quartz. *Quatern. Sci. Rev.* (*Quatern. Geochron.*) **20**, 701–704.

Erskine, W. D. & Livingstone E. A. (1999). In-channel benches: the role of floods in their formation and destruction on bedrock-confined rivers. In: *Varieties of Fluvial Form* (ed. by A. J. Miller & A. Gupta), 445–476. John Wiley & Sons, Chichester, UK.

Galbraith, R., Roberts, R. G., Laslett, G., Yoshida, H. & Olley, J. (1999) Optical dating of single and multiple grains of quartz from Jinmium Rock Shelter, Northern Australia. Part I: Experimental design and statistical models. *Archaeometry* **41**, 339–364.

Graf, W. L. (1988) *Fluvial Processes in Dryland Rivers*. Springer-Verlag, Berlin, Germany.

Gregory, S. V., Swanson, F. J., McKee, W. A. & Cummins, K. W. (1991) An ecosystem perspective of riparian zones. *Bioscience* **41**, 540–551.

Hilgers, A., Murray, A. S., Schlaak, N. & Radtke, U. (2001) Comparison of quartz OSL protocols using Lateglacial and Holocene dune sands from Brandenburg, Germany. *Quatern. Sci. Rev.* (*Quatern. Geochron.*) **20**, 731–736.

Junk, W. J., Bayley, P. B. & Sparks, R. E. (1989) The flood pulse concept in river-flood plain systems. In: *Proc. Int. Large River Symp.* (ed. by D. P. Dodge), 110–127. Canadian Special Publications in Fisheries and Aquatic Science, Ottawa, Canada.

Kilpatrick, F. A. & Barnes, H. H. (1964) Channel geometry of Piedmont streams as related to frequency of floods. *US Geol. Survey Prof. Paper* **422E**, 1–42.

Marron, D. C. (1992) Flood plain-storage of mine tailings in the Belle Fourche River system: A sediment budget approach. *Earth Surf. Processes Landf.* **17**, 675–685.

Mejdahl, V. (1979) Thermoluminescence dating: beta-dose attenuation in quartz grains. *Archaeometry* **21**, 61–72.

Miller, R. A., Troxell, J. & Leopolds, L. B. (1971) Hydrology of two small river basins in Pennsylvania before urbanisation. *US Geol. Survey Prof. Paper* **701A**, 1–18.

Murray, A. S. & Clemmensen, L. B. (2001) Luminescence dating of Holocene aeolian sand movement, Thy, Denmark. *Quatern. Sci. Rev.* (*Quatern. Geochron.*) **20**, 751–754.

Murray A. S. & Roberts R. G. (1998). Measurement of the equivalent dose in quartz using a regenerative-dose single aliquot protocol. *Radiat. Measurements* **29**, 503–515.

Murray, A. S. & Olley, J. M. (2002) Precision and accuracy in the optically stimulated luminescence dating of sedimentary quartz: a status review. *Geochronometria* **21**, 1–16.

Olley, J. M. & Caitcheon, G. G. (2000) The major element chemistry of sediments from the Darling-Barwon River and its tributaries: Implications for sediment and phosphorus sources. *Hydrol. Processes* **14**, 1159–1175.

Olley, J. M., Murray, A. S. & Roberts, R. G. (1996) The effects of disequilibria in the uranium and thorium decay chains on burial dose rates in fluvial sediments. *Quatern. Geochron.* **15**, 751–760.

Olley, J. M., Caitcheon, G. & Roberts, R. (1999) The origin of dose distributions in fluvial sediments, and the prospect of dating single grains of quartz from fluvial deposits using OSL. *Radiat. Measurements* **30**, 207–217.

Olley, J. M., Pietsch, T. & Roberts, R. G. (2004) Optical dating of Holocene sediments from a variety of geomorphic setting using single grains of quartz. *Geomorphol.* (in press).

Prescott, J. R. & Hutton, J. T. (1994) Cosmic ray contributions to dose rates for luminescence and ESR dating: large depths and long-term time variations. *Radiat. Measurements* **23**, 497–500.

Prescott, J. R. & Stephan, L. G. (1982). The contribution of cosmic radiation to the environmental dose for thermoluminescence dating: latitude, altitude and depth dependencies. *PACT* **6**, 17–25.

Radtke U., Janotta A., Hilgers A. & Murray A. S. (2001) The potential for OSL dating Lateglacial and Holocene dune sands with independent age control of the Laacher See tephra (12880a) at the Section 'Mainz-Gonsenheim'. *Quatern. Sci. Rev.* (*Quatern. Geochron.*) **20**, 719–724.

Roberts, R., Bird, M., Olley, J. M., Galbraith, R., Lawson, E., Laslett, G., Yoshida, H., Jones, R., Fullagar, R., Jacobsen, G. & Hua, Q. (1998) Optical and radiocarbon dating at Jinmium rock shelter in northern Australia. *Nature* **398**, 358–362.

Taylor, G. & Woodyer, K. D. (1978). Bank deposition in suspended load streams. In: *Fluvial Sedimentology* (ed. by A. D. Miall), 257–276. Canadian Society of Petroleum Geologists, Memoir 5.

Thoms, M. C. (2003) Flood plain-river ecosystems: lateral connections and the implications of human interference. *Geomorphol.* **56**(3–4), 335–350.

Thoms, M. C. & Sheldon, F. (1997) River channel complexity and ecosystem processes: the Barwon-Darling River, Australia. In: *Frontiers in Ecology: Building the Links* (ed. by N. Klomp & I. Lunt), 193–206. Elsevier, Oxford, UK.

Thoms, M. C. & Sheldon, F. (2000) Water resource development and hydrological change in a large dryland river: the Barwon-Darling River, Australia. *J. Hydrol.* **228**, 10–21.

Thoms, M. C., Beyer, P., & Rogers, K. R. (2004) Variability, complexity and diversity—The geomorphology of river ecosystems in dryland regions. In: *Changeable, Changed, Changing: The Ecology of Desert Rivers* (ed. by R. T. Kingsford), 26–42. Cambridge University Press, Cambridge, UK.

Thorp, J. H. & Delong, M. D. (1994) The riverine productivity model: an heuristic view of carbon sources and organic processing in large river ecosystems. *Oikos* **70**, 305–308.

Vannote, R. L., Minshall, G. W., Cummins, K. W., Sedell, J. R. & Cushing, C. E. (1980) The river continuum concept. *Can. J. Fisheries Aquatic Sci.* **37**, 130–137.

Walker, K. F., Sheldon, F. & Puckridge, J. T. (1997) A perspective on dryland river ecosystems. *Regulated Rivers: Research and Management* **11**, 85–104.

Webster, J. R., Covich, A. P., Tank, J. L. & Crockett, T. V. (1994) Retention of coarse organic particles in streams in the southern Appalachian Mountains. *J. North Am. Benthological Soc.* **13**, 140–150.

Woodyer, K. D. (1968) Bankfull frequency in rivers. *J. Hydrol.* **6**, 114–142.

Woodyer, K. D. (1975) Concave-bank benches on the Barwon River. *NSW Australian Geographer* **13**, 36–40.

Woodyer, K. D., Taylor, G. & Crook, K. A. W. (1979) Depositional processes along a very low-gradient, suspended-load stream: the Barwon River, New South Wales. *Sedimentary Geol.* **22**, 97–120.

Sediment Transfer through the Fluvial System (Proceedings of a symposium held in Moscow, August 2004).
IAHS Publ. 288, 2004

On sediment transport in the Łososina River in the Polish Carpathians

TADEUSZ BEDNARCZYK[1], ARTHUR RADECKI-PAWLIK[1], PRZEMYSŁAW BARAN[2] & EWA SŁOWIK-OPOKA[1]

1 *Department of Water Engineering, Agricultural University of Krakow, Al. Mickiewicza 24–28, 30-059 Krakow, Poland*
rmradeck@cyf-kr.edu.pl

2 *Department of Soil Mechanics and Earth Structures, Agricultural University of Krakow, Al. Mickiewicza 24–28, 30-059 Krakow, Poland*

Abstract The Łososina River is one of the Polish Carpathian mountain streams that crosses the south of the Beskid Wyspowy Mountains. It is mostly gravel-bed, it is flashy, experiences frequent flooding and often causes trouble for the local communities as far as spring floods are concerned. At the mouth of the Łososina River there is one of the biggest Polish Carpathian artificial lakes—the Żywiecki Water Reservoir (ŻWR). Since the Łososina River transports mostly gravel as the bed load to the ŻWR, in the early seventies it was partly canalized, especially in places where it passes the inhabited areas. The situation of the Łososina River before and after the engineering works is compared. Features such as changes in cross-section geometry, slope, granulometry and bed load transport balance were compared using archives and present-day studies. For the purpose of the study, old engineering projects and reports were used to find out the difference in the Łososina River behaviour.

Key words bed load transport; granulometry; mountain stream; Poland; water reservoir

INTRODUCTION

Economic development of a society is related to the ability to maximize the benefits and minimize the damage caused by rivers. Rivers very often adjust their cross section and their longitudinal profile through the process of downstream sediment transport (Yang, 1996). Generally one can recognize two types of sediments in rivers: bed load and suspended load. In mountain streams where the streambed consists mostly of gravel and coarse sands, bed load is reported to constitute in some extreme cases up to ~70% of total bed load (Selby, 1985). The problems caused by the sediment movement are especially dangerous when water reservoirs for flood protection and for water storage are constructed on such rivers because the sediment trapped in the reservoirs tends to fill them up, reducing their water capacity. To reduce this process (basically reducing the bed-load transport) many engineering works known as river training works are undertaken. This paper examines the sediment budget that was calculated before and after river training works on the Łososina River in the Polish Carpathians. At the mouth of the Łososina River the artificial lake (the Żywiecki Water Reservoir, ŻWR) was built. To undertake the calculations archival materials from river design offices were used with up-to-date measurements.

STUDY AREA

The Łososina River in the Polish part of Carpathian Mountains (Fig. 1) drains the Carpathian flysch. The stream is flashy and experiences frequent bed load movement. Its streambed

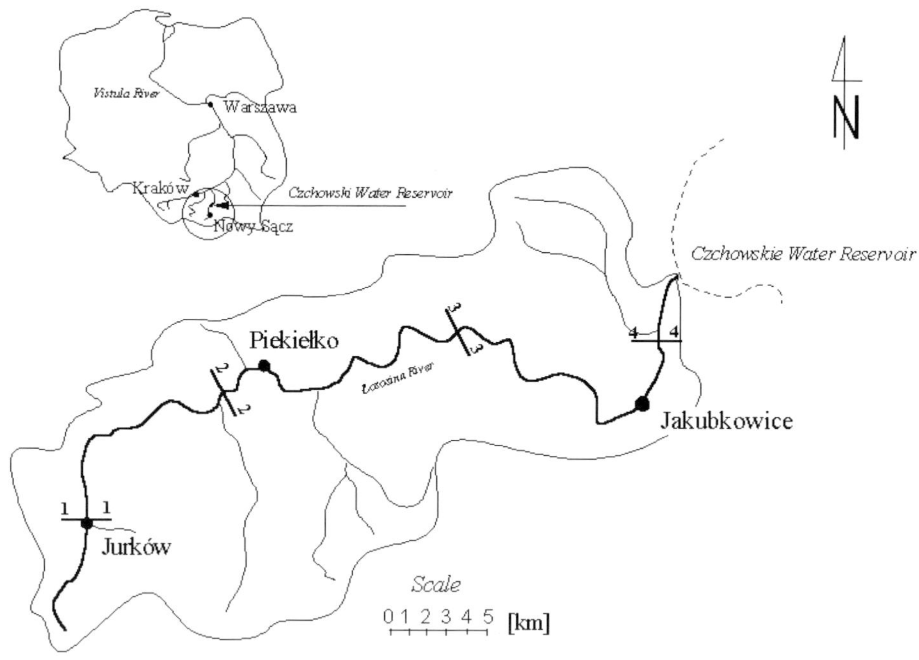

Fig. 1 Catchment study region with the detailed sketch of the research reach.

consists mostly of sandstone and mudstone bed load pebbles and cobbles forming a framework, the interstices of which are filled by a matrix of finer sediment. Suspended sediment load is small but its contributions to channel morphology were taken into consideration during sediment calculations. Many gravel river bed-forms, such as point and middle bars, occur within the investigated reaches of the Łososina River. Most gravel bed-forms can are observed at the riverbanks and within the river channel. After 1975, many river training works were performed along the Łososina channel to prevent bank erosion and to reduce the channel slope. The river cross sections were trained by building drop-hydraulic structures (to reduce slope) and by constructing gabions (stone-baskets along the banks—preventing bank erosion). These works were aimed at reducing the bed load transport along the Łososina and stopping its degradation after the river reservoir Czchów was constructed at the river mouth. The basic hydrological characteristics of the river are presented in the Table 1. All numbers refers to the river channel between cross sections 1-1 and 4-4 (see Fig. 1).

Table 1 Physical characteristics of sites investigated.

Precipitation (mm)	896	Max. stream depth D (m)	2.2
Catchment area (km^2)	410	W/D ratio	21.53
Max. catchment altitude (m a.s.l.)	760	Minimum annual discharge (m^3 s^{-1})	1.26
Min. catchment altitude (m a.s.l.)	241	Mean annual discharge(m^3 s^{-1})	4.78
Channel gradient (average within study area) (–)	0.0106	$Q_{50\%}$ (m^3 s^{-1})	48.63
Stream length L (km)	48.65	$Q_{3\%}$ (m^3 s^{-1})	196.41

METHODS

Sediment load was calculated at four cross sections (Fig. 1). The technique described by Church *et al.* (1987) was applied for sediment sampling. The surface sediment layer was identified as armoured rather than censored (Carling & Reader, 1981; Parker, 1990). Samples were collected from homogeneous bodies of sediment, so as not to combine them with the distinct surface material. The sieving analysis for coarse grains was carried out in the field by hand using round-mesh sieves (Michalik, 1990). Fine material was carefully collected and analysed in the laboratory.

For bed load transport calculations (especially for calculations of transport to the ŻWR) the Meyer-Peter-Muller (Meyer-Peter & Muller, 1948; Michalik, 1990) formula was used:

$$q_i = \left[\frac{\rho_w g h I - f_i g \Delta\rho d_i}{0.25 \rho_w^{\frac{1}{3}}} \right]^{1.5} p_i b \qquad (1)$$

where q_i is unit bed load transport, ρ_w ,and ρ_r are water and sediment density respectively (kg m^{-3}), g is acceleration due to gravity (m s^{-2}), h is water depth (m), I is slope, f_i is Shields shear stress value, $\Delta\rho = \rho_r - \rho_w$ (kg m^{-3}), d_i is sediment size (mm), p_i is percentage of the sediment fraction within the sediment probe, and b is active channel width (m).

Two computer models were used: SPAW_2003v.1.0 (Radecki-Pawlik & Baran, 2000) and SandCalc-1.2 (Wallingford, 1996). A special computer model called Shear_v.1.0 was developed for shear stress calculations (Radecki-Pawlik & Radecki-Pawlik, 2003). The land survey was done following methods described by Przewłocki (2000) and using a TOPCON AT-G7 gheodimeter. Hydrological calculations were made using the Punzet formulae (Punzet, 1972, 1981) using the WODA_2000_v.2.0 computer model (Radecki-Pawlik, 1985). Suspended sediment was measured at the bathymetric station in Jakubowice. Finally, the TransCalc model was used to calculate the sediment budget before and after the river training works on the Łososina River (DWE, 2003). This model calculates the sediment transport under a given threshold discharge (which is the competent flow for the beginning of the movement of bed load) using data such as calculated bed load transport, hydrology (in the case of this paper the discharges for the 12-year recurrence period) and the threshold of beginning of motion derived from the bed load calculations or measurements. Since the software was designed especially for this work a schematic of it is presented in Fig. 2.

All archival materials concerning the slopes of the river and granulometry of its bed before the river training works were provided by Nowy Sącz Municipality Authority, the State Geological Institute in Warsaw and River Water Authority in Krakow and Nowy Sącz.

RESULTS

All basic granulometric parameters calculated for each of the research cross sections and recognized are presented in Table 2. Tables 3 and 4 show the sediment transport data for Łososina before and after river training works. Table 5 presents changes of the unit bed load transport results for the Łososina River after regulation. Figure 3 presents the hydrological events before and after the river training works on the Łososina used in the TransCalc model to calculate the sediment budget along the Łososina.

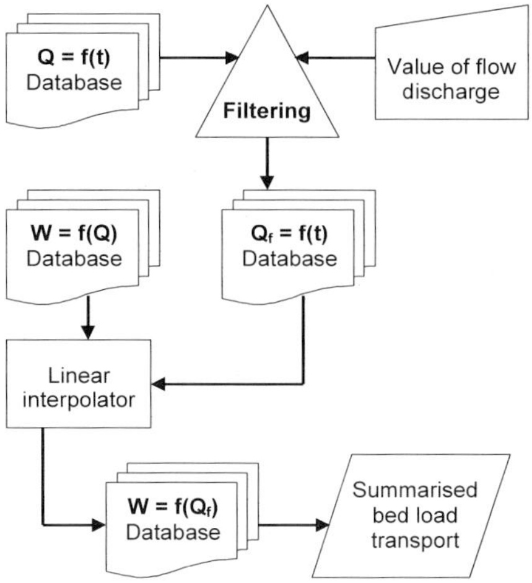

Fig. 2 Schematic of the TransCalc model.

Table 2 Characteristic grain-size diameters before and after the training works.

Sampling cross-section	Before – sediment diameter (mm)				After – sediment diameter (mm)			
	d_{16}	d_{50}	d_{84}	d_{90}	d_{16}	d_{50}	d_{84}	d_{90}
1-1	7	28	83	88	7	30	85	90
2-2	10	30	70	76	6	22	65	70
3-3	12	40	90	95	10	35	88	90
4-4	10	30	58	67	11	22	50	65

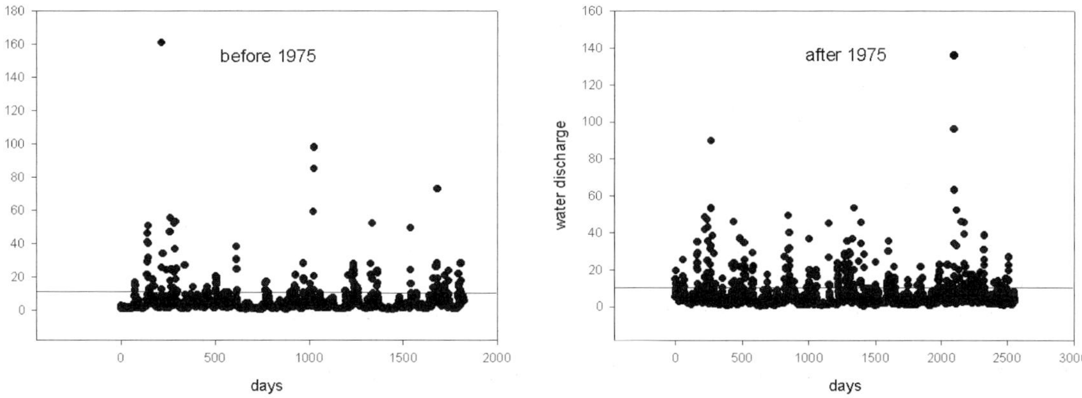

Fig. 3 Hydrology of the Łososina River with the threshold line (the beginning of motion for the sediment) above which the bed-load transport was calculated. The bed load values during those periods were calculated using the TransCalc model.

RECAPITULATION

The following conclusions can be drawn from the analysis of the data:

(a) The most important hydraulic parameter, which determines the value of the decreased shear stresses and bed load transport, is the slope of the river bed which was changed (here: reduced) by the river training works along the Łososina River.

Table 3 Unit bed load transport at the Łososina River—before the training works.

Water depth h (m)	Sampling cross-section 1-1 d_{50} = 28 (mm) Shear stress τ (N m^{-2})	Transport (m^3 s^{-1} mb^{-1})	Sampling cross-section 2-2 d_{50} = 30 (mm) Shear stress τ (N m^{-2})	Transport (m^3 s^{-1} mb^{-1})	Sampling cross-section 3-3 d_{50} = 40 (mm) Shear stress τ (N m^{-2})	Transport (m^3 s^{-1} mb^{-1})	Sampling cross-section 4-4 d_{50} = 30 (mm) Shear stress τ (N m^{-2})	Transport (m^3 s^{-1} mb^{-1})
0.6	No bed load transport							
0.7	21.39	0.00000424						
0.8	23.24	0.0000543						
0.9	28.07	0.0002978	No bed load transport					
1.0	29.28	0.000377			No bed load transport		No bed load transport	
1.1	32.38	0.000606						
1.2	34.67	0.000796						
1.3	43.76	0.001707	25.78	0.0000951				
1.4			28.24	0.0002186	33.47	0.0001043	22.65	0.0000014
1.5			32.28	0.0004829	39.36	0.0004541	27.42	0.0001736
1.6			35.60	0.000747	49.38	0.001345	32.57	0.000505
1.7	Max depth in cross-section 1.3 (m)		41.33	0.001287	52.34	0.001663	38.09	0.00097
1.8			43.98	0.001566	55.38	0.002012	43.98	0.001566
1.9			50.23	0.002295	58.49	0.002393	50.23	0.002295
2.0			53.09	0.00266	61.69	0.002805	53.09	0.00266
2.1			Max. depth in cross-section 2.0 (m)		Max. depth in cross-section 2.0 (m)		Max. depth in cross-section 2.0 (m)	

Table 4 Unit bed load transport at the Łososina River—after the training works.

Water depth h (m)	Sampling cross-section 1-1 d_{50} = 30 (mm) Shear stress τ (N m^{-2})	Transport (m^3 s^{-1} mb^{-1})	Sampling cross-section 2-2 d_{50} = 22 (mm) Shear stress τ (N m^{-2})	Transport (m^3 s^{-1} mb^{-1})	Sampling cross-section 3-3 d_{50} = 35 (mm) Shear stress τ (N m^{-2})	Transport (m^3 s^{-1} mb^{-1})	Sampling cross-section 4-4 d_{50} = 22 (mm) Shear stress τ (N m^{-2})	Transport (m^3 s^{-1} mb^{-1})
0.8	No bed load transport				No bed load transport			
0.9	22.85	0.0000039						
1.0	27.78	0.0001931			28.43	0.0000524	No bed load transport	
1.1	33.03	0.000539	No bed load transport		30.68	0.0001493		
1.2	35.36	0.000727			31.77	0.0002066		
1.3	44.63	0.001637			32.28	0.0002357	17.36	0.0000133
1.4	Max depth in cross-section 1.3 (m)				33.46	0.000307	18.27	0.0000381
1.5					44.77	0.001255	19.21	0.0000714
1.6			17.80	0.0000241	47.54	0.001546	19.95	0.0001022
1.7			24.21	0.0003381	50.38	0.001865	20.94	0.0001485
1.8			26.26	0.00048	53.31	0.002214	21.95	0.0002015
1.9			28.38	0.000645	56.30	0.002591	25.29	0.0004112
2.0			30.57	0.00083	59.38	0.002997	25.72	0.0004418
2.1			32.82	0.001037	Max depth in cross-section 2.0 (m)		Max depth in cross-section 2.0 (m)	
2.2			35.99	0.001353				
2.3			Max depth in cross-section 2.2 (m)					

Table 5 Budget of the unit bed load transport—Łososina River after regulation.

Unit bed load transport (m^3 s^{-1} mb^{-1})							
Depth h (m)	Sampling cross-section 1-1	Depth h (m)	Sampling cross-section 2-2	Depth h (m)	Sampling cross-section 3-3	Depth h (m)	Sampling cross-section 4-4
0.9	0.0002939	1.6	0.0007229	1.5	–0.000800	1.4	–0.0000367
1.0	0.0001839	1.7	0.0009489	1.6	0.000117	1.5	0.0001022
1.1	0.0000670	1.8	0.0010860	1.7	–0.000202	1.6	0.0004028
1.2	0.0000690	1.9	0.0016500	1.8	–0.000202	1.7	0.0008215
1.3	0.0000700	2.0	0.0018300	1.9	–0.000198	1.8	0.0013640
–	–	–	–	2.0	–0.000192	1.9	0.0018838
–	–	–	–	–	–	2.0	0.0022182
Total	0.000683	Total	0.00623	Total	–0.00147	Total	0.006750

(b) The greatest decrease in bed load transport was observed at cross section 4-4. This was the main aim and achievement of the river training works.

(c) The change of the bed load transport along the Łososina River in the research cross sections was as follows: along cross section 1-1 the unit bed load transport was bigger before the river training by about 0.683×10^{-3} m^2 s^{-1} (aggradation after river training works); along cross-section 2-2 the unit bed load transport was also bigger before the river training by about 6.23×10^{-3} m^2 s^{-1} (aggradation after river training works). Along cross-section 3-3 the unit bed load transport is bigger after the river training by about 1.47×10^{-3} m^2 s^{-1} (degradation of the river bed), and finally along cross-section 4-4 the unit bed load transport was bigger before the river training by about $q = 6.75 \times 10^{-3}$ m^2 s^{-1}. Along the whole river the unit bed load transport was larger before the river training by about $q = 6.7 \times 10^{-3}$ m^2 s^{-1}. In other words the river training reduced the bed load transport by that value.

(d) The river training works carried out along the Łososina River changed the bed load transport conditions within the whole river. The bed load transport before the river training works was 7180 t year^{-1} and it became 2279 t year^{-1} after the training works, whereas the suspended load was 70 464 t year^{-1} throughout that whole period. In other words the bed load was reduced from roughly 10% of suspended load down to 3%. It seems the river training works had a great influence on the bed load.

Acknowledgement The authors would like to thanks to the Nowy Sącz Municipality Authority for providing the archive design files and graphs connected with the Łososina River training works. The same thanks to the State Geological Institute in Warsaw and River Water Authority in Krakow and Nowy Sącz for the files and documents they provided.

REFERENCES

Carling, P. A. & Reader, N. A. (1981) Structure, composition and bulk properties of upland stream gravels. *Earth Surf. Processes Landf.* 7, 349–365.

Church, M. A., McLean, J. F. & Wolcot, J. F. (1987) River bed gravels: sampling and analysis. In: *Sediment Transport in Gravel-bed Rivers* (ed. by C. R. Thorne), 43–87. John Wiley & Sons, London, UK.

Department of Water Engineering (DWE) (2003) *TransCalc– a simple model for suspended sediment transport under the flooding condition.* Agricultural University of Kracow, Kracow, Poland.

Meyer-Peter, E. & Mueller R. (1948) Formulas for bedload transport. In: *Proc. of II Congress IAHR*, 39–64. Stockholm, Sweden.

Michalik, A. (1990) Badania intensywności transportu rumowiska wleczonego w rzekach karpackich (Bed-load transport investigations in some Polish Carpathians rivers) (in Polish). *Zesz. Nauk. AR Kraków seria Rozpr. Hab.* **138**, 115.

Parker, G. (1990) Surface-based bedload transport relation for gravel rivers. *J. Hydraul. Res.* **28**, 417–434.

Przewłocki, S. (2000) *Geodezja dla inżynierii środowiska* (Land survey for environmental engineers). PWN, Warszawa, Poland.

Punzet, J. (1972) *Empiryczne wyznaczenie przepływów maksymalnych o określonym prawdopodobieństwie pojawienia się w zlewniach karpackich dopływów* Wisły Empirical determination of maximum discharges in the Carpathian basin of the Vistula River) (in Polish). PIHM, Warszawa, Poland.

Punzet, J. (1981) *Empiryczne systemy oceny charakterystycznych przepływów rzek i potoków w karpackiej części dorzecza Wisły* (Empirical stream and river discharge assesment systems in Carpathian basin of the Vistula River). PIHM, Warszawa, Poland (in Polish)

Radecki-Pawlik, A. (1995) WODA_2000 - v. 2. 0 - A Simple Hydrological computer model to calculate the *t*-year flood. In: *Hydrological Processes in the Catchment* (ed. by B. Więzik) (Proc. Int. Conf. Kracow, Poland), 131–141. Institute of Water Engineering and Water Management, Kracow University of Technology, Kracow, Poland.

Radecki-Pawlik, A. & Baran P. (2000) Zastosowanie równania Parkera do obliczania intensywności transportu rumowiska wleczonego dla cieków podkarpackich (Using the Parker equation for bed-load transport calculations) (in Polish). *Zesz. Nauk. AR w Krakowie* **20**, 163–177.

Radecki-Pawlik, A. & Radecki-Pawlik, B. |(2003) SHEAR_v.1.0—shear stress calculations computer model. Dept. of Water Engineering, Agricultural University of Kraków, Kracow, Poland.

Selby, M. (1985) *Earth's Changing Surface. An Introduction to Geomorphology*. Oxford University Press—Clarendon Press, New York, USA.

Yang, C. (1996) *Sediment Transport—Theory and Practice*. McGraw-Hill, New York, USA.

Wallingford HR (1996) *SandCalc1_1—sediment transport computer model*. Wallingford, UK.

Sediment Transfer through the Fluvial System (Proceedings of a symposium held in Moscow, August 2004).
IAHS Publ. 288, 2004

Sedimentological assessment of the Tucuruí Reservoir (Tocantins River, Brazil)

NEWTON DE OLIVEIRA CARVALHO[1], ANTÔNIO RAIMUNDO SANTOS RIBEIRO COIMBRA[2], BRUNO LEONEL PAYOLLA[2], TARCÍSIO LUIZ COELHO DE CASTRO[3] & ANDERSON BRAGA MENDES[4]

1 *Sedimentology and Water Resources Adviser, Rua Conde de Baependi, 112 ap. 904, Flamengo, Cep 22231-140, Rio de Janeiro, RJ; Brazil*
newtonoc@openlink.com.br

2 *Centrais Elétricas do Norte do Brasil, ELETRONORTE. SCN, Quadra 6, Bloco, C Sala 501, Ed. Venencio 3000, Cep 70716-900, Brasília, DF, Brazil*

3 *Engevix Engenharia S/A, SCN Q. 04, Bloco B, 13° andar, Pétala D, Centro Empresarial Varig, Brazil*

4 *Engevix Engenharia S/A, Rua José Rodrigues dos Santos, 117, Bela Vista, Duque de Caxias, RJ, Brazil*

Abstract The results of a study on the sedimentation in Tucuruí Reservoir, located in the lower reaches of the Tocantins River, Brazil, are presented. Morphological information on reservoir sediment deposits and sediment transport data collected since dam closure in 1984 were utilized. These data were used to construct an empiricial model of reservoir sedimentation that allowed an assessment of the impact of land use change upstream Tucuruí Dam. Results of the model show there is not any short or medium-term problem regarding the development of sediment deposits at the water intake sill of the reservoir. Future work on topographic surveys in the river and sedimentological studies in the reservoir are recommended.

Key words Brazil; deposit heights at dam toe; erosion; reservoir; sediment; sediment distribution; Tocantins River; useful life

INTRODUCTION

Reservoir sedimentation is an important issue facing many water resource managers. Assessment of reservoir sedimentation requires information on discharges of sediment to the dam, calculation of sedimentation rates within the reservoir, and therefore computation of the time to reach the sill of water intakes—a computation of the useful life of water supply reservoirs. There are many large dams in Brazil and this paper reports on a study of sedimentation in the Tucuruí Reservoir, one of the largest artificial lakes in Brazil.

PRELIMINARY COMPUTATION

The Tucuruí Reservoir is located in the lower reaches of the Tocantins River, downstream from its confluence with the Araguaia River. It has a drainage area of 758 000 km^2, which represents 98.8% of the Tocantins River catchment. Thus, the dam has the capacity to trap most of the sediment load of the Tocantins River. The Tucuruí Reservoir began filling in September 1984 and since then there have been other reservoirs built upstream, namely: Serra da Mesa built in 1996 and Lajeado built in 2001, on the Tocantins River. Several other dams are planned for construction. All these structures will have an effect on the sediment load to the main Tucuruí Reservoir.

The first stage of the project was to evaluate the total sediment load to the reservoir. This involved the construction of sediment transport algorithms that related sediment discharge to river flows and therefore enabled the calculation of long term sediment loads to the reservoir. These data were then used to determine rates of sedimentation in the reservoir. The fundamental equations for reservoir sedimentation were:

$$S = D_{st}\, E_r / \gamma_{ap} = 356\, Q_{st}\, E_r / \gamma_{ap} \quad \text{and} \quad T = V_{res} / S \qquad (1)$$

where S is sediment volume retained by the reservoir (m^3 year^{-1}); D_{st} is annual average sediment discharge to the reservoir (t year^{-1}); E_r is sediment trapping efficiency in a reservoir (dimensionless); γ_{ap} is the gravity weight of the deposits (t m^{-3}); Q_{st} is average sediment discharge from the reservoir (t year^{-1}); T is sedimentation time (years); V_{res} is reservoir capacity (m^3).

It is pertinent to note that Q_{st} , D_{st} , E_r and γ_{ap} vary with time and sediment yields will also change over time because of increases in erosion potential in the reservoir catchment. The sediment trapping efficiency of the reservoir will decrease as sediment deposits increase and the gravity weight of the sediment deposits changes as a result of its compaction overtime. As the deposits become more significant, V_{res} decreases. The sediment trapping efficiency of the reservoir (E_r) was obtained through the Brune curve, whilst the gravity weight of the sediment deposits was computed via the Lara and Pemberton procedure (see Strand, 1974; ICOLD, 1989; Carvalho, 1994, 2000).

INCREASE IN THE SEDIMENT DISCHARGE

There has been significant land use changes in the Tocantins–Araguaia catchment associated with population increases between 3.5 and 8% per year. Land use changes have included increases in agricultural areas, deforestation, road and general construction. This has all contributed to a substantial increase in sediment yield from the catchment and subsequent increases in sediment discharges in the receiving rivers. Anecdotal evidence suggests commensurate in-channel and reservoir sedimentation causing drawbacks (severe flow events, etc). The rainfall runoff coefficients have increased over a longer time period at five out of seven monitoring stations in the catchment. It was also observed that the river flow at the dam site became greater, with the long-term average discharge between 1931 and 2000 increasing from 3.62 to 250 m^3 s^{-1}.

For this study, sediment discharge data for three stations in the catchment, with a record of about 20 years, were used. These data were assembled in 5-year periods, and from the flow and average yearly sediment discharge data, mass curves were traced (Fig. 1). Sediment yield rates were computed by using the angular coefficients of the straight lines. The coefficient of the first line of the mass curve provides r_1, whereas the second line gives r_2. E_c represents the increase/decrease of the phenomenon within a period, whereas R means the annual rate, according to the following equations:

$$E_c = (r_1 - r_2) / r_1 \quad \text{and} \quad (1 + R)^n = 1 + E_c \qquad (2)$$

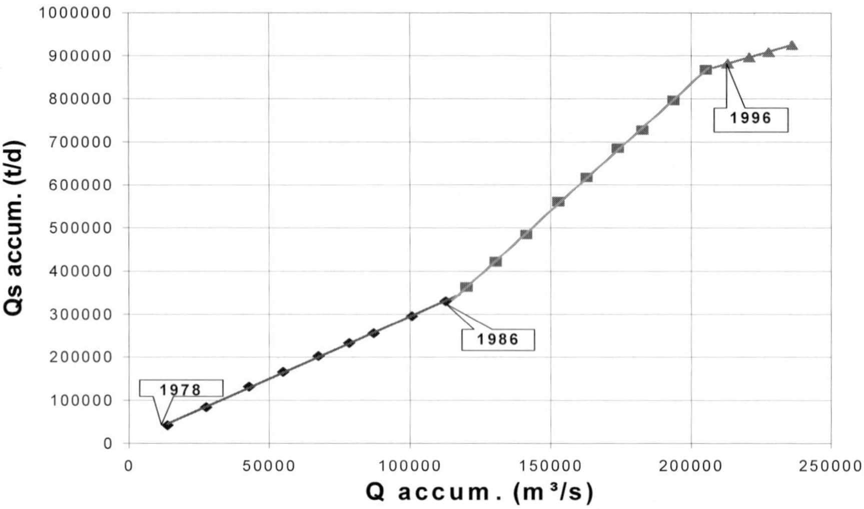

Fig. 1 Mass curve for the period between 1978–1999 (Tocantins River at Marabá station).

Table 1 Sediment discharge increase rate at stations on the Araguaia and Tocantins Rivers.

Code	Station	Period	Sediment yield rate (*R*) per year (%)	Drainage area (km^2)
2410 0000	Araguaia River at Cachoeira Grande	1977–1986	2.14	4504
		1982–1989	6.77	
		1987–1991	–10.95	
		1977–1991	1.04	
2905 0000	Tocantins River at Marabá	1978–1995	3.44	690 920
		1996–1999	2.85	
2910 0000	Itacaiúnas River at Fazenda Alegria	1979–1994	2.59	37 600

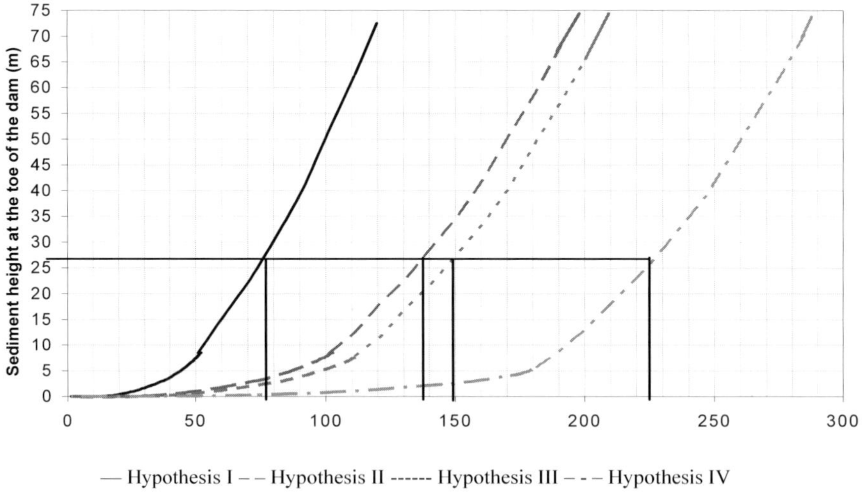

Fig. 2 Useful life of the plant for the four hypotheses. The horizontal line indicates the elevation of the sill of the water intake (27 m).

The results of the studies made for the three stations are presented in Table 1. The resultant reservoir sedimentation was assessed for four different scenarios—hypothesis I, not considering other reservoirs upstream; hypothesis II, considering the construction of Serra da Mesa dam upstream; hypothesis III, considering the existence of Serra da Mesa and Lajeado dams upstream; and hypothesis IV, considering Serra da Mesa, Lajeado, Santa Isabel and Serra Quebrada dams upstream Tucuruí dam (Fig. 2). From these scenarios it appears that there will be neither short nor medium-term sediment problems in the Tucuruí Reservoir if the sediment yield remains unchanged or presents the same rate through the years. However, it is necessary that sediment studies are made approximately every 10 years. Such studies include sediment discharge sampling, re-evaluation of the phenomena, topographic surveys, study on erosive processes of river banks and downstream from the dam, besides other studies aiming to verify the validation of the scenarios exhibited here.

REFERENCES

Carvalho, N. O., Jъnior, Naziano P. F., dos Santos, Paulo M. C., Lima & Jorge E. F. W. (2000) *Guia de Avaliação de Assoreamento de Reservatórios* (Reservoir Sedimentation Assessment Guideline). ANEEL/OMM/PNUD. Brasília, DF, Brazil.

Carvalho, N. O. (1994) *Hidrossedimentologia Prática* (Practical Hydro-sedimentology). CPRM & ELETROBRÁS, Rio de Janeiro, Brazil.

ENGEVIX-THEMAG, Consortium (2001) *Estudos Hidrossedimentológicos e Batimétricos no Reservatório da UHE Tucuruí—Relatório Final* (Hydro-sedimentological Studies on Tucuruí Plant Reservoir—Final Report). ELETRONORTE, Brasília, Brazil.

ICOLD (1989) *Sedimentation Control of Reservoirs*. ICOLD, Paris, France.

Strand, R. (1974) *Sedimentation. Appendix on Design of Small Dams*. Bureau of Reclamation, Washington DC, USA.

4 Modelling of Erosion and Deposition Processes

Soil erosion at the mesoscale: comparison of two erosion models for a pre-alpine Austrian basin

G. WOLKERSTORFER & P. STRAUSS
Federal Agency for Water Management, Institute for Land and Water Management Research, Pollnbergstrasse 1, A-3252 Petzenkirchen, Austria
peter.strauss@baw.at

Abstract In an attempt to get detailed information about amounts and spatial extents of soil erosion we conducted a study on sediment and water loads for the Ybbs River basin (1100 km^2), located in the pre-Alpine area of lower Austria. As the spatial validation of soil erosion and sediment yield at the mesoscale is almost impossible, we tried to gain knowledge about probable risk areas by application of completely different erosion models (MUSLE and MMF). We tried to evaluate whether they lead to a different pattern of risk areas and if they are comparable in terms of absolute values of soil loss. Measured flows were used to calibrate the two different erosion models in four sub-basins with markedly different land use. Differences in model results could be attributed to different methods of spatial aggregation. Both models overestimated sediment delivery to the river. Unrealistic parameter values for calculating transport capacity had to be used for calibration of sediment yields.

Key words drainage basin; mesoscale; MMF; soil erosion model; SWAT

INTRODUCTION

Non-point source pollution has become a serious concern in recent years. It is estimated, that 64–89% of the total nitrogen load and 41–80% of the total phosphorus load of the Danube River basin can be attributed to diffuse sources (Schreiber *et al.*, 2003). Phosphorus is a limiting factor for eutrophication in many inland rivers. As the main pathway of phosphorus transport into aquatic ecosystems is by erosion, approaches for better land management policy should include erosion models as a basis for sediment load estimation. To increase the understanding of processes and to improve the quantification of related fluxes at the mesoscale, the project *Nutrients Management in the Black Sea and its Impact on the Black Sea* (daNUbs) was launched.

Modelling of soil erosion for large basins is complicated by the fact that data availability is usually very limited and a spatially distributed validation is practically impossible. Generally, data requirements increase with the size of the drainage basin and on that score the accuracy of the model results decreases. A second difficulty of applying models is the level of uncertainty surrounding the model output. This arises from various sources such as difficulties during parameterization of the model; numerical errors; conceptual errors (Konikow & Bredehoeft, 1992); or simply the fact that modelling approaches are different. Hence, different models used for the same basin may produce different results (Svorin, 2003). Therefore, as a first step in the daNUbs project two soil erosion models with a completely different structure (SWAT and MMF) were compared in order to determine differences in modelling and results. Our major interest was whether they produce similar spatial patterns in terms of soil erosion rates and sediment yield into the river.

Table 1 Comparison of the different approaches of the SWAT and MMF models as used here.

Processes and key factors	SWAT	MMF
Surface runoff	Curve number (Mockus, 1972)	Daily rainfall exceeds soil moisture storage capacity (Kirkby, 1976)
Soil erosion	MUSLE (Williams & Berndt, 1972)	Total soil loss is compared to transport capacity (Meyer & Wischmeier, 1969)
Spatial disaggregation	Sub-basins ~15 km^2	Raster cells a 625 m^2
Connection between spatial units	No connection	Flow method of steepest ascend (Jenson & Domingue, 1988)

MODEL DESCRIPTION

Erosion rates were modelled using two erosion models, the MUSLE (Williams & Berndt, 1977) integrated into the SWAT model (Arnold *et al.*, 1998) and the MMF model (Morgan, 2001), incorporated into PCRaster (Wolkerstorfer, 2002). Because both models are already incorporated into GIS systems and require relatively few data, they seemed suitable for application to large basins. However, they completely differ in terms of their structure. Table 1 gives an overview of how the main processes in the two models are treated. Due to the different treatment of the main processes, it is necessary to use different input parameters. To describe the influence of plant cover on soil erosion, MUSLE for instance, uses the C-factor of the USLE (Wischmeier & Smith, 1978) while MMF requires canopy cover, ground cover, plant height and the USLE C-factor.

In general, there are two main methods of spatial disaggregation for the pre-processing procedure: either the use of a regular grid or the subdivision of a drainage basin into sub-areas or classes of sub-area that are assumed to be homogeneous in their hydrological response. In SWAT, a river basin may be partitioned into a number of sub-basins wherein the dominant land use and soil are used as a unique and homogenous value for the basin leading to a single result per sub-basin. MMF is implemented in a raster GIS wherein input parameter values and soil loss are calculated for each grid. Grid results are routed according to the topographic structure using the method of steepest ascent as described by Jenson & Domingue (1988).

THE CASE STUDY REGION

The River Ybbs is a tributary of the River Danube. The investigated area belongs to the northern limestone pre-Alpine area of Austria. Elevation ranges from 250 m to 1800 m a.s.l. Due to these differences in elevation climatic conditions are highly variable with mean annual precipitation between 650 mm and 2000 mm in the north and south of the region, respectively. Land use follows the pattern of precipitation with almost only forested land in the alpine area to intensively used agricultural land in the northern part of the Ybbs River basin (Fig. 1).

The databases used in this study consist of pre-existing maps and remote sensing data with a resolution of 25 m, field measurements and already existing data (Table 2).

CALIBRATION AND VALIDATION

As a basis for calculation and calibration of surface runoff the knowledge of the regional water balance is essential. Water balance calculations for the Ybbs River basin have been carried out by IHGW (2003) using the water balance model Difga2000 (Schwarze, 2001).

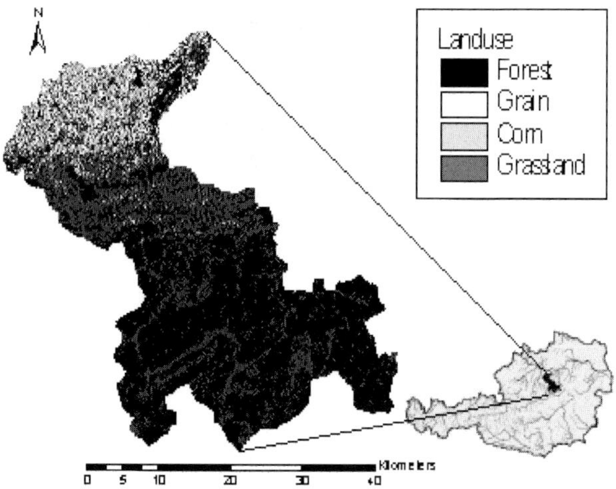

Characteristics of the study area*		
Basin area	km^2	1117
Precipitation	mm a^{-1}	1377
Slope	°	17
River discharge	mm a^{-1}	851
Share of arable land	%	16
Overland flow	%	36
Sediment load	kg ha^{-1}a^{-1}	610

* given values are long term means for the whole basin

Fig. 1 Land use in the Ybbs River basin.

Table 2 Origin and quality of data used for this study.

Data	Resolution	Source
Digital elevation model	25 m	Federal Office of Metrology and Surveying
Land use	30 m	Landsat-7 ETM+
Soil	1 : 25 000 m	Strauss & Wolkerstorfer (2004)
Climatic data	16 gauging stations (daily)	Hydrological Service NÖ
River measurements—flow	5 gauging stations (daily)	Hydrological Service NÖ
River measurements—sediment	3 gauging stations for flow proportional sampling	

This enabled separation of the total flow into slow groundwater flow, fast groundwater flow and direct flow. Direct flow is the surface or subsurface flow in the unsaturated zone. Direct flow rates were then used for the calibration of runoff for both soil erosion models. The drainage basin of the River Ybbs was divided into four sub-basins representing different land-use management areas. These monitoring points were used for calibration. In addition, river basin outlet data were used to validate calibrated results. An automatic calibration tool (van Griensven, 2002) could be applied to the SWAT model. Further details of the calibration for SWAT are described in IHGW (2003). Table 3 gives the calibration results for SWAT and MMF. The SWAT model overpredicted the mean flow conditions for most of the sub-basins. The reason for this is insufficient modelling of runoff events caused mainly by the lateral flow of the faster groundwater runoff (Schilling, 2003). Compared to mean flow conditions, low flow and high flow conditions are reproduced better.

The main parameter for calibrating the MMF model was the parameter "effective hydrological depth". The correlation coefficient between measured and predicted data for MMF is 0.93 which indicates that the model predicts values for surface runoff reasonably well. The different land-use management in the sub-areas is represented especially well. However, mean surface flow at the Ybbs basin outlet is overpredicted. Table 4 shows a comparison of values proposed in the original paper and those obtained after calibration. These values differ hugely. The Ybbs River basin is characterized by a strong gradient in

Table 3 Runoff calibration results for the different sub-basins (north to south) and the main outlet of the River Ybbs, percentage of total river discharge; simulation period 1991–1997.

Surface runoff (%)	River Ybbs Main outlet	Sub-basins Krenstetten Arable	Ybbsitz Grassland	Opponitz Forested area	Lunzois Forested area
Baseflow separation Difga	28.6	32.5	22.9	29.4	31.2
Soil erosion model MMF	32.0	32.5	22.4	29.4	31.5
Soil erosion model SWAT	37.0	56.0	39.0	37.0	23.0

Table 4 Calibration parameter "effective hydrological depth".

Vegetation	Values given by Morgan (2000) (m)	Calibrated Values (m)
Row crops	0.12	0.45
Mature forest	0.20	0.02
Cultivated grass	0.12	0.024

climate, land use, slope and geomorphology from south to north. Water flow follows this pattern. Therefore, the southern parts of the basin, which may be characterized as alpine areas, exhibit high water flow rates but the land is almost exclusively covered with forest or grassland. Direct flow rates given by Difga (2000) also include the quick subsurface flow in the unsaturated zone and this flow path is of particular importance for alpine areas with steep slopes and shallow soils. On the other hand, MMF deals only with surface runoff leading to an incompatibility between model structures. However, the consequences of these high "surface" flow rates for erosion estimation are less than expected due to the dense ground cover of these areas.

MODEL COMPARISON

After calibration of surface runoff we calculated soil erosion rates using both models. Best guess estimates for the different parameters were used, based on different sources of information (model proposals, measured values, literature). To make the models comparable, the results of MMF were averaged on the same sub-basin level (73 sub-basins based on the geomorphological characteristics of the study area) as used by SWAT. The results of this comparison of MMF and MUSLE for the period 1991–1997 (Fig. 2) indicate a general agreement on soil loss risk estimation. The results of both models reflect the land-use pattern with low erosion rates in the alpine areas and higher erosion rates in areas with more intense agricultural land use. However, SWAT exhibits a tendency to estimate higher soil losses compared to MMF for those sub-basins with a higher soil loss risk. This can be confirmed by the slope value of the linear regression between the results of both models which is 0.6 (1 indicates perfect agreement). This compares well with results for the USLE that demonstrate a general overestimation of model predictions at higher soil loss risks (Risse *et al.*, 1993; Strauss & Klaghofer, 2004). For particular sub-basins, considerable variation in results between the models occurs. For areas with low erosion rates calculated by SWAT and higher erosion rates calculated by MMF, this may be explained by the fact that SWAT calculates single input values for each sub-basin. In heterogeneous areas with very different land use intensities, SWAT uses those land-use parameters with the greatest spatial extension. Therefore, small areas with a high erosion risk may be neglected, whereas MMF uses all grid values of a sub-basin for calculation of average soil loss.

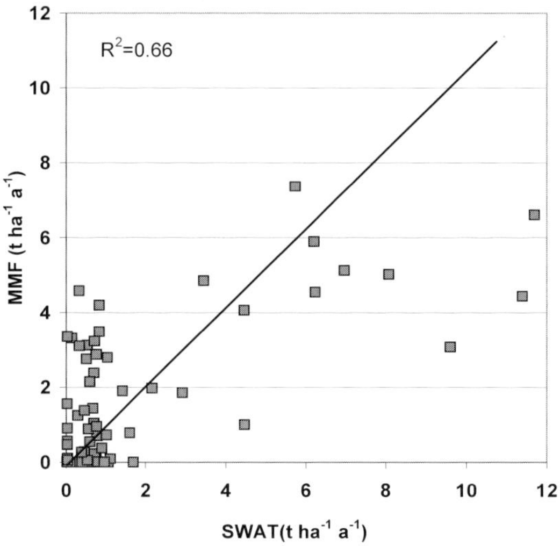

Fig. 2 Comparison of calculated soil loss (t ha^{-1} year^{-1}) using SWAT and MMF for 73 sub-basins of the Ybbs River basin.

A further model comparison was to calculate sediment concentrations by dividing soil loss of the different sub-basins by surface runoff. This resulted in a better correlation between both models (R^2 = 0.71). In addition, the slope value of the regression between MMF and SWAT was not different from 1.

In a second evaluation step we compared calculated soil loss rates to sediment loads measured at the outlet of three sub-basins. Table 5 demonstrates major differences between results calculated with erosion models and measured sediment yields especially in sub-basins with dominantly agricultural land use. Usually, the differences between *on land* erosion rates and *in river* sediment loads are taken into consideration by using sediment delivery ratios. In the case of SWAT, it is stated that due to the inclusion of an explicit runoff term into the erosion equations, delivery ratios are not required and calculated soil losses are equal to sediment input into the river (Arnold *et al.*, 1998). The huge differences therefore could only be explained by retention in the river itself. However, field investigation in the Ybbs River basin did not confirm such large amounts of retention. We therefore conclude that redistribution of soil inside the sub-basins constitutes the majority of soil erosion. This is confirmed by work of Martínez-Casanovas *et al.* (2001) who found soil loss retention of more than 50% already at the field scale, and Strauss & Peinsitt (2002) who mapped soil redistribution rates of a small basin of more than 800 t compared to sediment losses of about 20 t leaving the same basin.

Table 5 Comparison of sediment yields at three river gauging points representing different land use, measured and calculated in t ha^{-1} year^{-1}.

	Measured sediment yield	SWAT predicted	MMF predicted
Opponitz (mainly forested)	0.4	0.5	0.5
Krenstetten (arable land)	0.4	6.2	5.0
Greimpersdorf (river outlet)	0.7	2.7	1.8

To take the retention in the field and in the river into account, different ways of modelling exist: a first attempt was to calibrate MMF for sediment yield of the Ybbs River at the three sub-basins with measured sediment concentrations. Evaluation of the input parameters which are responsible for the soil loss calculation has shown that the transport capacity equation is the limiting process in the soil loss calculation. Transport capacity in MMF is determined by slope angle, surface runoff and land-use cover (C-factor in the USLE). The transport capacity equation appears to be most sensitive to slope angle. Due to this, sediment yields were too high from gentle slopes. As surface runoff has already been calibrated and slope angle cannot be changed, the only parameter to calibrate transport capacity is land-use cover. Calibrating MMF resulted in modelled sediment yields which were similar to measured values. However, the necessary changes of the land-use cover lead to unrealistic values for the C-factor. It was for instance necessary to change the C-factor values for corn from 0.43 to 0.01. Similarly, values for cereals had to be changed from 0.1 to 0.005.

A second possible way of taking the difference between measured sediment yield in the river and predicted soil erosion into account is to develop empirical sediment enrichment ratios. This will be the next step of evaluation.

Acknowledgements We are grateful to the European Commission which funded this work within the daNUBS project (EVK1-CT-2000-00051).

REFERENCES

Arnold, J. G., Srinivasan, R., Muttiah, R. S. & Williams, F. R. (1998) Large area hydrologic modelling and assessment. Part I: model development. *J. Am. Water Resour. Assoc.* **34**, 73–89.

IHGW (2003) *Water balance calculation for the case study regions in Austria, Hungary and Romania*. Deliverable D 1.1, Institute of Hydraulics, Hydrology and Water Resources Management, TU Vienna, Austria.

Jenson, S. K. & Domingue, J. O. (1988) Extracting topographic structure from digital elevation data for geographic information system analysis. *Photogram. Engng Remote Sens.* **54**(11), 1593–1600.

Kirkby, M. J. (1976) Hydrological slope models; the influence of climate. In: *Geomorphology and Climate* (ed. by E. Derbyshire), 247–267. Wiley, Chichester, UK.

Konikow, L. F. & Bredehoeft, J. D. (1992) Ground-water models cannot be validated. *Adv. Water Resour.* **15**, 75–83.

Martínez-Casasnovas, J. A., Ramos, M. C. & Ribes-Dasi, M. (2001) Soil erosion caused by extreme rainfall events: mapping and quantification in agricultural plots from very detailed digital elevation models. *Geoderma* **105**, 125–140.

Meyer, L. D. & Wischmeier, W. H. (1969) Mathematical simulation of the process of soil erosion by water. *Trans. ASAE* **12**, 754–758.

Morgan, R. P. C. (2001) A simple approach to soil loss prediction: a revised Morgan-Morgan-Finney model. *Catena* **44**, 305–322.

Mockus, V. (1972) Estimation of direct runoff from storm rainfall. In: *National Engineering Handbook*, 10.110.22. US Dept. Agric. Soil Conserv. Series. Washington DC, USA.

Risse, L. M., Nearing, M. A., Nicks, A. D. & Laflen, J. M. (1993) Error assessment in the Universal Soil Loss Equation. *Soil Sci. Soc. Am. J.* **57**, 825–833.

Schreiber, H. L. Th., Constantinescu, I., Cvitanic, D., Drumea, D., Jabucar, S., Juran, B. Pataki, S., Snishko, S., Zessner, M. & Behrendt, H. (2003) Harmonised inventory of point and diffuse emissions of nitrogen and phosphorus for a transboundary river basin. *Research Report* 200 22 232, Federal Environmental Agency, Berlin, Germany.

Schwarze, R. (2001) *Methodological Fundamentals of DIFGA*. Dresden, Germany.

Strauss, P. & Klaghofer, E. (2004) Scale considerations for the estimation of processes and effects of soil erosion in Austria. In: *Proc. of the OECD Expert meeting on Soil Erosion and Soil Biodiversity Indicators* (Rome, March 2003) (in press).

Strauss, P. & Peinsitt, A. (2002) Erosive rain in March 2002 and its consequences on two agricultural based microcatchments. *Tagungsband ALVA*, 259–261.

Strauss, P. & Wolkerstorfer, G. (2004) Erosion risk for a watershed—Database and comparison of two erosion models. *Mitt. d. Österr. Bodenkundl. Ges.* (in press).

Svorin, J. (2003) A test of three soil erosion models incorporated into a geographical information system. *Hydrol. Processes* **17**, 967–977.

van Griensven, A. & Francos, A. (2002) Sensitivity analysis and auto-calibration of an integral dynamic model for river water quality. *Water Sci. Technol.* **45**(5), 325–332.

Williams, J. R. & Berndt, H. D. (1977) Sediment yield prediction based on watershed hydrology. *Trans. ASAE* **20**(6), 1100–1104.

Wischmeier, W. H. & Smith, D. D. (1978) *Predicting Rainfall Erosion Losses*. USDA Handbook no. 537.

Wolkerstorfer, G. (2002) Comparison of two soil erosion models, Morgan-Morgan-Finney MMF and SWAT, for the Ybbs watershed. Master Thesis, University Vienna, Vienna, Austria.

Probability distribution function approach in stochastic modelling of soil erosion

ALEKSEY SIDORCHUK[1], ALISTAIR SMITH[2] & VLADIMIR NIKORA[2]

1 *Landcare Research, Private Bag 11052, Palmerston North, New Zealand*
sidorchuka@landcareresearch.co.nz

2 *National Institute of Water and Atmospheric Research (NIWA), PO Box 8602, Christchurch, New Zealand*

Abstract Stochastic modelling of soil erosion is based on calculation of the probability of soil particle detachment, which is the probability of excess of driving forces above resistance forces. These probability calculations require the probability distribution functions (PDFs) for the main hydrodynamic and soil structure characteristics, estimated experimentally or theoretically. The field of hydrodynamic forces (flow velocities and pressure distribution though space and time) is calculated with Large Eddy Simulation. Soil structure is estimated in terms of Kolmogorov's probabilistic approach to soil failure and aggregation. The PDF approach explicitly describes the process of soil erosion and gives a theoretical explanation of the great diversity in empirical relationships between erosion rate and main erosion factors.

Key words large eddy simulation; probability of detachment; probabilistic soil failure; soil erosion; stochastic modelling

INTRODUCTION

In spite of its major significance for strategic estimates and predictions related to many aspects of human activity, water erosion theory for cohesive soils is still largely undeveloped. For many years, efforts have mainly focused on the development of the empirical predictive relationships, based on data collected in areas with different climatic and land-use conditions (Merritt *et al.*, 2003). The most successful example is the so-called Universal Soil Loss Equation (Wischmeier & Smith, 1965). More recently, erosion models that address causative aspects have appeared, providing strong competition for the purely empirical models. An important step in this development was a paper by Foster & Meyer (1972), in which the sediment-budget approach to erosion modelling was suggested and developed. However, these models are still semi-theoretical or semi-empirical, as simplified stream power (or bed shear stress) relationships are used to describe such complicated phenomenon as the rate of erosion, while the whole complexity of soil resistance to erosion is expressed by simplistic erodibility coefficients.

Purely empirical and semi-empirical models do not promise much progress in soil erosion predictions and simulations. A new generation of theoretical erosion models is urgently needed that can account for the stochastic nature of soil erosion, based on mechanistic representations of the key physical processes. Recent achievements in deterministic-stochastic hydrodynamics of shallow rough-bed flows (Nikora *et al.*, 2001) make the development of such an approach feasible. Here we present a stochastic concept first, and then describe potential modelling approaches, which should provide necessary parameterization for bulk stochastic models and also give a deeper insight into erosion processes.

STOCHASTIC CONCEPT IN SOIL EROSION MODELLING

The rate of soil erosion can be estimated in two main ways (Sidorchuk, 2004): by multiplication of sediment concentration by soil particle vertical velocity (velocity–concentration); and by spatiotemporal averaging of unstable sediment particle volume on the time period for detachment (double–averaging). Within the first approach, two stochastic variables are required to calculate the rate of soil aggregates detachment *DER*: unstable aggregate concentration in the bed surface layer C_Δ and mean vertical velocity of unstable aggregates $U_\uparrow$:

$$DER = C_\Delta U_\uparrow \tag{1}$$

Bed concentration of unstable soil aggregates of a given size

The bed concentration of unstable aggregates is the ratio of the volume V_u of unstable aggregates and the whole volume V of the bed surface layer: $C_\Delta = V_u/V$. The volume of unstable aggregates can be written as the product of the number of unstable aggregates N and the mean unstable aggregate volume V_a:$V_u = NV_a$. The volume of aggregates in a surface layer can be presented as the product of the number of aggregates M, exposed to the flow on the unit area, and their mean volume V_{sm}:$V = MV_{sm}$. Therefore the concentration of unstable aggregates is:

$$C_\Delta = NV_a/MV_{sm} \tag{2}$$

The ratio N/M is the probability (P_{DER}) of soil aggregate detachment, and the ratio V_a/V_{sm} is a measure k_D of those soil aggregates' relative size. Therefore:

$$C_\Delta = k_D P_{DER} \tag{3}$$

An equation of this type was proposed by H. Einstein (1937), and is of main significance in the stochastic approach to erosion calculation. As sediment concentration appears to be proportional to the probability of detachment, the main goal of a stochastic methodology in soil erosion is to estimate this probability. The main method is to find the parameters of the probabilistic field of driving and resistance forces. Then, the probability of soil aggregate detachment can be found with the use of the condition of soil aggregate instability on the flow bed.

Soil aggregate instability

Soil aggregate detachment occurs because driving hydrodynamic forces exceed gravitational, hydrodynamic and geo-mechanical resistance and stabilizing forces. The main driving forces are form drag force (F_{fd}), wave drag force (F_{wd}), lift force (F_l), negative turbulent dynamic pressure (F_{dp}), pore water pressure (F_{pw}), and tangent component of submerged weight (F_{wt}). Resistance and stabilizing forces are normal components of submerged weight (F_{wn}), static pressure (F_{sp}), and positive turbulent dynamic pressure (F_{dp}). Mirtskhoulava (1988) and Lawrence (2000) showed that:

$$F_{fd} = C_R \rho S_d \frac{U^2}{2} \tag{4}$$

$$F_{wd} = \rho S_d C_{Rw} \frac{k_e D U^2}{2d} \qquad (5)$$

$$F_l = C_y \rho S_a \frac{U^2}{2} \qquad (6)$$

$$F_{dp} = 3.5\lambda\rho S_b \frac{U_m^2}{2} \qquad (7)$$

$$F_{pw} = g\rho S_p z_p \qquad (8)$$

$$F_{wt} = V_a(\rho_s - \rho) g \sin\beta \qquad (9)$$

$$F_{wn} = V_a(\rho_s - \rho) g \cos\beta \qquad (10)$$

$$F_{sp} = g\rho S_b d \qquad (11)$$

where C_R is the coefficient of drag resistance; C_y is the coefficient of uplift; C_s is the coefficient of static drag; C_{Rw} is the coefficient of wave drag; U is the actual near-bed flow velocity, and U_m is its mean (time averaged) value; λ is the coefficient of hydraulic resistance; S_d is the cross-sectional area of the soil aggregate, perpendicular to the flow; V_a is the volume of the soil aggregate; S_a is the cross-sectional area of the soil aggregate, parallel to the flow (vertical projection); S_b is the area of the soil aggregate that is attached to other aggregates; S_p is the area of pores; D is the aggregate diameter; z_p is capillary pressure height; β is the angle of flow bed local inclination; k_e is the exposure of a soil aggregate and d is water depth.

Finally, there is a complex system of geo-mechanical and electro-chemical forces, defined as soil cohesion (F_c). This is a reactive force; its magnitude and direction are determined by the sum of all the above-listed active forces. Its maximum magnitude is:

$$F_c = C_0 S_b \qquad (12)$$

where C_0 is soil cohesion.

Detachment occurs when the sum of the driving forces is larger than the sum of the resistance forces. For simplification, only normal components of the forces are analysed further. We define $\Theta_\uparrow$ as the inertial force that results from the force balance, normalized by $1/2\rho S_a C_y$.

$$\Theta_\uparrow = U^2 + k_{pw} z_p \frac{S_b}{S_a} \mp k_{dp} \lambda \frac{S_b}{S_a} U_m^2 - k_w D \frac{(\rho_s - \rho)}{\rho} - k_{sp} d \frac{S_b}{S_a} - k_c \frac{C_0}{\rho} \frac{S_b}{S_a} > 0 \qquad (13)$$

Driving and resistance forces are stochastic variables and, consequently, the function $\Theta_\uparrow$—the condition of instability—has some stochastic distribution (within a spatial/temporal "window" at the flow bed surface) with the PDF p_Θ. The PDF of the function of stochastic variables can usually be calculated when the PDFs of those stochastic variables are defined.

The probability of the detachment of the aggregate P_{DER} is the sum of p_Θ for all positive values of $\Theta_\uparrow$:

$$P_{DER} = \int_0^\infty p_\Theta d\Theta \qquad (14)$$

The vertical velocity of soil aggregates

The vertical velocity of soil aggregates is the second component of the formula (1) for the detachment rate calculation. The acceleration along the vertical co-ordinate z at the moment of an aggregate detachment can be derived from the second Newton law, written for the normal component of forces (see equation 13) and aggregate acceleration:

$$V_a \frac{\rho_s}{2} \frac{\partial U_\uparrow^2}{\partial z} = \frac{1}{2} \rho S_a C_y \Theta_\uparrow \tag{15}$$

In a bed layer with thickness D, an aggregate accelerates from zero velocity to its maximum value, $U_{\uparrow\max}$. The soil integrity $I_s = S_b/S_a$ decreases from maximum I_{s0} to zero within the bed surface layer (at the distance equal to D).

$$I_s = I_{s0} - \frac{I_{s0}}{D} z \tag{16}$$

The integral of (15) with (13) and (16) gives a parabolic expression for actual vertical velocity of an aggregate in a bed layer:

$$\begin{aligned} U_\uparrow^2(z) = {} & \frac{\rho C_y}{\rho_s D}\left(U^2 + k_{pw} z_p I_{s0} \mp k_{dp} \lambda I_{s0} U_m^2 - k_c \frac{C_0}{\rho} I_{s0} - k_{sp} d I_{s0} - k_w \frac{(\rho_s - \rho)}{\rho} D \right) z \\ & + \frac{\rho C_y}{2\rho_s D}\left(k_{pw} z_p \mp k_{dp} \lambda U_m^2 + k_{sp} d + k_c \frac{C_0}{\rho} \right) \frac{I_{s0}}{D} z^2 \end{aligned} \tag{17}$$

Averaged in the bed layer, the vertical aggregate velocity $U_{\uparrow m}$ can be easily calculated from (17), not presented here because of the great length of the expression.

In the field of random forces the vertical velocity for an aggregate is a random variable with PDF $p_{U\uparrow}$. Its mean value:

$$\overline{U_\uparrow} = \int_0^\infty p_{U\uparrow} U_{\uparrow m} dU_\uparrow \tag{18}$$

is combined with P_{DER} (14) to give the expression for the aggregate detachment rate calculation (1). These calculations require probability distributions of hydrodynamic and soil characteristics, which can be estimated both experimentally and theoretically.

MECHANISM-BASED APPROACHES IN NUMERICAL MODELLING OF PDF

Hydrodynamic characteristics

To underpin and complete the probabilistic approach for modelling erosion and sedimentation processes, we require a physically based model. This mechanistic hydrodynamic model will provide a sound representation of the fluid flow in sedimentary environments. Shallow flow and changeable rough surfaces are characteristic of such environments. To be consistent with our objective of considering fundamental principles of turbulence hydrodynamics and soil physics, simple configurations of erosion should be addressed first, with the intent that the modelling can be developed to incorporate higher degrees of complexity and additional factors. In particular, we intend to model four situations, in which we consider: flow, flow and rain (as a significant source of energy to the system), flow and erosion, flow and rain and erosion, respectively. The magnitudes of the forces (4) to (12) and, therefore, probability of detachment, are different for these four situations. All scenarios involve modelling flow past complex boundary conditions, which is therefore an important criterion for potential models. The fluid model must therefore be able to accommodate a high level of spatial complexity. Further, there is a dynamic fluid–solid interaction that takes place in these environments: the pressure gradients and shear stresses generated by the fluid flow causes the surface to erode, which in turn affects the flow structure. This interplay may or may not be incorporated into the hydrodynamic model, but it stands as a criterion in consideration of potential models.

The area of computational fluid dynamics (CFD) has advanced markedly in recent years, driven in part by advances in computational technology of solving the governing Navier-Stokes equations for incompressible flow. There are many and various approaches for solving the Navier-Stokes equations. The first is to solve them directly for specific boundary and initial conditions. This is an ideal approach; the only potential errors in this Direct Numerical Simulation (DNS) method are the ones introduced by the numerical scheme. That is, accuracy is highly dependent on the grid system used and level of spatial-temporal resolution, limiting the approach to simple geometries. But, because it solves the Navier-Stokes equations directly, it gives explicit instantaneous velocities. This is extremely useful for purposes of erosion modelling, in which the entire distributions of velocity values and pressure forces are needed. Due to the physical complexity associated with rough and changing solid surfaces, however, it is unrealistic to employ this approach for erosion modelling.

Secondly, there is Large Eddy Simulation (LES), which filters all instantaneous variables so that they operate at the level of grid resolution or larger, thereby reproducing only the large-scale flow structure. In particular, a turbulent viscosity value is used, which encompasses the range of all viscous forces below the grid resolution scale. This approach gives instantaneous velocity and pressure values that are spatially averaged at the scale of grid cell width. In consideration of the erosion problem, we note that this scale must not be significantly larger than the scale of soil aggregates.

As a third approach, decomposing the instantaneous flow into mean and fluctuating elements and then averaging gives rise to the Reynolds Averaged Navier-Stokes (RANS) equations. This introduces an additional Reynolds stress term, so that some closure model relating this stress to the mean flow is required. The RANS approach is more widely applicable than DNS, in terms of adapting to complex boundary conditions, but depends on the modelling assumptions inherent in the closure scheme. Furthermore, time-averaged velocity profiles are not useful in the context of erosion modelling, since it is primarily the extremes of the pressure distribution that cause soil detachment.

The LES has proven to be a flexible tool for erosion modelling, since it can accommodate complex solid boundaries adequately, it gives full distributions of velocity and pressure, and is compatible with a variety of methods for incorporating sediment dynamics. This modelling method will provide extensive information on velocity and sediment fields, which are needed to underpin the stochastic concept. There are many possibilities for, and difficulties with, implementing the LES method. The key aspect is the representation of the complicated rough boundary, for which there are many approaches, depending on the desired modelling scale. Nonetheless, our methodology for incorporating the results from LES simulations into the stochastic models is straightforward. The output from the LES will be in the form of time series data of velocity and pressure, calculated at each point on the grid. PDFs can then be extracted from spatially averaged time series data by assigning velocity and pressure data to a finite set of bins, and normalizing the frequency at which the data fall into each bin. Other useful statistical constructs, such as structure functions, can also be extracted from the LES time series data.

Soil structure modelling

Soil structure is the spatial/temporal distribution of soil physical characteristics within a soil body. One of these characteristics is the size (linear, by the area; volumetric, by the weight) of soil particles and aggregates. Distribution of soil particles and aggregates by size is described with PDFs, and more recently by fractal dimensions (FDs). These distributions change in time due to fragmentation of soil aggregates or aggregation of soil aggregates and particles. Nevertheless, there are quite a few main types of PDF, estimated empirically and associated with all variety of soils in different conditions. There is the logarithmically Normal distribution, and the Rosin-Rammler relation and power-law distribution, associated with the fractal approach (Perfect *et al.*, 1993). Only the logarithmically Normal distribution has theoretical basis (Kolmogorov, 1941). This work described the process of random failure of soil particles, when the probability of fragmentation of a particle to some number of parts was scale-invariant, and the result was asymptotically logarithmically Normal.

The Kolmogorov-type algorithm of soil particles failure can be simulated numerically, and in numerical experiments the assumption of scale independence of fragmentation can be avoided. These experiments with different relationships between probability of failure and particle size show a great stability of result. The logarithmically Normal distribution of soil particles is valid in a broad range of scenarios of fragmentation. This distribution is asymptotic, but is developed within a first few steps of simulation. Each type of fragmentation process is characterized by specific rates of mean size decrease and particle size variability increase.

NUMERICAL EXPERIMENTS

Numerical experiments were undertaken to show the general advantages of the proposed stochastic approach for soil erosion calculation. To investigate the most important soil erosion factors, the proposed approach was simplified: not all driving and resistance forces were included in the aggregate instability inequality; we only considered lift, gravity and cohesion forces. The stochastic variables in this inequality are assumed to be independent; this allows using the expressions for calculating the probability of the sum and product of independent stochastic variables. Three main types of probability distribution functions for hydrodynamic and soil characteristics were used: Normal, logarithmically Normal and Gamma distribution. The input data consisted of mean bed velocity U_m, mean soil cohesion C_{0m}, mean soil integrity I_s, mean aggregate diameter D_m, and standard deviations for all those variables. Numerical experiments were carried out to analyse the influence of these four

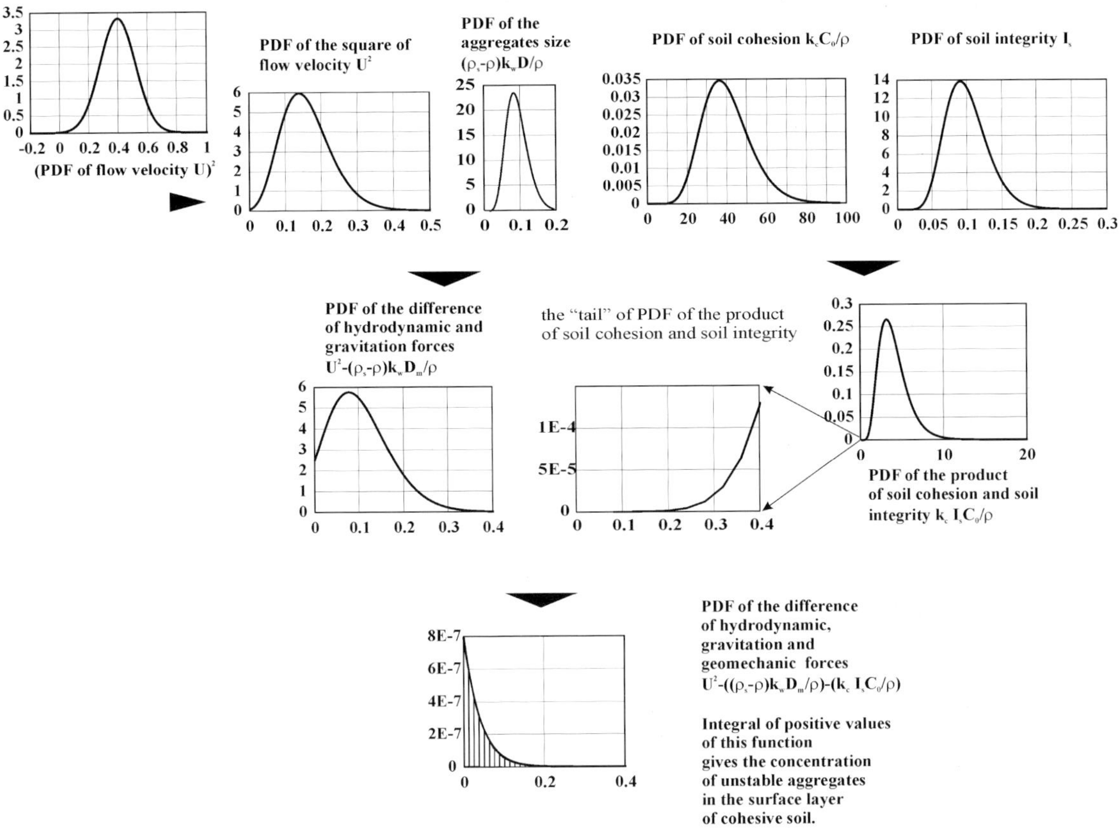

Fig. 1 The order of soil aggregate concentration calculation with the stochastic model.

stochastic factors on the detachment rate. The range of mean flow bed velocity was 0.1–2.2 m s^{-1}, the range of mean cohesion was 1–30 kPa, mean soil integrity ranged from 0.1 to 4, aggregate mean size in the natural soil varied from 1 to 10 mm; and standard deviations for all PDFs varied from 0.1 to 1.0–2.0 times the mean value. The sequence of calculation of sediment concentration of unstable aggregates from PDFs of driving and resistance forces is shown in Fig. 1. The same order is used to obtain the mean vertical velocity and, finally, the detachment rate.

RESULTS AND DISCUSSION

The following main phenomena were observed (Fig. 2):

(a) The increase in erosion rate with flow velocity cannot be described with an often-used simple power function with *a priori* known exponent n: $DER \sim U^n$. Calculations show that, when velocities are relatively low, the detachment rate increases more rapidly than in relatively high velocities. A similar effect was described by Nearing *et al*. (1997) on the basis of empirical soil erosion measurements. In this investigation the phenomenon was underpinned theoretically.

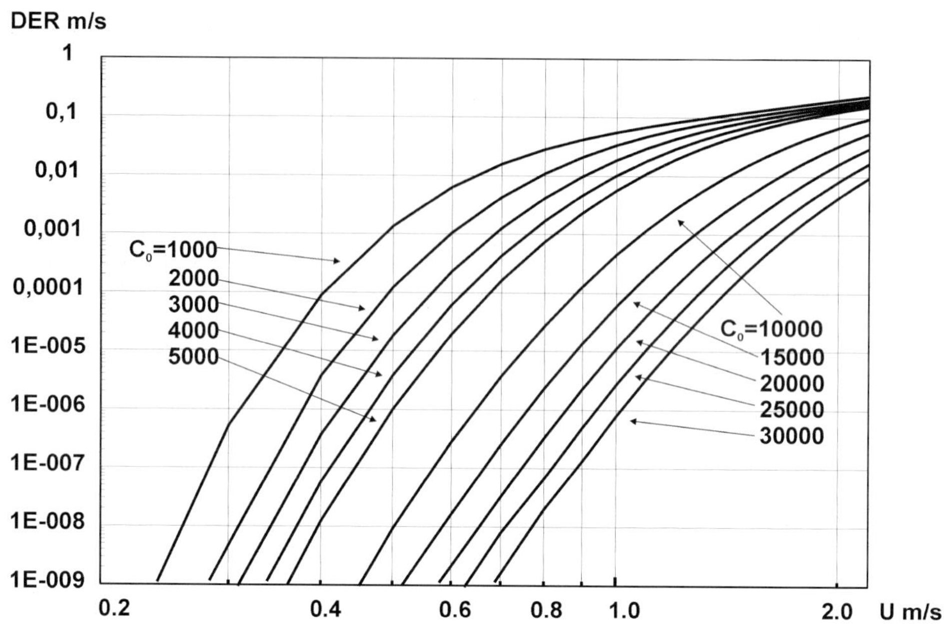

Fig. 2 Relationship between detachment rate (*DER*) and the mean flow velocity (*U*). The third variable is soil cohesion. The calculations were performed with soil integrity 0.1; mean aggregates size 2 mm, variability coefficients for these stochastic variables 0.3.

(b) The analysis of relationships between the hydraulic characteristics of the flow (actual flow velocity), the geo-mechanical properties of the soil (aggregates size, cohesion and integrity), and the soil aggregates detachment rate makes possible an explanation of the difference in relationship types between detachment rate and flow velocity (shear stress, stream power) for different soils. This difference is caused by the relative energy of the flow: the ratio between driving and resistance forces, as well as by the spatial/temporal variability of these forces. In high flow velocities, when driving forces significantly exceed stabilizing forces, the rate of erosion increase with flow velocity is relatively low. The influence of the variability of soil properties (cohesion, aggregate size, and soil integrity) is also less important in determining the soil erosion rate of relatively high flow energy. With low flow velocities and with driving forces only slightly exceeding the stabilizing forces, erosion rates increased rapidly with flow velocity, and all soil properties became sufficient for erosion rate estimation (see Fig. 2 for the influence of soil cohesion, other soil properties give the same effect).

The stochastic erosion models are third-generation models, accepting empirical statistical models (USLE-type) as first-generation models, and shear stress-based models (WEPP-type) as second-generation models. In the new model the relationship between soil detachment rate and the factors of erosion (flow and soil characteristics) is not obtained in advance from some empirical data. They are calculated within the model from the information about PDFs of driving and stabilizing forces with the use of basic equations and are different for the different combinations of erosion factors. Therefore third-generation models promise more precise soil erosion prediction due to more accurate description of soil erosion mechanics, but they require better information about flow and soil.

Acknowledgements This research is funded by the Marsden Fund administered by the Royal Society of New Zealand (grant LCR-203). The comments of Dr H. Middelkoop were very valuable and were incorporated into the text.

REFERENCES

Einstein, H. A. (1937) Bedload transport as a probability problem. PhD Thesis, ETH Zurich, 1937 (Translation by W. W. Sayre in *Sedimentation*, Appendix C, ed. by H. -W. Shen, C1-C105, Colorado State Univ., Fort Collins, 1972) (in German).

Foster, G. R. & Meyer, L. D. (1972) A closed-form soil erosion equation for upland areas. In: *Sedimentation* (ed. by H. -W. Shen), 12.2–12.9. Water Resources Publication, Fort Collins, Colorado, USA.

Kolmogorov, A. N. (1941) O logariflicheski-normal'nom zakone raspredeleniya razmerov chastits pri droblenii (About a logarithmically normal distribution of particles during the fragmentation). *Doclady Akademii Nauk SSSR* **31**(2), 99–101 (in Russian)

Lawrence, D. S. L. (2000) Hydraulic resistance in overland flow during partial and marginal surface inundation: experimental observations and modelling. *Water Resour. Res.* **36**(8), 2381–2393.

Merritt, W. S., Letcher, R. A. & Jakeman, A. J. (2003) A review of erosion and sediment transport models. *Environmental Modelling & Software,* **18**(8–9), 761–799.

Mirtskhoulava, T. E. (1988) *Principy fiziki i mehaniki erosii rusel* (Principles of Physics and Mechanics of Channel Erosion). Gidrometeoizdat, Leningrad, USSR (in Russian).

Nikora, V. I., Goring, D. G., McEwan, I. & Griffiths, G. (2001) Spatially-averaged open-channel flow over a rough bed. *J. Hydraul. Engng ASCE* **127**(2), 123–133.

Nearing, M. A., Norton, L. D., Bulgakov, D. A., Larionov, G. A., West, L. T. & Dontsova, K. (1997) Hydraulics and erosion in eroding rills. *Water Resour. Res.* **33**, 865–876.

Perfect, E, Kay, B. D., Ferguson, J. A., da Silva, A. P. & Denholm, K. A. (1993) Comparison of functions for characterizing the dry aggregate size distribution of tilled soil. *Soil Till. Res.* **28**, 123–139.

Sidorchuk, A. (2004) Stochastic modelling of erosion and deposition in cohesive soils. *Hydrol. Processes* (in press).

Wischmeier, W. H. & Smith, D. D. (1965) Predicting rainfall-erosion losses from cropland east of the rocky mountains—a guide for selection practices for soil and water conservation. *USDA Agricultural Handbook 282*. US Government Printing Office, Washington, DC, USA.

Sediment Transfer through the Fluvial System (Proceedings of a symposium held in Moscow, August 2004).
IAHS Publ. 288, 2004

A model of rill erosion by snowmelt

YURI P. SUKHANOVSKI[1], VALERY V. DEMIDOV[2] & GREGOR OLLESCH[3]

1 *The All Russian Research Institute of Agronomy and Soil Erosion Control, Karl-Marx-Sts. 70B, 305021 Kursk, Russia*
soil-er@kursknet.ru

2 *Institute of Basic Biological Problems, Russian Academy of Sciences, 142290 Pushchino, Moscow Region, Russia*

3 *UFZ-Center for Environment Research, Department of Soil Science, Brueckst. 3A, D-39114 Magdeburg, Germany*

Abstract Measurements of rill flow profiles, water discharges, sediment concentration, temperature of water, soil and air were conducted during spring snowmelt events on an experimental station located 100 km south of Moscow, Russia. The results indicate that: (a) the rill profiles have, as a rule, a triangular form; (b) the side-wall slope of a rill is close to the natural slope for non-frozen soils and depends on the water discharge; and (c) in general, the thawing of the soil surface occurs faster, than the soil particle detachment. As the knowledge of frozen soil erosion mechanics is limited, a number of assumptions have to be made for the model design. In detail, the Snow Melt Erosion Model (SMEM) includes the Chezy–Manning's equation, the Goncharov's equation to calculate bottom flow velocity, the Mirtskhulava's equation for estimation of soil particle detachment and the Kuznetsov's equation for critical bottom flow velocity. The model is tested with 7 years of data from two runoff plots located in the Central-Chernozem Zone of Russia (soil type is chernozem).

Key words rill erosion; snowmelt; erosion model; Russia

INTRODUCTION

Results of erosion studies in northern, central and eastern Europe indicate that the erosion rate during snow melt events can reach or even exceed the rainfall erosion rate. Rill formation is the fundamental erosion process during winter conditions. Understanding the nature of snowmelt erosion processes is essential for solving both the on-site and off-site problems and to deduce recommendations for management practices. Predictive modelling is an important tool in evaluating alternative technologies.

Recently, well known but often unadapted empirical equations have been applied for the assessment of soil losses during snowmelt periods (Wischmeier & Smith, 1978; Cheboterev *et al.*, 1979; Surmach, 1979; Edwards *et al.*, 1998). The main problem in the design of a snowmelt erosion model is connected to the characterization of soil detachment processes by snowmelt overland flow is one of the major problems to be solved. Additionally, the formation of a rill net by snowmelt overland flow is an open question. A physically based equation for particle detachment at frozen soil conditions was developed on the basis of laboratory experiments (Kuznetsov *et al.*, 1999, 2001; Kuznetsov & Demidov, 2002). However, extensive investigations have to be conducted to define the values of a number of relevant parameters. This partly restricts its application. The purpose of this paper is the presentation of a physically reasonable model for snowmelt rill erosion on hillslopes with a minimum of input parameters.

DATA AND MODEL

Rill profiles

The field station of the Institute of Basic Biological Problems (Russian Academy of Sciences) is located 100 km to the south of Moscow. The main agricultural practice of the predominant grey forest soils is autumn ploughing to a depth of 20–22 cm and winter wheat cultivation. The measurement of the following rill characteristics was conducted: water discharge, concentration of sediments, and cross section of a water flow. In addition to snow characteristics, air and soil temperature were measured. The results indicate that: (a) the rill profiles have in general a triangular form; (b) the slope of the side-wall of a rill depends on the water discharge and is close to the natural slope for non-frozen soils; (c) in general, the thawing of the soil surface occurs faster than the soil particle detachment. Also Gatto (2000) observed triangular rill profiles. These findings are of particular importance for the model development.

During 4 years of investigation 75 rill profiles for ploughed soils and 23 profiles for soils under winter wheat and correlated runoff characteristics were measured. Figure 1 presents representative cross sections of a rill on a fallow plot for different discharge values. The typical triangular shape of the rill cross section clearly indicates that rill incision is not limited by a frozen soil layer. Statistical analysis of the observed data show that the tangent of the angle α of the bank slope (Fig. 2) can be described with the following empirical relationship:

$$Tg(\alpha) = Tg(\alpha_{max}) - [Tg(\alpha_{max}) - Tg(\alpha_{min})] \exp(-\beta Q) \quad (1)$$

where Tg is tangent; Q is water discharge (l s^{-1}); β is stationary value (s l^{-1}); α_{min} and α_{max} are potential minimum and maximal angle α of the bank slope, respectively. The values of $Tg(\alpha_{min})$ of equation (1) and the average weighted relative deviation ($\varepsilon_{a.w}$) as measures for accuracy differ for fallow and winter wheat (Table 1). The received values $Tg(\alpha)$ are close to those that are recommended for amelioration of earthen channels.

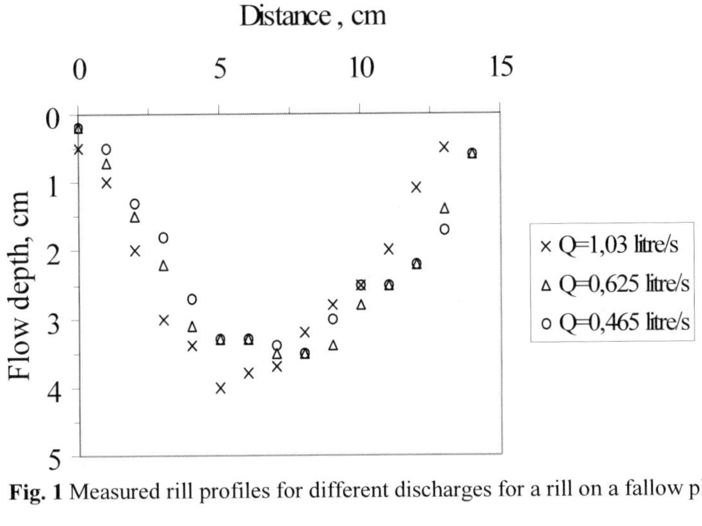

Fig. 1 Measured rill profiles for different discharges for a rill on a fallow plot, 31 March 1999.

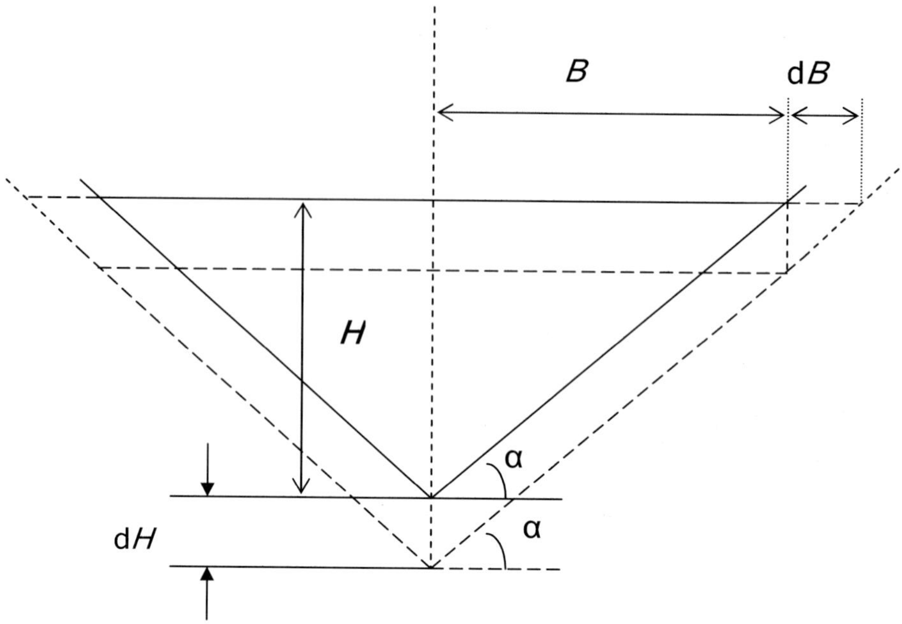

Fig. 2 Profiles of a rill at time t and t + dt.

Table 1 Parameters used in equation (1).

Parameter	Autumn ploughing	Winter wheat
$Tg\ (\alpha_{max})$	0.6	0.6
$Tg\ (\alpha_{min})$	0.1	0.2
$\beta\ (s\ l^{-1})$	3.5	3.5
$\varepsilon_{a.w}$ (%)	14.4	23.6

Soil detachment

The following assumptions are considered for a runoff plot with small length L and with the gradient i: (a) one rill is formed per plot; (b) the outlet discharge is known and the input of water in a rill normalized per unit length one will be identical for the entire plot; (c) for a small time increment the water discharge does not change practically; (d) the cross section of a water flow is determined by the water discharge at any time and at any distance from the top of a plot; (e) the soil particle detachment takes place for unfrozen soil conditions which are characterized by a minimum of coalescent force between soil particles; and (f) all detached soil particles are transported by water flow.

Let us consider profiles for a rill at time t and t + dt, where dt is small increment of time. In Fig. 2 for time t the profile is shown by a solid line, and for time t + dt the profile is shown by a dashed line. As a result of erosion, the water flow was lowered by a quantity dH (m). The increment of cross sectional area of rill to within the small value dH will be evaluated:

$$dS_{rill} = 2\ h\ \mathrm{Ctg}(\alpha)\ dH \quad \mathrm{m}^2 \qquad (2)$$

where *Ctg* is cotangent; *h* is depth of flow (m). Division of both parts of equation (2) with d*t* results in:

$$\frac{dS_{rill}}{dt} = 2hCtg(\alpha)\frac{dH}{dt} \quad m^2\ s^{-1} \tag{3}$$

Further:

$$\frac{dH}{dt} = \frac{q}{\rho_{soil}} \tag{4}$$

where q is the intensity of soil erosion (kg m^{-2} s^{-1}); and ρ_{soil} is the soil density (kg m^{-3}). From equations (3) and (4) follows that:

$$\frac{dS_{rill}(t,x)}{dt} = 2h(t,x)Ctg[\alpha(t,x)]\frac{q(t,x)}{\rho_{soil}} \tag{5}$$

where t is time (s); and x is the distance from the top of runoff plot (m). For any interval of time ($T_2 - T_1$) the volume of the rill will increase:

$$Volume = \int_{T_1}^{T_2} dt \int_0^L \frac{dS_{rill}(t,x)}{dt} dx \quad m^3 \tag{6}$$

For this interval of time the rill erosion will be equal, thus:

$$RillErosion = \rho_{soil}\ \text{Volume} \quad \text{kg} \tag{7}$$

Further we use a series of simple equations:
—water discharge:

$$Q\,(t, x) = [QL(t) / L]\, x \tag{8}$$

where L is length of a plot (m); $QL\,(t)$ is outlet discharge (m^3 s^{-1});
—the cross sectional area of water flow:

$$S_f = h^2 / \text{Tg}\,(\alpha) \quad m^2 \tag{9}$$

—hydraulic radius for the triangular form of the channel:

$$R = h\ \text{Cos}\,(\alpha) / 2 \quad m \tag{10}$$

where Cos is cosine;
—Chezy-Manning equation:

$$V = R^{2/3}\, i^{1/2} / n \tag{11}$$

where V is flow velocity (m s^{-1}), i is channel slope (dimensionless), n is Manning's coefficient;
—water discharge:

$$Q(t, x) = S_f V \quad m^3\ s^{-1} \tag{12}$$

From the equations (9) to (12), it follows that :

$$h(t, x) = 2^{1/4}\, Q(t, x)^{3/8}\, [n\ Tg(\alpha)]^{3/8}\ [i^{-3/16}\ \text{Cos}^{-1/4}\,(\alpha)] \tag{13}$$

Thus, knowing the outlet discharge $QL\ (t)$, it is possible to apply equation (8) to calculate

$Q(t, x)$, and with equation (1) it would be possible to calculate an angle $\alpha(t, x)$. Further, with equation (13) it is possible to calculate depth of a flow $h(t, x)$, which enters equation (5).

For the dimension of q we use the method from Mirtskhulava for unfrozen soils (Mirtskhulava, 1970, 2000):

$$q = 1.1 \times 10^{-6} \omega D_{wsp} \rho_{paricle} \left(\frac{V_{\Delta}^2}{V_{\Delta,cr1}^2} - 1 \right) \quad \text{kg m}^{-2}\ \text{s}^{-1} \tag{14}$$

where ω = 10 s^{-1} is the frequency of pulsations of water flow; D_{wsp} is the average diameter of the water stable aggregates (m); $\rho_{particle}$ is the density of aggregates (kg m^{-3}); V_{Δ} is the bottom velocity (m s^{-1}); $V_{\Delta,cr1}$ is the first critical bottom velocity (m s^{-1}). If V_{Δ} is less than $V_{\Delta,cr1}$ then q will be zero. Goncharov's equation is applied for the calculation of bottom flow velocity (Goncharov, 1962):

$$V_{\Delta} = 1.25\ V / \text{Log}_{10}\ (6.15\ h / \Delta) \quad \text{m s}^{-1} \tag{15}$$

where Log_{10} is logarithm; Δ is the roughness of bottom rill (m). The bottom roughness can be expressed through the diameter of soil particles (Kuznetsov, 1981):

$$\Delta = 0.7\ D_{wsp} \tag{16}$$

Hence, at a known water discharge $Q\ (t, x)$ it is possible to use equations (10), (11), (13), (15) and (16) to calculate the velocity of bottom flow V_{Δ} for any instance of time t and for any distance x from the top of a plot. The second critical velocity will be estimated without consideration of coalescence between particles following Kuznetsov (1981):

$$V_{\Delta,cr2} = 1.55 \sqrt{\frac{m_1 m_2 g}{\rho_{\text{water}} n_1} (1 - P) D_{wsp} (\rho_{\text{mineral}} - \rho_{\text{water}})} \tag{17}$$

where $V_{\Delta,cr2}$ is the second critical bottom velocity for fallow plots (m s^{-1}); for rill erosion m_1 = 1.4, m_2 = 1.0 and n_1 =2.3; g = 9.81 m s^{-2} is gravitational acceleration; P is the porosity of soil particles, (dimensionless); ρ_{water} is the density of water (kg m^{-3}); $\rho_{mineral}$ is the density of mineral (kg m^{-3}). The dependence between critical velocities is according to results from Mirtskhulava (1970):

$$V_{\Delta,cr2} = 1.4\ V_{\Delta,cr1} \tag{18}$$

With information on the physical characteristic of a soil, it is possible to calculate the quantity of $V_{\Delta,cr1}$ by using equations (17) and (18), which enters into equation (14). Thus, it is possible to calculate the intensity of erosion at the bottom of a rill $q(t, x)$ and quantify dS_{rill}/dt with a defined quantity of water discharge $Q(t, x)$. The integration in formula (6) gives the rill volume and equation (7) allows the quantity of rill erosion to be estimated.

MODEL RESULTS

The first application of the snowmelt rill erosion model (SMEM) was conducted with data that were received from two runoff plots of the Niznedevitsk water-balance station (the Voronezh region, Russia). Both plots have identical length and width (100 m × 20 m), a slope of 5.5% and a northern exposure. The local chernozem soil has the following characteristics: ρ_{soil} = 0.91 g cm^{-3}, $\rho_{mineral}$ = 2.58 g cm^{-3}, D_{wsp} = 0.5 mm, P = 0.408. Table 2

Table 2 The agricultural management of the runoff plots.

Year	Plot 9	Plot 10
1962	Winter wheat	Fallow
1963	Winter wheat	Fallow
1964	Fallow	Fallow
1965	Fallow	Fallow
1966	Fallow	There were no measurements
1968	Fallow	Fallow
1969	Fallow	Fallow

presents the crop rotation during the years that were selected for modelling. Measurements of the water discharges $Q_{L,1}$, $Q_{L,2}$,... $Q_{L,N}$ and the sediment concentrations $C_{L,1}$, $C_{L,2}$... $C_{L,N}$ (where N is the number of measurements) were conducted during the daytime at instants of time t_1, t_2... t_N. The measurement data were used to calculate soil losses using the following equation:

$$SoilLosses = \sum_{i=2}^{N} \frac{1}{2}\left(Q_{L,i-1}C_{L,i-1} + Q_{L,i}C_{L,i}\right)\left(t_i - t_{i-1}\right) \qquad (19)$$

To estimate the soil loss the length of each plot was divided into 10 equal parts (x_1, x_2... x_{10}, where x_j is distance from the top of a plot). According to data from Mirzkhulava, the first critical bottom velocity for winter grain $V_{\Delta,cr1,\text{Grain}} = 1.5\ V_{\Delta,cr1,\text{Fallow}}$ (Mirzkhulava, 2000). For each distance x_j and instant of time t_i (when the measurements were done) the values dS_{rill}/dt (equation 5) are calculated. Further, the values *Volume* (equation 6) and *RillErosion* (equation 7) were calculated for each day by applying a numerical integration. The estimated soil erosion varies between 32.6 kg year^{-1} for winter wheat of plot 9 and 66.5 kg year^{-1} for the fallow plot 10 kg year^{-1}. The deviation range was between 22.4% and 36.2%. In general, an overestimation of the model results compared to the measurement data can be observed (Fig. 3). Further analysis of the results shows that the average weighed relative deviation ($\varepsilon_{a.w}$) for rill erosion per day is $\varepsilon_{a.w} = 102\%$. The accuracy of the model increases for a longer period of 1 year to $\varepsilon_{a.w} = 62\%$.

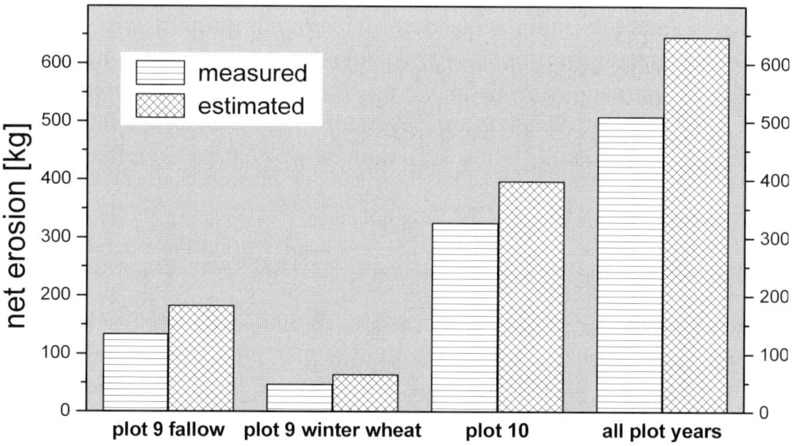

Fig. 3 Measured and modelled erosion from the two erosion plots.

CONCLUSIONS

The application of the snowmelt rill erosion model (SMEM) achieves results that are close to the data from erosion plot experiments. The accuracy of the modelling results increase with an increase in the modelling period. Hence, the basic assumptions can be used to develop a snowmelt rill erosion model that might be applied for the calculation of soil losses for longer periods. Further testing and analysis of parameter sensitivity has to be done to apply the model for single events and on a catchment scale.

REFERENCES

Cheboterev, A. I., Karauschev, A. V., Bogolubova, I. V., Bobrovitskay, N. N., Serpik, B. I. & Tumanovskay, S. M. (eds) (1979) *The Instruction on Calculation of the Hydrological Characteristics at Designing of Erosion-Protection Measures on European Territory USSR*. GIMIZ Press, Leningrad, Russia (in Russian).

Edwards, L., Richter, G., Bernsdorf, B., Schmidt, R. -G. & Burney, J. (1998) Measurement of rill erosion by snowmelt on potato fields under rotation in Prince Edward Island (Canada). *Can. J. Soil Sci.* **78**, 449–458.

Gatto, L. W. (2000) Soil freeze-thaw-induced changes to a simulated rill: potential impacts on soil erosion. *Geomorphology* **32**, 147–160.

Goncharov, V. N. (1962) *Dinamika Ruslovyh potokov (Dynamics of bed of streams).* GIMIZ Press, Leningrad, Russia (in Russian).

Kuznetsov, M. S. (1981) *Erosion-Preventive Stability of Soils*. Moscow University Press, Moscow, Russia (in Russian).

Kuznetsov, M. S. & Demidov, V. V. (2002) *Eroziya pochv lesostepnoi zony Centralnoi Rossii: modelirovanie, predotvrashenie i ecologicheskie posledstviya* (Soil Erosion of Forest Steppe Zone of Central Russia: Modelling, Prevention and Ecological Sequels). POLTEKS Press, Moscow, Russia (in Russian).

Kuznetsov, M. S., Demidov, V. V. & Gendugov, V. M. (2001) Opyt modelirovaniya erozii pochv pri snegotayanii (Experience of modelling of soil erosion at snowmelt). *Pochvovedenie* **8**, 1009–1014 (in Russian).

Kuznetsov, M. S., Gendugov, V. M. & Kosonozkin, V. I. (1999) Zakony erosionogo vozdeistviya potoka na taluyu pochvu (Laws erosive of action of a flow on thaw soil). *Pochvovedenie* **11**, 1393–1399 (in Russian).

Mirtskhulava, Ts. E. (1970) *Inzenernye metody rascheta i prognozirovaniya erozii pochv (Engineering Methods of Calculation and Forecast of Waters Erosion)*. Kolos Press, Moscow, Russia (in Russian).

Mirtskhulava, Ts. E. (2000) *Vodnaya eroziya pochv* (Water Erosion of Soils). MEZNIEREBA Press, Tbilisi, Georgia (in Russian).

Surmach, G. P. (1979) Opyt pascheta poter pochvy dlya stroitelstva compleksa protivoerozionnyh meropriyatii (Experience of soil losses calculation for build-up of the complex of anti-erosion measures) *Pochvovedenie* **4**, 92–104 (in Russian).

Wischmeier, W. H. & Smith, D. D. (1978) *Predicting Rainfall Erosion Losses*. Agricultural handbook 537. Washington, USA.

Hydrophysical model of soil erosion: a basic equation and influence of bed load and suspended sediment on soil detachment by shallow water flow

G. A. LARIONOV, N. G. DOBROVOLSKAYA, Z. P. KIRYUKHINA, S. F. KRASNOV & L. F. LITVIN
Laboratory for Soil Erosion and Fluvial Processes, Faculty of Geography, Moscow State University, Leninskye Gory 119992, GSP-2, Moscow, Russia
river@river.geogr.msu.su

Abstract The hydrophysical model is based on the three premises. According to the first and the second premises soil detachment is proportional to the cubed flow velocity in general. Within the range where flow velocity is higher than 0.4 u_0 and less than 1.6 u_0 the detachment of soil particles has a probabilistic nature. The experimental data are described well by the equation formed from these premises. The third premise runs as follows: the detached soil particles, which move in the water course of an overland flow as bed load and suspended sediment, in some way influence the detachment of new soil particles. The influence of natural and artificial bed load produced from different materials on wash out of chernozem samples are described satisfactorily by the exponential function. The suspended sediment corks up the pores of soil. Therefore it leads to a strengthening of the cohesion forces between soil particles and aggregates and consequently to a decrease of erodibility.

Key words bed load; erosion modelling; influence; soil detachment; soil erosion; suspended sediment

INTRODUCTION

The analyses of well known erosion models show that all of them contradict the field and experimental data (Larionov *et al.*, 2003). According to numerous plot studies soil loss is proportional to the sine of slope angle and slope length raised to an exponent which is significantly less than one. The equations of soil detachment written as a function of slope and slope length show the following. According to the models which are based on the assumption that soil detachment is the function of actual shear stress expressed either as shear stress minus the critical value (Foster, 1982) or the ratio of squared flow velocity to squared critical value minus one (Mirtskhulava, 1970), the soil detachment rate is proportional to the sine of slope angle and slope length, both raised to 0.666. Models which rely on a stream power to characterize the capability of flow to detach soil (Rose, 1985; Hairsine, 1992) indicate that soil loss is proportional to the sine of slope angle and slope length.

Nearing (1991) proposed a probabilistic model of soil detachment. One of the aims of the model was to overcome an apparent inconsistency in terms of the orders of magnitude differences between soil strength and flow shear stresses. According to his data soil strength is of the order of kPa, while flow shear stresses are of the order of Pa. It is known that local shear stresses associated with turbulent burst-events are much greater than the average flow shear stresses. The average shear stress of the burst-events is 150 times greater than the average flow shear stress. This fact is only one part of the explanation. Nearing (1991) suggests that the detachment process is determined by the overlapping tails of two distributions (burst-event shear stresses and soil particle resistance to detachment). The final equation of soil detachment written as a function of slope and slope length is a nonlinear function. The detachment rate is proportional to the sine of slope raised to 1.333 and slope length raised to 0.333.

The model developed by Gendugov *et al.* (1997) proceeds from the assumption that flow velocities below the critical value also produce soil detachment. According to this model soil detachment is proportional to the sine of slope angle raised to 0.88–0.58 and slope length raised to 0.56–0.44.

Analysis shows that the above mentioned models are partly or completely contrary to the facts. The models of Foster (1982), Mirtskhylava (1970) and Gendugov *et al.* (1997) underestimate both the slope steepness and slope length factors. The model of Rose (1985) greatly overestimates the slope length factor. The model proposed by Nearing (1991) overestimates the slope steepness factor.

In order to overcome the discrepancy between the field soil loss data collected at North Caucasus and modelled data using the second edition of USLE (Larionov & Krasnov, 1993), a new soil detachment model was developed. The new model contains a probabilistic element and is based on the most general laws of nature. The model describes existing flume experiment data from literature well and was confirmed by the experiments designed to study soil detachment by high velocity flow. The aim of this paper is to demonstrate the main features of the basic equation and to quantify the influence of bed load and suspended sediment on soil detachment.

THE EQUATION OF SOIL DETACHMENT

Parameterization and validation

The main equation of soil detachment is deduced from two premises: (a) erosion is the work of water flow, which is responsible for the detachment; (b) particles are detached by the flow streams if instantaneous flow velocities exceed the threshold. The simple mathematical construction based on the first premise leads to the conclusion that soil detachment is proportional to the power of flow:

$$D_r \propto k_r \gamma u^3 \qquad (1)$$

where D_r is the rill detachment rate, k_r is the rill erodibility of soil, γ is the specific weight of water, and u is the average flow velocity. It should be mentioned that in the case of shallow flow, water power is proportional to the cubed average flow velocity.

According to the second assumption the detachment of soil particles has to have a stochastic nature if the average flow velocity is close to the critical value. In that case detachment events depend on overlapping of two probability distributions. The first is the distribution of instantaneous flow velocity; the second is the distribution of resistance of soil particles to detachment. Probability of events falling within a certain limit is usually determined from the special tables but this is inconvenient for our case. It is easier to use the cumulative curve of probability expressed by a logistic function. Then the probability density function (P_u) of instantaneous flow velocity may be approximately expressed as:

$$P_u = \left[1 + 10^{a(1-u/u_0)}\right]^{-1} \qquad (2)$$

where u is the average flow velocity, u_0 is the critical velocity and a is the coefficient depended on the instantaneous flow velocity dispersion. According to measurements of instantaneous flow velocity (Mirtskhulava, 1967) coefficient a equals approximately 4. Conveniently, the probability of soil particle tensile stress should be described by a logistic equation with the same variables as in the case of flow velocity. Because the stress exerted on soil particles by water flow is proportional to flow velocity raised to the second power, the average relative value of soil tensile strength may be expressed as squared critical flow velocity. Then the probability (P_s) of soil particle tensile strength may be written as:

$$P_s = \left[1 + 10^{b(1-u^2/u_0^2)}\right]^{-1} \qquad (3)$$

where b is the coefficient depended on the dispersion of soil particles tensile strength. Other variables are already mentioned above.

Recalling equation 1 soil detachment (D_r) by clear water should have a form:

$$D_r = k_r \gamma u^3 \left[1 + 10^{a(1-u/u_0)}\right]^{-1} \left[1 + 10^{b(1-u^2/u_0^2)}\right]^{-1} \qquad (4)$$

Equation (4) of soil detachment by clear water flow was adjusted and verified using existing literature experimental data (Kuznetsov & Glazunov, 1985; Nearing *et al*., 1981; Larionov & Krasnov, 1997) obtained under laboratory conditions over a wide range of flow velocities (0.21–1.97 m s^{-1}). According to the consequence of the first and the second premises the relationship between the soil detachment and the cubed flow velocity should be linear if $u \geq 1.6u_0$. This is confirmed by Fig. 1. It also indicates that detachment takes place in the zone where flow velocity is much less than the critical magnitude. In this case the relationship between soil detachment and cubed flow velocity is linear too. The two segments of straight line are connected by the *s*-shaped line. This part of the graph corresponds to the zone where the soil detachment process has a probabilistic nature. This peculiarity of the relationship between soil detachment and cubed flow velocity is indicated by equation (4). It should be added by the term designated to describe the soil detachment in the zone where flow velocity is less than the critical value. Results of parameterization and validation of equation (4) adding the above mentioned term are presented in Table 1 (Fig. 1). The flow velocity at the height of roughness elements and average flow velocity at the near-bed 1 cm layer were used as variables for the soil detachment equation. The results were good enough in both cases. The coefficients of determination are between 0.98 and 0.99. This is slightly higher than described by Nearing *et al.* (1991). However, from a practical point of view the flow velocity at the near bed of 1 cm layer is much more reasonable than flow velocity at the height of roughness elements. Firstly, the magnitude of erodibility, k_{1r} and k_{2r}, is proportional to the tensile strength of soils. This in turn is inversely proportional to the weighted average soil aggregate diameter according to the data of six soil materials used by Nearing *et al.* (1991). The flow velocity at the height of roughness elements does not indicate such regularity. Secondly, for all sieved samples the magnitude of the critical velocity, in the case of using the flow velocity in the near-bed 1 cm layer, is almost the same, contrary to the flow velocity at the height of roughness elements. Thirdly, in the case of flow velocity at the near bed layer the constant value of coefficient b can be applied for soils with aggregates of uniform size. The equation of soil detachment by clear shallow water flow is written in the form:

$$D_r = 10^{-6} \gamma u_s^3 \{k_{1r}\left[1+10^{-4(1-u_s/u_0)}\right]^{-1} + k_{2r}\left[1+10^{4(1-u_s/u_0)}\right]^{-1}\left[1+10^{b(1-u_s^2/u_0^2)}\right]^{-1} \qquad (5)$$

where D_r is the soil loss per unit area and unit time (g m^{-2} s^{-1}); γ is the specific weight of water, (kg m^3); u_s is the flow velocity at the 1-cm near bottom layer, (m s^{-1}); k_{1r} and k_{2r} are the soil rill erodibility (m^{-1} s^2); a and b are dimensionless coefficients.

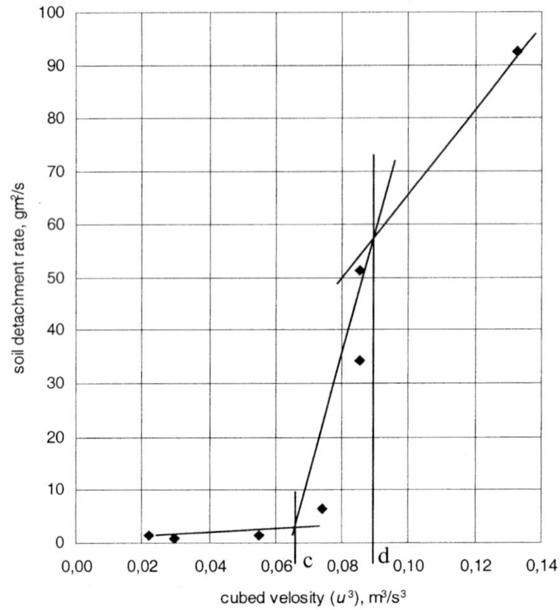

Fig. 1 Relationship between cubed flow velocity and soil detachment rate.

Table 1 The magnitude of parameters and coefficients of equation (3).

Parameters	SOIL							
	Russell			Paulding			Chernozem	Loam
	Aggregate size. mm							
	0.47	1.022	2.065	0.596	1.428	2.719	2.4	1.5
Flow velocity calculated at the height of roughness elements								
U_0. m s^{-1}	0.33	0.427	0.536	0.357	0.49	0.578	0.374	–*
K_1. m^{-1} s^2	50	30	45	40	18	16	0.015	–*
K_2. m^{-1} s^2	1500	850	1250	700	590	980	62	67
A	4	04	4	4	4	4	4	4
B	10	22	18	10	5	10	2	2
C	–4	–4	–4	–4	–4	–4	–4	–4
R^2	0.976	0.989	0.996	0.990	0.986	0.992	0.984	0.900
Average error. %	24.19	17.4	19.49	8.24	16.38	7.86	7.84	15.12
Average flow velocity at 1 cm near bottom layer								
U_0. m s^{-1}	0.770	0.772	0.770	0.772	0.773	0.767	0.53	–*
k_1. m^{-1} s^2	2.8	5.7	18.0	4.0	5.0	8.0	6.2	–*
k_2. m^{-1} s^2	115	138	440	69	132	410	29	17
A	4	4	4	4	4	4	4	4
B	14	14	14	14	14	14	2	2
C	–4	–4	–4	–4	–4	–4	–4	–4
R^2	0.980	0.984	0.994	0.991	0.994	0.996	0.981	0.898
Average error %	26.52	22.27	19.98	10.88	17.56	21.10	7.47	26.52

* The tests were carried out under flow velocities which were much greater than u_0.

The erodibility k_{1r} works if flow velocity in the standard layer is less than 0.4 u_0. The logistic expression in the first rectangular brackets in equation (5) turn to zero if flow velocity $\geq u_0$. The value of a equals 4 for the upland flows as it is shown above. The coefficient b depends on the dispersion tensile strength of soil particles or aggregates. Coefficient b equals 14 for soils of uniform aggregate size. The coefficient becomes 2 for ploughed and undisturbed soil. The threshold velocity (u_0) can be calculated taking c and d from Fig. 2:

$$u_0 = \sqrt[3]{\frac{c+d}{2}} \qquad (6)$$

The average flow velocity (u_s) in the near bed layer of 1 cm may be estimated by the equation of Izbashas & Haldre (1959) which is proposed for rough bed flows. It has the form:

$$u_s = u\,D^{-1.333} \qquad (7)$$

where D is the flow depth (cm); u is the average flow velocity (m s^{-1}).

Within the framework of this paper it is impossible to give the exhaustive answer on the question, why the cohesion forces which are initially of order of K_p, under the energy of water flow decrease so greatly that the maximum values of the fluctuating bed shear stresses which are of the order of the first 100 Pa (Nearing *et al.*, 1991) attain the capacity to produce soil detachment. Probably this may be conditioned by two causes: The first one is believed to be due to the weakening of the cohesion forces of soil particles and due to the hydration of soil contacted water flow and ions exchange between the surface layer of soil and water. This is in conformity with the fact that the slow process of soil detachment takes place under flow velocities below the critical value. It may be possible that in that case the cohesive forces in the upper layer of soil become negligibly small and the water flow acts only against the force of gravity.

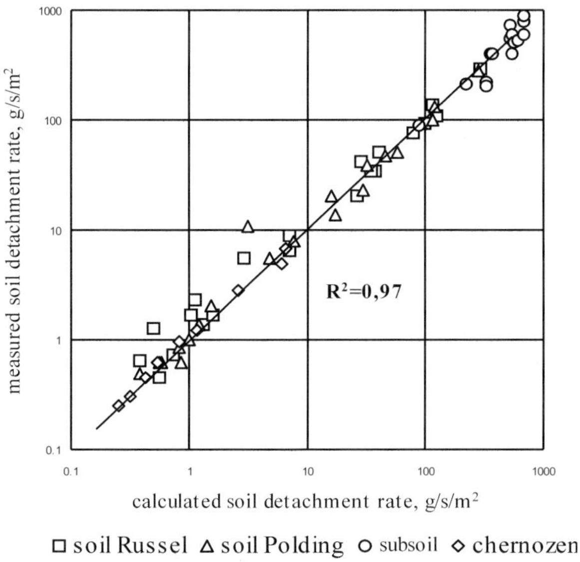

Fig. 2 Predicted detachment rates using equation (5) *vs* measured detachment rate.

If flow velocity is strong enough to impart to soil particles protruded into water reciprocal movement (trembling) the fatigue-limit of soil particle can be overcome by pulsating shear stresses which magnitudes are much less than tensile strength. Mirtskhulava (1970) was the first who emphasized the role of fatigue destroying in the process of soil detaching.

Influence of bed load and suspended sediment on soil detachment

The simple mathematical construction based on the third premise shows that bed load should decrease the detachment capacity of water flow. Furthermore the magnitude of influence is strongly affected by soil aggregate stability. The best known erosion models take the influence of sediment load on the process of soil detachment into account. According to Foster (1982) the detachment of soil particles and its further transportation are produced by the free energy of water flow. The quantity of energy required for transportation of sediment is much less than for detachment. Thereafter the free energy of water flow is used in the first place on sediment transport. Rose (1985) suggests that the main reason of the decrease of the soil detachment rate along the rill is due to deposition the flow bed by the settled soil particles. But Merten *et al.* (2001) have shown experimentally that the magnitude of sediment which influences the soil detachment is less than predicted by Foster. With regard to the suggestion of Rose (1985) the sediment layer at the bed of rills is usually absent. Thus it is widely accepted that sediment influences soil detachment but there is no common opinion on the reason and mechanisms of this process.

Experiments were carried out in order to understand and quantify the influence of bed load and suspended sediment on soil detachment. The circular type of flume was used for this study. Their length and width are 2069 cm and 20 cm, respectively. The volume of the water circulated in the flume is approximately 40 l. Pieces of porolon (5 × 5 × 5 mm), rubber (3 × 3 × 3 mm) and rubber coated copper conductor (3 mm in diameter, 3.5–4.5 mm in length) were used as soft bed load. Gravel of 1–2 mm size was used as material for the hard bed load. The suspended sediment was imitated by silty loam. The soil used in this study for the detachment tests, is a loamy chernozem dried and sieved to obtain the 1.5–2 mm size fraction. Pre-wetted soils for the detachment tests were formed in containers of 7–2 cm cross section under constant pressure to obtain the desired density (1.1–1.3 g cm^{-3} in bed load experiments and 1.1; 1.2 and 1.3 g cm^{-3} in suspended sediment experiments). The samples of soil were placed in the pan with thin (approximately 0.5 cm) layer of water to saturate the soil from the bottom. The saturated samples were placed into the chamber for 10–12 h before testing.

The sample was mounted in the flume so that the surface of the sample was flash with the flume bed. The sample squeezed out from the container during the experiment so that the soil surface remained at the level of the flow bed. Each treatment was replicated 5–10 times.

The results of the bed load experiments indicate that the influence of the soft bed load particles and gravel are absolutely different. The bed load produced from soft material significantly decreases the soil detachment rate. The gravel particles strongly increase the soil detachment. The influence of bed load on the soil detachment rate can be expressed by an exponential function

$$D_{rl} = D_r e^{ax_h + bx_s} \qquad (8)$$

where D_{rl} is the soil detachment rate produced by the sediment loaded water flow; D_r is the soil detachment rate produced by clear flow; a and b are the coefficients which describe the influence the soft and hard bed load the soil detachment; x_h and x_s are the amounts of hard and soft particles transported as bed load above the flow bottom, units bed load particles over 1 m^2 of flow bed. Equation (8) indicates satisfactory results (Fig. 3). The magnitude of coefficients is shown in Table 2. The influence of hard particles on soil detachment was justified by the corrosion effect. This confirmed by multitude elongated craters which can be easily seen by the naked eye on the surface of the treated soil sample. The reason for the influence of soft bed load particles on the soil detachment process is not so clear. Two suggestions can be made. On the one hand the soft particles protect the soil from scouring of water flow at the point where particles contact the soil surface. On the other hand bed load reduces the turbulence of water flow and thus diminishes the magnitude of instantaneous shear stresses. The experiments can not give clear answers to this problem.

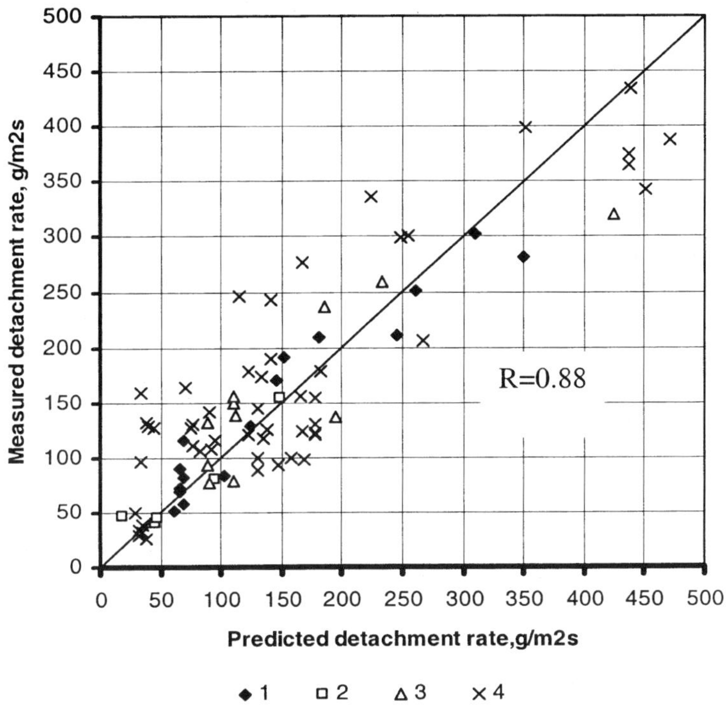

Fig. 3 Soil detachment rate predicted by equation (8) *vs* measured soil detachment rate. Bed load material: (1) porolon; (2) rubber; (3) rubber coated conductor; (4) quartz gravel.

Table 2 Influence of bed load on soil detachment.

Material	Bulk density, g cm^{-3}	Coefficient of Eq. (8)
Porolon	≈1.00	–0.00063
Rubber	1.21	–0.00048
Rubber coated conductor	3.36	–0.00016
Quartz gravel	2.65	0.00034

Table 3 Influence of suspended sediment on the soil erodibility.

Bulk density, g cm^{-3}	1.1				1.2				1.3			
Sediment concentration, g l^{-1}	0	6	12	24	0	6	12	24	0	6	12	24
Without infiltration												
Erodibility	243.1	135.4	108.1	–	123.5	74.5	67.8	–	10.8	11.9	5.3	–
Infiltration takes place												
Erodibility	370.8	85.38	58.2	57.1	126.1	43.3	25.9	17.8	15.4	7.6	3.9	1.8

The suspended sediment leads to an apparent decrease of the soil detachment rate (Table 3) which is believed to be due to the siltation of inter-aggregate pores. The light mass of sediment particles between dark aggregates can be seen easily on the break of treated samples. The soil samples with predominant point contact connections between aggregates turn into the more homogeneous soil body wherein tensile strength is sufficiently higher compared to the same soil with unsilted pore space. Thereafter the erodibility increases during the process of pore siltation. The degree of pore space siltation depends on the sediment concentration and the water infiltration rate. It explains why the decrease of detachment is higher in the tests that conducted under free infiltration of water and relatively high sediment concentration.

CONCLUSION

The proposed soil detachment model based on the three self evident premises show that soil detachment is proportional to cubed average flow velocity if it meets the condition $u \geq 1.6u_0$ and the probabilistic function is in the domain where flow velocities ranged from $u = 0.4u_0$ up to $u = 1.6u_0$. If flow velocity is less than $0.4u_0$ the rate of soil detachment is small, but proportional to the cubed velocity of flow too. The model explains well existing literature data of soil detachment..

The bed load effects strongly on the soil detachment rate but the mechanics of influence of soft bed load material is not yet clear. The suspended sediment decreases the soil erodibility due to siltation of the interaggregate porous space. The study should contribute to a better understanding of the mechanisms of the influence of sediment on soil detachment and a further quantification of this process.

Acknowledgments Financial support of RFBR (project 03-05-64822) and Leading Scientific Schools program grant (NS-1443.203.5) are gratefully acknowledged.

REFERENCES

Gendugov, V. M., Kuznetsov, M. S., Khalilov, M. S. & Ivanyta, C. C. (1997) Novyi podhod k ocenke vliyaniya erozii na pochvy (A new approach to evaluation of flow erosion effect on soil). *Vestnik Mosk. Un-ta. Cer. 17. Pochvovedenie* **2**, 37–41 (in Russian).

Foster, G. R. (1982) Modeling the erosion process. In: *Hydrologic Modeling of Small Watersheds* (ed. by C. T. Haan *et al.*). *ASAE Monogr.* **5**, 296–380.

Izbash, S. V. & Haldre, Kh. Y. (1959) *Gidravlika perekrytiya rusel rek* (The hydraulics of river damming). Gosenergoizdat, Moskva-Leningrad, USSR (in Russian).

Hairsine, P. B. & Rose, C. W. (1992) Modeling water erosion due to overland flow using physical principles: rill flow. *Water Resour. Res.* **28**, 245–250.

Kuznetsov, M. S. & Glazunov, G. P. (1985) *Eroziya pochv* (Soil Erosion). Izdatel'stvo Mosk. Universiteta, USSR (in Russian).

Larionov, G. A. & Krasnov, S. F. (1997) Gydrofizicheskaya koncepciya erozii pochv (A hydrophysical concept of soil erosion). *Pochvovedenie* **5**, 85–93.

Larionov, G. A., Dobrovol'skaya, N. G., Krasnov, S. F. & Yuan, Lu B. (2003) Novoe uravnenie factora rel'efa dlya statisticheskih modelei vodnoi erozii (New equation of relief factor for statistic models of water erosion). *Pochvovedenie* **10**, 1239–1247 (in Russian).

Merten, G. H., Nearing, M. A. & Borges, A. L. O. (2001) Effect of sediment load on soil detachment and deposition in rills. *J. Soil Sci. Am.* **65**, 861–868.

Mirtskhulava, Ts. E. (1967) *Ochistka kanalov i metod ocenki ikh stabil'nosti* (Scouring of channels and method of evaluation of its stability). Moscow, Kolos, USSR (in Russian).

Mirtskhulava, Ts. E. (1970) *Inzhenernye metody rascheta i prognoza vodnoi erozii* (Engineering Methods of Soil Loss Prediction). Gidrometeoizdat, Moscow, USSR (in Russian).

Nearing, M. A. (1991) A probabilistic model of soil detachment by shallow turbulent flow. *Trans. of ASAE.* v. 34, **1**. 81–85.

Nearing, M. A., Bradford, J. M. & Parcker, S. C. (1991) Soil detachment by shallow flow at low slopes. *Soil Sci. Soc. Am. J.* **55**. 339–344.

Rose, C. W. (1985) Development of soil erosion and deposition models. *Adv. Soil Sci.* **2**, 1–163.

Sediment Transfer through the Fluvial System (Proceedings of the Moscow Symposium, August 2004).
IAHS Publ. 288, 2004

Spatial modelling of debris flows in an alpine drainage basin

VOLKER WICHMANN & MICHAEL BECHT
KU Eichstaett-Ingolstadt, Ostenstrasse 18, D-85072 Eichstaett, Germany
volker.wichmann@ku-eichstaett.de

Abstract A model to describe the spatial distribution of torrent bed type debris flows in alpine drainage basins was developed and validated by field measurements. In addition to the determination of debris flow initiation sites, process pathway and erosion and deposition zones were identified. Potential process initiation sites were derived from channel slope, upslope contributing area and a material contributing area. Process pathway and travel distance were modelled by a grid-based "random walk" in conjunction with a 2-parameter friction model and Monte Carlo simulation. Erosion and deposition zones were derived by threshold functions of channel gradient and modelled velocity. A high magnitude rainstorm event in 2002 in the Lahnenwiesgraben drainage basin was used to validate the model. It is suggested that the model output may be used for the investigation of spatially distributed sediment transfers and for natural hazard zonation.

Key words debris flows; deposition; erosion; friction model; natural hazard zonation; Northern Limestone Alps; random walk; sediment transfer; travel distance

INTRODUCTION

Debris flows are common features in mountain drainage basins and they play a decisive role in sediment routing. They act as a sediment transport link between hillslopes and channels and thus have an important impact on the sediment budget. Debris flows influence the spatial and temporal distribution of sediments in alluvial channels, either because they deposit sediments in channels or because the deposits provide a source for increased sediment transport further downstream. Therefore, knowledge of the role of debris flow routing will aid in the construction of sediment budgets, and in interpreting the measurements of sediment transport in mountainous basins (Benda & Dunne, 1987).

Beside slope inclination and vegetation, the availability of water and the physical properties of the debris are most important for the initiation of debris flows. In a downslope direction, the flow increases in volume by entraining additional sediments, water and organic debris, and tends to become more destructive with travel distance. Debris flows can affect regions far from the initial failure sites, and in inhabited areas they usually cause damage to infrastructure and land resources, and may even claim lives. Therefore, methods for predicting initiation sites must be complemented by methods for predicting erosion and deposition areas (Benda & Cundy, 1990).

Our work forms part of the project sediment cascades in alpine geosystems (SEDAG), aiming at quantifying and modelling the sediment budget of mountain drainage basins (Haas *et al.*, 2004). In this paper, we describe models to identify process initiation sites, process pathway and run-out distance of debris flows (including lateral spreading and identification of erosion and deposition areas). The model output may be used for the investigation of spatially distributed sediment transfers and for natural hazard zonation. Field observations of initiation, erosion and deposition of debris flows in the Lahnenwiesgraben drainage basin after a high magnitude rainstorm event in the year 2002 are used to validate the models.

STUDY AREA AND DATA

The Lahnenwiesgraben drainage basin with an area of 16.6 km^2 is located in the Northern Limestone Alps, Bavaria, Germany and drains into the River Loisach. A wide range of fluvial and gravitational processes occur within the basin, including slope type and torrent bed type debris flows. For a more detailed description of the natural conditions see Haas *et al.* (2004).

In the early morning of 21 June 2002, a high magnitude rainstorm event triggered several slope type debris flows. Most channels remained stable, but a further event in the early afternoon caused failures in most of the steep torrents. These torrent bed type debris flows destroyed the forest road at several sites while large volumes of debris were transported and deposited. In this paper we focus upon torrent bed type debris flows, methods for modelling slope type debris flows are described elsewhere (Wichmann *et al.*, 2002; Wichmann & Becht, 2003, 2004).

In order to calibrate and validate the models, several debris flows were mapped in the field after the event (erosion and deposition heights and volumes). The consequences of the event for fluvial transport rates are discussed in Haas *et al.* (2004).

The model version presented here only requires data that can be derived directly from a digital elevation model (DEM). The DEM of the drainage basin was interpolated with ARC/INFO's Topogrid command at a resolution of 5 m. Elevation information was obtained from photogrammetric contour data (scale 1:10 000, contour interval 20 m, © Bayer. Landesvermessungsamt München (Bavarian Ordnance Survey), http://www.bayern.de/-vermessung, Az: VM-DLZ-LB-0628). Interpolation errors are relatively small and are mostly encountered in regions with steep rock walls (>60°). Due to forest cover, some elevation accuracy is already lost in the original data. Further problems arise from the fact that parts of the forest road had not been built when the elevation data was obtained. Thus, the forest road alignment is not represented in the DEM.

METHODS

The models described here were calibrated by empirical functions and field observations. We developed the models as module libraries (coded in C++) for a new GIS called SAGA (Böhner *et al.*, 2003). The determination of debris flow initiation sites is possible with the knowledge of the type and characteristics of the starting zone. Initiation sites of torrent bed type debris flows were derived in several steps (Zimmermann *et al.*, 1997; Wichmann & Becht, 2003, 2004). The channel network was extracted from the DEM by thresholds of upslope contributing area and plan curvature. The FD8 algorithm of Freeman (1991) in combination with a threshold for single flow direction was used to derive the upslope contributing area of each cell. Grid cells with a flow accumulation value higher than 2500 m^2 and a concave plan curvature (Zevenbergen & Thorne, 1987) less –0.02 were classified as channel cells. The thresholds were found by comparing the resulting grids with the channel network visible on orthophotos and topographic maps.

The failure mechanisms are complex, but beside high discharges there must be at least enough material available in the channel. On this account, a material contributing area was delineated to specify the amount of sediment input from the hillslopes (Zimmermann *et al.*, 1997). Starting from the channel network, a special algorithm recursively searches uphill until a maximum distance to the channel network (250 m) or a minimum slope gradient in flow direction (20°) is reached. This material contributing area may be weighted in relation to its sediment production. The weights (values between zero and one) can be chosen to reflect the natural conditions, e.g. vegetation cover, weathering rates and delivering process. As this information is rather difficult to obtain and has not been available up to now, we manage without this feature. This results in a maximum material contributing area for each channel cell (worst case scenario). Possible initiation sites were extracted from the channel network by combining empirically derived thresholds of upslope contributing area, slope and material contributing area. We used a minimum material contributing area of 10 000 m^2 and applied a threshold function derived by Zimmermann *et al.* (1997) for debris flows in Switzerland: $s = 0.32a^{-0.2}$, where s is the local channel gradient in percent and a is the upslope contributing area in km^2 (with increasing discharge downstream, lower channel gradients suffice for debris flow initiation). The last step is to use a filter along the channel network to remove redundant starting cells. The resulting grid shows the spatial distribution of potential debris flow initiation sites within the drainage basin and is used as input for modelling process pathway and travel distance (Fig. 1).

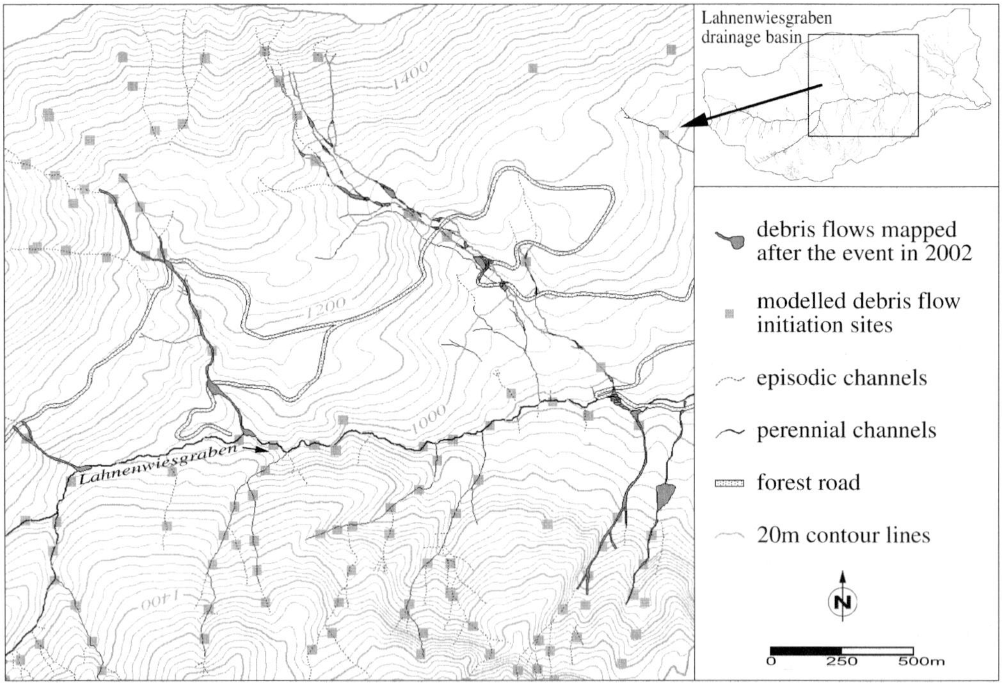

Fig. 1 Debris flows mapped after the event in 2002 and modelled debris flow initiation sites. The magnified map shows the central part of the Lahnenwiesgraben drainage basin.

The downstream movement of debris flows was modelled by a grid-based "random walk" (Price, 1976; Gamma, 2000) in conjunction with a 2-parameter friction model (PCM model, Perla *et al.*, 1980) and Monte Carlo simulation. Wichmann & Becht (2003) provide a more detailed description of the model. The "random walk" is a combination of single and multiple flow direction algorithms and can be adjusted by three calibration parameters (slope threshold for lateral spreading, parameter for divergent flow, persistence factor). On steep slopes near to the slope threshold, only neighbours with high gradients are allowed in addition to the steepest descent. In flat regions almost all lower neighbours are potential flow path cells and the tendency for divergent flow is increased. A higher weighting of the previous flow direction (persistence factor) is used to reduce abrupt changes in flow direction. A general tendency towards the steepest descent is achieved as the transition probabilities are weighted by slope (Gamma, 2000). We used a slope threshold of 20°, a divergence factor of 1.3, a persistence factor of 1.5 and calculated 1000 iterations from each initiation site. The parameters of the "random walk" were calibrated by field observations and are comparable with values used by Gamma (2000) for debris flows in Switzerland. Along with the identification of the next cell in the flow path, the local change in velocity is calculated with the PCM model. The model was originally developed for snow avalanches, but has also been applied to debris flows (Rickenmann, 1990; Gamma, 2000; Wichmann *et al.*, 2002; Wichmann & Becht, 2003, 2004). It is assumed that the motion is mainly governed by a sliding friction coefficient (μ) and a mass-to-drag ratio (*M/D*). We used a constant *M/D* of 75 m along the flow path and a spatially distributed friction coefficient μ derived from an estimating function of upslope contributing area a ($\mu = 0.13a^{-0.25}$; Gamma, 2000) to account for different rheology with higher discharges. A minimum threshold for μ is set to 0.045 and a maximum threshold to 0.3 (Gamma, 2000). It is possible to use different functions for different scenarios (Gamma, 2000; Wichmann & Becht, 2004). Here, we focus upon maximum run-out distances.

To model the sediment transfer throughout the basin, erosion and deposition zones need to be identified. Benda & Cundy (1990) developed an empirical model to estimate erosion volumes and deposition sites of debris flows from channel gradient and the junction angle between the contributing and receiving channel. We used threshold functions of channel gradient and/or modelled velocity to determine erosion and deposition zones and relative volumes assuming transport-limited conditions. The model may be run in two modes:

(a) Relative erosion and deposition heights are determined along the process pathway and specified (e.g. measured) volumes are scaled accordingly after model calculations. This mode may be used for stochastic forcing of sediment supply and routing as proposed by Benda & Dunne (1997).

(b) The volume of the initial failure is specified and further erosion depths along the process pathway are calculated using the slope and velocity thresholds. The entrained material is used for deposition and deposited material influences the following runs of the Monte Carlo simulation. Thus the filling of sinks and barriers and the plugging of the channel can be simulated.

It is possible to use a combination of slope and velocity thresholds to minimize artefacts resulting from the usage of one threshold alone (e.g. there is no material deposited in flat parts of the profile if the velocity is still high; deposition = min {height(velocity); height(slope)}; erosion = max {height(velocity); height(slope)}). The curves in Fig. 2 are used for calculating relative erosion and deposition heights along the process pathway. The

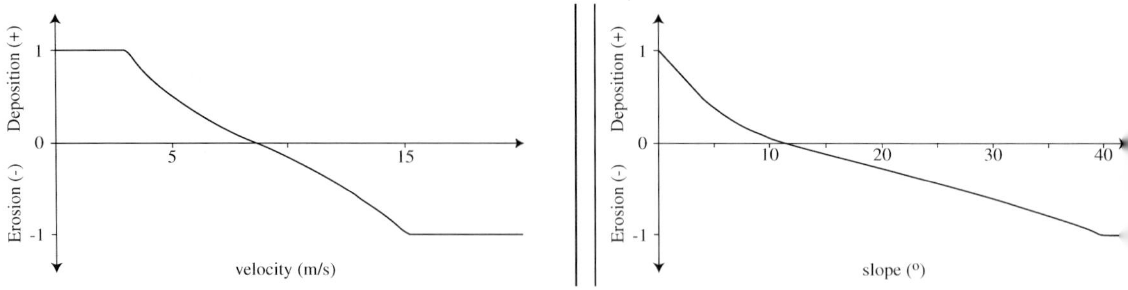

Fig. 2 Estimating functions of velocity and slope used for calculating erosion and deposition heights along the process pathway.

values, calibrated by field observations, do not necessarily correspond to physical reality as velocity is modelled and slope depends on grid resolution.

RESULTS

The examples presented here were calculated with the model settings discussed in the previous section. Figure 1 shows some of the debris flows mapped after the event in 2002, together with the modelled debris flow initiation sites. As we performed no weighting of the material contributing area, nearly all steeper channels are prone to debris flow initiation. This worst-case scenario reproduced all torrent bed type debris flows of the event in 2002. Figure 3(a) shows the Herrentischgraben sub-basin in greater detail. The pattern is quite complex as some of the debris flows unite downstream. Figure 3(b) shows the modelling

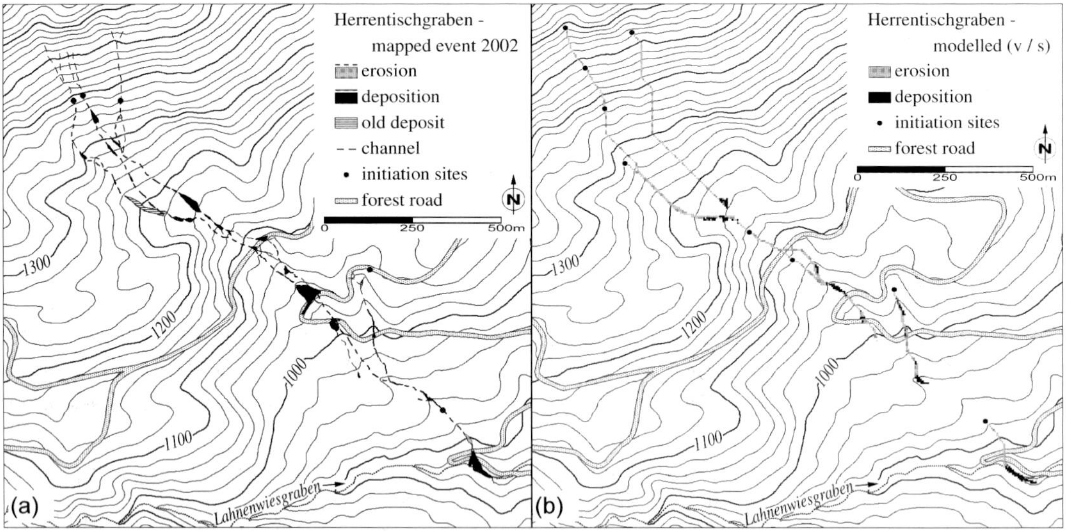

Fig. 3 Maps of (a) observed and (b) modelled debris flows in the Herrentischgraben sub-basin (Lahnenwiesgraben). Erosion and deposition zones are modelled by a combination of velocity and slope thresholds.

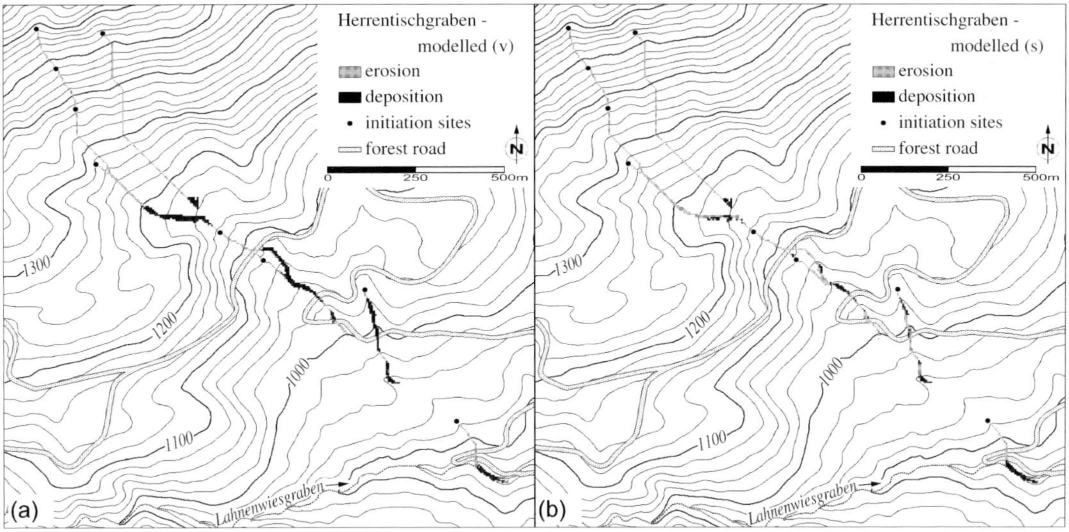

Fig. 4 Maps of modelled debris flows in the Herrentischgraben sub-basin (Lahnenwiesgraben). Erosion and deposition zones are modelled by (a) velocity and (b) slope thresholds.

results obtained by the usage of a combination of velocity and slope thresholds. The initiation sites match quite well with the observed ones, only the highest site mapped in Fig. 3(a) is missing. Besides that, the corresponding channel is not represented in the DEM very well. This debris flow may be classified as slope type debris flow (see the corresponding deposition area just below the initiation site). Process pathways and travel distances resemble the mapped ones, the exact location is not reproduced due to inaccuracies of the elevation data (nearly the whole sub-basin is covered by forest). These inaccuracies become most apparent by comparing erosion and deposition zones. As the forest road alignment is not included in the DEM, the extent of the deposits on the road was modelled too small (no lateral spreading and no deposition because of high slope gradients). The results obtained by using velocity and slope thresholds alone are shown in Fig. 4(a,b). Using the velocity threshold function to predict erosion and deposition zones resulted in a slight overestimation of depositional areas. The usage of the slope threshold reproduced the observed pattern quite well, only the extent of some depositional areas was underestimated.

CONCLUSIONS

Both predicted process area and erosion and deposition zones resemble the pattern of the observed event. This is remarkable as only data obtained from a DEM was used for modelling. The good results may be supported by mainly transport-limited conditions during the event in 2002.

The model can be used by engineers and resource planners to recognize and zone hazard areas. In addition, the model can be used in geomorphic process studies to assist in predicting erosion and deposition zones. The accuracy of the DEM has a major impact on the results and further investigations with different data and resolutions is needed to extend and validate the model.

Acknowledgements This study is funded by the German Science Foundation (DFG, Bonn), which is gratefully acknowledged by the authors. The comments of an anonymous reviewer on the paper are highly appreciated.

REFERENCES

Benda, L. & Cundy, T. (1990) Predicting deposition of debris flows in mountain channels. *Can. Geotech. J.* **27**, 409–417.

Benda, L. & Dunne, T. (1987) Sediment routing by debris flows. In: *Erosion and Sedimentation in the Pacific Rim* (ed. by R. L. Beschta, T. Blinn, G. E. Ice & F. J. Swanson) (Proc. Corvallis Symp., August 1987), 213–223. IAHS Publ. 165. IAHS Press, Wallingford, UK.

Benda, L. & Dunne, T. (1997) Stochastic forcing of sediment supply to channel networks from landsliding and debris flows. *Water Resour. Res.* **33**(12), 2849–2863.

Böhner, J., Conrad, O., Köthe, R. & Ringeler, A. (2003) System for an automated Geographical Analysis. http://134.76.76.30

Freeman, T. G. (1991) Calculating catchment area with divergent flow based on a regular grid. *Computers & Geosciences* **17**(3), 413–422.

Gamma, P. (2000) *dfwalk*—Ein Murgang-Simulationsprogramm zur Gefahrenzonierung (*dfwalk*—a debris flow simulation program for natural hazard zonation). *Geographica Bernensia* **G66**, Bern, Switzerland (in German).

Haas, F., Heckmann, T., Wichmann, V. & Becht, M. (2004) Change of fluvial sediment transport rates after a high magnitude debris flow event in a drainage basin in the Northern Limestone Alps, Germany. In: *Sediment Transfer through the Fluvial System* (ed. by V. Golosov, V. Belyaev & D. E. Walling) (Proc. Moscow Symp., August 2004). IAHS Publ. 288. IAHS Press, Wallingford, UK (this volume).

Perla, R., Cheng, T. T. & McClung, D. M. (1980) A two-parameter model of snow-avalanche motion. *J. Glaciol.* **26**(94), 197–207.

Price, W. E. (1976): A random-walk simulation model of alluvial-fan deposition. In: *Random Processes in Geology* (ed. by D. F. Merriam), 55–62. Springer, New York, USA.

Rickenmann, D. (1990) Debris flows 1987 in Switzerland: modelling and fluvial sediment transport. In: *Hydrology in Mountainous Regions. II – Artificial Reservoirs; Water and Slopes* (ed. by R. O. Sinninger & M. Monbaron) (Proc. Lausanne Symp., August 1990), 371–378. IAHS Publ. 194. IAHS Press, Wallingford, UK.

Wichmann, V. & Becht, M. (2004) Modelling of geomorphic processes in an alpine catchment. In: *GeoDynamics* (Conf. Proc. GeoComputation 2003, Southampton) (ed. by P. Atkinson, G. Foody, S. Darby & F. Wu) (in press).

Wichmann, V. & Becht, M. (2004) Modellierung geomorphologischer Prozesse zur Abschätzung von Gefahrenpotentialen (Modelling of geomorphic processes for natural hazard zonation) (in German). *Z. Geomorph. N. F.*, Suppl. (submitted).

Wichmann, V., Mittelsten Scheid, T. & Becht, M. (2002) Gefahrenpotential durch Muren: Möglichkeiten und Grenzen einer Quantifizierung (Hazard potential of debris flows: Chance and limit of a quantification). *Trierer Geographische Studien* **25,** 131–142 (in German).

Zevenbergen, L. W. & Thorne, C. R. (1987) Quantitative analysis of land surface topography. *Earth Surf. Processes Landf.* **12**, 47–56.

Zimmermann, M., Mani, P., Gamma, P., Gsteiger, P., Heiniger, O. & Hunziker, G. (1997) Murganggefahr und Klimaänderung—ein GIS-basierter Ansatz (Debris flow hazard and climate change—a GIS based approach) Schlussbericht NFP 31, Zurich, Switzerland (in German with English summary).

Computation of the regime configuration of a meandering stream

D. TERMINI[1] **& M. S. YALIN**[2]

1 *Dipartimento di Ingegneria idraulica ed Applicazioni Ambientali, Facoltà di Ingegneria, Università di Palermo, Viale delle Scienze, I-90128, Italy*
dony@idra.unipa.it

2 *Department of Civil Engineering, Queen's University, Kingston, Ontario, Canada*

Abstract Regime channel formation is a delicate adaptation to the imposed environmental conditions compatible with flow and sediment transport mechanisms. The present paper concerns the computation of the regime configuration of a meandering stream. It is supposed that the channel develops until it reaches the state of "final thermodynamic equilibrium", where the ratio of the kinetic energy of the flow to its cross-sectional potential energy is minimum. An optimization procedure that allows the analysis of the regime channel formation is presented and it is checked using a real case.

Key words predictive model; regime configuration; river's plane-form evolution

INTRODUCTION

Natural rivers tend to adjust their plane shape and longitudinal profile in order to assume a configuration compatible with the changing hydraulic and man-made constraints. The final configuration, called "regime configuration", represents the equilibrium geometry self-formed by the alluvial stream. The prediction of the regime channel configuration has generated a lot of interest amongst scientists and engineers. In technical terms the approach consists of the identification of the independent variables (often called controlling variables) and of the dependent variables (channel width, water depth, channel slope, meandering or braiding pattern). The independent variables, which are imposed upon the river by nature or by man, maintain their constant values through the duration of the regime development and they can be considered as the cause of the evolution process; the dependent variables vary during the channel evolution in order to obtain a value compatible with the imposed conditions. In selecting the independent variables for the prediction analysis, it is necessary to take into account the time scale of the process under study. In long-term processes, as considered in the present work, the mechanisms controlling the planimetric channel evolution become significant (Chang, 1988; Di Silvio, 1994): the water discharge and the sediment discharge can be considered as independent variables and the channel characteristics (width, depth and slope) as dependent variables. The planimetric evolution of the channel follows as effect of the regime-formation process that could take place in different ways: in some cases the stream is meandering, in others it is braiding (Yalin & da Silva, 2001).

There is still no unanimity on the question of which process is mainly responsible for the initiation and development of meandering stream. In particular, two major theories can be considered: the stability theories (Engelund & Skovgaard, 1973; Callander, 1978; Parker *et al.*, 1982; Zolezzi & Seminara, 2001) which consider the meandering pattern as a consequence of an unstable response of the river to some perturbations; the rational theories or extreme methods (Yang & Song, 1979; White *et al*., 1982; Davies & Sutherland, 1983; Yang, 1984), which are motivated by the conviction that the regime configuration develops because an energy related physical quantity tends to acquire its minimum value. In the latter case the regime configuration is revealed by solving a system of equations determining the dependent variables (width, depth and slope). Within the rational theories various physical quantities reaching the minimum value have been considered: the rate of energy dissipation (Yang & Song, 1979), the transport rate (White *et al.*, 1982), the unit stream power (Yang, 1984), the stream power (Chang, 1988), etc.

Recently Yalin & da Silva (2000, 2001) demonstrated, on the basis of thermodynamics considerations, that the channel develops until it reaches the state of "final thermodynamic equilibrium", where the ratio of the kinetic energy of the flow to its cross-sectional potential energy (as implied by the flow Froude number) is minimum. In the present work, this last approach is applied and a computational procedure to determine the regime channel formation is presented.

THE SYSTEM OF EQUATIONS

In accordance with the rational theories, the regime characteristics (channel width, water depth, channel bed slope) of an alluvial stream of constant flow rate, Q, can be estimated by solving a system of equations formed by:

—the resistance equation, that can be written as:

$$Q = Bhc\sqrt{gSh} \tag{1}$$

where c is the friction factor; g is the acceleration due to gravity; B is channel width; h is the water depth, S is the bed slope;

—the transport equation, written in the following general form:

$$Q_s = D_{50} u_* \phi_{Q_s}(X, Y, Z) \tag{2}$$

where Q_s is the sediment transport rate; u_* is the shear flow velocity; D_{50} is the representative sediment diameter; ϕ_{Q_S} is in general a function of the non-dimensional parameters $X = \dfrac{u_* D_{50}}{\nu}$, $Y = \dfrac{\rho u_*^2}{\gamma_s D_{50}}$, $Z = \dfrac{h}{D_{50}}$, where ρ is water density; γ_s is specific weight of water; ν is kinematic viscosity, and it has to be specified on the basis of the relationship used to express the transport rate (Einstein, 1942; Meyer-Peter; 1949; Yalin, 1963).

—the optimization condition of a quantity related to the stream energy. As a rule, it is written as :

$$\text{Minimize } O.F. = \varphi(V, V_c)$$

$$V = (B, h, S, \ldots); \quad V_c = (Q, Q_s, \gamma_s, \gamma, \rho, D_{50}, \ldots) \tag{3}$$

where V is the decision variables vector, V_c is the controlling variables vector. The optimization condition (3) expresses the condition to which the river system tends spontaneously.

The equations (1) and (2) can be treated as constraint equations of the optimization problem (3). The equations (1)–(3) system is solved and the regime values of the variables B, h and S are estimated. Because it is not easy to evaluate the controlling variable Q_s, Yalin & da Silva (2001) proposed to replace equation (2) by the following equation:

$$B = f_1\, f_2 \sqrt{\frac{Q}{u_{*,cr}}} \tag{4}$$

$$f_1 = 0.639 \left[\frac{\gamma_s D_{50}^3}{\rho \nu^2} \right]^{0.3} \tag{5}$$

if the term in square brackets is <15;

$$f_1 = 1.42 \text{ otherwise } f_2 = \left[0.2\, b \left(1 - e^{-0.35\left|\bar{c} - 27.5\right|^{1.2}} \right) + 1.2 \right] \frac{c}{\bar{c}} \tag{6}$$

with $b = 1$ if $\bar{c} \geq 27.5$; $b = -1$ if $\bar{c} < 27.5$

where $u_{*,cr}$ is the critical shear flow velocity, $\bar{c}$ is the friction factor with flat bed.

In accordance with previous experimental observations (Leopold & Wolman, 1957; Ackers, 1964), equations (4)÷(6) were derived by assuming that the variation of channel width essentially takes place during a very short time period (say *tb*) in comparison with the entire duration of the regime channel formation (*tR*); thus, no substantial adjustment in channel width occurs during the time interval (*tR*–*tb*).

THERMODYNAMIC CONSIDERATIONS

The second law of thermodynamic establishes that the entropy, ζ, of an isolated system tends to increase with the passage of time. If a river system is considered as a closed sub-system (Yalin & da Silva, 2001), it tends to evolve in order to increase its entropy. Thus, it is:

$$\frac{d\zeta}{d\theta} \geq 0 \tag{7}$$

where $\theta = t/t_R$ represents the no-dimensional time.

On the other hand, in accordance with Gibb's equation, the increment of the system entropy is related to the increment of the system internal energy U:

$$T \frac{d\zeta}{d\theta} = \frac{dU}{d\theta} \tag{8}$$

where T is the absolute temperature.

In general, a body of mass M and volume W (control volume of the river system) receives both mechanical energy by the work of forces to which it is subjected and thermal energy through the contour surface. The contribution to the system of these forms of energy is equal to the variation in time both of the kinetic energy and of the internal energy of the system. If a river system is considered, as aforementioned, a closed sub-system, it is supposed that the net heat and the work time rates exchanged between the stream and the boundaries is null. Thus, the variation in time (i.e. during the regime development) of the sum of the kinetic and internal energy has to be equal to zero:

$$\frac{d}{d\theta}\left(U+\frac{\rho}{2}Wu^2\right)=0 \tag{9}$$

where u is the average flow velocity.

Consequently, the increment of the system entropy corresponds to a decrement of the kinetic energy and, because W and ρ are positive quantities, the decrement of the kinetic energy coincides with the decrement in average flow velocity.

Thus, the kinetic energy (and thus the average flow velocity) represents the physical quantity that tends to reach its minimum value in time, producing the channel regime formation. On the other hand, from the constraint equation (1), for a wide alluvial stream, it is:

$$\frac{u^2}{gh}=c^2S \tag{10}$$

where the term on right hand side represents the Froude number F_r. Thus, replacing equation (10) in (9) and taking into account equation (8), after simple passages (Yalin & da Silva, 2001) it can be concluded that, during a time period $t_b \le t \le t_R$, the increment of the entropy (equation (7)) is necessarily accompanied by a decrement of the stream Froude number.

SOLUTION OF THE OPTIMIZATION PROBLEM

On the basis of the thermodynamic considerations reported in previous section, it is assumed that the regime channel formation has to be related to the minimization of the stream Froude number, that can be estimated by equation (10). Thus, the objective function of the optimization problem (3) has to express the minimization of the product (c^2S).

For given geometric characteristics of the sediment on the bed, the friction factor c depends on the hydraulic characteristics of the flow and it can be estimated as (Engelund, 1966; Yalin, 1992):

$$c=\sqrt{1\Big/\left[\frac{1}{\bar{c}^{2}}+\frac{1}{2h}\left(\delta_r{}^2\Lambda_r+\delta_d{}^2\Lambda_d\right)\right]} \qquad \bar{c}=\frac{1}{\kappa}\ln\left(\frac{h}{k_s}\right)+B_s \tag{11}$$

where κ is the Von Karman's constant (assumed equal to 0.4); k_s is the representative bed roughness (assumed equal to $2D_{50}$); B_s is a function of the Reynolds shear number $\left(\text{Re}_*=\frac{u_*D_{50}}{\nu}\right)$, δ_r and δ_d represent respectively the ripples and dunes steepness; Δ_r and Δ_d represent respectively the ripples and dunes height. These last four terms (δ_r, δ_d, Δ_r, Δ_d) depend on three non-dimensional parameters: $Z=\frac{h}{D_{50}}$, $\xi=\left(\frac{\gamma_s D_{50}^3}{\rho\nu^2}\right)^{1/3}$ and $\eta=\frac{\tau}{\tau_{cr}}=\frac{u_*^2}{u_{*,cr}^2}$.

Thus, it can be synthetically written (Yalin, 1992) that:

$$\delta_r=\delta_r(\eta,Z),\ \ \Lambda_r=\Lambda_r(\eta,\xi,Z)\ \ \delta_d=\delta_d(\eta,\xi,Z),\ \ \Lambda_d=\Lambda_d(\eta,\xi,Z) \tag{12}$$

The parameter ξ depends on the physical characteristics of water and sediment, that are supposed constant during the regime channel evolution; the parameter Z depends on the water depth h, that is a state variable of the considered optimization problem; the parameter η represents the ratio of the average shear stress to the critical shear stress and, thus, it is related to the stream energetic properties; thus, η is considered the decision variable of the optimization problem.

The bed slope, S, varies in time and it can also be expressed as a function of the variable η. Taking into account that $u_*^2 = g\,h\,S$, it is:

$$S(\eta) = \eta \frac{u_{*,cr}^2}{gh} = \eta \left[\frac{u_{*,cr}^2}{gD_{50}} \right] \frac{D_{50}}{h} \qquad (13)$$

The term in square brackets, multiplied by the ratio γ/γ_s (γ = specific weight for water), represents the no-dimensional number $Y_{cr} = \frac{\gamma u_{*,cr}^2}{\gamma_s g D_{50}}$. Thus (13) becomes:

$$S(\eta) = \eta \frac{\gamma_s}{\gamma Z} Y_{cr} \qquad (14)$$

by taking into account equation (1), it is: $N = \frac{Q}{BD_{50}u_{*,cr}} = \frac{h}{D_{50}} c \frac{\sqrt{gSh}}{u_{*,cr}} = Z\,c\,\sqrt{\eta}$ and the equation (14) can be rewritten as:

$$S(\eta) = \frac{\gamma_s}{\gamma} Y_{cr} \frac{c(\eta)\eta^{2/3}}{N} \qquad (15)$$

Finally, the optimization problem (3) is expressed as:

$$\text{Minimize FO} = \left[\frac{\gamma_s}{\gamma N} Y_{cr} \eta^{2/3} c(\eta)^2 \right] \qquad (16)$$

subjected to constraints (1) and (4)÷(6). The function $c(\eta)$ is determined by (11).

In gravel-bed channels the evolution process stops when $\eta = 1$ (i.e. $\tau = \tau_{cr}$). In sand bed channels the regime configuration could be reached for $\eta>1$, because of the presence of bed forms. Thus, the following constraint has also been considered:

$$\eta \geq 1 \qquad (17)$$

The critical mobility number Y_{cr} can be estimated as a function of the parameter ξ (Yalin & da Silva, 2001), that depends on the physical characteristics of the granular material and of the fluid, both assumed constant during the regime channel formation. Thus, on the plane Fr-η, the objective function (16) is a curve of parameter N.

APPLICATION

The optimization problem (1)–(4)÷(16)–(17) is nonlinear and it has been solved with the help of the commercial code MINOS (Murtagh & Saunders, 1983). The solution of the optimization problem allows to estimate the regime values of the channel width, water depth and longitudinal bed slope, for a given bed material and for a given flow rate. The procedure is iterative and it gives as result the values of pairs (F_r–η) at each i-th stage of the channel evolution. At $t = 0$ (initial condition) both the characteristics of the "initial" channel (B_o, h_o, S_o) and the physical characteristics of the bed material and of the fluid have to be known. Thus, the parameters ξ and Y_{cr} can be estimated and the critical shear velocity is calculated as:

$$u_{*,cr} = \left(\frac{\gamma_s D_{50}}{\rho}\right)^{1/2} Y_{cr}^{1/2} \tag{18}$$

The no-dimensional parameter N_i, at i-th evolution stage, is determined and the minimum value of the Froude number ($F_{r,i}$) and the corresponding flow intensity η_i are obtained as solution of the optimization problem (16). Then, the corresponding flow depth is evaluated as $h_i = \left(\dfrac{Q^2}{gB_i^2 F_{r,i}}\right)^{1/3}$ and the new channel width is calculated by (4)÷(6). The procedure stops when the values of the channel width estimated for two following iterations are almost the same.

Finally, the regime bed slope is calculated as:

$$S_R = \eta_R \frac{\gamma_s D_{50} Y_{cr}}{\gamma h_R} \tag{19}$$

The flow chart of the proposed procedure is reported in Fig. 1.

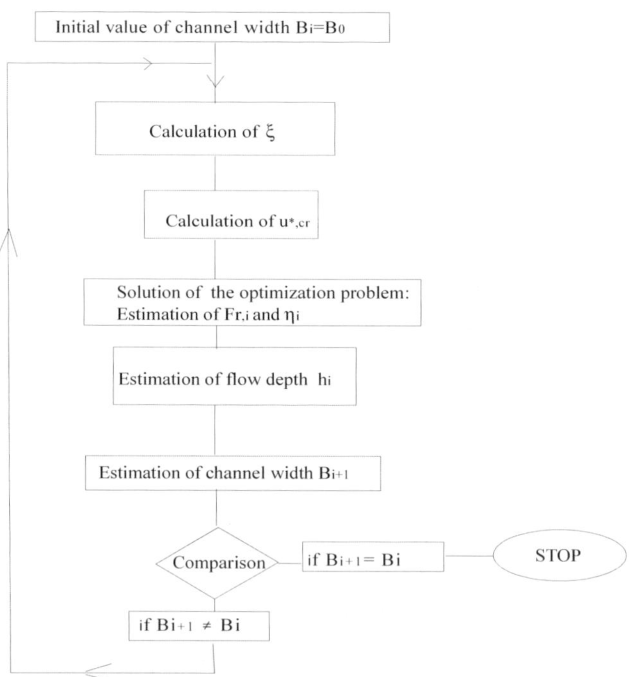

Fig. 1 Flow chart of the proposed procedure.

It is assumed that the channel plane shape follows the sine-generated curve (Langbein & Leopold, 1966). Thus, it is:

$$\vartheta = \vartheta_o \cos\left(2\pi \frac{l}{L_m}\right) \tag{20}$$

where l is the longitudinal abscissa along the channel centreline; L_m is the meander length; ϑ_o is the deflection angle at $l = 0$; ϑ is the deflection angle at the abscissa l.

Thus, the regime plane configuration of the channel is evaluated by (20), taking into account that L_m and the meander wave length (Λ) are interrelated by ϑ_o as (Yalin, 1992):

$$\frac{L_m}{\Lambda} = \frac{1}{J(\vartheta_o)}\left(= \frac{S_o}{S_R}\right) \tag{21}$$

where $J(\vartheta_o)$ is the Bessel function of the first kind and zero-th order.

The procedure has been applied in order to evaluate the regime configuration of Peace River in Alberta (data found in Yalin & da Silva, 2001). The initial channel width (B_o) is 619.2 m, the initial water depth (h_o) is 9.33 m and the initial bed slope $(S)_o$ is 0.0084%; the water discharge is equal to 9905 $m^3 s^{-1}$ and the median size of bed material (D_{50}) is 0.31 mm.

The regime values of channel width, water depth and longitudinal bed slope estimated as solution of problem (1)–(6) and (16)–(17) are: B_R= 797.7 m; h_R = 18.6 m; S_R = 0.0000339. These values have been obtained for $\eta_R = 29.64$ and $Fr_{,R} = 2.7 \times 10^{-3}$ The regime (at $\theta = 1$) planimetric channel configuration is characterized by a value of the deflection angle $\vartheta_o = 96.6°$, and it has been determined by (20). In Fig. 2 the configurations of the channel axis estimated at three evolution stages are reported.

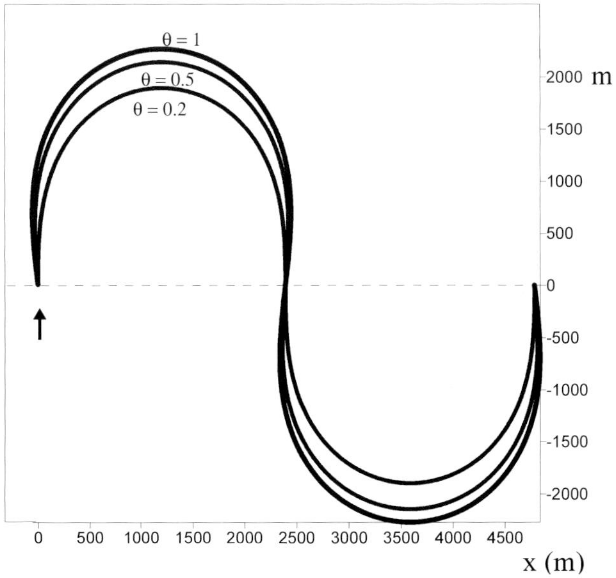

Fig. 2 Estimated planimetric channel configurations.

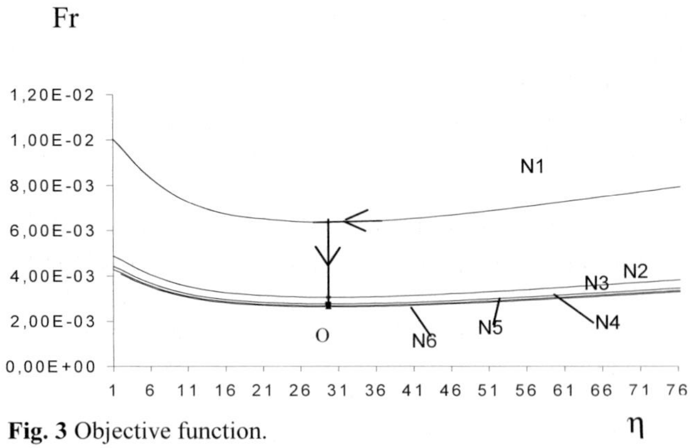

Fig. 3 Objective function.

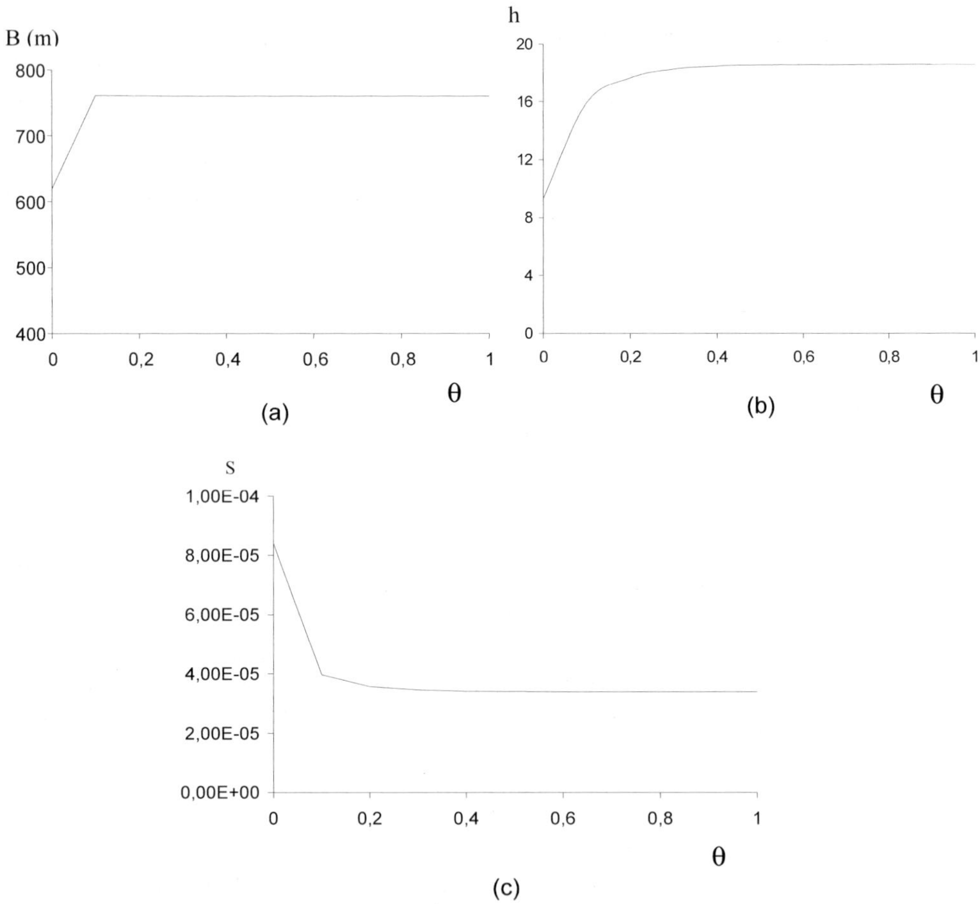

Fig. 4 State variables calculated at each time step: (a) channel width; (b) water depth; (c) channel bed slope.

In Fig. 3 the pairs of Fr–η estimated at each evolution stage are reported. The parameter of each curve is the value of N estimated at the corresponding evolution stage; the point O identifies the final regime condition.

The channel width, the water depth and the channel bed slope estimated at each evolution stage are reported in Fig. 4(a–c), respectively. Both the channel width and the water depth increase during the channel evolution, while the longitudinal channel bed slope decreases. As Fig. 4(a) shows, the variation of the channel width is significant in the early evolution stages ($\theta < 0.1$); it is almost equal to the final regime value already for $\theta = 0.1$.

Furthermore, Fig. 4(b) and (c) show that three variation phases can be identified both for the water depth and for the longitudinal bed slope: a first phase ($\theta \leq 0.1$) where both the water depth and the longitudinal bed slope vary rapidly; a second phase ($0.1 \leq \theta \leq 0.5$) where gradual variations occur, and a third phase ($\theta > 0.5$) where very little variation takes place until the variables reach the corresponding regime values.

CONCLUSION

In this paper a computational procedure for the prediction of the regime channel formation is presented. The procedure solves an optimization constrained problem where the objective function expresses the minimization of the stream Froude number, as recently suggested by one of the authors. The proposed computational procedure allows the evaluation of the planimetric configuration of the channel at each evolution stage until the regime configuration is reached. The application to a real case highlights the procedure capability.

REFERENCES

Ackers, P. (1964) Experiments on small streams in alluvium. *J. Hydr. Div., ASCE* **90**(4), 4–37.

Callender, R. A. (1978) River meandering. *Ann. Rev. Fluid Mechanics* **10**, 129-158.

Chang, H. H. (1988) On the cause of river meandering. In: *International Conference on River Regime* (ed. by W. R. White), 83–93. John Wiley & Sons, Chichester, UK.

Davies, T. H. R. & Sutherland, A. J. (1983) Extremal hypothesis for river behaviour. *Water Resour. Res.* **19**(1), 141–148.

Di Silvio, G. (1994) Trasporto solido e modellamento di fondo (Sediment transport and bed modelling). In: *XXIV Convegno di Idraulica e Costruzioni Idrauliche*. Napoli – 20/22 Settembre, CUEN s.r.l. Napoli , Italy, 323–355.

Einstein, H. A. (1942) Formulae for the transportation of bed load. *Trans. ASCE* **107**, 561–597.

Engelund, F. (1966) Hydraulic resistance of alluvial streams. *J. Hydr. Div., ASCE* **92**(2), 315–326.

Engelund, F. & Skovgaard, O. (1973) On the origin of meandering and braiding in alluvial streams. *J. Fluid Mech.* **57**(2), 289–302.

Langbein, W. B. & Leopold, L. B. (1966) River meanders-theory of minimum variance. *US Geol. Survey Prof. Paper* **422H**, 1–15.

Leopold, L. B. & Wolmann, M. G. (1957) River channel patterns: braided, meandering and straight. *US Geol. Survey Prof. Paper* **282B**, 39–85.

Meyer, P. E. (1949) *Formulas for Bed-Load Transport*. Int. Ass. Hydr. Res., Stockolm, Sweden.

Murtagh, B. A. & Saunders, M. A. (1983) MINOS 5.0 User's Guide. Systems Optimization Laboratory. Stanford University, USA.

Parker, G., Sawai, K. & Ikeda, S. (1982) Bend theory of river meanders. 2. Nonlinear deformation of finite-amplitude bends. *J. Fluid Mech.* **115**, 303–314.

White, W. R., Bettess, R. & Paris, E. (1982) Analytical approach to river regime. *J. Hydrol. Div., ASCE* **108**(10), 1179–1193.

Yalin, M. S. (1963) An expression for bed-load transportation. *Proc. ASCE* **89**(3), 221–250.

Yalin, M. S. (1992) *River Mechanics*. Pergamon Press, Oxford, UK.

Yalin, M. S. & Ferreira da Silva, A. (2000) Computation of regime channel characteristics on thermodynamics basis. *J. Hydrol. Res.* **38**(1), 57–63.

Yalin, M. S. & Ferreira da Silva, A. (2001) *Fluvial Processes*. IAHR Monograph, Delft, The Netherlands.

Yang, C. T. (1984) Unit stream power equation for gravel. *J. Hydrol. Engng*., ASCE **110**(12), 1783–1798.

Yang, C. T. & Song, C. S. (1979) Theory of minimum rate of energy dissipation. *J. Hydrol. Div., ASCE* **105**(7), 769–784.

Zolezzi, G. & Seminara, G. (2001) Downstream and upstream influence in river meandering. Part I—general theory and application development. *J. Fluid Mech.* **438**, 183–211.

Sediment Transfer through the Fluvial System (Proceedings of a symposium held in Moscow, August 2004).
IAHS Publ. 288, 2004

Using computer modelling for regulation of sediment transport under hydraulic structures on a large river

A. A. ZAITSEV[1], V. V. BELIKOV[2] & A. N. MILITEEV[2]

1 *Laboratory for Soil Erosion and Fluvial Processes, Faculty of Geography, Moscow State University, Leninskye Gory, GSP-2, Moscow 119992, Russia*
zaitsev@river.geogr.msu.su

2 *OAO "NIIES", B. 7a, Stroitelniy proezd, Moscow 123362, Russia*

Abstract Mathematical models based on shallow water equations and equations of bed deformations are proposed for use in designing hydraulic structures and channel straightening for large rivers in northeast Siberia (Russia). The objective of the simulation was to evaluate short time changes of channels under man-made and natural impact. The methods of hydrodynamic models application for studying of river relief change are described, and examples of simulations are demonstrated.

Key words channel deformations; computer model; equation of deformation; equations of unstratified shallow water; hydraulic structures

INTRODUCTION

The economic problems are related to questions of the planning and construction of small hydraulic structures such as water intakes, river harbours, underwater pipelines, bridges, spring water levies, bank protection dams, etc. Most projects are purposed for the regulation of water flow and sediment transport, along rather short river sections. In some cases it is required to induce the sedimentation near hydraulic constructions, for instance to protect the coast from erosion. However, in other cases the sedimentation is an undesirable factor, e.g. when it is necessary to stabilize the balance of alluvium around water intakes. Navigable rivers need to control configuration and variation of bed forms, even without any hydraulic constructions.

Forty years ago the preference was given to large flow regulation structures, the design of which was associated with widely practiced physical modelling. Application of physical models for designing of small hydraulic engineering constructions is unfairly expensive. Usually empirical methods are applied for calculation of construction parameters. The operation of such structures revealed the imperfection of the design technology, which was based on very rough calculations and methods of the similarity. As a consequence, the errors in designing resulted in quick destruction and failures in operation of small hydraulic structures. For design of different variants of the engineering constructions and artificial influence on the river, it is possible to use the computer models, which should be based on equations of unstratified shallow water and equation of deformation.

Simulations made by such models determine trends of channel deformations and allow the consideration of the required number of alternatives for the arrangement of hydraulic structures with regard to their interaction with the river channel under natural and hypothetical conditions. The technological sequence worked out by the authors was tested for large rivers in the northeast of Russia and offered the solution of a number of problems related to the national economy.

COMPUTER MODELLING IN FLUVIAL HYDRAULIC PRACTICE

Before starting computer model development the following information on the subject is required:

(a) to carry out the analysis of natural controls of river channel development, including endogenic (geological structure and tectonics) and eczogenic (climatic conditions, runoff, etc.) factors;
(b) to execute channel analysis;
(c) to determine initial (morphologic data) and boundary (hydrologic data) conditions;
(d) to decide about the location and configuration of hydraulic structures in the course of design, the provision of permanent control of their technical condition, and the elaboration of alternatives of repair recommendations based on the results of the channel deformation study.

At the following stage it is required to collect the data for modelling. In the present work the river bathymetry data was collected by hydrographic survey of the study reach. Improvement of ultrasonic methods of sounding has led to the development of up-to-date echo sounders, side-scan sonars, seismoacoustic profilographs (Zaitsev & Egorov, 1999). The spatial location of measured point was defined using navigation satellites of NAVSTAR (USA) and GLONASS (Russia) systems. The survey procedure underwent continuous improvement and was gradually put into practice. Standard and specialized software was used to plot graphs and layouts of channels on the basis of the observation results.

The flood plain topography data was obtained by vectorization of detailed topography maps. High-altitude marks of a ground and a water table were determined by means of geodesy systems and satellite positioning. Numeric data arrays were used for developing a digital model of channel configuration. The network nodes for the modelling area were automatically generated on a regular and non-regular grid by special software (Zajtsev *et al.*, 2003). As a result a numerical model of a relief was obtained.

The computer model consists of system of equations describing the stimulated phenomenon, numeric scheme of their discretization, computer program, and information on a particular object. The model should adequately reveal the flow structure at the studied site. The use of computer models in further calculations is based on solving a system of equations (Saint-Venant equations) and an elementary equation, which are used to calculate bed deformations (Grishanin, 1979; Lyatkher & Militeev, 1981). For rivers the 2-D system of equations is usually applied:

$$\frac{\partial q_i}{\partial t} + \frac{\partial q_j U_i}{\partial x_j} + gh\frac{\partial \zeta}{\partial x_i} = \tau_{n_i} + \tau_{w_i} \qquad (1)$$

$$\frac{\partial \zeta}{\partial t} + \frac{\partial q_i}{\partial x_i} = q_{ïð} \qquad (2)$$

$$(1-p)\frac{\partial z}{\partial t} + \frac{\partial U_1 sh}{\partial x_1} + \frac{\partial U_2 sh}{\partial x_2} = 0 \qquad (3)$$

(use for all the equations the indices *i* and *j*).

Where i = 1,2; j = *1*, are postscripts for Cartesian coordinates x_1, x_2; *t*-time; $\vec{q} = (q_1, q_2)$, vector of unit discharges of water (m^2 s^{-1}); *h*, depth; $\vec{U} = (U_1, U_2)$ –

vector of averaged velocity; g, gravity acceleration; ζ, elevation of water table; $\vec{\tau}_n = (\tau_{n_1}, \tau_{n_1})$, $\vec{\tau}_w = (\tau_{w_1}, \tau_{w_1})$, shear stress accordingly at the bottom and on a free surface of water (wind tension); z, elevation of bottom; p, porosity of sediment; S, equilibrium sediment concentration.

Shear stress at the bottom and on a free surface can be expressed as:

$$\bar{\tau}_n = -\lambda_n |\bar{U}| \bar{U}, \qquad \lambda_n = g n_n^2 h^{-1/3} \tag{4}$$

$$\bar{\tau}_w = -\lambda_w |\overline{W}| \overline{W}, \qquad \lambda_w = 2.8 \cdot 10^{-7} \tag{5}$$

where $\overline{W}$ is the vector of wind velocity; λ_n, λ_w are the hydraulic resistance factors respectively at the bottom and on a free surface of water; n is Manning's roughness factor.

For calculation of sediment discharge the most acceptable Bagnold's formulae could be simplified for practical application:

$$S = 0.42 \frac{U^2}{gh} \left(0.24 + 0.01 \frac{U}{\omega} \right) \tag{6}$$

Where ω is fall velocity of bottom sediment particles. Equation (6) demonstrates that sediment concentration and yield depended on flow velocity and can strongly vary in non-uniform and non-stationary flows. At present, more complicated models, which take into account the degree of flow saturation with sediments and the noncolinearity of the sediment flux and averaged flow direction, are also used in calculations (Belikov & Volchenkov, 1985; Militeev & Bazarov, 1999). However, the sediment mass conservation equation gives quite reasonable results in solving most practical problems, particularly if wide channels of large rivers are considered.

Effective numerical procedures for solving shallow-water equations (Militeev *et al.*, 1983; Militeev, 1999; Belikov *et al.*, 2001), which have been developed throughout recent years, permit the calculation of bed deformations over a short time period (years or less). Systems of equations are numerically solved using different methods on rectangular and triangular grids. The channel bed configuration at the initial moment is taken as the initial condition for calculation. Boundary conditions are specified in the form of water discharges or levels at liquid borders; the condition characterizing the absence of flow is specified as solid borders (river banks, man-made structures); the sediment discharge is specified at inlet borders. As a consequence, the flow velocity field, the field of water surface levels, water discharges through specified cross sections, and the "new" bed configuration can be determined.

COMPUTER MODELS OF LARGE RIVER REACHES

One-dimensional model describes a process on long sections of narrow flows. Catastrophic deformations of bottom relief of the large rivers are not observed in nature. Such deformations are probably possible after the break of dams of large reservoirs. Usually one-dimensional (1-D) models are used for calculation of the water table. We have applied 2-D and 1-D models to flooding calculation for Lensk—the town on the Lena River where the great ice jam occurred in 2001. Calculations have shown that the water level sharply rises (up to 6.5 m) when ice movement stops. The factor of roughness under ice cover increased from 0.024 to 0.045. When the ice dam reached a height of 4 m, 40 km upstream, the water level only rose by 0.2 m (Fig. 1).

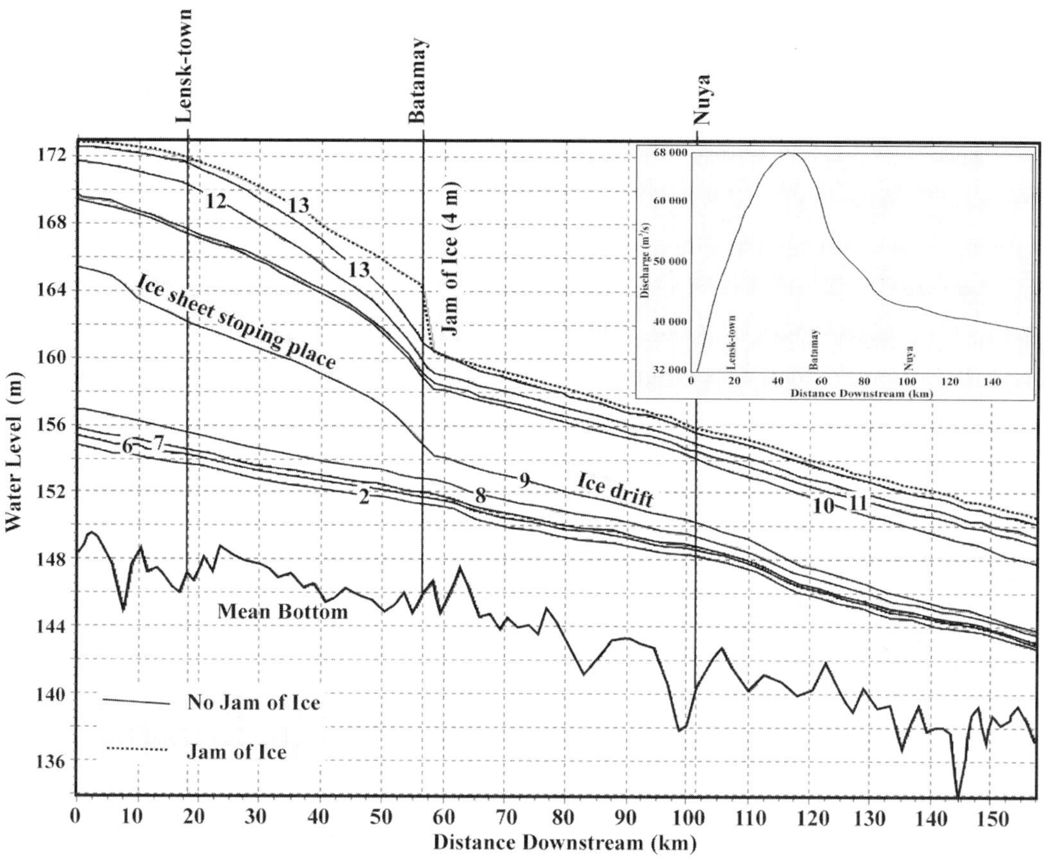

Fig. 1 Simulation of water surface profile upstream of the ice jam at the middle reaches of the Lena River.

Two-dimensional equations of shallow water are accepted for modelling in the particular case of large rivers. The computer 2-D models are used to decide more difficult problems, in the cases of: (a) low water and bankfull flows in singlethread channels; (b) anabranched rivers, (c) confluence of a river with its tributary; and (d) overflowing a flood plain.

(a) Simulation was executed for elaboration of the bank protection strategy in the area of the town of Vilyuisk at the Vilui River. The length of the modelled river section was 15 km, and the mean river width was about 1 km. During the low-flow period, the main stream moves away from the right bank, where a large sand shoal is being formed within the town area; this shoal pushes the stream away at low water. During the flood period, the coastal shoal becomes submerged and the midstream rushes to the right to the terrace, where the town is located. The urban area has been flooded for many days. For >150 years the flow washed out a strip of 150 m wide. Many dwelling houses and buildings in the historical centre of the town were ruined.

In recent years, the bank erosion has become more intensive as a result of the formation of a gentle meander near the town. The concave bank of the meander attracts the stream during the flood period. Moreover, the abovementioned sand shoal gradually shifts downstream and its terrace-protecting effect decreases. During low-flow periods, water discharges are insignificant (about 1000 $m^3 s^{-1}$). However, during flood periods, the water discharge increases up to 10 000 $m^3 s^{-1}$ and more. At this time, water level rises up to 12 m and flow velocity reaches 2 m s^{-1}. The river bed is mainly composed by sand of particle size 0.28 mm. Slopes of the terrace where the town is located are constituted by homogeneous sands of mean particle size 0.22 mm. Therefore, bank erosion and bed deformations are considerable here. At the depth of about 4 m from the low-flow level, the riverbed is underlain by bedrock; that contributes to intensive horizontal channel deformations. In addition the large tributary (the Tyung River) enters the Vilyui River from the left bank side 8 km upstream of the town. The tributary noticeably influences the flow kinematic structure. The runoff volume of the Tyung River, during the open channel period, makes 10–40% of the Vilyui runoff volume. The hydrographs of the Vilui and the Tyng Rivers are often asynchronous.

It was suggested that a series of relatively short wing dykes should be constructed. Their number and length were also determined using mathematical modelling. Fragments of the velocity field around dykes in the course of modelling are presented in Fig. 2. For example, the construction of two short dykes (about 100 m long) did not provide the complete protection of the bank against erosion, both in the period of increased low flow (Fig. 2(a)) and during the flood period (Fig. 2(b)). The construction of four dykes causes the reduction of flow velocities along the bank to the nonscouring values. Dykes are of different length and directed at the angle of about 85° to the flow direction. In the periods of increased low flow (Fig. 2(c)) and flood (Fig. 2(d)), bank protection should prevent bank erosion. The calculations for different combinations of the tributary and main river discharges as well as for real hydrographs of low-flow, high-flow and medium-flow years resulted in obtaining of the "new" bed configuration. The conclusions made are supported by the results of the followed observations.

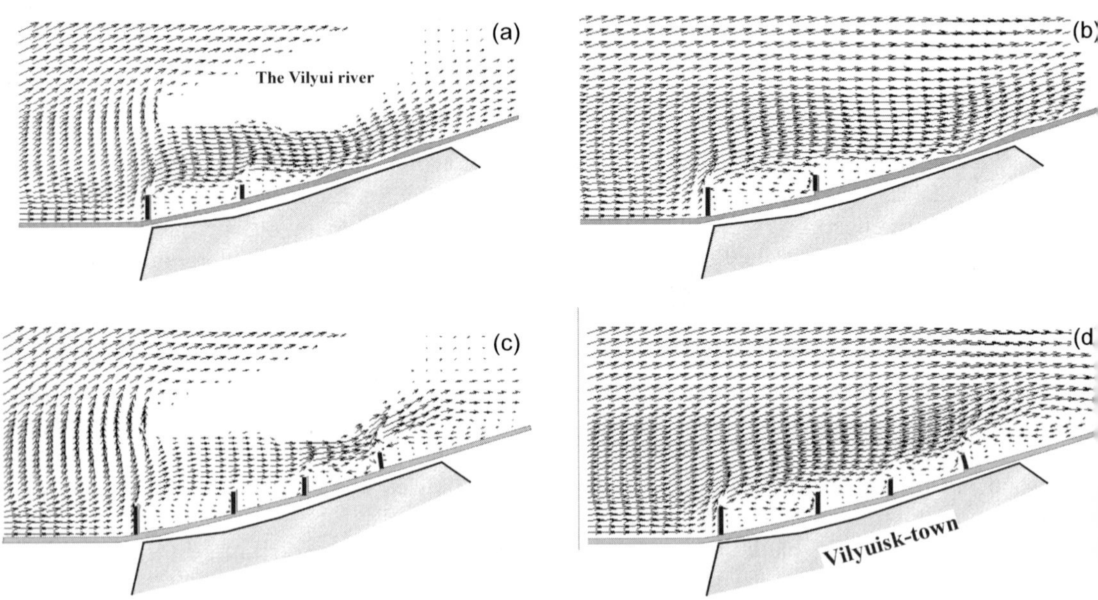

Fig. 2 Simulation of the Vilui River flow around bank protection structures at the town of Viluisk.

(b) The first mathematical model of the Lena River was successfully used in 1989 when it was applied for the project of flood plain and old channel aggradation for the construction of dwelling houses in a new district of the city of Yakutsk. The aggraded areas are located at the bank of the secondary branch Gorodskaya Protoka, anabranched from the main channel at approximately 10 km upstream of the city and deriving 4–7% of the Lena runoff. The ground for the aggradated area was taken from the branch bottom. The volume of annually withdrawn sand made up several million tons and in some years this volume exceeded the total yield of bed load of the entire river. The sand extraction caused an increase of water discharge in the branch and decrease of water level that should result in activization of the branch. It was necessary to evaluate consequences of large-scale sand withdrawals and the amount of sand that could be withdrawn from the channel of Gorodskaya Protoka without any adverse effects. Mathematical models were also used to work out recommendations for designing bank protection structures; these models made it possible to understand the mechanism of flow interaction with hydraulic structures and to choose the model parameters. The example of calculation of bottom relief transformation is demonstrated in Fig. 3.

(c) A computer model of a multibranch river section at the confluence of the Lena and the Aldan rivers was developed in 1999. The model was based on the bathymetrical map plotted in compliance with the results of field surveys carried out in 1996. The size of the object covered >600 km^2. The model was verified using a series of measured water discharges. The most complicated stage of the model development was related to editing the data on the channel configuration, because the data array was of a huge dimension. As a consequence, the detailed

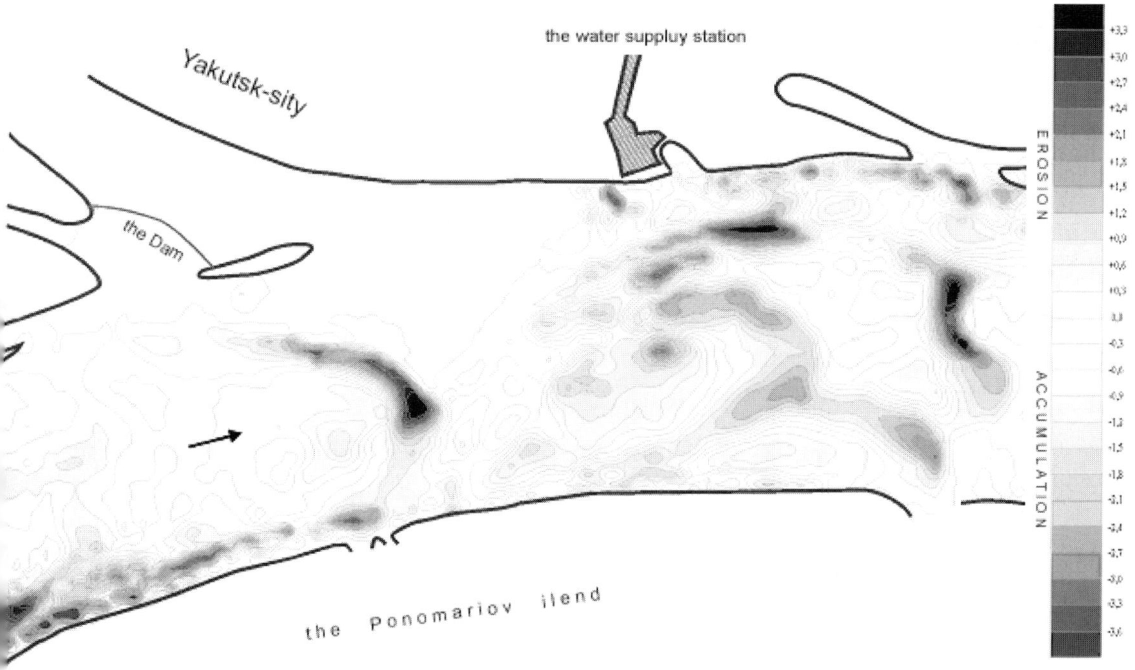

Fig. 3 Simulation of bed relief transformation of the Lena River at the City of Yakutsk during one hydrological year (fragment).

information on the bed configuration made it possible to plot a specific irregular triangular grid with the number of elements exceeding 20 000. The grid condensing was made at places of detailed calculation of deformations and around numerous small islands. Computer modelling permitted the solution of one of the most widespread and important problems—the distribution of water discharges among branches and arms at different water levels. Modelling of channel deformations made it possible to reveal the area of erosional and accumulative flow activity amidst the river section. Any significant deformations were not recorded in the Aldan River channel composed of pebbles (unlike the Lena River channel where sands dominate). Forecast of channel deformations were made.

(d) The flood plain at a period of the flooding had a serious influence upon the river channel process. Therefore the relief of the riverbed and flood plains should be included in composition of the models. There are many settlements along the section of the Lena River near Yakutsk where many problems connected with fluvial processes occur. So the computer model of the river section of 40 km length was constructed in 2002. The relief of the channel and valley bottom was described by 500 000 points. For this section of the river the flow kinematics and channel deformations were calculated. A fragment of calculation of the river bottom deformation is shown in Fig. 4.

The three-dimensional hydrodynamic model considers streams with small sediment load, allowing not to take it into account in equations of water movement. Influence of deposits is only reflected by the change of bottom elevation that causes change of stream parameters. This assumption causes the pressure to be close to hydrostatic. Movement of liquid is described by 3-D equations of shallow water. This model could be applied for the following conditions: (a) vertical acceleration in liquid is much less than gravity acceleration, therefore distribution of pressure on a vertical can be assumed as hydrostatic; (b) change of tangents of pressure in the plan is much less than in a vertical direction.

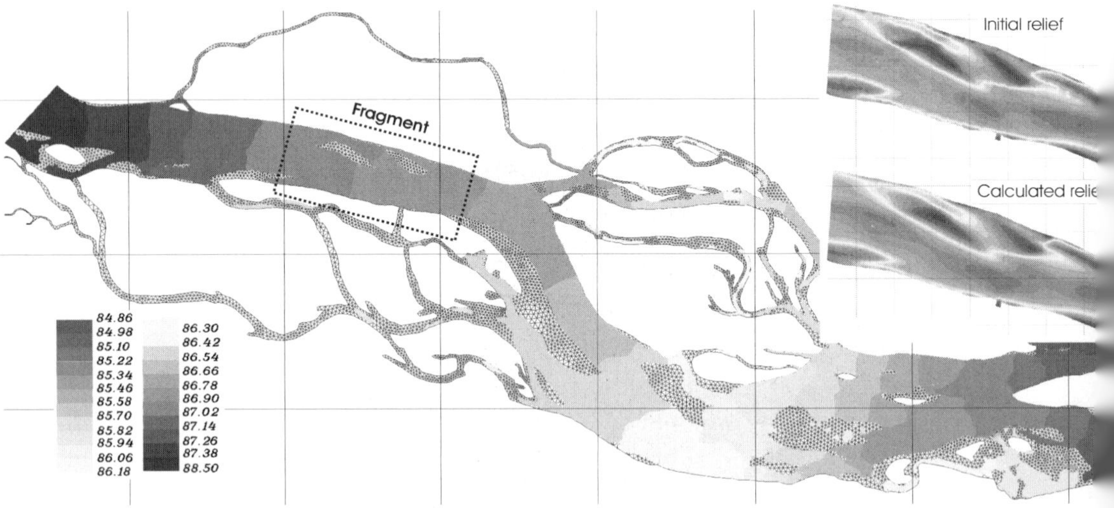

Fig. 4 Simulation of mean annual parameters of channel processes of the Lena River for 3 years.

Transport of deposits occurs both in a suspension, and in the benthonic layers. The discharge of suspended deposits of various fractions can be presented as:

$$\vec{q}_{s\,m} = \int_{Z_d}^{Z_s} \vec{U} c_m \mathrm{d}z \tag{7}$$

where c_m is the volumetric concentration of fraction m of deposits. Wilson's formula was used for bed load:

$$\vec{q}_b = \alpha_b \frac{1}{g} \sqrt{|\vec{\tau}_b|} \vec{\tau}_b \tag{8}$$

The 3-D model simulates moving of a sandy bars that allows to solve problems connected with parameters of hydraulic engineering constructions (Fig. 5).

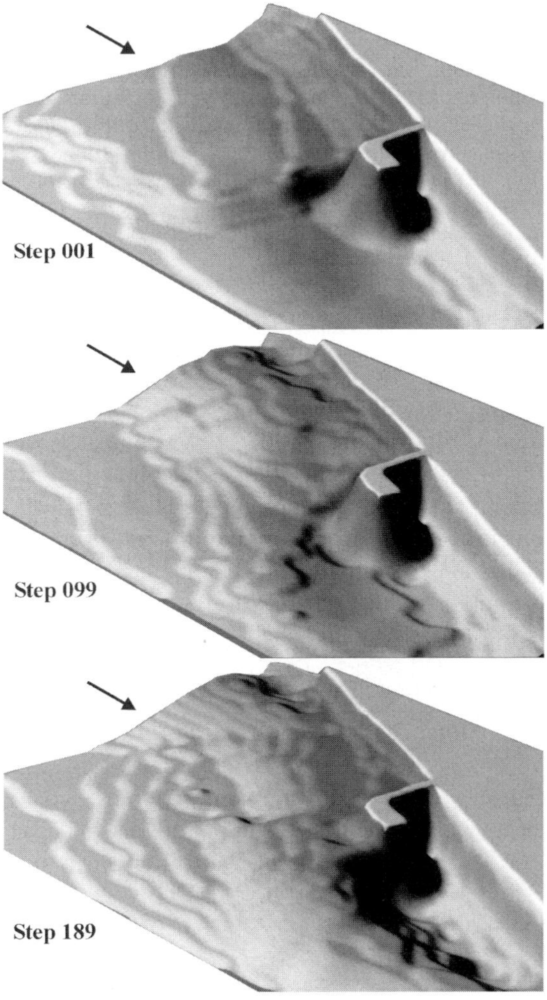

Fig. 5 Simulation of sandy bar movement near water intake.

CONCLUSIONS

Computer modelling allows study ofthe regime of natural channel deformations and impact of hydraulic structures on the flow. The objectives of computer modelling are as follows:

(a) analysis of the river flow structure and bed deformations under natural conditions at different water discharges;
(b) calculation of the parameters of water flow and sediments in a stream with hydraulic structures to choose the optimum arrangement of these structures.

The maximum available information on channel configuration is used to construct a mathematical model of fluviomorphological processes. Initial parametric characteristics are achieved by collecting a database for the object under study. Main and secondary factors of channel formation, as well as methods of organizing further observations over fluviomorphological processes, are determined after introducing the model into practice and analyses of results. The application of channel deformation equation in the computer model allows the definition of zones of accumulation and erosion. The use of real or hypothetical hydrographs makes it possible to forecast channel development and record the rates of its natural transformation with regard to the maximum possible number of the most essential factors.

REFERENCES

Belikov, V. V. & Volchenkov, G. Ya. (1985) *Gidravlika dorozhnykh vodopropusknykh sooruzhenii* (Hydraulics of Road Water Conduits). SPI, Saratov, Russia (in Russian).

Belikov, V. V., Zaitsev, A. A. & Militeev, A. N. (2001) Numerical modeling of flow kinematics in a segment of inerodable channel. *Vodn. Resur.* **6**, 640–648.

Grishanin, K. V.(1979) *Dinamika ruslovykh potokov* (Channel Flow Dynamics). Gidrometiozdat, Leningrad, Russia (in Russian).

Lyatkher, V. M. & Militeev, A. N. (1981) Hydraulic investigation by digital methods (Hydraulic studies in energy and water facilities) *Vodn. Resur.* **3**, 36–41 (in Russian).

Militeev, A. N. (1983). Gidravlicheskie issledovaniya v energetike i vodnom khozyaistve (Hydraulic study in energy and water industries). In: *Collection of Works of Hydroprogekt Hydraulic Studies in Power Engineering and Water Management* (ed. by A. M. Prudovskii), 67–80. Energiya, Moscow, Russia (in Russian).

Militeev, A. N. & Bazarov, D. R. (1999) Gidravlicheskie issledovaniya chislennuimy metodamy (Hydraulic studies by numerical methods). *Vodn. Resur.* **l**, 22–29 (in Russian).

Militeev, A. N., Petrov, P. G. & Belikov, V. V. (1989) Paket prikladnih programm dliy rascheta techeniy gidkosty v otkruituih potokah (package of the applied programs for calculation of the current to liquids in opened flowpackage of the applied programs for calculation of the current to liquids in opened flow). In: *Dinamikai termika rek, ozer, vodokhranilishch i okrainnykh morei.* Moscow, IVPAN SSSR **2**, 85–87 (in Russian).

Zaitsev, A. A. & Egorov, V. F. (1999) Primenenie programmno-apparatnogo compleksa I sputnikovogo coordinirovaniay pri provedenii iziskanii na vodnuih putiah (Using the software-instrument complex and satellite positioning for investigation the prospecting on water ways). In: *Sovremennoe sostoyanievodnykh putei i problemy ruslovykh protsessov* (ed. by R. S. Chalov), 113–124. Mosk. Gos. University, Moscow, Russia (in Russian).

Zajtsev A. A., Il'jasov A. K., Savel'ev R. A., Belikov V. V. & Militeev, A. N. (2003) Compiuternuie modeli ruslovuih processov (Computer models of channel processes). Problems of channel processes. *Trudy Academii vodohozaistvennyh problems* **9**, 89–105 (in Russian).

Movement of bed forms and sediment yield of rivers

NIKOLAY I. ALEKSEEVSKIY
Department of Geography, Moscow State University, Leninskie Gory, GSP-2, 11992 Moscow, Russia
n_alex@hydro.geogr.msu.su

Abstract Factors influencing the characteristics of suspended load yield and bed load yield were studied. Estimation methods of bed load yield for unstudied rivers, which are mainly composed of sandy bed deposits, were substantiated. Correlation between suspended load yield and bed load yield for the rivers which flow in different natural conditions was analysed.

Key words bed load; channel relief; sediment yield; suspended load

INTRODUCTION

Transportation of water, bed and suspended (wash) load, chemical substances, biological substances and heat, is constantly taking place along rivers. Oscillations in matter streams along rivers and in time, cause many dangerous processes influencing the safety of population, economics, and conditions of the existence for aquatic biota. In particular, their trend and intensity mainly depends on sediment yield. Sediment yield (W) includes two components: suspended load W_R and bed load W_G, i.e. $W = W_R + W_G$. Suspended load is composed of small fractions (mud, clay, dust, sand). The bed load component also includes fractions of gravel, pebbles and boulders. In different natural conditions the correlation between the components of sediment yield is rather variable and affects the character of bed deformations.

FACTORS OF SUSPENDED LOAD VARIATION

Suspended load yield depends on the mean long-term discharge Q_o and sediment concentration S, i.e. $W_R = pSQ_0$ where p is the duration of a year (3.15×10^{-6} s). Sediment concentration depends on the zonal and azonal natural factors, anthropogenic impact on river basin and the channel. When the drainage area is the same the increase of water turbidity is observed from humid to arid areas, from mountains to plains. It causes complicated spatial changes of suspended load yield W_R, mean long-term discharge Q_o and sediment concentration in rivers S. In similar natural conditions ($S = const$) suspended load yield mainly depends on the stream order (N) where N is the function of the number of the tributaries (L) with a length of 10–15 km. When L = 1, 2–3, 4–7, 8–15, 16–31, ... 512–1023, 16384–32767, then N = 1, 2, 3, 4, 5 ... 10 ... 15 (Alekseevskiy, 1998). The greater N is the larger the drainage basin and the values Q_o and W_R accordingly. This regularity is quite often broken, especially on large rivers crossing several natural zones.

Characteristics of suspended load yield undergo long-term, annual, seasonal and daily oscillations. It is a result of the specific rhythm of denudation products inflow from the river basin. Depending on the combination of the water balance components of the territory and the variability of natural conditions of the drainage basin, more or less washing products come into the rivers that determine the variation of water runoff, and suspended load yield.

The analysis of spatial and temporal variability of suspended load shows that a large number of natural processes exist, which influence the value and conditions of mineral particles movement. For some period of time:

$$W_R = \sum_i^n W_{Ri} = W_{R1} + W_{R2} + \ldots + W_{Rn} \quad (1)$$

where W_{Ri} is the impact of certain natural processes into the total suspended load yield. The genetic role of these processes is seen after transformation of the equation (1):

$$1 = \frac{W_{R1}}{W_R} + \frac{W_{R2}}{W_R} + \ldots + \frac{W_{Rn}}{W_R} \quad (2)$$

When the expression of the type $W_{R\gamma}/W_R$ could be named as the coefficient of the genetic value $\alpha_\gamma = W_{R\gamma}/W_R$ and consequently for some period of time $\sum_i \alpha_i = 1$. The change of α_γ is the result of the specific way of sediment origin and temporal averaging of W_R. When $\Delta t \to \infty$ all sources of sediment yield change for a certain section of the river can be taken into account using equation (1). The value of separate sources can depend on certain natural conditions. Differences in frequency of appearing of such conditions can lead to the situation when the period Δt is completed, γ process hasn't yet effected W_R, i.e. $a_\gamma = 0$. In the river mouths sediment yield decreases up to zero and consequently the coefficient of genetic value of accumulation processes $\alpha_A = -1$. *Vice versa*, in the upper part of the slopes where the washing out of soil is the main source of sediment yield the coefficient $\alpha_c = 1$. If natural processes for Δt contribute the increase of W_R: $0 \le \alpha_\gamma \le 1$. In general case: $-1 \le \alpha_\gamma \le 1$.

FEATURES OF BED LOAD TRANSPORT

The movement of individual particles depends on the hydraulic characteristics of streams and particle size of alluvial deposits. The greater the stream velocity is the higher the probability of movement for sandy, gravel and other coarse particles on the bottom. When the stream velocity $v < v_H$ where v_H is the critical velocity, they are stable and bed load discharge $G = 0$. If $v > v_H$ then $G > 0$. Depending on the ratio between v and v_H bed load particles either roll on the bottom surface or saltate. Both forms can coexist in time and can form a cycle of the process when the smooth and bed form phases of bed relief change each other successively. The increase of v causes transition from the stable state of the particles in the surface layer of sediments to saltation and formation of the movable elements of bed relief. As a result the structure of bed load transport is changed since the total discharge $G = G_1 + G_2$ where G_1 is bed load caused by shifting of relatively small bed forms (of the length l_g) and G_2 is bed load due to particle movement in the form of larger elements of bed relief and saltation length $l > l_g$. The value $G_1 \to 0$ and $G \approx G_2$ in the process of washing out of the smaller bed form. The process is repeated (Goncharov, 1962) and when G_2 decreases larger bed forms are formed, resulting in the increase of G_3 and so on. As a result, a complicated relation between the total discharge and bed load discharge G_3 appears due to the form shifting being in close correspondence with hydraulic condition of the stream. The total discharge of bed load is greater than G_i where i is the index of the bed relief form type. According to Rossinsky & Debolsky (1980) $G_i = 0.7G$. The coefficient depends on the mean diameter of bed deposition and the relative depth of the channel h/b_p, where b_p is the width of the stream.

Variation of the conditions of the flow and river deposits interaction causes the change of form parameters. The increase of water discharge and stream velocity leads to the increase of their length and height. When the levels become lower the new elements of bed relief do not always manage to reflect the whole cycle of the opposite changes. Small forms corresponding to hydraulic conditions of low water and large bed forms preserved from the previous high water regime can occur in bed relief at the same time (Fig. 1). Bed relief forms of different size move both actively and passively (Znamenskaya, 1968). The active movement is connected with particle transport at the distance $l = l_{gi}$. The passive movement is characterized by the movement of large forms by shifting of smaller forms along their surface. This creates a complicated structure of bed load transport in different water regime phases, when different kinds of bed relief elements move by active and/or a passive shifting mechanism. The longer the bed form, the lower the probability of its active shifting and the higher "the weight" of its passive shifting. In contrast, the smallest forms only migrate by active shifting. The forms of intermediate size undergo active and passive shifting from time to time.

A STRUCTURE OF BOTTOM RELIEF

Side bars (point bars) are the largest bed forms (A type) in river channel relief structure (Fig. 1). Their length is in proportion to the stream order, the channel width (Popov, 1965). The macroform is distinguished by the typical change of the bottom slope (μ): (i) $\mu = 0$ at the beginning of the macroform; (ii) $\mu < 0$ at the upper part of the form; (iii) $\mu = 0$ at its crest; $\mu > 0$ at the lower part of the form, $\mu = 0$ at the beginning of the next macroform. In a similar way, forms of smaller linear size could be distinguished. The method, based on the

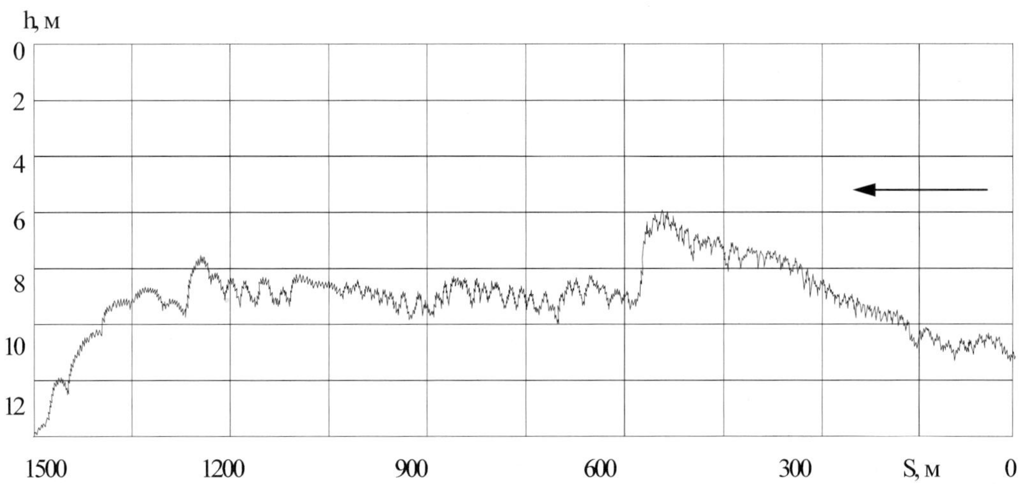

Fig. 1 Hierarchy of bed forms (S, the distance; h, depth of the stream).

double change of sign of bed slope along the form is repeated until the form of the smallest length is determined.

The forms of five types exist in the hierarchy of bottom forms despite the river size. To escape wrong interpretation of the terms we suggest that a letter index of the forms should be used: the largest form A coincides with the length of macroform (point bar), which depends on the channel width and the stream order *N*. One or two forms of Б type can exist on the surface of this macroform. В type forms develop on the surface of Б type form and finally Г and Д forms are situated on their surface. The smallest Д-form type (microform) has the length $l_g \leq h$ where *h* is the stream depth.

The height of different type forms is a perpendicular to the form bottom from its crest (full form height h_g) or the distance from the bottom of the given type to the bottom of the next form (Fig. 2). For example, the proper height of *A*-form (h_{gs}) is equal to the thickness of the deposit layer lower the bottom of Б form type. Between $(h_{gs})_i$ and $(h_g)_i$ there is a connection:

$$(h_{gs})_i = \beta_i\,(\mathrm{h_g})_i \tag{3}$$

Where β_i is the adaptation coefficient for the *i* form type. It is accordingly equal to 0.55; 0.31 and 0/39 for A, Б and В forms type. The proper and full height for Г and Д forms coincide, for them $\beta = 1$ since the shifting of the smallest bed forms is quick and does not effect the difference between the full and proper height of Г form type.

There is a certain correspondence between form parameters of every type and the stream order. The treatment of natural data shows a close statistical correlation ($\sigma > 0.75$) between $(h_g)_i$ and *N,* where *N* is the order of the river. The height of the relief forms of all types, which are in accordance with the hydraulic regime of the stream, is proportional to *N* (Table 1). The increase of *N* results in the nonlinear increase of the height of bed forms of one type in accordance with the equation:

$$(h_g)_{ij} = a_{ij}N^{bij} \tag{4}$$

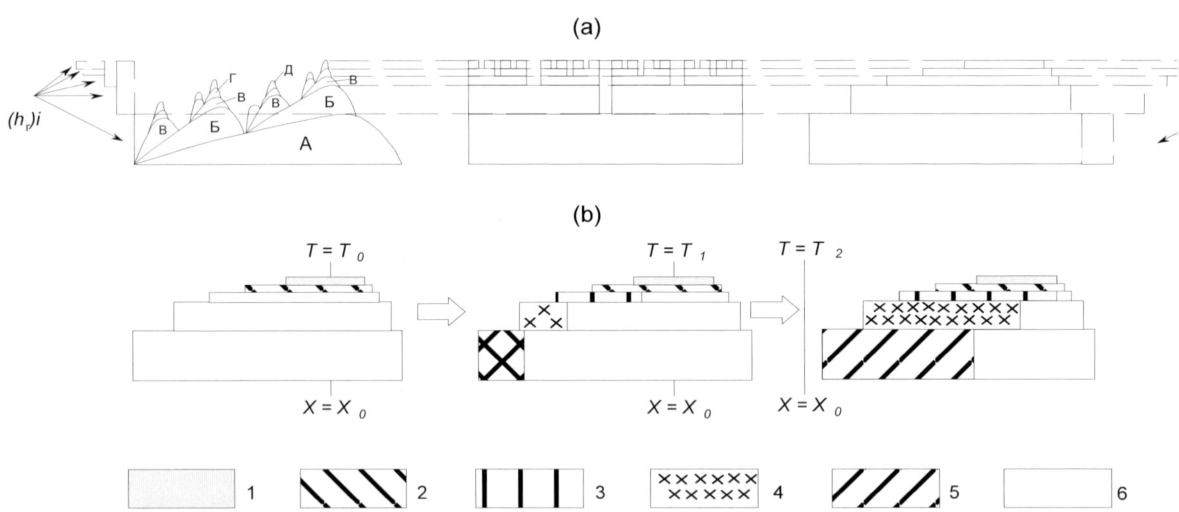

Fig. 2 The structure of relief (a) and the scheme of active (1,2,3,4,5) and passive (b) migration of bedforms. T_0, T_1, T_2, X_0 – time and spatial coordinates.

Table 1 Parameters of equation (4).

Water season	Parameter	Type of forms, *i*				
		А	Б	В	Г	Д
High water ($j = 1$)	*a*	0.046	0.018	0.032	0.023	0.005
	в	1.95	2.0	1.5	1.4	1.9
Low water ($j = 2$)	*a*	0.046	0.018	0.008	0.005	0.003
	в	1.95	2.0	2.0	2.0	2.0

where a_{ij}, b_{ij} are the empirical parameters, j is the index of water season.

The forms of bed relief move along the stream at some velocity C_g. Analysis of shifting of А, Б, В, Г and Д forms brings to the conclusion about the seasonal character of the relation between shifting velocity (C_g) and the river order N. The variability of relative shifting velocities of forms of different type depends on the stream order (Fig. 3). For one type of forms the shifting velocity increases, for other it decreases, when N increases. For A and B forms the dependence between the variables is of increasing character and for Б, Г and Д forms of decreasing character. In general case they look like:

$$\frac{C_{u_{ij}}}{v_j} = r_{ij} N^{p_{ij}} \tag{5}$$

The coefficients r_{ij} and p_{ij} characterize the influence of form type and water season phase on this form type mobility. In low water the greater part of large forms is moving passively due to active shifting of small forms. Observations on many rivers show that only Д and Г forms undergo such a shifting. The velocity of this process for the forms of both types is the same, hence:

$$C_{g_{Г,Д}} = 600 v_m N^{-2.0} \tag{6}$$

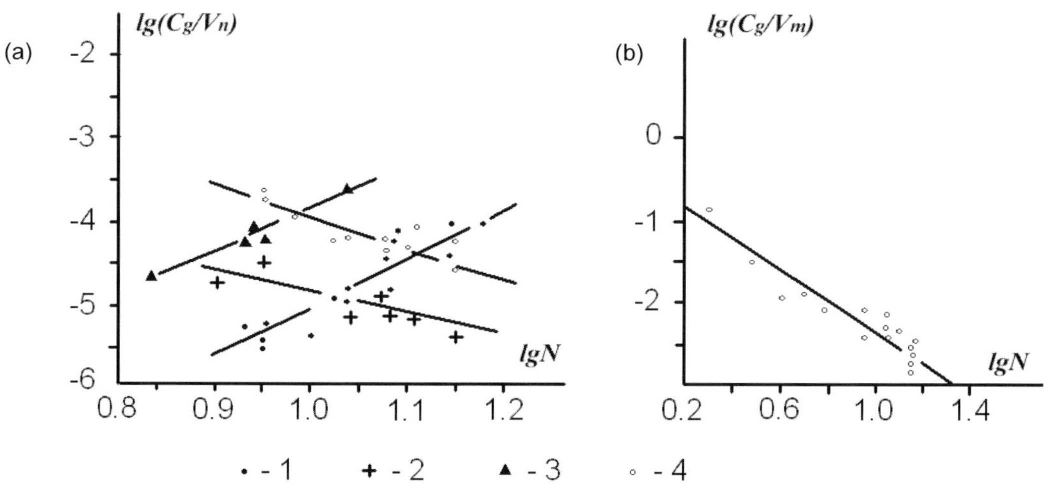

Fig. 3 The change of relative rate of different types of bed forms migration (А (1), Б (2), В (3), Г and Д (4)) for the period of high (($lg(C_g/V_n)$, (a) and low (($lg(C_g/V_m)$), (b) water.

The increase of the river order and the decrease of the mean velocity of the stream in low water is accompanied by the decrease of the $C_{g_{Г,Д}}$ (m day^{-1}) which is in agreement with the known quantitative conclusion: in large rivers the bottom forms move slower (Gryshanin, 1979). The dependence (6) shows a good approximation to natural data.

In high runoff the active movement of Г and Д forms is subjected to the similar in structure dependence:

$$C_{g_{Г.Д}} = 2.2 \cdot 10^{4} v_n N^{-3.5} \tag{7}$$

where v_n (m s^{-1}) is the mean velocity of the stream in the period of high water. In this season larger forms shift more actively as well. To simplify the process it can be assumed that in the period of high runoff the whole structure of the bed relief is preserved and the "washing out" processes of small forms are included into the velocity value of their shifting. This hypothesis is proved by the observations on the Niger River (Sidorchuk, 1983) and the Kama River (Ivanov, 1984). Its application leads accordingly to obtaining the equations to calculate the velocity of active shifting (m day^{-1}) of В, Б and А forms:

$$C_{g_В} = 3.5 \cdot 10^{-4} \cdot v_n \cdot N^{-4.6} \tag{8}$$

$$C_{g_Б} = 3.1 \cdot 10^{2} \cdot v_n \cdot N^{-2.6} \tag{9}$$

$$C_{g_А} = 1.1 \cdot 10^{-6} \cdot v_n \cdot N^{-5.4} \tag{10}$$

The character of relation between the variables in equations (8)–(10) demonstrates various changes of different types of form mobility, when the stream order increases (Fig. 3a). The increase of v_n and N results in the decrease of the intensity of the active shifting of Г, Д and Б forms and in the increase of В and А forms of shifting velocity. The reason for such a change of the velocity of the active shifting of different types of bottom relief forms lies in the change of the effect of individual forms in bottom relief structure on the formation of the total bed load transport when moving from small rivers to large ones (Alekseevskiy, 1998).

BED LOAD YIELD CALCULATION

Bed load discharge G (kg day^{-1}) depends on the height of the form h_g (m) and its shifting velocity C_g (m day^{-1}) within the active width of the bed b_p (where $v > v_н$) when the forms of one type move actively. The value G depends on the density of river deposits (δ_0) and the parameters of the element bottom relief form k = 0,6, hence:

$$G_g = k\delta_0 b_p (C_g h_g) \tag{11}$$

Following the equation (11) some forms $n = b_p/b_g$ of the same width b_g, length and height move through the river cross section at C_g velocity. The greater the longitudinal and transverse velocity gradient the greater this schematization differs from the natural conditions of bed load transport. In particular, the probability of the active movement of the forms of other type, represented in the hierarchy of bed relief forms, becomes higher. At low water (Fig. 2, $T = T_0$) the movement of the active forms causes the passive dynamics of greater forms, which do not have their own development mechanism. It explains the coincidence of the G values, calculated according to the equation (11) using the parameters of the active and passive forms.

The situation changes radically during the flood, when the particles move at the distance $l > l_g$. Under this, the conditions for the active evolution of the forms of coordinated types are created. Their participation in the total transport of bed load needs the improvement of the equation (11), the estimation of the impact of active forms of different types on the total bottom deformation $|\Delta Z|$, caused by the shifting of the whole hierarchy forms for a rather long period of time. During the active movement of all forms ($T = T_1$, $T = T_2$) this impact is proportional to the proper height of forms (Fig. 2). Taking into account equation (1) the active layer of river deposit is the sum of individual values of $h_{gs,i}$:

$$|Z| = \sum_{i=1}^{M} \beta_i h_{gi} \quad (12)$$

here M is the number of forms of different type, crossing the calculated gauge for the time interval Δt. The volume of the transported material connected with i form type migration and individual deformation of the deposit surface ΔZ_i may be calculated according equation:

$$W_i = k \cdot b_{pa} \cdot l_{gi} \cdot |\Delta Z_i| \quad (13)$$

where k is the form coefficient, l_{gi} is the length of i type forms. Sediment discharge, taken as a mean value Δt, corresponds to the volume W_i:

$$G_i = k \cdot b_{pa} \cdot \sigma_0 \cdot C_{gi} \cdot |\Delta Z_i| \quad (14)$$

where the velocity of form shifting $C_{gi} = l_{gi} / \Delta t$. In case of the active development of the whole form complex the total bed load discharge may be determined from following equation:

$$G_i = k \cdot b_{pa} \cdot \sigma_0 \cdot \sum_{i}^{M} \beta h_g C_g)_i \quad (15)$$

Bed load discharge G characterizes the mean value of particle transport for T time, when the macroform (form A) is shifted at full length. At some period of time individual sediment discharge connected with the shifting of i type forms do not coincide with G since they make up only a part of the total discharge. Only for the time T the longitudinal coordination of the erosion and accumulation volumes and the sediment discharge, obtained using the form parameters, approaches its real value.

The mean annual and long-term volume of bed load yield may be calculated from equation:

$$W_G = W_{mg} + W_{ng} = (365 - T_n) \cdot \sum_{i=1}^{2} G_i + T_n \cdot \sum_{i=1}^{5} G_i \quad (16)$$

where T_n is the duration (in days) of high water. The individual values of bed load yield G_i (kg day^{-1}) are calculated according to the equation (11) using the dependences to calculate the height (Table 1) and the velocity of shifting of forms (equations 6–10).

RIVER RUNOFF AND BED LOAD YIELD CORRELATION

It is usually assumed that bed load discharge does not make up more than 1% of the suspended sediment discharge (when the mean diameter of the deposits is $d < 0.25$ mm) and

Table 2 Suspended sediment and bed load discharge ratio upstream and within the branched section of the Lena River near Yakutsk during the flood in 1989.

Data	Q (m^3 s$^{-1)}$	Upstream the branched section			Within branched section		
		R (kg s^{-1})	G (kg s^{-1})	$G/(R+G)$ (%)	$\sum R$ (kg s^{-1})	$\sum G$ (kg s^{-1})	$\sum G/(\sum R+\sum G)$ (%)
06 June	34 000	2400	1320	35.5	2400	500	17.2
23 June	24 600	1740	850	32.8	2060	470	18.6

in the case of prevailing in pebble and boulder sediments and when $d/h \geq 0.01$ the ratio R/G increases up to 5–10% (Karasev, 1975). On plain rivers (Kopaliani,1985) the ratio of bed load discharge to river sediment discharge $R/G = \delta = R/(R+G)100$ as a rule makes up 1–3% (rarely 10%). In mountain streams the value varies from 15 to 70%. It depends on the natural conditions of R formation and on the water volume of the rivers Q (Chalov *et al.*, 2000). Bed load yield is maximum for the rivers and their parts, where the suspended yield is relatively small. On some rivers bed load yield predominates (δ = 90%) in river sediments. The smaller the river is the greater the bed load yield and the value δ.

The correlation between the suspended sediment and bed load discharge changes in the branched section of the Lena River (Table 2). According to the observations upstream the branched section at large water discharge bed load discharge makes up 32.8–35.5% of the total river sediment discharge. In the branched part of the river the structure of sediment transport changes under the water runoff distribution. In river branches the relative impact of bed load is not higher than 19% of the total river sediment discharge. The main part of bed load discharge during the flood enlarges the volume of river deposits. Specific discharge of bed load (G/B) in the arms of the branched section within different development tendency is the function of the Reinolds number: $G/B = n(Vh)^m$ where Vh is the mean velocity and the depth of the stream; $n \sim v^{-m}$. Here v is the kinematic viscosity at water temperature 18°C, m is the empirical parameter. The corresponding values n for the developing, stable and disappearing (dying) arms in the middle course of the Ob River are equal to 2.75×10^{-4}; 1.8×10^{-2}; 2.9×10^{-2}, and the parameter m are 3.5; 1.26 and 1.0. The ratio $G/B = const$ is reached in these branches under different values of Vh. In the developing arms a greater increase of Vh is needed to move the mineral particles with the intensity of 0.03 kg s^{-1} m^{-1} in comparison with the stable and dying arms. When the value G decreases the value R and a part of coarse fraction of bed load discharge increases.

CONCLUSIONS

River sediment yield is an important component of the matter flux in different parts of the network. It is often considered to be practically equal to the suspended sediment yield. In a number of cases such a hypothesis is incorrect, it results in the errors in the determination of river sediment outflow into the reception reservoirs, the time of reservoir silting and the intensity of channel deformations. The study showed, that a part of the bed load yield could not be assumed as a constant value equal to 1–10% of the suspended sediment yield. It is changed in a big variety of values depending on the suspended sediment yield, the type and size of the river and other factors.

Acknowledgement Funding for this research was provided by RFBR (project N 03-05-64306) and Federal Program "Universities of Russia – fundamental research".

REFERENCES

Alekseevskiy, N. I. (1998) *Phormirovanie i dvigenie rechnyh nanosov* (River Sediment Formation and Movement). Izdatelstvo MSU, Moscow, Russia (in Russian).

Chalov, R. S., Lu Shuguan & Alekseevskiy, N. I. (2000) *Stok nanosov i ruslovye processy krupneyshih rek Rossii i Kitaja* (Sediment Yield and Channel Processes on Large Rivers of Russia and China). Izdatelstvo MSU, Moscow, Russia (in Russian).

Goncharov, V. N. (1962) *Dinamika ruslovyh potokov* (Water Streams Dynamic). Gidrometeoisdat, Leningrad, USSR (in Russian).

Gryshanin, K. V. (1979) *Dinamika ruslovyh potokov* (Water Streams Dynamic). Gidrometeoisdat, Leningrad, USSR (in Russian).

Ivanov, B. A. (1984) Ismerenie I raschety parametrov donnyh grjad, rashoda nanosov I rasmyva rusla r.Kama v polovodie v nignem bjephe Nignekamskoy GES (Estimation and calculation of bed forms parameters, bed load of Kama river and channel erosion in the low reach of the Nizhnekamsk Hydroelectric power station in the flood). In: *Trudy Gosydarstvennogo gidrologicheskogo Instituta* **318**, 63–91 (in Russian).

Karasev, I. F. (1975) *Ruslovye processy pri perebroske stoka* (Channel Processes Due Runoff Transfer). Gidrometeoisdat, Leningrad, USSR (in Russian).

Kopaliani, Z. D. (1985) O correlyacii pashodov vzveshennyh i vlekomyh nanosov (On the correlation of bed load and suspended load discharge in rivers). In: *Hydrophysical Processes in Rivers and Reservoirs,* Izdatelstvo Nauka, Moscow, USSR, 143–147 (in Russian).

Rossinskiy, K. I. & Debolskiy, B. K. (1980) *Rechnye nanosy* (River Sediments). Moscow, Izdatelstvo Nauka, USSR (in Russian).

Sidorchuk, A. Y. (1983) Morphologija i dinamika ruslovogo reliepha nignego Nigera (Morphology and bed relief dynamics of the low Niger). In: *Morphodynamics Problems*, 21–38. Izdatelstvo Nauka, Moscow, USSR (in Russian).

Znamenskaya, N. S. (1968) *Grjadovoe dvigenie nanosov* (Bed Forms Movement and Bed Load). Gidrometeoisdat, Leningrad, USSR. (in Russian).

Sediment Transfer through the Fluvial System (Proceedings of a symposium held in Moscow, August 2004). IAHS Publ. 288, 2004

A particle tracking method to simulate sediment transfer over flood plains

IVO THONON, KOR DE JONG, MARCEL VAN DER PERK & HANS MIDDELKOOP

Centre for Geo-ecological Research (ICG), Faculteit Geowetenschappen, Universiteit Utrecht, Postbus 80115, 3508 TC, Utrecht, The Netherlands
i.thonon@geog.uu.nl

Abstract Deposition of contaminated sediments during high discharges is an important aspect on lowland river flood plains. Quantification of deposition often proceeds through modelling. However, raster-based sedimentation models in particular suffer from numerical dispersion in the sediment-transfer routine. To simulate conveyance of sediments over flood plains without introducing numerical dispersion, we developed a sediment-transfer model in which we implemented the particle tracking method "Method of Characteristics". This raster-based model uses input data such as initial concentrations, water levels, flow velocities and dispersion coefficients. In the future, we will use this model as part of a flood plain sedimentation model.

Key words flood plain; Method of Characteristics; modelling; numerical dispersion; particle tracking; Rhine River; sedimentation; suspended sediment

INTRODUCTION

Flood plains sequester large amounts of sediments and associated pollutants (e.g. Walling *et al.*, 1998; Middelkoop & Asselman, 1998; Owens *et al.*, 1999). To simulate the deposition patterns of these sediment-associated pollutants, several workers have developed sedimentation models. Some of these models have been based on a finite-element approach (e.g. Stewart *et al.*, 1998; Siggers *et al.*, 1999; Hardy *et al.*, 2000); other models have been based on a finite-difference approach (e.g. Nicholas & Walling, 1997, 1998; Middelkoop & Van der Perk, 1998). On the one hand, the finite-element models have the advantage of employing a mesh that can be optimized for local situations (i.e. a finer mesh in environments with higher gradients and *vice versa*) so as to minimize numerical errors. On the other hand, these models have the disadvantage that their output cannot be imported directly into a raster GIS for further processing, whereas this is relatively easy using finite-difference or raster-based models. Raster-based models, however, suffer from numerical dispersion (i.e. dispersion as an unwanted artefact of the numerical modelling technique), which may lead to erroneous results. Here we propose particle tracking as a method to cope with this problem in raster-based modelling.

An example of a particle tracking method is the "Method of Characteristics" (MoC). According to Garder *et al.* (1964) this method minimizes numerical dispersion. Konikow & Bredehoeft (1978) used MoC to simulate solute transport in groundwater. We modified MoC as used by Konikow & Bredehoeft (1978) to be applicable for surface water and implemented it in PCRaster (Wesseling *et al.*, 1996), a tool that aids in numerical modelling of environmental processes using rasterized input maps.

The purpose of this paper is to describe the modified MoC model, discuss the input data needed and give an example of the output. A stand-alone hydrodynamic model calculated the input data—such as water levels, *x*- and *y*-flow velocities—and we used empirical formulae to calculate dispersion coefficients.

MODEL CONCEPT

Flow equation

Konikow & Bredehoeft (1978) have developed a complete groundwater model that solves both the flow and transport equation. We, however, only considered the transport equation in their model, since we derived stationary flow fields from another model. The adapted transport equation is:

$$\frac{\partial(Ch)}{\partial t} = \frac{\partial}{\partial x_i}\left(hE\frac{\partial C}{\partial x_j}\right) - \frac{\partial}{\partial x}(hCu) - C'S \qquad (1)$$

where C is suspended sediment concentration (*SSC*) (mg l^{-1}), t is time (s), x_i and x_j are Cartesian coordinates (m), h is water depth (m), E is hydrodynamic dispersion coefficient (m^2 s^{-1}), u is flow velocity (m s^{-1}), C' is *SSC* in a source or sink (mg l^{-1}), S is volume flux per unit surface area in a source or sink (m s^{-1}). The first term on the right-hand side is the dispersion term, the second one the convection term and the third one the source/sink term.

Konikow & Bredehoeft (1978) assumed that convective transport (transport of particles and solutes by flowing water) dominates solute transport and used a grid-based model. Under these circumstances, it is possible to define concentration changes for flowing reference particles that pass fixed nodes, rather than the conventional way of defining these changes for fixed reference nodes passed by flowing particles. MoC thus solves the transport equation taking flow lines as a basis instead of grid cells. Garder *et al.* (1964) called these flow lines "characteristic curves", hence the name Method of Characteristics. With this approach, equation (1) can be reduced to the following form:

$$\Delta C = \Delta t\left[\frac{1}{h}\frac{\partial}{\partial x_i}\left(hE\frac{\partial C}{\partial x_j}\right) + \frac{CS - C'S}{h}\right] \qquad (2)$$

Note that the convection term in equation (1) has disappeared in equation (2).

Particle tracking

The first step in the calculation procedure involves placing a number of particles in each cell of the model grid (Fig. 1). All particles inherit the initial concentration from the grid cells in the input map. During each subsequent time step, all particles move a distance of, at most, one grid cell according to their characteristic curves (the curved lines in Fig. 1). After each time step, the model calculates a temporal average concentration C^* for each node on the basis of the new spatial arrangement of particles. Then, the change in concentration (ΔC) due to hydrodynamic or mechanical dispersion, divergence of velocity, sources/sinks and change in water depth is calculated using equation (2). The present version of the model does not yet

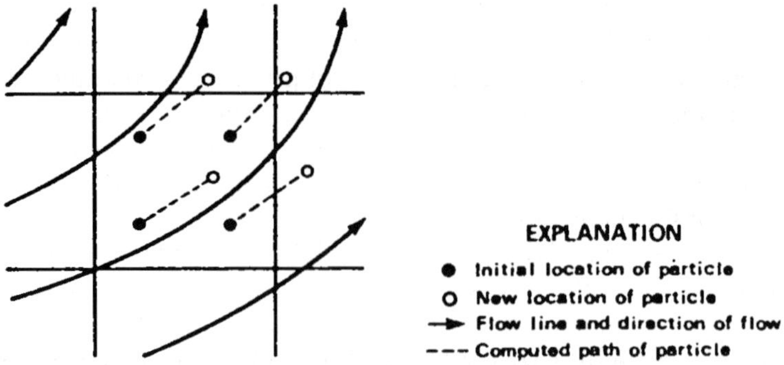

Fig. 1 The location of the particles in the raster cells and subsequent displacement by the model. Source: Konikow & Bredehoeft (1978).

consider deposition or resuspension of sediment, so the resulting concentration ($C^* \pm \Delta C$) is the new concentration at each node for the next time step. At the end of each time step, PCRaster provides the user with maps of this concentration.

Stability criteria and boundary conditions

The model treats the user-defined time step as the maximum time increment. If necessary, the model divides this time increment into smaller ones to ensure that particles cannot move more than one cell length at a time. The model bases this division on four stability criteria, involving the dispersivity values, water depth and flux, and flow velocities in the *x*- and *y*-direction, respectively.

The model uses two types of boundaries: no-flow boundaries and flux boundaries. The user has to define no-flow boundaries at the locations of river dikes or around high terrains in the flood plains that remain dry during high discharges. Next, the upstream and downstream boundaries of the study area have to be flux boundaries, with the flux being either constant or transient. The model considers upstream boundary cells as sources, while downstream boundary cells are considered to be sinks.

Model input

The model uses the following data as input maps: initial suspended sediment concentrations, flow velocities in *x*- and *y*-direction, water depth, flux, longitudinal and transverse dispersion coefficients. The size of the grid cell is a property of the input maps. The user also has to define the number of particles per cell (four, five or nine, while these numbers provide, at least for the first time step, a regular coverage of the grid cells) and the length of the time step.

EXAMPLE CALCULATION

We simulated the sediment dispersion for a river reach of the River Rhine during a period of high discharge that occurred in spring 2002, with a peak discharge of 8054 m^3 s^{-1} at the

Dutch-German border. In this simulation we used input maps with cell sizes of 50 × 50 m. We derived the input data in the following ways:

(a) Initial suspended sediment concentration, C' (mg l^{-1}): the *Infocentrum Binnenwateren* in Lelystad daily measures the *SSC* at Lobith (at the Dutch-German border). As initial *SSC* for all the flood plains in the model area we used the concentration at Lobith on the first day of the inundation: 30 mg l^{-1}. For the river channel, we adopted the average of the first five days of the inundation period, i.e., 53 mg l^{-1}.

(b) Water depth, h (m): we derived the water depth by subtracting the surface elevation from the water level. To define the surface elevation we constructed a Digital Elevation Model (DEM) from the base data provided by the *Adviesdienst voor Geo-informatie & ICT* (RWS-AGI) in Delft. RIZA-WST in Dordrecht calculated the water depth using the hydrodynamic model WAQUA (MX.Systems 2003). WAQUA solves the Saint-Venant equations in a curved grid, using as boundary conditions the river discharge at the upstream end and the water level at the downstream end of the modelled river stretch.

(c) Flow velocity, u (m s^{-1}): WAQUA provided the flow velocities in x- and y-direction (Fig. 2). From these values, we calculated the absolute flow velocities, which we later used in the calculation of the dispersion coefficients.

(d) Flux, S (m s^{-1}): we recalculated the river discharge entering and leaving the model area to a flux by dividing it by the cell area of the PCRaster grid.

(e) Transverse dispersion coefficient, E_T (m^2 s^{-1}): we used an equation proposed by Holley & Abrahams (1973):

$$E_T = c \cdot h \cdot u \tag{3}$$

(f) where E_T is the transverse dispersion coefficient (m^2 s^{-1}), c is a constant (–), h is the water depth (m) and u is the absolute velocity (m s^{-1}). Using tracer experiments in the model area on the Waal River, Holley & Abrahams (1973) found an approximate value of 0.03 for c. Since there are no data available for flood plains, we assumed this value to be valid for both the river channel and its flood plains. Still, the different character of the flood plains finds expression in lower E_T values because of their lower water depths and flow velocities.

(g) Longitudinal dispersion coefficient, E_L (m^2 s^{-1}): we used a formula provided by Fischer *et al.* (1985):

$$E_L = 0.011 \cdot u^2 \cdot w^2 / h \cdot u_* \tag{4}$$

where E_L is longitudinal dispersion coefficient (m^2 s^{-1}), 0.011 is an empirical constant, w is cell length (m) and u_* is shear velocity (m s^{-1}).

Model output

Figure 3 shows the model output for the middle reach of the Waal River, a distributary of the River Rhine. The flow is from right to left. The flood plain in the northeast of the model area receives a lot of suspended sediment, whereas the one in the south-central part receives far less because non-inundated areas block the inflow of river water. In the case of the flood plain in the northwest, the river water apparently cannot reach the far ends of it, leading to a pronounced north–south gradient in *SSC*.

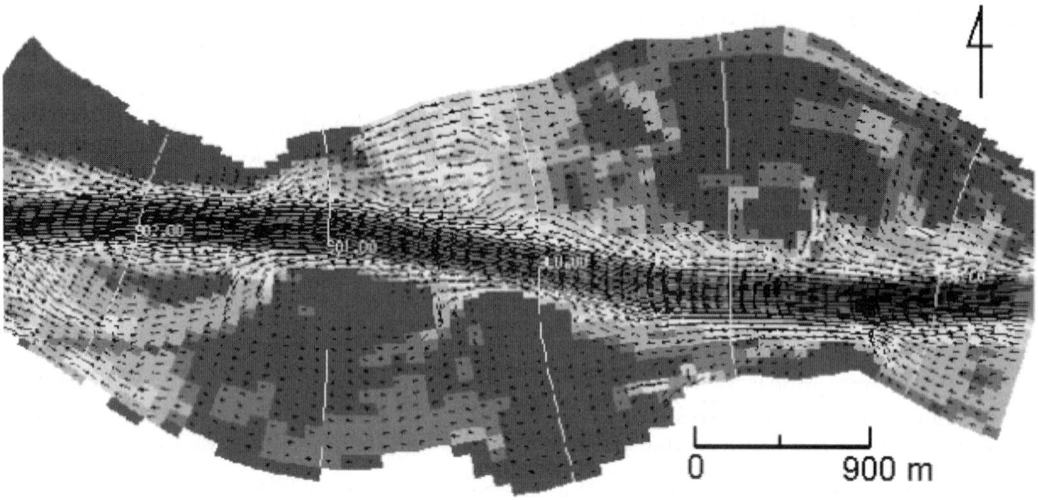

Fig. 2 Average flow velocities and flow directions in the study area as modelled by WAQUA. The length of the arrows indicates the relative size of the velocity. The lighter the colour, the higher the flow velocity. The channel is however dark due to the high number of arrows.

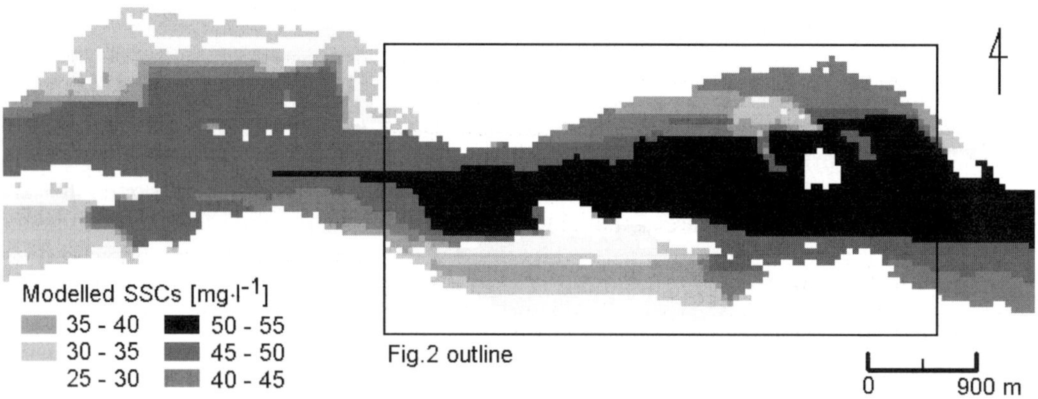

Fig. 3 Modelled suspended sediment concentrations for the middle reach of the Waal River. Flow is from right to left. The box denotes the area in Fig. 2.

DISCUSSION AND PERSPECTIVE

The implementation of the Method of Characteristics in a 2-D environmental modelling tool is promising and can aid in the prediction of the dispersion of contaminated sediments over flood plains. In the future, we will extend the model with a sedimentation module. We will calibrate this module with settling velocities reported in Thonon & Van der Perk (2003) and sedimentation data given in Middelkoop *et al.* (2003) and Thonon (2003).

Acknowledgements We acknowledge the support from Ton Visser and Dr. Gertjan Zwolsman (RIZA-WST, Dordrecht) for carrying out the WAQUA calculations and the stimulating discussions, respectively.

REFERENCES

Fischer, H. B., List, E. J., Koh, R. C. Y., Imberger, J. & Brooks, N. H. (1985) *Mixing in Inland and Coastal Waters*. Academic Press, Orlando, USA.

Garder, A. O., Peaceman, D. W. & Pozzi, A. L. Jr (1964) Numerical calculation of multidimensional miscible displacement by the method of characteristics. *Soc. Petrol. Engrs J.* **4**(1), 26–36.

Hardy, R. J., Bates, P. D. & Anderson, M. G. (2000) Modelling suspended sediment deposition on a fluvial floodplain using a two-dimensional dynamic finite element model. *J. Hydrol.* **229**, 202–218.

Holley, E. R. & Abraham, G. (1973) Field tests on transverse mixing in rivers. *J. Hydraul. Div. ASCE* **99**(HY12), 2313–2331.

Konikow, L. F. & Bredehoeft, J. D. (1978) Computer model of two-dimensional solute transport and dispersion in ground water. *Techniques of water-resources investigations of the United States Geological Survey.* Alexandria, Virginia, USA.

Middelkoop, H. & Asselman, N. E. M. (1998) Spatial variability of floodplain sedimentation at the event scale in the Rhine-Meuse delta, The Netherlands. *Earth Surf. Processes Landf.* **23**(6), 561–573.

Middelkoop, H. & Van der Perk, M. (1998) Modelling spatial patterns of overbank sedimentation on embanked floodplains. *Geogr. Ann.* **80A**(2), 95–109.

Middelkoop, H., Van der Perk, M. & Thonon, I. (2003) Herverontreiniging van uiterwaarden langs de Rijntakken met sediment-gebonden zware metalen (Recontamination of flood plains along the Rhine branches with sediment-associated heavy metals) (in Dutch). *ICG Rapport* 03/3. Faculteit Geowetenschappen, Universiteit Utrecht, Utrecht, The Netherlands.

MX.Systems (2003) Users's Guide WAQUA. *SIMONA-report* 92-10. MX.Systems/Ministerie van Verkeer & Waterstaat, Rijswijk/The Hague, The Netherlands.

Nicholas, A. P. & Walling, D. E. (1997) Modelling flood hydraulics and overbank deposition on river floodplains. *Earth Surf. Processes Landf.* **22**(1), 59–77.

Nicholas, A. P. & Walling, D. E. (1998) Numerical modelling of floodplain hydraulics and suspended sediment transport and deposition. *Hydrol. Processes* **12**(8), 1339–1355.

Owens, Ph. N., Walling, D. E. & Leeks, G. J. L. (1999) Deposition and storage of fine-grained sediment within the main channel system of the River Tweed, Scotland. *Earth Surf. Processes Landf.* **24**(12), 1061–1076.

Siggers, G. B., Bates, P. D., Anderson, M. G., Walling, D. E. & He, Q. (1999) A preliminary investigation of the integration of modeled floodplain hydraulics with estimates of overbank flood plain sedimentation derived from Pb-210 and Cs-137 measurements. *Earth Surf. Processes Landf.* **24**(3), 211–231.

Stewart, M. D., Bates, P. D., Price, D. A. & Burt, T. P. (1998) Modelling the spatial variability in floodplain soil contamination during flood events to improve chemical mass balance estimates. *Hydrol. Processes* **12**(8), 1233–1255.

Thonon, I. (2003) The effect of river rehabilitation projects on the sediment-associated heavy metal pollution of the Dutch floodplains. In: *Current themes in Dutch river research* (ed. by R. S. E. W. Leuven, A. G. van Os & P. H. Nienhuis) (Proc. NCR-days, November 2002), 142–143. NCR-publication no. 20-2003. NCR, Delft, The Netherlands.

Thonon, I. & Van der Perk, M. (2003) Measuring suspended sediment characteristics using a LISST-ST in an embanked flood plain of the River Rhine. In: *Erosion and Sediment Transport Measurement: Technological and Methodological Advances* (ed. by J. Bogen, T. Fergus & D. E. Walling) (Proc. Oslo Workshop, June 2002), 37–44. IAHS Publ. 283. IAHS Press, Wallingford, UK.

Walling, D. E., Owens, Ph. N. & Leeks, G. J. L. (1998) The role of channel and floodplain storage in the suspended sediment budget of the River Ouse, Yorkshire, UK. *Geomorphol.* **22**(3–4), 225–242.

Wesseling, C. G., Karsenberg, D., Van Deursen, W. P. A. & Burrough, P. A. (1996) Integrating dynamic environmental models in GIS: the development of a dynamic modelling language. *Transactions GIS* **1**(1), 40–48.

Sediment Transfer through the Fluvial System (Proceedings of a symposium held in Moscow, August 2004).
IAHS Publ. 288, 2004

Monitoring and modelling flow and suspended sediment transport processes in alluvial cutoffs

R. I. SUTTON, A. P. NICHOLAS & D. E. WALLING
Department of Geography, University of Exeter, Exeter EX4 4RJ, UK
r.i.sutton@exeter.ac.uk

Abstract Sedimentation within cutoffs is controlled by mean and turbulent flow structures that drive suspended sediment transport by advection and diffusion and control particle settling and resuspension processes. The relationship between these flow and sedimentation processes is poorly understood at present due to the difficulty of collecting synchronous instantaneous velocity and sediment concentration data above an undisturbed bed. These relationships are examined using a combined approach involving field monitoring and numerical modelling applied to a site on the River Culm, Devon, UK. Flow and suspended sediment time series were monitored using an array of Acoustic Doppler Velocimeters (ADVs), which provide synchronous instantaneous measurements of three-dimensional (3-D) velocity and sediment concentration at a frequency of 25 Hz. ADV data provide an estimate of the net sediment flux into the cutoff and allow the relationship between vertical turbulent sediment transport and momentum transport events to be examined. Results from a two-dimensional (2-D) finite volume numerical model are also presented and provide further insights into the stage-dependent nature of hydraulics at this site.

Key words alluvial cutoff; ADV; numerical model; suspended sediment

INTRODUCTION

Alluvial cutoffs are typically subject to very high rates of sediment deposition (Lewis & Lewin, 1983) due to their proximity to the river channel and high frequency of inundation (Nicholas & Walling, 1997). Transport of fine sediment into these features is driven by complex flow structures at the interface between the dead water in the cutoff and the flow of the main channel (Nicholas & McLelland, 1999). However, these processes are not well understood at present because little research has been conducted to examine flow and sedimentation within cutoffs, largely due to problems of collecting representative process data in natural rivers. Here we present some results from a study of three-dimensional (3-D) flow hydraulics and suspended sediment transport at a channel-cutoff interface on the River Culm, Devon, UK. Acoustic Doppler velocimeters (ADV) were used to monitor patterns of mean flow, turbulence and suspended sediment concentration simultaneously. Flow data are also compared with results from a two-dimensional (2-D) numerical model which further elucidates the stage-dependent nature of interface hydraulics at this site.

FIELD DATA COLLECTION

Velocity and sediment concentration data were collected at a transect located near the interface between the main channel and a cutoff area at a site on the River Culm, Devon, UK. Data were collected using an array of four ADVs at two separate discharges. A 10-m long, adjustable frame was built to allow easy deployment of the ADVs and eradicate any need for the researcher to enter the water, thus reducing disturbances to the flow and fine sediment. The ADVs were orientated looking into the flow and positioned in a vertical profile separated by gaps of 0.05 m at low discharge and 0.1 m at high discharge. A total of 80 measurements were obtained at 20 vertical profiles at low flow, compared to 64 measurements obtained at 10 vertical profiles at high flow.

The ADV transmits a short acoustic pulse at a frequency of 10 MHz, from a transmitter at the centre of the probe head, into the water column. Air bubbles and small suspended particles reflect acoustic energy back to three receivers focused on a sampling volume of ~0.05 m below the probe head. Velocities can be calculated from the Doppler-shifted phase of the reflected acoustic energy. At each measurement position ADV data were collected for ~4 minutes at a sampling rate of 25 Hz (6000 data points). Data were post-processed using the methods of McLelland & Nicholas (2000).

ADVs provide measurements of 3-D mean velocity and turbulence statistics. They can also be used to obtain simultaneous measurements of suspended sediment concentration since this is related to the intensity of backscatter monitored by the ADV. The probes used here were calibrated by collecting one 250-ml bottle of water from the head of each ADV, using a pump sampling system for each ADV measurement point. Mean suspended sediment concentrations determined for each bottle sample were used to derive a calibration relationship between backscatter intensity and sediment concentration for each probe. Once calibrated in this way the ADV can be used to examine sediment concentration time series and their relationship with turbulent flow structures (cf. Nikora & Goring, 2002).

NUMERICAL SIMULATION

A 2-D finite volume model was used to predict patterns of flow depth and depth-averaged velocity at the field site. The model was implemented using a regular grid, with a resolution of 0.5 m, representing the topography of the field site. The continuity of mass and momentum equations solved by the model can be written as follows:

$$\frac{\partial h}{\partial t}+\frac{\partial q}{\partial x}+\frac{\partial r}{\partial y}=0 \tag{1}$$

$$\frac{\partial q}{\partial t}+\frac{\partial (q^2/h)}{\partial x}+\frac{\partial (qr/h)}{\partial y}+\frac{g}{2}\frac{\partial (h^2)}{\partial x}+gh\frac{\partial z}{\partial x}-\frac{1}{\rho}\frac{\partial (h\tau_{xy})}{\partial y}-\frac{1}{\rho}\frac{\partial (h\tau_{xx})}{\partial x}+\frac{\tau_{bx}}{\rho}=0 \tag{2}$$

$$\frac{\partial r}{\partial t}+\frac{\partial (r^2/h)}{\partial y}+\frac{\partial (qr/h)}{\partial x}+\frac{g}{2}\frac{\partial (h^2)}{\partial y}+gh\frac{\partial z}{\partial y}-\frac{1}{\rho}\frac{\partial (h\tau_{yx})}{\partial x}-\frac{1}{\rho}\frac{\partial (h\tau_{yy})}{\partial y}+\frac{\tau_{by}}{\rho}=0 \tag{3}$$

where h is the flow depth (m), q and r are unit discharge ($m^2\ s^{-1}$) in the x and y directions respectively, t is time (s), z is bed elevation (m), g is acceleration due to gravity, ρ is the fluid density, τ_{xx}, τ_{yy}, τ_{xy} and τ_{yx} are turbulent stresses (calculated using a zero order eddy viscosity model), and τ_{bx} and τ_{yx} are the bed shear stresses (calculated as a quadratic function of velocity). For example, in the x direction:

$$\frac{\tau_{bx}}{\rho}=\frac{g}{(Ch)^2}q\sqrt{q^2+r^2} \tag{4}$$

where C is the Chezy friction coefficient. An optimum uniform value of C within the river channel was determined by comparing model predictions with depth and velocity data collected from 20 cross sections along the study reach.

FLOW HYDRAULICS

ADV data (Figs 1 and 2) and model results (Fig. 3) show broadly consistent changes in flow structures at the channel-cutoff interface as discharge rises. At low discharge (Fig. 1(a)) water enters the cutoff at its upstream end (point B) and returns to the channel where it meets the cutoff bank (point A). At high discharge (Fig. 1(b)) flow into the cutoff is restricted to a 2-m portion of the transect AB. Results shown in Fig. 2 (based on depth-averaged ADV data) illustrate changes in flow direction across the transect in greater detail. At low discharge (Fig. 2(a)) the interface can be divided into three broad regions. Zone I contains water flowing out of the cutoff and from B to A. Zones II and III are characterized by water

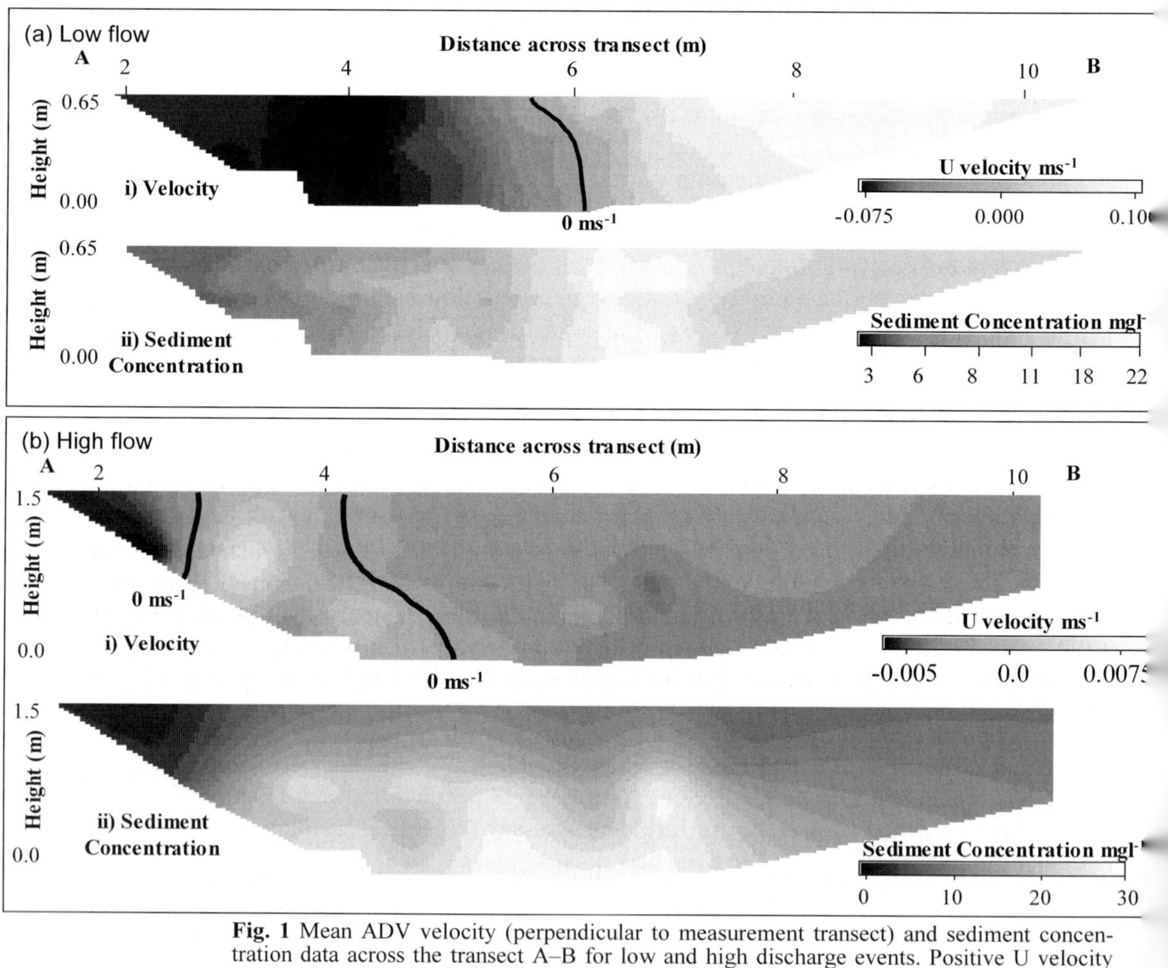

Fig. 1 Mean ADV velocity (perpendicular to measurement transect) and sediment concentration data across the transect A–B for low and high discharge events. Positive U velocity indicates flow into the cutoff.

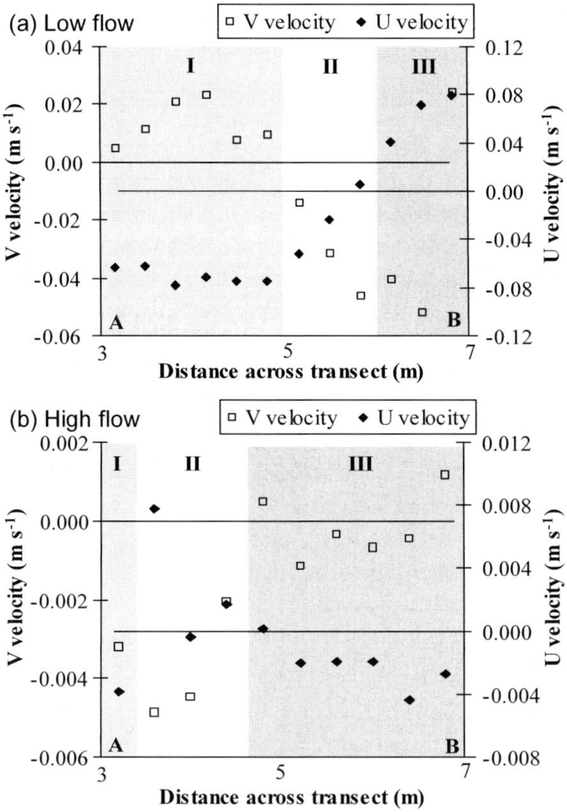

Fig. 2 Depth averaged ADV data obtained along transect AB. Highlighted regions indicate areas characterized by distinct flow structures. Positive U and V velocities indicate flow moving into the cutoff and from B to A, respectively.

moving from A to B, out of the cutoff in zone II and into it in zone III. This pattern of circulation is broadly equivalent to that shown in the model results at low discharge (Fig. 3), the main difference being that the circulation cell is more symmetrical in the model while in the field, flow out of the cutoff occupies a larger part of the channel-cutoff interface. At high discharge (Fig. 2(b)) the interface can again be split into three zones. Zone I represents a small recirculation cell near the bank at A. The fact that this feature is not predicted by the model may be due to the resolution of the model grid. Zone II contains water that is predominately entering the cutoff and flowing from A to B. In zone III, water is leaving the cutoff and moving weakly from A to B. It is also noticeable that velocities are generally an order of magnitude lower in the field at the higher discharge. Both the overall reduction in flow velocity and the circulation patterns observed in zones II and III are reproduced by the numerical model (Fig. 3).

Model results suggest that the significant reduction in velocity and the reversal in the direction of flow circulation at the channel-cutoff interface occur because the main channel flow moves away from the cutoff as discharge rises. At the low discharge, water is clearly flowing directly into the cutoff. In contrast, at bankfull discharge the high velocity core within the channel is displaced approximately 5 m to the right so that it bypasses the cutoff.

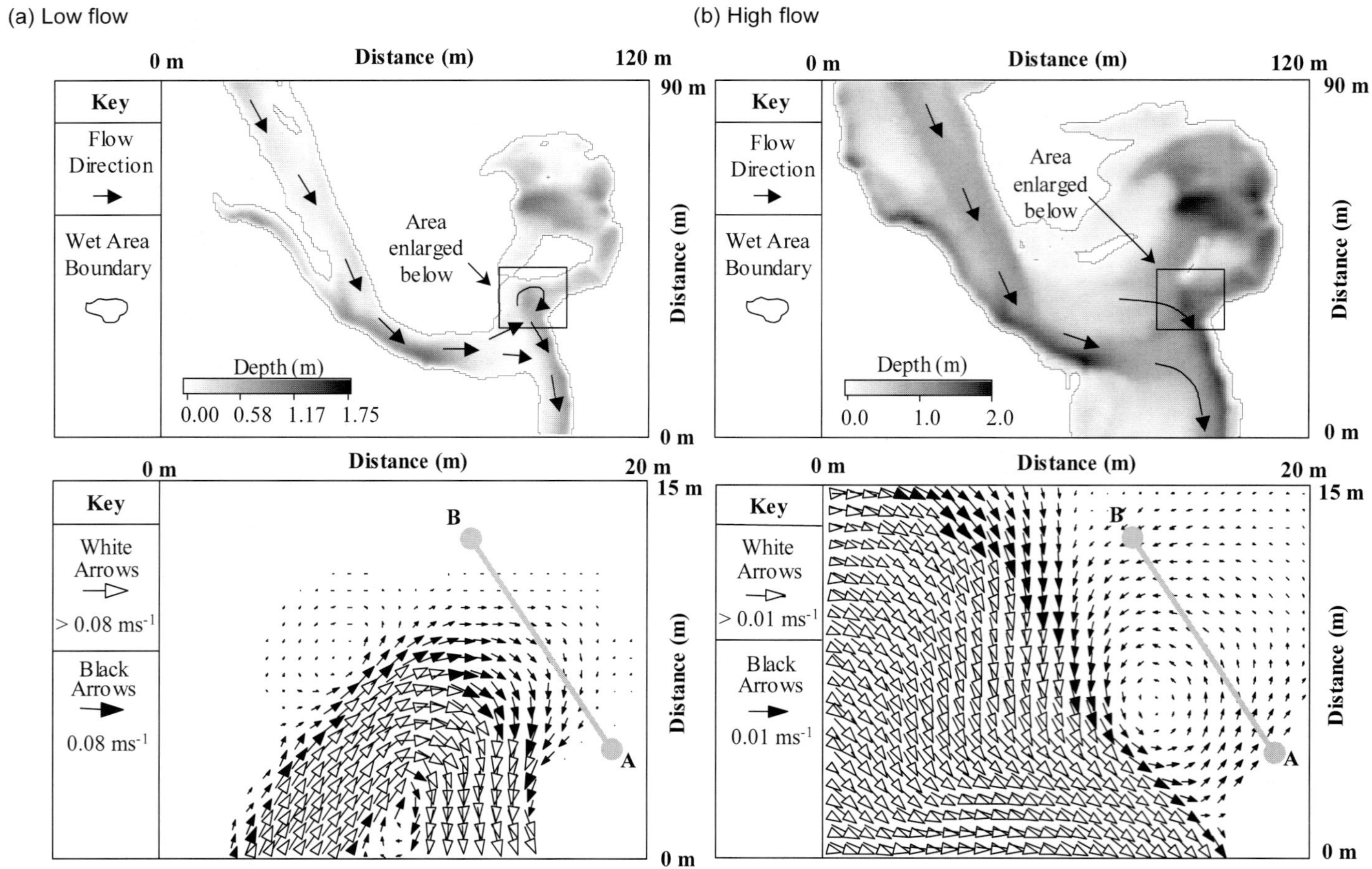

Fig. 3 Modelled flow depth within a 150 m reach of channel (upper diagrams) and velocity vectors at the channel-cutoff interface (lower diagrams).

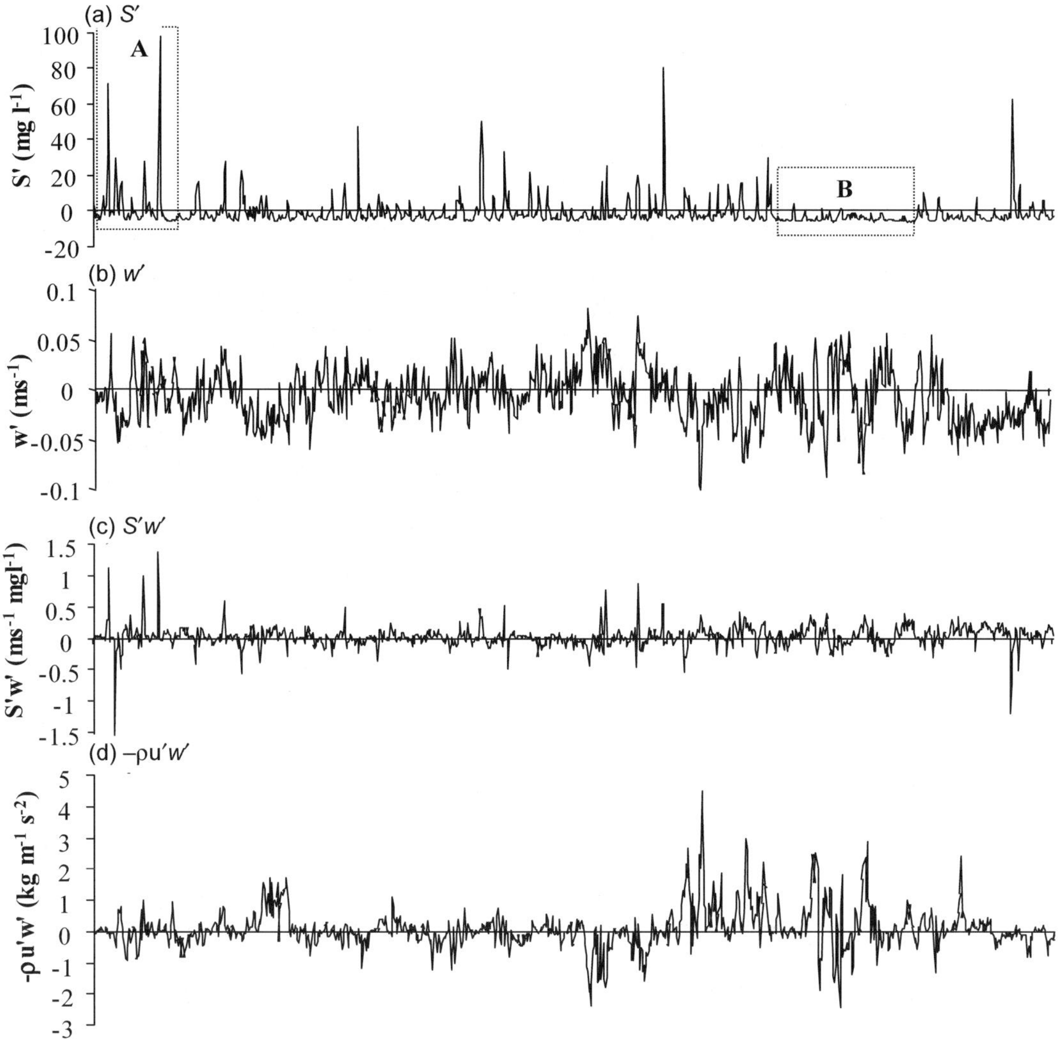

Fig. 4 Forty-second time series from one ADV located 5.5 m along the transect AB and 0.1 m above the bed during the low flow event.

entrance and a recirculation zone forms within this slack water area. The presence of this recirculation zone agrees well with other field studies (Nicholas & McLelland, 1999; Hankin *et al.*, 2001) and laboratory experiments (Kimura & Hosoda, 1997) that have identified such flow structures as common where channels are attached laterally to near-stationary water bodies.

SUSPENDED SEDIMENT TRANSPORT

Mean suspended sediment concentration estimates calculated using the ADV (Fig. 1) show clear systematic trends with peak concentrations occurring near the bed and in the centre of the channel-cutoff interface. Integration of the sediment flux across the interface at low discharge (using sediment concentration and velocity data) indicates that there is net sediment transport into the cutoff which is, presumably, balanced by deposition within this feature. The average deposition rate over the cutoff estimated in this way is in the order of 108 g m^2 day^{-1}, which is consistent with evidence of medium-term sedimentation rates on the River Culm derived from radionuclide studies (e.g. Nicholas & Walling, 1997).

Preliminary analysis of velocity and sediment concentration time series obtained using the ADV illustrate the potential for using this technique to investigate turbulent sediment transport within cutoff environments. Figure 4 shows 40 seconds of a single ADV time series obtained at low discharge at the 5.5 m point of the measurement transect. Upward turbulent sediment transport ($S'w' > 0$) can be linked with two phenomena. First, highly intermittent peaks in sediment concentration (S) that are associated with positive w' events, but which do not appear to be linked to significant Reynolds shear stress ($-\rho u'w'$) production (region A marked on Fig. 4(a) provides a good example of this). Second, periods of lower than average sediment concentration ($S' < 0$) where contributions to upward turbulent sediment transport are due to downward movement ($w' < 0$) of clean water (region B on Fig. 4(a)). These periods are sometimes associated with Reynolds shear stress production during sweep-type events.

SUMMARY

Results have been presented from an investigation of flow and suspended sediment transport at the interface between the main channel and a cutoff on the River Culm, Devon, UK. Comparisons between field data and model results are consistent and suggest that flow structures in such locations are likely to be strongly dependent on stage. At high flow the cutoff is occupied by a recirculation zone similar to that observed in previous field-based and experimental studies. However, differences in flow direction within the main channel prevent this recirculation pattern from forming at low stage. These hydraulic structures have important implications for the transport of fine sediment into and out of the cutoff by advective and diffusive mechanisms.

REFERENCES

Hankin, B. G., Hardy, R., Kettle, H. & Beven, K. J. (2001) Using CFD in a GLUE framework to model the flow and dispersion characteristics of a natural fluvial dead zone. *Earth Surf. Processes Landf.* **26**, 667–687.

Kimura, I. & Hosoda, T. (1997) Fundamental properties of flows in open channels with dead zone. *J. Hydraul. Engng* **123**, 98–107.

Lewis, G. W. & Lewin, J. (1983) Alluvial cutoffs in Wales and the Borderlands. *International Association of Sedimentologists, Special Publication* 6, 145–154.

McLelland, S. J. & Nicholas, A. P. (2000) A new method for evaluating errors in high-frequency ADV measurements. *Hydrol. Processes* **14**, 351–366.

Nicholas, A. P. & McLelland, S. J. (1999) Hydrodynamics of a floodplain recirculation zone investigated by field monitoring and numerical simulation. In: *Floodplains: Interdisciplinary Approaches* (ed. by S. B. Marriott & J. Alexander) , 15–26. Special Publications 163. Geological Society, London, UK.

Nicholas, A. P. & Walling, D. E. (1997) Modelling flood hydraulics and overbank deposition on river floodplains. In: *Earth Surf. Processes Landf.* **22**, 59–77.

Nikora, V. N. & Goring, D. G. (2002). Fluctuations of suspended sediment concentration and turbulent sediment fluxes in an open-channel flow. *J. Hydraul. Engng* **128**, 214–244.

5 Implications for Nutrient and Contaminant Transfer

Monitoring suspended sediment and associated trace element and nutrient fluxes in large river basins in the USA

ARTHUR J. HOROWITZ
US Geological Survey, Peachtree Business Centre, 3039 Amwiler Road, Atlanta, Georgia 30360, USA
horowitz@usgs.gov

Abstract In 1996, the US Geological Survey converted its occurrence and distribution-based National Stream Quality Accounting Network (NASQAN) to a national, flux-based water-quality monitoring programme. The main objective of the revised programme is to characterize large USA river basins by measuring the fluxes of selected constituents at critical nodes in various basins. Each NASQAN site was instrumented to determine daily discharge, but water and suspended sediment samples are collected no more than 12–15 times per year. Due to the limited sampling programme, annual suspended sediment fluxes were determined from site-specific sediment rating (transport) curves. As no significant relationship could be found between either discharge or suspended sediment concentration (SSC) and suspended sediment chemistry, trace element and nutrient fluxes are estimated using site-specific mean or median chemical levels determined from a number of samples collected over a period of years, and under a variety of flow conditions.

Key words large river basins; monitoring; nutrient fluxes; suspended sediment fluxes; trace element fluxes

INTRODUCTION

The United States Geological Survey's (USGS) National Stream Quality Accounting Network programme (NASQAN) began in 1973 and was designed to determine the occurrence and distribution of a wide variety of water quality parameters (Ficke & Hawkinson, 1975; OWQ, 1996). The programme operated more-or-less continuously using this approach, until 1995, and at its height, incorporated more than 500 sites. In 1994, in response to substantially diminishing resources, changes in data requirements, and to better integrate the programme with other USGS ambient water-quality monitoring programmes, NASQAN was redesigned (OWQ, 1996). Initially, the new programme, which began in late 1995 (the 1996 water year), had two data objectives: (a) to characterize large US river basins by measuring the mass flux of selected constituents at critical nodes in various systems; and (b) to determine the fluxes of a variety of chemical constituents to the coastal zone. An important secondary goal was to provide error estimates for all the calculated fluxes. After the first year, the second objective was dropped due to a lack of sufficient spatial coverage. Based on historical data, the revised NASQAN network should account for >80% of the annual movement of water within the conterminous USA, as well as the discharge of more than 260 Mt of suspended sediment, to the coastal zone (Meade & Parker, 1985).

SAMPLING DESIGN AND METHODS

The revised NASQAN programme sampling design was predicated on evaluating the movement of water and a variety of water-quality-associated parameters in large rivers, where individual point sources tend to have only relatively local impacts. As such, water and dissolved constituents tend to behave conservatively; hence, large river segments tend to be compositionally homogeneous. Thus, contributions from sizeable portions of a basin, and/or potential sources or sinks for a variety of chemical parameters, can be estimated by adding/subtracting the contributions/losses of large river segments. Hydrophilic substances tend to move through large basins along with the water itself. Hence, sampling the same segment of water (i.e. Lagrangian sampling) as it moves through the system should permit the delineation of concentration changes and, if the discharge or water velocity is known, travel time(s) for the segment may be predicted.

Contrariwise, suspended sediment does not behave conservatively; it moves in and out of suspension, and there are constant exchanges between the water column, the riverbed, and banks. Thus, the particles making up a "packet" of sediment, and their associated chemical constituents, continuously change as material moves in (deposition) and out (resuspension) of "storage" as the packet moves downstream. As such, the packet rarely retains its original composition, even over relatively short distances, and travel times for individual particles are unpredictable. Thus, suspended sediment and sediment-associated chemical constituents display much more marked spatial and temporal variability than dissolved constituents (e.g. Horowitz, 1995). As a result of this non-conservative behaviour, contributions from sizeable portions of a basin, and/or potential sources or sinks for a variety of sediment-associated chemical parameters, cannot be reasonably estimated by adding/subtracting the contributions/losses of large river segments. As a result of the behavioural differences between water and suspended sediment, a sampling programme designed to monitor dissolved fluxes may not be applicable to monitoring sediment and sediment-associated chemical fluxes. This dichotomy can create network design problems that may necessitate separate sampling schemes (both spatial and temporal) for water and suspended sediment.

Each revised NASQAN site was instrumented to determine mean daily discharge, but water and suspended sediment samples were collected no more than 12 to 15 times per year (OWQ, 1996). Sampling schedules were established to try to cover >80% of the typical range of annual flows at each site; however, sampling tended to be biased in favour of non-baseflow periods. As a result of the limited physical sampling programme, and the perception that at least a mean daily suspended sediment concentration (SSC) would be required to produce reasonable flux estimates (Walling & Webb, 1981, 1988; de Vries & Klavers, 1994; Horowitz, 1995; Phillips *et al.*, 1999), each site either had to be instrumented with some type of automatic sampling/measuring device (Horowitz, 1995), or a site-specific discharge-based regression equation had to be developed for predicting SSC (de Vries & Klavers, 1994; Phillips *et al.*, 1999). As no revised NASQAN site was instrumented with automatic sampling equipment or measuring devices, the only means of estimating mean-daily SSC was by developing site-specific discharge-based regression equations (rating curves).

ESTIMATING SUSPENDED SEDIMENT FLUXES AND ASSOCIATED ERRORS

For more than sixty years, in the absence of actual continuous or near-continuous SSC data, hydrologists have used rating (sediment transport) curves to estimate (predict) SSCs for flux calculations. Although there are more than 20 methods for developing rating curves, the most common is a power function (regression) that relates SSC to water discharge, with the discharge measurement constituting the independent variable (e.g. Phillips, *et al.*, 1999; Asselman, 2000). This requires the log-transformation of SSC and discharge data prior to the analysis. Comparisons of actual and predicted SSC, partially as a result of scatter about the regression line, as well as the conversion of results from log-space to arithmetic-space, indicate that rating curves can substantially underpredict actual concentrations (Walling & Webb, 1988; Asselman, 2000). To compensate, various method modifications have been applied; these include dividing the SSC–discharge data into seasonal or hydrological groupings, developing various correction factors, or using nonlinear regression equations (Duan, 1983; Ferguson, 1986; Walling & Webb, 1988; de Vries & Klavers, 1994; Phillips *et al.*, 1999; Asselman, 2000).

Within NASQAN, the Mississippi River at Thebes site is unique because it constitutes the only long-term, ongoing, daily SSC-measuring site in the network. As such, the data from this site are uniquely suited to evaluating such issues as sampling frequency, temporal resolution, and flux calculation/estimation errors. All calculations used in these evaluations are based on a 20-year data set covering water years (October–September) spanning 1981 to 2000.

The first evaluation entailed an examination of the temporal resolution, and associated errors, of estimated suspended sediment fluxes at the Thebes site covering the first 5 years (1996–2000) of the revised NASQAN programme. The actual flux for that 5-year period was 414 Mt (mega tonnes), whereas the predicted flux, using daily values was 404 Mt, a 3% underestimate. Despite this close agreement for the entire 5-year period, maximum errors in daily estimates of SSC ranged from –76% to +205%. The 5-year suspended sediment flux estimate using the approximately monthly NASQAN samples was 439 Mt, a 6% overestimate. The various errors associated with different levels of temporal resolution also were calculated for the same 5-year period; the errors tend to decline with increasing temporal resolution (Table 1). This accrues because the rating-curve approach underestimates highs and overestimates lows. Hence, the longer the period of interest, the greater the chance for the over- and underestimates to balance.

Table 1 Various levels of temporal resolution and their associated errors for the Mississippi River at the Thebes site for the five-year period 1996–2000.

Temporal Resolution	Maximum underestimate (relative %)	Maximum overestimate (relative %)	Average absolute error (relative %)
Daily	–76	205	26
Weekly	–63	157	24
Monthly	–43	58	19
Quarterly	–33	21	13
Yearly	–12	8	8

The effect of sampling frequency on the accuracy and associated errors of 5-year suspended sediment flux estimates also was investigated as part of the same evaluation. This entailed using the daily SSC values for the Thebes site and calculating a large number of rating curves to predict daily SSC values assuming different levels of sampling intensity.

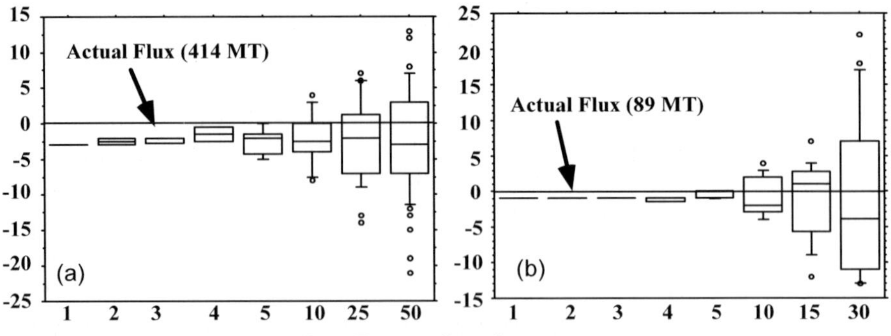

Fig. 1 Effect of sampling frequency on the errors associated with the estimation of suspended sediment fluxes over a 5-year (1996–2000 WY) period (a), and for a 1-year (1995 WY) period (b), for the Mississippi River at Thebes site.

The sampling frequencies evaluated in this way corresponded to: (a) once a day; (b) once every other day; (c) once every 3 days; (d) once every 4 days; (e) once every 5 days (weekly); (f) once every 10 days, (g) once every 25 days (monthly); and (h) once every 50 days (every other month). Not surprisingly the accuracy of the 5-year estimates decreased, and the size and range of the associated errors increased with decreasing sampling frequency (Fig. 1(a)). Interestingly, there was little difference between sampling frequencies ranging from 1- to 5-days. On the other hand, estimation errors from sampling frequencies on the order of once every two months (once every 50 days) were little compromized, and tended to fall within a range of ±20%. As the calculations were based solely on calendar distributions, they probably represent the maximum error likely to occur with this level of sampling frequency (Fig. 1(a)). If the same level of sampling (once every 50 days) were hydrologically distributed such as to encompass some 80 to 85% of the typical range of site-specific discharge, the associated estimation errors likely would be substantially less (e.g. Horowitz, 1995). The effect of sampling frequency on the accuracy and associated errors of annual suspended sediment flux estimates also were investigated concurrently (Fig. 1(b)). These evaluations covered high, median, and low flux years. The sampling frequencies evaluated in this way corresponded to: (a) once a day; (b) once every other day; (c) once every 3 days; (d) once every 4 days; (e) once every 5 days (weekly); (f) once every 10 days, (g) once every 15 days (fortnightly); and once every 30 days (every month). Note that as with the 5-year study, there is little difference between 1- and 5-day sampling. Further, even collecting a sample as infrequently as once a month only produced errors on the order of ±20%, regardless of the flow conditions (high, low, median). The same caveats apply to the annual study as to the 5-year study, hence, hydrologically-based sampling, as opposed to calendar-based sampling, is likely to produce substantially more accurate estimates.

The actual 20-year suspended sediment flux for the Thebes site for the period 1981 to 2000 was 2100 Mt. A single rating curve, using the entire 20-year data set, yielded an estimate of 2000 Mt, representing an error of <1%. This is a fairly standard approach for generating site-specific rating curves where long-term data are available, and is based on the assumption that all the data from the site are part of the same statistical population. Note that the annual errors associated with this single rating-curve approach can be significant (Fig. 2(a)). However, when individual annual rating curves are calculated for the same

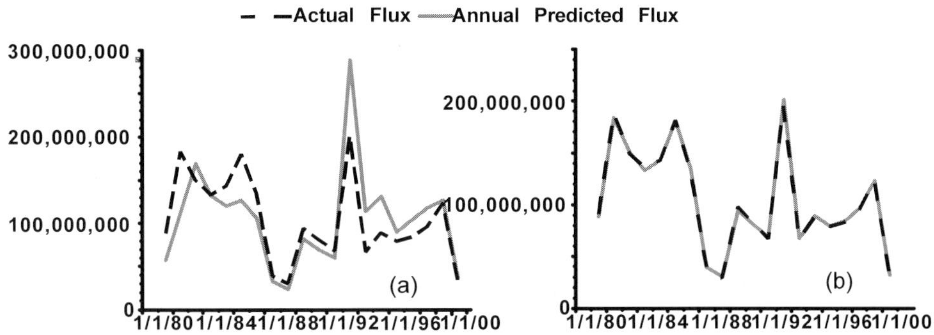

Fig. 2 Comparisons of annual fluxes calculated by using a 20-year rating curve (a), and by calculating individual annual rating curves (b), at the Mississippi River Thebes site for the period 1981–2000.

20-year period, it is apparent that the data are not part of the same statistical population. Some curves are linear whereas others are nonlinear (both concave and convex). Interestingly, the sum of the annual fluxes for the 20-year period now is 2100 Mt; however, the individual annual estimates are significantly closer to the actual fluxes (Fig. 2(b)). Hence, although the estimate of total flux does not substantially improve through the use of annual rating curves as opposed to a single rating curve, better annual estimates within the 20-year period can be obtained if individual calculations are used.

SEDIMENT-ASSOCIATED TRACE ELEMENT AND NUTRIENT FLUXES

The traditional, as well as the regulatory method of determining suspended sediment-associated trace element concentrations entails the collection and analysis of filtered (dissolved) and unfiltered (whole water) sample pairs, with subsequent subtraction of the associated concentrations of the former from the latter (Office of Water Data Coordination (OWDC), 1978, 1982). This approach is problematic because: (a) it does not provide total concentrations (with "total" being defined as ≥95% of the constituent concentration present); (b) the small sample masses typically collected can be affected by inhomogeneities; and (c) the small sample masses, combined with the dilution effects of the associated water in whole-water samples, can lead to analytical detection limit problems (e.g. Horowitz, 1995). Based on the foregoing, the revised NASQAN programme adopted a different approach. Large-volume (10 to 100 l), depth- and width-integrated isokinetic whole-water samples are collected such that aggregate suspended sediment masses would be between 1.00 and 1.25 g. All the large-volume whole-water samples were shipped to a central location (Atlanta, Georgia) for dewatering by flow-through centrifugation, and subsequent total trace element analysis.

Predicting suspended sediment-associated trace element concentrations using either discharge or SSC as the independent variable proved inadequate. However, site-specific intra- and interannual suspended sediment-associated trace element variations were markedly less (usually no more than a factor of two) than those for either discharge or SSC in all the NASQAN basins (Table 2). In fact, except where concentrations approach the reporting limit, the differences between site-specific interannual trace element means/medians typically do not exceed the errors (generally ±10%) associated with the analytical methods

Table 2 Annual mean and median values for discharge (*Q*), suspended sediment (S. Sed.), and sediment associated chemical concentrations for selected NASQAN sites (*n* is number of data points).

Site	Water Year	S. Sed. (mg kg^{-1})	*Q* (m^3 s^{-1})	Cu (mg kg^{-1})	Pb	Zn	Cd	P	As	Hg	Mn	Fe (wt. %)	Al	TOC
Mississippi River at Thebes, Illinois														
n	1996			16	16	16	16	16	16	10	16	16.0	17	13.0
Mean		289	6 723	23	28	97	0.6	1 000	8.5	0.05	760	2.5	5.5	3.1
Median		193	4 842	22	27	90	0.5	1 000	8.5	0.05	830	2.5	5.5	2.3
n	1997			15	15	15	15	15	15	15	15	15	15	15
Mean		288	7 504	19	24	89	0.5	850	8.8	0.04	980	2.4	5.4	1.8
Median	1998	221	6 428	18	22	89	0.5	850	8.7	0.03	1 000	2.4	5.5	1.8
n				17	17	17	17	17	17	13	17	17	17	17
Mean		354	8 181	20	25	93	0.5	850	9.0	0.03	1 000	2.6	5.5	1.9
Median		279	7 504	19	26	96	0.5	880	9.0	0.03	980	2.6	5.5	1.9
Willamette River at Portland, Oregon														
n				11	11	17	11	11	10	11	11	11	11	9
Mean	1996	58	1 584	88	40	160	0.4	2 100	8.3	0.11	2 400	5.7	8.0	4.4
Median		12	968	53	31	140	0.3	2 100	8.2	0.10	2 400	5.6	8.3	4.1
n				19	19	19	17	19	19	16	19	19	19	15
Mean	1997	51	1 628	52	24	130	0.3	1 700	8.6	0.07	1 900	5.8	8.6	2.8
Median		11	864	53	18	130	0.2	1 600	8.8	0.06	1 500	5.9	8.9	2.3
n				20	20	20	17	20	20	14	20	20	20	12
Mean	98	16	918	62	23	150	0.4	1 800	8.0	0.08	2 100	5.9	8.0	3.5
Median		10	702	60	23	140	0.3	1 800	8.0	0.08	1 800	6.1	8.1	3.7
Green River at Green River, Utah														
n				12	11	12	11	12	13	9	13	13	13	9
Mean	96	4 610	44	18	17	68	0.4	680	7.3	0.03	350	1.8	4.8	1.0
Median		197	34	19	16	69	0.4	700	7.2	0.04	330	1.7	4.6	1.2
n				13	13	13	12	13	13	9	13	13	13	12
Mean	97	2 745	105	18	18	78	0.5	670	6.5	0.03	360	2.1	5.3	0.8
Median		2 819	71	16	18	63	0.5	660	5.6	0.02	360	1.8	4.7	0.8
n				11	11	11	11	11	11	5	11	11	11	7
Mean	98	807	54	15	16	66	0.3	660	6.0	0.02	350	1.8	4.5	1.2
Median		636	33	15	16	66	0.3	660	6.0	0.02	340	1.8	4.2	1.2
Rio Grande at Foster Ranch, Texas														
n				15	15	15	13	15	15	15	15	15	15	11
Mean	96	1 120	25	14	19	82	0.3	700	8.0	0.29	440	2.2	5.4	1.2
Median		274	15	15	17	92	0.3	800	7.9	0.06	450	2.4	5.7	0.9
n				11	11	11	11	11	11	11	11	11	11	11
Mean	97	570	23	15	16	85	0.4	760	8.6	0.04	560	2.4	5.9	1.5
Median		272	15	15	17	81	0.4	780	8.5	0.04	450	2.5	6.1	1.1
n				10	10	10	10	10	10	10	10	10	10	10
Mean	98	328	15	14	17	89	0.4	750	9.0	0.05	480	2.6	6.2	1.3
Median		242	14	14	16	91	0.4	800	9.0	0.04	480	2.4	6.1	1.2

used to determine them (Table 2). With the failure of the regression model approach, this relative lack of variability provided one of the few means of estimating annual sediment-associated trace element fluxes, through the use of derived mean/median concentrations. Potential errors associated with the use of this approach ranged from <1% to as much as 75% and were determined by comparing the differences between the actual as opposed to the calculated (based on the selected mean/median concentrations) summed instantaneous daily fluxes for actual NASQAN samples.

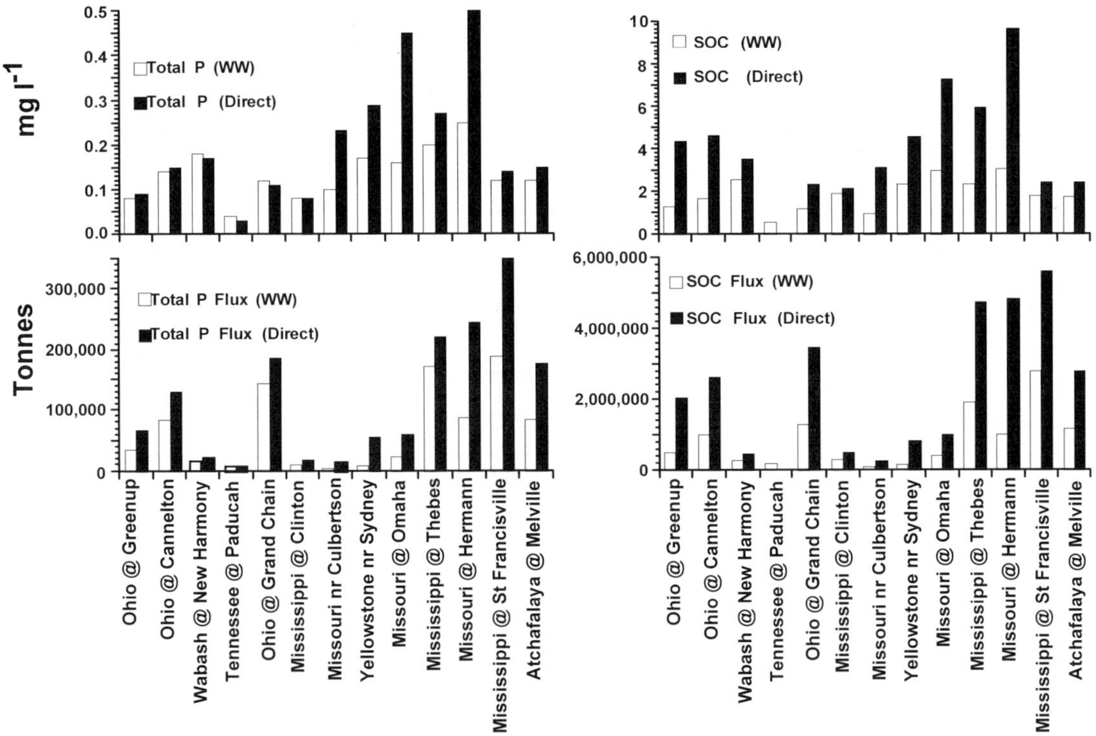

Fig. 3 A comparison of suspended sediment-associated average total phosphorus and organic carbon concentrations and cumulative (1996–1998 Water Years) suspended sediment-associated fluxes for the Mississippi River Basin sites indicating the general pattern of underestimates associated with the traditional/regulatory paired whole-water/filtered water approach, relative to direct determinations made on dewatered suspended sediment.

As part of the revised NASQAN program, suspended sediment-associated P as well as organic carbon (SOC) were determined both by the traditional/regulatory paired whole-water/filtered-water approach, and directly on dewatered suspended sediment. The latter method consistently generated significantly higher (by factors ranging from 1.5- to 10-fold) concentrations/fluxes for both constituents (Fig. 3). The apparent underestimates associated with the traditional/regulatory paired whole-water/filtered-water approach seem to result from a combination of sampling and analytical factors, with the latter likely to be more significant. These consistent underestimates may be important factors when addressing such issues as nutrient transport, eutrophication, algal blooms, and coastal/estuarine productivity.

The determination of suspended sediment, as well as total trace element, total phosphorus (P), and total organic carbon (TOC) fluxes is one of the major goals of the revised NASQAN programme. Unfortunately, as a result of a large number of censored (less than the detection limit) dissolved concentrations, total fluxes for a number of constituents (e.g. Ag, Pb, Cd, Cr, Co, V, Be, As, Sb and Hg) could not be estimated. The majority (≥70%) of the Cu, Zn, Cr, Ni, Ba, P, As, Fe, Mn and Al are transported in association with suspended sediment; Sr transport seems dominated by the dissolved phase, whereas the transport of Li and TOC seem to be divided equally between both phases (Fig. 4).

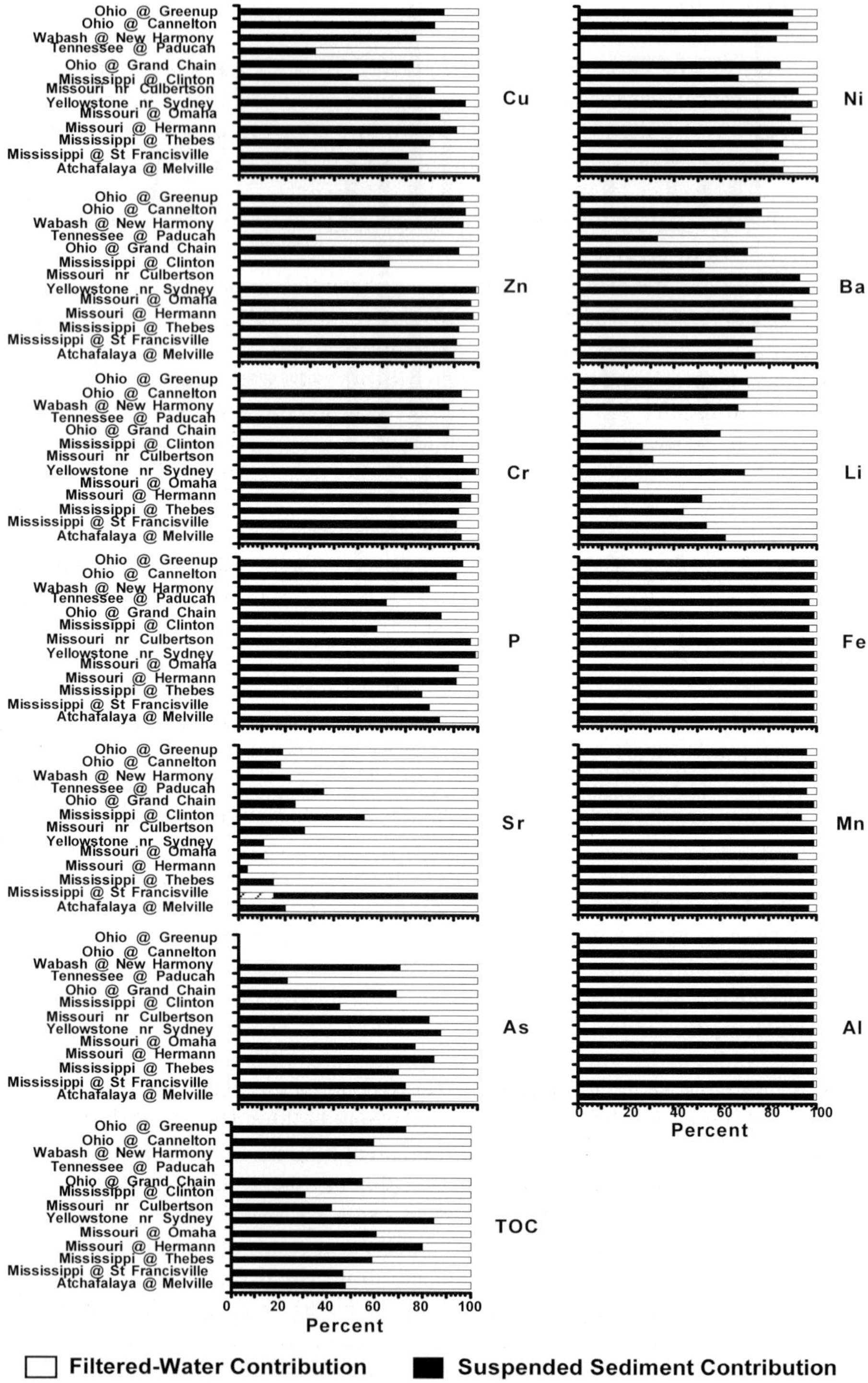

Fig. 4 Partitioning of various trace elements and nutrients between the dissolved (filtered water) and solid phases for the sites in the Mississippi River Basin.

CONCLUSIONS

(a) In large river basins, water and dissolved constituents tend to behave conservatively, whereas suspended sediment and solid phase-associated constituents tend to behave non-conservatively; therefore, different sampling schemes may be required for each type of media to adequately monitor suspended sediment, trace element, and nutrient fluxes.

(b) It appears that at least for large rivers, sediment rating curves can be used to generate reasonably accurate (≤15–20%) suspended sediment flux estimates for quarterly timeframes or greater.

(c) Sampling frequencies can exercise a substantial impact on the accuracy of sediment rating curve-generated flux estimates; however, good estimates (errors of ≤±20%) can be obtained from relatively infrequent samples. This is especially true if the samples are collected on a hydrological as opposed to a calendar-basis; good 5-year flux estimates appear to require as few as six samples per year, whereas good annual estimates appear to require as few as 12 samples per year.

(d) Site-specific intra- and interannual variations in suspended sediment associated-trace element concentrations tended to be much smaller (usually less than a factor of two) than those for discharge or SSC (typically more than an order of magnitude). As there are no strong interrelations between discharge or SSC, and suspended sediment-associated trace element chemistry, the use of site-specific mean/median concentrations provides one of the only means of estimating annual sediment-associated trace element and nutrient fluxes.

(e) The concentrations, and hence the annual fluxes, for suspended sediment-associated P and organic carbon, determined from the direct analyses of dewatered suspended sediment samples were markedly higher than those determined using the more traditional/regulatory paired whole-water/filtered-water approach; this could be important for such issues as eutrophication, algal blooms, and coastal productivity.

(f) At least in the NASQAN basins, the majority (typically ≥70%) of Cu, Zn, Cr, Ni, Ba, P, As, Fe, Mn and Al are transported in association with suspended sediment; in contrast, Sr fluxes seem to be dominated by the dissolved fraction, whereas the transport of Li and TOC appear to be divided about equally between both phases.

REFERENCES

Asselman, N. E. M. (2000) Fitting and interpretation of sediment rating curves. *J. Hydrol.* **234**, 228–248.

de Vries, A. & Klavers, H. C. (1994) Riverine fluxes of pollutants: monitoring strategy first, calculation methods second. *European J. Water Pollution Control.* **4**, 12–17.

Duan, N. (1983) "Smearing estimate": a nonparametric retransformation method. *J. Am. Statist. Assoc.* **78**, 605–610.

Ferguson, R. I. (1986) River loads underestimated by rating curves. *Water Resour. Res.* **22**, 74–76.

Ficke, J. F. & Hawkinson, R. O. (1975) *The National Stream Quality Accounting Network (NASQAN)—Some Questions and Answers*. US Geological Survey Circular 719, US Government Printing Office, Washington DC, USA.

Horowitz, A. J. (1995) *The Use of Suspended Sediment and Associated Trace Elements in Water Quality Studies*. IAHS Special Publication 4. IAHS Press, Wallingford, UK.

Meade, R. H. & Parker, R. S. (1985) Sediment in rivers of the United States. In: *National Water Summary 1984, US Geological Survey Water-Supply Paper 2275*, 49–60. US Government Printing Office, Washington DC, USA.

Office of Water Data Coordination (1978) *National Handbook of Recommended Methods for Water Data Acquisition*, Chapter 3: Sediment, 3-17 – 3-27. US Geological Survey, Reston, Virginia, USA.

Office of Water Data Coordination (1982) *National Handbook of Recommended Methods for Water Data Acquisition*, Chapter 5: Chemical Quality, 5-10–5-11. US Geological Survey, Reston, Virginia, USA.

Office of Water Quality (1996) *Redesign Plan for the National Stream Quality Accounting Network.* US Geological Survey, Reston, Virginia, USA.

Phillips, J. M., Webb, B. W., Walling, D. E. & Leeks, G. J. L. (1999) Estimating the suspended sediment loads of rivers in the LOIS study area using infrequent samples. *Hydrol. Processes* **13**, 1035–1050.

Walling, D. E. & Webb, B. W. (1981) The reliability of suspended sediment load data. In: *Erosion and Sediment Transport Measurement* (Proc. Symp., Florence, June 1981), 177–194. IAHS Publ. 133. IAHS Press, Wallingford, UK.

Walling, D. E. & Webb, B. W. (1988) The reliability of rating curve estimates of suspended sediment yield: some further comments. In: *Sediment Budgets* (ed. by M. P. Bordas & D. E. Walling) (Proc. Porto Alegre Symposium, December 1988), 337–350. IAHS Publ. 174. IAHS Press, Wallingford, UK.

Sediment–contaminant interactions and transport: a new perspective

IAN G. DROPPO & GARY G. LEPPARD
NWRI, Environment Canada, PO Box 5050, Burlington, Ontario L7R 4A6, Canada
ian.droppo@ec.gc.ca

Abstract This paper investigates the sediment structural controls over the transport and compartmentalization of Cd, Cu and Pb within flocculated sediments. The compartmentalization of contaminants within flocculated sediment particles demonstrates the complex underlying biogeochemical controls of sediment contaminant interactions. The aims and objectives of the paper are to demonstrate (a) how the structure of suspended sediment particles will influence the transfer of metals within aquatic systems and (b) how the internal structure/composition of the particles promotes compartmentalization of metals in a selective manner.

Key words bacteria; Cd; clay; Cu; EPS; floc; flocculation; metals; Pb

INTRODUCTION

Sediment transport and sediment contaminant interactions are often diagnosed using bulk sediment/contaminant loads. Such loading estimates are useful for the identification of basin disturbances, trend assessments, receiving water body loadings (infilling rates) and for the general management of water systems. Our understanding of how the sediment structure influences the transport of this sediment and how and where contaminants are compartmentalized within the sediment particles (flocs) is, however, poorly understood. This paper reports results from various image analysis techniques, microscopy and sequential metal extraction for the assessment of how sediment structure influences sediment and contaminant transport and provides insight into the complex interaction of metals with various components of flocculated sediment particles. A better understanding of these relationships will further our ability to predict the source, fate, and effect of sediments and contaminants within aquatic systems, thus facilitating more effective management decisions.

METHODS

Sample sites

Suspended and bed sediment samples were collected from Hamilton Harbour near the outfall of Red Hill Creek/Windermere Basin (RHWB) and Grindstone Creek (GC). The geology of both creeks is similar (Ongley, 1974). The primary difference is that GC is primarily a rural catchment while RHWB drains a large area of urban and industrial land. In addition RHWB is also the receiving water body for the main Hamilton sewage treatment plant (STP) and receives multiple combined sewer overflows during storm conditions. As such the organic content of RHWB is substantially higher and has a characteristic sewage odour. Five additional sites, distributed throughout the harbour, were used for bed sediment collection.

Sediment sampling and analysis

Recently deposited surficial fine-grained laminae (SFGL) (Droppo & Stone, 1994) were collected by scraping the top few mm of sediment from a 50 cm^2 box core. These bulk samples were used for sequential extraction metals analysis following the method of Tessier & Campbell (1988). This method allows for the assessment of the partitioning of metals according to the categories exchangeable, bound to carbonates, bound to Fe/Mn oxides, bound to organics and held within the crystalline structure of the sediment. These fractions can be viewed as a gradient of decreasing bioavailability, with the residual fraction not being bioavailable. In our assessment, exchangeable and carbonate fractions were combined.

Suspended floc samples for structural and compositional analysis were sampled following the method of Droppo *et al.* (1997) which incorporates the use of plankton chambers. Disturbance of the flocs is minimized by the plankton chambers functioning as both the sampling instrument and the analytical cell for laboratory observation and measurement.

A 12 l water/suspended sediment sample was used for determination of floc settling velocity, density and porosity following the settling image analysis technique described in Droppo *et al.* (1997). Further sub-samples were used for correlative microscopic analysis following the methods of Liss *et al.* (1996). Correlative microscopy here entails the use of conventional optical (COM), environmental scanning electron (ESEM) and transmission electron microscopy (TEM) being applied to the same sample (Leppard *et al.*, 2003). The TEM was equipped with energy dispersive spectroscopy (EDS) for the identification of metal associations with submicrometer entities of the floc. The reader is referred to the above publications for detailed descriptions of these techniques.

RESULTS AND DISCUSSION

Flocs were characterized, with regard to principal structural, chemical, and microbiological components and properties, in relation to metal contaminant–particle/colloid associations and floc settling as a surrogate of sediment transport.

Principal floc structure and composition—gross and fine

The structure of flocs is discussed here to illustrate their complex nature and the multiple binding sites for metals. Figure 1 illustrates the substantive differences between sample site flocs with representative micrographs for COM (Fig. 1(a),(b)), ESEM (Fig. 1(c),(d)), and TEM (Fig. 1(e),(f)) analysis. It is evident that both floc samples possess irregular shaped particles with open matrices composed of both organic and inorganic particles (Fig. 1). RHWB flocs were much more organic and diffuse in appearance compared to the GC flocs which were composed of more inorganic matter (Fig. 1(a),(b)). Many of the RHWB flocs had an amorphous organic coating (Fig. 1(c)), compared to the more mineral contact of the GC flocs (Fig. 1(d)). Loss-on-ignition (LOI) analysis of respective bed sediments also revealed that the RHWB sediment had an organic content 3 times higher than the GC sediments (Table 1). The internal structure of the both floc samples (Fig. 1(e),(f)) can be described as tortuous. Multiple convoluted channels are defined by the constituent particles of clay minerals, bacterial cells, residues of bacterial cell walls, granules, higher level organisms such as diatoms and the colloidal particles of extracellular polymeric substances (EPS) manifested as fibrillar material (Fig. 1(e),(f)). The TEM micrographs illustrate the large degree of EPS associated with the floc matrix and pores for both floc samples, but to a higher degree within the RHWB flocs. This EPS increases the cohesiveness of the floc thus promoting further flocculation, provided that shear conditions are favorable. It maintains a high porosity within both flocs and with the EPS fibrils being colloidal in size they induce high surface tensions and water retention. In addition, EPS is believed to be largely responsible for floc strength, with EPS densities observed to be the greatest in the centre of the floc and diffusing outwards (Droppo *et al.*, 1997; Liao *et al.*, 2002). Such an EPS

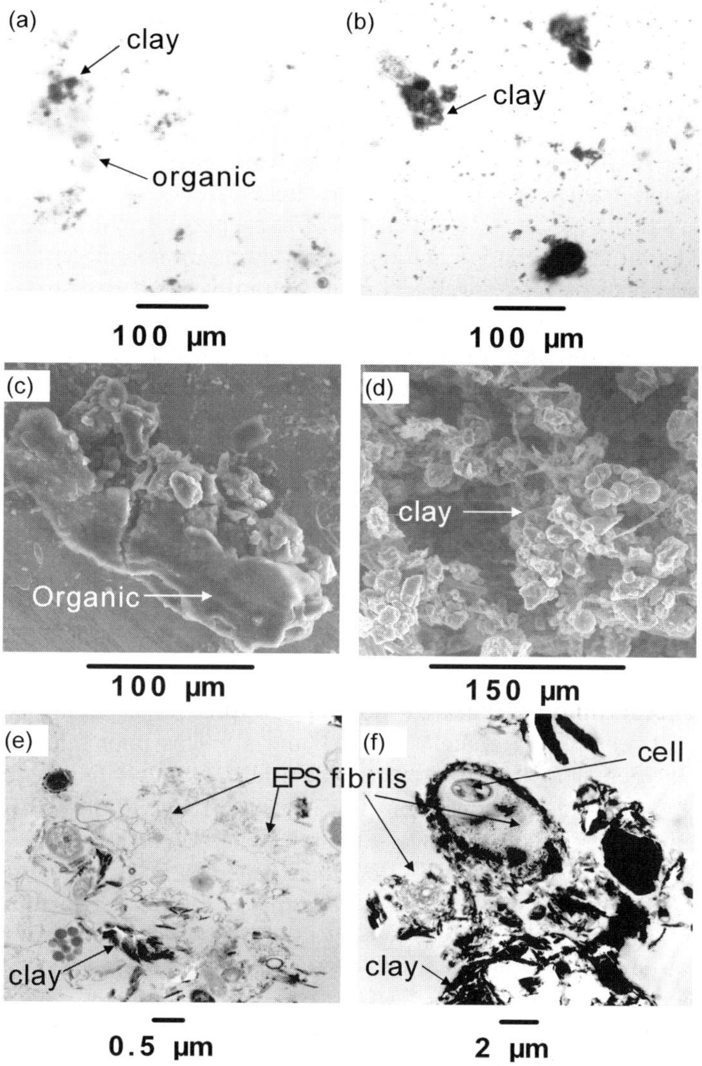

Fig. 1 Representative micrographs of RHWB (a, c and e) and GC (b, d and f) sample site flocs using three different microscopes: COM (a and b), ESEM (c and d) and TEM (e and f).

Table 1 Characteristics of flocs and bed sediment samples collected at stations.

	Grindstone Creek	Red Hill Creek–Windermere Basin
LOI (%)	7.6	23.0
% Sand	2.3	3.1
% Silt	67.1	61.5
% Clay	30.6	35.4
Mineralogy	Quartz, Calcite/Dolomite, Feldspars & Goethite	Quartz, Calcite/Dolomite, Feldspars & Goethite
Dominant clay mineral	Mica>Kaolinite>Chlorite	Mica>Kaolinite>Chlorite
D_{50} sonicated (μm)	6.1	7.8
d_{50} floc (μm)	119.0	188.0
Mean settling velocity (mm s^{-1})	3.08 ± 1.92	1.75 ± 0.56
Mean floc density (g cm^{-1})	1.12 ± 0.06	1.05 ± 0.04
Mean porosity (%)	81.7 ± 9.7	92.6 ± 7.1

± standard deviation; Greater than 5000 particles analysed for particle size distributions, 102 particles analysed for settling velocity, density and porosity.

gradient is suggestive of a strong core with a high potential of surficial floc growth in addition to chemical transport within the floc being likely related to diffusional and electrochemical gradients (due to high density EPS micro pore structure) (Droppo, 2001).

Both sites had similar mineralogy and inorganic grain size characteristics (Table 1) with floc size by-volume distributions negatively skewed and by-number distributions positively skewed. As such, the larger floc sizes represent the majority of the mass for the samples but the majority of the particles were small in size. The RHWB site showed a much larger floc size distribution (Table 1) which is indicative of flocs having a high organic content (Droppo, 2001). Similar floc sizes were found by Droppo *et al.* (2002) for combined sewer flocs.

Floc structural influence on sediment/contaminant transport

As the majority of sediment is transported as flocs (Droppo, 2001), contaminant transport will be dictated by the size, structure, composition and density characteristics of the flocs themselves. GC flocs had settling velocities which ranged between 1 and 10 mm s^{-1} and showed the common relationship of increasing settling velocity with floc size (r^2 = 0.68). The RHWB sediments, however, showed no relationship between these two variables with all flocs settling between approximately 3 and 1 mm s^{-1}. This poor relationship and lower settling velocity is related to the structure of the RHWB flocs being highly organic and diffuse in nature promoting a consistent low density range (0.002–0.12 g cm^{-3}) and minimizing the effects of size on settling. Although the RHWB flocs were generally larger (Table 1), more than half of the RHWB flocs had densities an order of magnitude below those of the GC flocs (density range 0.04–0.14 g cm^{-3}). A strong inverse relationship was observed between floc size and density ($r^2 > 70\%$) for both sample sites. As such, one would expect that contaminants associated with RHWB would be transported further than those from GC. Studies of Hamilton Harbour have shown that the entire Windermere arm of the harbour is contaminated with metals, PCBs and PAHs (Zeman & Patterson, 2003). It is likely that this contaminant distribution is controlled, in part, by floc structural influences on sediment/contaminant transport.

Floc structural influence on floc contaminant characteristics

The complexity of a floc's structural, chemical and microbial components and properties will play havoc on any attempt to provide a universal relationship of contaminants to some gross sediment particle property (e.g. size). Unlike traditional mono-structure sediments, flocs have multiple binding sites (Droppo, 2001) which are not static, but which may continually vary the internal metal composition/phase or concentration. Nevertheless, it is evident from the above structural synopsis that some general insights are possible. Clay particles will be active receptors of contaminants; however, it is likely the colloidal fraction of bacterial cells, residues of bacterial cell walls, granules (likely Fe oxides), higher level organisms (such as diatoms) and EPS fibril material will be the dominant surfaces controlling metal behaviour within the flocs due to their small size, relatively great unit surface area, and therefore high reactivity. This will be particularly true for the RHWB flocs which are much higher in organic content.

Recent interdisciplinary studies on flocs support the idea that EPS composition is an important component whose surface properties control hydrophobicity and surface charge at the floc-water interface, while bound water can be a function of floc size and total EPS (Liao *et al.*, 2001, 2002). These properties, as influenced by EPS, all play a role in the flocculation and contaminant interactions of cohesive sediments. Flocs have also been observed to possess anaerobic zones within the core of the floc where EPS densities are usually the highest and oxygen may be depleted (Simon *et al.*, 2002). Such gradients in redox potential may result in the fixation or mobilization of metals within the flocs. Generally an anaerobic zone will promote metal uptake. Ph gradients set up within a floc will also affect the migration of metal ions within the floc matrix through diffusional processes (Warren & Haack, 2001).

Metal composition—bulk analysis and EDS analysis

Bulk metal analysis provides little insight into the factors that control the transport and interaction of contaminants with/within the sediment particles. Traditional geochemical relationships have demonstrated that as particle size decreases, the potential for contaminant uptake increases. This relationship is, however, based on individual particles and does not reflect the true structure by which sediment and contaminants are transported (i.e. as flocs). While classical metal binding to individual clay particles prior to flocculation occurs, metals may be transformed, relocated within the floc matrix, or added to by the diffusion and electrochemical processes operating within the floc. These processes, in conjunction with microbial metabolism, chemical precipitation and co-precipitation, dissolution and volatilization, are what dictate the dynamic metal concentrations within the floc (see Warren & Haack (2001) for a comprehensive review of biogeochemical controls on metal behaviour).

While it is not feasible to fully understand the complex relationship of floc-metal interactions, sequential extraction (even with its limitations, Tessier & Campbell, 1988) does provide insight into the relative partitioning of metals within the flocs (although still with a bulk sample). A preliminary survey of Hamilton Harbour SFGL (including SFGL from the two study sites), suggest that both Cd and Pb were primarily bound within Fe/Mn oxides

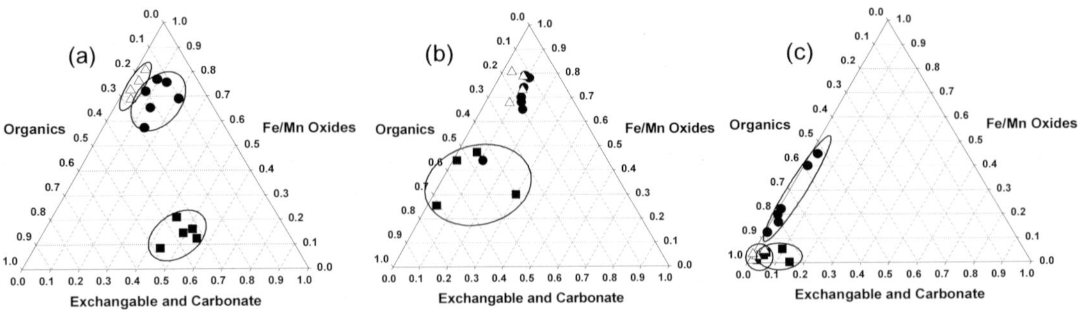

Fig. 2 Ternary plots of (a) Cd, (b) Pb and (c) Cu for labile fractions only within SFGL (●), street dust (■) and CSO (Δ) sediments (Residual fraction has been removed, exchangeable and bound to carbonate phase has been combined and the fractions adjusted to 100%).

(60–80%) (Fig. 2). Similar findings have been found for urban sediments and urban impacted rivers by Stone & Marsalek (1996), Droppo *et al.* (1998) and Sutherland & Tack (2000). Fe/Mn oxides are often manifested as precipitated crystalline forms or coatings and often dominate the geochemical reactivity of the sediment component. Microbial activity can further promote Fe/Mn oxides within flocs. These forms of oxides are often smaller and more diffuse than chemical precipitates and as such possess extensive surface areas for metal uptake (Warren & Haack, 2001). The presence of Fe/Mn oxides is substantiated by goethite being present as one of the dominant minerals (Table 1). Goethite is a transformed stable version of ferrihydrite; a common Fe oxide within freshwater (Warren & Haack, 2001). Fig. 2 also plots the data for CSOs and street dust collected within the Kenilworth sewershed which contributes to the RHWB sample site from a previous study (e.g. Droppo *et al.*, 1998; Irvine *et al.*, 1999). Comparison of the overall Hamilton Harbour SFGL data shows that the CSO samples contained similar ratios of the metal phases while the street dust was quite different. This suggests that the sanitary portion of the CSO and three STP discharges may be a significant contributor to the metal characteristics of Hamilton Harbour SFGL. The CSO will deliver street dust to the harbour during washoff and overflow events, but its ratio signature does not correspond with the SFGL samples. Only for Pb does the street dust appear to make a contribution to the SFGL. However, this is only one sample and further work will be required to resolve this issue.

Copper analysis revealed a common association with the organic fraction of the sediment (55–90%) (Fig. 2). This organic fraction is heterogeneous and can include viable or non-viable bacteria, humic and fulvic acids and the colloidal fraction contributed by the extensive EPS network of the flocs and bacteria. Organics are effective scavengers of metals due to their large surface areas and multiple surface functional groups. Small organic molecules are generally more labile and taken individually may have a lower residence time in aquatic systems (Warren & Haack, 2001). However, organic molecules often are adsorbed onto mineral surfaces and when incorporated in flocs, these organic materials contribute to floc building and influence the behaviour of metals (organic coatings can inhibit or enhance metal binding – Warren & Haack, 2001). Bacteria themselves possess surface functional groups and can sorb metal species to their surface although this process could be complicated by the EPS secreted by the bacterium or other substances dissolved in the water (Warren & Haack, 2001; Chatellier *et al.*, 2004). It would appear (Fig. 2) that the phases of

Cu are possibly controlled more by fluvial and erosional processes than by the CSOs or street dust as the data plots in a different area of the graph (although the RHWB sample was closest to the CSO and street dust data points).

Recent assessment of individual binding sites within industrial wastewater flocs using TEM and EDS has revealed considerable insight (Leppard *et al.*, 2003). However, within the present study, only in very limited cases was Cu, Cd or Pb found to be attached to any component of the floc, even though high concentrations were observed with sequential extraction analysis. This lack of data generated by EDS may be related to: (a) concentrations not being high enough for detection to consistently show binding to one specific site, and/or (b) the small sample volume analysed for TEM missing important metal attachment sites. In the Leppard *et al.* (2003) study, although concentrations of metals were orders of magnitude higher than in this study, Pb was found to be preferentially bound to polyphosphate granules. It is possible that a similar relationship is at play for natural sediments, but that concentrations of metal ions on any one surface are not high enough for detection.

CONCLUSIONS

(a) Flocculation is a dynamic process which influences the transport and contaminant-binding behavior of suspended sediment.
(b) Floc characteristics and settling behaviors were very different between the RHWB and GC sites due to significant structural differences (primarily organic differences).
(c) Colloidal fractions are important binding sites for metals due to their large reactive surface area and therefore high reactivity.
(d) Cd and Pb were preferentially bound to Fe/Mn oxides while Cu was preferentially bound to organic material. Concentrations were likely not elevated enough for metal spectrum analysis with EDS.

Acknowledgement The authors thank M. Reid and C. Jaskot for their imaging work.

REFERENCES

Chatellier, X., West, M. M., Rose, J., Fortin, D., Leppard, G. G. & Ferris, F. G. (2004) Characterization of iron-oxides formed by oxidation of ferrous ions in the presence of various bacterial species and inorganic ligands. *Geomicrobiol. J.* (in press).

Droppo, I. G. (2001) Rethinking what constitutes suspended sediment. *Hydrol. Processes* **15**, 1551–1564.

Droppo, I. G. & Stone, M. (1994) In-channel surficial fine-grained sediment laminae (Part I): Physical characteristics and formational processes. *Hydrol. Processes* **8**, 101–111.

Droppo, I. G., Leppard, G. G., Flannigan, D. T. & Liss, S. N. (1997) The freshwater floc: A functional relationship of water and organic and inorganic floc constituents affecting suspended sediment properties. *Water Air & Soil Pollut.* **99**, 43–53.

Droppo, I. G., Irvine, K. N., Murphy, T. P. & Jaskot, C. (1998) Fractionated metals in street dust of a mixed land use sewershed, Hamilton, Ontario. In: *Hydrology in a Changing Environment*, Volume 3 (ed. by H. Wheater & C. Kirby), 383–394. John Wiley & Sons Ltd., New York, USA.

Droppo, I. G., Irvine, K. N. & Jaskot, C. (2002) Flocculation of cohesive sediments in the urban continuum: Implications for stormwater management. *Environ. Technol.* **23**, 27–41.

Irvine, K. N., Droppo, I. G. & Murphy, T. P. (1999) Contaminant levels and loading estimates: The first step in identifying restoration options for a Great Lakes Area of Concern. In: *Aquatic Restoration in Canada* (ed. by T. Murphy & M. Munawar), 51–68. Backhuys Publishers, The Netherlands.

Leppard, G. G., Droppo, I. G., West, M. & Liss, S. N. (2003) Compartmentalization of metals within the diverse colloidal matrices comprising activated sludge microbial flocs. *J. Environ. Qual.* **32**, 2100–2108.

Liao, B. Q., Allen, D. G., Droppo, I. G., Leppard, G. G. & Liss, S. N. (2001) Surface properties of activated sludge and their role in bioflocculation and settling: extracellular polymeric substances, hydrophobicity and surface charge. *Water Res.* **35**, 339–350.

Liao, B. Q., Allen, D. G., Leppard, G. G., Droppo, I. G. & Liss, S. N. (2002) Interparticle interactions affecting the stability of sludge flocs. *J. Coll. Interf. Sci.* **249**, 372–380.

Liss, S. N., Droppo, I. G., Flannigan, D. T. & Leppard, G. G. (1996) Floc architecture in wastewater and natural riverine systems. *Environ. Sci. Technol.* **30**, 680–686.

Ongley E. D. (1974) *Hydrophysical Characteristics of Great Lakes Tributary Drainage, Canada.* Pollution and Land Use Activities Reference Group (PLUARG), IJC, Windsor, Canada.

Simon, M., Grossart, H. -P., Schweitzer, B. & Ploug. H. (2002) Microbial ecology of organic aggregates in aquatic systems. *Aquat. Microb. Ecol.* **28**, 175–211.

Stone, M. & Marsalek, J. (1996) Trace metal composition and speciation in street sediment: Sault Ste. Marie, Canada. *Water Air & Soil Pollut.* **87**, 149–169.

Sutherland, R. A. & Tack, F. M. G. (2000) Metal phase associations in soils from an urban watershed, Honolulu, Hawaii. *Sci. Total Environ.* **256**, 103–113.

Tessier, A. & Campbell, P. G. C. (1988) Partitioning of trace metals in sediments, In: *Metal Speciation: Theory, Analysis and Application* (ed. by J. R. Kramer & H. E. Allen), 183–199. Lewis, New York, USA.

Warren, L. A. & Haack, E. A. (2001) Biogeochemical controls on metal behaviour in freshwater environments. *Earth Sci. Rev.* **54**, 261–320.

Zeman, A. J. & Patterson, T. S. (2003) *Sediment survey of Windermere Arm, Hamilton Harbour*. Environment Canada, Burlington, Ontario, Canada, 1–12. NWRI Cont. no. 03-171

Transport and retention of copper fungicides in vineyards

MARCEL VAN DER PERK, VICTOR JETTEN, ERIK HESKES, MARJA SEGERS & INGRID WIJNTJENS

Department of Physical Geography, Utrecht University, PO Box 80115, 3508 TC Utrecht, The Netherlands

m.vanderperk@geog.uu.nl

Abstract Copper fungicides, for example Bordeaux mixture, are frequently used in vineyards to protect young grapes against mildew. The copper accumulates in the topsoil layer where it is adsorbed to clay minerals and soil organic matter. Part of the copper may be transported across the vineyard by surface runoff and soil erosion, which could then enter the fluvial system. To investigate the transport and retention of copper in vineyards, detailed mapping of both the lateral and vertical spatial distribution of copper in vineyard soils was performed in two vineyards located in the Ouvèze catchment in southeastern France. The vertical distributions of copper concentrations over the soil profile, and the spatial distribution patterns of copper deposition densities enabled us to identify and assess the redistribution mechanisms and transfer rates of sediment-associated copper within the vineyard fields.

Key words copper; deposition; erosion; France; heavy metals; spatial variation; vineyards

INTRODUCTION

Copper (Cu) fungicides, such as Bordeaux mixture (a mixture of lime and copper sulphate), are frequently used in viticulture to protect young grapes against mildew. Most Cu accumulates in the upper layer of the soil (Deluisa *et al.*, 1996), where is binds to soil organic matter and ferric oxides in the clay fraction (Parat *et al.*, 2002). Soil erosion processes may lead to Cu redistribution within the vineyards and Cu export to the fluvial system (Ribolzi *et al.*, 2002). Depending on the downstream physico-chemical conditions, this Cu may lead to, or contribute to, toxicity to aquatic life.

This study aims to quantify long-term copper transport within and from vineyards due to soil erosion and deposition. Detailed mapping of both the vertical and lateral spatial distribution of copper was carried out in two vineyards (Bigaude and Madeleine) located near Malaucène in the Ouvèze catchment in southeastern France. Table 1 lists the main geomorphological features of these vineyards. The subsoil of both vineyards consists of Pliocenic calcareous loamy sands. The loss on ignition content of the topsoil ranges between 1.5% and 3%. The Bigaude vineyard contains 35 vine rows of approximately 60 to 80 years age and the Madeleine vineyard contains 28 vine rows of about 40 years age (Heskes *et al.*, 2003).

SOIL SAMPLING AND ANALYSIS

During summer 2001, soil samples from both vineyards were taken using a stainless steel corer with a diameter of 70 mm. To explore short range variation of Cu concentrations, the topsoil (0–5 cm) in each vineyard was sampled every 10 m along a transect parallel to the

Table 1 Geomorphological features of the vineyards studied.

Feature	Bigaude	Madeleine
Surface area (ha)	0.99	0.39
Maximum width (m)	89	70
Maximum length (m)	120	91
Maximum altitude difference (m)	4.4	4.3
Mean slope (%)	3.2	4.4

vine rows. Each sampling location was subdivided into four sublocations, viz. below the grapevine (about 10 cm from the vine stem), in the rill along the vine row (about 25 cm), in the wheel track (about 60 cm), and in the middle between the vine rows (about 125 cm). In total 48 topsoil samples were collected from the Bigaude vineyard, and 40 topsoil samples from the Madeleine vineyard. To investigate the vertical distribution pattern of Cu, five soil profiles in the Bigaude vineyard and four soil profiles in the Madeleine were sampled along the same transects. Bulk samples were taken at depths of 0–1, 1–5, 5–10, 10–15, 15–20, 20–40 and 40–60 cm.

To explore the spatial distribution of Cu inventories in both vineyards, bulk samples were collected up to a depth of 30–60 cm below the soil surface using a stratified random sampling scheme (Heskes *et al.*, 2003). Thirty-five bulk samples were collected from the Bigaude vineyard and 30 bulk samples from the Madeleine vineyard. In addition, 12 bulk samples were collected along the transect in the Bigaude vineyard and 10 bulk samples along the transect in the Madeleine vineyard. The sampling depth depended on topographical position and was based on a pilot field study on the vertical distribution in soil using a field Cu test (Merck© Photometric Copper Test Kit Cat. no. 1.14767.0001). This pilot study showed that a sampling depth of 30 cm at the top of the slopes, increasing to 60 cm in the downslope direction, is sufficient to sample the majority of Cu present in the soil profile. To minimize the effect of short range spatial variation, at each sampling location three subsamples were collected at an interval of about 0.3 m perpendicular to the vine rows, and were bulked manually to one composite sample in a plastic bucket in the field. The bulk sample was weighed to determine soil bulk density. After homogenization, an aliquot of about 600 g was taken from the bulked sample to reduce sample weight. In the laboratory, these portions of the samples were dried and weighed. The Cu content was determined by decomposing about 0.5 g of sample with 2 mol l^{-2} HNO_3 in a microwave oven. The Cu concentration in the extract was measured by inductively coupled plasma atomic emission spectrometry (ICP-AES).

SHORT-RANGE VARIATION

Figure 1 shows the Cu concentrations in the topsoil along the transects in the Bigaude and Madeleine vineyards at the various sublocations perpendicular to the vine row. In the Bigaude vineyard, the Cu concentrations increase generally from approximately 100 mg kg^{-1} at the top of the slope to nearly 250 mg kg^{-1} at the foot of the slope. In the Madeleine vineyard, a similar trend can be observed, but the concentration levels are lower (approximately 40 mg kg^{-1} and 150 mg kg^{-1}, respectively). The Cu concentrations in the rill,

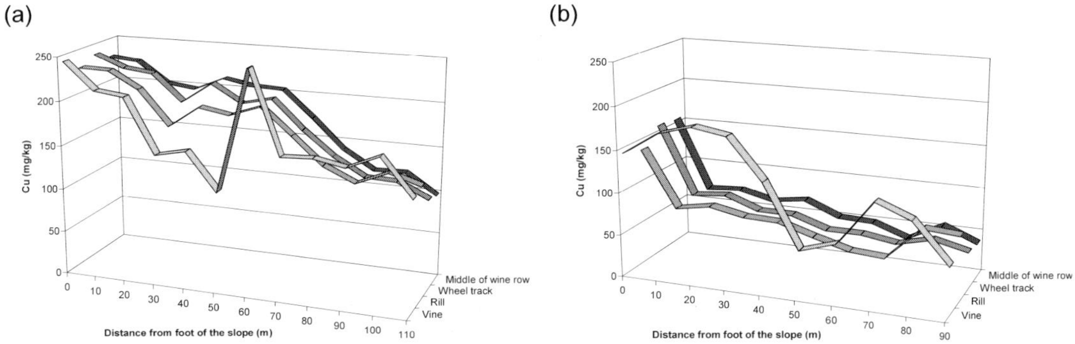

Fig. 1 Short range variation of Cu concentrations in the topsoil along a transect in: (a) the Bigaude vineyard, and (b) the Madeleine vineyard.

wheel track, and the middle of the wine row follow are almost equal along the transects in both vineyards. However, the Cu concentrations in the topsoil below the grapevines show much more variation. In the Madeleine vineyard, the Cu concentrations below the vines are generally larger than farther away from the vines.

VERTICAL VARIATION

The vertical Cu distribution patterns of the upper and lower profile along the transects in the Bigaude and Madeleine vineyards are shown in Fig. 2. These vertical distributions show that the majority of Cu is concentrated in the top 20 cm of the soil profile. In the Bigaude vineyard, the Cu concentrations below the 20 cm are close to the background concentrations of about 6.8 mg kg^{-1} (average Cu concentration of the deepest samples). In the Madeleine vineyard, the vertical distribution is more variable. Only in the upper vineyard profiles do the Cu concentrations decrease close to the background concentration (minimum concentration = 5.9 mg kg^{-1}). The Cu concentrations at 40 and 60 cm depth in the two lower soil profiles remain elevated. Figure 2 also shows that in both vineyards, the Cu concentrations in the topsoil exhibit an increasing trend from the top to the foot of the slope.

WITHIN-FIELD SPATIAL VARIATION

General

The Cu concentration values (mg kg^{-1}) of the bulk samples were converted into Cu inventories (kg m^{-2}) using soil bulk density and sampling depth. To obtain the anthropogenic Cu enrichment in the soil profile, the Cu inventories were corrected for the local background Cu concentrations mentioned above. Table 2 presents the basic statistics of anthropogenic Cu enrichment in the Bigaude and Madeleine vineyards subdivided into stratified random samples and the transect samples. These statistics show that the soil of the Bigaude vineyard contains more Cu than the soil of the Madeleine vineyard. The distributions of the Cu inventories are positively skewed.

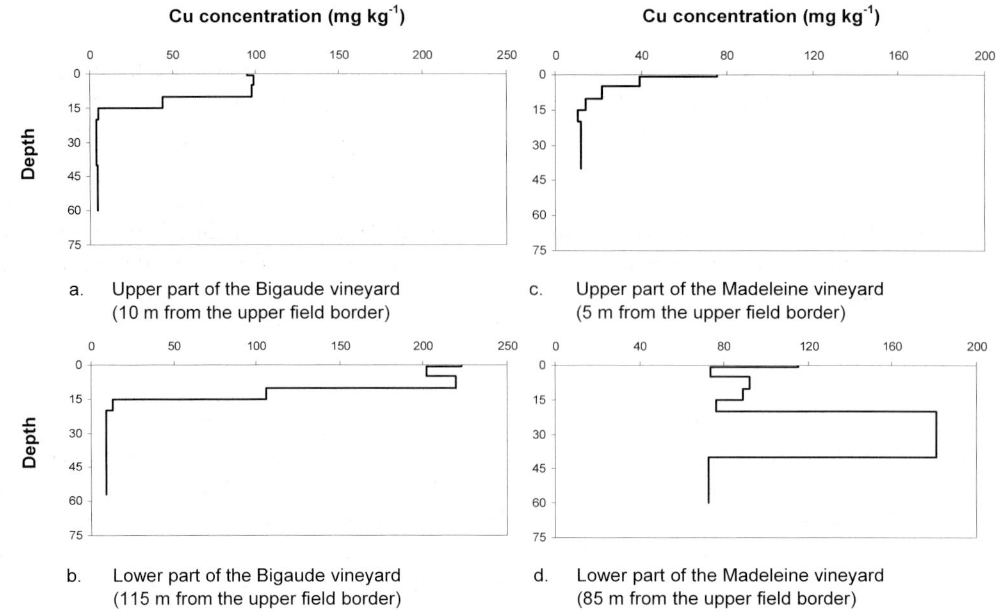

Fig. 2 Short range variation of Cu concentrations in the topsoil along a transect in: (a) the Bigaude vineyard, and (b) the Madeleine vineyard.

Table 2 Statistics of Cu inventories in the vineyard soils.

	Bigaude Stratified random	Transect	Total	Madeleine Stratified random	Transect	Total
N	35	12	47	30	10	40
Mean (kg m^{-2})	0.0369	0.0558	0.0418	0.0302	0.0344	0.0313
Median (kg m^{-2})	0.0322	0.0502	0.0332	0.0198	0.0223	0.0200
St. dev. (kg m^{-2})	0.0167	0.0314	0.0226	0.0266	0.0320	0.0277
Skewness	2.70	1.82	2.49	2.09	1.49	1.83

Relationship between Cu inventories and soil erosion

The relationship between soil erosion and Cu inventories in vineyard soil was explored using a straightforward GIS-based sediment budget index model based on a rill erosion model described by Govers *et al.* (1993). The basic map for analysis of erosion and deposition patterns was comprised of a 1 m × 1 m resolution digital elevation model (DEM) from which the necessary input for the sediment budget index model was derived.

A relative measure for potential soil erosion was derived using:

$$E_r = L^n S^m \tag{1}$$

where E_r is the relative potential erosion rate per unit area, L is the slope length (m), S is the sine of the slope gradient, and n and m are exponents. The sediment transport capacity (Tr) on a given location in the catchment was considered to be proportional to the potential for erosion (Govers *et al.*, 1993):

$$Tr = g_t E_r \tag{2}$$

where g_t is the transport capacity coefficient. The amount of eroded material per grid cell is transported downstream over the drainage network, as long as it does not exceed the transport capacity. The surplus is deposited. The *accucapacitystate* operator in the PCRaster GIS (PCRaster 2001) was used to obtain the amount of deposited material given the drainage network, the available eroded and transported material from the upstream area, and the transport capacity. The relative sediment budget index D_r is calculated by subtracting the eroded material from the deposited material. A positive value for D_r means deposition and negative value erosion.

The Cu inventory was predicted using the regression equation:

$$Cu = A + B \times D_r + \varepsilon \tag{3}$$

where Cu is the Cu inventory (kg m^{-2}), A and B are regression parameters, and ε is the prediction error with $E(\varepsilon) = 0$. The exponents n and m, the transport capacity coefficient g_t, and the regression parameters A and B were simultaneously estimated for both vineyards by means of least squares fitting using the PEST package (Doherty, 2003), a tool for automatic calibration. The lower and upper parameter bounds for the exponents n and m and coefficient g_t were *a priori* set to the values reported in the literature. The *a priori* and calibrated parameter values are listed in Table 3. Table 3 also lists the R^2 between the predicted and observed Cu inventories.

Figure 3 shows the spatial distribution of the Cu inventories predicted using equation 3. This figure demonstrates that the Cu inventories are relatively small in the upper part of the vineyard and relatively large in the lower part.

Table 3 *A priori* and calibrated parameter values.

Parameter	Initial estimation: Lower limit	Upper limit	Calibrated values: Bigaude	Madeleine
N	0.4	0.8	0.8	0.4
M	1.2	1.5	1.2	1.2
g_t	40	90	41.8	56.6
A	0	0.2	0.0421	0.0343
B	0	0.2	0.025	0.184
R^2			0.319	0.556

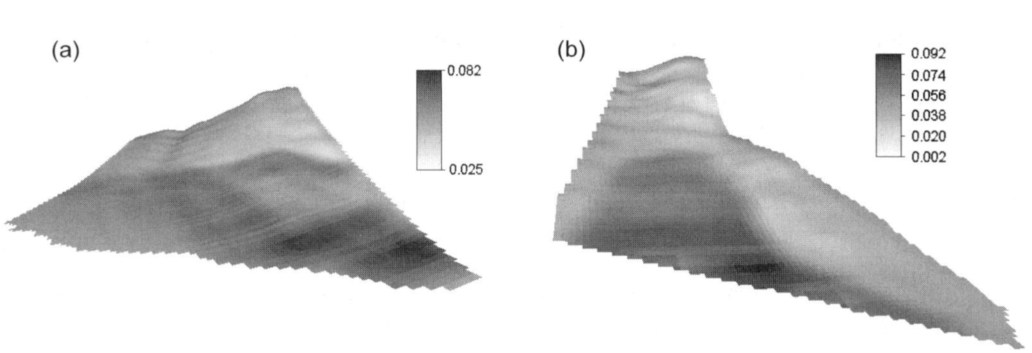

Fig. 3 Spatial distribution of Cu inventories (kg m^{-2}) in: (a) the Bigaude vineyard and (b) the Madeleine vineyard.

ASSESSMENT OF CU TRANSPORT

The vertical distribution of Cu in the soil profiles demonstrates that the Cu is fixed in the upper layer of the soil. This implies that leaching of Cu to deeper horizons is of minor importance. The results from the different sampling campaigns all show that there is a general increasing trend of Cu concentrations in the topsoil and total Cu inventories from the top of the slope to the foot of the slope. These trends suggest that soil erosion and deposition is the principal redistribution mechanism of Cu in the vineyards studied. The soil erosion and deposition processes, as estimated using a straightforward soil redistribution model, account for 32% of the variation in Cu inventories in the Bigaude vineyard and 56% of the variation in Cu inventories in the Madeleine vineyard. The portion of variation that is not explained by soil erosion and distribution may be attributed to spatial differences in the amount of Cu application and accumulation, errors in the model predictions of the relative sediment budget index D_r, and variation due to possible incomplete homogenization of the subsamples.

Parameter A in the regression equation (3) represents the anthropogenic Cu enrichment that would be present in the vineyards if erosion and deposition were absent. Comparison with the mean Cu inventories of the stratified random sampling, which represent the actual Cu enrichment in both vineyards, shows that the actual Cu enrichment is less than the Cu enrichment in the case of no erosion and deposition processes. This implies that there has been a net loss of Cu from the vineyards due to net soil erosion. The time-integrated Cu loss due to soil erosion amounts to 0.0052 kg m^{-2} for the Bigaude vineyard and 0.0041 kg m^{-2} for the Madeleine vineyard. Given the respective ages of both vineyards, 70 years and 40 years, the average net Cu loss rate due to net soil erosion amounts to 0.74 kg ha^{-1} $year^{-1}$ for the Bigaude vineyard and 1.02 kg ha^{-1} $year^{-1}$ for the Madeleine vineyard. Comparison with local Cu application rates is not possible because this information is lacking for the vineyards studied, but the Cu losses found in this study are about one order of magnitude lower than the average Cu input rate reported by Moolenaar & Beltrami (1998) for Italian vineyards (3–16 kg ha^{-1} $year^{-1}$). The greater Cu loss rate from the Madeleine vineyard compared to that from the Bigaude vineyard is probably due to the steeper average slope of the Madeleine vineyard (see Table 1).

CONCLUSIONS

Both the Cu concentrations in the topsoil and Cu inventories in soil in the Bigaude and Madeleine vineyards show a general increasing trend in the downslope direction. The simple soil erosion–deposition model enabled us to estimate the time-integrated Cu losses from the vineyards. Whether the Cu lost is deposited at or near the field borders or transferred into the fluvial system requires further research.

Acknowledgements The authors wish to thank personnel of the Laboratory of Physical Geography for constructing the sampling devices and for carrying out the Cu analyses on the ICP-AES. The authors also thank Mrs P. Scheijen for her contribution to field sampling and laboratory analyses.

REFERENCES

Deluisa, A., Giandon, P., Aichner, M., Bortolami, P., Bruna, L., Lupetti, A., Nardelli, F. & Stringari, G. (1996) Copper pollution in Italian vineyard soils. *Commun. Soil Sci. Plant Anal.* **27**, 1537–1548.

Doherty, J. (2003) *PEST 7.2; Model-independent Parameter Estimation; Manual.* Watermark Numerical Computing, Brisbane, Australia.

Govers, G., Quine, T. A. & Walling, D. E. (1993) The effects of water erosion and tillage movement on hillslope profile development: A comparison of field observations and model results. In: *Farm Land Erosion in Temperate Plains and Hills* (ed. by S. Wicherek), 285–300. Elsevier, Amsterdam, The Netherlands.

Heskes, E., Segers, M. & Wijntjens, I. (2003) Transport of sediment-associated copper on vineyards in the Ouvèze River basin. MSc Thesis, Department of Physical Geography, Utrecht University, The Netherlands.

Moolenaar, S. W. & Beltrami, P. (1998) Heavy metal balances of an Italian soil affected by sewage sludge and Bordeaux mixture applications. *J. Environ. Qual.* **27**, 828–835.

Parat, C., Chaussod, R., Lévéque, J., Dousset, S. & Andreux, F. (2002) The relationship between copper accumulated in vineyard calcareous soils and soil organic matter and iron. *Eur. J. Soil Sci.* **53**, 663–669.

PCRaster (2001) *PCRaster manual.* Department of Physical Geography, Utrecht University, The Netherlands.

Ribolzi, O., Valles, V., Gomez, L. & Voltz, M. (2002) Speciation and origin of particulate copper in runoff water from a Mediterranean vineyard catchment. *Environ. Pollut.* **117**, 261–271.

Sediment Transfer through the Fluvial System (Proceedings of a symposium held in Moscow, August 2004).
IAHS Publ. 288, 2004

The role of soil phosphorus in controlling sediment-associated phosphorus transfers in river catchments

PHILIP N. OWENS & LYNDA K. DEEKS
National Soil Resources Institute, Cranfield University, North Wyke Research Station, Okehampton, Devon EX20 2SB, UK
philip.owens@bbsrc.ac.uk

Abstract Many of the models used to determine phosphorus (P) transfers within river catchments use soil-P concentrations as one of the main input data sets. Such soil-P values are often based on bulked soil cores. To test the validity of using such data, depth profiles of total-P and Olsen-P were determined for arable and grassland fields in the catchment of the Hampshire Avon, UK. For the soils under grassland, maximum soil-P values occur at or near the surface and decrease with depth. For soils under arable, total-P values are broadly constant within the plough layer, whereas for Olsen-P maximum values are located at or near the surface, below which values are broadly constant to the plough depth. These findings have important implications for the use of bulked soil-P data to represent the concentrations of P at the soil surface, particularly for modelling P transfers associated with soil erosion and overland flow.

Key words fluvial sediment; models; NSI database; phosphorus; PSYCHIC; soil profiles

INTRODUCTION

Phosphorus (P) is often cited as the main nutrient causing eutrophication in rivers and lakes (Correll, 1998). In most catchments, much of the phosphorus being delivered to, and transported by, rivers is associated with fine-grained sediment (Russell *et al.*, 1998; Owens & Walling, 2002). In order to help reduce the extent of eutrophication in river catchments, there has been a concerted effort in many countries to reduce the delivery of P to rivers. To date, most effort has focused on reducing the delivery of P from point sources such as sewage treatment works (STWs), and available evidence suggests that in some cases this may be working. Thus Owens & Walling (2003) used flood plain deposits to reconstruct changing P transfers through rivers in Yorkshire, UK, over the last approx. 100 years. They demonstrated that for a river receiving inputs from STWs and other point sources, the P content of fluvial sediment had decreased over the last approx. 10 years, whereas there was no similar evidence of a decrease in P levels in rivers draining rural catchments that received primarily non-point, or diffuse, sources of P. In catchments where there has been a reasonable amount of success in controlling point sources of P, increasing attention is focusing on reducing P delivery to rivers from diffuse sources.

Because of the inherent complexities associated with trying to understand and control the delivery of P to surface waters from diffuse sources in all but the simplest (and usually smallest) catchments, a variety of models have been developed recently in order to predict the extent and spatial distribution of P transfers within land and river systems. In the UK, such models include the Export Coefficient Model (Johnes, 1996), the P-Expert System Model (Harrod & Fraser, 1999) and the Phosphorus Indicators Tool (Heathwaite *et al.*,

2003). A new model currently being developed by the authors and colleagues to estimate the delivery of P from land to rivers in the UK is the PSYCHIC (Phosphorus and Sediment Yield Characterization) model. Many of these models, including PSYCHIC, and other similar models worldwide, require input values of the P content of soils. In most catchments, a large proportion of the diffuse P delivered to waters is via surface pathways and is associated with soil erosion and sediment transfers by overland flow (Haygarth *et al.*, 1998; Harrod & Fraser, 1999; Heathwaite *et al.*, 2003). Consequently, accurate values of the P content of the uppermost layers of topsoil (i.e. upper few centimetres) is crucial for these models to be able to determine P delivery to rivers and lakes via surface pathways. However, to date, many models have used lumped soil-P data. Thus, for example, the soil-P input data for the Phosphorus Indicators Tool and the PSYCHIC models are based on the National Soil Inventory (NSI) database, which represents bulked values for the 0–15 cm topsoil layer for samples collected at 5 km grid intervals throughout England and Wales between 1978 and 1983 (McGrath & Loveland, 1992). In consequence, the soil-P values used in these models represent the average P content of soil within the upper 15 cm of the soil profile. Studies (e.g. Haygarth *et al.*, 1998; Withers *et al.*, 2001) have, however, shown that in uncultivated soils there tends to be a decrease in soil-P concentrations with depth, with maximum values at, or close to, the surface. Although the depth profile of P in cultivated soils is expected to be more homogeneous (due to soil mixing during tillage), it is still uncertain how representative a bulked value for 0–15 cm depth is of P concentrations in the upper few centimetres.

This paper presents depth profiles of P (both total and Olsen) in agricultural fields for different soil types in one of the PSYCHIC study catchments, and examines the effect of these depth profiles on the values of soil-P to be used in the PSYCHIC model for determining (particulate) P delivery to rivers in this catchment.

STUDY AREA AND METHODS

The Hampshire Avon is a large catchment (~1700 km^2) located in the counties of Wiltshire, Dorset and Hampshire in southern England. The upper catchment is characterized by rolling chalklands and sheltered river valleys and includes the arable landscape of Salisbury Plain. For the catchment as a whole, land use is dominated by arable cropping (33%) and permanent grassland (26%), with temporary grassland, woodland and lowland heath occupying 19, 9 and 3%, respectively.

Detailed soil-P profiles were determined for four dominant soil types within the Hampshire Avon catchment (Table 1). Soil samples were taken in both arable (for all four dominant soil types) and permanent grassland (for two of the soil types) fields. For two of the soil types (the typical brown calcareous earth and the typical argillic brown earth) it was difficult to find good examples of agricultural soil profiles, without any history of disturbance in the last few decades, and consequently grassland soil-P profiles were not determined for these soil types. To prevent profile compaction, which can occur using augers, a soil pit was dug to expose the profile face. Soil samples were collected by excavating into the profile face delineating samples in 1 cm depth increments down to 8 cm and then in 2 cm increments down to 20 cm (grassland) and 30 cm (arable).

Soil samples were air-dried, broken up and passed through a 2 mm sieve. Plant available P (Olsen-P) was determined by the Olsen's bicarbonate extractable P fraction method as described by Olsen *et al.* (1954). For total-P the air-dried soil was milled to a fine powder

Table 1 Description of the four soils examined in the Hampshire Avon catchment.

Site	Soil subgroup†		Soil description†	Surface area of catchment (%)
	Number	Name		
Knoyle Down	3.43	Brown rendzina	Shallow well-drained calcareous silty soils	32
Mere Down	3.41	Humic rendzina	Shallow, mostly humose, well drained calcareous soils	14
Wilsford	5.11	Typical brown calcareous earth	Well-drained calcareous fine silty soils	10
Sunnyhill	5.71	Typical argillic brown earth	Well-drained fine and coarse loamy glauconitic soils	9

†Based on Avery (1980).

(<1 mm diameter) and the sodium hydroxide fusion method described by Smith & Bain (1982) was used to determine total phosphate. In both cases, the Murphy & Riley (1962) method was used to determine the Mo-reactive P in the extract.

RESULTS

Figure 1 shows plots of the depth profiles for both total-P and Olsen-P for the four arable and the two grassland fields. In the case of the depth profiles of P in the two soils under grassland (Knoyle Down and Mere Down) there are clear decreases in both total-P and Olsen-P with depth. For Knoyle Down, maximum values are at the soil surface (0–1 cm depth increment), whereas for Mere Down the maximum values of soil-P are in the 1–2 cm increment. These depth profiles are consistent with expectations for undisturbed grassland soils, with both the surface maxima and the marked decreases of soil-P with depth reflecting inputs of P to the soil surface due to additions of P-based fertilizers, manures and other artificial inputs, and the return of P in dead plants and storage in soil organic matter (Haygarth *et al.*, 1998). As expected, the decrease in values with depth is most pronounced with Olsen-P for both soil types.

In the case of the depth profiles of soil-P for the samples collected from the four arable fields there are some important trends. For the four profiles of total-P there is a trend of broadly consistent values from the surface to the depth of ploughing, which lies between 20 and 25 cm depth for these fields. Unlike the total-P profiles for the grassland fields, where maximum values are located at, or near, the soil surface, maximum values of total-P for the arable fields are not necessarily located near the soil surface and appear to be located anywhere within the plough layer. For the four profiles of Olsen-P, there are more complex depth distributions, with maximum values at, or near, the soil surface and with subsequent decreases in values with depth, which in turn are superimposed on a pattern of broadly constant values with depth. In other words, the effect of tillage processes on mixing the soil and associated P is still evident (particularly in the case of Sunnyhill) but there is also a trend of maximum values at the surface and marked decreases in Olsen-P values with depth, particularly within the upper 5–10 cm. The surface peak in Olsen-P values is consistent with the application of artificial sources of P (such as fertilizers) to the soil before tillage processes have mixed them thoroughly within the plough depth (i.e. between consecutive ploughing events).

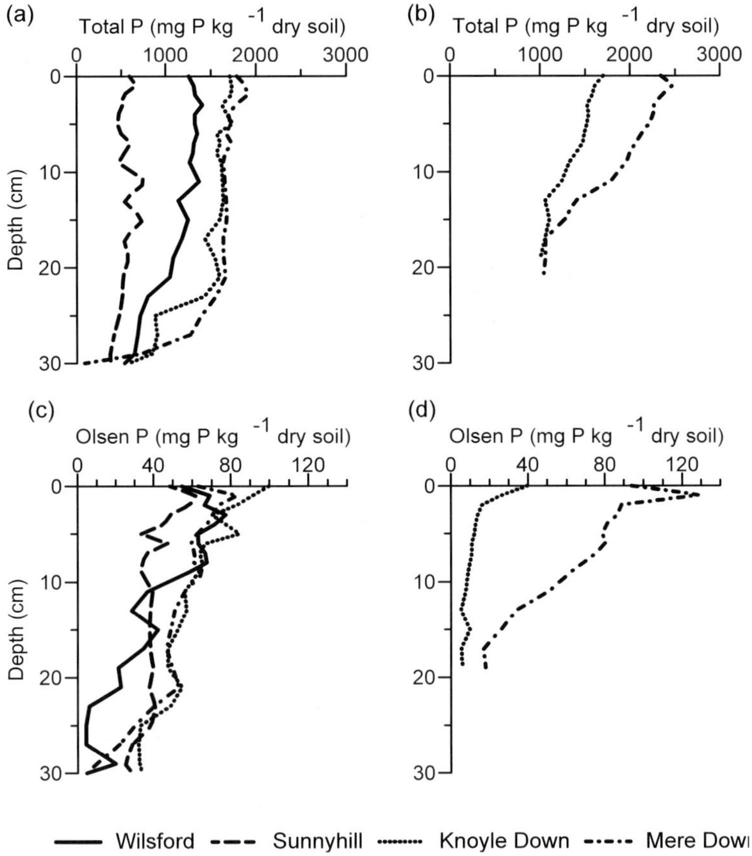

Fig. 1 The depth distribution of total-P in: (a) arable and (b) grassland soils, and Olsen-P in (c) arable and (d) grassland soils for key sites.

DISCUSSION

Tables 2 and 3 present summary statistics of soil-P for the soils under arable and grassland. For total-P (Table 2), values for the soil surface (expressed here as the average for the upper 2 cm) are in most cases greater than average values for the 0–15 cm depth. The depth 0–2 cm has been chosen to represent that layer of soil that is susceptible to rill and inter-rill erosion, and thus relevant to the modelling of P transfers associated with surface processes. The difference between values for the surface layers and the average for 0–15 cm depth are especially noticeable for the two grassland profiles, where surface and 0–15 cm values are 1658 and 1428 mg kg^{-1} and 2409 and 2040 mg kg^{-1}, respectively. Thus, surface values of total-P are 230 and 369 mg kg^{-1} higher than the average for the 0–15 cm depth, for the Knoyle Down and Mere Down sites, respectively. In the case of the soils under arable, surface concentrations of total-P are generally similar (i.e. differences are within the range 18–109 mg kg^{-1}) to the average values for the 0–15 cm depth. Clearly, for the soils examined here, the use of average (bulked) values for the 0–15 cm depth underestimates the total-P concentration of soil within the upper 2 cm of the profile by 14–15% and 4–10% in the case of the soils under grassland and arable ($n = 3$), respectively.

Table 2 Total-P concentrations (mean (0–15 cm), maximum and average for upper 2 cm of the profile) for the different soils under arable and grassland.

Site	Soil type	Land use	Total-P (mg P kg^{-1} dry soil) Mean	Max	0–2 cm
Knoyle Down	Brown rendzina	Arable	1648	1733	1724
		Grassland	1428	1704	1658
Mere Down	Humic rendzina	Arable	1730	1904	1839
		Grassland	2040	2475	2409
Wilsford	Typical brown calcareous earth	Arable	1305	1408	1287
Sunnyhill	Typical argillic brown earth	Arable	568	670	632

Table 3 Olsen-P concentrations (mean (0–15 cm), maximum and average for upper 2 cm of the profile) for the different soils under arable and grassland.

Site	Soil type	Land use	Olsen-P (mg P kg^{-1} dry soil) Mean	Max	0–2 cm
Knoyle Down	Brown rendzina	Arable	71.1	99.6	96
		Grassland	14.0	39.7	33
Mere Down	Humic rendzina	Arable	63.3	83.1	73
		Grassland	73.5	128.7	112
Wilsford	Typical brown calcareous earth	Arable	58.8	77.3	62
Sunnyhill	Typical argillic brown earth	Arable	43.3	61.9	55

In the case of Olsen-P (Table 3), there is a similar pattern. For soils under both grassland and arable, values of Olsen-P are greater in the 0–2 cm depth increment than the average value for 0–15 cm depth. As with total-P (Table 2) these difference are most pronounced for

Table 4 Values of total-P and Olsen-P based on NSI database (bulked 0–15 cm) for the four different soil types examined and also for all soils under arable and permanent grassland in England and Wales. Values relate to the national database except for the values specific to the Hampshire Avon (based on NSI sites within this catchment).

Soil type	Land use	N	Total-P (mg kg^{-1})	Olsen-P(mg l^{-1})†
Brown rendzina (3.43)	Arable	115	1157 (1054)‡	30 (23)
	Grassland	18	1476 (1238)	33 (15)
	All	168	1184 (1078)	28 (21)
Humic rendzina (3.41)	Arable	3	1397 (1479)	40 (35)
	Grassland	9	1453 (1548)	18 (11)
	All	20	1480 (1456)	23 (15)
Typical brown calcareous earth (5.11)	Arable	130	1027 (942)	33 (26)
	Grassland	33	1130 (1110)	21 (15)
	All	205	1077 (1023)	31 (24)
Typical argillic brown earth (5.71)	Arable	101	768 (680)	37 (30)
	Grassland	56	812 (753)	24 (14)
	All	255	767 (680)	33 (24)
All—Hampshire Avon	Arable	23	1236 (1237)	25 (20)
All—Hampshire Avon	Grassland	18	1343 (1388)	19 (15)
All—England and Wales	Arable	1888	830 (734)	32 (26)
All—England and Wales	Grassland	1560	939 (847)	24 (17)

†It is assumed that mg l^{-1} approximately equates to mg kg^{-1} for this database.
‡Values are means with medians in brackets.

the two grassland profiles, where surface values of Olsen-P are 19 and 39 mg kg^{-1} greater than average values for the 0–15 cm depth, for the Knoyle Down and Mere Down sites, respectively. For the soils under arable, surface concentrations of Olsen-P are between 3 and 25 mg kg^{-1} greater than average values for 0–15 cm depth.

It is useful to compare the values described above with values based on the NSI data set for the four soil types examined (Table 4). Table 4 illustrates the type of data that is commonly used in models to determine the extent and spatial variation of P transfers within river systems. Typically, soil-P input values are classified either according to soil type (i.e. subgroup) or by land use (i.e. arable or grassland). It is clear that the surface (0–2 cm) concentrations of total-P (Table 2) and Olsen-P (Table 3) for the four soil types in the Hampshire Avon catchment are considerably greater than equivalent values for the same soils based on the full national NSI database, especially in the case of Olsen-P.

CONCLUSIONS

The depth profiles of total-P and Olsen-P clearly demonstrate that there are variations in soil-P concentrations with depth in uncultivated grassland soils, particularly in the case of Olsen-P. For soils under arable, values of total-P are broadly constant within the plough layer. Concentrations of Olsen-P, however, exhibit maximum values near the soil surface reflecting inputs of fertilizers and manures to the soil surface between consecutive ploughing events. For 11 out of 12 soil-P profiles, the concentrations of soil-P in the surface layer (0–2 cm) are greater (in one instance >100%) than the average concentration for the 0–15 cm depth. This has important implications for the use of soil-P data based on bulked 0–15 cm soil cores for modelling surface transfers of P in river systems.

Acknowledgements We are grateful to DEFRA for funding this work, to Paul Withers (ADAS) for his role in leading the PSYCHIC project, to IGER-North Wyke for allowing us to use their laboratories and to Phil Haygarth (IGER) for encouraging us to publish these results.

REFERENCES

Avery, B. W. (1980) *Soil Classification for England and Wales (Higher Categories)*. Soil Survey Technical Monograph 14, National Soil Resources Institute, Silsoe, UK.

Correll, D. L. (1998) The role of phosphorus in the eutrophication of receiving waters: a review. *J. Environ. Qual.* **27**, 261–266.

Harrod, T. R. & Fraser, A. I. (1999) *Soil Erosion and Phosphorus Loss from Agriculture*. Final report to MAFF, Project NT1014. SSLRC (NSRI), Cranfield University, Okehampton, UK.

Haygarth, P. M., Lepworth, L. & Jarvis, S. C. (1998) Forms of phosphorus transfer in hydrological pathways from soil under grazed grassland. *Europ. J. Soil Sci.* **49**, 65–72.

Heathwaite, A. L., Fraser, A. I., Johnes, P. J., Hutchins, M., Lord, E. & Butterfield, D. (2003) The Phosphorus Indicators Tool: a simple model of diffuse P loss from agricultural land to water. *Soil Use Manage.* **19**, 1–11.

Johnes, P. J. (1996) Evaluation and management of the impact of land use change on the N and P load delivered to surface waters: the export coefficient modelling approach. *J. Hydrol.* **183**, 323–349.

McGrath, S. P. & Loveland, P. J. (1992) *The Soil Geochemical Atlas of England and Wales*. Blackie, London, UK.

Murphy, J. & Riley, J. P. (1962) A modified single solution method for the determination of phosphate in natural waters. *Anal. Chim. Acta* **27**, 31–36.

Olsen, S. R., Cole, C. V., Watanbe, F. S. & Dean, L. A. (1954) *Estimation of Available Phosphorus in Soils by Extraction with Sodium Bicarbonate*. USDA Circular 939, USDA, Washington, DC, USA.

Owens, P. N. & Walling, D. E. (2002) The phosphorus content of fluvial sediment in rural and industrialized river basins. *Water Res.* **36**, 685–701.

Owens, P. N. & Walling, D. E. (2003) Temporal changes in the metal and phosphorus content of suspended sediment transported by Yorkshire rivers, UK over the last 100 years, as recorded by overbank flood plain deposits. *Hydrobiologia* **494**, 185–191.

Russell, M. A., Walling, D. E., Webb, B. W. & Bearne, R. (1998) The composition of nutrient fluxes from contrasting UK river basins. *Hydrol. Processes* **12**, 1461–1482.

Smith, B. F. L. & Bain, D. C. (1982) A sodium hydroxide fusion method for the determination of total phosphate in soils. *Commun. Soil Sci. Plant Anal.* **13**, 185–190.

Withers, P. J. A., Edwards, A. C. & Foy, R. H. (2001) Phosphorus cycling in agriculture and implications for phosphorus loss from soil. *Soil Use Manage.* **17**, 139–149.

Soil erosion, nutrient migration and surface water pollution in Russia

LEONID F. LITVIN & ZOYA P. KIRYUKHINA
Laboratory of Soil Erosion and Channel Processes, Moscow State University, GSP-2, 119992 Vorob'evy Gory, Moscow, Russia
river@river.geogr.msu.su

Abstract The evaluation of soil erosion-originated transport of nutrients (C, N, P, K) from arable slopes is presented in this paper. The results are based on the map "Erosion-affected land in Russia" (scale 1:1 500 000) and agrochemical data. Annual N, P and K losses (in gross forms) are estimated to be 1.4, 1.1 and 11.8 × 10^6 t, respectively. Soluble forms of P (phosphorus) and K (potassium) are exported at the rate of 53 and 68 × 10^3 t, respectively. Average annual soil loss from arable land of Russia is about 560 × 10^6 t, including 26.3 × 10^6 t of humus. Landscape zones and, most importantly, arable land topography, are major factors controlling regional variability of nutrient losses. Conditional P concentration is directly proportional to the amount of its erosion-originated loss and is inversely related to runoff layer. The P conditional concentration (C_{cP}) calculations for all drainage basins of large Russian rivers have been used for to assess the ecological risk of surface water pollution. The North-Caucasian rivers appear to be the most endangered from this point of view.

Key words conditional concentration; delivery from slopes; erosional losses; gross content and soluble forms; human-accelerated erosion; landscape zones; nutrients; pollution; Russia; surface water

INTRODUCTION

Soil erosion is the main agent of lateral material transport on anthropogenically disturbed land. Its intensity in temperate climate zones is 2–3 times higher on arable slopes than under natural vegetation. The total mass of soil being annually washed off agricultural land in Russia is estimated at about 560 × 10^6 t (Litvin, 2002).

Human-accelerated soil erosion has important implications for landscape ecological conditions. Firstly, nutrient (N, P, K) export from slopes decreases soil fertility. Secondly, surface waters are polluted by mineral and organic substances held by transported sediment (sometimes toxic or radioactive). Human-accelerated soil erosion is the main source of river suspended sediment in plain areas. This conclusion is confirmed by an increase of specific suspended sediment yield by 5–8 times in intensively used drainage basins (Hudson, 1971; Dedkov & Mozzherin, 1984; Razgulin, 1993). This paper aims to assess the amount and intensity of nutrient losses from arable slopes in various natural zones and to evaluate the potential risk of surface water pollution in Russia.

METHOD

This study has been based on a cartographic method—the overlay of erosion intensity maps on maps of soil types. Annual soil erosion rates on arable land have been obtained from the map "Erosion-affected land in Russia" (scale 1:1 500 000) (Litvin, 2002). Average annual rates (t ha^{-1} $year^{-1}$) shown on that map were calculated using the modified models USLE (USA) and GGI (State Hydrological Institute, St. Petersburg, Russia) which were adapted for North Eurasian conditions (Larionov, 1993). The point-statistical method of slope morphometry measurement (70–80 × 10^3 points) has allowed us to evaluate variability of soil loss rates on large parcels of arable and pasture land. Adequacy of the modelling has been confirmed by field measurements in key areas. For chernozem soils of European Russia, the error term does not exceed 3–60%. For soddy-podzolic soils, erosion rates may have been overestimated by 20–30% (Litvin, 2002).

Soil nutrient content is closely related to humus content and texture. For most chemical elements there are typical stable ratios between soluble and insoluble forms (for example, 1:10–1:15 for phosphorus). This makes it possible to use soil type maps for territorial evaluation of soil nutrient content.

Amounts and ratios between different forms of nutrients exported from slopes depend on a number of factors such as soil loss intensity, contribution of rainfall and snowmelt erosion, degree of particle size selectivity, type of crop rotations, etc. Nevertheless, analysis of measurements of soil erosion and nutrient output from erosion plots (Azhigirov *et al.*, 1987; Zhilko *et al.*, 1999) showed that it is reliable to employ estimates of average annual erosion rates for large-scale evaluation of nutrient transport (Litvin & Kiryukhina, 2004). Nutrient fluxes from slopes were estimated combining information on average soil erosion rates, soil type and topsoil nutrient content for various soil types, derived from sources explained for each study point.

Risk of river water pollution is determined by the amount of a pollutant, its toxicity and concentration. To assess the potential concentration of erosion-derived pollutants in river waters, a value of conditional concentration C_c (mg l^{-1}) has been employed:

$$C_c = 10 \cdot A \cdot S \cdot P / H \qquad (1)$$

where A is soil erosion rate (t ha^{-1} $year^{-1}$); S is percentage of arable land relatively to total drainage area (%); P is relative pollutant content (gross forms) in the plough layer (%); H is average annual runoff depth (mm). Conditional concentration characterizes gross amount of a pollutant in a unit volume of eroded soil. To derive real concentrations of pollutant in surface waters from that value it should be multiplied by the ratio of soluble and insoluble forms.

It is known that the proportion of slope-derived sediment reaching permanent water objects is inversely related to the drainage basin area. Most of the sediment is deposited in small valleys of ephemeral streams, small lakes, ponds and reservoirs, as well as on slope sections covered with natural vegetation. The general form of the relationship between drainage area and sediment yield reduction is similar for the USA and European Russia territories:

$$K_n = aF^{-0.2} \qquad (2)$$

where K_n is the sediment delivery ratio; F is the drainage basin area (United States Soil Conservation Service, 1971; Sidorchuk, 1995). The coefficient a reflects the influence of a number of factors and varies regionally. For example, for the Dnieper and the Volga River basins it equals 0.25, and for the Dnestr and the Don River basins it equals 0.75 (Sidorchuk, 1995). Sediment redeposition on slopes and within small catchments can potentially be evaluated using models such as CREAMS or WEPP (Kinsel, 1980; Laflen *et al.*, 1991). However, there is no database available for large-scale evaluation in regions of Northern Eurasia. For medium-sized plain rivers of European Russia K_n approximately equals 0.06–0.1.

Soil phosphorus (P) can serve as an indicator of the migration and transfers of soil particle-associated pollutants, such as heavy metals and radionuclides. Phosphorus itself plays a leading role in the eutrophication of water bodies. Agricultural land is a major source of this P, estimated to contribute up to 60–90% of its total surface water content (Khrisanov & Osipov, 1993; Kudeyarova, 1993).

RESULTS AND DISCUSSION

Natural and anthropogenic factors of soil erosion vary significantly between different landscape zones of Russia. Most important for our study are percentage of arable land and typical nutrient content and composition in zonal soils. These differences determine zonal variability of C, N, P and K migration from arable land (Table 1). Differences in rates of N, P and K export in various landscapes are not significant. For humus it is more noticeable, though still lower than zonal and regional variation of its soil content (for example, 2–3 times between Chernozem and Soddy-podzolic soils). The lower intensity of P and K losses in Siberia is a consequence of the limited extent of soil erosion on the West-Siberian Lowland. In general, for landscape zones nutrient export intensity is predetermined by spatial distribution of erosion. The latter is dependent on azonal factors—topography and regional land use patterns.

Losses of humus, N, P and, to some extent K in absolute terms, are substantially higher in landscape zones dominated by Chernozem soils than in those with Soddy-podzolic soils. The European steppe and forest-steppe zones lose 3.2 times more nutrients, and Siberian 4.6 times more, than the taiga forest zones of the same territories (Table 1). The main reasons for this are both higher nutrient content in Chernozems and the larger percentage of arable land. On average 199 kg of humus, 11 kg of N, 8.3 kg of P and 89 kg of K (gross) are lost annually from 1 ha of arable land in Russia.

Table 1 Annual erosional losses of soil, humus and nutrients from arable land of different landscape zones of Russia.

Regions and landscape zones	Eroded soil volume ($\times 10^3$ t)	Losses of humus and gross forms of nutrients							
		Rate (kg ha^{-1})				Volume ($\times 10^3$ t)			
		Humus	N	P	K	Humus	N	P	K
European Russia	436 137	198	11.7	8.7	96.2	18 674	1100	823	9056
Forest, including:	140 089	162	11.7	10.2	129.5	3379	246	213	2707
(1) North and middle Taiga	15 141	156	10.3	9.2	139.1	303	20	18	270
(2) Southern Taiga	124 948	162	11.9	10.3	128.5	3076	226	195	2437
Forest-steppe	148 772	215	12.3	8.3	94.0	6948	394	268	3039
Steppe	147 276	204	11.2	8.4	81.0	8347	460	342	3310
Siberia	130 833	202	9.3	7.1	72.3	7692	356	271	2757
Forest	28 407	208	12.7	8.5	103.0	1223	75	48	572
Forest-steppe	53 092	216	9.4	6.9	66.6	3812	165	122	1175
Steppe	49 334	178	7.8	6.6	67.7	2657	116	98	1010
Entire Russia	566 970	199	10.9	8.3	89.3	26 366	1456	1094	11 800

Table 2 Annual erosional losses of soil, humus and nutrients from arable land of different economic regions of Russia.

Economic regions	Soil loss rate (t ha^{-1})	Eroded soil volume (× 10^3 t)	Humus and nutrient content in suspended sediment (upper figure × 10^3 t; lower figure kg ha^{-1})					
			Gross				Soluble	
			Humus	N	P	K	P	K
Northern	6.3	8207.0	164.14 125.9	10.7 8.2	9.8 7.5	147.7 113.6	0.9 0.68	1.1 0.76
North-Western	4.6	8447.1	169.0 91.4	11.8 6.3	11.0 6.0	152.0 82.0	1.1 0.60	1.2 0.65
Volga-Vyatka	8.4	63 472.2	1777.2 235.2	120.6 16.0	95.2 12.6	1206.0 159.6	6.5 0.86	7.7 1.0
Central	5.0	73 486.9	1984.1 134.8	132.3 9.0	110.2 7.5	1396.3 95.0	7.6 0.51	7.7 0.52
Central-Chernozem	3.1	33 874.6	2405.0 220.2	115.2 10.5	67.7 6.2	745.2 68.2	3.0 0.28	3.7 0.34
Povolzhskiy	2.5	62 588.3	3004.3 120.7	181.5 7.3	112.7 4.5	1314.4 53.0	5.6 0.23	7.5 0.30
North-Caucasian	6.5	105 938.9	6144.5 376.9	339.0 20.8	264.8 16.8	2436.6 149.5	7.8 0.48	12.0 0.73
Ural	4.0	89 012.8	4895.7 220.0	267.0 12.0	186.9 8.4	1958.3 88.0	6.2 0.27	10.4 0.47
Western-Siberian	1.3	26 355.0	1408.6 69.5	65.0 3.2	60.6 3.0	553.5 27.3	3.1 0.16	3.6 0.18
Eastern-Siberian	8.1	77 926.1	3762.1 391.0	163.6 17.0	148.0 15.4	1558.5 162.0	8.9 0.93	10.7 1.12
Far East[1]	6.0	16 931.3	638.4 226.2	47.4 16.8	25.4 9.0	338.6 120.0	2.0 0.70	2.3 0.81
Entire Russia[2]	4.3	566 240.2	26353 199.1	1454 11.0	1092 8.3	11807 89.7	53 0.39	68 0.51

[1]In the Far East region, arable land of Amur, Khabarovsk and Primorskiy regions is also included;
[2]Without Kaliningrad region.

In order to develop and validate national-scale ecological and soil conservation programmes it is necessary to have information on nutrient export from slopes within administrative or economic territorial units (Table 2). Estimations have shown that the migration rates of P and K in soluble forms from arable land are relatively low, both in individual regions and for the whole of Russia. However, gross annual losses for the whole of Russia (52.7×10^3 t P and 67.9×10^3 t K) are economically serious, especially taking into account the high prices of artificial chemical fertilizers.

Average annual rates of nutrient migration vary regionally: for N and P from 7 to 12 kg ha^{-1}, for K from 6.5 to 14.7 kg ha^{-1}. Lower rates are observed in the Central-Chernozem and Volga regions where arable Chernozems prevail. Minimum values are found for Siberia as soil erosion there is only active locally in the far southeast (Table 2). Maximum migration rates are characteristic for the North-Caucasian region where high erosion rates are combined with relatively high nutrient content in the topsoil. Average annual rates of humus loss from arable land in Russia vary from 70 to 390 kg ha^{-1}. Total export of organic matter from slopes under agricultural land use exceeds 26×10^6 t (Table 2).

Results of our calculations of spatial distribution and rates of nutrient migration for European Russia are in satisfactory agreement with the results of other studies based on independent methods (Podgorny & Butenko, 1991; Kashtanov & Yavtushenko, 1997).

Table 3 Characteristics of P delivery into surface water in different landscape zones.

Landscape zones and topography	Number of drainage basins	Gross soil loss on arable land (t ha^{-1} $year^{-1}$)	Specific sediment yield from arable land (t ha^{-1} $year^{-1}$)	Annual runoff layer (mm)	C_{cP} (mg l^{-1})
Forest	**86**	**6.7**	**2.4**	**225**	**1.4**
Lowland	73	6.4	2.1	228	1.1
Upland	13	8.8	4.8	202	3.2
Forest-steppe	**37**	**4.6**	**3.2**	**150**	**4.1**
Lowland	21	3.4	1.8	166	2.3
Upland	16	6.2	5.0	126	6.6
Steppe	**33**	**3.6**	**3.0**	**57**	**10.2**
Lowland	24	2.2	1.8	50	8.2
Upland	9	6.5	5.6	72	13.0

RISK OF SURFACE WATER POLLUTION

Assuming that phosphorus can be used as an indicator of environmental behaviour of many other pollutants exhibiting similar geochemical properties, we have employed the estimated conditional concentration of P (C_{cP}) as an indicator of surface water pollution from erosional sources. The spatial pattern of such pollution is mainly determined by variations in river runoff, soil erosion rate and pollutant concentration in soils (Table 3). Within European Russia these factors, as well as land use characteristics, change according to latitudinal zones. The value of C_{cP} exhibits a similar tendency of variation, with significant additional influence of azonal geomorphic factors. Surface water in drainage basins of the forest-steppe and steppe zones are characterized by the highest C_{cP} values. Water bodies and watercourses of these zones are at largest risk of eutrophication and disappearance. Upland areas are at increased risk of surface water pollution in all landscape zones as more ranged terrain favours erosional transport of nutrients from slopes to river network.

Zonal characteristics can be used to evaluate the potential pollution of small and medium rivers by P. For large rivers an individual approach is preferred (Table 4) as such rivers cross contrasting landscapes with different natural and land use conditions. This causes significant variation of pollution within drainage basins. Major factors controlling the P delivery are the same as for smaller rivers, though the extent of agricultural land becomes dominant for northern European Russia and southeastern Siberia.

The Volga River basin is characterized by the largest internal variation of P pollution. The average soil erosion rate for that area is 6.4 t ha^{-1} $year^{-1}$, specific sediment yield is 2.8 t ha^{-1} $year^{-1}$, and C_{cP} is 2.0 mg l^{-1}. The upper part of the Volga River basin is located within the forest zone. The values of C_{cP} in the largest Volga tributaries in forested area vary from 0.1 to 2.5 mg l^{-1}. The steppe part of the basin is characterized by higher values of C_{cP} (2.6–7.5 mg l^{-1}) due to lower surface runoff, despite relatively low erosion rates (1.5–3.9 t ha^{-1} $year^{-1}$). In the lower part of the basin within the Caspian Lowland erosion-derived P pollution is almost absent.

The North-Caucasian river basins (Manych, Kalaus, Kuma, Terek) have maximum values of erosion-derived P pollution in Russia (up to 44 mg l^{-1}). It is associated with a combination of high erosion rates and low surface runoff. The Kuban River basin located in the same region has much lower average C_{cP} due to large unpolluted water runoff from its mountainous tributaries. The Western Siberia rivers have low potential P pollution. Only the Kuznetsky Alatau and Salair Ridge piedmonts are characterized by an increase in erosion

Table 4 Potential pollution of river water by erosion-derived P in Russia.

Large rivers / Lakes / Seas	Number of drainage basins	Gross soil loss on arable land (t ha^{-1} $year^{-1}$)	Specific sediment yield from arable land (t ha^{-1} $year^{-1}$)	Annual runoff layer (mm)	C_{cP} (mg l^{-1})
Finskiy Bay	2	3.0	0.3	260	0.1
Ladoga Lake	6	3.3	0.2	287	0.1
Ilmen Lake	7	5.2	1.9	246	0.9
Zap. Dvina River	9	5.0	1..5	233	0.7
Sev. Dvina River	5	7.8	0.9	265	0.5
Dnieper River	9	6.3	4.2	158	3.6
Volga River	85	6.4	2.8	196	2.0
Oka River	17	6.7	3.8	166	3.6
Kama River	10	8.2	3.8	192	2.6
Don River	34	4.0	3.4	85	6.6
Severskiy Donets River	4	7.5	6.3	95	11.8
Azov Sea	2	5.2	4.7	30	25.4
Kuban River	5	8.1	4.7	220	4.0
Terek River	2	13.5	10.3	99	22.0
Kuma River	4	10.9	9.1	68	30.0
Ural River	4	1.9	1.2	72	2.7
Ob River	39	2.6	1.0	126	1.9
Irtysh River	12	0.4	0.1	40	0.5
Enisei River[1]	16	6.6	1.9	117	3.2
Lena River[1]	3	18.2	2.8	126	4.0
Baikal Lake[1]	3	15.8	3.2	43	6.0
Amur River[1]	10	11.5	1.7	66	3.8

[1]Within agricultural zone only.

rate and C_{cP} (up to 5–10 t ha^{-1} $year^{-1}$ and 5.5 mg l^{-1}, respectively). The latter is also relatively high in agriculturally developed parts of Eastern Siberia and the Far East (Table 4).

Real concentrations of soluble P in river water is much lower than the estimated C_{cP}, as the latter includes gross P forms delivered from slopes. When accounting for soluble forms only (normally not exceeding 10% of gross value) and making corrections for the sediment delivery ratio, estimates become closer. In the river water of agriculturally developed drainage basins, mineral P concentrations are in the range of 0.002 to 1.145 mg l^{-1} and gross P values are 0.007 to 3.2 mg l^{-1} (Savenko & Zakharova, 1997).

However, absolute values of pollutant concentration do not allow an evaluation of ecological risks. It is necessary in this case to establish "ecological threshold" values of concentration. As regards P, it has been found that water vegetation does not develop when its concentration is below 0.01 mg l^{-1}. Its growth begins in a range of concentrations from 0.01 to 0.025 mg l^{-1}, and it flourishes under higher concentrations—from 0.09 to 1.8 mg l^{-1} (Kudeyarova, 1993). Thus, we can establish the first "ecological threshold" value for P as 0.01 mg l^{-1} (C_{cP} about 1 mg l^{-1}). The second ("crisis") value will be 0.2 mg l^{-1} (C_{cP} about 20 mg l^{-1}, and it is associated with flourishing of water vegetation. Considering these values, we have distinguished five types of areas within the agricultural zones of Russia, according to their risk of water pollution by soil erosion-derived P (Fig. 1). These are based on C_{cP} estimations discussed above.

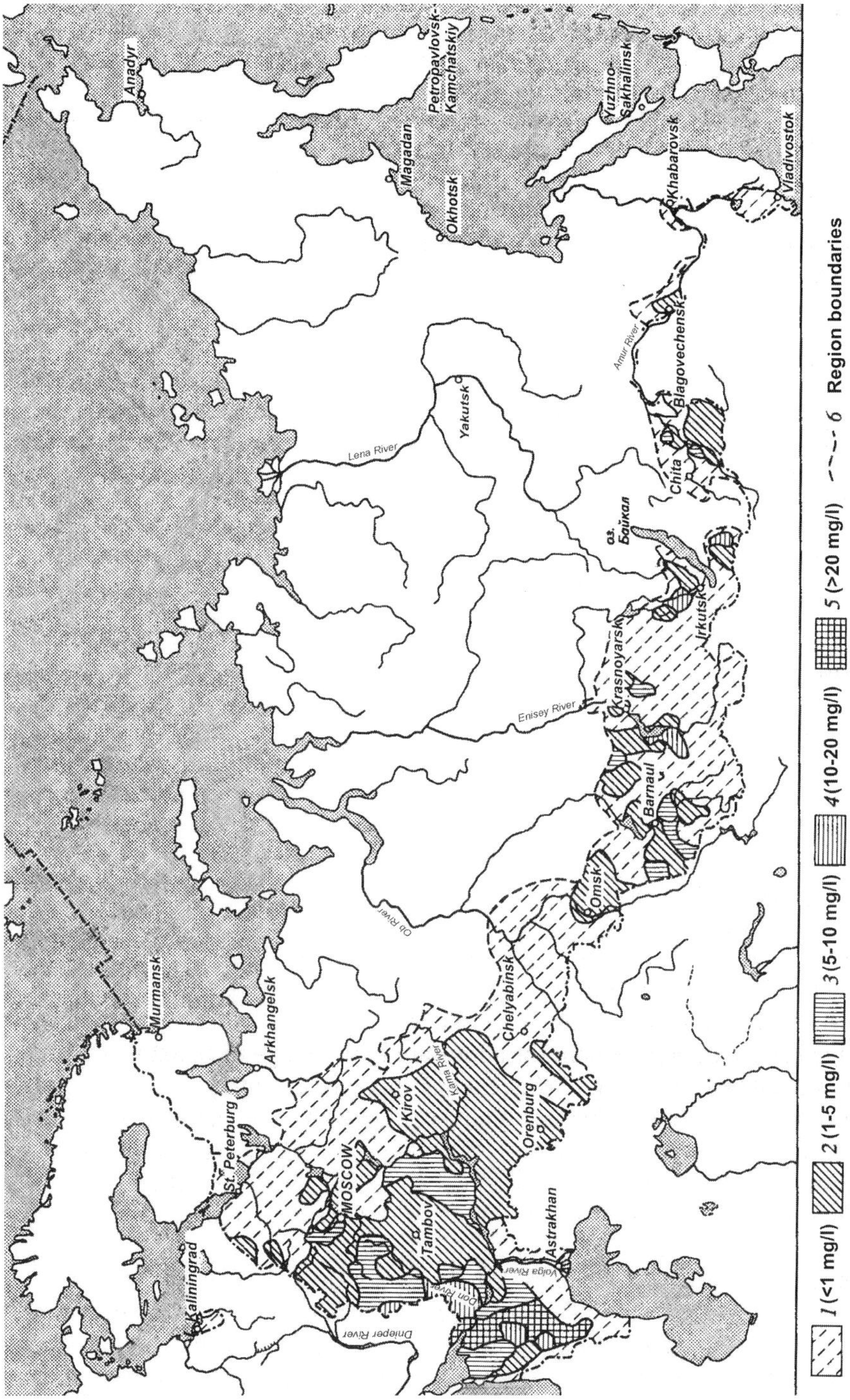

Fig. 1 Potential surface water pollution by erosion-derived P in the agricultural zone of Russia. Five regions are distinguished according to estimated C_{cP} value.

CONCLUSION

Erosion remains the main factor of soil degradation and surface water pollution by nutrients on agricultural land in Russia. Total output of gross forms of N, P and K from arable slopes is 2–3 times higher than its input to fields with organic and mineral fertilizers. Hence, erosion causes substantial economic losses. Annual losses of soluble (readily available for cultural plants) forms of N, P and K comprise about 8% of their mass in mineral fertilizers. Soil erosion is also the main agent of surface water pollution by P.

Regional variation of nutrient erosional transport intensity mainly depends on soil erosion rates. River water pollution is also controlled by variation of specific surface runoff.

Acknowledgement This research has been made possible thanks to financial support of the program of the Russian leading scientific schools support (project NS 1443.2003.5).

REFERENCES

Azhigirov, A. A., Golosov, V. N., Dobrovolskaya, N. G., Belotserkovskiy, M. Yu., Zharkova, Yu. G., Kalinina, V. R., Kiryukhina, Z. P., Krasnov, S. F., Larionov, G. A., Litvin, L. F., Petrov, V. N., Samodurova, L. S., Veretennikova, M. V., Zorina, E. F., Lyubimov, B. P. & Nikolskaya, I. I. (1987) *Investigations of Water Runoff and Sediment Yield from Slope Catchments of the Protva River basin.* Report VINITI No. 6389-B 87. Moscow University, Moscow, Russia (in Russian).

Dedkov, A. P. & Mozzherin, V. I. (1984) *Erosion and Sediment Yield on Earth.* Kazan University Publ., Kazan, Russia (in Russian).

Hudson, N. W. (1971) *Soil Conservation.* Batsford, London, UK.

Instruction for Hydrological Calculations for Design of Erosion Countermeasures in the European USSR (1979) Gidrometeoizdat Publ., Leningrad, Russia (in Russian).

Kashtanov, A. N. & Yavtushenko, V. E. (1997) *Agroecology of Soils on Slopes.* Kolos Publ. Moscow, Russia (in Russian).

Khrisanov, N. I. & Osipov, G. K. (1993) *Water Bodies Eutrophication Management.* Gidrometeoizdat Publ., Saint Petersburg, Russia (in Russian).

Knisel, W. G. (1980) *CREAMS: A Field-Scale Model for Chemicals, Runoff and Agricultural Management Systems.* USDA Conservation Res. Report 26. USDA-ARS, Washington, DC, USA.

Kudeyarova, A. Yu. (1993) *Pedogeochemistry of Ortho- and Polyphosphates Under Use of Fertilizers.* Nauka Publ., Moscow, Russia (in Russian).

Laflen, J. M., Lane, L. J. & Foster, G. R. (1991) WEPP: A new generation of erosion prediction technology. *J. Soil Water Conser.* **46**, 34–38.

Larionov, G. A. (1993) *Erosion and Deflation of Soils.* Moscow University Publ., Moscow, Russia (in Russian).

Litvin, L. F. (2002) *Geography of Soil Erosion on Agricultural Land of Russia.* IKC Akademkniga Publ., Moscow, Russia (in Russian).

Litvin, L. F. & Kiryukhina, Z. P. (2004) Erosional migration of nutrients and surface water pollution. In: *Soil Erosion and Fluvial Processes*, vol. 14. Moscow University Publ., Moscow, Russia (in Russian) (in press).

Podgorny, V. K. & Butenko, L. G. (1991) Erosion and ecological problems of agriculture in the Central-Chernozem region. In: *Basics of Landscape Ecology-Oriented Contour-Meliorated Agriculture in the Steppe Zone.* vol. 1, 17–19 (in Russian).

Razgulin, S. M. (1993) Nitrogen and phosphorus in water of the Rybinsk reservoir. *Vodnye Resursy* **20**(1), 24–31 (in Russian).

Savenko, V. S. & Zakharova, E. A. (1997) Phosphorus in water of upper parts of hydrological network. *Vodnye Resursy* **24**(2), 243–249 (in Russian).

Sidorchuk, A. Yu. (1995) Erosion-accumulation processes on the Russian Plain and problems of small river siltation. In: *Proc. Academy Water Industry Sci.* **1**, 74–83 (in Russian).

United States Soil Conservation Service (1971) *Sediment Sources, Yields and Delivery Ratios.* SCS National Engineering Handbook (Section 3, Chapter 6). United States Departament of Agriculture, Washington, DC, USA.

Zhilko, V. V., Zhukova, I. I., Chernysh, A. F., Tsybulka, N. N. & Tishuk, L. A. (1999) Humus and nutrient losses from Soddy-podzolic soils of Belorussia caused by water erosion. *Agrohimiya* **10**, 41–46 (in Russian).

Hydrological and biological event based variability in the fine-grained sediment structure of a small undisturbed catchment

J. L. MCCONNACHIE & E. L. PETTICREW
Geography Program, University of Northern British Columbia, 3333 University Way, Prince George, British Columbia V2N 4Z9, Canada
ellen@unbc.ca

Abstract In 2001 a study in an undisturbed, highly productive salmon bearing stream (O'Ne-eil Creek) was undertaken in the northern interior of British Columbia, Canada. The aim was to determine if the structure and settling characteristics of the aggregated sediments varied significantly over the different hydrological and biological regimes of an open-water season. Sampling regimes included the hydrological conditions of the rising limb of spring freshet, falling limb flows of snowmelt and summer rainstorms, while biological conditions included sampling during spawning and die-back of 13 580 sockeye salmon. A negative relationship between shear stress and effective particle size diameter, D_{50} is evident for the hydrological regimes, while the largest flocs occur when live and dead salmon are present in the stream. Increased organic matter quality and biological resuspension of gravel-stored material attributed to spawning salmon, reflect their physical and chemical influence on the change in size and density of aggregated fine sediment in this stream system.

Key words aggregation; Canada; effective density; floc factor; flocculation; image analysis; organic matter quality; shear stress; spawning salmon; suspended sediment

INTRODUCTION

Downstream transport of fine-grained sediment (<63 μm diameter) in river channels is assumed to be efficient, according to conservative hydrodynamic theory, such that yield models should accurately predict sediment fluxes from catchments. However, field studies indicate conveyance losses in longitudinal fine sediment flux attributed to storage on flood plains as well as within channel beds (Owens *et al.*, 1999), which renders yield models inadequate. This difference is largely due to the misconception that particles are transported as single grains with very slow settling rates.

Recent research indicates that suspended sediment is commonly transported in a flocculated form (e.g. Petticrew, 1996, 1998; Droppo *et al.*, 1997; Woodward *et al.*, 2002), whereby single sediment grains (inorganic and/or organic) combine together to form larger units. The composite particles have been further categorized by Petticrew & Droppo (2000), who visually identified two populations: (a) fast settling, compact aggregates, and (b) slow settling, amorphous flocs.

Particle structure is a major factor in regulating the behaviour of suspended material in aquatic environments (Nicholas & Walling, 1996). Flocculation alters the hydrodynamic properties of sediment by changing particle density, shape, porosity and composition through increases in effective size and internal complexity of particles (Petticrew & Biickert, 1998; Droppo, 2001). Petticrew & Biickert (1998) suggest that these alterations to the suspended particle population should increase storage of fine-grained sediment in gravel beds compared to what is currently predicted by hydrodynamic models.

While the magnitude and duration of fine-grained sediment storage depends mainly on sediment supply and hydrological conditions, the stored load may comprise a significant portion of a system's annual sediment budget (Owens *et al.*, 1999). However, even small quantities of fine-grained sediment stored on or in the gravel bed for short (weeks) periods could be significant for streams that support salmon-spawning habitat because of the potential for decreased oxygen flow through interstitial spaces that house salmon eggs, and degradation and/or loss of aquatic habitat for food species (Newcombe & MacDonald, 1991). The purpose of this study is to evaluate the temporal variability in suspended fine sediment structure and settling properties over a full field season of hydrological (snowmelt and rain storms) and biological (salmon spawning and die-off) events in an undisturbed salmon stream.

METHODS

Field site and sampling design

The O'Ne-eil drainage basin is a small (75 km^2), relatively undisturbed (one approximately 20-year old access road) system at the most northerly extent (55°N, 125°50′W) of the Fraser River catchment in British Columbia, Canada (Petticrew, 1996). The main channel is about 20 km in length with the lower 2 km exhibiting excellent conditions for sockeye salmon (*Oncorhynchus nerka*) spawning. A 10-m riffle in the main channel, approximately 1500 m upstream of the mouth, was sampled during the period of 18 May to 21 August 2001. Samples were collected to evaluate the seasonal changes in fine sediment structure and morphology over a range of hydrological and biological events (Fig. 1(a)). Hydrological events included: (a) the rising limb of the freshet, (b) the falling limb of snowmelt, and (c) five rain events sampled at a range of times relative to the start of each storm (displayed as points in Fig. 1(a)). Biologically important periods were: (a) active spawning upon the return of migrating salmon (return date 21 July), and (b) the post-spawn die-off. The spawn regime includes the period of active digging of redds (nests for eggs) as well as the start of post-reproductive die-off, while post-spawn is devoid of live fish but incorporates a greater presence of decaying fish carcasses. In this year, 13 580 sockeye salmon returned to the lower reaches of O'Ne-eil Creek to spawn with an approximate density of 0.7 fish m^{-2}.

Field sampling

The seasonal hydrograph was reconstructed from stage height data logged by the Department of Fisheries and Oceans at a gauge within the study reach. Bottle samples were removed at the thalweg for (a) suspended particulate and organic matter concentrations (SPM and SOM), (b) carbon and nitrogen content, and (c) absolute particle size distributions (APSD). Composite 20 L samples were collected for effective (*in situ*) particle size and settling analysis. Samples were withdrawn near the water's surface and thus do not reflect depth integrated values. Shear velocity and stress was calculated as per Gordon *et al.* (1992) from

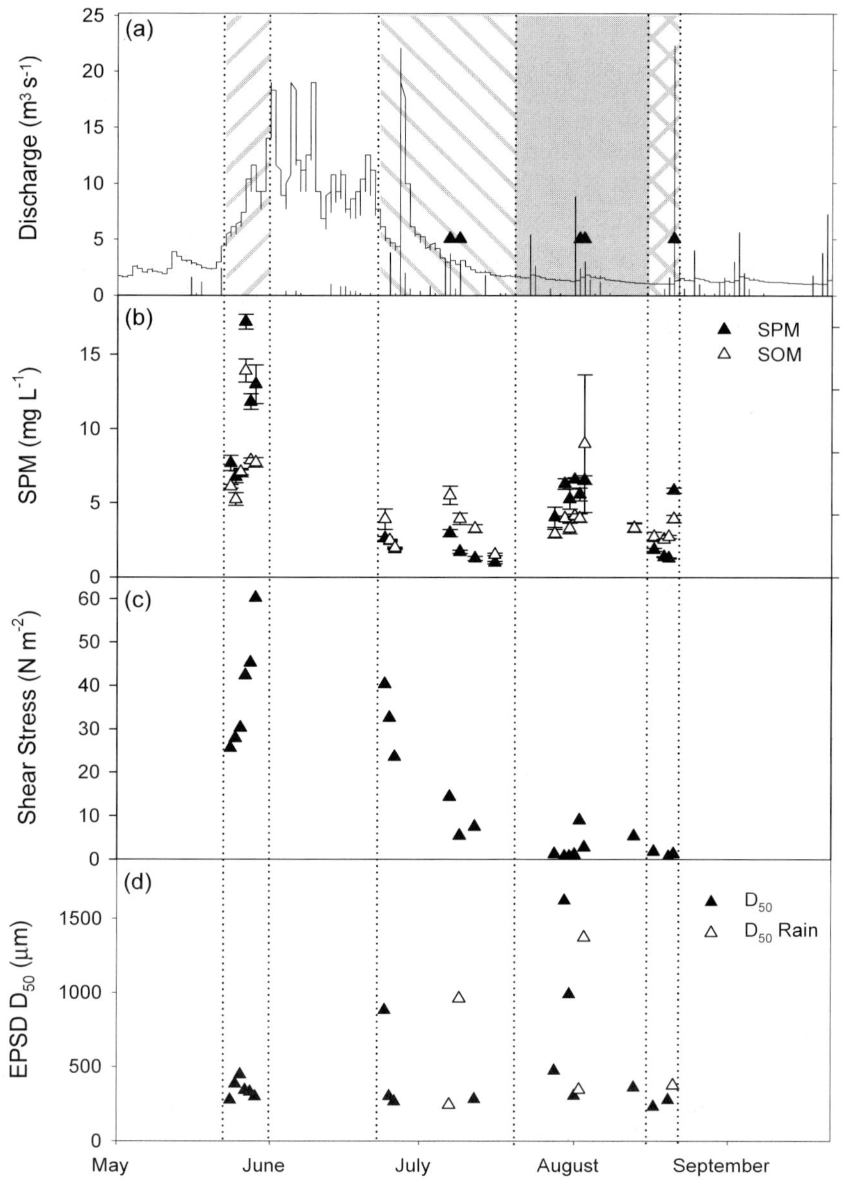

Fig. 1 (a) Discharge and precipitation over the 2001 field season for O'Ne-eil Creek in northern British Columbia, Canada. Sampled rain events are noted with triangles. Forward diagonal hatching denotes the rising snowmelt response, backward hatching indicates falling water levels after snowmelt, presence of live spawners is displayed as grey, and the post-spawn period is crosshatched. (b) SPM and SOM (mg l^{-1}) values from triplicate samples of suspended sediment. Error bars represent ± 1 SE. (c) Shear stress (N m^{-2}) calculated from vertical velocity profiles. (d) D_{50} particle diameters calculated from cumulative percent volume effective particle size distributions (EPSD).

vertical profiles of velocities measured at 5-cm depth intervals with a Swoffer Model 2100 current meter.

Laboratory analysis

Water samples were filtered through triplicate pre-combusted and pre-weighed 47 mm diameter, 0.7 μm pore size glass-fibre filters for gravimetric determination of SPM and SOM. A second set of these filters was freeze-dried and analysed for carbon and nitrogen content using a Carlo Erba NC2500 elemental analyser with an AS 128 autosampler. C:N ratios were calculated from the resultant percentages. Sediment was also filtered for analysis using a Coulter Multisizer (detection range of <1 to 111 μm) to size and count the inorganic constituent particles in the manner of Petticrew (1998). All particles for each sample were grouped into classes defined by their equivalent spherical diameters as per Kranck *et al.* (1993) yielding APSD.

Effective particle size distributions (EPSD) were collected from analysis of images obtained from experiments using a Plexiglas settling tube (cf. Petticrew & Droppo, 2000). A charge-coupled device (CCD), with a resolution of 512 × 512 pixels, interfaced with an Intel-based PC running Northern Exposure (Empix Imaging, Mississauga, Ontario, Canada) was used to capture sequential images of particles as they settled due to gravity. A procedure change occurred mid-June, where the analog CCD was replaced by a Retiga 1300 digital CCD (resolution 1280 × 1024 pixels). At the same time the software was upgraded to Empix's Northern Eclipse. These two changes were not found to bias the particle image collection. Dimensions (e.g. diameter, area, perimeter, and shape) of 500–1500 particles for each sample date were measured and recorded using the Northern Eclipse package (detection minimum of 42 μm). Population particle distributions were obtained in the same manner as APSD. D_{50} particle sizes were determined from linear interpolation of cumulative percent volume data for both EPSD and APSD, and the ratio of the two was calculated as the floc factor (degree of particle aggregation). Settling velocities were determined from the distance travelled by the particle divided by the time interval between images, and particle densities were calculated from these measurements as per Namer & Ganczarczyk (1993).

RESULTS

The hydrograph is shown along with the results of the suspended particulate matter, suspended organic matter, shear stress and the median size of the suspended EPSD (D_{50}) (Fig. 1). As expected, the SPM and SOM increase on the rising limb of the snowmelt flood and decrease on the falling limbs of the large floods. The EPSD D_{50} also rises at the start of the snowmelt, but only for three consecutive days after which it begins to decrease. Shear stresses increase consistently over this six day period, indicating that higher shear conditions are able to entrain more material but potentially disaggregate, or inhibit the aggregation of particles. Alternately, the channel and/or terrestrial supply of aggregates could be smaller in size during the latter part of the rising limb of snowmelt.

Of interest in Fig. 1 is the statistically significant elevated values of SPM during the period of active spawn that decrease when no live spawners are present in the reach. The EPSD D_{50} also reaches its highest values at this time. Soulsby *et al.* (2001) noted that female fish dig spawning redds to bury eggs exerting enough force on the bed to move the gravels. During this process, finer material stored in the gravels is resuspended; a considerable disturbance when hundreds of salmon are active simultaneously. The data in Fig. 1(b) reflect this biotic resuspension of sediment. The enlarged D_{50} is presumed to be a result of particle resuspension without the accompanying high shear stresses as seen during higher discharges. Although the intermittent shear stresses exerted by salmon activity may also cause particle break up, it does not appear to affect the suspended sediment structure to the extent of the hydrological events.

A contributing factor causing increased EPSD D_{50} during this period could be the introduction of marine-derived organic matter. It is well-established that nutrients, especially nitrogen, released during salmon carcass decay can enhance microbial growth in freshwater ecosystems (Wold & Hershey, 1999) and a connection between microbial activity and flocculation has been suggested by many researchers. For example, Droppo (2001) found flocs coated liberally with extracellular polymeric substances (EPS) typically derived from bacteria. These sticky fibrils act like a glue that improves particle binding capabilities. Thus, it is a possibility that salmon presence has two important implications for flocculation in freshwater systems: (a) particle resuspension, and (b) enhanced particle binding potential.

Three rising limb events comprising early snowmelt, a summer storm and a storm during active spawn are compared in Fig. 2. The increasing D_{50} and floc factor over the season is associated with decreases in shear velocities and C:N ratios. This trend of decreasing floc size with increasing shear is also evident in the larger data set (Fig. 1(c) and (d)) for the hydrological events. Alternately, the largest flocs tend to occur during summer rains and/or in low shears (<10 N m^{-2}) when fish are present, potentially reflecting differences not just in the quantity but also in the quality of organic matter. The seasonal decrease observed in C:N ratios from rising limb events (Fig. 2) reflects the increasing quality of the organic matter in the suspended sediment as low C:N ratios imply higher quality OM for bacterial consumption (Bouillon *et al.*, 2000). These low shear environments, combined with abundant nutrients for bacterial productivity, generate large suspended flocs.

Figure 3 depicts 1186 individual suspended particles, which were characterized by image analysis for size, shape and settling velocity allowing for the calculation of particle density. The five different regimes are plotted separately, and, while the majority of particles

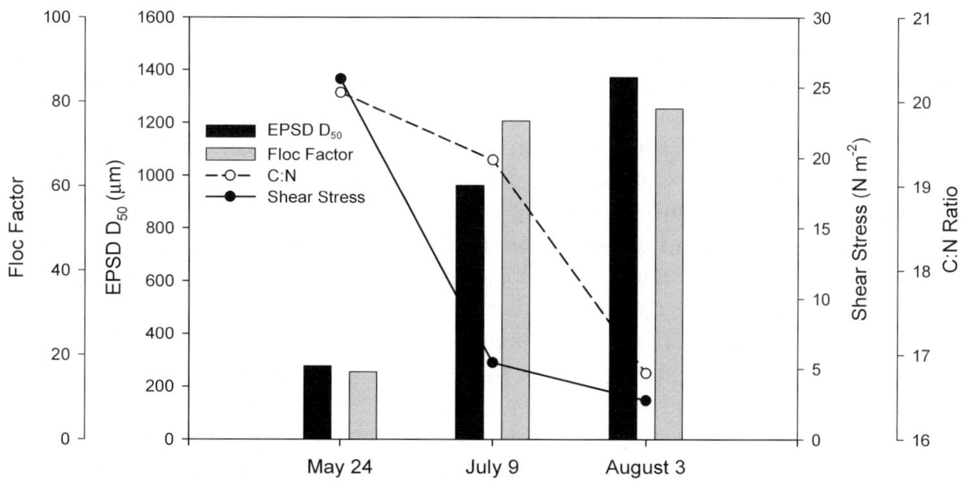

Fig. 2 EPSD D_{50}, floc factor, shear stress, and C:N ratio for three sample days representing rising limb events on the hydrograph for the O'Ne-eil Creek.

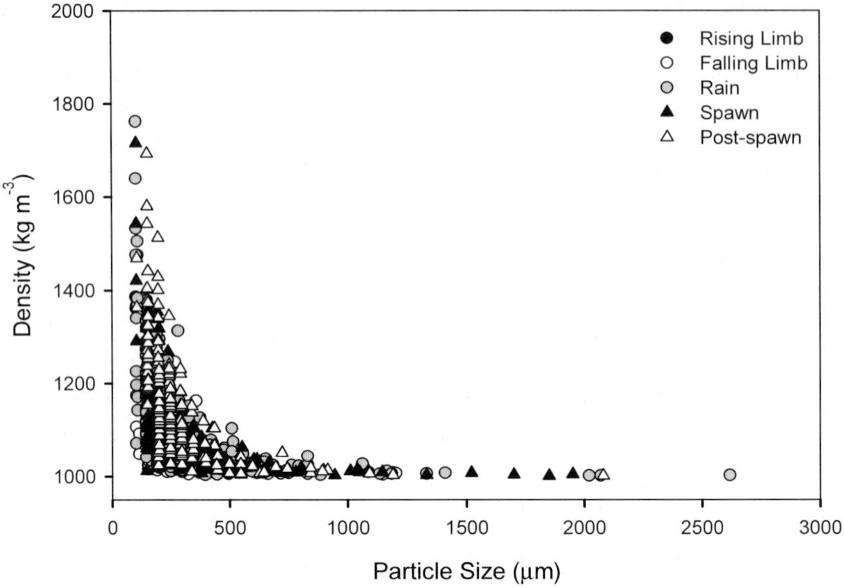

Fig. 3 Effective particle density by diameter differentiated by hydrological and biological response types. An exponential relationship is observed with the majority of the large, low density particles occurring during active salmon spawn and a high number of small, high density particles exhibited post-spawn.

are <500 µm and have densities lower than 1200 kg m^{-3} (or less than an excess density of 0.2 g cm^{-3}), particles in the spawn, post-spawn and rain events exhibit the widest range of particle structure and behaviour. These regimes have the greatest proportion of large, low density particles >1000 µm, as well as the highest proportion of most dense particles (>1400 or in excess of 0.4 g cm^{-3}) which tend to be <300 µm. Large, low density or small, very dense particles are not found in the rising or falling limb samples. A statistical analysis of the full settling data set indicates that the rising limb of the snowmelt flood exhibits significantly smaller particles, while post-spawn consists of the fastest settling and densest particles (all $p < 0.01$). This implies that the potential for gravel-storage of fines is highest when decaying carcasses provide abundant organic material to the stream, which tends to occur when stream flows are low. In fact, Petticrew & Arocena (2003) report increased proportions of gravel stored inorganics, <63 µm, during the die-off and following the decay of all the fish in this same riffle for 2001. Gravel stored fines were at their lowest proportions at mid-spawn when the fish would have both a physical (resuspension) and chemical (decaying OM) effect on the sediment. Infiltration rates from July through September (2001) in this spawning reach were relatively low (<50 g m^2 day^{-1}), but did exhibit a 39% increase following the die-off and rotting of the fish carcasses. As this increase coincides with the period when salmon eggs are incubating, the importance of the timing and the quality of inorganic and organic sediment delivery to the stream on fine particle morphology and behaviour is apparent.

The presence of spawning and decaying fish plays a substantial physical and chemical role in fine sediment dynamics of productive streams. Therefore further examination of the relationship between fine sediment transfers and organic matter source and quality is warranted.

Acknowledgements Thanks are extended to A. Ullrich, R. McConnachie, L. Chamberlist, and T. Waqar for field and laboratory assistance. Hydrological data and field facilities were donated by the Canadian Department of Fisheries and Oceans and Canadian Forest Products. Funding for this project was received from the Northern Land Use Institute and Fisheries Renewal British Columbia. Both authors also received funding for this project from the Natural Science and Engineering Research Council (NSERC) of Canada.

REFERENCES

Bouillon, S., Chandra Mohan, P., Sreenivas, N. & Dehairs, F. (2000) Sources of suspended organic matter and selective feeding by zooplankton in an estuarine mangrove ecosystem as traced by stable isotopes. *Maine. Ecol. Prog. Series* **208**, 79–92.

Droppo, I. G. (2001) Rethinking what constitutes suspended sediment. *Hydrol. Processes* **15**, 1551–1564.

Droppo, I. G., Leppard, G. G., Flannigan, D. T. & Liss, S. N. (1997) The freshwater floc: a functional relationship of water and organic and inorganic floc constituents affecting suspended sediment properties. *Water, Air and Soil Pollution* **99**, 43–45.

Gordon, N. D., McMahon, T. A. & Finlayson, B. L. (1992) *Stream Hydrology: An Introduction for Ecologists*. John Wiley & Sons, Chichester, UK.

Kranck, K., Petticrew, E. L., Milligan, T. G. & Droppo, I. G. (1993) *In situ* particle size distributions resulting from flocculation of suspended sediment. *Coastal and Estuarine Study Series* **42**, 60–74.

Namer, J. & Ganczarczyk, J. J. (1993) Settling properties of digested sludge particle aggregates. *Water Res.* **27**, 1285–1294.

Newcombe, C. P. & MacDonald, D. D. (1991) Effects of suspended sediment on aquatic ecosystems. *North Am. J. Fish. Management* **11**, 72–82.

Nicholas, A. P. & Walling, D. E. (1996) The significance of particle aggregation in the overbank deposition of suspended sediment on river floodplains. *J. Hydrol.* **186**, 275–293.

Owens, P. N., Walling, D. E. & Leeks, G. J. L. (1999) Deposition and storage of fine-grained sediment within the main channel system of the River Tweed, Scotland. *Earth Surf. Processes Landf.* **24**, 1061–1076.

Petticrew, E. L. (1996) Sediment aggregation and transport in northern interior British Columbia streams. In: *Erosion and Sediment Yield: Global and Regional Perspectives* (ed. by. D. E. Walling & B. W. Webb) (Proc. Exeter Symp., July 1996), 313–319. IAHS Publ. 236. IAHS Press, Wallingford, UK.

Petticrew, E. L. (1998) Influence of aggregation on storage of fine-grained sediments in salmon-bearing streams. In: *Proceedings of the Forest-Fish Conference: Land Management Practices Affecting Aquatic Ecosystems* (Tech. coords. M. K. Brewin & D. M. A. Monita), 241–247. AB, Calgary, Canada.

Petticrew, E. L. & Arocena, J. M. (2003) Organic matter composition of gravel-stored sediments from salmon bearing streams. *Hydrobiologia* **494**, 17–24.

Petticrew, E. L. & Biickert, S. L. (1998) Characterization of sediment transport and storage in the upstream portion of the Fraser River (British Columbia, Canada). In: *Modelling Soil Erosion, Sediment Transport and Closely Related Hydrological Processes* (ed. by W. Summer, E. Klaghofer & W. Zhang) (Proc. Vienna Symp., July 1998), 383–391. IAHS Publ. 249. IAHS Press, Wallingford, UK.

Petticrew, E. L. & Droppo, I. G. (2000) The morphology and settling characteristics of fine-grained sediment from a selection of Canadian rivers. In: *Contributions to the Hydrological Programme-V by Canadian Experts*, 111–126. Tech. Documents in Hydrology no. 33. UNESCO, Paris, France.

Soulsby, C., Youngson, A. F., Moir, H. J. & Malcolm, I. A. (2001) Fine sediment influence on salmonid spawning habitat in a lowland agricultural stream: a preliminary assessment. *Science Total Environ.* **265**, 295-307.

Wold, A. K. F. & Hershey, A. E. (1999) Effects of salmon carcass decomposition on biofilm growth and wood decomposition. *Can. J. Fish. Aquat. Sci.* **56**, 767–773.

Woodward, J. C., Porter, P. R., Lowe, A. T., Walling, D. E. & Evans, A. J. (2002) Composite suspended sediment particles and flocculation in glacial meltwaters: preliminary evidence from Alpine and Himalayan basins. *Hydrol. Processes* **16**, 1735–1744.

Sediment Transfer through the Fluvial System (Proceedings of the Moscow Symposium, August 2004).
IAHS Publ. 288, 2004

Transport of phosphorus, wash load and suspended sediment in the River Varde Å in southwest Jutland, Denmark

HANS THODSEN, BENT HASHOLT & MORTEN PEJRUP
Institute of Geography, University of Copenhagen, Østervoldgade 10, DK1350 København K, Denmark
ht@geogr.ku.dk

Abstract Total phosphorus (TP) concentrations, suspended sediment concentrations (SSC) and wash load have been measured at three river monitoring stations in the River Varde Å system since 1998. This provides the possibility of studying the link between SSC and wash load and concentrations of TP. Transport rates of TP, suspended sediment and wash load at the three stations, calculated using rating curves, indicate the dependence of TP transport on the transport of suspended sediment and wash load. Two stations are located on tributaries flowing upstream of the third station located at a weir at the end of a small impoundment. Transport rates at the upstream stations were 57% higher for suspended sediment and 27% higher for wash load than at the downstream station, while transport of TP was the same. This indicates that phosphorus is transported adhered to the finest grain size fractions that do not deposit in the impoundment.

Key words Denmark; phosphorus transport; rating curve; river system; River Varde Å; sediment transport; suspended sediment; wash load

INTRODUCTION

The transport of phosphorus has been studied in the central part of the River Varde Å system, through a monitoring programme started in 1998. The transport of suspended sediment and wash load has been studied at the same locations since 1969 (Hasholt & Madeyski, 1998). The monitoring programme gives the possibility of studying the connections between transport of phosphorus and the transport of water, suspended sediment and wash load. It is well known that phosphorus, to some extent, is transported adhered to suspended particles, e.g. Kronvang *et al.* (1997). It could also be argued that phosphorus is transported adhered to the finest particles in transportation because of electrical surface bounding, and therefore will be transported by wash load rather than suspended sediment.

STUDY AREA

The River Varde Å system is located in southwest Denmark (Fig. 1). The catchment area is 1090 km^2 and the average water discharge at the river mouth is 16.00 m^3 s^{-1} (Ovesen *et al.*, 2000) (Table 1). The river is an alluvial lowland river; it has three main tributaries, of which two are considered here, the River Grindsted Å and the River Ansager Å (Fig. 1). The monitoring stations are located in the central part of the river system, where the River Grindsted Å and the River Ansager Å meet and become the River Varde Å. Shortly downstream of the confluence the River Varde Å is split into the River Gammel (Old) Varde Å and the Ansager, at a dammed site, where the water is directed to the River Gammel

Table 1 Mean annual discharge and catchment areas for three river stations in the River Varde Å system.

	River Ansager Å Lavborg Bridge	River Varde/Grindsted Å Mølby Bro	Ansager Canal Ansager Weir
Discharge mean annual ($m^3 s^{-1}$)	2.04*	3.39	5.19
Catchment area (km^2)	131*	238†	439‡

*(Ovesen *et al.*, 2000) †(Hasholt, 1983) ‡(Hasholt, 1972).

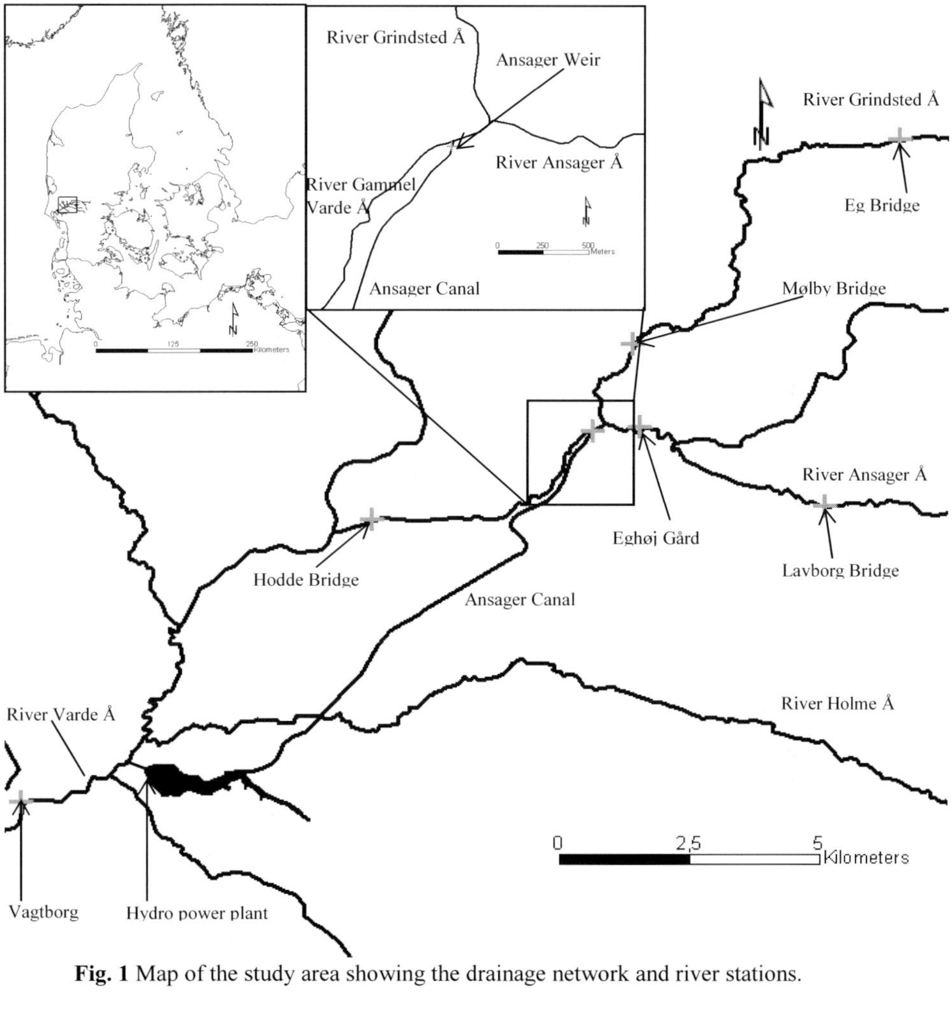

Fig. 1 Map of the study area showing the drainage network and river stations.

Varde Å and the Ansager Canal over two weirs. There is a small impoundment upstream of the dam. One monitoring station is located on the River Ansager Å at Lavborg Bridge and another is located on the River Grindsted Å at Mølby Bridge (Fig. 1). The third monitoring station is located downstream of the confluence at the weir between the impoundment and the Ansager Canal. Most of the water from the River Varde Å enters the Ansager Canal which feeds a hydropower plant. A minor part of the water is still supplied to the natural stream channel, the River Gammel Varde Å (Fig. 1).

Discharges at the Ansager Weir on the Ansager Canal are calculated using *Q*/*Q* relations between the stations at Eg Bridge and Mølby Bridge on the River Grindsted Å and at Lavborg Bridge and Enghøj Gård on the River Ansager Å (Fig. 1). Then the following expression is used $Q_{\text{Mølby Bridge}} + Q_{\text{Enghøj Gård}} - Q_{\text{Hodde Bridge}} = Q_{\text{Ansager Weir}}$, (Fig. 1). Discharge from the Ansager canal at the Ansager Weir are only calculated for the period 5 May 1994–31 December 2002 because of a lack of data from the previous period. The method of calculating the discharge at Mølby Bridge is given above. Discharge at Lavborg Bridge is obtained from a permanent gauging station at that location. Mean annual precipitation in the catchment between 1 January 1990 and 31 January 2001 is about 875 mm year^{-1}. Geomorphologically the catchment mainly consists of sandy outwash plains from the Weichsel glaciations, and older sandy moraine landscapes from the Sahle glaciations, the most easterly parts of the catchment consists of younger moraines from the Weichsel glaciations (Smed, 1978).

METHOD

Samples of suspended sediment have been collected using a depth integrating water sampler (Nilson, 1969) at all stations approximately once a month since 1969 (Hasholt & Madeyski, 1998). Data from 1983–1984 are missing. During 1987 additional samples were collected on days with high discharges. Samples of suspended sediment collected just beneath the water surface using the depth integrating water sampler, have been collected since 1980 at the same time as the suspended sediment samples: these samples provide an estimate of wash load. Since November 1990 water samples were also collected 1–3 times a day using an ISCO 2700 automatic water sampler (ISCO, 1988) at Lavborg Bridge on the River Ansager Å. The ISCO samples water approximately 50 cm above the bed.

In the laboratory total suspended sediment concentrations (SSC) and wash load were determined in two ways. Before 1975 sediment concentrations were determined by filtering water samples through OOH paper filters with a retention diameter of 1–2 μm. Since 1975 Whatman GF/F filters with a retention diameter of 0.7 μm have been used. Filters were dried for at least 3 h at 60°C; after drying filters were weighed after resting for 20 min at room temperature. Water samples were analysed for total phosphorus concentrations using a standard photometric method (Dansk standard, 1985)

A rating curve approach for modelling the transport of suspended sediment as a function of discharge was used, because more sophisticated approaches such as the rising/falling limb approach and the duration curve approach did not significantly improve the modelling results. Suspended sediment concentration rating curves were calculated for the three monitoring stations as a function of water discharge. At the River Ansager Å rating curves were calculated for both manually and automatically collected samples. The difference in transport rates calculated on the basis of manually and automatically collected samples were used to calculate a correction factor for correcting the transport rates from the manually collected samples. The correction factor was calculated by adjusting the mean annual transport given by the rating curve to the mean annual transport given by the automatically sampled data, at the monitoring station at River Ansager Å, Lavborg Bridge. Total phosphorus concentrations were modelled using three kinds of rating curves: those derived by using water discharge, SSC and wash load as the predictors.

RESULTS AND DISCUSSION

Transport of suspended sediment, wash load and phosphorus are modelled and the dependence of phosphorus transport to SSC and wash load concentrations is described.

Suspended sediment

Rating curves for the total SSC, measured by the depth integrating sampler and the ISCO automatic sampler as a function of discharge, were calculated. The rating curves are power functions fitted by a standard least square technique (Table 2). Mean annual transport rates of suspended sediment were calculated using the rating curve functions for the manually collected samples and water discharge for the period 1 January 1995 to 31 December 2002 (Table 2). Estimates of suspended sediment transport obtained by rating curves tend to underestimate the transport in comparison with estimates made by high frequency sampling (Ferguson, 1986; Phillips *et al.*, 1999; Asselman, 2000). At the Ansager Å River station measurements were made 1–3 times a day by the ISCO sampler in the period determining suspended sediment transport. The transport rate estimated by the high frequency sampling method is calculated using mean daily water discharges. A correction factor is applied to the mean annual transport determined by the rating curve approach, giving the corrected mean annual transport. Correction factor:1285 t $year^{-1}$/1100 t $year^{-1}$ = 1.17. The correction factor is multiplied by the mean annual transport thereby giving the mean annual corrected transport, (Table 2). The correction factor was used at the three monitoring stations.

At the River Ansager Å station suspended sediment yields were previously found to be 7 t km^{-2} $year^{-1}$ (Hasholt & Madeyski, 1998). The transport rates indicate that the area between the two upstream stations and the Ansager Canal station is a deposition area for suspended sediment. A part of the suspended sediment transported in that part of the River Varde Å upstream of Ansager Canal is transported to the River Gammel Varde Å. No direct measurements of this transport exists but it is expected to be much smaller than the transport to the Ansager Canal because of a much smaller discharge to the River Gammel Varde Å. Deposition of suspended sediment would be expected at the impounding upstream of the weirs of the Ansager Canal and the River Gammel Varde Å. Approximately 57% of the

Table 2 Rating curves of total suspended sediment concentration (SSC) as a function of dischange (Q), based on depth integrated samples and samples collected automatically by an ISCO sampler. Mean annual transport rates and corrected mean annual transport rates.

River & station	Number of samples (*n*)	R^2	Function	Mean annual transport (t $year^{-1}$)	(t km^{-2} $year^{-1}$)	Mean annual corrected transport (t $year^{-1}$)	(t km^{-2} $year^{-1}$)
River Ansager Å, Lavborg Bridge	390	0.30	SSC = 5.488 $Q^{1.256}$	1100	8.4	1285	9.8
River Grindsted Å, Mølby Bridge	397	0.24	SSC = 5.471 $Q^{1.140}$	2800	11.8	3270	13.7
River Ansager Canal, Ansager Weir	104	0.36	SSC = 3.861 $Q^{0.528}$	1660	3.8	1940	4.4
River Ansager Å, Lavborg Bridge ISCO	5153	0.30	SSC = 5.321 $Q^{0.965}$	1285	9.8	–	–

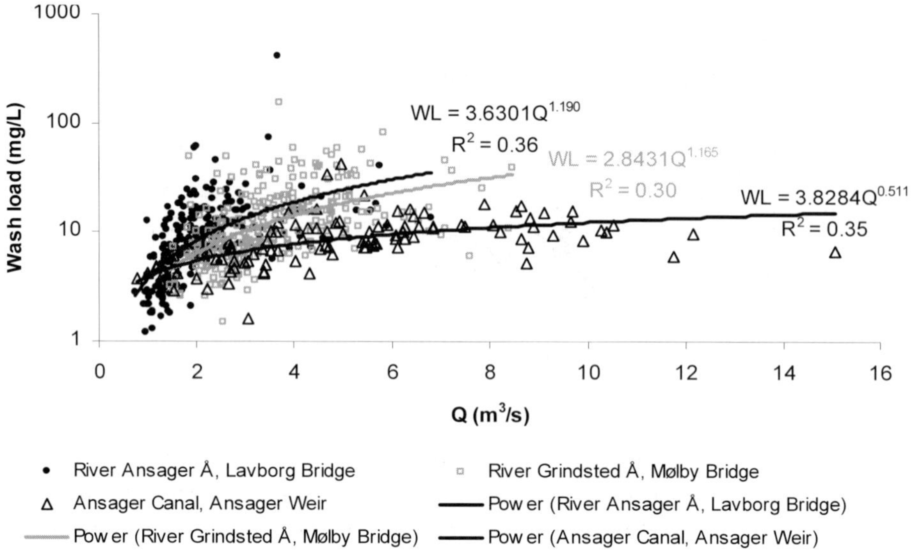

Fig. 2 Rating curves of concentrations of wash load as a function of water discharge (Q), at three stations at the River Varde Å system. WL is wash load concentration.

sediment transported through the two upstream stations is not transported through the station at the Ansager Canal (Table 2).

Wash load

Wash load concentrations are collected from all three monitoring stations. Rating curves have been calculated for the three monitoring stations (Fig. 2).

The rating curves given in Fig. 2, and discharge from the period 1 January 1995 to 31 December 2002 were used to calculate mean annual transport rates which are as follows:

River Ansager Å at Lavborg Bridge, 675 t year^{-1}, 5.2 t km^{-2} year^{-1};
River Grindsted Å at Mølby Bridge, 1500 t year^{-1}, 6.3 t km^{-2} year^{-1};
Ansager Canal at Ansager Weir, 1590 t year^{-1}, 3.6 t km^{-2} year^{-1}.

The sum of the transport rates of the two upstream stations on the River Ansager Å and the River Grindsted Å are approximately 27% higher than at the downstream station on the Ansager Canal. This also indicates that sediment is deposited between the stations. For suspended sediment 57% is deposited between the stations, because it holds fractions of coarser grains from suspended bed load, which is deposited in the impounded water.

Phosphorous

Measurements of total phosphorus concentrations have been made at the three stations. At the River Grindsted Å and the Ansager Canal measurements have been made since May 1998 and at the River Ansager Å since the beginning of 1999. Samples have been collected at the same dates as sediment samples. Rating curves have been made for the relationship

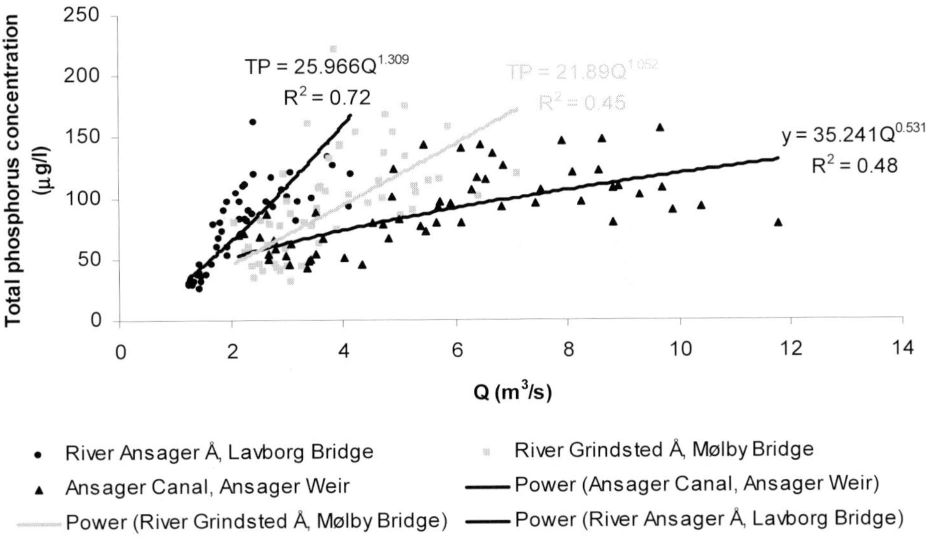

Fig. 3 Rating curves of Total Phosphorus concentrations (TP) as a function of water discharge (Q) at three stations at the River Varde Å system.

Table 3 Total phosphorous concentrations (P_{total}) as a function of suspended sediment concentrations (SSC) and wash load (WL).

River	Samples *n*	SSC R^2	Function	Wash load concentration R^2	Function
River Ansager Å, Lavborg Bridge	47	0.82	P_{total}=18.185 $SSC^{0.623}$	0.86	P_{total}=15.236 $WL^{0.75}$
River Grindsted Å, Mølby Bridge	53	0.58	P_{total}=23.048 $SSC^{0.497}$	0.69	P_{total}=39.603 $WL^{0.51}$
Ansager Canal, Ansager Weir	57	0.86	P_{total}=20.003 $SSC^{0.659}$	0.83	P_{total}=21.303 $WL^{0.64}$

between total phosphorus and discharge (Fig. 3). Relatively large differences are seen in the constants of the three power functions (Fig. 3). The two upstream stations have higher power values, reflecting a higher degree of response to changes in discharge, but also reflecting the fact that the discharge is higher at the downstream station while the total phosphorus concentration is in-between the values of the two upstream stations. Total phosphorus concentrations are better correlated to discharges at the station on the River Ansager Å, than the other two stations.

It has been shown that phosphorus is, to a large extent, transported through fluvial systems adhered to suspended particles (Kronvang *et al.*, 1997). Sediment sampled just beneath the water surface has a smaller grain size than depth integrated samples because the latter samples contain a part of the suspended bed load. Phosphorus is adhered to the smallest grain sizes because of electrical bounding. Phosphorus may then be transported with wash load rather than with total suspended sediment. The influence of the suspended sediment and wash load on total phosphorus concentrations are seen in Table 3.

On average wash load has a better correlation with total phosphorus than suspended sediment, but no clear trend is seen (Table 3). The small difference between the two functions for the Ansager Canal possibly reflects the deposition of the coarsest suspended particles in the impoundment just upstream of the weir (Table 3). The effect of this is that the transports, measured as suspended sediment and as wash load over the weir, to a great extent consists of the same particle size fractions.

Rating curves predicting total phosphorus concentrations can be developed from water discharge, SSC and wash load. Mean annual transport rates of phosphorus are calculated using water discharge rating curves for the three stations. For comparison, a mean annual transport rate is calculated for River Ansager Å at Lavborg Bridge, by a rating curve using SSC, measured by the automatic ISCO sampler, as the input value. Mean annual transport rates of phosphorus derived by water discharge rating curves are:

River Ansager Å at Lavborg Bridge, 5.5 t $year^{-1}$, 42.0 kg km^{-2} $year^{-1}$;
River Grindsted Å at Mølby Bridge, 9.8 t $year^{-1}$, 41.2 kg km^{-2} $year^{-1}$;
Ansager Canal at Ansager Weir, 15.2 t $year^{-1}$, 34.6 kg km^{-2} $year^{-1}$.

Mean annual transport rates derived by SSC rating curve are:

River Ansager Å at Lavborg Bridge, 9.0 t $year^{-1}$, 68.7 kg km^{-2} $year^{-1}$.

Transport of total phosphorus calculated by water discharge indicates that phosphorus is not accumulated between the two upstream stations and the Ansager Canal to the same extent as suspended sediment and wash load. The transport through the upstream stations are 9.8 t $year^{-1}$ + 5.5 t $year^{-1}$ = 15.3 t $year^{-1}$ and 15.2 t $year^{-1}$ through the station at the Ansager Canal. Comparison between phosphorus transport rates at the River Ansager Å at Lavborg Bridge, calculated by water discharge and SSC indicates that the transport rate calculated by water discharge is underestimated. The transport rate calculated by SSC is approximately 64% higher than the transport rate calculated by water discharge. The calculated values indicate that phosphorus is not deposited in the area.

Comparisons can be made for the transport of phosphorus to measurements of transport made by the Ribe Amt County (Jepsen *et al.*, 2003). For the River Varde Å at Vagtborg (Fig. 1) the Ribe Amt County had a transport rate of 36.787 t $year^{-1}$ for 2002, and, with the catchment area at Vagtborg being 814 km^2, this gives 45.2 kg km^{-2} (Fig. 1). At the River Grindsted Å the equivalent number is 44.8 kg km^{-2}; for the River Ansager Å it is 46.7 kg km^{-2} and at the Ansager Canal it is 39.0 kg km^{-2}. All values are seen to be of approximately the same size, indicating that the observed transport rates are of the right magnitude. It is also indicated that using SSC, which has a much higher R^2 value with total phosphorus concentrations than that found for water discharge, could produce a considerably different and perhaps more reliable estimate of transport rates.

It is seen by the above, that the transport of phosphorus is less influenced by deposition in the impoundment upstream of the Ansager Weir than the transport of wash load and especially suspended sediment. This indicates that phosphorus is transported with only the finest sediment particles, which are not deposited in the impounding area. Modelling of phosphorus concentrations using SSC is therefore more site-specific than using wash load. It could, therefore, be argued that wash load is more appropriate in modelling phosphorus concentrations in river systems than SSC, because SSC is highly influenced by the location of deposition areas.

CONCLUSIONS

It can be concluded that phosphorus is transported adhered to fine grain size particles in the wash load transport fraction. It may also be concluded that the transport of phosphorus is not affected by the passage of the river flow through an area of impounded water. The transport rates of suspended sediment is reduced by 57% and the transport of wash load is reduced by 27%, by passing through the impounded area. It was found that total phosphorus concentrations are well correlated with SSC and wash load, and poorly correlated with water discharge. Transport rates of total phosphorus calculated using SSC as the predictor at the River Ansager Å gives a 64% higher transport rate than using water discharge as the predictor. If possible, SSC or wash load concentrations should be used instead of water discharge as the predictor of total phosphorus concentrations.

Acknowledgements The authors wish to thank, Kirsten Simonsen at Skallingen Laboratoriet in Esbjerg for collecting water samples and laboratory work, Søren Vinsløv at the County of Ribe Amt and Niels Bering Ovesen at NERI for providing discharge data.

REFERENCES

Asselman, N. E. M. (2000) Fitting and interpretation of sediment rating curves. *J. Hydrol.* **234**, 228–248.

Danish Standard (1985) DS 292:1985. Data sheet (in Danish).

Ferguson, R. I. (1986) River loads underestimated by rating curves. *Water Resour. Res.* **22**, 74–76.

Hasholt, B. (1972) Materialevandringen og sedimentationen i Karlsgårde Sø og dens tilløb (The material transport and the rate of sedimentation in the lake Karlsgårde Sø with affluents). *Geografisk Tidsskrift* **71**, 54–90 (in Danish with English summary).

Hasholt, B. (1983) Dissolved and particulate load in Danish water courses. In: *Dissolved Loads of Rivers and Surface Water Quantity/Quality Relationships* (ed. by B. W. Webb) (Proc. Hamburg Symp., 1983). 255–264. IAHS Publ. 141. IAHS Press, Wallingford, UK

Hasholt, B. & Madeyski, M. (1998) Suspended sediment yield from the Ansager A River basin, Denmark. *Geografisk Tidsskrift* **98**, 38–45.

ISCO (1988) *Instruction Manual, Model 2700 Sampler*. ISCO, Inc., Nebraska, USA.

Jepsen, E. O., Andersen, H. T. & Hammerstrøm, M. (2003) *Vandløb og kilder 2002 (Watercourses and sources 2002).* Ribe Amt. (County of Ribe Amt) (In Danish).

Kronvang, B., Laubel, A. & Grant, R. (1997) Suspended sediment and particulate phosphorus transport and delivery pathways in an arable catchment, Gelbaek Stream, Denmark. *Hydrol. Processes* **11**, 627–642.

Nilsson, B. (1969) *Development of a Depth-Integrating Water Sampler*. UNGI Report.

Ovesen, N. B., Iversen, H. L., Larsen, S. E., Müller-Wohlfeil, D.-I., Svendsen, L. M., Blicher, A. S. & Jensen, P. M. (2000) *Afstrømningsforhold i danske vandløb (Runoff characteristics of Danish watercourses).* Miljø- og Energiministeriet, Danmarks Miljøundersøgelser. Faglig rapport fra DMU nr. 340. (Ministry of Energy & Environment. NERI Tech. report no. 340) (in Danish with English summary).

Phillips, J. M., Webb, B. W., Walling, D. E., and Leeks, G. J. L. (1999) Estimating the suspended sediment loads of rivers in the LOIS study area using infrequent samples. *Hydrol. Processes* **13**, 1035–1050.

Smed, P. (1978) *Landskabskort over Danmark, blad 2* (Geomorphological map of Denmark, sheet 2). Geografforlaget.

Sediment Transfer through the Fluvial System (Proceedings of a symposium held in Moscow, August 2004).
IAHS Publ. 288, 2004

Nitrogen content of suspended matter in the Kam Tin River, Hong Kong

M. R. PEART
Department of Geography, University of Hong Kong, Hong Kong
mrpeart@hkucc.hku.hk

Abstract There is little information about the physical and chemical properties of suspended matter in the streams and rivers of Hong Kong. Observations have been made in the Kam Tin drainage basin of the nitrogen content of suspended matter near the settlement of Kam Tin in the developed lowland area of the basin. The median %N of the suspended matter was 4.27 ($n = 53$) and 1.7 ($n = 58$) for stable flow and storm flow samples, respectively. Two small upland streams have been sampled and they afford an interesting contrast to the lowland site. At the KARC stream, where secondary woodland dominates, stable flow and storm flow sediment samples have median %N values of 1.01 ($n = 49$) and 0.97 ($n = 154$), respectively. A second small undisturbed upland stream (RDH) has a storm period median %N value of 0.61 for suspended matter where $n = 38$. The data show evidence of spatial variation in the nitrogen content of suspended matter in the Kam Tin basin and enrichment compared to soil and regolith materials. Comparison of N in suspended matter to fresh leaves reveals depletion of N in suspended sediment with the exception of samples collected under stable flow conditions at Kam Tin.

Key words enrichment; Kam Tin basin, Hong Kong; nitrogen; suspended matter

INTRODUCTION

Traditionally suspended sediment has been regarded as a physical pollutant in the riverine environment (e.g. Guy & Ferguson, 1970) and it has been recognized that it may be an important control upon geomorphological and biological processes. For example, Ongley (1996) highlights the impact of sediment on turbidity that affects the penetration of sunlight into the water. This may reduce food chain production and limit, or prohibit the growth of algae and aquatic plants. Other researchers who note the impact upon aquatic ecology include Shiel & Walker (1984) and Walling & Webb (1992).

It is now well established that, in addition to its physical effects, suspended sediment plays an important role in the transport and biogeochemical cycling of nutrients and other contaminants in the aquatic system (e.g. Ongley *et al.*, 1981; Kronvang, 1992; Gao *et al.*, 2003). The importance of particulate elemental transfer is well evidence by Meybeck (1982) who reported that particulate phosphorus represents 95% of P transported by rivers whilst 55% of river-borne nitrogen is associated with particulates.

Whilst observations on suspended sediment transport and loads in the streams and rivers of Hong Kong have been undertaken (e.g. Lam, 1978) with expansion to include sedimentation (e.g. Port Works Division, 1988), relatively few observations have been made on the physical and chemical properties of suspended matter. However, Dudgeon (1984) and Peart (1993, 2003) provide some information on the physical and chemical properties of suspended sediment. This lack of observation in Hong Kong is paralleled in the rivers of South China with the work of Gao *et al.* (2002, 2003) being a notable exception.

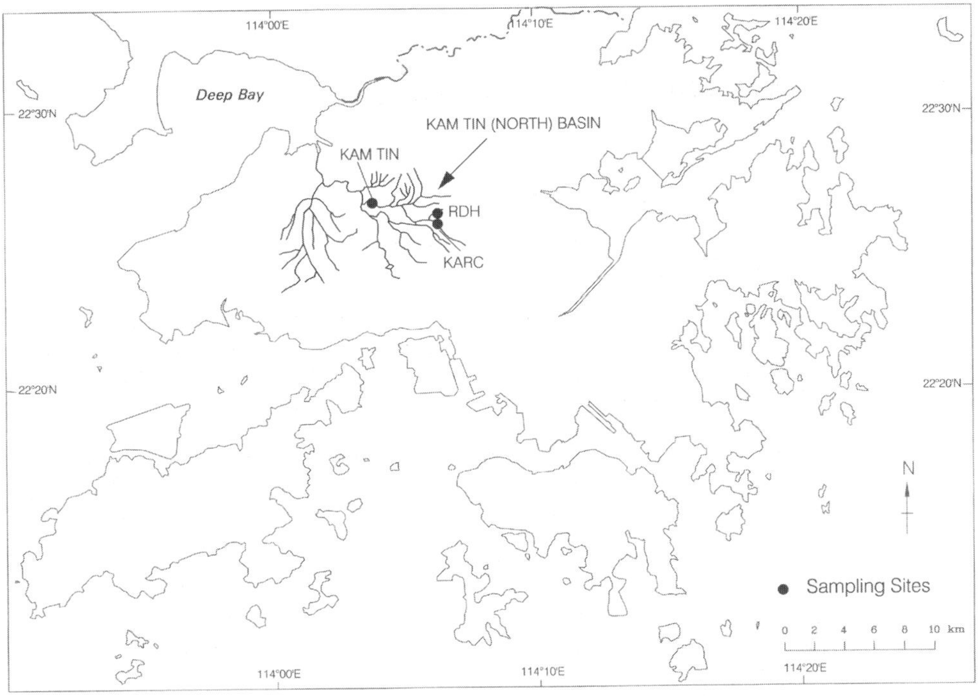

Fig. 1 Location of the sampling sites.

STUDY BASIN AND METHODS

The Kam Tin basin is located in the North West New Territories and it drains the slopes of Kai Keung Leng, Tai To Yan and the western slopes of Tai Mo Shan. The solid geology of the basin is dominated by granodiorite along with volcanic rocks of Mesozoic age. Topographically the basin consists of steep uplands and a contrasting lowland alluvial plain. The steep upland slopes are well vegetated with grass, fern, scrublands and woodland. Agriculture has dominated the lowland plain and remains important. However, an increasing area is occupied by housing (high and low density), open storage and some light industrial uses. Most of the houses and other buildings are not connected to a public sewer system. Near Kam Tin, at the now defunct Water Supplies Department gauging station, the basin has an area of 11.75 km^2. A flood improvement scheme involving the training of some lowland sections of the river has been undertaken.

The drainage system consists of two main branches, the Kam Tin (north) and Kam Tin (south) systems which meet near Yuen Long and ultimately discharge westwards into Deep Bay. A sampling programme was undertaken to monitor the nitrogen (N) content of suspended matter at three points in the drainage basin (Fig. 1), one of which is in the developed lowlands on the Kam Tin (North) River near the settlement of Kam Tin. Two sampling sites were established in the undisturbed uplands. The first site is a small (around 0.1 km^2), steep basin with a predominantly woodland and tall scrub vegetation cover at the Kadoorie Agricultural Research Centre (KARC). The second is a small and steep catchment, but with a grass/fern/low shrub vegetative cover, called RDH. This small stream may become dry in the winter whilst that at KARC always has water present.

Only the Kam Tin and KARC sites were been sampled under steady flow conditions; large volume water samples were collected due to the generally low suspended sediment concentrations that occur under such flow conditions. At all three sites storm-period sampling was undertaken using sample volumes of 400 ml or more with volume dependent upon prevailing sediment concentration. At Kam Tin, most samples were collected during the period 1998 to 2000. Sampling of sediment quality began at KARC and RDH in 1998. For the RDH basin most samples have been collected since 2001. The suspended sediment was separated by filtration using GF/C filter papers that were not pre-ignited because they are composed of 100% borosilicate and are binder free. The filter papers were air dried and the sediment disaggregated before analysis. Suspended matter was not pre-treated to remove carbonates prior to analysis.

Soil samples were collected to allow comparisons with the suspended matter. Samples were obtained from four landslide scars, one of which was just below the outlet of the KARC basin, and the other three being in the headwaters and adjacent slopes of the RDH basin. Landslide scars were sampled because they afford a convenient means to sample the full range of soil and regolith materials. Samples were collected so as to represent the organic A horizon, with sampling typically being of the top 0–20 cm. Material from deeper down the profile was collected to represent sub-soil/regolith materials. The landslides were typical of those in the uplands of Hong Kong being quite shallow. Fresh leaf material from a number of shrub and tree species was also collected during both summer and winter seasons from the KARC drainage basin. A total of 10 alluvial channel bank material samples were collected from two active meander bends on the Kam Tin River just above the settlement of Kam Tin. The soil, regolith, channel bank material and leaves were returned to the laboratory where they were air dried prior to analysis. All samples were passed through a 250 micron sieve before analysis. Total N, along with total carbon (C) was determined on the suspended sediment, soil, regolith, channel bank and leaf material using a Perkin Elmer 2400 elemental analyser. All values are the means of at least two samples.

During the period 1961–1990 Hong Kong had an average annual rainfall of 2214.3 mm of which over 80% fell in the period from April to September inclusive. A rain gauge located in the Kam Tin basin at the KARC recorded an average annual rainfall of 2747.5 mm for 1998–2001 inclusive, of which 87% fell in the summer wet season. Runoff is similarly concentrated in the wet season. For the period 1961–1990 mean monthly air temperature peaked in July at 28.8°C with the minimum occurring in January at 15.8°C.

RESULTS

Table 1 presents mean and median values for total N, along with mean values for total C and their weight ratio for the three study sites in the Kam Tin drainage basin. The presence of N in the suspended matter, and therefore, its transport in particulate form is shown in Table 1. Median %N values range from 1.01 to 4.27 under stable runoff and 0.61 to 1.70 during storm flow conditions. At the two stations for which both storm and stable flow samples are available (Kam Tin and KARC) the N content of suspended matter is lowest under storm flow runoff although this contrast is most pronounced at Kam Tin. Some evidence of a relationship between sample location and N content in the suspended matter can be detected. For example, under stable flow conditions the suspended matter N content is highest at the lowland site with the mean %N value being ~4 times that of the upland KARC site. Under storm flow conditions, the highest mean nitrogen content is again exhibited at the downstream lowland site, with %N content at Kam Tin being 2.0 and 3.2 times that for the KARC and RDH basins respectively. Table 1 also shows considerable variability of N content in suspended matter under storm flow runoff conditions at all sites, with coefficients of variation ranging from 50 to 101%. In contrast, under stable flow conditions the downstream site at Kam Tin exhibits a comparatively low coefficient of variation for N content of 18.5%. However, in the upland stream at KARC much more variability is apparent under stable runoff conditions.

The mean C content of suspended matter ranges from 12.6 to 31.8% under stable runoff, and from 6.95 to 13.9% during storm flow. A pronounced spatial contrast exists under stable runoff but is less apparent under storm flow conditions, with the lowland site at Kam Tin always exhibiting the highest amount of C in the suspended solids.

Tables 2 and 3 present data on N, C, and their ratio, for the potential source materials of the suspended sediment in the study basin. Soil and regolith materials are characterized by Table 2 whilst a range of plant leaf materials are presented in Table 3. Comparison of the data in Tables 1 and 2 reveals considerable enrichment of the suspended matter at both lowland and upland sites in comparison to soil and subsoil/regolith samples. For example, a comparison of suspended matter collected during storm flow conditions to topsoil material

Table 1 N, C and C/N ratio of suspended matter at the sampling sites.

Site	Hydrological sample	Size	Median N (%)	Mean N (%)	Coeff. variation N (%)	Mean C (%)	Mean C/N ratio
Kam Tin	Stable flow	4.27	4.22	18.5	31.83	7.62	53
	Storm flow	1.70	1.94	65.5	13.90	7.32	58
KARC	Stable flow	1.01	1.05	96.2	12.60	12.36	49
	Storm flow	0.97	0.96	101.0	12.65	13.21	154
RDH	Storm flow	0.61	0.60	50.0	6.95	11.5	38

Table 2 Soil/regolith sample N, C and C/N ratio.

	Median N(%)	Mean N (%)	Mean C (%)	Mean C/N ratio	Sample size *n*
Topsoil	0.2	0.24	2.86	13.05	9
Subsoil regolith	0.02	0.05	0.34	15.95	11

Table 3 Mean N, C and C/N ratios for fresh leaves.

Vegetation type	Species	N (%)	C (%)	C/N	Sample size *n*
Tree	*Bridelia monoica*	3.03	44.76	14.81	6
	Helicia cochinchinensis	1.73	41.33	23.87	5
	Stereulia nobilis	2.33	43.63	19.58	4
	Hibiscus tiliaceus	2.31	42.97	18.74	5
	Psychotria rubra	2.40	43.99	18.81	4
Average for all samples		2.39			
Shrub	*Rhodymyrtus tomentosa*	1.66	47.64	30.72	4
	Litsea rotundifolia	2.43	47.07	20.41	4
Average for all samples		2.05			

using mean values reveals enrichment ratios for N of 8.1, 4.0 and 2.5 for Kam Tin, KARC and RDH respectively. The equivalent storm flow enrichment ratios for N using subsoil/regolith samples as source materials are 38.8, 19.2 and 12 respectively for Kam Tin, KARC and RDH. Table 3 reveals that the leaf material has N contents that are rather higher than those observed in the suspended matter collected under storm flow conditions, especially for the two upland sites. Using a mean N content of fresh leaf matter from trees gives N enrichment ratios at the Kam Tin, KARC and RDH sites of 0.81, 0.40 and 0.25, respectively, for storm-period samples, whilst shrubs give enrichment ratios of 0.95, 0.47 and 0.29. These values of less than 1 indicate a loss of N in suspended matter compared to the source materials. Comparison of tree and shrub leaf materials under stable runoff conditions to the suspended matter reveals a contrast between the upland site (KARC) and the lowland sampling station, with the former evidence of depletion of N whilst the latter exhibits considerable enrichment (enrichment ratios = 1.77 and 2.05 for tree and shrub leaf materials, respectively). In terms of channel bank material, analysis identified two types of sample, those containing no N and those with only a small amount, ranging from 0.06 to 0.08% N. Considerable enrichment occurs in the suspended matter of the Kam Tin River as documented in Table 1, in comparison to channel margin sediments.

DISCUSSION

The data presented in Table 1 confirms the presence of N in the suspended matter of the Kam Tin drainage basin. This recognition may be timely given the recent occurrence of red tides in Hong Kong coastal waters (e.g. Hodgkiss & Ho, 1997) and the release of nutrients from sediment documented by Hu *et al.* (2001). The mean particulate organic nitrogen (PON) content of suspended matter was reported by Meybeck (1982) as typically between 0.1 and 1.3%. The values for the two upland sites are within this range but the downstream site, Kam Tin, exceeds these values, especially under stable runoff conditions. Meybeck (1982) cites the Sopchoppy River in the USA as having a PON value of 4.04% and suggests that in that river organic material constitutes much of the suspended matter. Interestingly, Table 1 indicates that the Kam Tin River has a much higher C level in comparison to the upland KARC site under stable runoff conditions and this might help account for the high N levels downstream at Kam Tin. As noted in the previous section, spatial variation of N occurs in the Kam Tin River basin. Few other researchers have observed the spatial variation of N in suspended matter in a single drainage system.

Comparison of the fresh leaf material to the suspended matter is also instructive in that it reveals that, at the upland sites (KARC and RDH), and at Kam Tin under storm flow conditions, the suspended matter contains less N than the potential source material, be it trees or shrubs. This suggests that if the ultimate source of the N is allochthonous organic matter, which Dudgeon (1983) revealed to be important in some streams in Hong Kong, it must have undergone some decomposition. The lower C/N ratios of the suspended matter, and the much decreased C content, relative to the fresh leaf material, are also suggestive of this. Decomposition of leaf material in streams and rivers is widely reported in the literature (e.g. Lopez *et al.*, 2001). The enrichment of N in the suspended matter at the Kam Tin sampling station under stable flow conditions, coupled with its mean C/N ratio of around 7.6 is suggestive of very different influences at this site under these flow conditions. Olley (2002) suggests that a C/N ratio of 5–6 may be indicative of algae or phyto plankton and this may be relevant to the Kam Tin site, especially under stable runoff. The lowering of the percentage C and N in the suspended matter at the Kam Tin site during storm events may reflect an input from channel bank materials and such a process has been observed in the field.

The enrichment of suspended sediment in comparison to potential source material, such as soil and regolith, has not been reported previously for Hong Kong. The enrichment reported for the study basin shows evidence of spatial variation with the greatest enrichment being associated with the lowland Kam Tin site. A number of investigators have, however, reported that eroded sediments are enriched in N as compared to the soils from which they are derived (e.g. Johansen *et al.*, 2001; Avnimelech & McHenry, 1984). It should be noted, however, that erosion processes are very scale dependent (e.g. Walling, 1983; Johansen *et al.*, 2001) and, therefore, comparison between studies is difficult and the contrast between the small upstream basins and the larger downstream station may reflect the size of the drainage basins. There may also be a need to consider the lowland agricultural soils, for the data in Table 2 refers to undisturbed upland soils. Nevertheless, it is important to recognize that suspended matter may be enriched in N in comparison to soil/regolith source materials for this indicates selective erosion and transportation which may give rise to degraded soils. Peart (1993) mentions the possibility of selective erosion in Hong Kong.

CONCLUSION

N has been shown to be present in the suspended matter at three sampling stations in the Kam Tin drainage basin. Evidence of spatial variation has been presented. The presence of N in suspended matter indicates that water quality monitoring in Hong Kong may need to consider this aspect, and that N export in particulate matter may need to be considered in studies of soil degradation.

Acknowledgements Funding from the Dr Stephen S. F. Hui Trust Fund, the Lau Wong Fat Fund and CRCG is gratefully acknowledged.

REFERENCES

Avnimelech, Y. & McHenry, J. R. (1984) Enrichment of transported sediments with organic carbon, nutrients and clay. *Soil Sci. Soc. Am. J.* **48**, 259–266.

Dudgeon, D. (1983) Preliminary measurements of primary production and community respiration in a forest stream in Hong Kong. *Arch. Hydrobiol.* **98**, 287–298.

Dudgeon, D. (1984) Seasonal and long-term changes in the hydrology of the Lam Tsuen River, New Territories, Hong Kong, with special reference to benthic macroinvertebrate distribution and abundance. *Arch. Hydrobiol. Suppl.* **69**, 55–129.

Gao, Q. Z., Tao, Z., Shen, C. D., Sun, Y. M., Yi, W. X. & Xing, C. P. (2002) Riverine organic carbon in the Xijiang River (South China): seasonal variation in content and flux budget. *Environ. Geol.* **41**, 826–832.

Gao, Q. Z., Tao, Z., Xie, M., Cui, K. & Zeng, F. (2003) Effects of hydrological processes on the chemical composition of riverine suspended sediment in the Zhujiang River, China. *Hydrol. Processes* **17**, 2365–2373.

Guy, H. P. & Ferguson, G. E. (1970) Stream sediment: an environmental problem. *J. Soil Water Conserv.* **25**(6), 217–221.

Hodgkiss, I. J. & Ho, K. C. (1997) Are changes in N:P ratios in coastal waters the key to increased red tide blooms? *Hydrobiologia* **352**, 141–147.

Hu, W. F., Lo, W., Chua, H., Sin, S. N. & Yu, P. H. F. (2001) Nutrient release and sediment oxygen demand in a eutrophic land-locked embayment in Hong Kong. *Environ. Int.* **26**, 369–375.

Johansen, M. P., Hakonson, T. E, Ward, F. W., Simanton, J. R. & Stone, J. J. (2001) Hydrologic response and radionuclide transport following fire at semiarid sites. *J. Environ. Quality* **30**, 2010–2017.

Kronvang, B. (1992) The export of particulate matter, particulate phosphorus and dissolved phosphorus from two agricultural river basins: implications on estimating the non-point pollution load. *Water Res.* **26**, 1347–1358.

Lam, K. C. (1978) Soil erosion, suspended sediment and solute production in three Hong Kong catchments. *J. Tropical Geography* **47**, 51–62.

Lopez, E. S., Pardo, I. & Nuria Felpeto (2001) Seasonal differences in green leaf breakdown and nutrient content of deciduous and evergreen tree species and grass in a granitic headwater stream. *Hydrobiologia* **464**, 51–61.

Meybeck, M. (1982) Carbon, nitrogen and phosphorus transport by world rivers. *Am. J. Sci.*, **282**, 401–450.

Olley, J. M. (2002) Organic carbon supply to a large lowland river and implications for aquatic ecosystems. In: *The Structure Function and Management Implications of Fluvial Sedimentary Systems* (ed. by F. J. Dyer, M. C. Thomas & J. M. Olley), 27–33. IAHS Publ. 276. IAHS Press, Wallingford, UK.

Ongley, E. D. (1996) *Control of Water Pollution from Agriculture*. Irrigation and Drainage Paper 55. FAO, Rome, Italy.

Ongley, E. D., Bynoe, M. C. & Percival, J. B. (1981) Physical and geochemical characteristics of suspended solids, Wilton Creek, Ontario. *Can. J. Earth Sci.* **18**, 1365–1379.

Peart, M. R. (1993) Using sediment properties as natural tracers for sediment source: two case studies from Hong Kong. In: *Tracers in Hydrology* (ed. by N. E. Peters, E. Hoehn, Ch. Leibundgut, N. Tase & D. E. Walling), 313–381. IAHS Publ. 215. IAHS Press, Wallingford, UK.

Peart, M. R. (2003) Carbon and nitrogen content of suspended matter in a Hong Kong drainage basin. *Hydrobiologia* **494**, 215–222.

Port Works Division (1988) *The Sedimentation Study of Various Man-Made River Channels in the Territory*. Civil Engineering Services Department, Government of Hong Kong, Hong Kong.

Shiel, R. J. & Walker, K. F. (1984) Zooplankton of regulated and unregulated rivers: the Murray-Darling System, Australia. In: *Regulated Rivers* (ed. by A. Lillehammer & S. J. Saltviet), 263–270. University of Oslo Press, Oslo, Norway.

Walling, D. E. (1983) The sediment delivery problem. *J. Hydrol.* **65**, 209–237.

Walling, D. E. & Webb, B. W. (1992) Water quality: physical characteristics. In: *The Rivers Handbook* 1 (ed. by P. Calow & G. E. Petts), 48–72. Blackwell Scientific, Oxford, UK.

Spatial distribution of particulate phosphorus forms in the Slave River Delta, Northwest Territories, Canada

MICHAEL STONE
School of Planning and Department of Geography, University of Waterloo, Waterloo, Ontario N2L 3G1, Canada
mstone@fes.uwaterloo.ca

Abstract Deltaic landforms are important sinks for a portion of the suspended sediment flux transported by many large river systems in the world. The trapping and subsequent deposition of fine-grained (< 63 μm) sediment by rooted aquatic plants has implications for the distribution of sediment-associated nutrients and contaminants in these ecologically diverse ecosystems. The present study examines the spatial distribution of chemically defined particulate phosphorus (P) forms [nonapatite inorganic P (NAIP), apatite inorganic P (AP), organic P (OP)] in a remote northern Canadian delta where sediment accumulation rates are influenced by delta morphology and plant assemblages. Sediment samples were collected in three morphological zones (outer, mid, apex) of the delta. The median total particulate P concentration was 329.2 $\mu g\ P\ g^{-1} \pm 50\ \mu g\ P\ g^{-1}$. NAIP represented approximately 68% of particulate P in the delta but total particulate P concentrations are considerably lower than reported in the literature for mid-latitude river and lake environments. Macrophytes influence the spatial distribution of particulate P forms in the delta by influencing the patterns of sediment deposition and grain size.

Key words bioavailability; particulate phosphorus; sediment geochemistry; Slave River delta

INTRODUCTION

Deltas of large rivers are among the most productive and environmentally sensitive ecosystems in northern Canada. These deltas include the Mackenzie River delta, the Slave River delta and the Peace-Athabasca delta. The Slave River delta, which is located at the confluence of the Slave River and Great Slave Lake, supports a highly productive and diverse aquatic ecosystem and is ecologically important at the local, regional and international scale (Thompson *et al.*, 1979; Tripp *et al.*, 1981). The people of Fort Resolution rely on local harvests from the Slave River delta to provide income and a supply of country food (Geddes, 1981). The productivity of this landform is related to the supply of sediment-associated nutrients derived from upstream sources that deposit in channels as well as perched and riparian habitats of the delta. Morphology of the delta is controlled by a series of complex geomorphic and hydraulic factors that influence sediment erosion and deposition (Vanderburg & Smith, 1988; Milburn & Prowse, 1998). Aquatic macrophytes play an important role in establishing and stabilizing deltaic features (English *et al.*, 1997).

The geochemical cycling and distribution of particulate phosphorus has been well documented, particularly in the context of mid-latitude lakes (Bird, 1986; Boers *et al.*, 1993) and rivers (Miller *et al.*, 1982; Stone & English, 1993; Stone *et al.*, 1991) that receive higher inputs of nutrient from point and non-point sources. However, very little is known about the source, chemical speciation and spatial distribution of particulate phosphorus in remote northern rivers and its role in nutrient cycling within these ecologically diverse northern habitats (Ruttenberg & Goni, 1997). Given the importance of vegetation for sediment trapping and its potential for nutrient uptake (Kairesalo & Matilainen, 1994; Schulz *et al.*, 2003; Vargo *et al.*, 1998), this study examines the distribution of particulate phosphorus forms (NAIP, AP, OP) in depositional zones within three morphological areas of the Slave River delta. The distribution of particulate P forms is related to delta morphology as well as the geochemical composition and grain size characteristics of sediment.

METHODS

Study area

The Slave River Delta is located in the Northwest Territories at the confluence of the Slave River with Great Slave Lake (Fig. 1). Three morphological areas of the delta (outer, mid and apex) are distinguished by increasing levee height, botany and geomorphology (English *et al.* 1997). The outer delta is a large flat marsh, which serves as an important staging, breeding and feeding ground for migratory bird species. *Equisetum fluviatile* and other emergent species of aquatic vegetation attract a diverse range of wildlife including muskrat

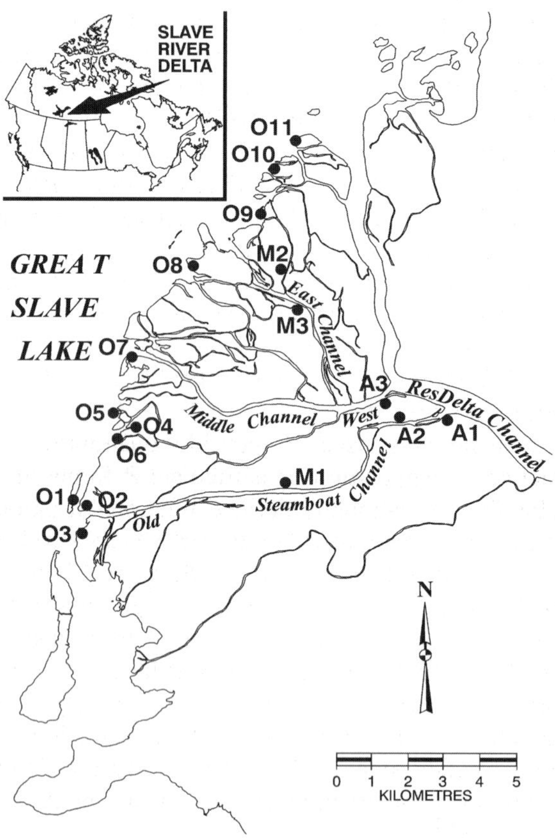

Fig. 1 Study area and sampling stations.

and moose. Cleavage bar islands are the principal deposition features in the outer delta where emergent plant species such as *Equisetum fluviatile*, *Carex rostra* or *Carex aquatilis* invade and enhance sediment trapping (English, 1984). The mid delta is a transitional area between the water-dominated landscape of the outer delta and the elevated, relatively dry apex area. The mid delta represents approximately 45% of the active delta and average levee height is approximately 1 m above summer lake water levels. This portion of the delta supports plant species adapted to a mesic environment including *Equisetum arvense*, *Cornus Stolonifera*, *Salix* spp., *Alnus tenuifolia* and *Populus balsamifera*. The *Alnus–Salix* plant assemblage is by far the most representative in the mid delta. In the apex, the average elevation of levees above summer water levels is approximately 2 m. Flood frequency in the apex is approximately 35 years (English, 1984). About 6% of the apex is classified as aquatic and most of these areas are elevated and cut-off from the Slave River. A significant portion of this zone has reached the climax forest stage of *Picea glauca* that is underlain by permafrost.

Sample collection and chemical analysis

Surface sediment samples (top 10 cm) were collected from depositional features in 17 locations across the delta (Fig. 1), including 11 outer delta sites (O1–O11), three mid delta sites (M1–M3) and three apex sites (A1–A3). Surface sediment samples were stored for 1 month at 4°C before analysis. The grain size distribution was determined with a combination of sieve and hydrometer techniques and the organic matter content was determined by loss on ignition (Environment Canada, 1979).

The concentration of major elements (SiO_2, Al_2O_3, Fe_2O_3, CaO, MgO, K_2O, Na_2O, TiO_2, MnO and P_2O_5) was determined using a Phillips PW1480 XRF spectrometer in the Geology Department at McMaster University, Hamilton, Ontario, Canada. The practical detection limit of the instrument is 50 ppm. Precision of the analysis was determined by analysing three pellets made from a homogenized sediment sample. Accuracy of the analysis was verified by running Canadian Reference Standards and comparing the results with the stated reference values for major elements. The mineralogical composition of sediments was investigated by powder X-ray diffraction using a Cu-target with a Ni-filter. Mineralogy of the sediment is reported as percent dry weight.

A sequential extraction scheme was used to provide quantitative and qualitative assessment of the relative fractional composition of particulate P forms in delta sediment. The fractionation scheme yields five operationally defined fractions of particulate P (Stone & English, 1993). In order of extraction the particulate P fractions are: (a) loosely sorbed P; (b) reductant soluble P; (c) reactive P sorbed to metal oxides; (d) P bound to carbonates, apatite P and P released by the dissolution of oxides; and (e) non-reactive organic P extractable in hot (85°C) NaOH. The first three fractions combined are considered bioavailable and referred to collectively as non-apatite inorganic P (NAIP). Fractions four and five are referred to as apatite P (AP) and organic P (OP), respectively. The OP fraction is potentially available after mineralization (Bostrom *et al.*, 1988). Extractions were performed in triplicate. After centrifugation, extracts were analysed on a Technicon Autoanalyzer using the stannous chloride ammonium molybdate method (Environment Canada, 1979). Detection limit of the analytical method is 1 $\mu g\ l^{-1}$.

RESULTS AND DISCUSSION

Mineralogy and major element composition

Sediment mineralogy is remarkably consistent in the three morphological zones of the delta. Sediments are primarily composed of quartz (60–82%) with lesser proportions of calcite (2–6%), feldspars (8–15%), micas (3–7%), montmorillonite (2–9%), dolomite/ankerite (2–4%) and kaolinite (3–5%). Differences in sediment mineralogy are related to variation in the textural composition of the samples. Fine grained deposits typically have less quartz but higher percentages of feldspar and montmorillonite.

The concentrations of major elements are relatively uniform across the delta (Table 1). The grain size characteristics of deposited sediment in the delta are controlled by sediment supply and transport dynamics which are related to variable energy conditions and water level fluctuations (Vanderburg & Smith, 1988). In addition, floral succession plays a significant role in establishing and stabilizing deltaic features such as lateral accretion bars and cleavage bar islands (English, 1984). The major element composition was influenced by grain size (median diameter, D_{50}). The D_{50} was positively correlated with SiO_2 (0.95, $P = 0.01$), CaO (0.76, $P = 0.01$) and Na_2O (0.93, $P = 0.01$) but inversely correlated with Al_2O_3 (–0.96, $P = 0.01$), Fe_2O_3 (–0.88, $P = 0.01$) and MnO (–0.77, $P = 0.01$). Similar relationships between grain size and sediment chemistry have been previously reported (Stone & Mudroch, 1989; Stone & English, 1993). The data show that concentrations of Al_2O_3, Fe_2O_3, and MnO increase with decreasing grain size, which is related to the abundance of clay minerals present in fine grained sedimentary deposits. Conversely, coarse grained deposits contain higher concentrations of SiO_2.

Particulate phosphorus forms

Previous studies report that total particulate P concentrations in a variety of calcareous and non-calcareous lakes range from 580 to 7000 μg P g^{-1} (Williams *et al.*, 1976; Peterson *et al.*, 1988; Ostrofsky & McGee, 1991; White & Stone, 1996). In contrast, the median total particulate P concentration in the Slave River delta is 329.2 ± 50.3 μg P g^{-1} and the range

Table 1 Concentration of major elements in morphological zones of the delta (percent dry weight).

	Outer Delta			Mid delta			Apex		
	Mean	Range	CV (%)	Mean	Range	CV (%)	Mean	Range	CV (%)
SiO_2	74.3	69.4–77.3	3.8	72.9	71.1–76.2	3.89	72.8	67.7–80.5	5.41
Al_2O_3	10.9	9.1–14	16.67	11.1	8.4–13	21.73	11.8	6.9–15.8	23.59
Fe_2O_3	3.5	2.9–4.8	20.47	3.8	3.2–4.2	14.37	3.9	2.2–5.6	28.5
MgO	2.3	2.1–2.7	9.81	2.3	1.9–2.6	16.23	2.4	1.8–2.8	11.26
CaO	4.4	3.9–4.7	7.38	4.8	4.3–5.6	14.58	4.2	2.8–5.6	20.53
Na_2O	0.53	0.41–0.63	15.81	0.49	0.44–0.57	14.29	0.49	0.34–0.72	23.87
K_2O	2.1	1.8–2.5	12.82	2.1	1.8–2.3	12.18	2.2	1.5–2.8	18.44
TiO_2	0.71	0.58–0.86	14.94	0.62	0.41–0.76	29.74	0.7	0.5–0.88	18.05
MnO	0.059	0.043–0.084	26.05	0.047	0.025–0.074	82.67	0.057	0.031–0.11	41.2
P_2O_5	0.3	0.27–0.32	6.09	0.29	0.27–0.32	9.12	0.29	0.27–0.32	5.34

varied from 195 μg P g $^{-1}$ at Site O3 to 390 μg P g $^{-1}$ at Site O1. The Peace and Athabasca rivers erode parent materials from northern Alberta that provides an upstream source of sediment to the Slave River. The eroded parent materials contain approximately 440 μg P g^{-1}, which is lower than the average crustal abundance of 1050 μg Pg^{-1} (Spiers *et al.*, 1981). Because phosphorus levels are low in parent source materials, particulate P concentrations in the delta are also low. Higher particulate P levels in mid-latitude rivers and lakes cited above are due to high deposition rates of atmospheric P as well as increased loading due to point and diffuse source inputs from agricultural and urban sources (Wall *et al.*, 1982).

The distribution of particulate P forms by sample location is shown in Fig. 2. The median concentration of each particulate P form is NAIP (227.7 ± 49.1 μg P g^{-1}), AP (38.4 ± 7.2 μg P g^{-1}) and OP (65.8 ± 16.7 μg P g^{-1}). The average concentrations of NAIP in the apex, mid and outer delta are 241.8 μg P g^{-1}, 222.2 μg P g^{-1} and 205.7 μg P g^{-1}, respectively. While total particulate P levels in sediment from the Slave River delta are much lower than values reported in the literature, NAIP represents 68% ± 8 of the total particulate P. This value is higher than the 40 to 55% NAIP observed in calcareous lakes (White & Stone, 1996) and indicates while total particulate P levels are low; a significant fraction of the particulate P in sediment of the Slave River delta is potentially bioavailable. In a study of nutrient composition of sediments in the Mackenzie River delta, Ruttenberg & Goni (1997) reported the total P:inorganic P ratio was 0.77. They attributed the higher proportion of inorganic P to low weathering rates of particulate matter in arctic environments.

The bioavailability of sediment-associated P for plant uptake depends upon several interrelated physical, chemical and biological factors (Bird, 1986). Grain size is one factor that

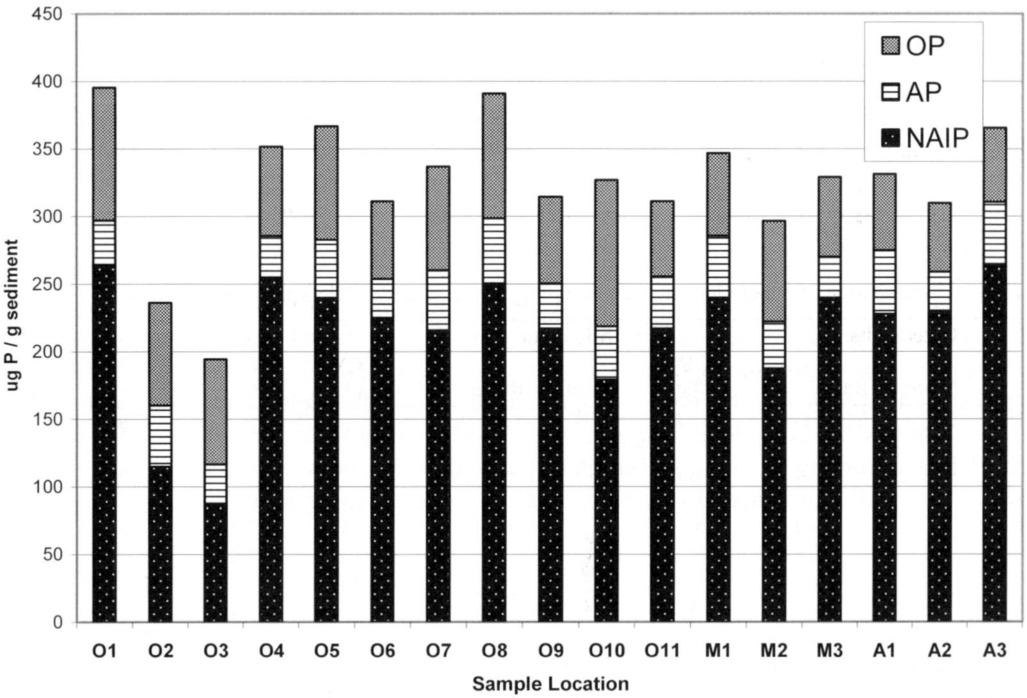

Fig. 2 Distribution of particulate phosphorus forms in the Slave River delta.

Table 2 Slave River water quality data at Fort Smith, NWT (1982–2002).

Variable	*n*	Mean	Range
Suspended solids (mg l^{-1})	59	245	18–2800
pH	64	7.92	7.3–8.4
TP (μg P l^{-1})	61	237	22–4400
Total Fe (μg P l^{-1})	63	4786	8–24700
Total Mn (μg P l^{-1})	41	86	6–482

strongly influences the major element composition and distribution of particulate phosphorus forms in aquatic sediment (Stone & Mudroch, 1989; Stone & English, 1993). The NAIP fraction was positively correlated with Al_2O_3 (0.68, $P = 0.05$), Fe_2O_3 (0.78, $P = 0.01$) and inversely correlated with SiO_2 (–0.70, $P = 0.05$) and suggests that P is primarily bound to metal oxy-hydroxide surfaces of the sediment (Stone & Mudroch, 1989). The mean dissolved P concentration in the Slave River at Fort Resolution is low (5μg P l^{-1}) and levels of total Fe and total Mn are relatively high (Table 2). In rivers with low calcium levels, phosphate concentrations are regulated by the solubility of a solid-solution of ferric-hydroxide in colloidal suspensions or suspended sediments (Fox, 1993). High concentrations of total Fe and low levels of dissolved P suggest that ferric-hydroxide coatings on colloidal suspensions and particulate matter is the dominant factor controlling the kinetics of phosphate in the Slave River.

Much of the sand and coarse-silt fraction of the Slave River is transported through Res Delta channel directly into Great Slave Lake (English *et al.*, 1997). However, finer grained materials enter delta distributaries and a portion of these materials are trapped by vegetation in deposits such as lateral accretion bars, mid-channel bars, channel entrance bars and levees. Results of the present study show that higher particulate P concentrations were found in finer-grained deposits that are stabilized by macrophytes. Further studies are required to estimate nutrient loading to the delta and to quantify the availability of particulate P for plant uptake.

CONCLUSION

The Slave River delta is a remote and environmentally sensitive landform in northern Canada. The present study is the first to quantify the spatial distribution of particulate phosphorus in morphological regions of the delta. The bioavailable NAIP fraction represented approximately 68% of particulate P in the delta but total particulate P concentrations are considerably lower than reported in the literature for mid-latitude river and lake environments. The distribution of particulate P is controlled by sediment grain size which was partly influenced by floral succession and delta morphology.

Acknowledgements Craig Smith assisted with the sample collection and conducted the phosphorus speciation analysis. Water quality data for the Slave River was provided by Wayne Starling, Water Resources Office, DIAND, Fort Smith, NWT.

REFERENCES

Bird, G. A. (1986) *Phosphorus Dynamics in Great Lakes Ecosystems*. Environment Canada, Ottawa, Canada.

Boers, P., Cappenberg, T. & van Raaphorst, W. (eds) (1993) Proceedings of the third international workshop on phosphorus in sediments. *Hydrobiologia* **253**, 1–376.

Bostrom, B., Persson, G. & Bromberg, B. (1988) Bioavailbility of different phosphorus forms in freshwater systems. *Hydrobiologia* **170**, 133–135.

English, M. (1984) Implications of upstream impoundment on the natural ecology and environment of the Slave River delta, NWT. In: *Northern Ecology and Resource Management* (ed. by R. Olson), 11–336. University of Alberta Press, Alberta, Canada.

English, M. C., Stone, M. & Hill, B. (1997) Geomorphological and plant assemblage change in the Slave River Delta, NWT resulting from Peace River Impoundment. *Hydrol. Processes* **11**, 1707–1724.

Environment Canada (1979) *Analytical Methods Manual*. Ottawa, Canada.

Fox, L. (1993) The chemistry of aquatic phosphate: inorganic processes in rivers. *Hydrobiologia* **253**, 1–16.

Geddes, F. (1981) *Part E. Productivity and Habitat Selection of Muskrats in the Slave River delta*. Mackenzie River Basin Study Report Supplement, Mackenzie River Basin Committee.

Kairesalo, T. & Matilainen, T. (1994) Phosphorus fluctuation in water and deposition into sediment with an emergent macrophyte stand. *Hydrobiologia* **275**, 285–292.

Milburn, D. & Prowse, T. (1998) Sediment bound contaminants in a remote northern delta. *Nordic Hydrol.* **29**(4/5), 397–414.

Miller, M., Robinson, J., Coote, D., Spires, A. & Draper, D. (1982) Agriculture and water quality in the Great Lakes basin: III Phosphorus. *J. Environ. Qual.* **2**, 487–493.

Ostrofsky, M. & McGee, G. (1991) Spatial variation in the distribution of phosphorus species of in the surficial sediments of Canadohta Lake, Pennsylvania: Implications for internal phosphorus loading estimates. *Can. J. Fish. Aquat. Sci.* **48**, 233–237.

Peterson, K., Bostrom, B. & Jacobsen, O. (1988) Phosphorus in sediment speciation and analysis. *Hydrobiologia* **170**, 91–101.

Ruttenberg, K. & Goni, M. (1997) Phosphorus distribution, C:N:P ratios and $\delta^{13}C_{oc}$ in arctic, temperate and tropical coastal sediments: tools for characterizing bulk sedimentary organic matter. *Mar. Geol.* **139**, 123–145.

Spiers, G., Dudas, M. & Turchenek, L. (1989) The chemical and mineralogical composition of soil parent material in northern Alberta. *Can. J. Soil Sci.* **69**, 721–737.

Stone, M. & English, M. (1993) Geochemical composition, phosphorus speciation and mass transport characteristics of fine-grained sediment in two Lake Erie tributaries. *Hydrobiologia* **253**, 17–29.

Stone, M. & Mudroch, A. (1989) The effect of particle size, chemistry and mineralogy of river sediments on phosphate adsorption. *Environ. Technol. Lett.* **10**, 501–510.

Stone, M., English, M. C. & Mulamoottil, G. (1991) Sediment and nutrient transport dynamics in two Lake Erie tributaries: A numerical model. *Hydrol. Processes* **5**, 371–382.

Schulz, M., Kozerski, H., Pluntke, T. & Rinke, K. (2003) The influence of macrophytes on sedimentation and nutrient retention in lower River Spree (Germany). *Water Res.* **37**, 569–578.

Thompson, R., Quinlan, R. & Ambrock, K. (1981) *Fisheries studies in the Slave River Delta, NWT, Final report*. Mackenzie River Basin Study Report Supplement, Mackenzie River Basin Committee.

Tripp, D., McCart, R., Saunders, R. & Hughes, G. (1981) *Assessment of Migratory Bird Resources in the Slave River Delta*. Mackenzie River Basin Study Report Supplement, Mackenzie River Basin Committee.

Vanderburg, S. & Smith, D. (1988) Slave River Delta: geomorphology, sedimentology and holocene reconstruction. *Can. J. Earth Sci.* **25**, 1990–2004.

Vargo, S., Neely, R. & Kirkwood, S. (1998) Emergent plant decomposition and sedimentation: response to sediments of varying texture, phosphorus content and frequency of deposition. *Environ. Exper. Biol.* **40**, 43–58.

Wall, G., Dickinson, T. & Van Vliet, J. (1982) Agriculture and water quality in the Canadian Great Lakes Basin: II Fluvial sediments. *J. Environ. Qual.* **11**, 482–486.

White, A. & Stone, M. (1996) Spatial variation and distribution of phosphorus forms in surficial sediments of two Canadian Shield lakes. *Can. Geogr.* **40**(3), 258–265.

Williams, J., Murphy, T. & Mayer, T. (1976) Rates and accumulation of phosphorus forms in Lake Erie sediments. *J. Fish. Res. Bd. Can.* **33**, 430–439.

Sediment Transfer through the Fluvial System (Proceedings of a symposium held in Moscow, August 2004).
IAHS Publ. 288, 2004

Implications of nutrient and soil transfer with runoff in the northeastern region of India

U. C. SHARMA[1] & VIKAS SHARMA[2]

1 *CNRM, VPO Tarore 181133, District Jammu, Jammu & Kashmir, India*
ucsharma2@rediffmail.com

2 *S. K. University of Agricultural Sciences & Technology, Jammu & Kashmir, India*

Abstract The northeastern region of India is predominantly hilly with an average annual precipitation of 2500 mm. The Brahmaputra River and the Barak River are two major rivers in the region with an annual runoff of 537.2 km^3 and 59.8 km^3, respectively. Prevalence of shifting cultivation over an area of 14 660 km^2 is responsible for an annual transfer of 88.3×10^6 t soil and 10.65, 0.337 and 6.05×10^3 t available N, P_2O_5 and K_2O, respectively. Soil and nutrient transfer depends on the amount and intensity of rainfall, slope, vegetation cover, soil texture and human interference. The major sinks for sediment and nutrients are the sea, responsible for an intake of 51.4% of the total transferred quantity, followed by river systems (14.4%), and seasonal streams (11.0%). A major sink for the forest litter and humus transferred from the hill slopes is valley land adjacent to the hills, which retain about 50–65% of them, making the soil highly fertile.

Key words implications; India; nutrient and soil transfer; rain water; sinks

INTRODUCTION

The northeastern region of India, covering an area of 255 090 km^2, lies between latitudes 21°57′ and 29°28′ N and longitudes 89°40′ and 97°25′E (Fig. 1). The region is endowed with natural resources of soil, water and vegetation, but their indiscriminate use has rendered these resources in a fragile state (Sharma, 2003). About 600×10^6 t soil is lost every year through erosion, and land degradation occurs via other means (Borthakur, 1992). This huge quantity of soil, along with runoff, takes away large amounts of absorbed and dissolved crop nutrients to different sinks. Besides other factors, mismanagement of water resources has been the major cause of low productivity in the region. There is an annual loss of 88.3×10^6 t soil and 10.06, 0.37 and 6.05×10^3 t of available N, P_2O_5 and K_2O, respectively from shifting cultivation in the region due to soil erosion during the rainy season (Sharma & Prasad, 1995). The people are attached socio-culturally to the practice of shifting cultivation involving deforestation and burning of the biomass before land preparation. With the present rate of increase in population of the region (2.47%), the situation may become alarming if the use of existing water resources is not suitably planned, developed and conserved (Sharma, 1997). Though it is a gigantic task to estimate the amounts of soil and nutrients deposited through erosion and other means in the various sinks, an attempt has been made to delineate the possible sinks for eroded soil and other constituents for the northeastern region of India. Further studies are suggested to refine the estimated values of the transported fluvial load deposited en-route and at its final destination, the sea.

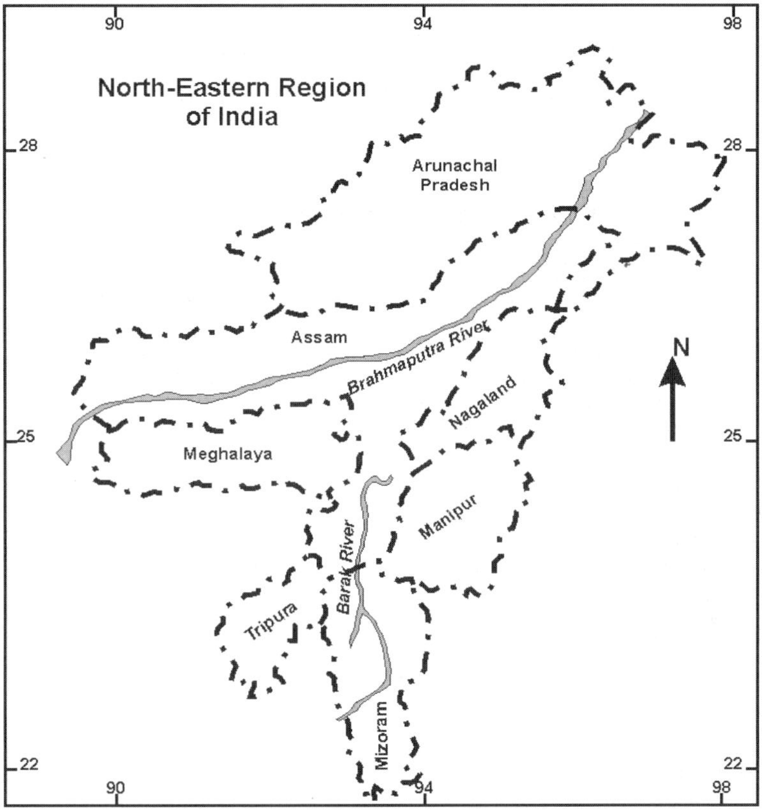

Fig. 1 Study area location.

METHODOLOGY

The methodology for estimating the possible deposition of eroded soil and associated nutrients from the northeastern region of India is primarily based on determining the loss of soil and its deposition at different places by quantifying the soil/sediment and nutrient load in runoff. Soil loss from erosion was estimated through representative gauges from shifting cultivation and other cultivated areas as well as from non-agricultural land. It may be difficult to justify the exact soil and nutrient loss from an area of 255 090 km^2 of northeastern region of India. However, total soil loss was estimated as a product of soil loss from the representative areas of the watersheds studied under different situations, extrapolated to the total area of the region, taking into consideration slope, vegetation, rainfall and soil texture. Deposition of eroded soil in the sediment sinks was estimated by analysing the water samples for dissolved elements. Total amount of elements was determined by multiplying with quantity of water collected in different sinks. The sediment load carried to the sea was estimated by subtracting the sediment load in different sinks (except sea) from the total sediments lost from the northeastern region. Eight to ten samples were taken from each sink from around the region for estimation. Nutrient content of the soil was determined by standard procedures as outlined by Jackson (1967).

Table 1 Water availability in major rivers of northeastern region of India.

River	Drainage (km^2)	Average annual runoff (Mm^3)
Brahmputra	194 413	537 240
Barak	78 150	59 800

Table 2 Aquatic resources of northeastern region of India.

Type	Area
River length (km)	19 548
Reservoirs (ha)	51 697
Lakes (ha)	91 265
Tanks (ha)	40 706

DISCUSSION

Sources of water

The northeastern region has two major river basins viz. Brahmaputra and Barak, draining 194 413 km^2 and 78 150 km^2 respectively, with an annual runoff of 537.2 and 59.8 km^3, respectively (Table 1). Total river length in the region is 19 548 km (Table 2). The region also has a number of other streams and tributaries. Due to a high silt load entering from the catchments and northern tributaries, the Brahmputra River has a tendency to change its course, resulting in bank cutting and thereby causing additional loss of land and generating further particulate and nutrient load. Most of the surface water in the region is confined to rivers, which are in a highly dynamic state due to the high gradient, and they transfer a high load of soil and nutrients to various sinks. The region receives 510 km^3 of water annually as rainfall. Most of the rains are received from May to October, which is also the period during which maximum soil and nutrient displacement takes place. Reservoirs, lakes and tanks account for about 183 668 ha of the area.

Floods

It is important to mention the floods in the north-eastern region of India where their occurrence is a common feature every year during the rainy season. The scale, as well as the frequency, of the floods has been increasing year after year (Borthakur, 1992). The main cause of floods in the Brahmaputra basin is the heavy silt load brought from hill slopes due to erosion because of heavy rainfall in the catchment area and deforestation, which continues unabated. About 35 840 km^2 area of the region is prone to floods while an area of about 3760 km^2 is affected every year (Table 3). On average, the flood water contains 2250 mg l^{-1} of sediment (Table 4) (Fig.1). It also contains 10.6, 3.2, 13.5 and 23.6 mg l^{-1} of NO_3-N, P-PO_4, K_2O and other nutrients, respectively (Table 4).

Sinks for soil and nutrients

The possible sinks for the soil and nutrients displaced from various parts of the region are the major riverbeds, streams, flood affected areas, valleys, areas of temporary water storage after

Table 3 Area affected by floods in northeastern region of India (km^2).

Sub-region	Area prone to floods	Annual flood affected area
I	290	12
II	4100	151
III	22450	1721
IV	9000	1876
Total	35840	3760

Table 4 Sediment and nutrient concentration range in flood water (dissolved phase).

Soil / nutrients	mg l^{-1}
Soil (sediment)	1500–3000
NO_3-N	6.4–25.8
P-PO_4	2.3–8.5
K_2O	15.4–33.8
Other nutrients	21.5–51.6

the floods recede, lakes and reservoirs including ponds, and the sea. The total displacement of soil, nitrogen, phosphorus, potassium, manganese, zinc, calcium and magnesium has been estimated at 601 million tonnes (soil) and 685.8, 99.8, 511.1, 22.6, 14.0, 57.1 and 43.0 thousand tonnes of nutrients and trace elements, respectively (Table 5). As sinks, the rivers, flood prone area, streams, valleys, temporary water storage areas after floods, lakes and reservoirs, and the sea account for about 4.6, 7.6, 11.0, 9.6, 5.0, 1.0 and 51.2% of the sediment load of soil as against 11.4, 7.1, 10.6, 8.5, 4.0, 1.1 and 57.3% of the nutrient load, respectively (Table 5).

Implications

The transfer of the soil and nutrient load, along with runoff from rainfall has many implications for the resource base and environment in the region. Due to heavy soil erosion because of rains, there has been large scale land and environment degradation. The prevalence of shifting cultivation over an annual area of 386.9 thousand ha has resulted in

Table 5 Sinks for the soil nutrients and trace elements transferred in northeastern region.

Sink	Soil	Nutrient/elements (10^3 t)						
	(10^6 t)	N	P	K	Mn	Zn	Ca	Mg
River	86	77.5	11.6	57.0	2.5	1.9	7.7	5.6
Flood area	46	48.7	7.5	36.5	1.6	0.9	3.6	2.6
Streams	66	72.0	10.3	53.1	2.3	1.8	7.0	5.5
Valleys	58	57.6	8.9	42.9	1.8	1.4	5.5	4.3
Temporary water storage	30	27.4	4.9	20.0	0.8	0.6	2.3	1.9
Lakes & Reservoir	6	7.5	1.2	5.4	0.2	0.2	0.7	0.6
Sea	309	395.1	55.4	296.2	13.4	7.2	30.3	22.5
Total	601	685.8	99.8	511.1	22.6	14.0	57.1	43.0

Table 6 Land degradation in northeastern region of India.

State of the region	Degraded land (10^3 ha)
Arunachal Pradesh	2854
Assam	2999
Manipur	734
Meghalaya	1102
Mizoram	610
Nagaland	482
Tripura	279
Total	8860

deforestation because the practice involves burning of the forest biomass. About 8.86×10^6 ha land has been degraded due to soil erosion and other degradation problems in the northeastern region of India (Borthakur, 1992) (Table 6) (Fig 1). Continuous degradation of land has affected the water resources, which may lead to crisis of enormous magnitude. Planning for management of land and water resources, therefore, assumes great significance. An increase in population putting more pressure on water and land resources has also contributed to environmental degradation. Pollution by sewage and other domestic products, poisonous industrial effluents, agricultural chemicals etc. has increased. Eutrophication of water bodies with agricultural runoff has caused drastic changes in the biomass. Water pollution has been caused by an increase in the concentration of chlorides, sulphates, nitrate and aluminium, possibly because of high aluminium content in the soil. High runoff during the rainy season from May to October has resulted in perpetual floods (Table 3). More than 95% of the soils in the region are acidic in nature, which is the result of leaching of salts due to heavy rainfall in the region that is 2450 mm per annum, on average. Soil degradation and mismanagement have had an impact on crop productivity, which is quite poor and the region is still in deficit in food grains.

Nevertheless, there has been a positive effect of runoff in the immediate vicinity of the hill slopes as a lot of forest litter and humus from the forests becomes deposited in the near valley areas. This has increased the soil fertility in the limited adjoining areas to the forests and crop productivity is higher in these areas even without the application of inorganic fertilizers.

CONCLUSIONS

Transport of large load of soil and sediment associated nutrients in the runoff and its deposition in various sinks is a matter of concern for the northeastern region of India. Deforestation and denudation of basins has resulted in water scarcity because the natural water cycle has been upset. The runoff during rains goes untapped from denuded hill slopes instead of infiltrating to recharge aquifers. There has been large scale land and environmental degradation in the region. There is an urgent need to conserve and manage the rain water judiciously to reduce runoff. Sustainable solutions to water problems and judicious use of water resources require a radical policy at the government level as well as relevant institutional reforms to create awareness among the people.

REFERENCES

Borthakur, D. N. (1992) *Agriculture in the Northeastern Region.* Bee Cee Publishers, Guwahati, Assam, India.

Jackson, M. L. (1967) *Soil Chemical Analysis.* Prentice Hall of India, New Delhi, India.

Sharma, U. C. (1997) Food security in the northeast. In: *Perspective for Planning and Development in Northeast India* (ed. by R. C. Sundryal, Uma Shankar & T. C. Upreti), 197–212. G. B. Pant Institute of Himalayan Environment and Development, Almarah, India.

Sharma, U. C. (2003) Impact of population growth and climate change on the quantity and quality of water resources in the northeast of India. In: *Water Resources Systems—Hydrological Risks, Management and Development* (ed. by G. Bloschl, S. Franks, M. Kumagi, K. Musiake & D. Rosbjerg) (Proc. Sapporo Symp., July 2003), 349–357. IAHS Publ. 281. IAHS Press, Wallingford, UK.

Sharma, U. C. & Prasad, R. N. (1995) Socio-economic aspect of acid soil management and alternate land use systems for northeastern states of India. In: *Plant–Soil Interactions at Low pH* (ed. by R. A. Date, N. J. Grudon, G. E. Rayment & M. E. Roberts), 689–695. Kluwer, Dordrecht, The Netherlands.

Key word index

IAHS Publications on Surface Water Topics

World Catalogue of Maximum Observed Floods

compiled by Reg Herschy

This revised and updated edition of the flood catalogue, includes data for 120 countries:

- the location of the flood flow observation sites, and information about the drainage basins (e.g. climate regime, mean annual precipitation and mean annual discharge, if known);
- the maximum instantaneous discharge during the events listed in 1, plus ancillary information about the event (e.g. the antecedent rainfall, flood duration);
- for sites in some countries, series of annual maximum instantaneous discharge data are also available.

UNESCO IHP and WMO provided financial support for preparation of this publication

Publ. 284 (*2003*) ISBN 1-901502-47-3; 320 pp., £80.00

Weather Radar Information and Distributed Hydrological Modelling

Y. Tachikawa, B. E. Vieux, K. P. Georgakakos & E. Nakakita

The combined use of radar hydrology with distributed hydrological modelling creates powerful numerical tools for forecasting storms, floods and related natural disasters. Such tools, when used with historical data, can also contribute rich information for water resources assessment. Much attention has been focused on these developments. However, fundamental issues of model development associated with the existence of uncertainties or errors in radar rainfall estimation and prediction, and in distributed model parameters, still remain unresolved; clarification as to how prediction accuracy will be improved is essential. Also, suitable model resolution and required forcing resolution might be a function of catchment scale, but the relationship between them is still unknown. In association with the scale issues, assessing in what environment a distributed model shows better performance than a lumped model, is also important. This volume comprehensively tackles these issues.

Publ.282 (2003) ISBN 1-901502-37-6; 324 + xii;pp.; £55.40

Regionalization in Hydrology

edited by Bernd Diekkrüger, Mike J. Kirkby & Ulrich Schröder

There is an increasing demand for the forecasting of water quantity and quality at all scales from the local (point) to the regional (meso and macro) scale, and so regionalization has become one of the major issues in hydrology.

This volume of 28 papers explores new mathematical and computational tools to describe and analyse the behaviour of hydrological systems at all relevant scales from the point to the global, for whole systems and for subsystems. It provides an overview of actual model and regionalization concepts, scaling of properties and processes, and modern software tools.

The German National Committee of the International Hydrological Programme of UNESCO and the WMO Operational Hydrology Programme provided financial sponsorship for this volume

Publ. 254 (1999) ISBN 1-901502-80-95; 260 + xii pp., £41.50

Integrated Methods in Catchment Hydrology—Tracer, Remote Sensing and New Hydrometric Techniques

edited by Chris Leibundgut, Jeff McDonnell & Gert Schultz

Tracer techniques and remote sensing have become important tools in catchment hydrology. While remote sensing is mainly used to address surface parameters, tracers have been useful for quantifying sources of streamflow, groundwater residence times, and surface and subsurface flow paths. However, little integration of tracer techniques with other techniques has been attempted. The papers in this volume deal specifically with integrated use of the available advanced techniques.

Publl. 258 (*1999*) ISBN 1-901502-01-5; 284 + xii pp.; £43.00

The FRIEND Series

Flow Regimes from International Experimental and Network Data

FRIEND, an initiative of the UNESCO International Hydrological Programme, brings together research and operational hydrologists, water users and policy makers. There is a network of scientists in FRIEND groups: Northern European, Alpine and Mediterranean, Southern Africa, West and Central Africa, Hindu-Kush Himalayan, Nile Basin, Asian Pacific and Caribbean.

FRIEND 2002—Regional Hydrology: Bridging the Gap between Research and Practice

edited by Henny A. J. van Lanen & Siegfried Demuth

The fourth FRIEND conference (Cape Town, March 2002) demonstrated how advances in hydrology can be used for the development of integrated river basin management. Examples are given of the incorporation of research results into operational hydrology and water resources planning, with numerical and statistical models and visualization techniques.

- Hydrological Data policy, international rivers, databases, real time, dissemination
- Managing Hydrological Risk floods, droughts
- Water Scarcity overexploitation and poverty reduction
- Sustaining Water-related Ecosystems definitions, methods
- Continental Hydrology regimes, water sharing, teleconnections, snow, ice, international basins

Publ. 274 (*2002*) ISBN 1-901502-81-3; 518 + x pp.; £73.50

FRIENDS in Hydrology *edited by Lars Roald, Kjell Nordseth & Karin Anker Hassel*

Publ. 187 (*1989*) ISBN 0-947571-27-2; 492 + xxiv pp.; now £10.00

FRIEND: Flow Regimes from International Experimental and Network Data *edited by P. Seuna, A. Gustard, N. W. Arnell & G. A. Cole*

Publ. 221 (*1994*) ISBN 0-947571-04-3, 526 + x pp.; £48.50

FRIEND'97—Regional Hydrology: Concepts and Models for Sustainable Water Resource Management *edited by A. G., S. Blazkova, M. Brilly, S. Demuth, J. Dixon, H. van Lanen, C. Llasat, S. Mkhandi & E. Servat*

Publ. 246 (*1997*) ISBN 1-901502-35-X; 364 + xii pp.; £50.00

The Ecohydrology of South American Rivers and Wetlands

edited by Michael E. McClain

An overview of ecohydrological processes operating in South America's most important aquatic systems. River reaches ranging from pristine to heavily impacted and processes operating in channels, wetlands, and riparian environments are considered. The Amazon, Orinoco, and Paraná, receive the greatest attention, but the condition of the Piracicaba (São Paulo) and the Paraíba do Sul (Rio de Janeiro) are also evaluated. The natural attenuation processes in these ecosystems stand to aid South America in achieving its goal for sustainable use of its resources.

Sponsored financially by the UNESCO Regional Bureau for Science in Europe (ROSTE)

Sp. Publ. 6 (*2002*) ISBN 1-901502-02-3; 210 + xiv pp.; £39.50

*More details at **www.iahs.info** including abstracts of all papers since 2000.*

Please send orders and enquiries about books to:

Mrs Jill Gash, IAHS Press, Centre for Ecology and Hydrology, Wallingford, Oxfordshire OX10 8BB, UK

tel: +44 1491 692442; fax: +44 1491 692448/692424;

jilly@iahs.demon.co.uk

IAHS members benefit from discounts on book prices: 25% for all members, and a subsidized 80% discount for members in financially disadvantaged countries on titles published before the current year; IAHS membership is free of charge.

All prices include delivery by surface mail. Payment may be made by credit card (VISA, MasterCard, EUROCARD).